A SCIENTIFIC GUIDE TO SURFACE MOUNT TECHNOLOGY

INTERGRAPH CORP.

NOV 29 1988

INFORMATION CENTER

A SCIENTIFIC GUIDE TO SURFACE MOUNT TECHNOLOGY

by
C. LEA

ELECTROCHEMICAL PUBLICATIONS LIMITED
1988

ELECTROCHEMICAL PUBLICATIONS LIMITED
8 Barns Street, Ayr, Scotland

©
Electrochemical Publications Limited
1988

All Rights Reserved
No part of this book may be reproduced in any form
without written authorisation from the publishers

ISBN 0 901150 22 3

Typesetting by Brian Robinson, Buckingham
Printed in Great Britain by The Anchor Press Ltd
and bound by Wm Brendon & Son Ltd,
both of Tiptree, Essex

PREFACE

Surface mounting of electronic components is a novel technology, a rapidly evolving technology. The speed of writing and publishing a book on SMT cannot, at present, keep abreast of the innovative developments that are taking place, and yet there is a need for a guidebook delineating a path through the current technology, illustrating the developments of SMT and pointing the way forward, giving sufficient data and scientific understanding for the technical aspects of managerial decisions to be made. Thus I have aimed for a text that is not a handbook listing equipment to be superseded before the publication date, or giving a comprehensive selection of components and their footprints, but a book that offers underpinning scientific concepts and data in the hope that the reader will find the subject both fascinating and applicable to any future developments by the industry.

Although this book is my view of the scientific background to SMT, illustrated by examples from the current practical standing of the technology, most of the contents is the work of others, a fact accorded to by the citation of over 500 references to the scientific and technical literature. The book can only have been written with the help of others, through conversations, reading their published work and using their data. I have attempted here to review the literature and paint a balanced, rationalised picture, bringing together all aspects of SMT in a unified volume. Insofar as I have succeeded in this, I owe thanks to many people too numerous to name individually, especially my friends and colleagues in the electronics assembly industry, members of the Soldering Science & Technology Club, and those on whose published work I have drawn. I am particularly indebted to a few who have unknowingly kindled my interest and lit my path into several specific areas and who may well recognise their scientific influence on these parts of the text: Reinard Klein Wassink (Philips) on the metal wetting and wave soldering; Hector Steen (Multicore Solders) on the metallurgy of solders; Werner Engelmaier (AT&T Bell) on low cycle fatigue; and Brian Ellis (Protonique) on cleaning of electronic assemblies. I am indebted also to Robert Willis (GEC Technical Directorate) who found time to read the proof manuscript and make many suggestions.

Additionally, it is with pleasure that I acknowledge the contribution of my erstwhile colleague, Ernie Hondros, whose drive and encouragement initiated both my interest in metal wetting and the soldering work at this laboratory. Special thanks are due to my colleagues Martin Seah, whose scientific insight into technical problems is always an invaluable cornerstone of our work, and Doreen Tilbrook, without whose sterling assistance in the organisation of my paperwork, this manuscript would have been much delayed. I would like to thank my family for enabling me to spend much of my time over 18 months in the writing, and

Lorna Cullen of Electrochemical Publications, whose encouragement from the start and attention to detail during the proof stages have been invaluable.

November 1987

Colin Lea
National Physical Laboratory
Teddington, Middlesex

CONTENTS

PREFACE v

CHAPTER ONE
Overview 1

1.1	A New Technology	1
1.1.1	Types of Surface Mounting Components	2
1.1.2	New Substrate Materials	2
1.1.3	The Assembly Process	2
1.2	Advantages of Surface Mounting	4
1.3	Critical Issues in SMT	4
1.4	The Future for SMT	5

CHAPTER TWO
Components 5

2.1	Introduction	6
2.2	Towards Miniaturisation	6
2.2.1	Use of Board Area	6
2.2.2	Move away from DIL Packages	8
2.3	Surface Mounting Semiconductor Packages	9
2.3.1	SOIC Packages	9
2.3.2	SOT Packages	11
2.3.3	Cylindrical Diode Packages	13
2.3.4	Leadless Ceramic Chip Carriers	13
2.3.5	Plastic Leaded Chip Carriers	15
2.3.6	Flatpacks and Quad Packs	17
2.3.7	Land Grid Arrays	18
2.3.8	Modified Through-hole Components	18
2.3.9	Sockets	18
2.4	Bare Chips	20
2.4.1	Wire Bonding	20
2.4.2	Tape Automated Bonding	20
2.4.3	Flip-chip	22
2.5	Comparison of IC Packaging Options	22
2.5.1	Board Area Requirement	22
2.5.2	Conductor Lengths	25
2.6	IC Package Construction and Materials	26
2.6.1	Plastic Packages	27

2.6.2		Ceramic Packages	28
2.6.3		Leadframes	29
2.7		Passive Components	29
2.7.1		Capacitors	30
2.7.1.1		Ceramic Chip Capacitors	30
2.7.1.2		Cylindrical Ceramic Capacitors	33
2.7.1.3		Plastic Capacitors	34
2.7.1.4		Tantalum Chip Capacitors	34
2.7.1.5		Aluminium Electrolytic Capacitors	35
2.7.2		Resistors	36
2.7.3		Other Passive Components	37
2.8		Thermal Characteristics of SMCs	38
2.8.1		Temperature-dependent Effects	39
2.8.2		Thermal Parameters	39
2.8.3		Temperature Coefficient of Expansion	41
2.9		Component Packing	42
2.9.1		Bulk	42
2.9.2		Tape	42
2.9.3		Magazine	44

CHAPTER THREE
Substrates   45

3.1		Introduction	45
3.2		Compatibility of Thermal Expansion	45
3.2.1		Matched TCE	46
3.2.2		Compliant Joint	47
3.2.3		Compliant Substrate	47
3.3.		Thermal Management	47
3.3.1		Heat Dissipation by Conduction	48
3.3.2		Heat Dissipation by Convection	50
3.3.3		Heat Dissipation by Radiation	51
3.3.4		Relative Rôles of Heat Transfer Mechanisms	51
3.4		Interconnection Density	52
3.4.1		Interconnection Requirements	52
3.4.2		Blind, Buried and Thermal Vias	53
3.5		Types of Substrate and Interconnection	55
3.5.1		Organic Substrates	55
3.5.1.1		Phenolic Laminates	56
3.5.1.2		Epoxy Laminates	57
3.5.1.3		Polyimide Laminates	58
3.5.1.4		Other Organic Laminates	59
3.5.1.5		The Copper Foil	59
3.5.1.6		The Glass Cloth	59
3.5.1.7		Comparative Properties of Organic Laminates	60
3.5.1.8		PCB Fabrication with Organic Laminate Substrates	60
3.5.2		Matched TCE Substrates: Ceramics	63
3.5.2.1		Thin Film Circuitry	65
3.5.2.2		Thick Film Circuitry	66
3.5.3		Matched TCE Substrates: Laminates	71
3.5.4		Metal Core Substrates	72
3.5.4.1		Porcelain Enamelled Steel	72
3.5.4.2		Organic Metal-core Substrates	74

3.5.4.3	Multiwire	81
3.5.5	Compliant Substrates	82
3.5.6.	Polymer Thick Film Circuits	83
3.5.7	Comparison of Substrate-interconnection Systems	84

CHAPTER FOUR
Design and Assembly — 88

4.1	Assembly Variations	88
4.1.1	SMCs Single-sided	88
4.1.2	SMCs Double-sided	90
4.1.3	Mixed Assembly, SMCs Single-sided	91
4.1.4	Mixed Assembly, SMCs Double-sided	92
4.2	PCB Layout	93
4.2.1	Factors Affecting PCB Layout	93
4.2.2	Design Constraints from the Placement Operation	94
4.2.3	Design Constraints from the Soldering Operation	94
4.2.4	Design Constraints from the Testing Operation	96
4.2.5	Preferred Component Layouts	97
4.2.6	Conductor Routing	99
4.3	Component Footprints	100
4.3.1	Placement Accuracy	101
4.3.2	Typical Footprints	103
4.3.3	Solder Masks	108
4.3.4	Conductor Fan-outs	109
4.3.5	Thermal Mounting	110
4.4	Computer-aided Design	110
4.5	Component Placement	111
4.5.1	Hand Assembly	111
4.5.2	Sequential and Simultaneous Placement	111
4.5.3	Selection Criteria for Placement Machines	114
4.5.4	Placement Machine Functions	115
4.6	Chip-on-board Assembly	117
4.6.1	Die Bonding	117
4.6.2	Wire Bonding	117
4.6.3	Tape-automated Bonding	119
4.6.4	Flip-chips	120
4.6.5	Beam Leads	120
4.6.6	Encapsulation	120

CHAPTER FIVE
Wave Soldering — 121

5.1	Introduction	121
5.2	Temporary Adhesive Bonding	121
5.2.1	Application of Adhesive	122
5.2.2	Curing the Adhesive	124
5.3	Fluxing the Board	126
5.3.1	Application of the Flux	126
5.3.1.1	Foam Fluxing	127
5.3.1.2	Wave Fluxing	128
5.3.1.3	Spray Fluxing	128
5.3.2	Monitoring Flux Density	128

5.4	Preheating the Board	130
5.5	The Solder Wave	132
5.5.1	Wave Generation	133
5.5.2	The Soldering Process	134
5.5.3	Wave Shape	135
5.5.4	Shadowing	137
5.5.5	Jet Wave	138
5.5.6	Dual Wave	139
5.6	The Solder Alloy	140
5.6.1	The Choice of Alloy	140
5.6.2	Impurities in the Solder	141
5.6.3	Oxidation of the Solder	142
5.7	Oil in the Solder Wave	143
5.7.1	Reduction of Solder Bridges	143
5.7.2	Types of Oil	144
5.7.3	Application of the Oil	144
5.7.4	Colophony Addition	144
5.8	Air Knife	145
5.9	Circuit Board Finish	146
5.10	Wave Soldering Defects	148
5.10.1	Inserted Components	148
5.10.2	Surface Mounted Components	149
5.10.2.1	Avoidance of Bridges by Pattern Design	151

CHAPTER SIX
Solder Pastes — 154

6.1	Introduction	154
6.2	The Solder Powder	154
6.2.1	Solder Particle Shape	154
6.2.2	Solder Particle Size	156
6.2.3	Oxide Content	159
6.2.4	Solids Content	160
6.3.	The Flux in the Paste	162
6.3.1	Fluxes Soluble in Organic Liquids	162
6.3.2	Water Soluble Fluxes	164
6.4	The Solder Alloy	165
6.4.1	Reflow Temperature	166
6.4.2	Physical Properties	168
6.4.3	Metallurgical Systems	168
6.4.3.1	Tin-lead Systems	168
6.4.3.2	Tin-lead-silver Systems	168
6.4.3.3	Tin-silver Systems	169
6.4.3.4	Tin-antimony and Tin-lead-antimony Systems	169
6.4.3.5	Lead-indium and Tin-indium Systems	169
6.4.3.6	Tin-bismuth and Tin-lead-bismuth Systems	169
6.4.4	Metal Purity	170
6.5	Application of the Paste	172
6.5.1	Screen Printing	172
6.5.2	Stencil Application	177
6.5.3	Syringe Dispensing	178
6.5.4	Pin Transfer	179

6.6	Paste Rheology	179
6.6.1	Viscosity	179
6.6.2	Measurement of Paste Viscosity	181
6.6.3	Screenability	183
6.6.4	Slump	183
6.6.5	Tackiness	184
6.7	Drying the Solder Paste	186
6.7.1	Storage after Drying	188
6.8	Handling and Storage of Solder Pastes	189
6.9	Voids in the Solder Fillets	190
6.9.1	Flux Activation	191
6.9.2	Solvent	191
6.9.3	Pre-heating	192
6.10	Solder Balls	192
6.10.1	Solder Balling Test	193
6.11	Solder Paste Design Appraisal	194

CHAPTER SEVEN
Reflow Soldering Using Radiant Heating — 196

7.1	Introduction	196
7.2	Radiation Heat Transfer	197
7.2.1	The Electromagnetic Spectrum	197
7.2.2	Basic Definitions	197
7.2.3	Planck's Law	199
7.2.4	Wien's Law	201
7.2.5	Stefan-Boltzmann Law	203
7.2.6	Kirchoff's Law	205
7.2.7	Lambert's Law—Diffuse Emission	206
7.2.8	Heat Transfer by Diffuse Radiation	206
7.2.9	Heat Transfer Coefficient	208
7.2.10	Penetration of Radiation	209
7.3	Infra-red Energy Sources	210
7.3.1	Tungsten Filament Sources	211
7.3.2	Nichrome Filament Sources	217
7.3.3	Area Emission Sources	217
7.4	Workpiece Characteristics	219
7.4.1	Absorptivity of Materials	219
7.4.2	Surface Roughness and Directional Emissivity	221
7.4.3	Thermal Degradation	221
7.4.4	Geometrical Effects	223
7.5	The Gaseous Environment	224
7.5.1	Air	224
7.5.2	Nitrogen	225
7.5.3	Hydrogen-nitrogen	225
7.6	Infra-red Processing	225
7.6.1	Infra-red Character	226
7.6.2	Furnace Design	226
7.6.3	Estimate of Heating Requirements	228
7.6.4	Heating Cycles	229
7.6.5	Instrumentation and Control	232

CHAPTER EIGHT
Solder Reflow by Fluid Heat Transfer — 234

8.1	Introduction	234
8.2	Heat Transfer from a Fluid	235
8.2.1	Heat Transfer Coefficient	235
8.2.2	Heat Flow in the Workpiece	236
8.2.3	Evaluation of Heat Transfer	238
8.2.4	Condensation Heat Transfer	241
8.2.4.1	Vertical and Inclined Surfaces	242
8.2.4.2	Horizontal Surfaces	244
8.2.4.3	Cooling Tubes	246
8.2.4.4	Workpiece Surface Condition	247
8.3	Properties of the Heat Transfer Fluid	247
8.4	Vapour Phase Soldering	248
8.4.1	The Soldering System	248
8.4.2	The Primary Fluid	251
8.4.2.1	Fluid Chemistry	251
8.4.2.2	Heat Transfer Properties	253
8.4.2.3	Solder Reflow Properties	254
8.4.2.4	Operating Power Requirements	255
8.4.2.5	Rosin Flux Solubility	258
8.4.2.6	Thermal Degradation—Toxicity	260
8.4.2.7	Thermal Degradation—Corrosivity	262
8.4.2.8	Consumption of Fluid	263
8.4.3	The Secondary Fluid	266
8.4.4	Production Vapour Phase Systems	266
8.4.4.1	Heating	268
8.4.4.2	Condensing	269
8.4.4.3	Secondary Injection	269
8.4.4.4	Fluid Filtration	271
8.4.4.5	Cycling Process Control	272
8.5	Liquid Phase Soldering	273
8.5.1	Comparison of Heat Transfer in Liquid and Vapour	275
8.5.2	Practical Tests of Liquid Phase Soldering	276

CHAPTER NINE
Other Attachment Methods — 278

9.1	Introduction	278
9.2	Hot-plate Solder Reflow	278
9.2.1	Heat Conduction Through the Workpiece	279
9.3	Local Conductive Heating	279
9.3.1	Single-lead Soldering	279
9.3.2	Multiple-lead Soldering	281
9.3.3	Collet Soldering	282
9.3.4	Resistance Soldering	283
9.3.5	Local Heat Conduction in a Workpiece	283
9.4	Local Hot Gas Soldering	285
9.5	Laser Soldering	286
9.5.1	The Laser	286
9.5.1.1	The Physical Mechanism	286
9.5.1.2	The Nd:YAG Laser	291
9.5.1.3	The Carbon Dioxide Gas Laser	291

9.5.2	Selecting the Laser for SMT	292
9.5.2.1	Efficiency of Heating	292
9.5.2.2	Power Levels and Pulse Duration	294
9.5.2.3	Damage to the Board	295
9.5.2.4	Fluxless Soldering	296
9.5.2.5	Multiple Beam Laser Soldering	296
9.5.3	The Laser-soldering System	296
9.5.4	Practical Soldering with a Laser	298
9.5.4.1	Application of Solder	298
9.5.4.2	Laser Beam Angle	298
9.5.5	Joint Metallurgy	298
9.5.6	Intelligent Laser Soldering	299
9.5.7	Desoldering and Repair by Laser	301
9.5.8	Summary of Advantages of Laser Soldering	301
9.5.9	Laser Safety	302
9.6	Assembly with Adhesives	303
9.6.1	Conductive Adhesives	303
9.6.2	Application of the Adhesive	305
9.6.3	Adhesive Curing	306
9.6.4	Strength of Conductive Adhesive Joints	306

CHAPTER TEN
Solderability 308

10.1	Introduction	308
10.2	Wettability	308
10.2.1	Speed and Degree of Wetting	308
10.2.2	Surface Tension	309
10.2.3	Thermodynamics of Wetting	311
10.2.4	Liquid Meniscus Shapes	313
10.3	Wetting by Solder	316
10.3.1	Effect of Solder Alloy	316
10.3.2	The Rôle of Surface Composition	317
10.3.3	Effect of Surface Roughness	320
10.3.4	Hysteresis of Wetting	322
10.3.5	Degrees of Wetting	322
10.3.6	The Phenomenon of De-wetting	323
10.3.7	The Need for a Flux	324
10.4	Time-variant Changes in Wettability	326
10.4.1	The Rôle of Solderable Coatings	327
10.4.2	The Ageing Process	328
10.4.3	Intermetallic Compound Growth	329
10.4.3.1	Cooling and Solidification	331
10.4.3.2	Intermetallic Phases	333
10.4.4	Growth of Intermetallic Phases in Contact with Solid Sn or Sn-Pb	333
10.4.4.1	Copper Substrate	334
10.4.4.2	Other Substrates	336
10.4.5	Growth of Intermetallic Phases in Contact with Liquid Sn or Sn-Pb	337
10.4.6	Effect of the Intermetallic Layer on Solderability	337
10.4.7	Dissolution of Terminations in Molten Solder	338
10.4.8	The Oxidation and Corrosion of Solderable Surfaces	338
10.4.9	Accelerated Ageing Treatments	339
10.4.9.1	Relative Humidity	344
10.4.9.2	The Relevance of Accelerated Ageing Tests	345
10.5	The Assessment of Solderability	345
10.5.1	Solder Dip Method	346

xiv   Contents

10.5.2		Area-of-spread Test	348
10.5.3		Meniscus Shape Method	352
10.5.4		The Wetting Balance	353
10.5.4.1		Theoretical Wetting Force	354
10.5.4.2		Interpretation of Wetting Balance Curves	355
10.5.4.3		Wetting Balance for Surface Mounting Components	360
10.5.4.4		Scanning Mode Wetting Balance	361
10.5.5		The Globule Balance	363
10.5.5.1		Globule Size	363
10.5.5.2		Specimen-solder Contact	364
10.5.5.3		Restriction of Solder Rise	365
10.5.5.4		Thermal Response of Globule Block	367
10.5.5.5		Comparison between Globule Balance and Wetting Balance	368
10.5.6		Rotary Dip Method	369
10.6		Surface of Standard Solderability	370
10.7		Movement of Components During Soldering	371
10.7.1		Floating and Swimming of Components	372
10.7.2		Tombstoning of Components	373

CHAPTER ELEVEN
The Solder Fillet   378

11.1		Introduction	378
11.2		Metallurgy of the Solder	378
11.2.1		Tin	378
11.2.2		Tin-lead Alloys	379
11.2.3		Strength Properties of Solder	380
11.2.4		The Phase Diagram	386
11.2.5		Diffusion Reactions	391
11.2.6		Cooling Curves	393
11.2.7		Alloying and Impurity Elements	394
11.3		Fatigue in Solder Joints	396
11.3.1		Prediction of Fatigue Life	396
11.3.2		Origin of Fatigue in Solder Joints	399
11.3.3		Fatigue Mechanisms of Solder	403
11.3.4		Fatigue Life of Leadless Ceramic Chip Carriers	407
11.3.5		Effect of Frequency and Hold Time	413
11.3.6		High Frequency Fatigue	414
11.3.7		Effect of Solder Fillet Geometry	417
11.3.8		Effect of Substrate Material	421
11.3.9		Effect of Solder Alloy	423
11.3.10		Effect of Joint Microstructure	424
11.3.11		Effect of Temperature	426
11.3.12		Effect of Test Conditions	426
11.3.13		Choice of Failure Criterion	427
11.3.14		Validity of Fatigue-life Predictions	428
11.3.15		Fatigue Conditions of Solder Joints in Service	432

CHAPTER TWELVE
Post-assembly Operations   435

12.1		Cleaning	435
12.1.1		To Clean or not to Clean?	435
12.1.2		Effects of Contamination	436
12.1.2.1		Corrosion	436
12.1.2.2		Leakage Currents	437
12.1.2.3		Coating De-bonding	438

12.1.2.4		White Residues	438
12.1.2.5		Insulating Contact Surfaces	438
12.1.3		Solubility	438
12.1.3.1		Dissolution	439
12.1.3.2		Solubilisation	440
12.1.3.3		Flux Solubility	440
12.1.3.4		Metal Abietates	442
12.1.3.5		Solubility Parameter Theory	442
12.1.3.6		Solvent Temperature	444
12.1.3.7		Surface Wetting by Solvents	445
12.1.3.8		Capillary Penetration	446
12.1.3.9		Shear Stress Cleaning	448
12.1.4		Cleaning with Organic Solvents	449
12.1.4.1		Azeotropic Systems	449
12.1.4.2		Chlorinated Solvents	450
12.1.4.3		Fluorinated Solvents	451
12.1.5		Aqueous Cleaning	452
12.1.5.1		Water-soluble Flux Removal	453
12.1.5.2		Aqueous Removal of Rosin Flux	454
12.1.6		Cleaning Techniques	454
12.1.6.1		Liquid Solvent Cleaning	454
12.1.6.2		Vapour Solvent Cleaning	455
12.1.6.3		Spray Solvent Cleaning	456
12.1.6.4		Ultrasonic Agitation	458
12.1.6.5		Water Cleaning	459
12.1.7		Cleaning Considerations Specific to Surface Mounted Assemblies	460
12.1.8		Measurement of Cleanliness	461
12.1.8.1		Techniques for Contamination Assessment	461
12.1.8.2		Levels of Ionic Contaminants	462
12.1.8.3		Levels of Non-ionic Contaminants	464
12.1.8.4		Insulation Resistance Tests	465
12.1.8.5		Permissible Ionic Contamination	466
12.2		Inspection of Soldering Quality	466
12.2.1		Visual Inspection	467
12.2.1.1		Classification of Defects	467
12.2.1.2		Chip Component Solder Fillets	470
12.2.1.3		Leaded Component Solder Fillets	471
12.2.1.4		Leadless Chip Carrier Solder Fillets	472
12.2.2		X-ray Inspection of Solder Joints	472
12.2.3		Laser Inspection of Solder Joints	474
12.2.4		Scanning Acoustic Microscopy	477
12.2.5		Evaluation of Solder Joint Inspection Methods	478
12.3		Post-assembly Testing	480
12.3.1		A Testing Strategy	480
12.3.2		Test Efficiency	481
12.3.3		In-circuit Testing	481
12.4		Rework and Repair	483
12.4.1		Reworking Solder Fillets	483
12.4.2		Replacing Components	484
12.4.2.1		Heated Collet	484
12.4.2.2		Hot Gas	485
12.5		Protective Coatings	485
12.5.1		Protective Concepts	486
12.5.1.1		Conformal Coatings	486
12.5.1.2		Non-conformal Surface Coatings	486
12.5.1.3		Potted Coatings	486
12.5.2		Methods of Application	487
12.5.3		Types of Coatings	487

CHAPTER THIRTEEN
Quality and Reliability 489

13.1	Introduction	489
13.2	Reliability Behaviour	489
13.3	Reliability Functions	490
13.3.1	Random Failures	491
13.3.2	Wear-out Failures	492
13.4	Accelerated Assessment of Reliability	496
13.4.1	Ageing Mechanisms	497
13.4.2	Thermal Acceleration	497
13.4.3	Temperature-sensitive Parameters	499
13.4.4	Electrical Acceleration	500
13.4.5	Damp Heat Acceleration	500
13.5	Practical Reliability	502
13.5.1	Component Reliability	503
13.5.2	Assembly Reliability	504
13.5.3	Zero-hour Quality	504

CHAPTER FOURTEEN
Economics and Trends 506

14.1	SMT Growth	506
14.1.1	Growth of Infrastructure	506
14.1.2	Growth of Technology	507
14.2	Economics of SMT	508
14.2.1	Manufacturing Costs	509
14.2.2	Assembled System Costs	510
14.2.3	Increased Sales Potential	511
14.2.4	The Economic Decision	511
14.2.5	Introducing SMT	512
14.3	Trends	513
14.3.1	Surface Mounting Components	513
14.3.2	Surface Mounting Technology in Europe	514
14.3.3	Surface Mounting Technology in Japan	515
14.3.4	Surface Mounting Technology in USA	516
14.4	The Challenge of Surface Mounting Technology	518

APPENDIX 519

REFERENCES 535

AUTHOR INDEX 554

SUBJECT INDEX 559

Chapter 1

OVERVIEW

1.1 A NEW TECHNOLOGY

Electronic production worldwide is undergoing a revolutionary change in both components and manufacturing techniques.[1] Surface mounted assembly is a totally new kind of automated electronic assembly technology that uses a totally different kind of device package. The new device is connected mechanically and electrically to the surface of the interconnecting printed circuit board as opposed to conventional insertion assembly in which the component leads are inserted into holes in the board. In most cases, surface mounting can potentially result in smaller, less expensive electronic assemblies with an improved performance.

The conventional insertion methods of assembling electronic components have virtually reached their limits as far as improvements in cost, weight, volume and reliability are concerned. This situation is so because of the need to drill a hole in the substrate board for every component lead, because of the universal 2·54 mm (0·1 inch) interconnection grid in use, and because of the entrenched position of the ubiquitous plastic dual-in-line package for almost all integrated circuits. It required a sweeping change of thought to make the move to mounting components on boards rather than through them because, with this seemingly trivial change have also come changes in component packaging shapes and sizes, changes to fully automated component handling, new component attachment methods using screen printed solder pastes or conductive adhesives, innovative changes in board design algorithms and changes in the requirements of post-assembly cleaning, testing and inspection.

The three primary goals of the majority of electronic assemblies are reduced size, reduced cost and increased reliability. Surface mounting technology (SMT) helps meet all these goals. It provides new components and new techniques for their assembly that can result in size reductions of over 40% and assembly cost reductions of up to 50%. The performance of a surface mounted assembly is superior to that of a conventional assembly particularly at high frequencies, and the reliability is at least as good. The technology is applicable to all equipment, be it military, high reliability, industrial, commercial or consumer.

In addition to the advantages gained by the electronic assemblies through the implementation of surface mounting, there are benefits to be accrued in the shipping and warehousing of components and assemblies, and in the requirements of manufacturing space and equipment.

1.1.1 Types of Surface Mounting Components

Electronic components have been surface mounted on to ceramic substrates for many years by the hybrid microelectronics industry, using printed film circuits. The feature of surface mounting technology that differentiates it is the use of conventional printed circuit boards, single-sided, double-sided or multilayer, and the consequent available substrate size increase. The investment of capital, time and expertise in SMT has enabled many component types to be developed in addition to those already used by the hybrid industry.

Resistors and some capacitors in hybrid assemblies form part of the thick-film screen printed circuit, but for surface mounted assemblies small leadless components are used. Discrete transistors for surface mounting use packages developed for hybrid microelectronics. Integrated circuits for surface mounting are available in a variety of packages, some inherited from the hybrid industry, some developed in the USA for high performance military applications and some developed in Europe and Japan for lower cost consumer applications. In all cases these devices offer lower noise and improved frequency response resulting from shorter path lengths.[2] All the surface mounting components have an increased suitability for automated handling, testing and assembly. They can be delivered to the pick-up and placement head of an assembly machine in standardised tape or cartridge formats.

1.1.2 New Substrate Materials

The essence of SMT is that electronic components are mounted on standard printed circuit boards, normally manufactured for high-technology applications from copper-clad fibreglass-epoxy laminate. Notwithstanding, new substrates have been developed specifically to meet requirements arising from the implementation of the new technology.

Large leadless components soldered directly on to a normal PCB do not have the necessary compliance, available with a lead, to absorb strains and avoid damage and failure, during temperature changes. For this reason substrates have been developed that are themselves flexible or have a temperature coefficient of expansion that is closer to that of the components.[3]

A second need for new substrate types is to enable adequate dissipation of heat from the circuit. Surface mounting enables higher component densities to be achieved, which leads to greater heat dissipation problems. For surface mounted assemblies a PCB with a metal core is frequently used, to which components can be heat-sinked directly.

Interconnection boards used for surface mounting are not limited to drillable materials since it is not necessary to provide holes through the board. For smaller interconnection boards, more expensive materials such as ceramic and porcelainised steel can be used. The use of such materials with the new SMT components is blurring the historical division between the hybrid microelectronics industry and the surface mounting manufacturing industry.

1.1.3 The Assembly Process

Surface mounting components can be joined to the PCB using a conducting adhesive, but the vast majority of assemblies are soldered. There are two basically different approaches to soldering, wave soldering and reflow soldering.[4] For wave soldering each component is glued to the board using a tiny spot of adhesive which is then cured. The board is then turned over and, with the components on the underside, fluxed and wave soldered. The components pass wholesale through the solder. The reflow soldering route involves the use of a solder paste which is a mixture of tiny solder particles in a binder containing flux. The paste is screen printed on to the unpopulated PCB and the components are placed. The tackiness of the paste restrains them from moving while the whole is heated to a temperature typically 30-50°C above the melting point of the solder alloy in the paste. There are several ways of imparting the necessary heat, such as a hot plate, infra-red radiation, infra-red laser and immersion in a hot fluid. The choice depends on the type of substrate and components, the design of the board and the production run.

It is often necessary or judicious to use both insertion and surface mounting components on one board. In such situations reflow of a solder paste is inappropriate and wave soldering must be used. These assemblies are referred to as mixed technology. Figure 1.1 gives an indication of the penetration of surface mounting and mixed technology into the electronic assembly market.[5]

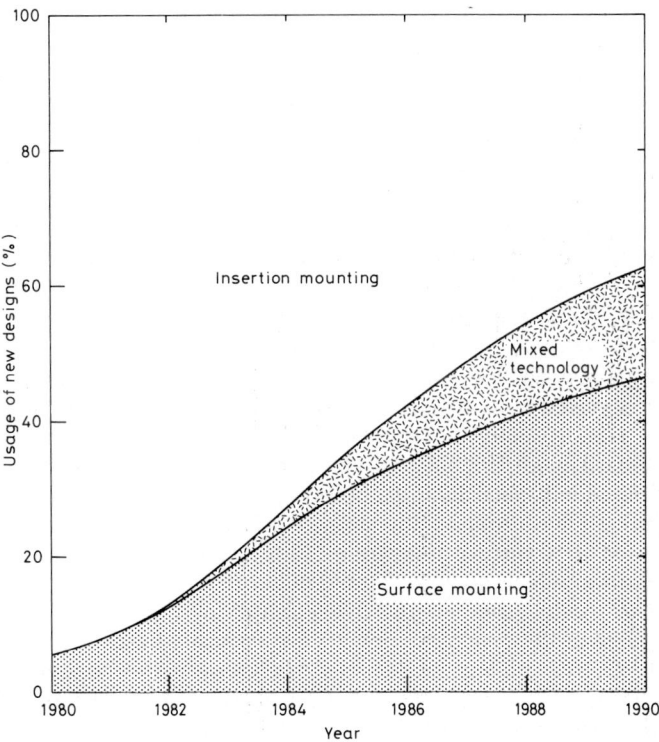

Fig. 1.1 The penetration of surface mounting technology into the electronics assembly industry over the decade 1980–1990.[5]

1.2 ADVANTAGES OF SURFACE MOUNTING

The most apparent advantage of surface mounting technology is space savings, either reducing the size and weight of a particular electronic assembly or increasing the number of functions within a fixed size. Even with the smaller surface mounting components, it is not always possible to reduce the size of the circuit board because the conductors require a certain board area and multilayering is not always practical. Generally, however, surface mounting does result in smaller and fewer boards to complete a system. Surface mounting almost always reduces the weight of the complete circuit board.

Potentially, surface mounting can reduce the cost of manufacturing a system. Fewer printed circuit boards are used, resulting in fewer interconnections and connectors, lower testing costs, smaller enclosures with less cooling apparatus, and low shipping and storage costs.

The more compact surface-mounted printed circuit board has the capability of performing better at high speed because the interconnection paths between the active IC chips are shorter and, on average, have less inductance and capacitance. Also, the impedance of short conductors can be more easily controlled. In most cases, reducing the number of printed circuit boards and connectors increases the reliability of the system.

1.3 CRITICAL ISSUES IN SMT

No manufacturing technology can be implemented without progression up a learning curve. Surface mounting is not yet a widely used technology and therefore is not as easily implemented as insertion mounting. The technological and management infrastructure is not fully developed: the pool of expertise on which to draw is limited. Additionally, fewer components are, at present, offered in surface mounting format than in a conventional format intended for insertion mounting. Even if components and equipment exist for surface mounting, the selection is limited and may not exactly fit the user's needs.

Because of the relative newness of the technology, PCBs manufactured for surface mounting may cost considerably more than conventional boards. Because components are packed more densely, interconnection tracks must also be packed more densely. The difficulty of reliable manufacture as well as the testing and inspection of boards with fine lines accounts for the cost premium. In addition, the costs associated with redesigning a board for surface mounting may outweigh the savings realised by adopting the technology. Some components cannot be produced or used in a surface mounting form; these have to be attached in a secondary operation which adds to the overall cost.

There are also special cases where surface mounting restricts design flexibility or reliability. Most of the design restrictions are associated with the routing of the conductors and the very high track densities often required. A common feature in conventional boards is to run tracks between DIL terminations and under the package. This is not always possible with the four sided, narrow-pitch chip carriers.

Technically, there remain some issues that are still being resolved. Notable among these issues is the sparsity of long-term reliability data, and the contentious problems of post-assembly cleaning and testing.[6]

1.4 THE FUTURE FOR SMT

The technology and the infrastructure associated with surface mounting are still evolving and eventually many of the existing disadvantageous issues will be eliminated. As more components specifically designed for surface mounting become available, their cost will fall. These components and the automated manufacturing equipment will become more multi-sourced. The pool of manufacturing expertise will grow as more companies get their surface mounting facilities in place, and as the distinction between surface mounting on PCBs and hybrid circuits gradually disappears.

Designers of surface mounted circuits will become increasingly aware of the advantages of the technology and will learn to design around its shortcomings. New VLSI packages are continually entering the market and the associated new designs will result in a very rapid growth of SMT and a gradual ousting of insertion mounting.

The standard FR-4 epoxy-fibreglass laminate will probably retain its supremacy for high-technology assemblies, but with high quality ultra-thin copper to aid in the manufacture of PCBs with finer lines. New board materials will continue to impact the market for special applications. Any existing cost premium of surface mounting components and PCBs will gradually disappear with the increase in demand.

Surface mounting technology has been heralded by some of its advocates as a revolutionary step in electronics, comparable to the introduction of the transistor and the integrated circuit. Others see SMT in rather less dramatic light. What is certain is that SMT is another significant step forward in the development of the electronics industry, bringing together the three distinct industry trends of reduction in size, high reliability and automated production.

Chapter 2

COMPONENTS

2.1 INTRODUCTION

The predominant packaging requirement for solid state devices originally arose when transistors were incorporated into high reliability military and telecommunications applications. This was the need for hermeticity to prevent junction leakage and degradation of transistor gain, caused by moisture and contamination. The need for high reliability led to silicon planar technology which protected the transistor junctions from contaminants. The integrated circuit (IC) was the natural development of this technology and this led to the requirement for an ever increasing number of lead-outs from the component package. A proliferation of multi-leaded packages arose in the 1960s, but the lack of standardisation and the difficulty of assembly soon left only two major contenders, the dual-in-line (DIL) package and the flatpack. Originally, these both had leads at 0·1 inch pitch along two opposite edges. The flatpack was smaller, but its leads, which were nominally planar with the package, tended to spring away from the circuit board during soldering, such that special tools became a requirement and repair was difficult. The leads of the DIL were inserted into plated-through-holes in the circuit board and were therefore self-retaining during any pre-soldering handling. The development of automated insertion equipment made possible fast and reliable assembly of PCBs and established the DIL as the accepted IC package.

2.2 TOWARDS MINIATURISATION

2.2.1 Use of Board Area

The most cost-effective means of placing more functions on a circuit board is to increase the complexity of the individual devices. The average price of an integrated circuit has fallen year by year, whilst the cost per area of the PCB is increasing, and there is therefore a strong economic case for not only increasing the complexity of devices but also increasing their density on the PCB. The entrenchment of the DIL and the plated-through-hole PCB has greatly curtailed this possibility. This is reflected by the fact that the cost per soldered interconnection has remained virtually static for 15 years compared with the expected 20% cost reduction per doubling of manufactured volume when using established technologies.

Enormous advances in IC technology to produce very large scale integration (VLSI) have eliminated, in some cases, the need for external connections, but in general the complexity and versatility of ICs have led to a steady introduction, and seemingly a continuing requirement, for higher lead-out packages.[7] Eventually, as levels of integration continue to increase and circuits become faster, allowing a greater degree of multiplexing on device leads, the demand for ever higher lead-counts will ease. The upward trend is set to continue for some while, however, with packages now available having in excess of 200 leadouts.

Figure 2.1 illustrates the trend in IC package lead count as a function of circuit complexity, exemplified by memories, microprocessors and gate arrays. Figure 2.2 illustrates the trend as a function of time. This latter trend is a function of both the practicalities of manufacture and the economics of manufacture and use. In Figure 2.2 the lower boundary of the shaded area is the lead count limit of packages commonly available, whilst the upper boundary line represents the limit for more specialised devices. It is clear that lead-count requirements for existing

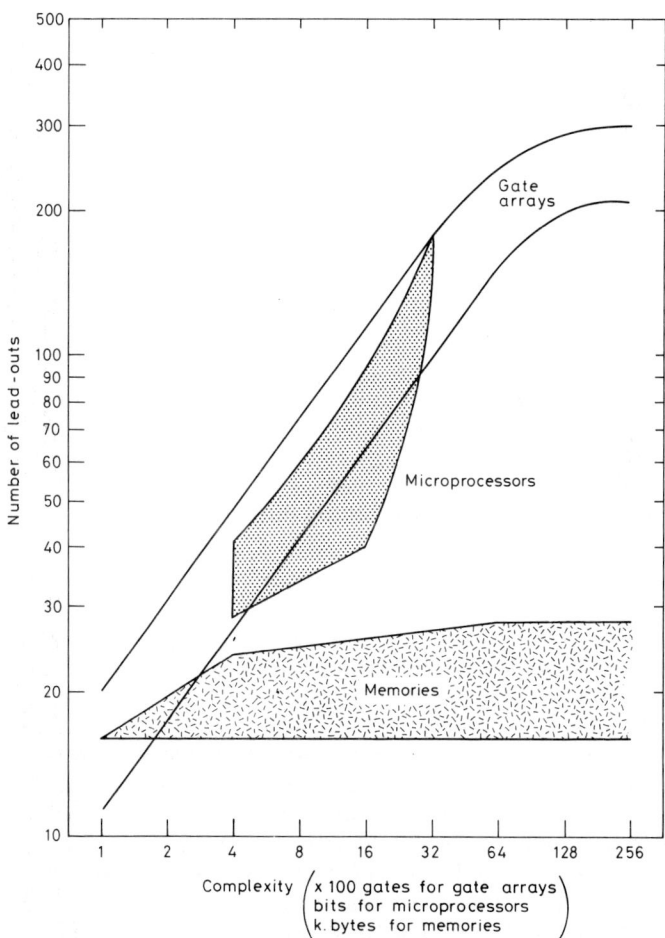

Fig. 2.1 As the complexity of active components increases so does the number of lead-outs from the package. This is illustrated for gate arrays, microprocessors and memories.[7]

products in common usage, such as 32-bit microprocessors and large gate arrays, cannot be met satisfactorily by the 2·54 mm (0·1 inch) lead pitch DIL packages.

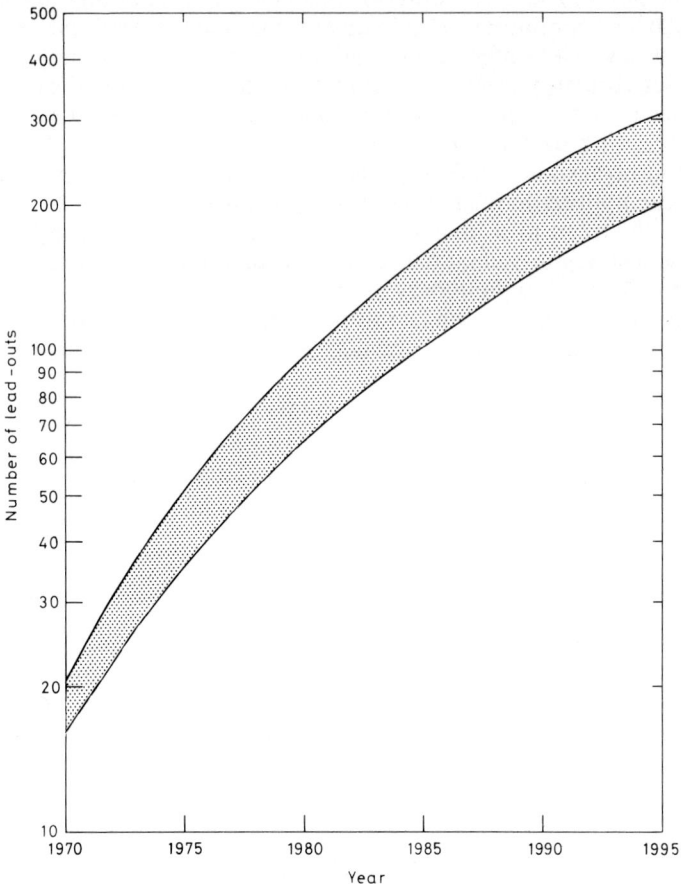

Fig. 2.2 There is a trend of increasing lead-outs from component packages year by year.[7]

2.2.2 Move Away from Dual-in-line Packages

In the mid 1980s the plastic dual-in-line package (DIL) accounted for 80% of all integrated circuits used in the electronics assembly industry. It had maintained itself in this overwhelming position for nearly 20 years. However, the DIL package has a number of disadvantages which began to become apparent as the pressure increased, and continues to increase, for higher lead-count devices.

For high lead-count devices the DIL package starts to become relatively expensive because of excessive size and material use to allow the pins to be brought out in a double row at a 2·54 mm (0·1 inch) pitch. Additionally, this format results in long internal lead lengths for the pins towards the ends of the package with consequent higher resistances and inter-lead capacitances, limiting the performance of the device. The package occupies more board area than is necessary, since the technology of manufacturing printed circuit boards has advanced so that track widths and separations can be reduced to accommodate much smaller lead pitches. The DIL, in its larger sizes, becomes progressively difficult to handle robotically and to insert automatically into the plated-through-holes of the PCB. In addition to all these problems, it is the mechanical properties of the package which have physically limited the size to which the DIL package can grow. For the plastic DIL format the maximum number of lead-outs is usually 68, limited by the danger of the moulded body warping. For the ceramic DIL type the limit is usually 48 above which the relatively brittle glass seals start to give problems with hermeticity.

Figure 2.3 shows a typical design plan of a 64 pin DIL and demonstrates how inefficient is the use of board area, using this device format.

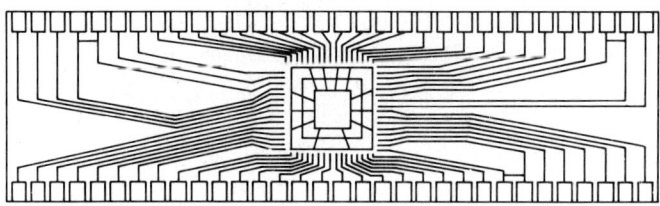

Fig. 2.3 Schematic of a typical lead-out pattern of a 64-pin DIL package, demonstrating the inefficient use of board area for the size of the active chip.

2.3 SURFACE MOUNTING SEMICONDUCTOR PACKAGES

Electronic components whose leads are positioned and soldered on a printed circuit board rather than in holes through it are termed surface mounting components (SMC) or (with less regard for grammar) surface mount components. After they have been soldered to the substrate they are referred to as surface mounted components. The abbreviation SMD for surface mounting device covers all types of components and is a registered service mark of the North American Phillips Corporation.

As the technology of surface mounting has developed, a range of packaging types has emerged, some by transposition from the hybrid microelectronics industry and some by development in their own right.[8]

2.3.1 SOIC Packages

The small outline integrated circuit (SOIC, or simply SO) and the small outline transistor (SOT) packages have a longer history of use than other surface mounting devices.[9] The SO package was developed in Europe in the mid-1970s particularly for the emerging electronic watch market.

The SOIC is a plastic package, currently available in 6, 8, 10, 14, and 16 pin versions with a body width of 4 mm, and in 16, 20, 24 and 28 pin versions with a wider body of 7·6 mm. The flattened leads are on standard 1·27 mm (0·05 inch) centres and are formed outwards in an inverted 'gull wing' fashion, so that the tips of the leads lie in contact with the PCB, as shown in Figure 2.4. The shapes of the SOIC range are shown in Figure 2.5 and the maximum dimensions given in Table 2.1. They may vary very slightly from one manufacturer to another except for the lead pitch which must be 1·27 mm. Beside the nine SO package sizes there are also two Very Small Outline (VSO) encapsulations with 40 and 56 leads at a pitch of 0·762 mm (0·03 inch). Particulars of these are also given in Table 2.1.

The low lead-count SOICs require less than half the area of the DIL equivalents and weigh only one tenth as much.

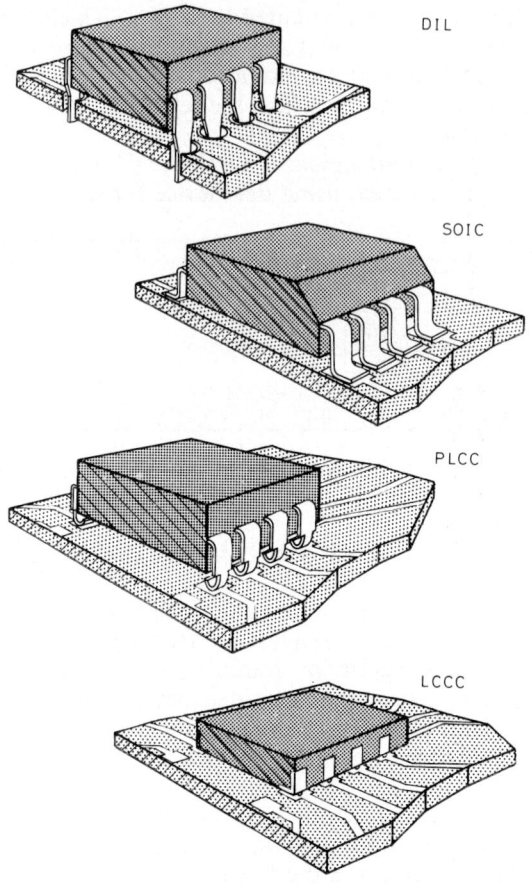

Fig. 2.4 Schematic illustrations of the various types of terminations encountered in semiconductor packages: the dual-in-line, the small outline integrated circuit, the plastic leaded chip carrier and the leadless ceramic chip carrier.

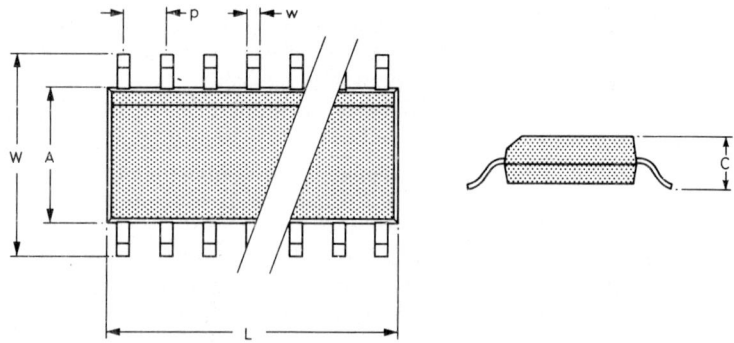

Fig. 2.5 Design of the SOIC range of packages. Dimensions are given in Table 2.1.

Table 2.1
Dimensions of SOIC Packages (mm)

Type	Leads	Lead Pitch [p]	Lead Width [w]	Maximum Length [L]	Maximum Width of Body [A]	Maximum Width of Device [W]	Typical Height [C]
SO-6	6	1·27	0·4	3·75	4·0	6·2	1·6
SO-8	8	1·27	0·4	5·00	4·0	6·2	1·6
SO-10	10	1·27	0·4	6·25	4·0	6·2	1·6
SO-14	14	1·27	0·4	8·75	4·0	6·2	1·6
SO-16	16	1·27	0·4	10·00	4·0	6·2	1·6
SO-16L	16	1·27	0·5	10·50	7·6	10·65	2·6
SO-20	20	1·27	0·5	13·00	7·6	10·65	2·6
SO-24	24	1·27	0·5	15·60	7·6	10·65	2·6
SO-28	28	1·27	0·5	18·10	7·6	10·65	2·6
VSO-40	40	0·762	0·35	16·00	7·6	12·80	2·6
VSO-56	56	0·762	0·35	22·00	11·1	15·80	2·6

The SOIC package is more useful than the chip carrier (see Section 2.3.5) for low pin-count integrated circuits, since the chip carrier, with pins on all four sides, has a practical lower limit of 16 pins. Additionally, having a dual-in-line configuration, the SOIC package is often easier to design into a high-density layout and if the leads are run parallel to the conveyor direction on a wave soldering machine, there are fewer problems with shadowing of the wave from the solderable surfaces than is the case with chip carriers, as discussed in Chapter 5. The SOIC package however can be successfully mixed with chip carriers and, on a mixed-technology board of insertion and onsertion, with DIL packages.

2.3.2 SOT Packages

Discrete semiconductors are available in standard small outline (SO), leaded packages for surface mounting. Originally designed for use in hybrid microelectronics assembly, the Small Outline Transistor (SOT) packages are used for discrete transistors and diodes.[10] The most common packages are the SOT-23 and the SOT-89 (now renamed in the JEDEC standard as TO-236 and TO-243 respectively). The construction of a typical SOT-23 is shown in Figure 2.6 with the standard dimensions of the package. As shown, the package has three leads, two along one edge and a third in the centre of the opposite edge. The leads are formed in a 'gull wing' fashion akin to those on the SOIC. The power handling capabilities of the SOT-23 make it suitable for most general purpose and RF transistors, typically used in audio and VHF amplifiers. The SOT-23 is also used for diodes and in practice is suitable for almost any semiconductor having a chip size of less than about 0·75 mm square.

Semiconductors on larger chips, up to about 1·5 mm square, are packaged in the SOT-89 format, shown schematically in Figure 2.7. Its three leads are all along the same edge of the package but the centre one extends across the bottom to improve the thermal conductivity.

As a guide, the SOT-23 can dissipate up to some 200 milliwatts in free air at 25°C whereas the SOT-89 can handle up to 500 milliwatts under the same conditions. New techniques of manufacture in combination with suitable

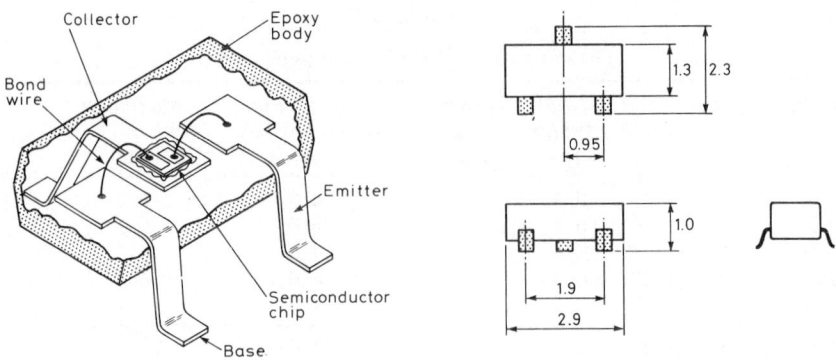

Fig. 2.6 Design of the SOT-23 package.

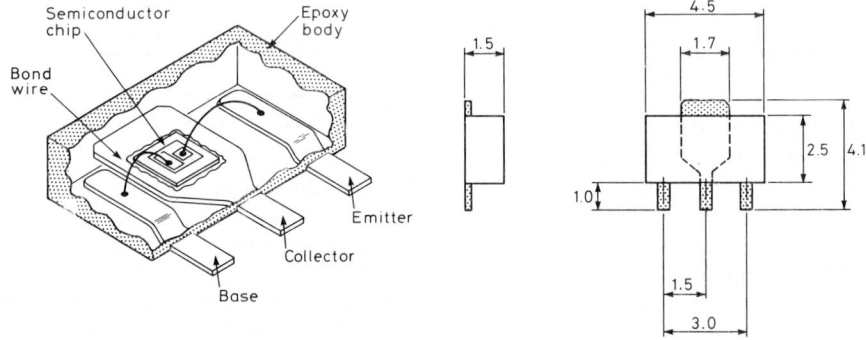

Fig. 2.7 Design of the SOT-89 package.

substrates are enabling these figures to be increased to 350 milliwatts and 1 watt respectively.

A power surface-mounting package, the SOT-194, has been developed to allow a derated power dissipation of up to 4 watts when used with a suitable heat dissipating interconnection substrate.[11] The dimensions of the SOT-194 are given in Figure 2.8. It can be used to encapsulate all standard discrete semiconductor die sizes up to 3 mm.

A four-lead SOT package, the SOT-143, is used for dual gate devices, whose outline drawing is shown in Figure 2.9.

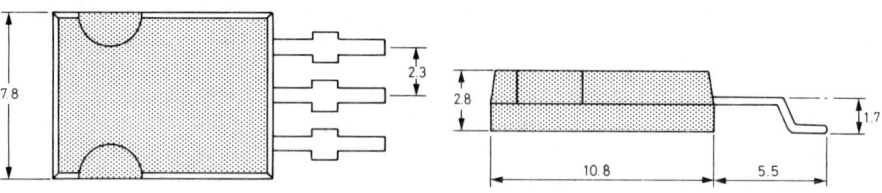

Fig. 2.8 Design of the high-power SOT-194 package.

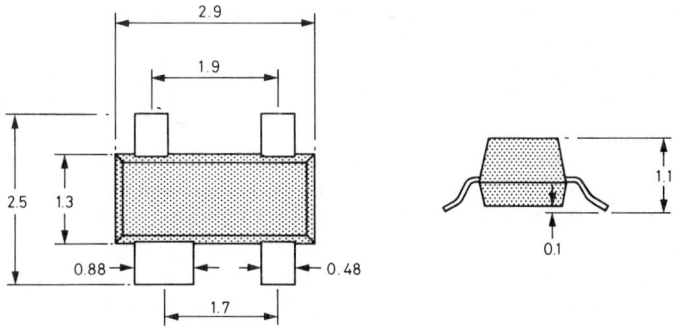

Fig. 2.9 Design of the four leaded SOT-143 package.

2.3.3 Cylindrical Diode Packages

Discrete diode components are frequently packaged in SOT-23 encapsulations, but one of the three contacts is then redundant. There are several two-terminal packages developed especially for diodes, the two most popular both being cylindrical, the SOD-80 and the MELF diode.

The small outline diode (SOD-80) encapsulation is less costly than the SOT-23, requires less board space and weighs less. Typical dimensions of the SOD-80 are given in Figure 2.10. It is generally of a hermetically sealed glass-to-metal structure. It is specifically designed for small diode chips and is usually limited to a power dissipation of 250 mW. When more power handling capability is required by the circuit design, a larger cylindrical encapsulation is used. It is often referred to as a 'MELF' diode because its dimensions, shown in Figure 2.10, are similar to the MELF (metal electrode face bonded) capacitors, described in Section 2.7.1.2. With adequate cooling, power dissipation can be as high as 2 watts. Note that the SOD-80 is sometimes referred to as the MiniMELF.

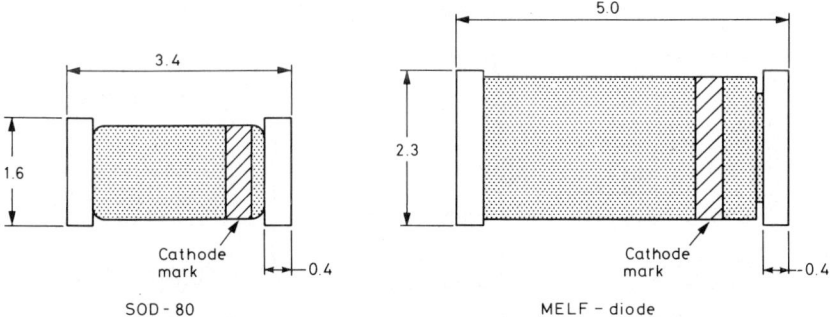

Fig. 2.10 Dimensions of the two common types of cylindrical diode packages.

2.3.4 Leadless Ceramic Chip Carriers

The term chip carrier refers to a range of IC packages that are square (or a nearly square rectangle) with their terminations brought out on all four sides.

The leadless ceramic chip carrier (LCCC) can be envisaged as the useful active centre of a hermetic DIL, with all the leads and excess packaging material discarded. The leadless ceramic package is suitable for direct attachment by soldering or for attachment by sockets with added leads. Since it is constructed of the same materials and in the same manner as hermetic DILs it is at least as reliable.

Leadless ceramic chip carriers are constructed in a variety of ways that are dictated by the end product use and the cost of manufacture.[12] The principle of the construction is that the IC chip is bonded to a ceramic base and connections are made with fine wires to metallisations that are brought out to solderable contact pads as shown in Figure 2.11. For the most demanding applications, where cost is not a major constraint, a three-layer construction is used with a flat gold plated lid sealed using a gold-tin solder preform. Glass sealing, using a pre-glassed ceramic lid with a three layer chip carrier, results in some cost saving. The use of a single layer chip carrier having a pre-glassed cavity and a ceramic cup-shaped lid gives a device of about half the price. Even further economies are obtained, if there is no need for hermeticity, by encapsulating the device on its ceramic base, in epoxy resin.

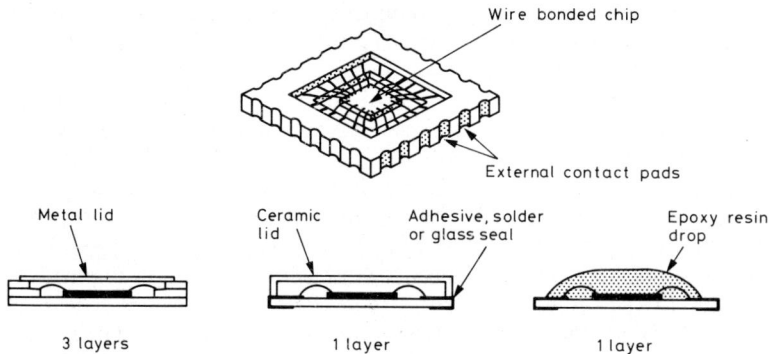

Fig. 2.11 Showing the construction of the leadless ceramic chip carrier (LCCC), with the lid removed (at top) and with three types of enclosure.

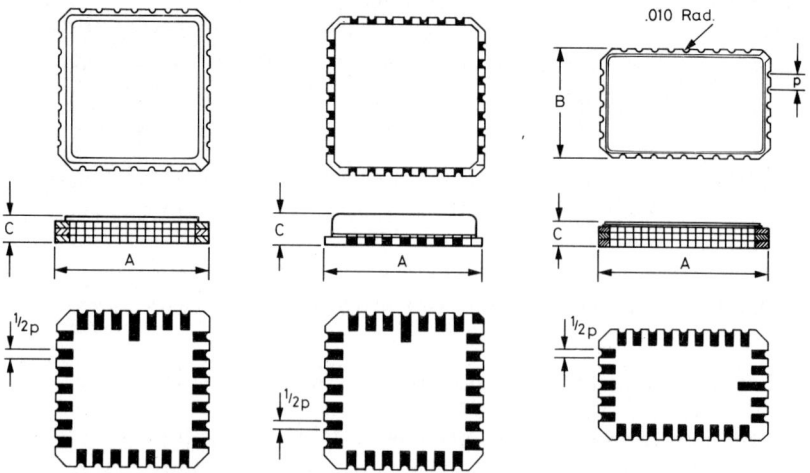

Fig. 2.12 The definition of dimensions of leadless ceramic chip carriers, as given in Table 2.2.

Since chip carriers require no holes to be drilled through the substrate board, the pitch of the contact pads can be less than the 2·54 mm (0·10 inch) of the DIL. In 1980 the square outline chip carriers received approval and are now standardised on a 1·27 mm (0·05 inch) pitch. Rectangular standard versions are now also available, examples of which are given in Figure 2.12.

Leadless ceramic chip carriers are commonly available in 18, 20, 28, 32, 44, 52, 68, 84, 100, 124 and 156 terminal versions. The dimensions and formats of the smaller sizes are given in Table 2.2, with reference to Figure 2.12. The component height depends on the construction of the package, but is typically 1·5–2·0 mm. In all cases the pitch between terminations is 1·27 mm and the shape and sizes of the solderable castellations are very similar from one supplier to another.

Table 2.2

Dimensions of LCCC Packages (mm)

Leads	Format	Pad Pitch [p]	Maximum Dimensions [A × B]
20	5 × 5	1·27	9·1
28	7 × 7	1·27	11·6
44	11 × 11	1·27	16·8
52	13 × 13	1·27	19·3
68	17 × 17	1·27	24·4
84	21 × 21	1·27	29·6
18	5 × 4	1·27	10·9 × 7·5
28	9 × 5	1·27	14·1 × 9·0
32	9 × 7	1·27	14·1 × 11·6

Compared with the DIL package the signal paths of chip carriers are shorter and of more uniform length, so that resistances and inductances are lower. Because the ceramic chip carrier is flat, thin and rugged, without leads, it is very amenable to automatic robotic assembly. However, the intrinsic cost of this type of sealed package has led to the development and the use of a low cost ceramic or plastic base for the IC, protected with a plastic encapsulation, and the direct equivalent of the plastic DIL, namely a post-moulded plastic chip carrier.

2.3.5 Plastic Leaded Chip Carriers

There are now available a wide range of low cost, moulded plastic leaded chip carriers (PLCCs) which have gained approval and hence are now accepted packages.[13] PLCCs are manufactured by fully automated high volume processes. They have leads which are intended to be sufficiently compliant to accommodate any thermal expansion mismatch between the component and the PCB. Whilst some PLCCs are produced with the 'gull-wing' profile for the leads, the majority of manufacturers have chosen 'J' leads that are folded underneath the package as shown in Figure 2.13. This facilitates robotic handling during automatic assembly of circuit boards.

The leads of all PLCCs are on 1·27 mm (0·05 inch) pitch and the packages are available in the same sizes and formats as the leadless ceramic chip carriers. Typical dimensions of the more popular sizes of PLCCs are shown in Table 2.3.

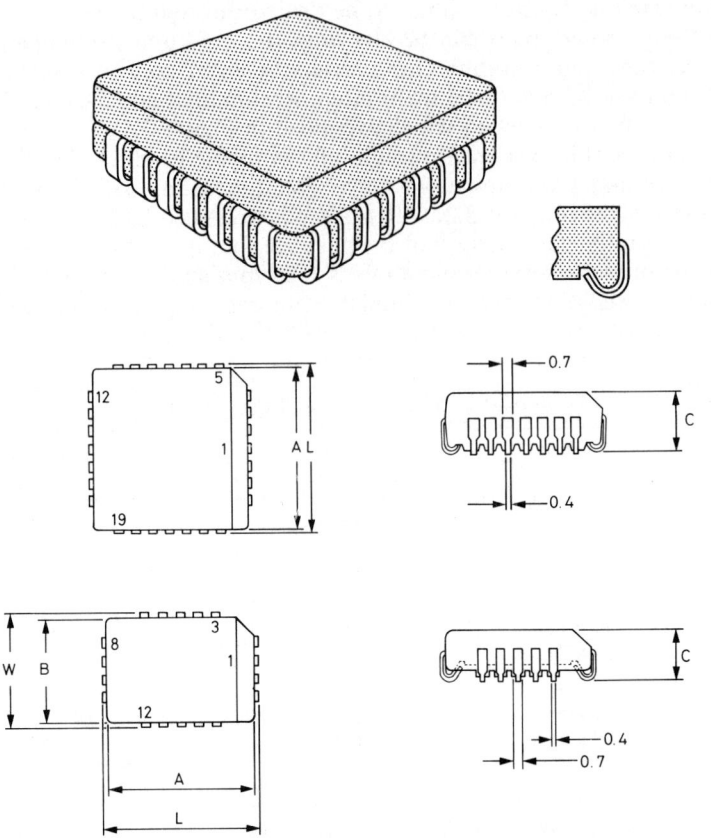

Fig. 2.13 The definition of dimensions of J-leaded PLCCs, as given in Table 2.3.

Table 2.3

Dimensions of PLCC Packages (mm)

Leads	Format	Lead Pitch [p]	Maximum Body Dimensions [A × B]	Maximum Device Dimensions [L × W]	Typical Height [C]
20	5×5	1·27	9·1	10·1	3·5–4·7
28	7×7	1·27	11·6	12·6	3·5–4·7
44	11×11	1·27	16·8	17·8	3·5–4·7
52	13×13	1·27	19·3	20·3	3·5–4·7
68	17×17	1·27	24·4	25·4	3·5–4·7
18	5×4	1·27	10·9× 7·5	11·9× 8·5	3·5–4·7
28	9×5	1·27	14·1× 9·0	15·1×10·0	3·5–4·7
32	9×7	1·27	14·1×11·6	15·1×12·6	3·5–4·7

Because the 'J' leads are tucked under the component, the footprint and the required board area are smaller than the 'gull wing' leads used for SO and flat pack packages. The absence of protruding leads and solder joints does, however, present difficulties for the inspection and testing of circuits.

The leads of PLCCs are commonly of a copper alloy with a plated or hot-dipped tin-lead coating applied to help ensure good solderability.

2.3.6 Flatpacks and Quad Packs

As mentioned in the introduction to this chapter, the flatpack concept with its lead frame co-planar to the body of the package is as old as the DIL package. The original flatpack had leads emerging on two sides of the body, but in recent years not only have versions appeared with finer lead pitches but also with leads on all four sides, namely the quad pack. These are used extensively in Japan, but, because of lack of standardisation, their use elsewhere has been limited.

The quad pack is a high lead-count plastic package that is square or rectangular. At present, devices in the range 40-100 leads are readily available, with developments under way to increase this to 200. The leads are formed to a 'gull-wing' mode to bring the ends of the leads level with the bottom of the package as shown in Figure 2.14. The package size remains the same and so the pitch of the leads varies with the lead-count, being, for example, 1·0 mm on packages up to 64 leads, 0·8 mm on the 80-lead version and 0·65 mm on the 100 pin package, as shown in the outline drawings in Figure 2.14, and given in Table 2.4. The new 200 pin package will use a 0·55 mm pitch.

The examples given in Table 2.4 all have a body size of 20 mm × 14 mm. New proposals have been made to standardise quad flatpacks over a range of body sizes, based on increments of a factor $\sqrt{2}$. Thus the range of body sizes in millimetres will progress 14 × 14, 20 × 14, 20 × 20, 28 × 20, 28 × 28, and so on.

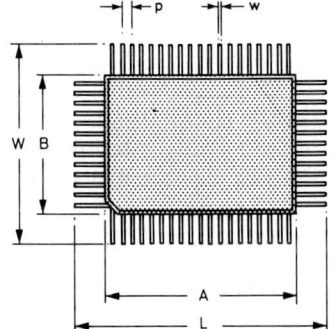
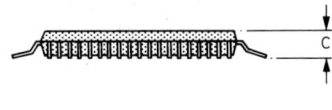

Fig. 2.14 Dimensions of a flatpack, as given in Table 2.4.

Table 2.4

Dimensions of Quad Flatpacks (mm)

Leads	Format	Lead Pitch [p]	Lead Width [w]	Body Dimensions [A × B]	Device Dimensions [L × W]	Height [C]
64	19 × 13	1·0	0·35	20 × 14	25·6 × 19·6	2·9
80	24 × 16	0·8	0·35	20 × 14	25·6 × 19·6	2·9
100	30 × 20	0·65	0·30	20 × 14	25·6 × 19·6	2·9

2.3.7 Land Grid Arrays

In an attempt to get many more pin-outs on insertion-mounting device packages, the pin grid array was developed in which pins emanate from an array on the underside of the package rather than just around its periphery[14]. The surface mounting version of the leaded pin grid array is the land grid array (LGA), whereby the pins are substituted by an array of solderable pads on the base. LGAs are available in various sizes, pad sizes and pad densities to cope with different lead-out arrangements[15]. Two common configurations are shown in Figure 2.15. Sockets are available for testing LGAs or can be used to convert to a leaded device.

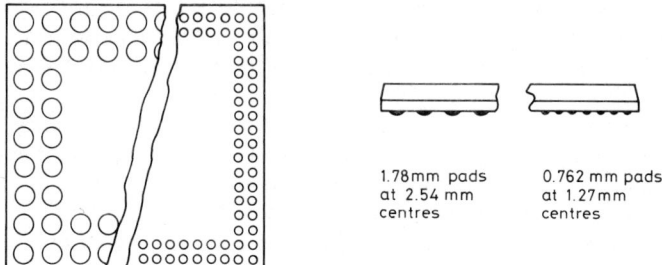

Fig. 2.15 The land grid array (LGA) is the surface mounting version of the pin grid array, with terminations in an array of pads across the underside of the component package.

2.3.8 Modified Through-hole Components

Surface mounting of components on to PCBs is a relatively new concept and the spectrum of devices in surface mounting formats is not complete. Many designers will want to change the physical layout and size of a circuit to allow it to be assembled by the surface mounting route without essentially altering the electronic design. This may require occasionally the use of devices which are unavailable in surface mounting packages.

Any leaded component can be readily converted for surface mounting by lead bending and cutting. Of all the through-hole components the most difficult to modify is the DIL integrated circuit package. Preset machines are available for automatically making the cuts and folds on the leads. The conversion is shown in Figure 2.16. Care must be taken to ensure the coplanarity of all the formed leads, especially if the component is to be reflow soldered with a paste rather than wave soldered. Alternatively the leads can simply be cropped and their ends butt soldered to the board. Butt joints are used widely in hybrid microelectronics, and on organic PCBs they normally have satisfactory thermal fatigue and vibration fatigue resistance.

2.3.9 Sockets

As well as the need to convert leaded components to surface mounting components, there is also a need to make the reciprocal conversion. It is often convenient for a designer of a conventional through-hole board to take advantage of the reduced size of the chip carrier compared with the DIL version of the same component or to use an IC that is available only in a surface mounting format. For such conversions, leaded sockets are available. The chip carrier is normally push fitted into a well, around the side of which are spring loaded contacts which align with the terminations of the chip carrier. Whilst the terminations on the chip carrier are normally at 1·27 mm centres, the pins on the underside of the

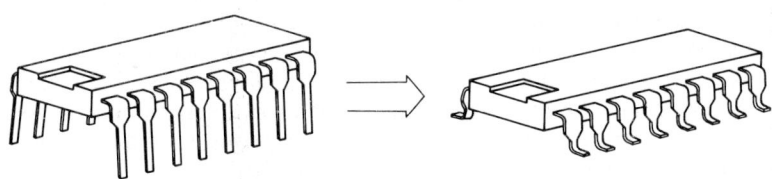

Fig. 2.16 Conventional leaded devices can be modified readily into a surface mounting format.

socket are normally at 2·54 mm centres corresponding to the standard through-hole board circuit array. It is thus necessary to have two rows of pins along each side of the socket, in the manner of a pin-grid-array. The use of through-hole sockets tends to complicate the circuit design by mixing 2·54 mm and 1·27 mm conductor grids around the device. This design problem is accentuated on multilayer boards. Sockets may also increase thermal management problems because of an increase in the thermal resistance between the device and the PCB.

The sockets are normally manufactured in plastic. Special slots are incorporated to enable the chip carrier to be easily removed without damage, using a tool. A socket also offers a convenient method of electrically testing leadless chip carrier devices before assembly.

Sockets are also very useful in situations where it is envisaged that a number of circuits will need to be interchanged during the design stage of an assembly, because of the ease of device removal and replacement with no need for solder rework. For this different requirement, sockets can be used with 1·27 mm contact centres both internally at the interface with the device and externally at the interface with the board. This type of socket is for surface mounting and simply replaces the chip carrier during assembly, although with a necessarily larger footprint. These sockets take much more strain on their solder fillets than other components, during the insertion and extraction of the device. Because of this, the footprint lands should be as large as is practical. Some sockets avoid this problem by allowing a screw fixing to the board prior to soldering.

Sockets are available in a great multiplicity of options and many factors should be considered. Many are designed with a stand-off height to ease the problem of thorough cleaning and to help with thermal dissipation by allowing air to circulate during operation. The contact plating materials and the contact configuration are of importance especially if multiple insertions are anticipated. The retention of the component in the socket may be either through the spring design of the contacts or, more robustly, with a clip-on cover plate. The force to insert the device will vary from about 3 kg for a small package to about 6 kg for a large package.

An important consideration point when using sockets is that, for many of them, the chip carrier can be physically inserted in four, and sometimes all, of the possible eight configurations. They may well offer visual aids for correct insertion but not any mechanical aid or safeguard.

2.4 BARE CHIPS

All of the IC packages described in Section 2.3 comprise a delicate semiconductor chip housed in a robust container, converting the microscopic contacts of the chip to contacts of a size convenient for interconnection. Often, however, the versatility of the chip carrier package necessarily leads to a redundancy of many of the terminations. Also the size of even the lowest profile chip carriers might be unacceptable for particular applications. For such cases the intermediate packaging stage can be omitted and the semiconductor chip mounted directly on to the PCB using essentially the same techniques as used for mounting it into a chip carrier. After such mounting it is necessary to encapsulate the assembly for protection of the delicate connections.

2.4.1 Wire Bonding

Wire bonding of a semiconductor chip to a PCB uses the same equipment and techniques as used to package a chip into a chip carrier. There are three basic types of wire bonding. The thermocompression concept uses a combination of heat (approximately 300°C) and compression to connect a wire to both the chip and the PCB conductor. There are in fact three techniques of thermocompression bonding used: ball, stitch and wedge, that are slightly different in their mechanical handling of the wire. The second concept is that of ultrasonic wire bonding which relies on the mechanical resonance of the interface to be bonded to absorb the energy needed to weld the wire to the land. An ultrasonic oscillator generates the resonant frequency. Thirdly, wire bonding can be accomplished thermosonically using a combination of heat, with the substrate at 120–150°C, and ultrasonic energy to provide the welding energy required. These wire bonding techniques are described more fully in Section 4.6.

Whichever wire bonding technique is used, the bare chip is glued to the board and the wire connections are made directly to pads on the interconnection circuit.

2.4.2 Tape Automated Bonding

Tape automated bonding (TAB) is a very convenient way of handling and mounting semiconductor chips directly on to PCBs. The delicate wire bond, with its necessity for sequential bonding of both ends of each lead, is completely dispensed with.

The TAB interconnection method consists of a process whereby a complete set of conductor fingers, designed to correspond to the bonding pads of a particular IC, and supported on a tape, is bonded in a single operation to the IC chip. The chips are then retained and stored in this form, on the tape. Electrical testing, if required before further assembly, can easily be performed because the outer ends of the fingers are electrically isolated and terminate in test pads, supported on the film, as shown in Figure 2.17. When required for assembly, the tape is transferred to the substrate, the outer pads are removed and all the free-ended leads soldered to the substrate circuit in one operation. Pulse soldering using a heated metal bracket is most suitable, but solder reflow by radiation, by vapour phase heat transfer or with a laser are all possible.

TAB was originally conceived as a rapid and robust alternative to wire bonding of chips to a lead frame, at a time when wire bonding was manual and therefore slow, but has now found a new niche for automatically mounting IC chips directly on PCBs.

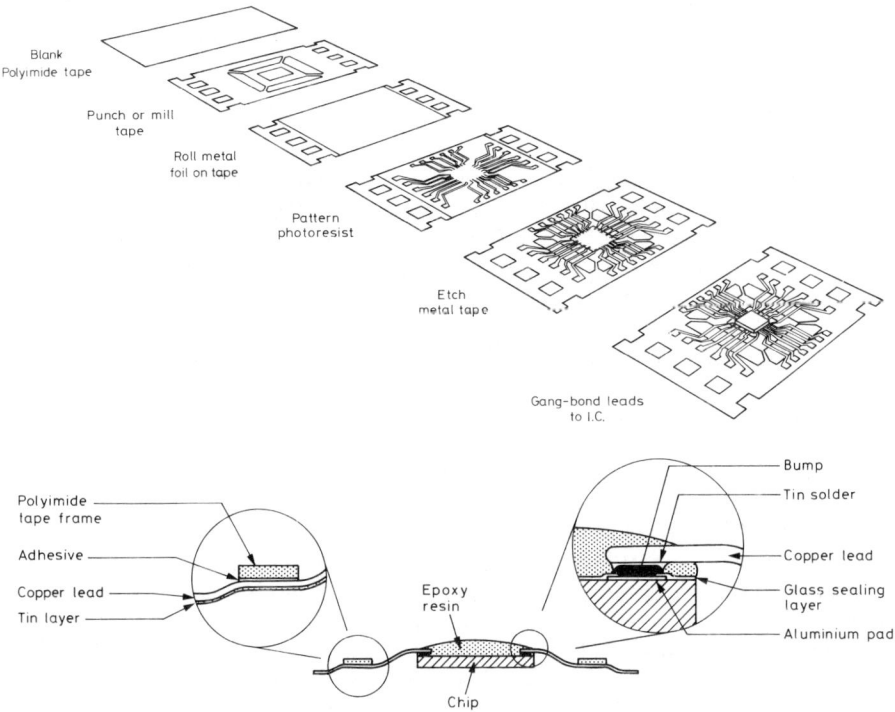

Fig. 2.17 Details of typical manufacturing steps and construction of a TAB component.

The robustness of TAB is demonstrated by a typical breaking strength of the bonds exceeding 50 g compared with around 4 g for wire bonds. The disadvantages of TAB are the cost of manufacture and the uniqueness of the tape artwork to each IC configuration. The advent of high speed automated wire bonding with its flexibility of rapid re-programming to cope with changes in IC configuration has reduced some of the attraction of TAB. It is nevertheless extensively employed in automated plastic packaging of ICs where the TAB bonding simply replaces the wire bond between the IC and the leadframe and the TAB user is also the IC manufacturer, and in situations in which the robustness and a very low profile (0·5 mm) of chip-on-board can be utilised, such as in intelligent identity cards.

The advantages offered to the electronic assembler by moving to surface mounting are secured further by the use of TAB. Because of the elimination of the intervening wire bonds, the board area occupied by TAB components can be considerably less than that occupied by the 1·27 mm (0·05 inch) pitch chip carriers.

During manufacture of the tapes the semiconductor chips are usually hermetically sealed by depositing a thin layer of glass over the whole surface of the chip. Contact holes are then etched through around the periphery to expose only the IC connection points. These holes are then closed by solderable metal 'bumps'. Initially the polyimide film has a bonded copper film which is pattern etched in a manner similar to the production of a PCB to form an array of copper foil fingers which are tin plated and soldered to the IC bumps. The IC is thus retained on the film.

TAB components are supplied on standard 8 mm, 16 mm and 35 mm tapes of the dimensions used in the film industry.

2.4.3 Flip-chip

The flip-chip is an integrated circuit chip supplied by the manufacturer with connection bumps, generally of solder, directly on the semiconductor surface. The device is 'flipped over' and reflow soldered upside down on to the substrate surface. Use of the flip-chip has been constrained by the perceived requirement of a rigid substrate such as ceramic, but it has been successfully used for COB.

2.5 COMPARISON OF IC PACKAGING OPTIONS

2.5.1 Board Area Requirement

The through-board mounting DIL package with 2·54 mm (0·10 inch) lead pitch starts to become very large above 20 leads and inefficient in terms of board area per lead. Shrink versions of the standard DIL using a 1·78 mm (0·07 inch) lead pitch were introduced for high lead-count packages, allowing, for example, a 64 pin device to require only a PCB area equal to that of a standard 42 pin DIL. A second DIL innovation for specific devices has been the skinny version, for which both the width and the height of the plastic package have been reduced. Skinny versions are very limited by the fact that a specially proportioned, long and thin die is needed requiring a high investment in new chip development. Another solution offered has been the quad-in-line package (QUIP) with leads emanating from the side of the package at 1·27 mm (0·05 inch) pitch, but staggered into two rows separated by 2·54 mm (0·1 inch).

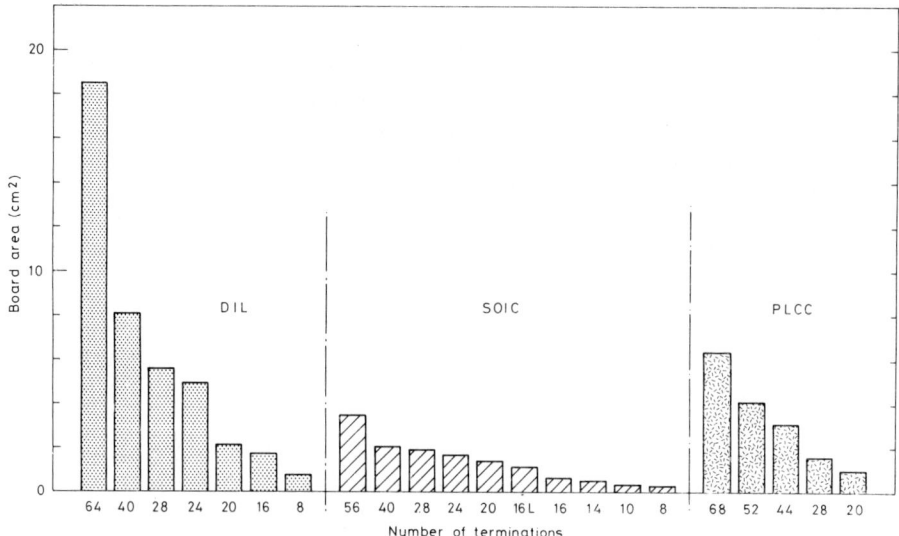

Fig. 2.18 The board area occupied by the most commonly used DIL, SOIC and PLCC packages.

In an attempt to increase the lead count of devices without sacrificing more board area, the pin grid array (PGA) was developed, originally for gate arrays in mainframe computers. These packages are square with pins on the underside in a square array usually at 2·54 mm (0·10 inch) centres. These devices suffer from the difficulty of full inspection of the soldered joints of the inner pins.

There is also the problem with shrink DILs, PGAs and other packages with many or close leads, that there is a limit on the number of holes that can be drilled in close proximity through a board while still maintaining reasonable mechanical strength.

Figure 2.18 compares the board area occupied by the most commonly used DIL, SOIC and chip carrier packages. In Figure 2.19 the board area is ratioed to the lead count for various types of IC package. It can be seen that the DILs, the shrink versions of the DIL and the PGAs form the upper limit of the board area per lead plot, over the entire range 0-200 leads. Surface mounting devices all fall well below the limit and the best choice of package is dictated by the lead-count requirement.

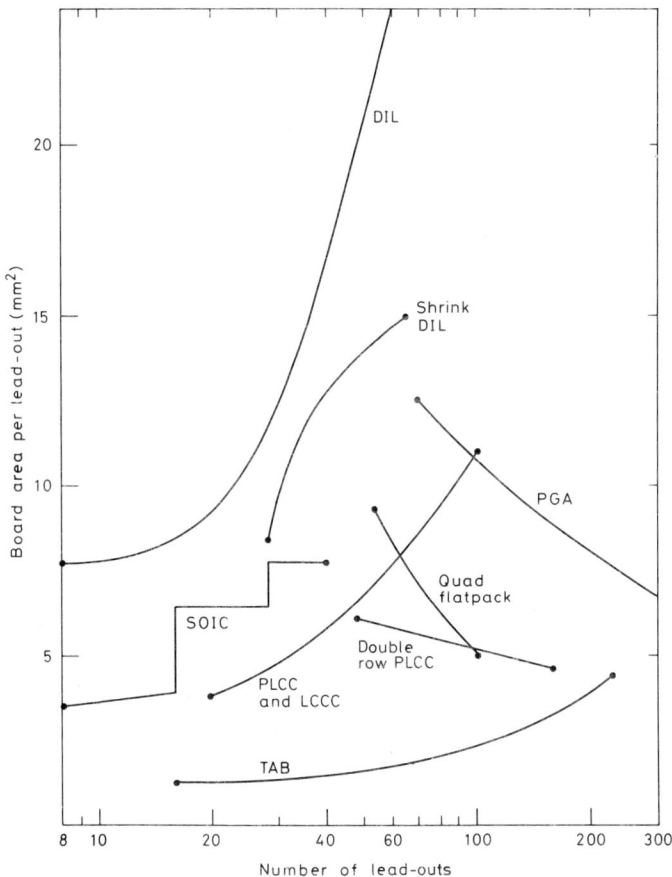

Fig. 2.19 The board area per lead-out for a variety of insertion and surface mounting package types.

For low lead counts, up to 16, the SOIC package is the most area-efficient. However, at 16 leads, the SOIC body width jumps from 4·0 mm to 7·6 mm, making the outline less area-efficient than the PLCCs and LCCCs. It is still less than half the figure for the DIL but nearly double that for the PLCC. For this reason SOIC packages in excess of 20 leads have not proved popular. Some improvement is gained by the VSO packages with their finer lead-pitch.

The line shown for PLCCs in Figure 2.19 corresponds to the 'J' lead version, with the leads tucked under the plastic body. This line is therefore very similar to that for LCCCs. The less common 'gull wing' leaded version of the PLCC requires more board area, adding about 1·9 mm² per lead.

Using 'J' leaded PLCCs, space savings can be achieved over the SOICs from 18 leads upwards and, for the whole range up to 100 leads, they offer considerable space savings compared with equivalent DILs. The 'J' leaded PLCC is the most area-efficient standard outline available.

An interesting package at present under development is the double row chip carrier. This outline has castellated edges and a lead pitch of 0·64 mm (0·025 inch). However, since alternative leads are offset from each other, some of the problems are reduced in providing a footprint for such a fine lead pitch. At

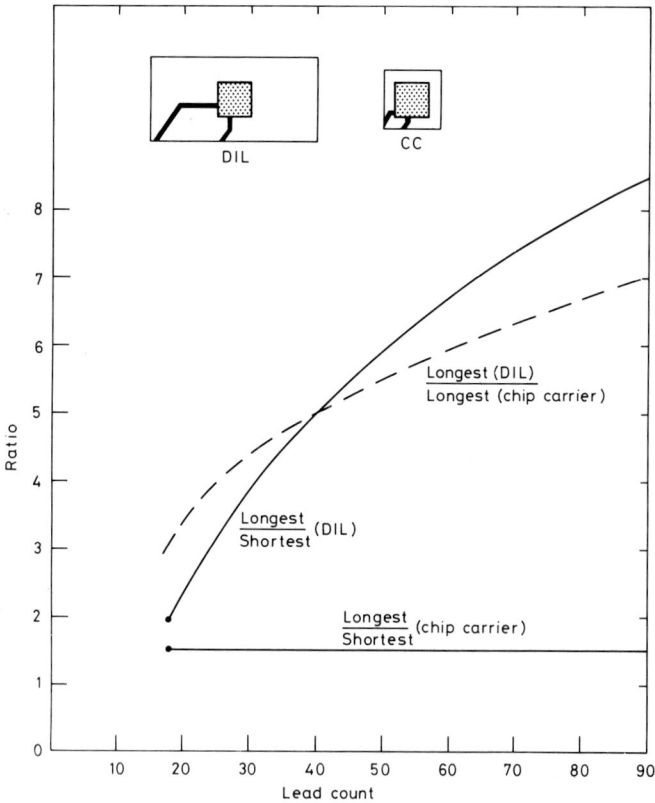

Fig. 2.20 The ratios of the longest to the shortest lead traces within DIL and chip carrier packages as a function of their size. Also shown is the ratio of the longest trace of a DIL to that of a chip carrier with the same number of lead-outs.

very high lead counts, the double row leaded chip carrier makes most efficient use of the board area, and will probably become an accepted standard.

SOIC packages are restricted to low lead counts by being leaded only on two sides. The flat plastic packages or quad packs have a similar outline to the SOICs but with leads on all four sides, using lead pitches from 1·0 mm down to 0·55 mm as the number of leads increases from 40 to 200. These packages give excellent utilisation of board area, but have not gained general market acceptance outside Japan because of their lack of standardisation.

2.5.2 Conductor Lengths

As well as the size advantages to be gained with the chip carrier format over the DIL format, there are also significant advantages in electrical performance arising from the more efficient packaging, and consequent lower and more uniform conductor lengths.

The ratio between the longest and shortest paths within the packages is compared in Figure 2.20. Because of the square form, the chip carriers offer advantages in a more uniform lead length. This is important when the timing between signals is critical, as in very high speed circuits. The overall shortening of the lead lengths improves the overall electrical performance by reducing signal lead resistance, and improved power distribution as well as speed. Particularly in high speed devices, the inter-conductor capacitance plays an important rôle. The

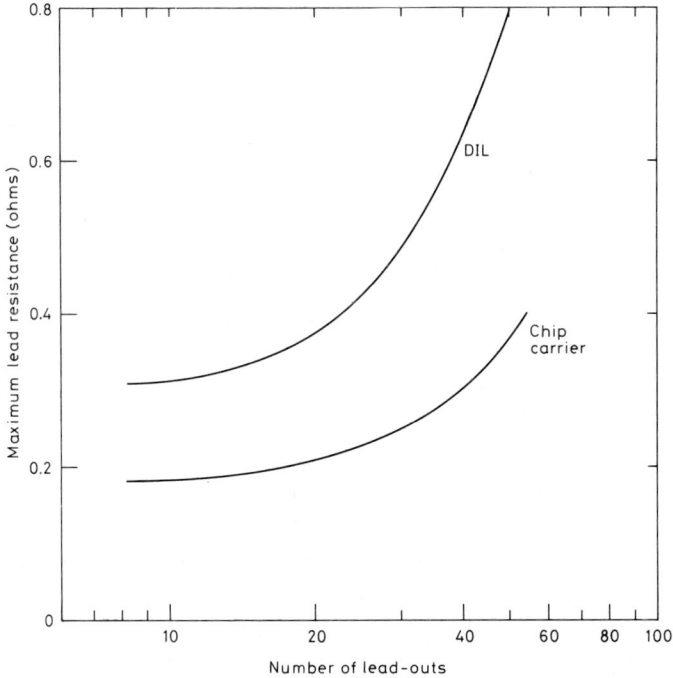

Fig. 2.21 The electrical resistance of the longest trace and lead within a DIL package compared with the equivalent chip carrier.[2]

maximum conductor resistance and the average inter-conductor capacitance, measured from the semiconductor chip through the wire bond and the leadframe, are shown respectively in Figures 2.21 and 2.22 for both DIL and chip carrier packages.

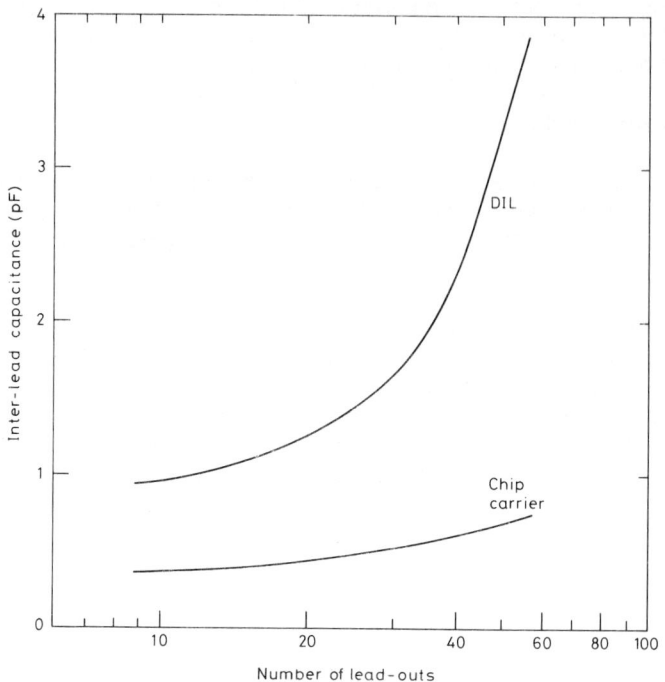

Fig. 2.22 The average inter-conductor capacitance measured for a DIL package, compared with its equivalent chip carrier package.[2]

Further advantage in electrical performance is gained by the reduction in path length between components on the circuit board, because of their smaller size and the finer pitch conductor grid used for surface mounted assemblies. A rough comparison of wiring requirements on a typical PCB can be made between a DIL, a chip carrier and a TAB package, as shown in Figure 2.23.

2.6 IC PACKAGE CONSTRUCTION AND MATERIALS

The basic functions of a package housing a semiconductor component are:

(i) to provide a means of interconnection from the semiconductor chip to the printed circuit board and hence to the outside world;
(ii) to protect the chip from the environment which may cause corrosion;
(iii) to aid dissipation of any generated heat.

Two classes of materials are used almost exclusively to make semiconductor packages, namely plastic and ceramic.

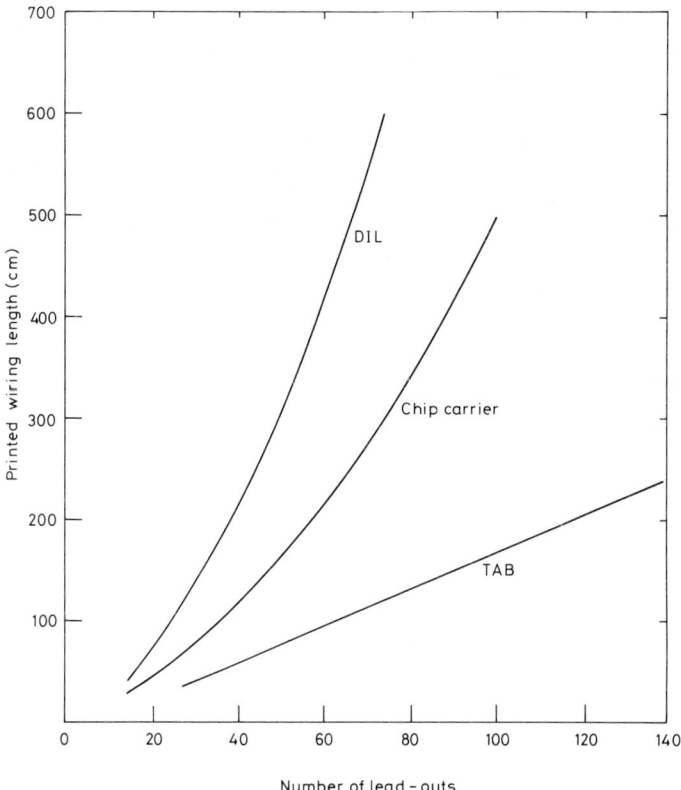

Fig. 2.23 The average length of PCB conductor track that must be associated with various device packages in order to interconnect them.

2.6.1 Plastic Packages

Developed primarily for commercial, industrial and consumer orientated products, chip carriers packaged in plastic are available in a wide range of sizes of several formats. All plastic packaged semiconductor components use thermosetting plastics, cured under extremely controlled conditions. The prime job of the plastic is that of protection and the major hindrance to the success of this task is moisture from the atmosphere. There exist two penetration routes for the moisture, first by tracking along the interface between the plastic and the leadframe and secondly by diffusion of moisture through the plastic itself. The first effect is exacerbated by the fact that the plastic tends to trap air around wire bonds and around lead frames, but modern techniques of controlled pre-treatment of the leadframe and of making more freely flowing plastics ensure the best possible watertight seal. The second route for moisture, directly through the plastic, cannot be avoided by altering the formulation of the plastic since the diffusivity of moisture is very insensitive to the plastic used. The plastic is relatively porous to moisture with a diffusion rate, at room temperature, of around 1 mm per year. Thus the thickness of the plastic directly affects its resistance to moisture ingress. Figure 2.24 gives the failure rate for ICs as a function of humidity of the surrounding air. The minor reduction in the mean time to failure of the flatpack compared with the DIL package is due to the reduction in plastic thickness protecting the device.

A recent development in plastic packages has arisen because of the discovery that the plastic is responsible for a large share of the alpha particle radiation that strikes the IC chip, giving rise to soft errors. This has led to the incorporation of alpha-absorbing coatings being used over the semiconductor[16] and also the development of plastics with minutely low alpha particle pollution levels.

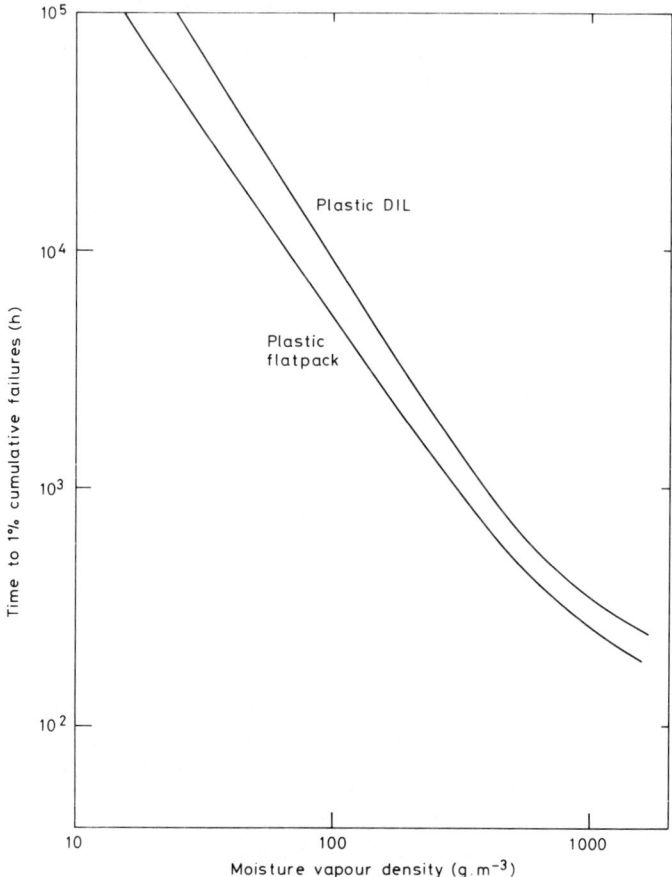

Fig. 2.24 The failure rate of plastic packaged ICs is related to the humidity of the surrounding environment.

2.6.2 Ceramic Packages

The cost of ceramic packaging compared with plastic equivalents has limited its application mainly to military and telecom uses when the reliability and performance outweigh the added cost. Generally speaking, ceramic engineering is able to produce complex cavities and interconnection between points with a degree of accuracy not possible with plastics.[17] As a result, ceramic packages are chosen for very complex devices, large-scale devices whose leadframes may not be mechanically stable enough for the plastic moulding process, for devices that require full protection from moisture, and for devices that require a lot of heat dissipation.

2.6.3 Leadframes

Traditionally, the leadframe on a DIL package is Alloy 42, chosen for its stability, rigidity and thermal characteristics. DIL leadframes are nearly always stamped out of sheet, and this alloy represents a good compromise between the requirements for the producer and the user of the component.

For DILs, the Alloy 42 provides the stiffness to withstand socket insertion and extraction forces and it is not as important that it conduct heat well because of the larger surface area compared with that of equivalent surface mounting packages.

The small size leadframes of surface mounting components often cannot be stamped with sufficient precision and are chemically etched due to the small geometries and complex details. The use of Alloy 42 is still possible, but, as the leads become smaller, the higher heat dissipation may dictate a change to an alloy with a higher thermal conductivity and a high copper content alloy is frequently used.

The diminutive nature of the smaller leadframes also means that any post-mould forming of the leads is a very delicate operation. It is difficult to perform any aggressive re-shaping without affecting the seal between the plastic and the leadframe.

2.7 PASSIVE COMPONENTS

To implement surface mounting assembly to its full potential all the components other than semiconductors must also be surface mounted. These include resistors and capacitors, inductors, potentiometers, connectors, etc. If this is not achieved, the product will compromise some of its size, weight and design benefits afforded by the use of surface mounting.

As with the active devices, it is not intended here to offer technical data, but rather to illustrate the general packaging properties of those components most commonly used.

Surface mounting capacitors and resistors usually have cubic dimensions and are customarily referred to as 'chips'. The designation 'chip' should only be used when confusion with semiconductor chips as used in integrated circuit technology can be excluded. Generally, chip components are easier to manufacture, are smaller and have a lower material content than their encapsulated leaded counterparts. Chip components also lend themselves readily to robotic pick-and-place handling and assembly, especially for vibratory feeds, having no leads to become bent or tangled. The most common chip components are resistors, capacitors and diodes, but every kind of two-terminal device is available in chip form, including chokes and crystals. The package most commonly used is rectangular in nature, but a cylindrical type is used in some instances, despite these being more difficult to handle robotically. Chip components may have their solderable terminations only on the end face, or on the top and bottom faces as well as the end, or on the sides in addition. Figure 2.25 illustrates these instances of one, three or five face metallisation for the end terminations. The choice of termination type is a matter of the manufacturing process. Where each component is dipped in a conductive ink, a five-face metallisation results, whereas, if the components are cut up as a final operation, one- or three-face metallisation results. Single-faced termination components exhibit significantly poorer assembly yields than the multi-faced ones.

Chip components are generally manufactured in set sizes which have become approved standards through common usage. A partial list of these is given in Table 2.5. The size designation arises from the length and width of the component expressed in hundredths of inches. Thus the 0805 size is approximately $0 \cdot 08 \times 0 \cdot 05$ inch and the 2220 is approximately $0 \cdot 22 \times 0 \cdot 20$ inch.

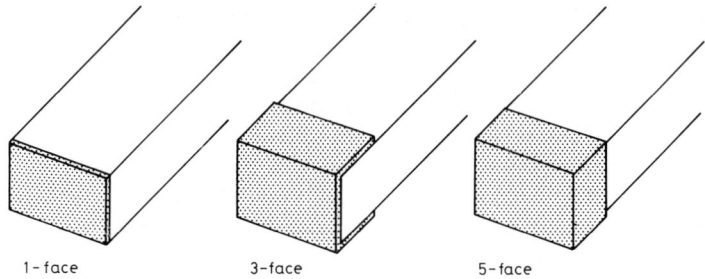

Fig. 2.25 Illustrating the difference between 1-face, 3-face and 5-face metallisations on chip components.

Table 2.5

Some Common Sizes of Surface Mounting Chip Components

Designation	Length (mm)	Width (mm)	Area (mm^2)	Height* (mm)	Volume* (mm^3)
0504	1·3±0·15	1·0 ±0·15	1·3	0·5 – 1·2	0·7 – 1·6
0805	2·0±0·15	1·25±0·15	2·5	0·5 – 1·3	1·3 – 3·3
1206	3·2±0·15	1·6 ±0·15	5·1	0·5 – 1·6	2·6 – 8·2
1210	3·1±0·2	2·5 ±0·2	8·0	0·5 – 1·9	4·0 – 15·0
1808	4·5±0·2	2·0 ±0·2	9·0	0·5 – 1·9	4·5 – 17·0
2220	5·7±0·2	5·0 ±0·2	28·5	0·5 – 1·9	14·0 – 54·0

*The height and therefore the component volume vary with component value.

2.7.1 Capacitors

Chip capacitors have had a long development period because of their requirement in the well established hybrid assembly technology. Whilst screen-printable film resistors are easily incorporated in hybrid circuits, only limited capacitor values are achievable by thick or thin film technologies and so there has long been a requirement for discrete surface mounting chip capacitors.

2.7.1.1 CERAMIC CHIP CAPACITORS

The type of chip capacitor that predominates because of its useful range is the multilayer ceramic construction. The basis of this structure is shown in Figure 2.26. In the manufacturing process, thick film capacitor electrodes are screen printed on to ceramic sheets using an interleaved pattern. These are stacked under pressure, dried, cut to size and sintered at a temperature around 1300°C. The electrodes must be of a metal with a melting point that is higher than the sintering temperature, and platinum (1774°C) or palladium (1552°C) are normally used. Metal terminations, commonly silver-palladium, are then added to the ends, each making contact with one set of electrodes, by a technique of screening or dipping followed by firing. The resultant capacitor is very rugged and the electrode system is totally enclosed and protected from the influence of the ambient atmosphere. Standard versions are able to withstand immersion in 250°C molten solder as well as high humidity, without the need for further encapsulation.

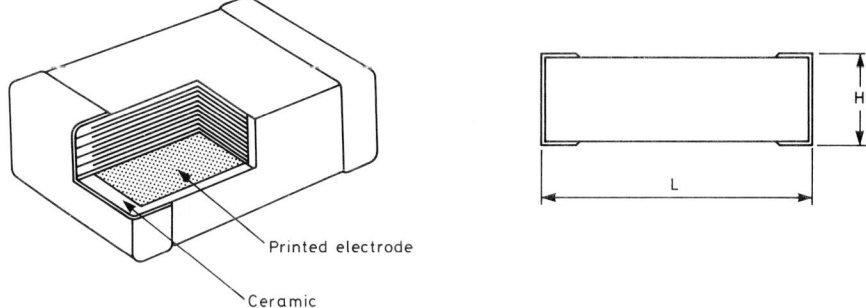

Fig. 2.26 The basic structure of the multilayer ceramic chip capacitor.

The ceramics used for multilayer capacitors fall into several classes of performance, concerning their dielectric constants, temperature variations and ageing effects. The most commonly used dielectric material is designated NPO (or COG). This has a dielectric constant in the range 30-150, providing capacitors up to 4·7 nF rated at 100 V, or up to 10 nF at 50 V. Further properties are given in Table 2.6 and the range in Figure 2.27. NPO capacitors have temperature compensating properties and high capacitance stability as shown in the data of Figure 2.28.

The second type, generally referred to as X7R, uses ferro-electric dielectrics with dielectric constants in the range 500-2000 to produce values in the range 470 pF to 1 μF. They have high capacitance per volume but their temperature coefficient is an order of magnitude greater than for NPO capacitors. Other characteristics such as the non-linear voltage-capacitance dependence restrict the use of X7R components to bypass and coupling capacitors where tolerances wider than ±5% are acceptable.

The third type of dielectric used for ceramic chip capacitors is termed Z5U. Tolerances of these capacitors are about −50%, +100% and hence are limited almost exclusively to decoupling applications for which this tolerance and a poor stability can be accepted as a compromise for the requirement of a high capacitance. The dielectric constant is usually in excess of 4000, enabling values up to 1·5 μF to be made.

Thus, the temperature coefficient of a multilayer chip capacitor is determined by the type and quality of the ceramic used. The temperature coefficient of the NPO capacitors is close to zero over the relevant temperature range, but for the others it can be either positive (capacitance increasing with rising temperature) or negative (capacitance decreasing with rising temperature). Temperature coefficients quoted in specifications are normally determined by measuring the component values C_{20} at 20°C and C_{80} at 80°C so that

$$TC = \frac{1}{60}\left(\frac{C_{20} - C_{80}}{C_{20}}\right) 10^6 \text{ ppm.K}^{-1}$$

which is the average value over that temperature range. In fact, as is clear from Figure 2.28, capacitance change is not linear with temperature and the quoted coefficient merely indicates a trend and the limits. If the coefficient is greater than 1000 ppm.K^{-1} it is often quoted as a %.K^{-1}.

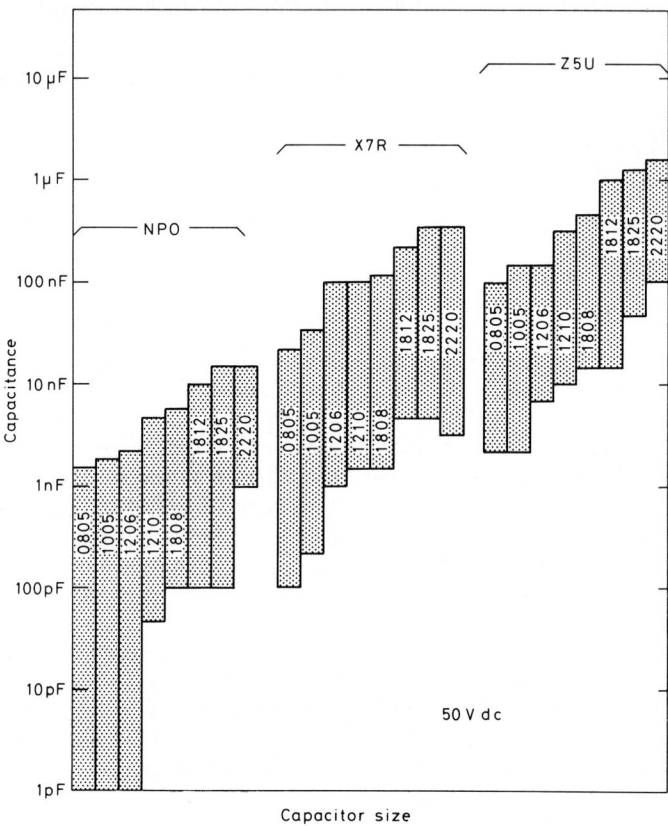

Fig. 2.27 The range of available multilayer ceramic chip capacitors, based on the three common types of ceramic used.

Table 2.6

Electrical Characteristics of Chip Capacitor Dielectrics

Classification	NPO(COG)	X7R	Z5U
Dielectric constant	30-150	500-2000	>4000
Operating temperature range:	−55 to +125°C	−55 to +125°C	+10 to +85°C
Capacitance range:	1 pF-10 nF	100 pF-200 nF	10 nF−1·5 μF
Tolerance	1-10%	3-20%	50-100%
Maximum voltage	50-200 V	50-200 V	50-200 V

Several solderable termination options are available for multilayer ceramic chip capacitors for which there is a trade-off between cost and retention of solderability of the metallisations. Good solderability is a critical process parameter in the quest for zero-defect assembly and, although the added cost in percentage terms per component may be high, the added cost in overall pricing of the product may be minimal or even zero.

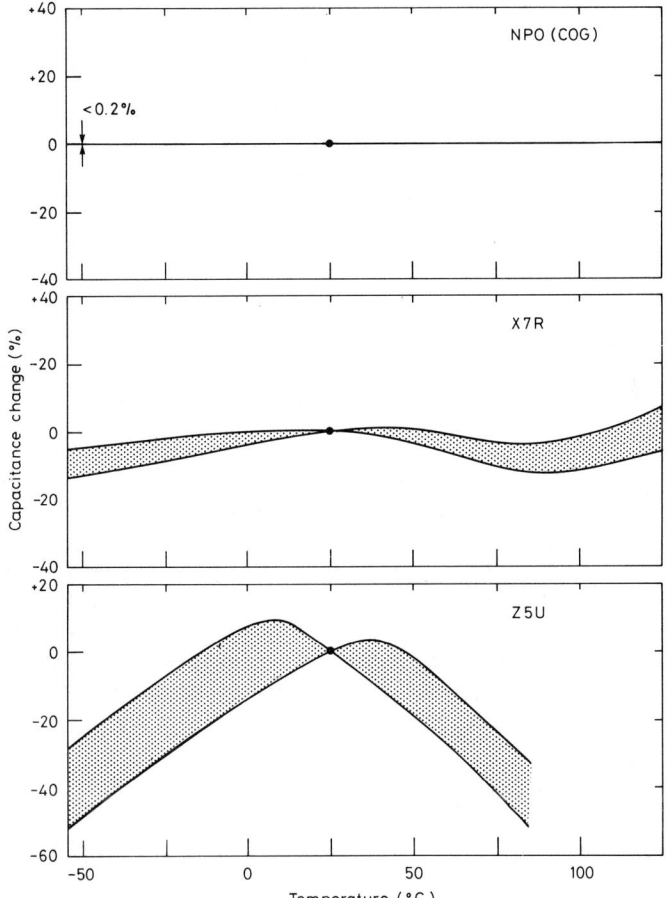

Fig. 2.28 The temperature stability of multilayer ceramic chip capacitors.

Two good types of termination for multilayer ceramic chip capacitors are shown in Figure 2.29. Contact is made to the capacitive layers by dipping in a silver or silver-palladium conductive ink which is fired on. A nickel barrier layer can then be deposited with a solderable outer coating. The nickel is present to prevent leaching of the silver into the molten solder during assembly.

2.7.1.2 CYLINDRICAL CERAMIC CAPACITORS

Metal electrode face-bonding (MELF) ceramic capacitors are cylindrical in form, consisting of a ceramic body with spiral copper metallisations from the end electrodes. MELF capacitors are usually constructed of ceramic with a high dielectric constant and consequently a poor temperature coefficient, for use particularly in de-coupling applications.

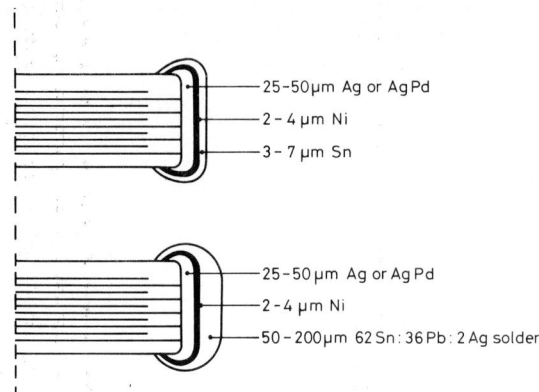

Fig. 2.29 Two typical end terminations frequently used for ceramic chip capacitors.

2.7.1.3 PLASTIC CAPACITORS

Metallised plastic capacitors offer a new cheap alternative to ceramic for non-critical applications. They are fabricated from sheets of flexible dielectric such as polyethylene terephthalate covered with a vacuum-deposited aluminium electrode. These metallised sheets are interleaved with their alternate edge-contact areas on opposite sides, and are then wound into a flattened cylindrical shape. The entire length of the outside edge of each metallisation is in contact with its own external contacts. The windings are then cut and encapsulated.

The capacitance values available are 1 nF to 0·47 μF at 50 V, in rectangular packages within the size range $5 \cdot 5 \times 4 \cdot 2 \times 2 \cdot 3$ mm to $7 \cdot 3 \times 7 \cdot 0 \times 4 \cdot 0$ mm.

2.7.1.4 TANTALUM CHIP CAPACITORS

Surface mounting capacitors with values greater than 1 μF are frequently required, especially in telecommunications applications. Solid tantalum capacitors are available in miniature chip form to meet this need, and generally cover the range 0·1 μF to 100 μF with working voltages up to 35 V, and tolerances down to ±5%.

Each capacitor consists of a near-rectangular anode of high-purity sintered tantalum with an electrolytically deposited tantalum oxide dielectric layer. The construction is encapsulated in a plastic moulding, as shown in Figure 2.30, giving the component very good mechanical and environmental properties.

Tantalum capacitors make highly efficient use of board area and component volume for high values of capacitance. These components are available in a large range of sizes from about 5 mm^3 to 200 mm^3, in several package shapes, as shown in Figure 2.30. The dimensions of the standard cuboid packages are given in Table 2.7. The tantalum capacitor is polar and therefore relatively sensitive to overvoltage.

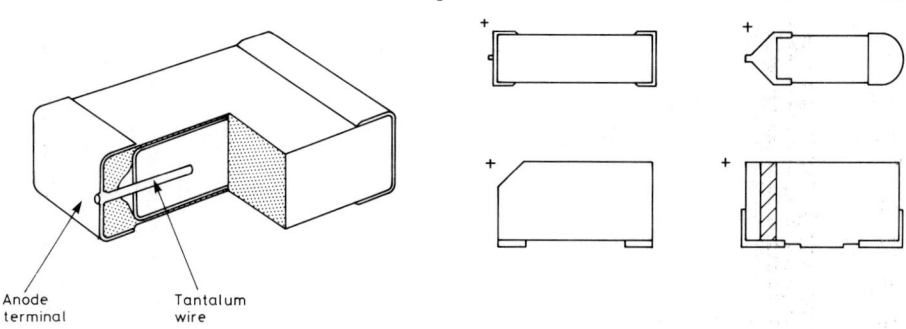

Fig. 2.30 Construction and polarity of tantalum capacitors for surface mounting.

Table 2.7

Dimensions of Cuboid Tantalum Chip Capacitors (mm)

Designation	Length [L] ±0·04	Width [W] ±0·04	Height [H] ±0·04	Metallisation Length
a	2·54	1·27	1·27	0·76
b	3·81	1·27	1·27	0·76
c	5·08	1·27	1·27	0·76
d	3·81	2·54	1·27	0·76
e	5·08	2·54	1·27	0·76
f	5·59	3·43	1·78	0·76
g	6·73	2·79	2·79	1·27
h	7·25	3·81	2·79	1·27

2.7.1.5 ALUMINIUM ELECTROLYTIC CAPACITORS

Some high value capacitors of the aluminium/wet electrolyte construction are available in surface mounting format, ranging from 0·1 μF upwards and rated at 6·3 or 50 V. This type of capacitor is manufactured by rolling aluminium foil sheets with paper separators and then encapsulating the roll with electrolyte in an aluminium cylinder. The whole is then encapsulated in plastic, as shown in Figure 2.31. Case sizes range from about $8 \times 4 \times 4$ mm to $12 \times 4 \times 4$ mm.

Electrolytic capacitors are, of course, polar and so care must be taken during assembly to ensure correct orientation.

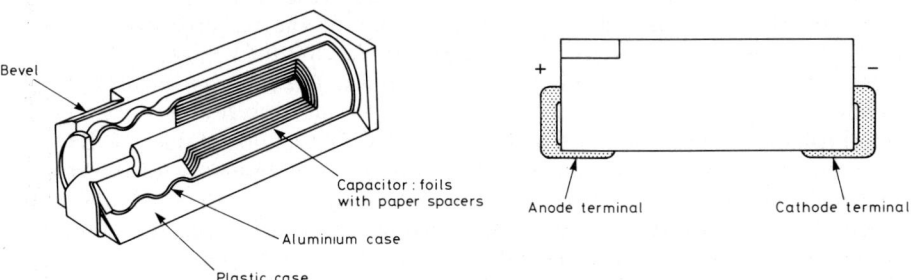

Fig. 2.31 Construction and polarity of an aluminium electrolytic capacitor for surface mounting.

2.7.2 Resistors

The resistance element of a chip resistor is a thick film[18] printed on to a high purity alumina substrate, as shown in Figure 2.32. The element is protected by a glass film and then solderable end terminations are incorporated. Such resistors typically range from 10 ohms to 2·2 megohms with the usual choice of tolerances in the range ±1% to ±20%.

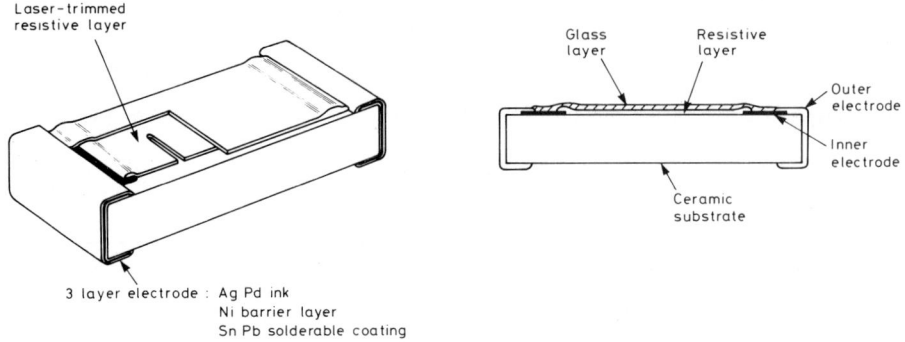

Fig. 2.32 The construction of a chip resistor.

Chip resistors use the same size notation as used for chip capacitors and given in Table 2.5, although only two sizes are commonly available, the 1206 with a height of 0·6 mm and the 0805 with a height of 0·5 mm. The precise sizes and tolerances depend on the source of the component. The 1206 package resistors have power ratings of ¼ W or ⅛ W over an operating range of approximately −40°C to +70°C, again depending on the source. The less common 0805 package is used for ¹⁄₁₆ W resistors.

The surface mounting resistor has the highest standards of size, quality and reliability and is a product of considerable ingenuity of design and manufacture.[19] The substrate is a high grade alumina ceramic sheet which is first scribed in two perpendicular directions to provide crack lines for breaking out the individual resistors at a later stage in the production. The internal metal electrodes, usually silver, sometimes gold, are printed across the appropriate cracks and fired on. The resistive element, normally based on ruthenium dioxide, is similarly printed between these electrodes and again baked. The resistive element is always of a value lower than that required so that computer-controlled laser trimming can be used on each resistor individually to trim to the precise value required; a small channel is burned out of the resistive element while monitoring the rising resistance.[20, 21] The central area is then glazed to cover over all the resistive element, followed by another bake. The primary substrate is then divided in one direction to give strips of side-by-side resistors. Nickel intermediate end electrodes are deposited across both complete sides of the strip, i.e., the ends of the individual resistors, making good contact with the existing internal electrodes. The strips are then fractured into individual chips and external contacts of tin-lead plated on, which are designed for good and retained solderability. The nickel intermediate layer serves the same purpose as it does on the chip capacitors, acting as a diffusion barrier layer to protect the silver electrode from the molten solder during soldering and during service life.

The temperature coefficient of this type of resistor is very low,[22] making it an extremely stable component. The resistance changes its value by no more than ±0·02% per degree celsius change in temperature.

Marking of surface mounting resistors to indicate their value and tolerance has not been standardised and is often not attempted at all. The chip is so small that a two or three digit code is most commonly used, whose significance is found from the manufacturer's data. However, identification is not generally important for the assembly of surface mounting chip components since they are usually transported and used in tape packing so that marking the tape and its spool is sufficient.

2.7.3 Other Passive Components

A comprehensive range of ancillary components in surface mounting format is available, including inductors, transformers, filters, crystals, thermistors, potentiometers, trimmer capacitors, switches and connectors.

Chip inductors are produced in cuboid shape using a magnetically shielded ferrite core, with very small weight and volume, potted in an epoxy resin, as shown schematically in Figure 2.33. Typically, chip inductors are some 3 mm cubed and available in the range 1-500 μH, the size depending on the current carrying capability required.

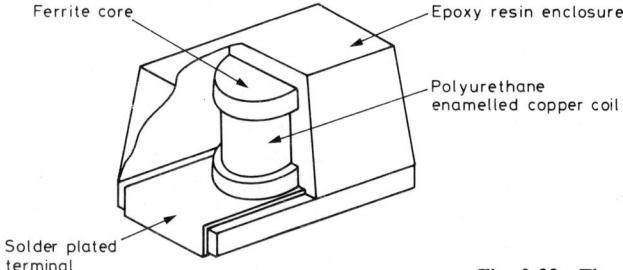

Fig. 2.33 The construction of a chip inductor.

Besides the miniature nature of the components, the important feature of all components is that they be capable of withstanding total immersion in molten solder for wave soldering, or total immersion in a condensing heat transfer fluid for vapour phase soldering, or irradiation with infra-red radiation during reflow. In addition, they must withstand stringent cleaning procedures with solvents after soldering. Thus the wiping arrangements on potentiometers and trimming capacitors, for example, must be close fitting to exclude entirely molten solder, condensing fluorocarbon fluid and cleaning solvents from the operational resistive film and the wiper.

Of all the component parts of a fully surface mounted assembly, the last to be the subject of specific surface mounting design requirements has been the edge connector.[23-28] The main problem with surface mounting a connector is the required mechanical strength of the solder joints. Additionally, the plastic parts must retain the required stiffness or flexibility during and after the solder immersion or solder reflow, and the temperature expansion mismatch with the substrate is important for such a relatively large component. Finally there is the not inconsiderable problem of robotic handling and assembly of such a component.

Design criteria for surface mounting edge connectors have been assessed. It has been found[25] that the required mechanical properties are best with a minimal solder stand-off between the connector terminations and the pads of the footprint, so that the leads are in contact with the board. Therefore, there should be some compliancy of the leads within the connector to absorb strains that inevitably arise. Coplanarity of the leads must be assured for good soldering quality and this can be accomplished in several ways depending on the product. For horizontal card connectors the contacts float in the housing cavities, sitting comfortably on the solder paste prior to reflow, then sinking, with some centring action, into the solder during reflow. For male connectors the pins can be stamped from a bi-metallic strip consisting of phosphor bronze and a soft compliant alloy that will provide leads that collapse against the solder pad with a minimal positive force. New plastics have been identified for the connectors to provide the required high temperature properties. These might be glass filled polyphenylene sulphide for edge connectors that must remain stiff, or glass filled polyetherimide for latched connectors that require a degree of flexibility.

2.8 THERMAL CHARACTERISTICS OF SMCs

Nearly all electronic components generate heat when they are in operation. This is true of both passive and active components, and includes the printed wiring tracks on the PCB. The generated heat raises the temperature of the component in which it originates, the temperature of the surface on which it is mounted and the temperature of the neighbouring components. Heat affects the circuitry in a number of detrimental ways, altering the electrical characteristics of semiconductors, the reactive properties of metals used in the devices and the mechanical properties of the PCB materials. For example, the interdiffusion of materials in contact and the growth rate of detrimental intermetallic compounds both increase exponentially with linear increase in temperature. Higher temperatures increase the activity of corrosive residues left on the board, perhaps by the flux. An elevated temperature can affect the reliability and the lifetime of an electronic assembly at a rate dependent on the time at the elevated temperature.

To prevent the generated heat raising the temperature to a point where it becomes a significant problem, the heat must be removed from each component and its vicinity. This can be achieved by conduction, convection or radiation or a combination of these.

In surface mounted assemblies, the heat generated is dissipated in smaller packages and concentrated in smaller areas than in conventional assemblies. This effect is counteracted by the fact that the thermal impedance paths are shorter, and made of an alloy that has a higher thermal conductivity, and therefore more efficient, conducting the heat from the surface mounted component to the PCB much faster. As a result, it is found in practice that the plastic leaded chip carrier (PLCC) is virtually equivalent to the plastic dual-in-line (DIL) in its heat dissipation, the equivalence becoming closer as the pin count increases. However, for surface mounted assemblies the heat absorption and dissipation of the PCB becomes much more important since it has to absorb a greater thermal flux. Some methods by which this greater heat absorption can be achieved are covered in

Chapter 3 on substrates for surface mounted assemblies. In this chapter are now discussed the thermal properties of the components themselves.

2.8.1 Temperature-dependent Effects

The dependence of the reliability of an electronic component on temperature is dealt with fully in Chapter 13. In brief, the rate of degradation R_T due to temperature effects is given by

$$R_T = K \exp\left(\frac{-Q_T}{kT}\right) \qquad (2.1)$$

where K is a constant, Q_T is the activation energy (joules), k is Boltzmann's constant ($1 \cdot 38 \ 10^{-23}$ J.K^{-1}) and T the temperature in kelvin. Different temperature-induced failure mechanisms have different activation energies,[29] but Equation (2.1) is valid for failure phenomena caused by diffusion of impurities, intermetallic growth, polarisation, ionic drift, electromigration, corrosion at a given humidity, etc. For many of the commonly occurring failure mechanisms the activation energy Q_T is around $1 \cdot 4 \ 10^{-19}$J(0·9 eV) which, using Equation (2.1), corresponds to a halving of the life of the component for every 6°C rise in temperature near room temperature or for every 10°C rise in temperature around 150°C.[30]

If the failure mechanism of the component is humidity dependent, a rise in the temperature may improve the reliability of the component[31] since it causes a lowering of the relative humidity and a corresponding reduction in the rate of degradation R_{TH} due to temperature and humidity combined[32]:

$$R_{TH} = K' \exp\left(Q_H(RH)^2 - \frac{Q_T}{kT}\right) \qquad (2.2)$$

where Q_H is the activation energy for the humidity degradation mechanism and RH is the relative humidity.

The sign and magnitude of R_{TH} are governed by the magnitude of the respective terms in Equation (2.2). In general, the degradation is reduced by a temperature increase when the temperature is relatively low and the humidity high.[33]

2.8.2 Thermal Parameters

For plastic packages, such as the PLCC, the SOIC and the flatpack, the most useful parameter that characterises the thermal performance of the device is the thermal resistance between the active junction and the ambient environment. This is designated θ_{JA}. For ceramic chip carriers the thermal performance is best characterised in terms of a parameter θ_{JC}, the thermal resistance between the active junction of the device and the case. This is chosen as the reference parameter because θ_{JA} for ceramic devices is very dependent upon the board material.[34]

The thermal resistance, measured in units of °C.W^{-1}, is a measure of the rate of change of junction temperature as a function of the power dissipated. A component with low thermal resistance is better able to conduct heat away from its active junction.

In Figure 2.34 the typical thermal resistance of SOICs and PLCCs is compared with that of DIL packages, the poorer performance of the surface mounting device resulting from its smaller size and greater concentration of heat. For the lower lead-count devices, θ_{JA} is some 25% greater than for equivalent DIL packages, but the difference decreases as the lead-count increases.

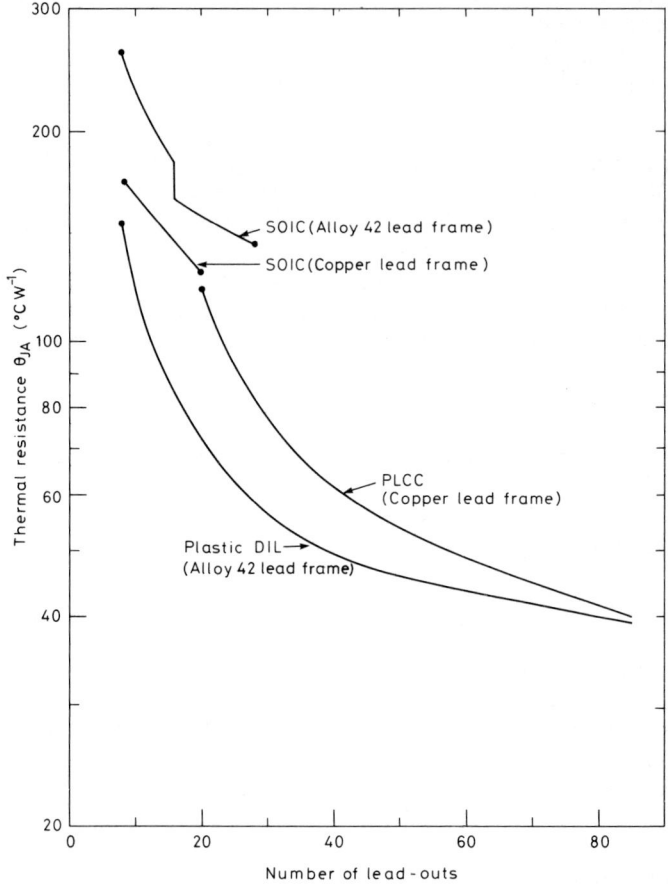

Fig. 2.34 The thermal resistance of SOICs and PLCCs compared with dual-in-line devices.[3]

For low lead-count devices the thermal performance of the SOIC package is superior to that of the PLCC. In Figure 2.34 the θ_{JA} for SOICs is compared with values for equivalent DILs constructed from the same moulding compound, and measured under identical conditions. A considerable reduction in thermal resistance can result from the use of a high copper alloy for the leadframe, which has a thermal conductivity of over 90% that of pure copper.

Leadless ceramic chip carriers (LCCCs) offer very good heat transfer characteristics since the ceramic has less thermal inertia than plastic. Ceramic packages have relatively low values of θ_{JC} as shown in Figure 2.35. These depend not only on the package size but on the die size also.

2.8.3 Temperature Coefficient of Expansion

Differences in the thermal coefficients of expansion (TCE) of the materials of a surface mounted assembly are very important to its reliability. This is because at one temperature there may be no stress at a point where two materials join, but at another temperature, if there is a differential between the TCEs, that same joint may well be under such considerable strain that the component parts fracture. Such situations occur where leadframes pass through plastic encapsulations and where component-solder-substrate joints are made.

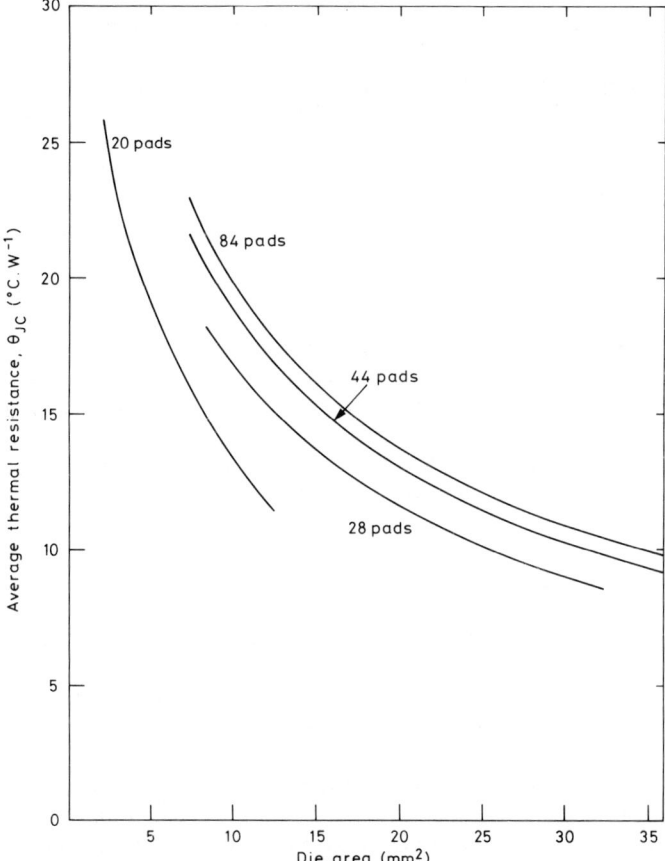

Fig. 2.35 The thermal resistance of LCCCs depends on the die area as well as the package size.[3]

Matching the TCEs at a joint is no guarantee of freedom from the problem because electronic components are their own heat sources so that the heating of a joint is not isothermal and one member will still expand more than the other. The subject of fatigue stresses arising in surface mounted assemblies as a result of different TCEs is dealt with in Chapter 11. The major problem arises from thermally induced cyclic stresses in the solder joints of the larger leadless components, i.e., higher pin-count LCCCs. The TCE of the ceramic used is dependent on its type and quality; a mean value of $6.4 \; 10^{-6} K^{-1}$ is normally taken for alumina ceramics.

2.9 COMPONENT PACKING

A major factor of the assembly of surface mounting electronic components is their ability to be handled robotically during all the necessary operations. The components must be delivered to the user in a format that is compatible with automatic assembly. The packing should preferably have all the following characteristics:

(i) low cost;
(ii) large quantity of components per packing unit;
(iii) variable number of components per packing unit possible;
(iv) small amount of packing per component;
(v) protective against transport and handling damage;
(vi) definite 3-dimensional orientation of the component in its packing;
(vii) compatible with automatic removal from packing, and different assembly machines;
(viii) protection against electrostatic charge;
(ix) limited number of packing sizes for a wide range of component types and sizes;
(x) standardised for all component suppliers.

Most surface mounting components can be supplied in three packing formats, bulk, tape and magazine. Some assembly machines will accept only one format but usually the choice is left to the user. Orderly packing, in the form of tape and magazines, has the advantage of virtually eliminating wrong or misplaced components. The disadvantages of the tape and magazine formats are increased cost and the finite limit to the number of components that can be loaded at one time.

2.9.1 Bulk

The simplest and lowest cost mode of supply of components is loose in bulk. Passive components and small semiconductors are suitable since almost no bending or entanglement of their leads can occur. The components are usually fed into a vibrating bowl hopper which shakes them into a common orientation suitable for the assembly machine to pick up and place on to the circuit board. This supply method permits a large quantity to be continuously available uninterrupted at the assembly machine, if required. The use of bulk loading does involve a danger of wrong components, because it is not mandatory, or sometimes even possible, to mark identifiers on to small SMCs, especially passive components, to indicate their value, tolerance and rating.

2.9.2 Tape

Many surface mounting component types are suitable for packing in tape, and it is the system most frequently used. The major benefit of the tape format is that it ensures safe keeping and virtually eliminates wrong component placement.

There are two main types of tape packing, shown in Figure 2.36, employing paper or card tape, or plastic blister tape, both with adhesive foils to retain the components. The paper version is cheaper but its use is restricted to those components whose thickness is less than the thickness of the paper tape, since holes, suitably sized for the components, are punched out of it. The paper or card cannot be made too thick or it becomes unmanageable on a spool. One millimetre thickness is a working limit. The paper tapes also can have a problem with abrasion giving rise to fine dust.

Paper tape

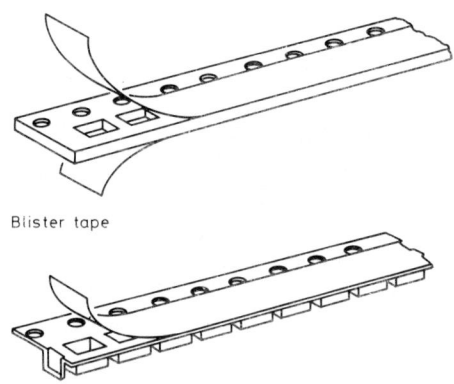

Blister tape

Fig. 2.36 The two types of tape packaging: paper or card and plastic blister tapes.

Blister tapes have preformed compartments corresponding to the component size. They are made either from plastic or a plastic-clad aluminium foil. When used for semiconductors which are sensitive to electrostatic charge, the tape has to have a degree of conductivity and must be suitably contacted to ground.

Both types of tape are aligned and fed using punched holes along one edge. The sizes and hole configurations are the subject of an IEC recommendation, to ensure that the tapes are accepted by all machines, as shown in Figure 2.37. The most used tapes are 8 and 12 mm wide, but 16 and 24 mm tapes are also standardised. The tape is wound on to reels with diameters of 178 mm (7·0 inches) or 330 mm (13·0 inches) depending on the number of components required.

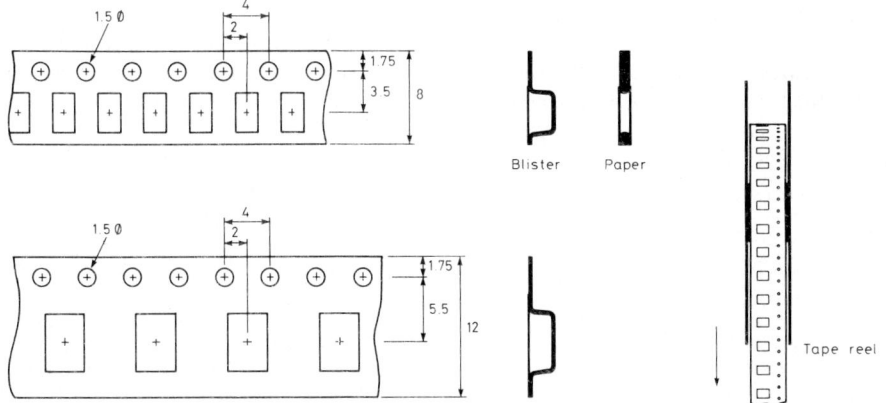

Fig. 2.37 The standard formats for 8 mm and 12 mm component tapes and the direction of feed from a spool.

2.9.3 Magazine

Several types of magazines are available as illustrated in Figure 2.38. These are the linear, the waffle, the stick and the stack. Magazines are normally specific to each assembly machine and have not been standardised. Stick and stack magazines are common and components can be supplied already packed in a magazine appropriate to a particular assembling machine. The linear and waffle magazines have been developed for components with protruding leads on all four sides, i.e., quad flat packs.

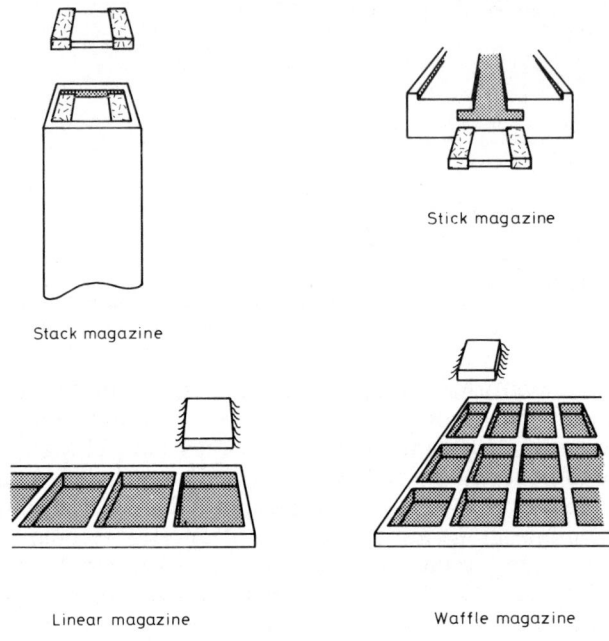

Fig. 2.38 An illustration of some of the types of component magazines that are used.

Magazines have the disadvantage that they can accommodate only a relatively small number of components, and therefore increase the operator time required to manage each placement machine.

Chapter 3

SUBSTRATES

3.1 INTRODUCTION

In any printed circuit electronic assembly the purpose of the substrate is to provide mechanical support and fixing for the components, to provide electrical interconnection between the terminations of the components and to assist in the dissipation of heat generated within the components.[35, 36] The mechanical rigidity is available from a wide range of board materials offering suitable properties of electrical insulation, thermal expansion, etc. The circuit interconnection tracks are either produced additively by printing on to the substrate, using for example conductive inks, or produced subtractively by chemical etching of a surface conductive foil. Thermal dissipation, if required, can be provided by using a metal board with an insulating surface on which the interconnections are made.

Three technical issues need to be addressed regarding the purposes of the substrate:

(i) the compatibility of the dimensional stability characteristics of the substrate with those of the components, in the absence of a compliant lead to relieve any strain arising through differential expansion;
(ii) the thermal conductivity and thermal dissipation properties of the substrate now that the heat generating components can be much more densely sited by the use of surface mounting;
(iii) the generation of sufficiently high interconnection densities on the substrate to satisfy the requirement of the very low board area available per component termination.

The choice of substrate for a particular assembly is made with regard to these three factors, as well as others as obvious as cost or size availability. It is the purpose of this chapter to discuss the surface mounting substrate options currently available, within a technical framework that will enable a similar appraisal to be made of future developments.

3.2 COMPATIBILITY OF THERMAL EXPANSION

With leaded components and through-hole mounting, the relative properties of the substrate and the component are not very critical because the two are fairly distant from each other and joined by a wire and solder fillet, both of which are relatively compliant and forgiving. With surface mounted components, however, there is intimate contact between the substrate and the component; sometimes they are glued, and with a minimal fillet of solder, insufficient to afford much compliancy.

In the general case, the system is under stress at room temperature. Usually, the component and the substrate are more rigid than the solder and so it is the solder that is the focus of the stress. The stress arises because the solder fillet has solidified at around 185°C and is more or less relieved of stress at that temperature assuming the cooling is homogeneous. As the assembly cools the various components contract by different relative amounts, determined by their respective temperature coefficients of expansion (TCE). A chip resistor or capacitor, made of alumina, has a TCE of about 6 ppm.K^{-1}. The substrate might well be a glass-epoxy composite with a TCE of around 16 ppm.K^{-1}. During the cooling process the induced strain and stresses are complex, but, since the substrate contracts 16.10^{-6} of its length for evey degree, whereas the component contracts only 6.10^{-6} of its length for every degree, when cooled to room temperature the component is in compression and the substrate in tension. Since normally the component and the substrate are more rigid than the solder, it is the solder that deforms in both shear and tension to relieve the strain. Whether or not the solder in the joint is able to withstand fracture at the level of strain imposed is dependent on a multitude of factors related to the size and metallurgical properties of the joint. These are discussed at length in Chapter 11. The joint will most frequently fail as a result of low cycle fatigue, that is a cyclical heating up and cooling down, usually over a prolonged time.

There are three ways in which this effect of differential expansion can be reduced:

(i) choosing construction materials with compatible TCEs;
(ii) increasing the ability of the joint to absorb the induced dimensional differences;
(iii) choosing a substrate that can absorb the induced dimensional differences.

These options are now briefly considered in turn.

3.2.1 Matched TCE

The first option, to match the TCE of the substrate to that of the component, has given rise to a number of suitable interconnection substrates with a TCE of around 6 ppm.K^{-1}, equivalent to that of the ceramic used for the manufacture of resistors, capacitors and leadless chip carriers. The most straightforward way of achieving TCE equivalence is to use an alumina ceramic substrate for surface mounting assembly. This is a long and well established technology commonly referred to as hybrid microelectronics assembly. In addition, a number of other substrates with TCEs close to 6 ppm.K^{-1} have been developed,[37] usually based on a metal core, since the TCE of a metal alloy can be readily adjusted metallurgically. The core is covered with an adherent, thin insulating layer which is constrained to the TCE of the metal. However, having a matched TCE only alleviates the problem; it does not remove it. This is because the unwanted stresses arise in practice not by isothermal heating of the assembly but by heat generated in the components themselves. During power-up of an assembly the temperature of the components rises faster than that of the substrate and so, even if the TCEs are exactly matched, a strain and therefore a stress is induced in the joints, until thermal equilibrium is achieved.

In summary, a substrate with a TCE matched to that of ceramic significantly reduces the rate of failure of solder joints by low cycle thermal fatigue in leadless surface mounted components. It does not however totally remove the stresses caused by differential expansion of the joint members nor therefore the thermal fatigue of the solder.

3.2.2 Compliant Joint

The second approach is to design the joint to absorb the thermally induced strain in the system arising from differential expansion. In conventional assembly the lead wire takes on this responsibility, being of a relatively ductile and thin nature, easily bending to fully relieve the induced strain. This is also true for leaded surface mounted components, such as the SO and PLCC packages or LCCCs in leaded sockets, which have leads that are long and thin enough to be sufficiently compliant to relieve the strain. It is the leadless components for which the problems arise since the wire is absent and the solder itself must absorb the strain. The ability of the solder joint to do this over a large number of heating up and cooling down cycles depends upon its shape and its metallurgical characteristics. The greater the stand-off of the component from the board, the more like a wire the solder joint becomes and the more able it is to withstand fatigue failure.[38] The means by which the stand-off can be increased are dealt with in Section 11.3.7.

3.2.3 Compliant Substrate

The third method by which the problems arising from differential thermal expansion can be minimised is to use a substrate that is compliant enough to absorb the strain induced. Several approaches are possible. The substrate can be fully flexible or it can be rigid with a flexible elastomer surface layer on which is the interconnection pattern of conductive tracks and pads. Another method is to incorporate local flexible heat sinks into the substrate for specific components.

3.3 THERMAL MANAGEMENT

There are a number of parameters of electronic components, both passive and active, that are seriously temperature sensitive and consequently their electrical performance and that of the assembly can be seriously affected by excess temperature.[39] The deterioration of performance as a function of increasing temperature results mainly from the fact that solid state diffusion processes across interfaces within the device or at terminations and solder joints increase their rates not linearly with temperature, but exponentially, thus enormously enhancing the observed effect of temperature upon performance. (This is discussed in Chapter 13.) Electronic components and assemblies are normally designed with this in mind to operate up to a certain temperature and then, at higher temperatures, are derated as a given function of the temperature. The thermal management of a circuit assembly is therefore often of prime importance. The problem of the dissipation of heat from a circuit has been exacerbated with surface mounting of components for two main reasons:

(i) The packaging of surface mounting components is smaller but the device might carry out the same functions as a DIL package, requiring the same power and the same dissipation of heat. Thus the increase in the density of packing gives rise to an increase in requirement for efficient heat dissipation.
(ii) The thermal path from the semiconductor chip to the printed circuit board and hence to the outside world has a significantly higher thermal resistance because the thermal conductors are perforce of smaller dimensions.

Heat transfer from its source at the component where power is dissipated, to the ambient air outside the finished product is effected by the three basic methods: conduction, convection and radiation.

3.3.1 Heat Dissipation by Conduction

Conduction is the flow of heat through an unequally heated body from a place of higher to one of lower temperature. The quantity of heat energy Q (joules) flowing in a thermal conductor between two planes normal to the direction, x, of heat flow, in an instantaneous time interval dt, is given by the rate of flow:

$$\frac{dQ}{dt} = -KA\frac{dT}{dx} \tag{3.1}$$

where $\frac{dT}{dx}$ is the temperature gradient in the heat flow direction, A (m^2) is the area of cross-section and K(J.s^{-1}.m^{-1}.K^{-1}) a constant of the system called the coefficient of thermal conductivity.

In the steady state of heat being supplied at one end of a thermal conductor and removed into a heat sink at the other, a loose analogy can be drawn to an electrical conductor of length ℓ, for which $\frac{dQ}{dt}$ is the current and ΔT the voltage, which leads to the concept of a thermal resistance

$$R_\theta = \frac{\ell}{KA} \tag{3.2}$$

with units (K.W^{-1}). The electrical resistivity r (ohms.m) gains equivalence with the 1/K, the inverse of the thermal conductivity.

With this in mind, the effect of thermally conducting components in a 'thermal circuit' can be summed using the same concepts of series and parallel resistances as in an electrical circuit. The heat flow through a multilayered solid in the steady state condition is illustrated schematically in Figure 3.1. The temperature T_4 can be calculated without calculating T_2 or T_3, by summing the thermal resistances:

$$R_\theta \text{ (total)} = \frac{1}{A}\left(\frac{\ell_1}{K_1} + \frac{\ell_2}{K_2} + \frac{\ell_3}{K_3}\right) \tag{3.3}$$

Similarly, in Figure 3.2 for a parallel heat flow pattern

$$\frac{1}{R_\theta(\text{total})} = A\left(\frac{K_1}{\ell_1} + \frac{K_2}{\ell_2} + \frac{K_3}{\ell_3}\right) \tag{3.4}$$

These types of one-dimensional summations provide a guide to the performance of different materials and different configurations, but the correspondence between thermal and electrical conductors is not rigorous because the former is physically able to dissipate heat energy three-dimensionally along its entire length, not only at its end. The rigorous approach requires the solution to Fourier's equation of heat propagation by conduction, with the appropriate boundary conditions. Fourier's equation gives the temperature T at time t and at a point with co-ordinates (x, y, z):

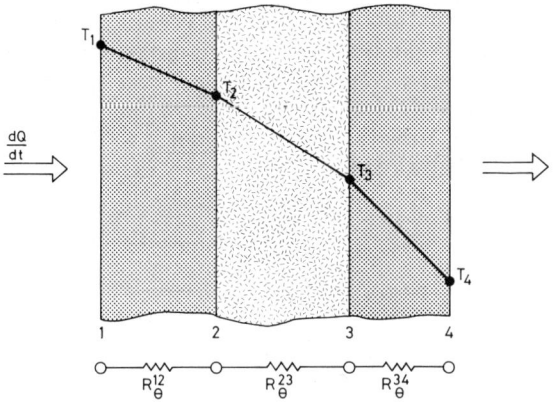

Fig. 3.1 Steady state heat flow through a multilayered solid, showing the equivalence to a series electrical resistor chain.

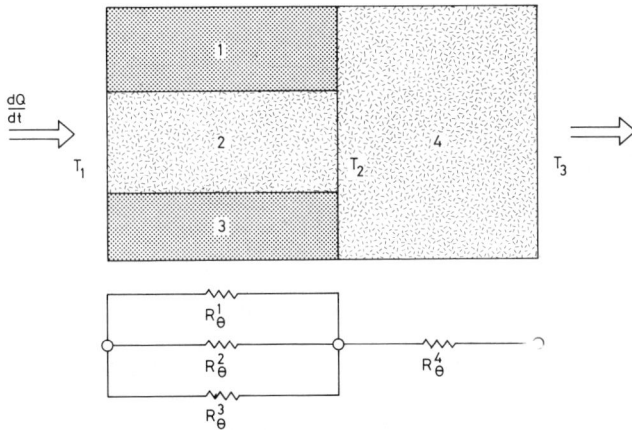

Fig. 3.2 Steady state heat flow through a multicomponent solid, illustrating the equivalence to a series-parallel electrical resistive network.

$$\frac{dT}{dt} = -\frac{K}{c\varrho}\left(\frac{d^2T}{dx^2} + \frac{d^2T}{dy^2} + \frac{d^2T}{dz^2}\right) \qquad (3.5)$$

where c is the specific heat (J.kg^{-1}.K^{-1}) and ϱ the density (kg.m^{-3}) of the conducting material. The solution of any problem in heat conduction must satisfy this equation. Specific solutions to Equation (3.5) can be found,[40] for example for a chip carrier lead being supplied with heat from the semiconductor chip and losing heat along its length, outside the package to the surrounding air as well as from its outer end by virtue of the heat sinking effect of the solder joint and interconnection track. Another example might be the three-dimensional heat flow in a submerged copper plane in a multilayer board with heat input at one or several points arising from thermal via heat sinks.

The practical effect of considering three-dimensional heat flow, comprising both the flow in the direction of temperature gradient and the lateral spreading, is, normally, by increasing the heat paths from the hotter areas to the cooler areas, to increase the heat flow and effectively reduce the thermal resistance below that calculated from the one-dimensional model given by Equation (3.2). Simplistically, but effectively, a spreading angle Ψ can be defined[41] as shown in Figure 3.3. The value of Ψ depends on the relative temperature gradients in the direction of the main heat flow from the heat source to the heat sink, and in the plane perpendicular to that direction.

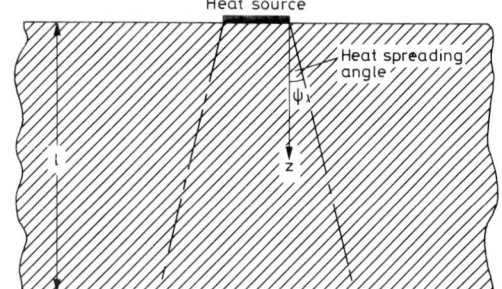

Fig. 3.3 Heat spreading due to three-dimensional heat flow.

3.3.2 Heat Dissipation by Convection

Transfer of heat by convection is less susceptible to mathematical treatment. In convection, heat is transported by the movement of a stream of fluid which, flowing along a temperature gradient, carries molecules of a higher energy content from points of high to points of low temperature. Convection is restricted to fluid systems and two classes may be distinguished:

(i) natural convection in which the movement occurs as a result of density changes produced by the heat;
(ii) forced convection in which the movement is maintained externally.

For electronic assemblies the fluid is normally air.

If an object is allowed to cool naturally the rate of loss of heat energy $\frac{dQ}{dt}$ (J.s^{-1}) is approximately proportional to the nth power of the excess temperature of the object over its surroundings. Whence from Equation (3.5)

$$\frac{dT}{dt} = -\frac{K_n}{c\varrho V}(T - T_f)^n \qquad (3.6)$$

where V is the volume of the object and T is temperature at time t. T_f is the temperature of the surrounding fluid.

If the object is cooled by forced convection the rate of loss of heat is given by an equation exactly analogous to Equation (3.6) but replacing K_n by K_p and n by p, where p differs from n.

The critical process that determines the value of n or p is the heat transfer across the boundary layer between the solid object and the fluid. Maybe 99% of the temperature differential occurs across the boundary layer only a few micrometres thick. If the thermal conductivity of the boundary layer is K and its thickness is δ, K/δ is defined as the heat transfer coefficient, h ($W.m^{-2}.K^{-1}$) of that interface, whose value depends on the physical properties of both the fluid and the solid surface. The equivalent of Equation (3.1) is then

$$\frac{dQ}{dt} = hA(T - T_f) \quad (3.7)$$

and the steady state thermal resistance is

$$R_\theta = \frac{1}{hA} \quad (3.8)$$

The heat transfer coefficient h, and consequently R_θ as well, are not fixed materials properties since the degree of roughness and angle of the solid with respect to gravity can affect markedly the heat transfer properties. Additionally turbulent flow of the fluid (air) augments the heat transfer by convection because of eddy currents which aid the flow of heat more than a streamline, laminar flow.

3.3.3 Heat Dissipation by Radiation

Whereas conduction and convection can take place only in a material medium, radiation can take place as a heat transfer mechanism in the absence of material. Thus radiant heat transfer becomes progressively more dominant than convection as the fluid density decreases, i.e., for the case of air being replaced by a vacuum. In addition, the rate of radiant heat transfer is proportional to the fourth power of the temperature (in kelvin) of the surface:

$$\frac{dQ}{dt} = \sigma A(\varepsilon_1 T_1^4 - \varepsilon_2 T_2^4) \quad (3.9)$$

where ε_1 and ε_2, and T_1 and T_2, are respectively the emissivities and the temperatures (in kelvin) of the emitting and absorbing surfaces. The constant of proportionality is the Stefan-Boltzmann constant σ (5·67 10^{-8} $W.m^{-2}.K^{-4}$). Because of the form of Equation (3.9) and the value of σ, radiant heat transfer rapidly becomes a dominant heat transfer process at high temperatures, but is generally not very significant as a mechanism for cooling electronic assemblies.

3.3.4 Relative Rôles of Heat Transfer Mechanisms

The relative magnitudes of the three heat transfer mechanisms for the thermal management of a surface mounted assembly depend on the materials properties, the temperature differentials and the geometry. Consider an idealised example shown in Figure 3.4 of a 1 cm square plate, 1 mm thick embedded in a perfect thermal insulator so that only one flat surface is exposed. The inner surface is measured at 70°C whilst the outer surface is 20°C. The rate of heat transfer at the outer surface of the plate is a measure of the efficiency of the cooling method to maintain this state. In (a) the plate is in intimate contact with a heat sink and the heat transfer mechanism is conduction. From Equation (3.1), the rate of heat flow is

$$\frac{dQ}{dt} = 5 \text{ K watts}$$

Thus for a copper plate ($K \approx 390$ J.s^{-1}.m^{-1}.K^{-1}), $\frac{dQ}{dt}$ is about 2000 W, whereas for alumina ($K \approx 25$ J.s^{-1}.m^{-1}.K^{-1}), $\frac{dQ}{dt}$ is about 125 W, and for epoxy-glass laminate ($K \approx 0.25$ J.s^{-1}.m^{-1}.K^{-1}), $\frac{dQ}{dt}$ is about 1 W.

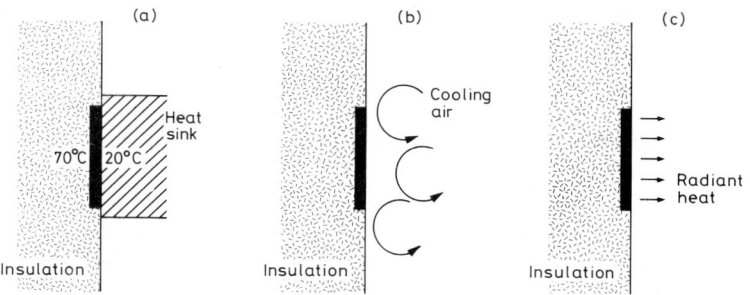

Fig. 3.4 A comparison of the efficiency of cooling by (a) conduction, (b) convection and (c) radiation. In each case the hot plate is 1 cm square and 1 mm thick with its surfaces at 70°C and 20°C.

The situation in Figure 3.4(b) is one of convective cooling. For natural convection by air, the heat transfer coefficient h is approximately 10 J.s^{-1}.m^{-2}.K^{-1} for metallic surfaces, so that, from Equation (3.7), $\frac{dQ}{dt}$ is only about 50 mW. If the convection is forced, the heat transfer coefficient might well increase by a factor 30, but $\frac{dQ}{dt}$ remains relatively low at about 1·5 W.

Heat transfer by radiation, indicated in Figure 3.4(c), is governed by the actual temperature rather than the temperature gradient. From Equation (3.9), with $T_1 = 70$°C, $T_2 = 20$°C and $A = 1$ cm, the radiant heat dissipation is less than 40 mW. If the temperature of the inner surface is increased to 150°C the radiant heat increases to about 140 mW, but it is the least significant heat transfer mechanism under normally experienced conditions of electronic assemblies.

3.4 INTERCONNECTION DENSITY

3.4.1 Interconnection Requirements

Although the conductor tracks on the substrate are considerably shorter per pin-out for surface mounted assemblies than for insertion mounted assemblies, the track density is dramatically higher because of the smaller spacings between the pin-outs and the more efficient use of the perimeter of each device for pin-outs. The conductor track length required for interconnections between components in an ideal model system is shown in Figure 3.5. This is the ideal or minimum density required.[42] More realistically the conductor routing efficiency is only about 50% and so the practical interconnection density required is twice these values. Thus, for a 44 pin LCCC a conductor track density of 50 cm/cm^2 is required. For a 68 pin LCCC this increases to 53 and for an 84 LCCC to 55 cm/cm^2.

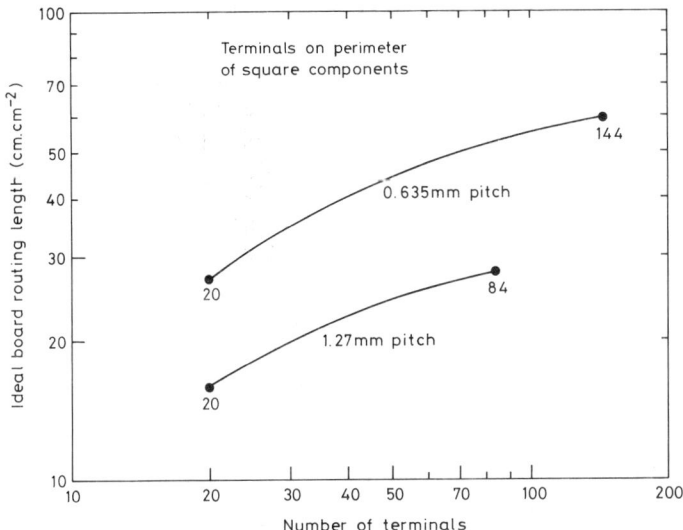

Fig. 3.5 The minimum interconnection track length required for square components, calculated assuming an ideal routing efficiency.[42]

A conductor track density of 50 cm/cm^2 in one conductor plane requires track widths and track spacings of 100 µm. This is right on the limit of present-day materials and processes. Using thin copper foils, of 12, 9 or even 5 µm thickness which have been carefully inspected for pinholes, it is possible, using FR-4 epoxy-glass copper clad laminate and standard production procedures, to achieve 125 µm lines and spaces. Therefore, to fully implement the spatial advantage of surface mounting it is normally necessary to use a multilayer structure of conductors. The multilayer technology is even more a requisite for the higher density pin-count, chip-on-board systems such as TAB, that require an interconnection density almost twice that of chip carriers. Figure 3.6 shows how the technology of high-resolution PCBs has progressed over 30 years, but, in this particular respect of line widths, the copper-clad laminate is now apparently approaching its achievable ultimate.

For screen printed conductor tracks using thick film inks, similar track widths and spaces of 100 µm are possible. On alumina substrates the thick film deposition process can be modified to incorporate a separate etching step, thus enabling finer line patterns down to 20 µm line widths and spacings to be achieved. This corresponds to an interconnection density of 250 cm/cm^2.

3.4.2 Blind, Buried and Thermal Vias

In electronic assembly parlance, vias are interconnections perpendicular to the plane of a circuit board and within the board, electrically or thermally joining

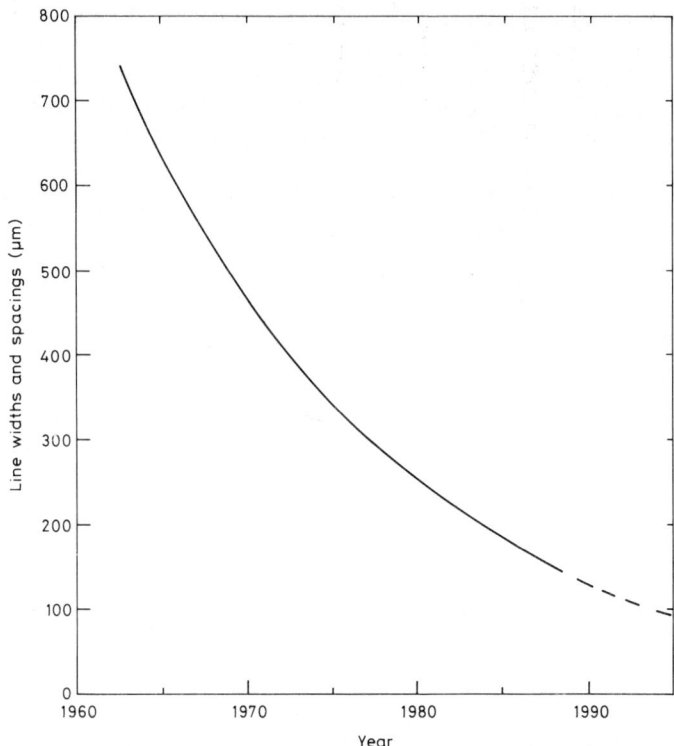

Fig. 3.6 Illustrating the steady reduction of track widths and spacings on PCBs achieved over 30 years, using copper-clad organic laminates.

conductor planes in the structure. There is no intention to insert component leads into vias. A *blind via* is an interconnection that connects the primary circuit side or the secondary underside to inner layers of the circuit board, but does not connect both outer layers. A *buried via* is an interconnection that connects only inner layers of the circuit board. A *thermal via* is an interconnection intended to conduct heat from a component heat source to a heat sink plane of the circuit structure. Schematic examples of vias are shown in Figure 3.7. Because of the required interconnection density on surface mounted substrates, if full benefit of the technology is to be gained, a multilayer structure is desirable and therefore vias are essential.

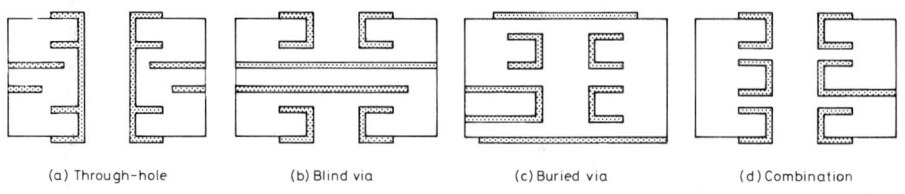

(a) Through-hole　　(b) Blind via　　(c) Buried via　　(d) Combination

Fig. 3.7 A schematic representation of through-board, blind and buried vias used for interconnecting tracks on multilayer PCBs.

A thermal via may contain a solid heat exchanger to remove heat from a high power component to a thermally conducting core, and the size of the via is dictated by the thermal management requirements. Blind and buried vias, however, rely on a plated copper barrel to provide an electrical connection. To produce buried vias, a sequential lamination method is required, using standard multilayer fabrication equipment. In this method, each copper-clad layer is treated as a separate circuit board with holes drilled and plated, first with electroless copper and then electrolytic copper. No z-axis controlled drilling is required with this fabrication method because all the vias are initially created as through-holes. The separate sub-layers are then laminated together and during the pressing ideally the vias completely fill with epoxy resin flowing from the prepregs. This eliminates air entrapment in the vias or the potential for solution entrapment that might occur in processes involving drilling and plating of vias after lamination.

In the manufacture of conventional PCBs, the drilling process is the most time consuming and by far the most expensive, representing around 50% of the production cost apart from the materials of the board. If vias are avoided, a surface mounting board can be produced without any drilling and it is therefore usually advantageous to sacrifice some of the size advantage gained by surface mounting to enable the circuit to be designed with only one conducting layer. If drilled holes are required, the need for finer tracks and high density interconnection demands smaller and smaller via holes. Vias are commonly drilled at 0·3 mm diameter. Since a depth-diameter aspect ratio of 5 is about as large as is practical, full-width boards can be drilled normally only one at a time.

Mass lamination is a technique used primarily to produce 4-layer multilayer boards with the inner layers as power and ground planes. The boards are produced as large panels, aimed to be cost-competitive with fine-line double-sided boards, which require more high level process control.

3.5 TYPES OF SUBSTRATE AND INTERCONNECTION

3.5.1 Organic Substrates

The format for substrates based on an organic resin, and used as a printed circuit interconnection system, comprises a sheet of laminated reinforced resin with an outer surface of copper foil, from which the conducting tracks are etched. The number of types of resin, reinforcement and lamination procedures used in the manufacture of such boards are many. The number of variations of manufacturing routes used to produce the finished PCB from the base material is enormous. It is almost the case that no two PCB houses use fully identical processes. Several books have been written on this subject alone.[43, 44] Here, only a brief outline of the most popular materials and processes involved will be given, specifically in relation to surface mounted assemblies.

Although the resins used are synthetic, they are based on natural products whose composition and properties may vary while enabling the laminate to retain its specification properties. The resin can be either thermoplastic or thermosetting, produced by means of a polymerisation process, although the latter is more usual. The choice of resins is potentially great, but the blend used is selected for its properties such as dielectric constant, dissipation factor, arc resistance, loss factor, absorption of water, mechanical properties of tensile, shear, flexural and impact strength, environmental resistance, flammability and self-extinguishing properties, heat resistance, adhesion to the reinforcing fabric and to the copper foil, machinability, dimensional stability, cost, and so on. The compromise of properties of the material used depends on the end-use of the product and its specification.

Over 90% of laminates for PCBs are produced using phenolic or epoxy resins. Others used in special applications and in pilot-scale plants are polyesters, melamines, silicones, polytetrafluoroethylene (PTFE) and polysulphones.

The cured resin on its own, whilst fulfilling the electrical requirements of the PCB, would not fulfil the mechanical properties, and an appropriate filler or reinforcement is incorporated, which is wet by the resin before curing. Reinforcing materials used include paper, glass cloth, glass mat, chopped glass, asbestos, nylon, and so on. Over the years of use, some combinations of resin and reinforcement have gained popularity and are classified by commonly used notations as given in Table 3.1, which are subject to specified properties (e.g., FR- indicates fire resistant).

Table 3.1

Common Types of Laminate for Rigid PCBs

Resin	Reinforcement		Classification
phenolic	paper	sheet	FR-2, X, XP, XX, XXP, XXX, XPC, etc.
	cotton	fabric	C, CE, L, LE
	asbestos	sheet	A
		fabric	AA
	glass	fibres	G-2
		cloth	G-3
	nylon	fibres	N-1
amino (melamine)	-	-	ES-1, ES-3
	glass	cloth	G-5, G-9
epoxy	paper	sheet	FR-3
	glass	cloth	G-10, G-11, FR-4, FR-5
alkyd (polyester)	glass	mat	GPO-1, GPO-2
silicone	glass	cloth	G-7

3.5.1.1 PHENOLIC LAMINATES

Phenolic resins are the most widely used because of their relatively low cost and good electrical properties (except arc resistance). Bakelite was the first laminate of this type and is still used today. Phenolic resins are made from phenol and formaldehyde. The chemistry is extremely complex but the main steps start with the formaldehyde reacting with the aromatic ring of the phenol to produce a mixture of benzyl alcohols, at the so-called 'A' stage of the polymerisation. As the reaction proceeds, more formaldehyde reacts and water is lost until it reaches a resol known as the 'B' stage. This is soluble in appropriate solvents for impregnation of the filler material. If the chemical nature of the filler is appropriate, as with the cellulose in paper for example, some chemical reaction takes place between the resol and the filler.

The filler is therefore perfectly embedded and the resin cures by establishing cross-links between its long molecules. The process involves dissolving the resol in a solvent, impregnating the cloth or paper and drying, at which stage the sheets are not tacky at room temperature and can be handled and cut to size. Then between 2 and 10 of these sheets are stacked, commonly with an outer copper foil sheet, between two heated plates in a laminating press. The combination of pressure and temperature causes the resin to flow to form a single laminated sheet.

Almost all phenolic laminates now are reinforced with paper, with for example a resin content of 35, 45 and 58% respectively for grades X, XX and XXX. The higher the resin content the harder the material; XXX cannot be punched and any holes must be drilled. Flexibilisers can be added, for example the code XXXP can be punched if heated to around 50–80°C and XXXPC can be cold-punched, usually at 25°C but only if less than about 1·5 mm thick.

3.5.1.2 EPOXY LAMINATES

Epoxy resins are produced from ethylene chlorohydrin (epichlorohydrin) and bisphenol-A, both of which are considerably more costly than the constituents of phenolic resins. The choice of epoxy resins is made because of their outstanding electrical, mechanical, chemical and thermal properties. FR-4 epoxy-fibreglass laminate is the standard for all high technology and professional electronic assemblies.

Some additives are used to modify the properties of epoxy resins. Phenolic novolacs for example improve the stiffness of the board during soldering. Up to 20% of the resin can be phenolic novolac, which increases the glass transition temperature T_g of the resin from about 120°C to 135°C. This is demonstrated[45] by measuring the temperature dependence of the z-axis TCE of laminates made with the resins, as shown in Figure 3.8. (The x,y axis expansion is constrained by the reinforcement fibres.) Trifunctional and tetrafunctional additives are used to add cross-linking and improve performance at higher temperatures. The T_g of these resins is again around 135°C. The T_g of an epoxy resin can be increased further to about 190°C by using vinyl phenols. These additives not only increase T_g but reduce the z-axis expansion at temperatures above T_g.

The 'B' stage resin is the result of reaction between the two constituents, and can be dissolved in appropriate solvents to impregnate the reinforcement, usually a woven fibre-glass cloth. The resultant sheet is dried and can then be handled and cut to size. In this state it is known as a 'prepreg'. Usually eight prepregs are laminated together, commonly with copper foil sheets on both sides to produce a copper-clad board 1·6 mm thick. During lamination the epoxy resin is cured by the heat, pressure and action of a catalyst in the resin formulation.

Cured epoxy resins are almost indestructible except by high temperatures and oxidising acids. Because of the resin cost, the reinforcement filler is also usually of high quality material. There are five grades of epoxy resin laminates in common use for PCB production. FR-3 is the only one reinforced with paper and contains fire retardants. It has good electrical characteristics, machinability and punchability, with reasonable dimensional stability and mechanical strength. The other four are reinforced with glass fibre cloths. G-10 and its flame retardant version FR-4 are the most used for professional boards. The mechanical properties and dimensional stability are provided by the glass cloth, but are anisotropic in the three axes because of the asymmetry of the cloth and the impregnation process. The flame retardancy of FR grades is achieved by substituting some functional groups of the resin by chlorine or bromine. G-11 and its flame retardant grade FR-5 are the highest quality and most expensive epoxy laminates. They are less easily machined and less punchable than G-10 and FR-4 but have better heat resistance, retaining about 50% of their flexural strength after an hour at 150°C.

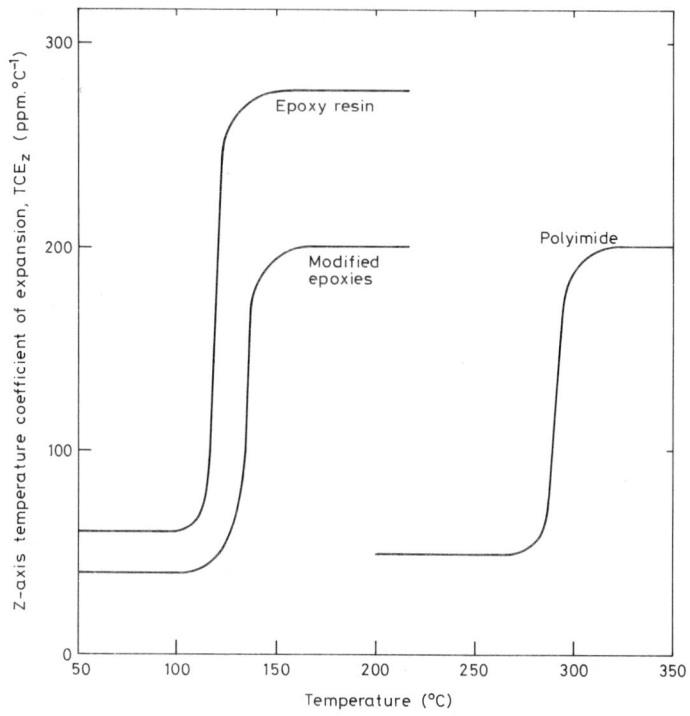

Fig. 3.8 The temperature dependence of the TCE in the z direction of laminate boards, manufactured with different resins, all demonstrating an abrupt increase at the glass transition temperature.[45]

3.5.1.3 POLYIMIDE LAMINATES

The polyimide resins[46] were developed for fibreglass reinforced laminates for manufacturing PCBs that have a glass transition temperature, T_g, higher than the soldering temperatures encountered. Typically T_g is around 290°C and so there is no increase in z-axis expansion at soldering temperatures, as illustrated in Figure 3.8. This property is of benefit to the integrity of through-hole plating which has the potential to crack during soldering when the z-axis expansion is too high.[47] Polyimide laminates are considerably more expensive than their epoxy equivalents. Additionally they are more difficult to bond to the copper foil as well as to nucleate electroless copper through the via holes. The advantages to be gained in plated-through-hole reliability however can be quite significant.

3.5.1.4 OTHER ORGANIC LAMINATES

Besides phenolic and epoxy resins, polyester, silicone resin and melamine laminates are also used for PCB substrates. Polyester laminates are relatively cheap materials but the problems of bow and twist after soldering may be considerable. Through-hole plating of vias is not successful and so they are used only for single-sided assemblies. Silicone resins have good resistance to chemicals and to heat and can be used up to 400°C. The electrical properties of silicone resins are good but the lamination and the adhesion of the copper foil is difficult. Melamine laminates are seldom used for electronic assemblies, except where the outstanding property of very high surface hardness is a benefit.

3.5.1.5 THE COPPER FOIL

Organic laminates are produced with one or both sides having an incorporated copper foil. The adhesion of the copper to the organic reinforced prepregs is achieved at the lamination stage.

Most copper-clad laminates are made using electrolytically produced foil, formed by plating on to a stainless steel drum slowly rotating in the liquid electrolyte. The side of the foil in contact with the drum is smooth and shiny whereas the other side is matt and granular. The granular side gives the foil adhesion to the resin of the laminate. The purity of the copper is usually about 99·5%.

The copper foil has to meet very strict requirements especially regarding pinholes which can lead to open circuits on fine-line conductor tracks. A high ductility is required to match the flexural properties of the laminate. The thickness of the copper foil specified on a laminate is commonly expressed in oz.ft^{-2}, the most commonly used being 1 oz.ft^{-2} which transposes to an average thickness of 35·5 μm depending on the density of the electrodeposit. For fine-line circuits, better resolution can be obtained with thinner foils, and laminates with foils as thin as 5 μm are in use.[48]

3.5.1.6 THE GLASS CLOTH

The glass fibres used to reinforce epoxy-fibreglass laminates are normally twisted into bundles to form a thread which is then woven into a cloth. The size of the individual fibres, the number of fibres per thread and the weave mesh, or threads per centimetre, are all carefully specified. Normally the mesh of the warp threads (running in the direction of the wound cloth length) is not the same as the mesh of the weft threads (running across the cloth).

Commonly, the glass fibres are 9·6 μm diameter. The threads consist either of 408 or 816 fibres, twisted one full turn about every 3-5 centimetres. The two most commonly employed glass cloths for PCB fabrication use either the 408-fibre thread for both warp and weft with a warp mesh of 17 threads per centimetre and a weft mesh of 13 threads per centimetre, or the 408-fibre thread at 17 per centimetre for the warp and the 816-fibre thread at 9 per centimetre for the weft.

The quality of an epoxy-glass laminate depends on the success of the resin impregnation between the threads and individual fibres. The glass fibres are usually treated with a wetting agent such as silicon hydride to improve the wetting and adhesion of the resin to the glass.

3.5.1.7 COMPARATIVE PROPERTIES OF ORGANIC LAMINATES

A brief guide to the relative properties of the copper-clad reinforced organic laminates suitable for PCB production is given in Table 3.2.

Table 3.2

Comparative Properties of Copper-clad Organic Laminates for PCBs

	XXXPC	FR-2	FR-3	G-10	G-11	FR-4	GPO-1
Density (g.cm^{-3})	1·28	1·30	1·45	1·75	1·75	1·85	1·5-1·9
TCE (ppm.K^{-1})							
in-plane:	11	11	13	10	10	11	15
through-plane:	12	12	15	15	14	15	21
Thermal conductivity (J.s^{-1}.m^{-1}.K^{-1})	0·24	0·24	0·23	0·26	0·25	0·25	–
Dielectric constant (1 MHz)	4·5	4·5	4·6	5·0	5·1	4·9	4·4
Dielectric strength (kV)	60-70	60-70	60-65	35-60	35-60	35-65	40
Tensile strength (MPa)							
in-plane:	92	88	83	280	280	280	70
through-plane:	71	66	63	235	235	235	83
Maximum water absorption (%)	0·8	0·8	0·75	0·35	0·35	0·35	1·0

3.5.1.8 PCB FABRICATION WITH ORGANIC LAMINATE SUBSTRATES

As already mentioned, the variations of manufacturing routes to progress from a sheet of copper-clad laminate to a PCB suitable for surface mounting components are legion, and beyond the scope of this book. The routes available to produce a PCB for surface mounting components are essentially those available for PCBs intended for insertion mounting. If a double-sided board is to be produced, the through-hole plating of vias replaces the through-hole plating of component lead holes.

The basic process for producing a double-sided through-hole connected PCB is called panel plating, the main steps of which are illustrated in Figure 3.9(a). In brief, the steps are drilling of via holes, cleaning, activation, deposition of electroless copper, electroplating of copper across the whole panel, laminating with photoresist, image transfer by exposure and wash, plating of tin-lead, removal of photoresist, etching of unprotected copper and reflow of the tin-lead coating.

A second manufacturing route, designed to prevent as far as possible, copper build-up on the areas from which it must subsequently be removed during etching, involves the selective plating of copper, and hence is known as the pattern plating process. The main steps involved are illustrated in Figure 3.9(b). They are drilling of via holes, cleaning, activation, deposition of electroless copper, lamination with photoresist, image transfer by exposure and wash, copper electroplating, tin-lead plating, removal of photoresist, etching of unprotected copper, and reflow of the tin-lead coating.

With the requirement of finer and finer interconnection geometries it is advantageous to apply a solder resist over all parts of the board except where

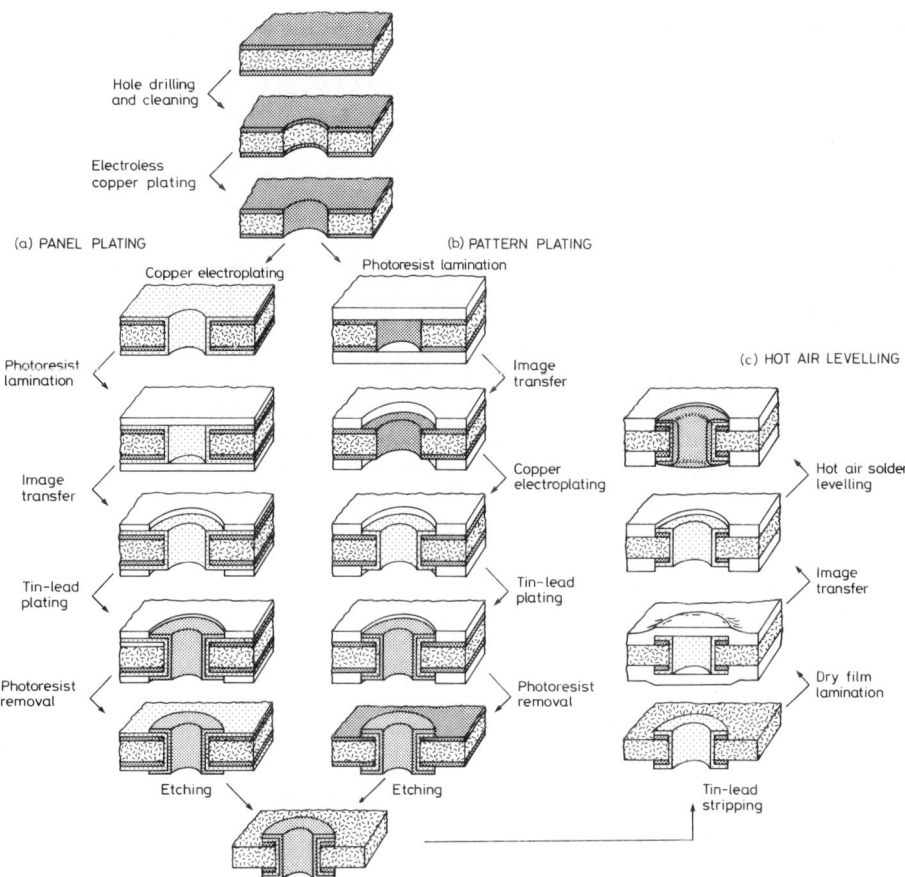

Fig. 3.9 The main production stages in the production of a double-sided PCB by (a) panel plating, (b) pattern plating and (c) hot air levelling.[43]

solder joints are to be made, in order to prevent solder bridging between tracks and to improve the overall cosmetic appearance of the board. There are three modes of applying the solder resists, wet ink, dry film or photoimageable ink.[49] The wet ink process involves screen printing on to the board area to be protected, and curing either by infra-red or ultra-violet radiation. In the dry film process a polymer film is applied and imaged with ultra-violet light, the uncured film being dissolved and washed away using a solvent.[50] Definition of the resist pattern is much better with dry film than wet ink, provided the dry film is not so thick that the ultra-violet light is scattered significantly in the film as it is being cured.[51] The third type of solder resist is applied by screen printing or curtain coater and then imaged and cured in the same manner as the dry film.[52]

Because of the widespread use of dry film solder resist, a new production route for PCBs was developed to overcome the cosmetic problem of the tin-lead coating of the conductor tracks under the solder mask wrinkling as it melts and then re-solidifies during the soldering assembly process. In this, the hot-air levelling production route, the dry film is applied over bare copper. One such process is illustrated in Figure 3.9(c): drilling of via holes, cleaning, activation, deposition of electroless copper, lamination of photoresist, image transfer by exposure and development, etching of unprotected copper with an etchant that does not attack the photoresist, removal of photoresist, application of solder resist followed by its exposure and development. It is then customary to apply a protective solderable coating to the bare copper by immersion vertically into a bath of molten solder. As the board is removed, hot air is blown perpendicularly to the board to level the solder and unplug the plated-through holes.[53-57]

On a surface mounting PCB the blind and buried vias can be produced as described in Section 3.3.2 and become filled with resin during lamination. The side-to-side vias can either be treated as normal through-plated holes, in which case they would become filled with solder if wave soldered, or they can be protectively tented over by the dry film. No problems of outgassing or of long-term reliability have been found to arise when vias are tented over in this way.

When it proves impossible to run all the interconnections between components on two planes, a multilayer board becomes necessary. Generally each board is separately laminated from specifically produced prepregs, making multilayer boards quite costly to produce. The simplest multilayer boards have 3 or 4 copper planes, the inner ones being power and ground planes and the outer ones being signal planes. Multilayer boards with up to 18 or even 24 layers are produced with very high acceptance rate.

If the multilayer board is to contain buried vias, these layers must be drilled first before lamination, as described in Section 3.4.2. The more usual

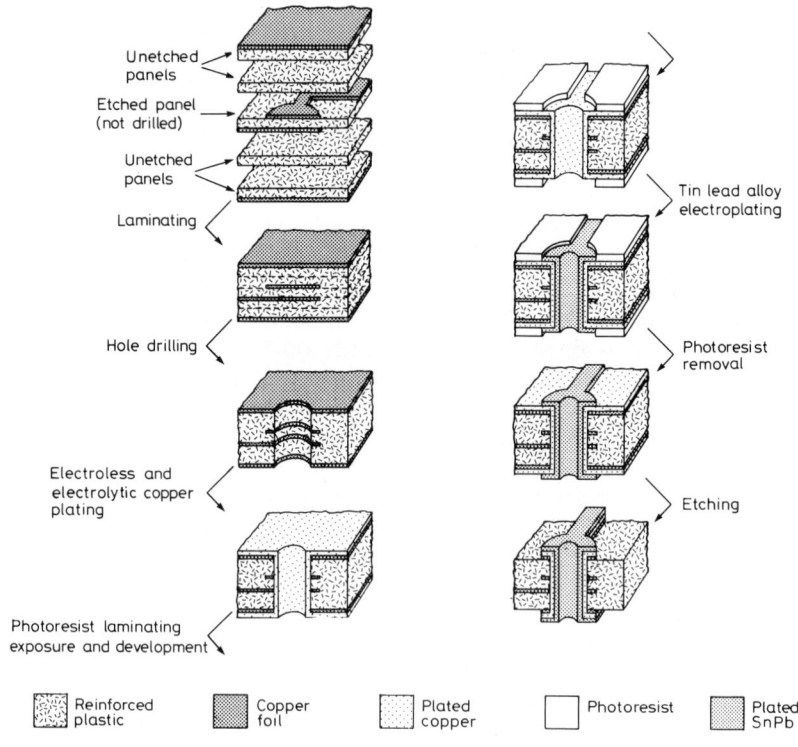

Fig. 3.10 The major steps in the manufacture of a multilayer PCB.[43]

manufacturing process, however, is to drill after lamination as illustrated in Figure 3.10. The process involves the lamination of etched and unetched panels. The resin used in these prepregs must have a greater flow than that used for double-sided laminates, in order to respond to the thickness variations of the etched internal copper planes. Contact between the copper planes is by way of the plating through the via holes.

3.5.2 Matched TCE Substrates: Ceramics

The use of ceramic substrates in hybrid microelectronics is a mature technology. The ceramic substrate can have the same TCE as the ceramic components and so much of the problem of differential expansion is alleviated. The substrate is usually glued to a heat distribution plate with a flexible adhesive as shown in Figure 3.11. The interconnection circuitry on the top surface is a combination of conductive, resistive and dielectric screen printed multilayers, fired on before soldering the components in place.

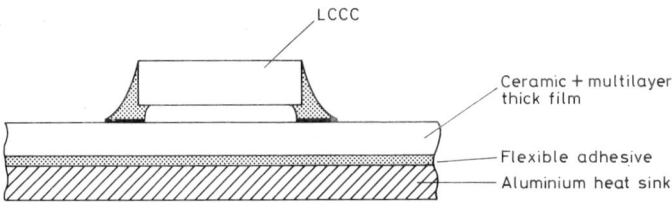

Fig. 3.11 Ceramic substrates are used to match the TCE of substrate and leadless components.

The conductors and resistors are defined by screen printing on the surface using thick film materials which are essentially mixtures of a conductive powder and a glass powder as an oxide loading agent in a resin vehicle. During firing, at about 900°C, the resin is burnt off and the glass bonds to the ceramic. Dielectric layers can likewise be printed over the top, to produce a multilayer structure.

An alternative technology to this 'thick film' approach is the 'thin film' approach in which the circuit patterns are laid down evaporatively in a vacuum. This is a time consuming and expensive process, although great precision of the interconnection pattern can be achieved.

Only two ceramics are commonly used at present, alumina (aluminium oxide)[58] and beryllia (beryllium oxide), the latter in cases where its very high thermal conductivity is essential. Two new materials are also gaining interest: aluminium nitride and a mixture of beryllia and silicon carbide.

The alumina substrates used are 96% for general purpose and 99·5% which is considerably more expensive and used when its superior mechanical strength or its excellent high frequency loss factor is required, as in microwave circuits. The material contains calcium and magnesium silicates in addition to alumina. The percentage figure does not relate directly to the purity but to the density of the sintered material.

Beryllia as a substrate material has electrical and mechanical properties equivalent to those of alumina, but an extraordinarily high thermal conductivity that is higher than aluminium metal. Beryllia is therefore used mainly for high power circuits. It does, however, have concomitant disadvantages notably of cost and toxicity. The latter problem is minimal if care is taken to control dust and use special precautions when laser or abrasive trimming the screen printed resistors.

A recent alternative to beryllia for high thermal conductivity without the health hazard is aluminium nitride. The composite material of beryllia and silicon carbide contains 20% beryllia and so retains its health hazard. Some physical properties of these materials are given in Table 3.3.

Table 3.3

Physical Properties of Ceramic Substrates

	96% Alumina	99·5% Alumina	99% Beryllia	Aluminium Nitride	Beryllia/ Silicon Carbide
Density (g.cm^{-3})	3·7	3·9	2·9	3·3	3·2
TCE [25-150°C] (ppm.K^{-1})	~6·4	~6·6	~5·0	5·6	
Thermal conductivity (J.s^{-1}.m^{-1}.K^{-1})	35	37	250	170	270
Flexural strength (kg.cm^{-2})	3200	4900	1900		
Dielectric strength (kV.mm^{-1})	8	9	14		
Resistivity [25°C] (Ω.cm)	7 10^{14}	7 10^{14}	10^{15}	10^{15}	>10^{13}
Dielectric loss [100 MHz]	0·0055	0·0008	0·0004		>0·05

The available sizes of ceramic substrates are generally limited not by mechanical strength but by the bow of the fired substrate.[59] In the hybrid industry 50 mm square is the common size, but with the intrusion of alternative substrates for surface mounting, such as epoxy-glass laminates, ceramics up to about 200 mm square have become available. Several circuits can be produced on one substrate and separated by scribing and cracking or by laser cutting.

Mounting and via holes are usually produced by laser but can also be diamond drilled or preformed during the pressing and sintering of the ceramic sheet, although shrinkage during this firing process is considerable and tight tolerances are not possible. The holes can be made conducting, to form vias between the surfaces of the substrate by either thin or thick film methods, which are the methods whereby the conducting and resistive tracks are also formed on the substrate surfaces.

The major disadvantages of using ceramic substrates are the limited size availability, the relative difficulty of machining and drilling, the cost and the limited electrical performance for high speed circuitry due to the high dielectric constant.

> **Matched TCE: Ceramic substrates**
>
> *Advantages:*
>
> > Same TCE as leadless ceramic components
> > Good thermal conductivity
> > Mature technology
> > Good process technology
> > Wide availability
> > Thick film circuitry cost competitive for small sizes
>
> *Disadvantages:*
>
> > Size limited to about 15 cm square due to warp and brittleness
> > Subject to fracture in mechanical or thermal shock
> > High dielectric constant
> > Limited rework possible
> > Considerably more expensive than epoxy fibreglass

3.5.2.1 THIN FILM CIRCUITRY

Thin film technology provided the precedent for printing circuits on to first glass and then ceramic substrates. It is a subtractive method of defining resistor and conductor patterns. The basic steps in the process (which may have a number of modifications in practice) are as follows. The substrate is cleaned and placed in an evaporator where residual surface contamination is removed by ion etching. A thin layer of resistive material, usually nichrome but sometimes tantalum nitride, is then evaporated, of a thickness that is sufficient for all the resistance values to be formed. If no resistors, but only conductors are to be used on the circuit, the nichrome is still deposited as a bonding coat, but only a few nanometres thick. A thin layer of copper is then deposited over the nichrome, and the substrate removed from the evaporator.

The following processing relies on the use of photoimaged resists to delineate first the copper tracks and then the resistors. Once the conductors have been outline masked with photoresist they are electroplated up to the required thickness, with copper or gold. The photoresist is then removed and the thin layer of copper chemically etched away from the unwanted areas, exposing the nichrome underlayer. If resistors are not required, further photolithographic and etching stages are used. The width and the length of the resistors determine their values since the thickness of nichrome evaporated is constant.

Nichrome has a resistivity of about 1 $\mu\Omega$.m. In Figure 3.12(a) a resistor or conductor is drawn schematically having a length ℓ and with a rectangular cross-section width w and thickness d. The resistance between the ends is

$$R = r\frac{\ell}{wd} \qquad (3.10)$$

where r is the resistivity. Now if $w = \ell$ as in Figure 3.12(b) the resistor (or conductor) is square and for a given thickness the resistance between two opposite edges can be described as R 'ohms per square', independent of the actual size of the square. This unit therefore is a combined measure of the resistivity of the material and its thickness. If the nichrome has a thickness in the range 0·1 to

1 μm the resistance lies in the range 1-10 ohms per square. For the copper, $r \approx 0.017$ μΩ.m, and the thickness is much greater since it has been built up electrolytically, so that its resistance is normally a few milliohms per square, i.e., a thousand times less than the nichrome.

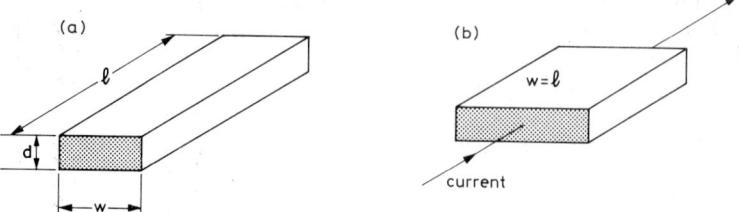

Fig. 3.12 Diagrams of a film resistor to illustrate the concept of resistivity and ohms-per-square, used in Equation (3.10).

Holes may be metallised to make the substrate double-sided, using the thin film process, but special care is needed to throw the evaporant into the holes to ensure success of the copper plating stage.

The advantages and disadvantages of the thin film technique are:

(i) Excellent line definition limited by the photolithographic equipment used; line widths below 50 μm are easily attainable.
(ii) For best line definition the photoresist should be thin and the plating, equal in thickness to the resist, will also be thin. If the resist or the plating is too thick, the tracks become non-rectangular in cross-section.
(iii) Thin film is a single layer process and, apart from through-hole plating, conductor cross-overs are not possible.
(iv) The nichrome is all of one thickness and therefore all the resistors have the same ohms per square, which may cause difficulties with resistor geometries. This can be overcome by using discrete chip resistors, surface mounted with the other components.
(v) Thin film circuits have excellent high frequency performance and are used extensively in low-loss type circuits requiring small and precise dimensions.

3.5.2.2 THICK FILM CIRCUITRY

Thick film printed circuits are a less expensive, more flexible method of forming conductor circuits with resistor elements on a substrate. The circuit is simply screen printed[60] on to the substrate using special inks developed as conductors, resistors and dielectrics (insulators). After printing each component the system is fired and hence the technology has been developed for, and is normally only applicable to, ceramic substrates.[61] The new technology of polymer thick film (Section 3.5.6) however is now available which uses low firing temperatures compatible with organic substrates.

The thick film inks for ceramic substrates are composed of metal and oxide powders, glass powders and a suitable organic carrier of resin and solvent.[62] After printing, the organic components are burnt out at a relatively low temperature, before the firing in the range 850-950°C to melt the glass and form a permanent bond. The bonding can be either by a mechanism involving the melting of the glass particles to form a frit which adheres to the ceramic, or a reactive mechanism whereby a metal-oxide bond is formed, or a mixture of the two. By overprinting conductors with dielectric, a complex multilayer structure can be built up, with a firing between each deposition.[63]

Conductors

Because multiple firings are required, the conductor systems have been confined to gold, silver, platinum and palladium, but, by firing in a nitrogen atmosphere, copper and nickel conductor systems have been developed and used successfully. The cost savings to be gained by using non-noble metals, however, is offset by the need for controlled atmosphere furnaces.

By far the most common conductor used in thick film circuits is palladium-silver fired in air.[64] It has now become virtually the standard for thick film conductors. The silver is chosen because of its high conductivity, the requirement for a noble metal because of the firing in air, and the low cost compared with other noble metals. The drawbacks to using silver are its high dissolution in molten solder, rapidly removing all the metallisation during soldering, and its propensity to migrate across a surface under the influence of an electric field. Both these problems are alleviated by adding palladium to the thick-film ink. The alternative metallisations that are quite commonly used are silver-platinum and gold, the latter being inapplicable for soldering, because of its high dissolution rate but used for gold-wire bonding of chips in the components manufacturing

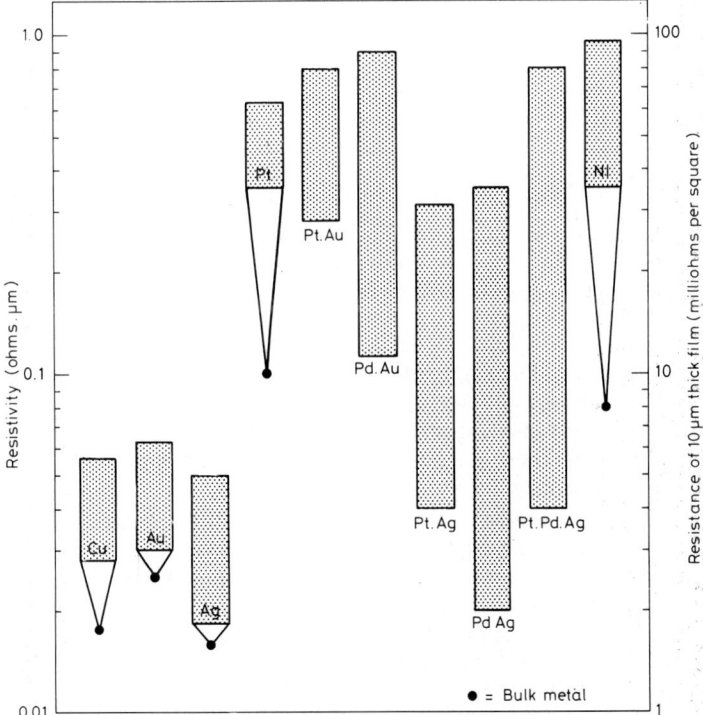

Fig. 3.13 The resistivities of various thick-film metallisations commonly used to print conductor tracks.

and the hybrid microelectronics sectors. As already mentioned, cost-saving copper and nickel systems fired in an inert atmosphere have been developed. The good conductive properties of copper together with its good soldering properties make it an ideal metallisation for multilayer circuits.

The properties of a printed conductor used for an interconnection system that are of importance are its resistivity, its adhesion to the substrate, its migration, its ageing characteristics and its solderability and dissolution in solder. These properties are not dependent alone on the metal alloy used, but on the binders and vehicles used, the particle size distribution, the firing temperature, the adhesion mechanism invoked, the storage conditions and so on.

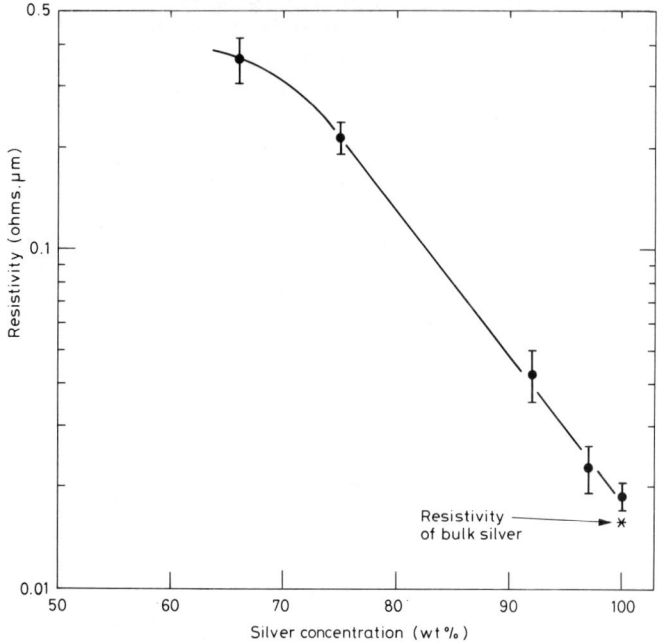

Fig. 3.14 The resistivity of Ag-Pd inks as a function of their Ag:Pd ratio.

Figure 3.13 shows how the resistivities of the various thick-film conductive metallisations compare. The range of resistivity for a particular metallisation can be quite large, depending on the formulation, firing temperature and bonding mechanism. Figure 3.14 illustrates how the resistivity of Ag-Pd inks is a function of the composition, all these particular inks being of the same family and fired under identical conditions.

The dissolution and leaching properties of thick film metallisations can be measured quantitatively. In Figure 3.15 data are given from a test involving printed lines 250 μm wide and 16 μm thick after firing, repeatedly dipped in a controlled manner into a solder bath at 215°C. The solder used in this instance is 62Sn:36Pb:2Ag, the silver addition intended to reduce the leaching effect. It is seen that an addition of 25% Pd into the silver does not significantly reduce the leaching but an addition of 35% Pd approximately halves the rate of degradation.

The thick film inks, besides being screen printed on to substrates to produce circuit interconnections, are also used in components, for example to produce the wire-bond to solderable pad connections in ceramic chip carriers. The inks are also used, with a different vehicle having a lower viscosity, for dipping chip components to produce the metallisations for soldering.

Some comparative attributes of the different thick film conductor systems are given in Table 3.4.

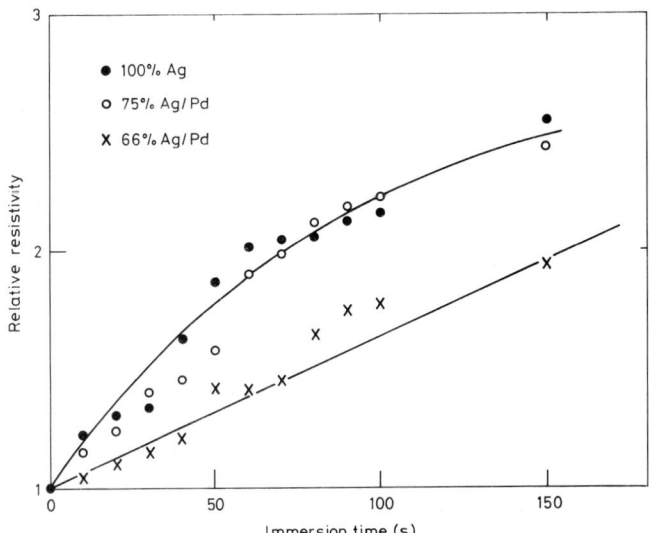

Fig. 3.15 The change in resistance of 16 μm thick fired deposits of silver and Ag-Pd inks, during immersion in 62/36/2 SnPbAg solder at 215°C.

Table 3.4
Thick Film Conductor Attributes

| System | Wire Bondability | | Solderability | Solder Leach | Corrosion |
	Gold	Aluminium		Resistance	Resistance
Au	excellent	excellent	good	very poor	excellent
Pt-Au	fair	fair	good	good	excellent
Pd-Au	fair	fair	good	good	excellent
Ag	good	poor	good	poor	very poor
Pt-Ag	good	poor	good	good	fair
Pd-Ag	good	good	good	good	fair
Pd-Pt-Ag	good	good	good	good	good
Cu	no good	fair	excellent	good	poor
Ni	no good	no good	poor	excellent	excellent

Resistors

Printed thick film resistors are a building component of hybrid microelectronics and of polymer thick film circuits and are not generally considered as part of surface mounting technology, but the distinction between these three interconnection and assembly methods is very blurred.

Resistive elements are of particular importance in hybrid circuits, enabling very high packing densities to be achieved, and also enabling the realisation of resistors with high stability and very close tolerance since each resistor is trimmed *in situ* to its required value. Thick film resistors cover a continuous range of values from 0·1 ohm to 1000 Mohm.

Most of the resistive inks available are based on ruthenium oxide (RuO_2) or bismuth ruthenate ($Bi_2Ru_2O_7$). Some new copper-compatible systems use tin oxide.[65] After printing and firing, trimming of the resistor is necessary. This used to be achieved by abrasive reductions of the resistor width using a narrow high pressure sand blaster but now most trimming is done by laser. A high intensity pulsed YAG laser under computer control is focused and directed to burn away a serpentine line in the resistor, thereby increasing its resistance until the required value is achieved. Tolerances below 0·1% are routinely achieved in high production manufacture.

Capacitors

Thick-film capacitors can be produced by printing a dielectric ink with one of a range of dielectric constants. Normally, dielectric pastes are used primarily as insulators to provide electrical isolation between crossing conductive tracks. Inks with dielectric constants up to about 100 can be used to produce low-value capacitors up to 50 pF.mm^{-2} of print area. Although inks with higher dielectric constant are available, up to 2000, these have a high loss factor and tend to result in a wide spread of values. For this reason, add-on chip capacitors are usually preferred.

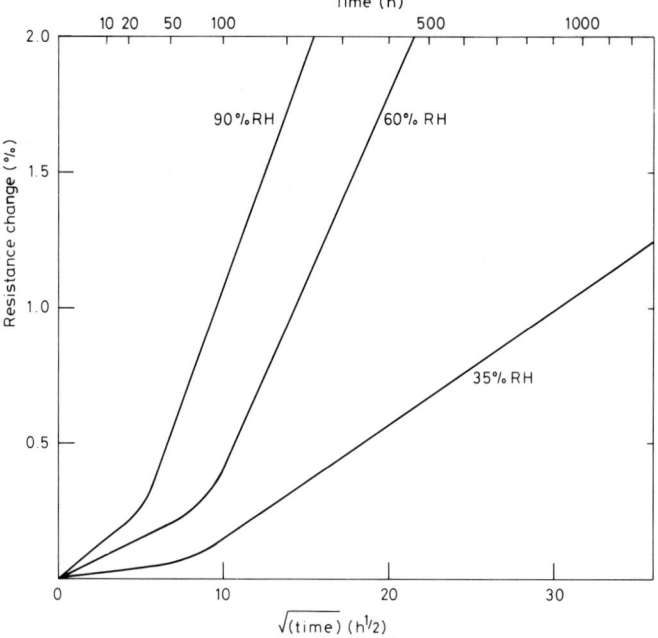

Fig. 3.16 The change in resistance of a particular thick film resistor, held at 110°C and at three levels of humidity, in an accelerated ageing test.[67]

Ageing of Thick Film Circuitry

The most consistent ageing behaviour of thick film components is a progressive increase in resistance as a function of square root time, demonstrating a diffusion mechanism. The change in resistance of a thick-film resistor for example is accelerated as the temperature increases and also as the humidity increases,[66, 67] as illustrated in Figure 3.16. Data of these type can be extrapolated, combining temperature, humidity and time to demonstrate that most thick film resistors would drift by less than 0·5% in 20 years, even in relatively stringent environmental conditions.

3.5.3 Matched TCE Substrates: Laminates

Organic composite laminates can also be manufactured with a TCE that is closely matched to that of ceramic. The approach is shown in Figure 3.17.[68] The most widely evaluated material is manufactured by replacing the glass fibres in epoxy-glass or polyimide-glass laminate with Kevlar*, an aramide fibre whose key property is a negative TCE achieved by its inherent physical properties and the method of its manufacture as fibres. The TCE of the laminated sheet can be tailored by suitable modification of the resin content. For surface mounting assembly a TCE of 6-7 ppm.K^{-1} is usual.

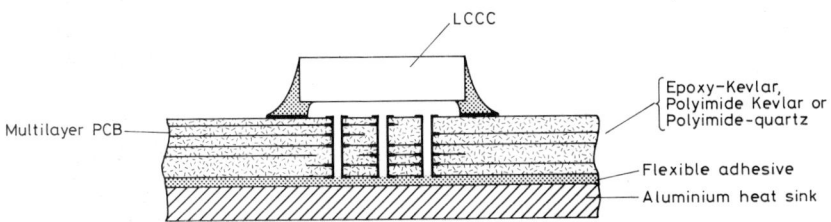

Fig. 3.17 Organic substrates can be expansion matched to ceramic leadless components by using Kevlar or quartz reinforcement.

The epoxy-Kevlar or polyimide-Kevlar laminates can be made in large sizes using the same processing equipment as used for the manufacture of normal epoxy-glass and polyimide-glass laminates.[69] They do however suffer from certain shortcomings, notably difficulty in drilling and machining, high moisture absorption, high cost and a tendency to form cosmetically deficient microcracks. The microcracks arise because of the large difference in TCE between the fibres and the resin in combination with relatively poor coupling of the fibres to the resin. The cracks originate where fibres cross over, spread around the fibre circumference and migrate to the surface of the laminate. The problem does not seem to be more than cosmetic, with no arising electrical failures reported, and it may be reduced through the development of more pliable resins and better bonding conditioners.

The relatively high moisture absorption is a more fundamental problem with Kevlar. The epoxy-Kevlar laminate on occasion exhibits internal delamination since the epoxy resin is not able to absorb much moisture from the Kevlar fibres. The polyimide resin is more obliging in this respect.

Another substitute for the glass fibres used in epoxy-glass and polyimide-glass laminates is quartz. Quartz (fused silica) fibres have a TCE of 0·54 ppm.K^{-1} and

---
*Kevlar is a registered trade name of DuPont.

a lower dielectric constant than glass. Again the laminate manufacture and the PCB processing are completely compatible with standard production lines. Because quartz is considerably harder than glass, drill life is significantly reduced.

The TCE of the laminate is tailored by controlling the resin content, but a compromise of properties is required.[70] For example, the TCE of polyimide-quartz in the required range for ceramic compatibility, 6-7 ppm.K^{-1}, is manufactured with a resin content of 37%. However, to make satisfactory multilayer circuit boards the prepregs require a higher resin content approaching 50% which increases the TCE to 11-13 ppm.K^{-1}. In addition, copper has a higher TCE of 17 ppm.K^{-1} and, since the system has a low modulus, the internal copper planes of a multilayer board also dictate, to some extent, the dimensional changes occurring with temperature. For a multilayer board using polyimide-quartz, the lowest TCE achievable is therefore only about 14 ppm.K^{-1}.

Matched TCE: Organic-fibre Laminates

Advantages:

    Potentially good match to TCE of ceramic
    TCE tailoring possible
    Large substrate size
    Processed with standard PCB production line

Disadvantages:

    More difficult to drill and rout
    Low modulus, so constrained by copper ground and power planes
    Expensive fibre material
    High water absorption (Kevlar)
    Cosmetic microcracking (Kevlar)

3.5.4 Metal Core Substrates

Metal core circuit boards have the potential of alleviating the problems of both differential thermal expansion and heat dissipation. Boards of this type consist of a rigid core or cores, or occasionally a mesh, normally metal, which dissipates the heat very efficiently and on to which heat sinks can contact.[71] The core has surface coatings of dielectric on which are the circuit tracks. If the core is designed to have a TCE matched to that of ceramic, the circuit tracks are constrained to expand and contract thermally also with this TCE, enabling large leadless ceramic components to be mounted.

3.5.4.1 PORCELAIN-ENAMELLED STEEL

Porcelain-on-steel substrates (PES) are a very common substitute for ceramic in surface mounted assemblies.[72] PES was originally used to overcome the disadvantages of size and cost of ceramic, and the technology of screen printing and firing porcelain ceramics of this type had been available for a long time. The steel composition is such that the TCE meets the requirement of matching to that of ceramic. The metal core also offers the availability of a good heat conductor and an integral ground plane. More recently porcelainised copper-clad Invar has been used for which the metal core can be tailored to provide a matched TCE.[73]

The enamel on steel must have a melting point high enough to allow for firing of thick film materials and low enough to minimise oxidation and warping of the steel during manufacture. Since enamels are relatively weak and fail in tension, the enamel is designed to have a lower TCE than the metal core in order to keep it in compression at the service temperatures of the electronic assembly. The lower expansion of the enamel compared with the steel holds only below about 450°C. Above this temperature the expansion of the enamel is greater than that of the steel, so that, during initial firing of the enamel and during the firing and subsequent cooling of thick film materials, the enamel undergoes an initial tensile stress but goes into compression below about 300°C, as shown in Figure 3.18. The faster the cooling rate during the firing of the thick films, the greater is the residual degree of compression in the enamel.[74]

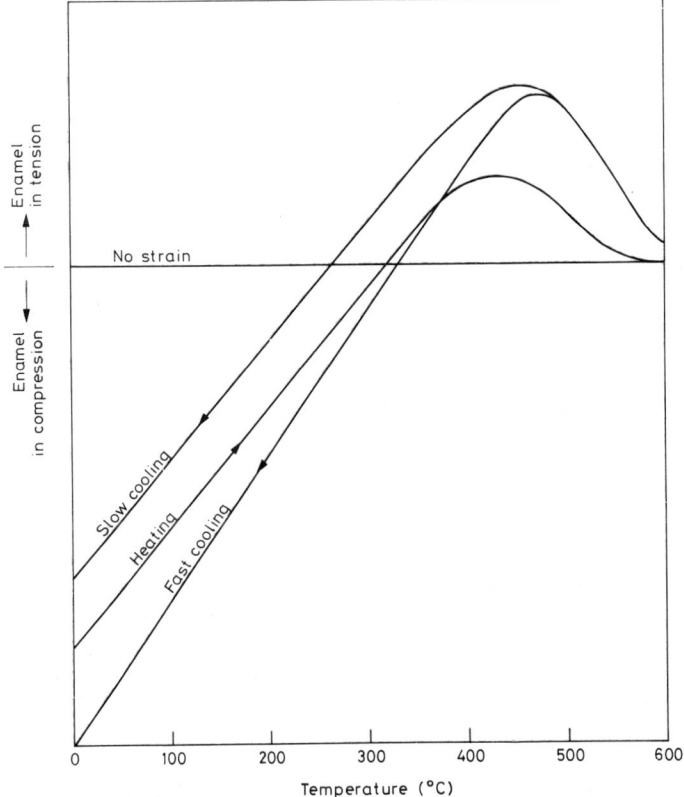

Fig. 3.18 The differential expansion between enamel and steel causes the enamel of a PES substrate to be in compression at working temperatures.[74]

Typically, for electronic grade substrates, extra low carbon steel (<0·003% C) has been found to minimise surface defects. The steel is chemically cleaned and pickled and plated with nickel. Glass powder is applied by dipping, spraying or electrophoretically, and fired for a few minutes at around 900°C.

Enamelled steel is both denser and stronger than other substrate materials.[75] The TCE of the steel is about 6-7 ppm.K^{-1}, matched to that of the alumina used for chip carriers and chip capacitors. The enamel surface layer is constrained to expand and contract in accord with the relatively massive steel core.

The effect of lowering the thermal resistance of the substrate by use of a metal core is to spread the heat laterally from its source very rapidly. The spreading angle Ψ is given approximately by

$$\tan \Psi \approx \frac{R_\theta(\text{through-board})}{R_\theta(\text{in-plane})} \tag{3.11}$$

A typical board might be constructed of a steel core, 0·6 mm thick, with a thermal conductivity 55 J.s^{-1}.m^{-1}.K^{-1} and an overlying porcelain layer 0·2 mm thick, thermal conductivity 1 J.s^{-1}.m^{-1}.K^{-1}. Thus, using Equations (3.3) and (3.4), per 1 cm square of surface:

$$R_\theta(\text{through-board}) = \left(\frac{6}{55} + \frac{2}{1}\right) = 2\cdot1\,°C/W$$

$$R_\theta(\text{in-plane}) = \left(\frac{55}{6} + \frac{1}{2}\right)^{-1} = 0\cdot1\,°C/W$$

so that $\Psi \approx 87°$.

This rough calculation demonstrates the significant difference between the through-board and in-plane thermal resistances and the effect that this has on the spreading of the heat. Figure 3.19 shows measurements of the surface temperature of a PES substrate in the vicinity of a square 2·5 mm × 2·5 mm powered-up resistor mounted on it. It is found that the spreading effect of the high conductivity core is reduced as the component, and consequently the heat source, becomes smaller.[76] The effect is negligible as the dimensions of the local heat source approach the thickness of the enamel.

3.5.4.2 ORGANIC METAL-CORE SUBSTRATES

The basic approach is to glue a thin single-sided PCB on each side of a metal sheet whose TCE has a value matched to that of ceramic, as shown in Figure 3.20. Because of the large difference in the TCE of the wiring assembly and the constraining core, the assembly must be double-sided to prevent warping. The board is assembled using a rigid adhesive to the core so that the interconnection system is bound to expand and contract as dictated by the metal core.[77]

This system has the advantage that it can use any standard PCB technology such as epoxy-glass or polyimide-glass, with surface or multilayer interconnections. The low TCE core can be fabricated as part of the standard process for multilayer boards. The core material can be any conductor with a suitable TCE, and Alloy 42, copper-clad molybdenum,[78] copper-clad graphite[79] and copper-clad Invar have all been used successfully.[80] Not only does the core constrain the board to the required TCE; it also aids significantly the thermal dissipation of the assembly. Heat sinks can be incorporated into the structure and glued with a thermally conducting glue to the chip carrier as shown in Figure 3.20.

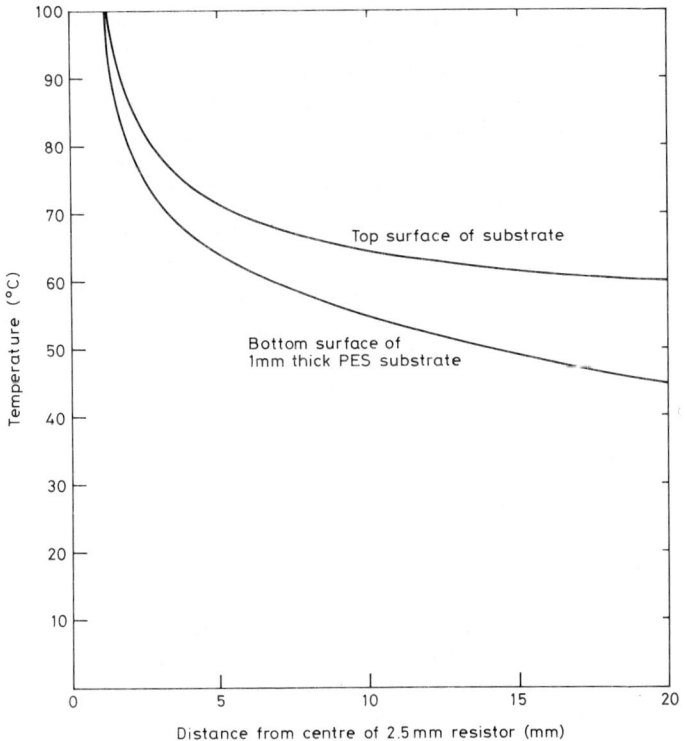

Fig. 3.19 Measurements of the surface temperature of a PES substrate in the vicinity of a 2·5 mm square powered-up resistor mounted on it.[76]

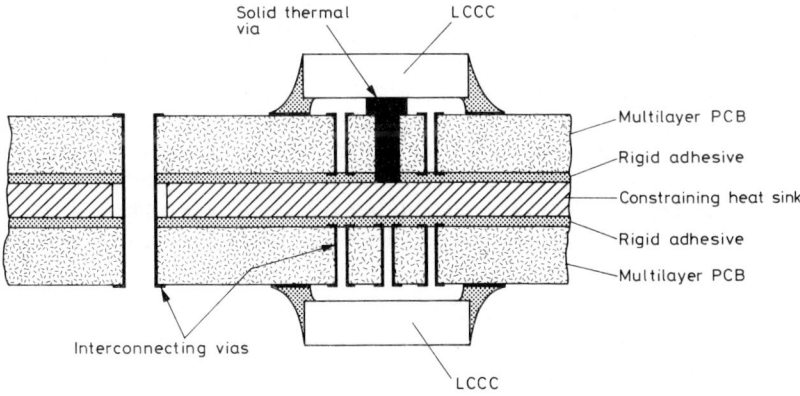

Fig. 3.20 The principles of a metal-core substrate: the substrate must be symmetrical on either side of the core. Heat sinks can be used to conduct heat from surface components to the metal core.

The material most commonly used for the core is copper-clad Invar.[81-83] The Invar is an alloy of 64% iron: 36% nickel, and for this application it is sandwiched between two layers of copper. The metal sheets are bonded together using the same process as used in making bi-metallic thermostat metals, which requires no brazing alloys or adhesives, but relies on solid-state diffusion welding to achieve a strong metallurgical bond between the layers. Figure 3.21 shows the dimensions of a typical constrained TCE multilayer structure, in this particular case with two cores which provide the power and the ground planes. The low TCE of the cores dominates the overall TCE of the board.

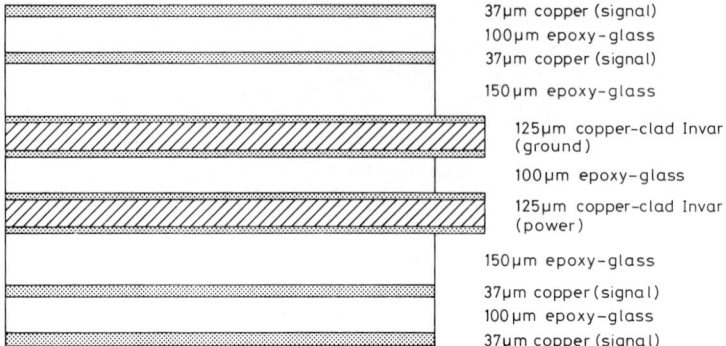

Fig. 3.21 The dimensions of a typical two-core multilayer board using copper-clad Invar to constrain the TCE.[3]

The designer has two variables to use: the ratio of copper to Invar and the ratio of copper-clad Invar to laminate.[84] The former determines the TCE and the thermal dissipation properties of the core whilst the latter determines the influence that the core has on the properties of the board. The copper provides the heat conductivity and the Invar provides the basis for a low TCE. To minimise weight and cost both the copper and Invar must be used effectively. If expansion constraint is a more significant factor than heat dissipation, the content of Invar should increase. Conversely, if heat dissipation is a more significant problem, the required amount of copper should be calculated first and then the amount of Invar to hold the expansion of the board to the acceptable level.

The TCE of copper is some 30 times greater than that of Invar and so an asymmetrical cladding would give rise to bi-metallic warpage. The designation of the copper-clad Invar is illustrated in Figure 3.22, with an overall thickness, followed by the percentage thicknesses of the copper-Invar-copper sandwich. The TCE of the sandwich and its thermal conductivity can be approximately given by mixing the respective values for copper alone and Invar alone, in proportion to

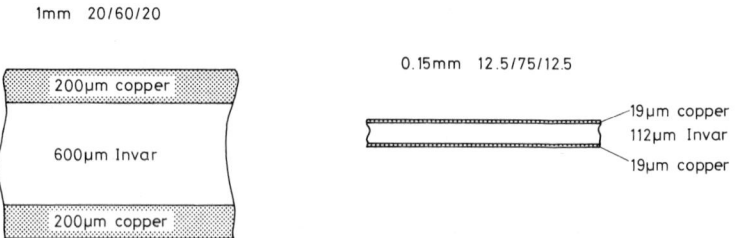

Fig. 3.22 The nomenclature of clad materials: the total thickness followed by the % thicknesses of the three layers.

their relative thicknesses. This is shown in Figure 3.23, illustrating the trade-off experienced between TCE and thermal conductivity. More rigorously, for a clad material, the effective TCE, α, is given by a function of the respective TCEs, α_1 and α_2 ($\alpha_2 > \alpha_1$), the respective elastic moduli, E_1 and E_2, the respective fractional volumes, V_1 and V_2 and the respective Poisson's ratios, v_1 and v_2:

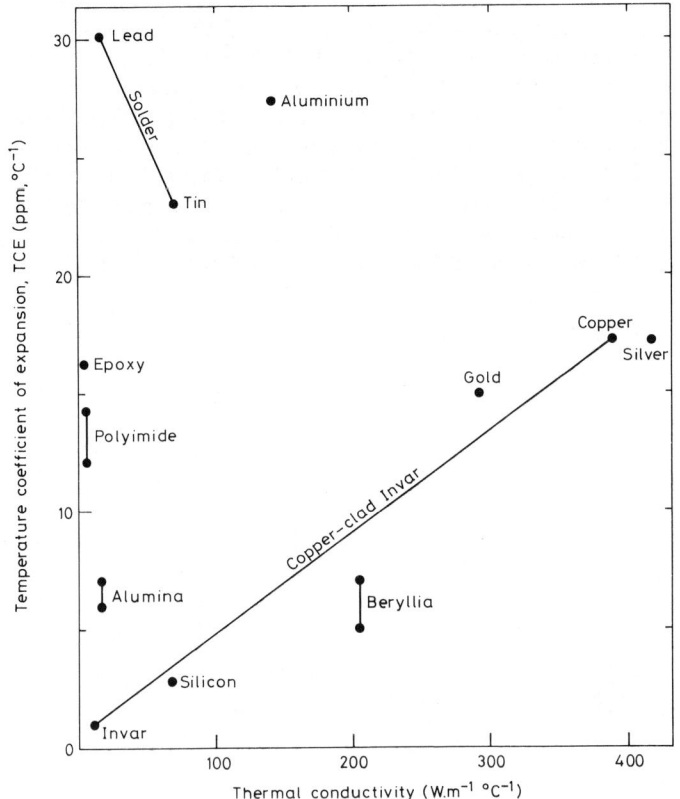

Fig. 3.23 Illustrating the trade-off between TCE and thermal conductivity for copper-clad Invar as well as other electronic assembly materials.

$$\alpha = \alpha_1 + \frac{(\alpha_2 - \alpha_1) V_1 E_1}{V_1 E_1 + V_2 E_2 \left(\frac{1-v_2}{1-v_1}\right)} \quad (3.12)$$

With an all-metal core, the difference in Poisson's effect for the two metals will be negligible since $v_1 \approx v_2$. Since both the TCEs and the elastic moduli of the materials vary with temperature, the calculated TCE is good only for one temperature.

Another factor to consider is that the thermal properties of the copper are affected by plastic-elastic transitions experienced during thermal cycling.[85] This occurs because the difference in TCEs of the copper and the Invar can stress the copper so much that its yield strength is exceeded. The yield stress σ_y of copper is 76 MPa and its Young's modulus E is $1 \cdot 2\ 10^5$ MPa, so that the yield strain

$\varepsilon_y = \sigma_y/E$ is 0·0006. This strain occurs whenever a thermal excursion of the clad material exceeds a critical range ΔT which can be calculated:

$$\varepsilon_y = \Delta T (\alpha_{CIC} - \alpha_C) \qquad (3.13)$$

Now for copper $\alpha_C = 17\ 10^{-6} K^{-1}$ and for 20/60/20 copper-clad Invar, as an example, $\alpha_{CIC} = 5 \cdot 2\ 10^{-6} K^{-1}$, which means that $\Delta T \approx 51°C$. Thus, whenever 20/60/20 copper-clad Invar is cycled more than 51°C above or below a stress-free state, the copper leaves the elastic regime and become plastic (Section 11.2.3). The Invar however remains elastic throughout the normal test cycling range ($-55°C$ to $+125°C$). This behaviour produces a hysteresis effect on the strain as the temperature is varied as shown in Figure 3.24. Since this phenomenon is driven by the relative strength of the copper and the Invar, it is a function of the cladding ratio; the greater the copper fraction, the wider is the hysteresis loop. As the copper fraction increases the composite expands at a higher rate and the temperature range over which the copper remains elastic is wider.

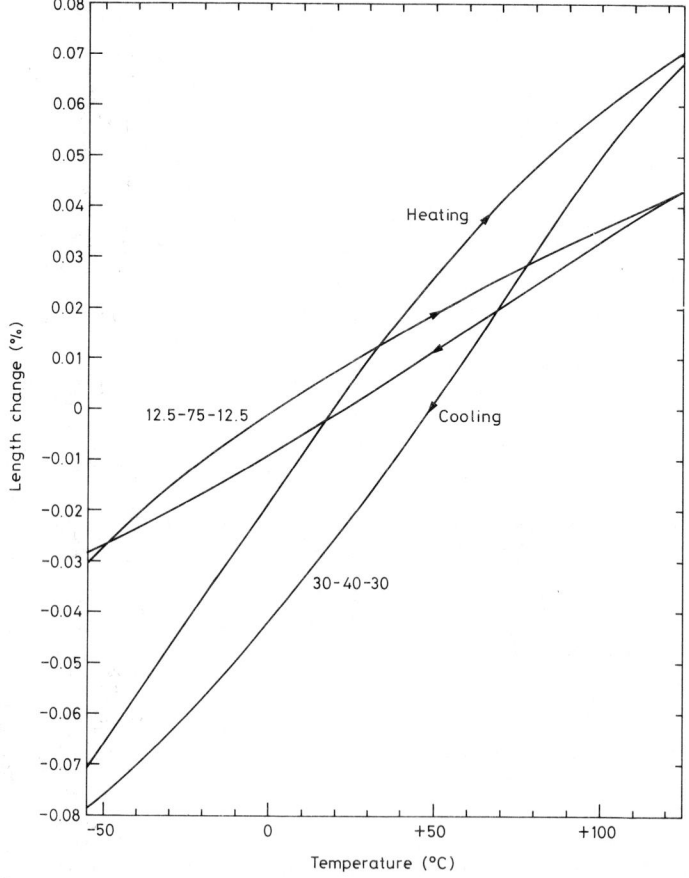

Fig. 3.24 The hysteresis effect of expansion of copper-clad Invar, caused by the copper entering the plastic regime; the hysteresis loop becomes wider as the copper content increases.[85]

The TCE is the slope of the strain-temperature curve of Figure 3.24 and it is clear that the hysteresis effect gives rise to a temperature dependence of TCE. Furthermore, at some temperatures, increasing the copper fraction increases the TCE and at others it decreases the TCE as shown in Figure 3.25. The range of TCE values within the test range $-55°C$ to $125°C$ increases as the copper fraction increases because of the widening of the hysteresis loop, as shown in Figure 3.26.

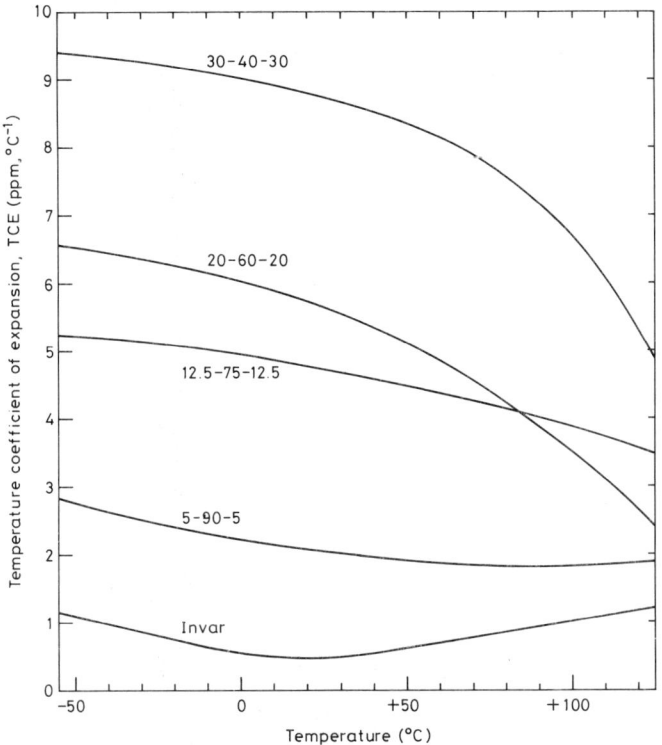

Fig. 3.25 Because of the hysteresis effect shown in Figure 3.24, the TCE of the clad material is temperature dependent.[85]

The dependence of thermal conductivity on the copper fraction of copper-clad Invar can be accurately calculated, as long as there is no significant interfacial thermal resistance between the layers. Because the clad material is obviously not isotropic, the thermal conductivity must be defined in both the in-plane (x-y) and through-board (z) directions. The thermal resistances are summed directly through the board but in the plane of the board the inverses of the thermal resistances are summed, as in Equations (3.3) and (3.4). Hence, if the respective volume fractions of copper and Invar are V_C and V_I, and the respective thermal conductivities are K_C and K_I, the thermal conductivity of the clad material is

$$K_{CIC} = V_C K_C + V_I K_I \quad \text{[in-plane, x-y]}$$

$$K_{CIC} = \left(\frac{V_C}{K_C} + \frac{V_I}{K_I}\right)^{-1} \quad \text{[through-plane, z]}$$

Taking the values of $K_I = 13 \cdot 8$ W.m^{-1}.K^{-1} and $K_C = 400$ W.m^{-1}.K^{-1} the thermal conductivities of the copper-Invar-copper sheet, K_{CIC}, calculated for room temperature, are shown in Figure 3.27. The thermal conductivities are only slightly temperature-dependent.

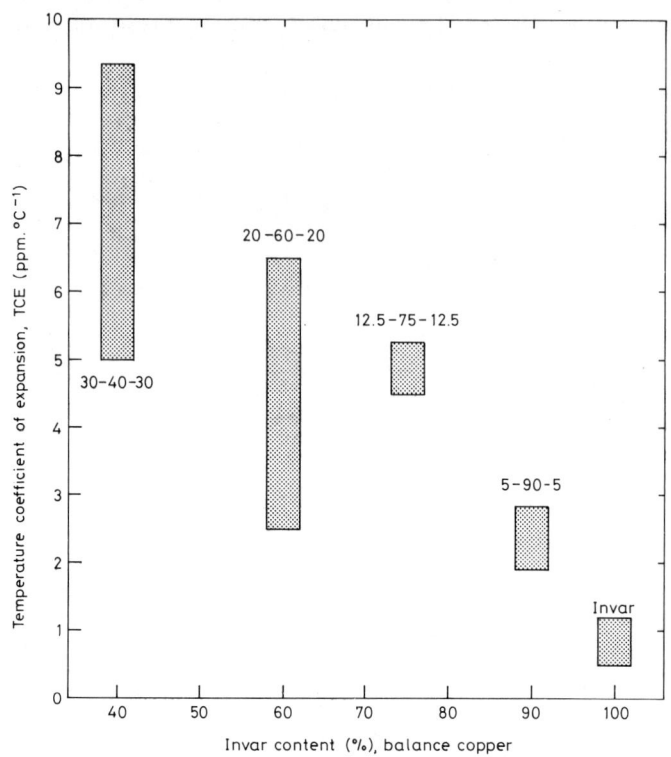

Fig. 3.26 The range of TCE values that can be measured for copper-clad Invar as a function of the composition. The range widens as the copper content increases because of the widening of the hysteresis loop.

Constrained TCE: Cored Substrates

Advantages:

    Readily matched and tailored TCE
    Standard PCB processing technology
    Uses standard epoxy or polyimide laminates
    Technology infrastructure available: widely used, lot of test data

Disadvantages:

    Increased weight with metal core
    Must be double-sided to avoid warping

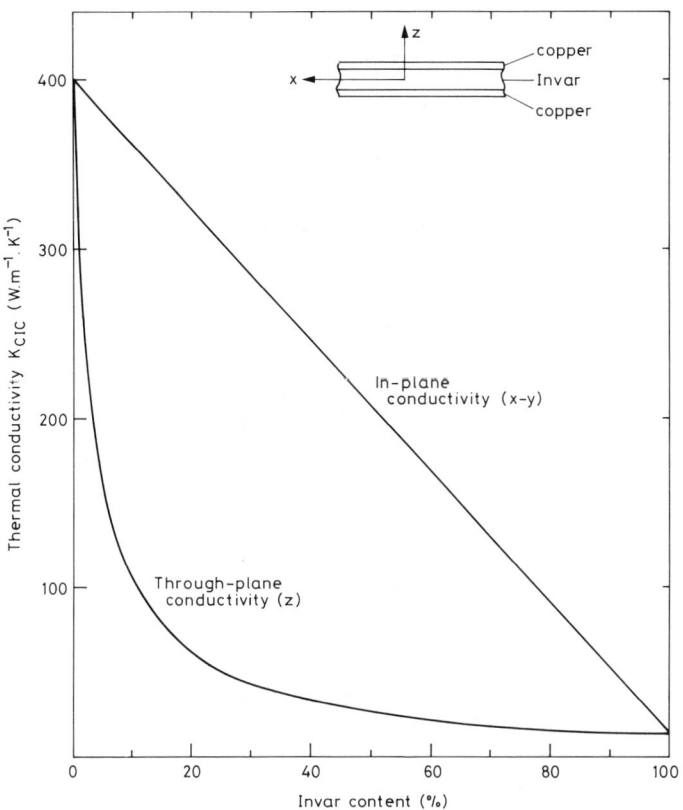

Fig. 3.27 The thermal conductivity of copper-clad Invar as a function of the composition, both in the plane of the laminations and through the laminations.

3.5.4.3 MULTIWIRE

Multiwire interconnection systems[86] comprise conductive tracks that are not etched from a thin surface foil but are insulated wires embedded in the prepregs of a laminated substrate. For a multilayer board the wires are typically 160 μm diameter copper which have the same conductive capability as conventional etched copper tracks 35 μm thick and 0·6 mm wide.[87] The wiring lengths and positions are organised by a dedicated computer. The wire ends are terminated or interconnected by plated-through holes, as shown in Figure 3.28.

This technique can be used to advantage for surface mounting assemblies,[88] possibly using wires as fine as 60 μm on a 300 μm grid, giving a maximum interconnection density, per double layer of wires in an x-y array, of 65 cm/cm^2. The wires can be bonded in a special adhesive to a normal epoxy-glass laminate with etched copper tracks, but more commonly in surface mounting this wiring approach is used in conjunction with a thermally matched metal core which supports and constrains the encapsulated wire array.[89]

If etched copper is used in addition to the wires, it usually provides the power or ground planes.[90] Electrical and mechanical connection between the wires, the tracks and the pads is by use of plated-through holes.

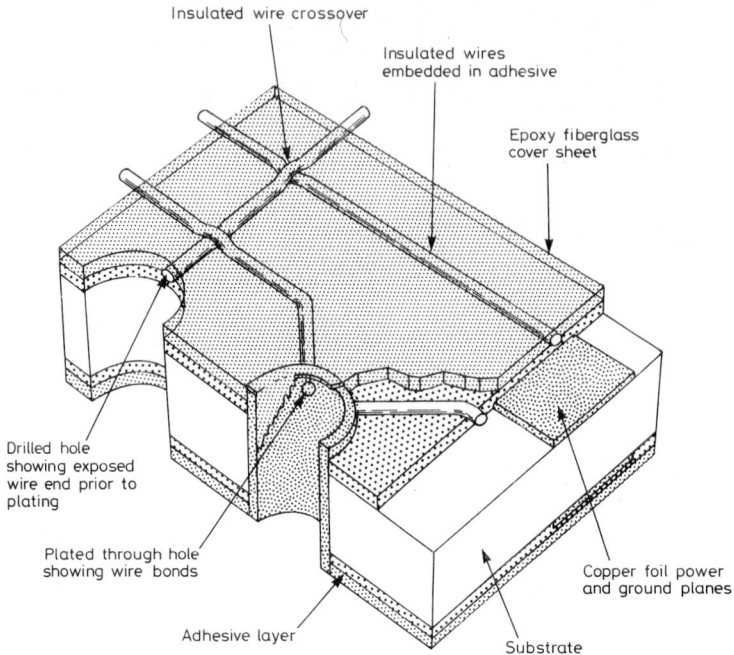

Fig. 3.28 The principle of construction of a multiwire interconnection substrate.

3.5.5 Compliant Substrates

The unconstrained approach to deal with the incompatibility of TCEs of the LCCC and its interconnection substrate relies on a flexible level between, or as part of, the two. The most obvious option is the use of compliant leads, usually in the form of a chip carrier socket, as discussed in Section 2.3.9. The nature of a lead, a relatively long and thin strip of a ductile metal, invariably offers adequate compliancy to negate detrimental effects of differences in thermal expansion. However, this approach can reduce the benefits of surface mounting, namely a low profile, high electronic speed and thermal dissipation. An alternative method is to use a 'solder lead' by artificially increasing the stand-off of the leadless component. The ways of doing this and the benefits accrued by so doing are discussed fully in Chapter 11. Yet another approach is to incorporate the compliance into the substrate. This can be done in one of two ways, either using a resilient layer on the top surface of the PCB or mounting components on flexible heat sinks within the substrate.[91] The first of these has found quite widespread use and acceptance for surface mounted assemblies incorporating LCCCs.

The flexible layer is incorporated on the laminate and the interconnection tracks are formed either additively or subtractively in a normal PCB manufacturing process. A number of proprietary laminates of this type have been produced and tested.[92, 93] A typical thickness of the compliant layer is 50 μm, over which there may be a copper foil as for a normal copper-clad laminate. The compliant layer can also be applied to multilayer boards. The selection of elastomer used is a compromise normally between the cost of the laminate and the temperature and environmental properties required. Two types are available based on polyacrylates and based on a nitride rubber such as acrylonitrile butadiene. The nitrile rubber is the elastomer most used and data of thermal fatigue of surface mounting solder joints using it show an increase by a factor of about 5 in the number of cycles to failure. For example the data for cycling large chip carriers, soldered with 60:40SnPb solder, over the temperature range $-55°C$ to $+125°C$, are given in Table 3.5.

Table 3.5

Benefit Gained in Thermal Fatigue by Using a Flexible Surface Layer

LCCC termination pads	Mean Number of Cycles to Failure ($-55°C$ to $+125°C$)			
	FR-4 SnPb	FR-4 SnPbAg	nitrile layer SnPb	acrylic layer SnPb
44	190	275	810	950
52	170	240	615	650
68	75	115	400	500

Most of the flexible surfaces show a large z axis expansion. No test data are available concerning the effect of this on the reliability of the plated-through-hole copper barrels for inter-plane vias in multilayer boards and thermal vias in all types of board.

Unconstrained TCE: Flexible Surface Layer

Advantages:

    Standard PCB processing
    Low cost and availability

Disadvantages:

    Proprietary products
    Limited test data, especially on PTH reliability

3.5.6 Polymer Thick Film Circuits

Polymer thick film technology[94-100] combines the attraction of versatility and economic material and production costs of thick film technology with standard PCB laminate materials such as fibre reinforced epoxy, phenolic, polyester and polyimide. When fully developed, the cost advantage and efficiency of production to be gained are very great.[101]

Conventional thick film inks cannot be used in conjunction with organic substrates because of the high firing temperatures required. New thick film conductors and resistors are now available[102, 103] that are based on polymers instead of glass, for which the 'firing' or curing temperature is only 150-300°C. Polymer thick film therefore allows the possibility of a much cheaper additive method of printed circuit fabrication, rather than the subtractive techniques of etching copper laminates. Resistors and cross-over dielectrics are printed in exactly the same way as for thick film circuits on ceramic,[104] to produce complex multilayer circuits. Fast curing systems permit continuous processing for large volume production.

The polymer curing after printing can be achieved by radiant heat[105, 106] (Chapter 7), by vapour phase condensation heat transfer (Chapter 8), or by immersion in a hot liquid (Chapter 8) using equipment similar to that used for reflowing solder paste. Radiant energy curing is usually performed in air whereas the fluids offer an inert environment. Generally the resistivity of the polymer conductors and resistors falls as the curing temperature is increased,[107] as shown in Figure 3.29, but the curing temperature profile is a compromise between the desire to fully cure the thick film ink while at the same time avoiding excessive degradation and loss of adhesion. In addition, the undercured deposits do not accept solder. In any case it is preferable to plate the solderable pads selectively with electroless nickel in order to avoid leaching of the silver from the conductor.[108]

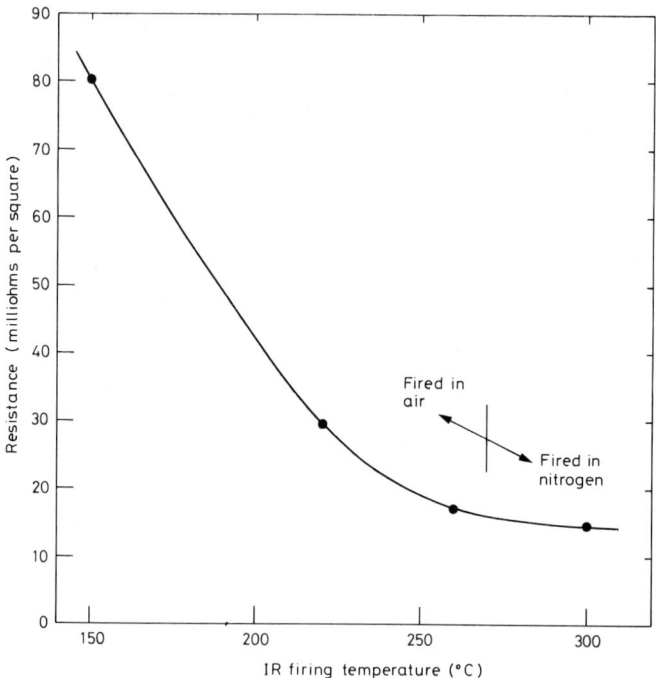

Fig. 3.29 An illustrative example of how the resistivity of polymer conductors and resistors falls as the curing temperature is increased.[107]

3.5.7 Comparison of Substrate-interconnection Systems

The fact that very many types of substrate and interconnection systems continue to be used for surface mounted assemblies is indicative that the many required ideal properties can only be compromised. The choice of system is dependent on the perceived importance of performance, material cost and production cost.

Three intrinsic materials properties are important: the thermal conductivity, the temperature coefficient of expansion and the dielectric constant. In addition each substrate-interconnection system is constrained physically by its variable size and interconnection density.

Thermal conductivity controls the success of the thermal management of an assembly, as has been discussed at length. The thermal conductivities of materials used or associated with substrate-interconnection systems for surface mounting are shown in Figure 3.30.

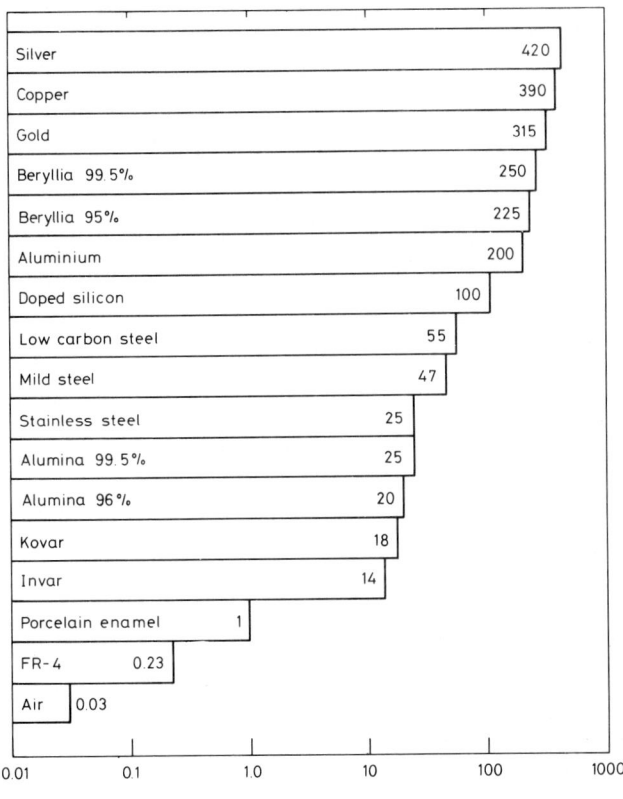

Fig. 3.30 The thermal conductivities of materials used, or associated with substrate-interconnection systems for surface mounting.

The differential expansion during temperature cycling is the major cause of solder joint failure in leadless surface mounted components. Some relevant TCE values are given in Figure 3.31.

The third property of importance is the dielectric constant. This is a measure of the degree to which the medium can resist the flow of charge and support an electrostatic stress. The dielectric constant of a material is the ratio of the capacitance of a capacitor with that material, to the capacitance of the same dimensions with the material replaced by a vacuum. A vacuum is the only perfect dielectric. The dielectric constant is a function of both temperature and frequency, and usually its value is only of importance for high frequency applications. Some values at 1 MHz are shown in Figure 3.32. Besides the dielectric constant, an insulator has two other principal properties; its dielectric loss, which is the amount of energy it dissipates as heat when placed in a varying electric field, and its dielectric strength which is the maximum potential gradient it can stand without breaking down.

Finally in Table 3.6 a range of properties of substrate-interconnection systems are compared.

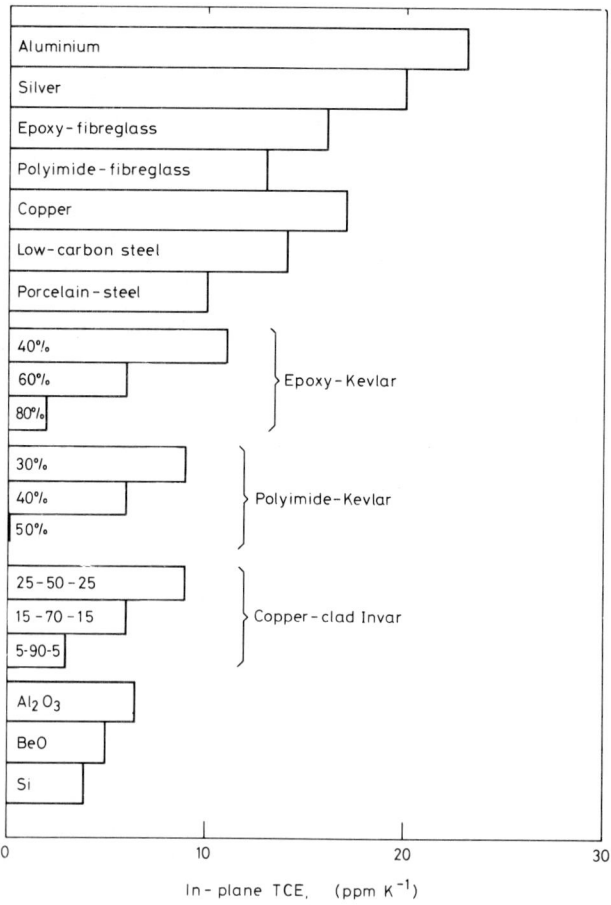

Fig. 3.31 Temperature coefficient of expansion (TCE) values of some materials used for interconnection substrates.

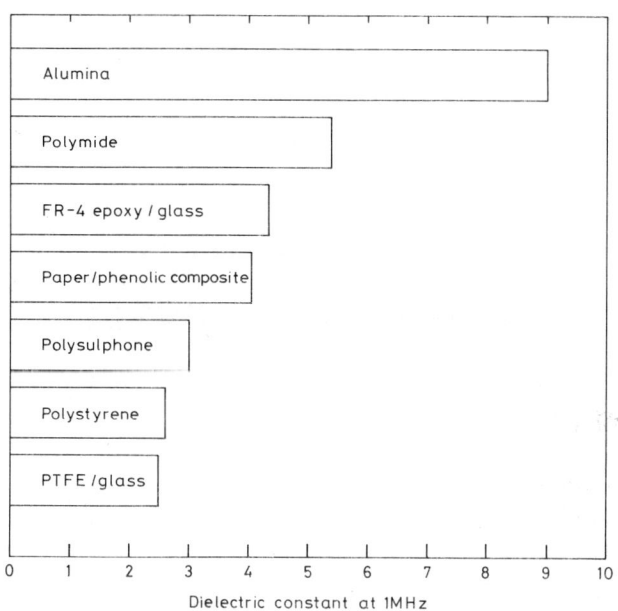

Fig. 3.32 Dielectric constant values at 1 MHz frequency, of substrate materials.

Table 3.6

Comparative Properties of Substrate-interconnection Systems

	Thin film on alumina	Thick film on alumina	Porcelain on steel	Epoxy-glass multilayer
Track widths and spaces (μm)	25	100	150	150
Minimum via diameter (μm)	-	200	200	300
Minimum via separation (μm)	-	750	750	2500
Maximum interconnection density per layer (cm/cm^2)	200	100	40	40
Number of layers	1	1-10	1-3	1-24
Normal substrate size maximum (cm)	15 × 15	15 × 15	30 × 30	60 × 60
Dielectric constant at 1 MHz	9·0	9·0	6·5	4·5
Capacitance (pF/cm^2)	-	350	60	30
Usable temperature (°C)	1000	800	650	110
TCE (ppm/°C)	6·4	6·4	10·3	16
Thermal conductivity (J.s^{-1}.m^{-1}.K^{-1})	20	20	1·0	0·23
Thermal resistance (°C.W^{-1}.cm^{-2})	0·005	0·02	0·04	1·7

Chapter 4

DESIGN AND ASSEMBLY

4.1 ASSEMBLY VARIATIONS

Surface mounting components are attached by solder or conductive adhesive to conductive tracks on a printed circuit board with no requirement for drilled holes through the board. Leaded components are located by insertion through holes drilled in the board and attached by solder either to a pad on one side of the board or, if the hole is plated-through, by a through-hole solder fillet. Whether or not surface mounting components are used exclusively or in combination with leaded components gives rise to several different board configurations and assembling procedures.

Figure 4.1 shows the PCB assembly variations possible with surface mounting components. Versions (a) and (b) use exclusively SMCs on one side of the board and on both sides respectively. If it is necessary or desirable to combine SMCs with leaded components, versions (c) and (d) are possible as examples of mixed assemblies. Versions (a) and (c) can be constructed using a single-sided PCB whereas for (b) and (d) a double-sided board is necessary.

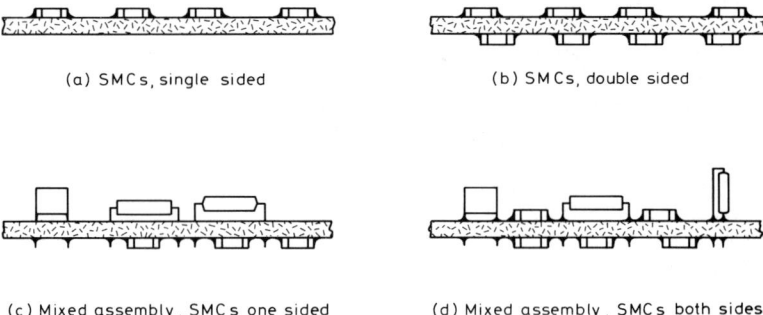

Fig. 4.1 Surface mounting variations: (a) SMCs single-sided, (b) SMCs double-sided, (c) mixed assembly, SMCs single-sided, (d) mixed assembly, SMCs both sides.

4.1.1 SMCs, Single-sided

Assemblies using SMCs exclusively, and on only one side of a substrate, can be produced in several ways, either by wave soldering or by reflowing a screen printed solder paste or, less commonly, by curing a conductive adhesive. These manufacturing processes are dealt with in detail in later chapters, but the separate assembly steps are listed here.

If the board is to be wave soldered, the components and all the solderable surfaces must be on the underside of the board, in contact with the wave. Thus all components physically pass through the wave and must therefore be of materials and construction designed to withstand such temperatures and thermal shock. To enable the board to be inverted all components are first glued to it. The glue can be applied either using a syringe or similar transfer mechanism, or by screen printing.

Thus, the steps for the wave soldered, single-sided, fully surface mounting assembly, as illustrated schematically in Figure 4.2, are: application of glue, placement of SMCs, cure glue, turn PCB over, flux and wave solder, followed by cleaning, inspection and testing.

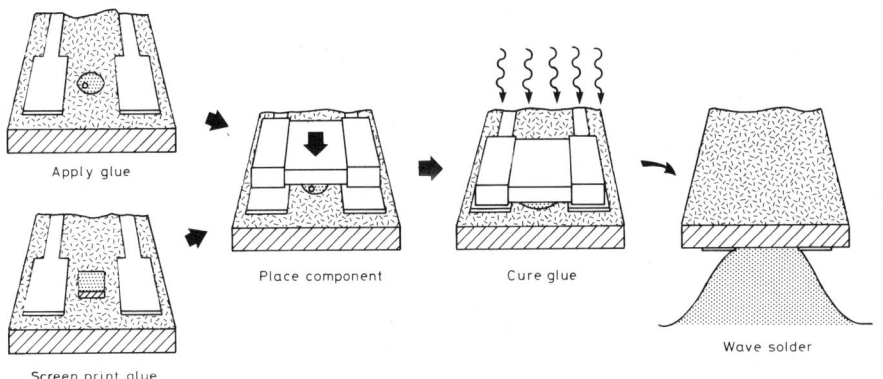

Fig. 4.2 The assembly steps for a wave soldered single-sided board.

For assemblies that are fully surface mounting, reflowed solder paste can be used to accomplish the soldering. The steps are shown in Figure 4.3; the solder paste is screen printed, the SMCs are placed and the whole is heated to melting and flow of the solder. The screen printed solder paste is tacky enough to retain the components without the temporary glueing necessary for wave soldering since the components are kept upperside of the board.

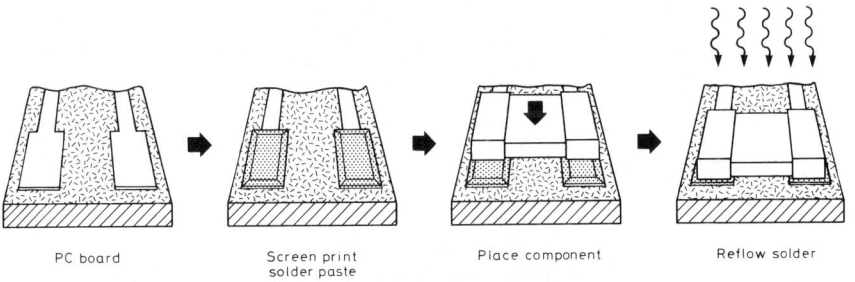

Fig. 4.3 The assembly steps for a reflow soldered single-sided board.

4.1.2 SMCs, Double-sided

One of the ways in which full advantage can be implemented of the potential space savings to be gained by surface mounting assembly is to mount the components on both sides of the printed circuit board, using through-plated vias to make the interconnections between the opposite sides. One procedure for double-sided SMD mounting is to screen print solder paste on to one side of the board, mount the SMCs and reflow the solder; then turn the board over, apply glue, place SMCs, cure glue, turn PCB over, flux and wave solder. In this method the reflowed solder on the board top surface may melt during passage over the solder wave but the components will not become dislodged because they are held aligned by surface tension forces of the molten solder. Other assembly variations for this type of board are possible, although this is the most satisfactory. Components are glued so that they do not fall off when the board is inverted as must occur when wave soldering. It is not possible to glue components when using a screen printed solder paste in order to invert the board for simultaneous

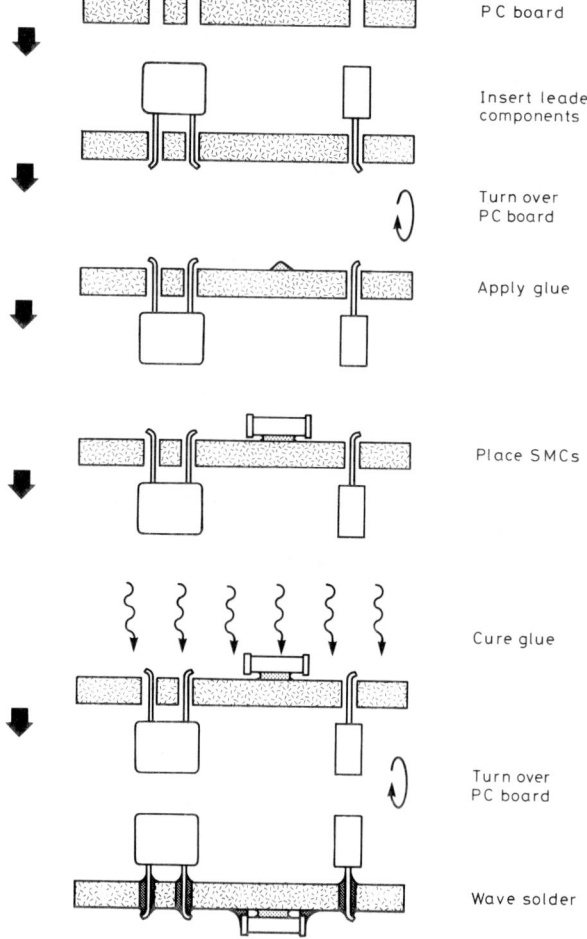

Fig. 4.4 The assembly of a mixed-technology board, placing the leaded components first.

reflow of both sides, because during reflow of solder paste there is a considerable volume reduction and the component must be free to pull down on to the pad and not be constrained by the glue.

4.1.3 Mixed Assembly, SMCs Single-sided

Again, for mixed assembly boards there are several variations of assembly procedure that can be used.[109] It is most common to place the leaded components first, as shown in Figure 4.4. The board is then turned over and glue applied. The SMCs are placed and the glue is cured. After a second board turn-over the assembly is fluxed and wave soldered.

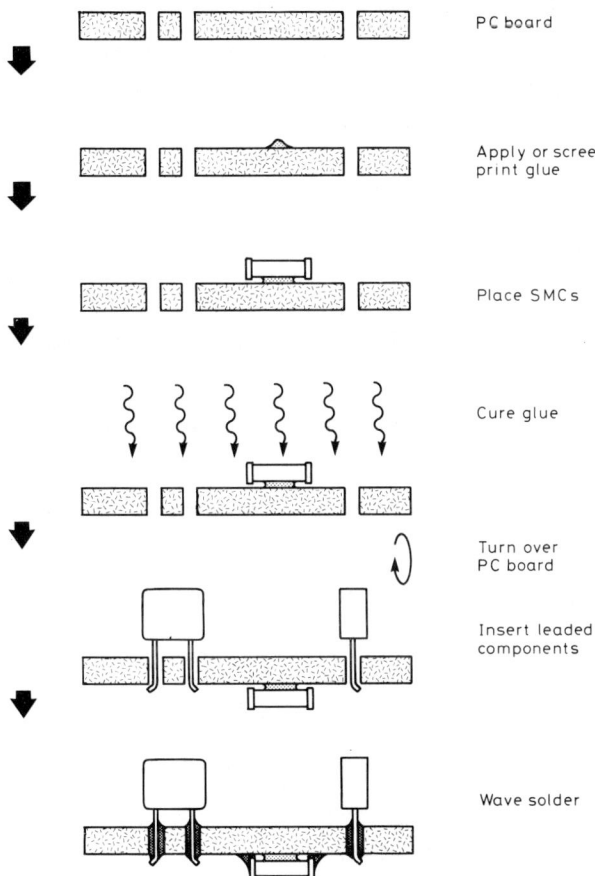

Fig. 4.5 The assembly of a mixed-technology board, placing the SMCs first.

The main variant of this assembly route is to place the SMCs before the leaded components, as shown in Figure 4.5. The glue is applied, SMCs are placed, the glue is cured, the board inverted, leaded components are inserted and the whole is fluxed and wave soldered. This method has the advantages that the glue can be screen printed, a much faster operation than a sequential syringe or pin-transfer method, and that there is only one board turn-over step. However, because the SMCs are already in place when the leaded components are to be inserted, vacant board space is required for the mounting tools of the insertion machines which are needed for the cutting and bending of the leads.

Assemblies with through-hole inserted components must be wave soldered; it is not possible to screen print or apply solder paste to successfully form a full through-hole solder fillet at reflow.

4.1.4 Mixed Assembly, SMCs Double-sided

The assembly procedure for double-sided surface mounting plus inserted components is as follows: screen printing of solder paste on top surface of board, placement of SMCs and reflow, insertion of leaded components taking precautions that the board can be turned over without losing them, application of glue to the other side of the board, placement of SMCs, curing of glue, turning over of the board for a second time, fluxing and wave soldering.

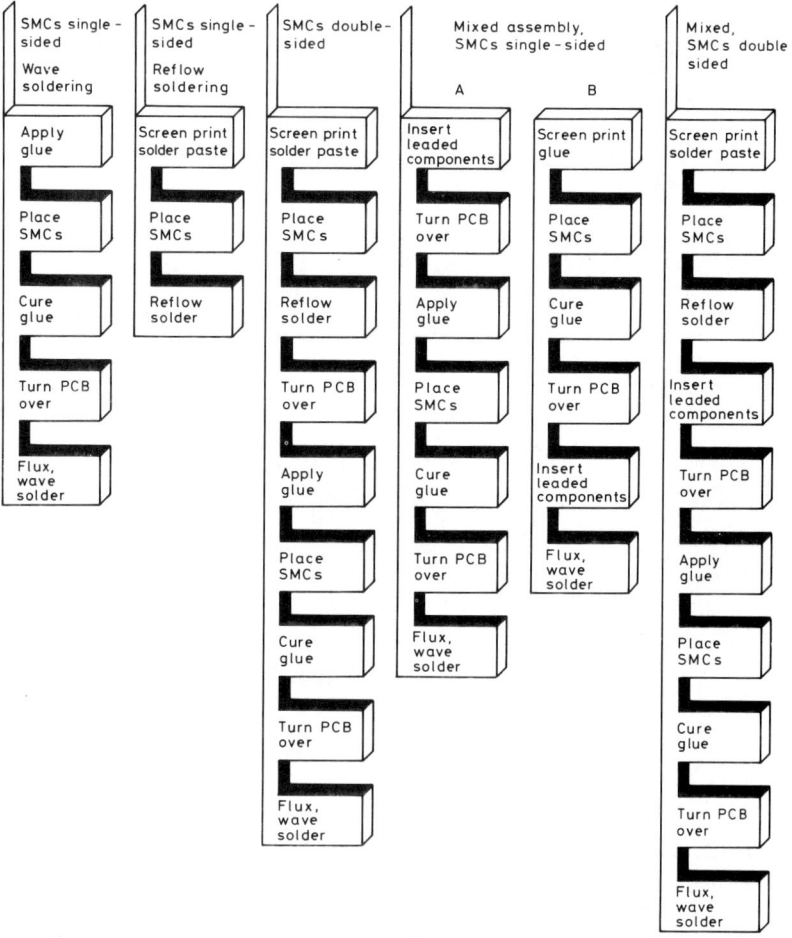

Fig. 4.6 The steps involved in the assembly of surface mounted boards, by the different routes.

A comparative flow chart of the possible assembly procedures for surface mounted assemblies is shown in Figure 4.6.

4.2 PCB LAYOUT

The design and layout of components, conductor tracks and solderable pads on a printed circuit board have to be fitted to the requirements of, for example, the components used, the function and environment of the electronic assembly, and the assembly method.[110]

4.2.1 Factors Affecting PCB Layout

The *component packages* used determine the choice of soldering method available and thus influence the PCB layout. Mixed assemblies, containing some through-board components, must be wave soldered. Certain SMCs cannot be wave soldered and a reflow method must be used. The lead pitch of the components defines the pad sizes, the spacings and their tolerances that are specified for the PCB as well as the required accuracy of placement of the components.

The *component placement equipment* that is to be used can impose restrictions on the circuit layout, since some machines require vacant space for placement tools and space for rotating components. These problems are largely avoided by the use of vacuum component pick-up heads rather than the tweezer or clamp styles.

The *soldering method* employed is also a major consideration in the design of the PCB artwork. In wave soldering the SMCs pass through the solder wave and any solder pads trailing behind the body of the component can be shadowed from the molten solder. Thus it is preferable to mount SO components parallel to the conveyor direction and to mount quad packs at 45° to the conveyor direction, in order to avoid or minimise shadowing. Also, certain minimum distances must be kept between adjacent solderable surfaces on the board to avoid solder bridged short circuits. Reflow soldering generally permits closer component spacing and smaller dimensioning of solder pads, although a dual solder wave with an air knife to eliminate bridges (Chapter 5) can produce results equivalent to reflow methods.

The *design method* is also influential in the PCB layout. The board may be designed either manually or by means of a CAD (computer-aided-design) system. Specific CAD systems require certain spacings and pad dimensions to be kept, which may constrain the design options.

The end-use *application* of the assembly may also require special design features of the PCB layout. This applies particularly to high frequency, high speed circuits or sub-circuits in an assembly.

The *specified quality* is another input into the design of a PCB interconnection system, because different quality levels may allow various trade-offs against solder pad sizes and closeness of components.

Finally, the electrical *testing requirements* may require special considerations such as test pads in the circuit and space between components adequate to allow test probes to be used.

4.2.2 Design Constraints from the Placement Operation

Due regard must be taken of a component placement machine that is intended to be used in the assembling. Apart from the very smallest prototype board, assembly of surface mounting components cannot be performed by hand, and a semi-automatic or fully automatic machine must be brought to bear, to cope with the component sizes, as well as the accuracy of placement and speed of placement required.

The speed of placement achieved in production is very dependent on the board design and a reduction of machine movements between placements will not only increase the overall assembly speed, giving rise to production time and cost reductions, but will also increase the service life of the machine. In order to design this concept into the PCB layout it is necessary to know if the placement machine is of a simultaneous or a sequential placement operation type and if it is a single or multiple headed type.

As mentioned in Section 4.1.3, the placement machinery will place further restrictions on the design if space has to be allowed for component holding devices. This is not normally the case for handling SMCs where a vacuum pick-up is invariably used, but in mixed assemblies where the insertion components are held mechanically.

Some placement machines may not be able to handle large devices and a second, normally slower, machine may need to be used. In such circumstances it is therefore advantageous to place all these larger components on only one side of the board so that only that side is slowed in its placement.

4.2.3 Design Constraints from the Soldering Operation

For optimum efficiency of design and reliability of assembly, the PCB layout should reflect the type of soldering that is intended to be used. Wave soldering must be used for mixed assemblies, but either wave or reflow soldering can be chosen for fully surface mounting assemblies. The design rules for wave and reflow soldering are different. For reflow soldering the components can be packed on the board more closely because the solder is already at the places where it is needed. Also reflow soldering is a symmetrical process and so the orientation of the components with respect to the board is not crucial. With wave soldering, the result of mounting components parallel or perpendicular to the conveyor movement across the wave can be very different. This applies both to solder skips

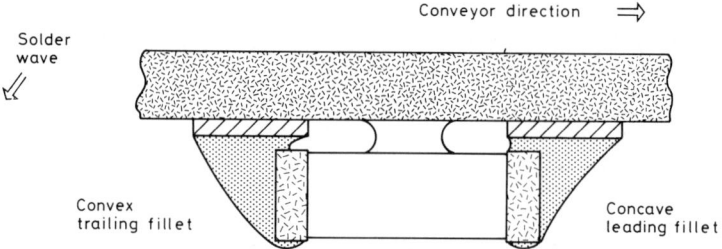

Fig. 4.7 When wave soldering a component mounted parallel to the conveyor direction, the solder fillet at the leading edge tends to be concave, while that at the trailing edge tends to be bulkier and more convex.

resulting from a shadowing effect of the body of the component and to solder bridges between terminations resulting from a surface tension effect. Additionally, because the wave soldering process is directional, the hydrostatic pressure, in combination with the surface tension forces at the point where the PCB leaves the wave, causes the solder joints at a leading edge of a component to be relatively concave whilst those at a trailing edge tend to be relatively bulkier and convex, as drawn in Figure 4.7.

In wave soldering, the joint is made primarily between the end face of the component metallisation and the solder pad as shown schematically in Figure 4.8, and the underside component metallisation plays little part in the bonding. Because of this the solder pads must extend well beyond the end of the component and be large enough to collect sufficient solder from the wave to make a satisfactory joint. In reflow soldering the solder is already on the pad and

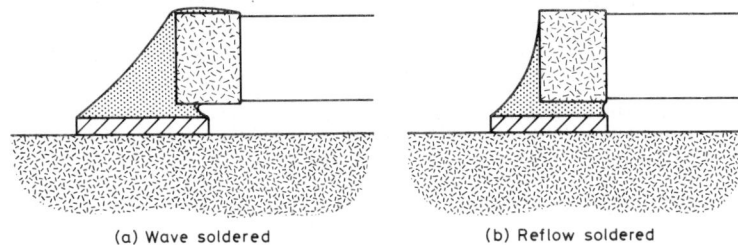

(a) Wave soldered   (b) Reflow soldered

Fig. 4.8 When wave soldering components, the end-face metallisation of the component and the pad extension on the board are of primary importance to the joint. When reflow soldering, however, the bottom-face metallisation and the under-component pad are more important.

the joint is made much more with the underside component metallisation, the end face playing a relatively minor rôle, as shown in Figure 4.8. The pads can therefore be made smaller since there is no need for them to extend far beyond the component to pick up solder from another source. Additionally in reflow soldering, because the solder is placed only where it is needed and in a precise amount, there is much less chance of solder bridging between pads or terminations, allowing a higher packing density of components on the board. However, the smaller the pads, the greater the accuracy that is required in the placement and the less the solder that can be screen printed since there is often a practical limit to the thickness of solder that can be applied in a one-pass screen print or else it will slump and spread beyond the pad edges.

With reflow soldering, components tend to move when the solder is molten (Section 10.7), driven by a thermodynamic need to minimise the surface energy of the system. This can be beneficial if the solderabilities of the component and the pad are good, acting as a self-aligning mechanism for the component on the pad, but it can also be detrimental if the conditions at each end of a two-terminal component are different. If the solderability of one metallisation or one pad is poor, or if one solder pad is larger or more heat-sinked than the other, the alignment of the component may become worse or the component may flip up on one end (Section 10.7.2). Thus, for reflow soldering, plated-through-holes or large areas of solder land that can act as solder drains should be separated from component solder pads by a short length of narrow track or some solder resist.

When wave soldering is used, the SMCs must be retained in position with a spot of non-conducting adhesive. This is one of the main problem areas because sufficient adhesive has to be applied to fill the gap between the PCB and the component. When the component is being placed, the adhesive of necessity spreads under pressure, but it is very important that it does not contaminate the solder pads or component terminations. The requirement is for a spot of adhesive of minimum diameter and maximum height. This can be achieved most satisfactorily by designing into the PCB layout a dummy copper bar or a conductor track routed under the component to reduce the height of the glue dot required, as shown in Figure 4.9.

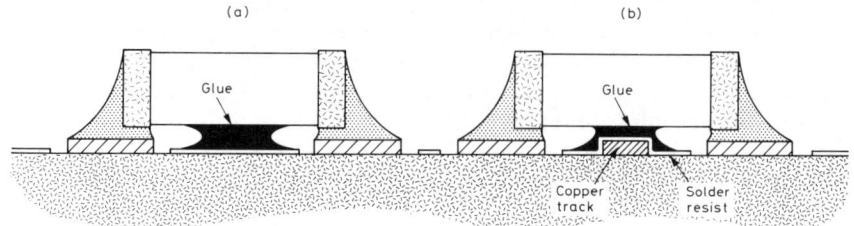

Fig. 4.9 To avoid a large amount of glue, which may spread and contaminate the solderable surfaces, a dummy or real copper track can be designed under the component body.

4.2.4 Design Constraints from the Testing Operation

It is advisable to be able to see the solder fillet on the end of a component in order to inspect the joint visually. For electrical testing, access to the terminations of SMCs in an easy and reliable way is a major problem.[111, 112] It is possible with leaded components such as SOICs, SOTs and flatpacks for a test probe on a lead to push the lead momentarily into contact with its pad and demonstrate a good joint even though the joint is, in reality, an open circuit. Another problem is that the test probe may be deflected across the surface of the tiny, and therefore steeply sloping, solder joint, causing damage. The scale of surface mounted assemblies is so small and the packing so dense that the pin sizes and tolerances of testing facilities may be quite inappropriate.

If a full electrical test of the assembled PCB is specified, the layout must take this into account, by including suitable pad extensions or dedicated pads as test points,[113] as illustrated in Figure 4.10. Such a requirement should be considered carefully because it can substantially increase the board size on a completely

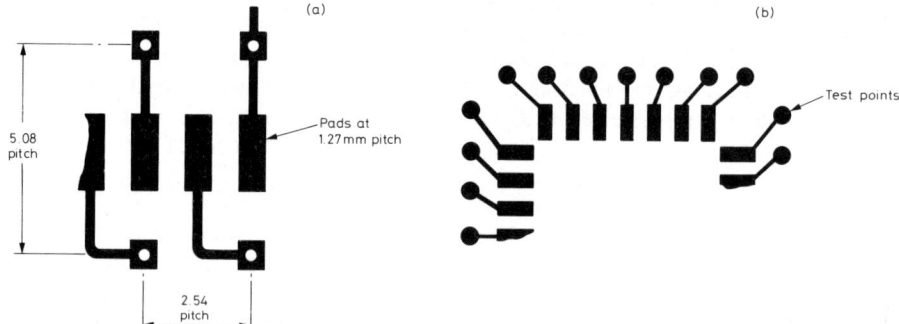

Fig. 4.10 If a full electrical test of the assembled PCB is required, the layout must include pad extensions or dedicated pads. It may be necessary to site these on a 0·1 inch grid.

surface mounted assembly.[114] With mixed assembly boards, the problem is not so critical, since the board area is not at such a premium. In Figure 4.10(a), the test points have been placed on a 2·54 mm grid although the component pads are at 1·27 mm pitch. Obviously, if this necessitates placing the test points under components, vias have to be used to access the test points from the other side of the board.

4.2.5 Preferred Component Layouts

The physical orientation and spacings of component packages are primarily dictated by the soldering method to be used. Reflow soldering has no directionality and so components can be placed at any orientation to each other with a proximity restricted only by the footprint pad size requirements. Wave soldering, however, is directional and requiring access to the pads by the solder wave. Thus both the relative orientation of components and their spacings are of prime importance to the reliability of the soldering process.

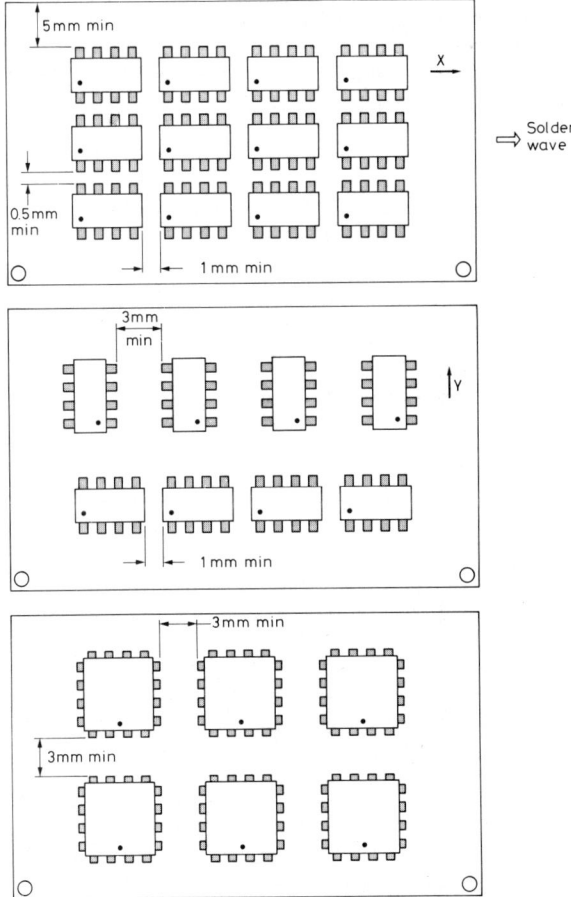

Fig. 4.11 Some illustrative examples of single component types of layout, giving minimum spacings between SOICs and between quad flatpacks, for successful wave soldering.

For a particular soldering process the highest packing density of components achievable is governed primarily by the component packaging format. For reflow soldering of SOICs, for example, the minimum separation of adjacent, parallel rows of pads can be as low as 300 μm. However, when assembling packages with J leads, such as PLCCs, it is important that this minimum separation be increased to several millimetres to allow visual solder-joint inspection and rework when necessary. The separation must usually be increased further if full testing is specified and test pads are incorporated. This restriction on the packing density can be overcome by using vias to enable all testing to be made on the reverse side of the board.

For wave soldering surface mounting boards, the minimum satisfactory separation of components is normally significantly greater and wave soldering should not be used when space reduction is the major consideration for adopting surface mounting. Comprehensive rules of layout design are not possible but some illustrative examples[115] of single type component layouts are shown in Figure 4.11 and of multi-type component layouts in Figure 4.12. The X-axis is that of the conveyor direction on a wave soldering machine. An X-axis layout will give significantly better soldering when using a wave, both in terms of equality of fillet shapes and in terms of solder skips. A minimum lateral spacing of 500 μm between the outer ends of solder pads is required for SOICs. The minimum spacing between the SOIC packages in the X direction needs to be about 1 mm to

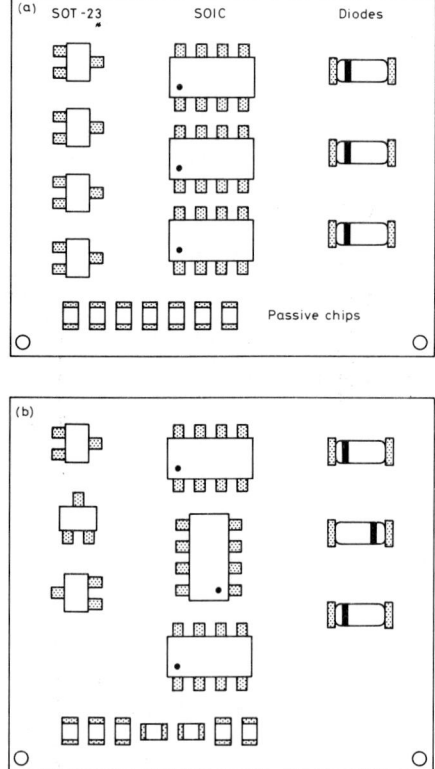

Fig. 4.12 A multi-component layout in (a) a preferred configuration with a minimum number of component axes and spans, and (b) a poor configuration involving many placement-head rotations.

take account of placement tolerances, component tolerances and extraneous plastic moulding material which sometimes extends from the component body.

The quad-flatpacks in Figure 4.11 have to be placed further apart in the X direction to avoid shadowing of the leads on the trailing edge of each component (Section 5.5.4). A space of 3 mm between the ends of the solder pads is shown. The shadowing effect for a flatpack is not so severe as for an SOIC or a PLCC because the height of the component body is less. If space permits, a preferred compromise for wave soldering 4-sided components is to mount them at 45° (Section 5.10.2.1) so that the differences attributable to leading edges, trailing edges and sides of components are minimised.

When considering the benefits of a high component packing density, and drawing up design guidelines, the following points must be considered sequentially in order of importance:

(i) Component packing formats; SOICs, SOTs, PLCCs
(ii) Intended method of soldering
(iii) Height of components when seated
(iv) Type and thickness of solder mask
(v) Direction of motion through solder wave
(vi) In-circuit test probe location needs
(vii) Visual inspection needs
(viii) Acceptable rework cost levels.

Figure 4.12 illustrates the use of axes, spans, and orientations in a layout. A preferred layout contains a minimum number of axes (in X direction), spans (in Y direction) and component polarity directions. This minimises machine placement movements such as head and table rotations, substantially improving productivity.

4.2.6 Conductor Routing

Due consideration of conductor routing on a surface mounting PCB can have a large effect on manufacturing yields and reliability. In low-density designs, based on a 2·54 mm grid, for insertion components it is possible and required normally to route only one conductor, 40 μm width between a pair of lands, as shown in Figure 4.13(a). In moderate density or 'two-track' circuit boards, the annular

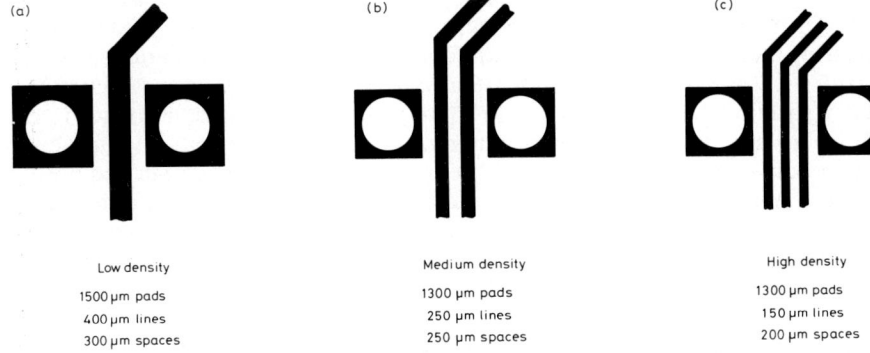

Fig. 4.13 Conductor routing on low-density, medium-density and high-density boards.

lands are reduced as shown in Figure 4.13(b), in order to use 250 μm lines and spaces. For high density or 'three-track' boards, as shown in Figure 4.13(c), three conductor tracks, 150 μm wide, pass between lands on a 2·54 mm grid.

Most surface mounting assemblies are now formatted on a 1·27 mm grid and it is possible to route only one conductor between pads, as illustrated in Figure 4.14(a). Even so, land widths that are commonly 0·9 mm wide allow only a 125 μm/125 μm track and spacing. Routing a conductor between lands therefore requires them if possible to be reduced to 0·65 mm width, allowing a 200 μm/200 μm track and spacing, as shown in Figure 4.14(b). Other lands can remain at 0·9 mm if no routing track is required.

The best method for routing a conductor track through a quad pack footprint is at the corners, by chamfering the corner pads as shown in Figure 4.14(c) to allow optimum spacing.

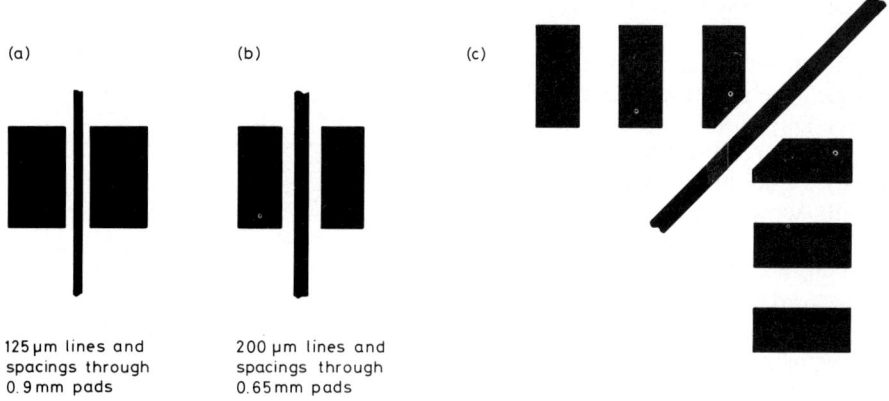

(a) 125 μm lines and spacings through 0.9 mm pads

(b) 200 μm lines and spacings through 0.65 mm pads

Fig. 4.14 (a) Component pad widths of 0·9 mm can be reduced to (b) 0·65 mm in order to ease the accommodation of an interpad conductor track. (c) Corner pads can be chamfered on a quad-pack to enable conductor routing.

4.3 COMPONENT FOOTPRINTS

The footprint on a PCB is the array of solderable pads associated with the attachment of a particular component. For a given component package and soldering method there is an optimum footprint design. Each solderable pad must be sufficiently large to accommodate misalignments of either the screen printed solder paste or the placement of the component but, on the other hand, it must not be so large that the spacings between the pads are liable to bridging by the molten solder. The shape and size of the pad in relation to the design and position of the component lead or pad, and the amount of solder present, must help to define the geometry of the solder fillet formed which, in turn, determines the thermal and mechanical properties of the joint.

Little controlled systematic experimental work on the design of footprints has been reported and recommended pad shapes and sizes are based mainly on *ad hoc* production experience. In this respect, the recommendations given here are not inviolable, but intended merely to act as guidelines.

4.3.1 Placement Accuracy

Besides the requirements of a footprint pad to be of a size commensurate with an ideal solder fillet geometry for mechanical strength and continued electrical integrity, consideration must be given to the range of relative placement positions between a component and its footprint. With respect to a 'true' position of a component termination and a solder pad, there is the positional accuracy (determined by the placement machine), the accuracy of position of the pad on the substrate (determined by the substrate manufacturing process) and the dimensional tolerances of the component itself (determined by the SMC manufacturer). Figure 4.15 shows how the placement accuracy is determined by these three factors. The worst situation is when all three errors combine cumulatively and this extreme must be taken as controlling the footprint design.

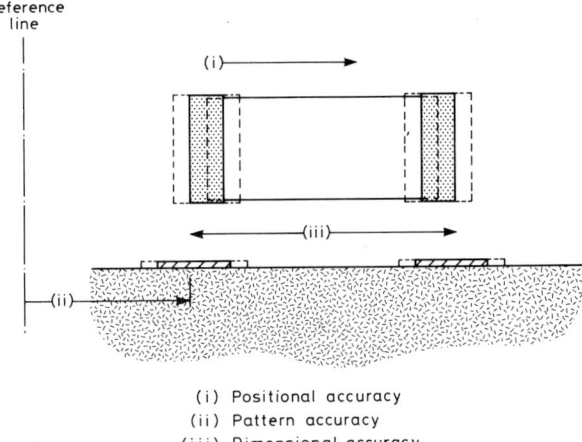

(i) Positional accuracy
(ii) Pattern accuracy
(iii) Dimensional accuracy

Fig. 4.15 Component placement accuracy is determined by three factors: the placement, the board and the component.

The simplest case is for a two terminal chip resistor or capacitor. The space between the solder pads needs to be as wide as possible to accommodate conductor tracks passing beneath the component, yet narrow enough to ensure adequate overlap with the metallisations on the component. As shown in Figure 4.16, the maximum distance between the pads, d_{max}, is related to the minimum

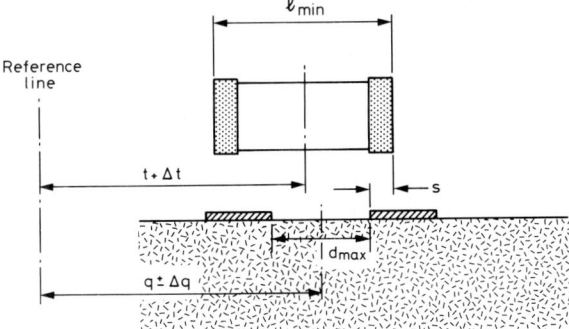

Fig. 4.16 Definition of the terms used in Equation (4.1) for the maximum separation of pads.

length of the component, ℓ_{min}, the positioning uncertainty, Δt, of the placement machine, the substrate positional tolerance, Δq, and the minimum overlap, s, of the component and each pad:

$$d_{max} = \ell_{min} - 2\Delta t - 2\Delta q - 2s \tag{4.1}$$

assuming that the nominal position, t, of the component and the nominal position, q, of the substrate pad are identical from some arbitrary reference point. Typically, values for these parameters are

placement accuracy	$\Delta t = \pm 0{\cdot}05$ mm
pattern accuracy	$\Delta q = \pm 0{\cdot}2$ mm
required overlap	$s = 0{\cdot}1$ mm

Thus, for a 1206 chip capacitor, for example, whose manufacturer quotes a length of 3·2±0·15 mm, ℓ_{min} is 3·05 mm and the maximum space under the component between the pads would be 2·35 mm, sufficient for three 250 μm parallel tracks with 250 μm spaces.

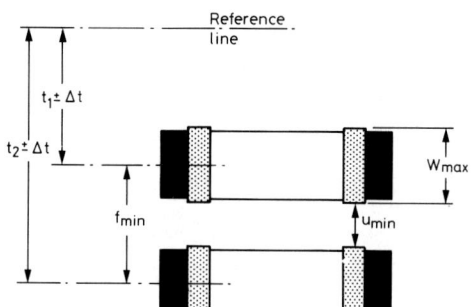

Fig. 4.17 Definition of the terms used in Equation (4.2) for the minimum transverse separation of footprints.

Cubic 0805, 1206, 1210	1	1	1	1	1
Cubic 1808	1.5	2	2	1	1
Cyl. SOD-80	1	1	1	2	1
Cyl. MELF	1.5	1.5	1.5	3	2
SOT-23	1.5	0.6	1.5	1.0	0.6
SOT-89	2	2	2	1	1

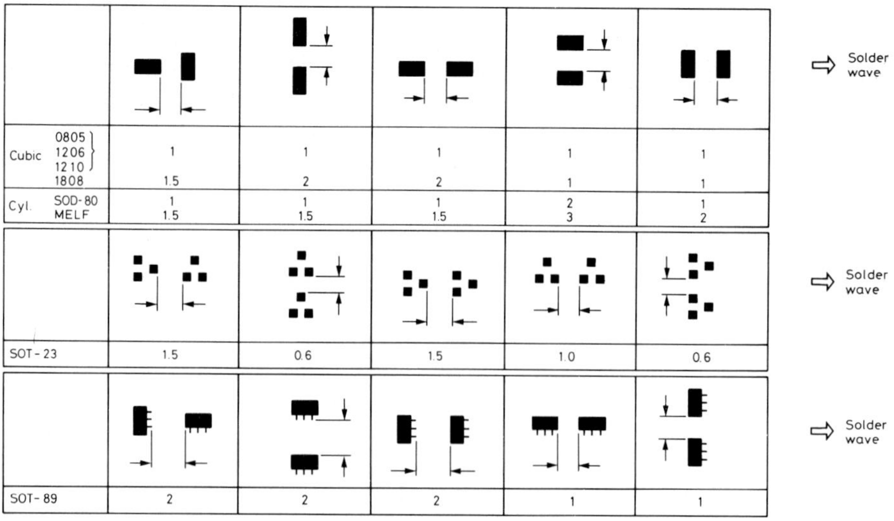

Fig. 4.18 Some examples of footprint separations (mm) for two-terminal components as well as SOT-23s and SOT-89s, when wave soldering.[109]

Design and Assembly

An equation, similar to Equation (4.1), can be constructed for the minimum distance f_{min} between the solder pads of parallel adjacent chip components, illustrated in Figure 4.17. In this case the maximum component width, W_{max}, and the minimum distance of component separation, u_{min}, necessary to avoid short circuit bridging when soldering, are important. Thus:

$$f_{min} = W_{max} + 2\Delta t + u_{min} \tag{4.2}$$

For reflow soldered assemblies, u_{min} can be very small, but for wave soldered assemblies, u_{min}, and therefore f_{min}, is based on experience for the particular components involved, their orientation to one another and their orientation to the solder wave. Examples are shown in Figure 4.18 for some cubic and some cylindrical chip components as well as SOT-23 and SOT-89 packages.

Both d_{max} and f_{min} should be slightly modified from the values calculated by Equations 4.1 and 4.2 to take account of rotational misplacement. The influence of component rotation is illustrated in Figure 4.19. The effective length of the component is increased by a factor $W \sin\Psi$ and the effective width is increased by a factor $L \sin\Psi$. In reflow soldering, components will tend to re-align with the footprint axis, so these corrections are strictly necessary only for wave soldering. The rotational placement accuracy Ψ is typically $\pm 3°$. This means that, for instance, the effective width of a $3 \cdot 2 \times 1 \cdot 6$ mm 1206 chip component can increase by 10% for a 3° rotational misalignment.

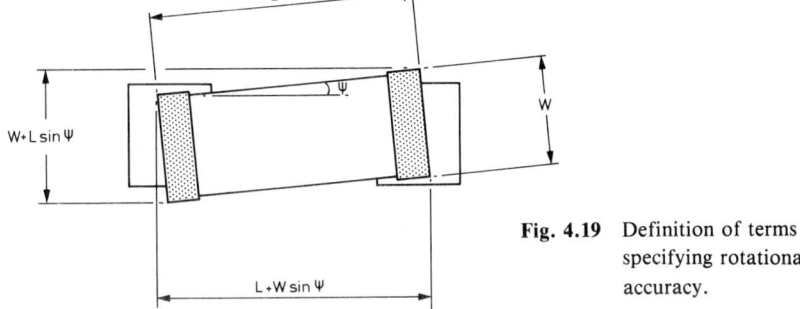

Fig. 4.19 Definition of terms involved in specifying rotational placement accuracy.

4.3.2 Typical Footprints

The basic guidelines given in the previous section can be used to design the footprint and the relative layout of footprints of any components. The optimum pad shapes and sizes are also a function of the soldering process used and the configuration of the component.[116]

The width of the pads is normally designed to be almost the same as the nominal width of the component metallisation, although this can be increased for components with small terminations that are relatively far apart such as the SOT-23 or chip components.[117]

For LCCCs with solderable castellations it is usual to add about 1 mm to the outside dimension in order to provide a termination area on the substrate for a solder fillet tapering up the side of the chip carrier castellation. A tested formula for designing LCCC footprints is given in Table 4.1, with reference to Figure 4.20.

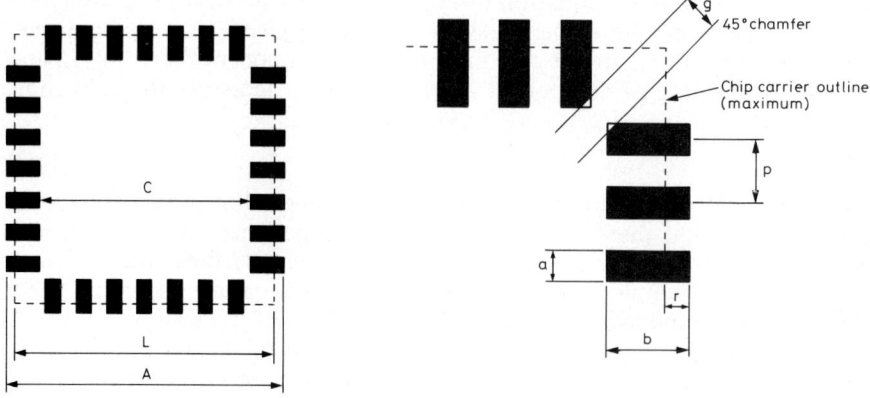

Fig. 4.20 Design guidelines for LCCC footprints (Table 4.1).

Table 4.1

Design of Footprints for 1·27 mm Pitch LCCCs (mm)

Chip carrier dimension	L
Maximum footprint dimension	A = L + 1
Inner footprint dimension	C = L − 2·5
Pad width	a = terminal width + 0·25
Pad length	b = 1·8
Pad extension	r = 0·5
Pad pitch	p = 1·27
Minimum corner separation	g = 0·5

Figure 4.21 with Table 4.2 gives the dimensions of footprints for SOICs. The recommended sizes may vary quite considerably from one source to another. This is because they have been optimised using one specific condition of the reflow soldering or of the wave soldering. Thus no footprint recommendation is mandatory, but a guideline based on the best available knowledge.

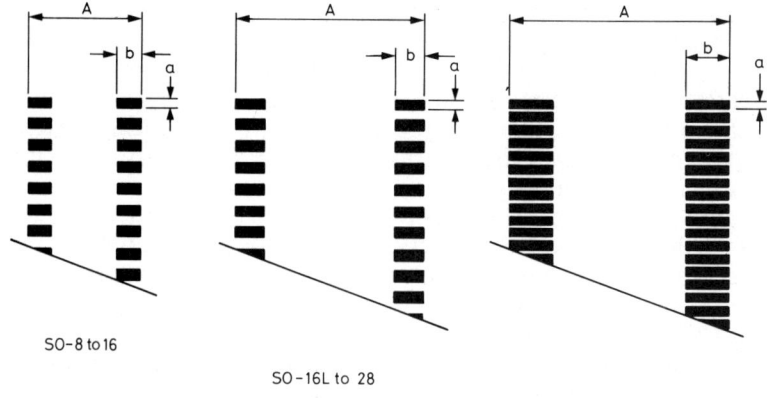

Fig. 4.21 Design guidelines for SOIC footprints (Table 4.2).

Table 4.2
Design of Footprints for SOIC Packages (mm)

Device Package	Pad Width [a]	Pad Pitch [p]	Pad Length [b]	Footprint Width [A]
SO-8 to SO-16	0·6	1·27	1·5	7·2
SO-16L to SO-28	0·6	1·27	1·7	11·6
VSO-40 to VSO-56	0·5	0·76	2·7	13·6

Figure 4.22 with Table 4.3 offers footprint dimensions for PLCCs. The outer dimension of the footprint is given by:

$$A = 1 \cdot 27N + 3 \cdot 8 \text{ mm}$$

when N is the number of pads along a side row of the footprint.

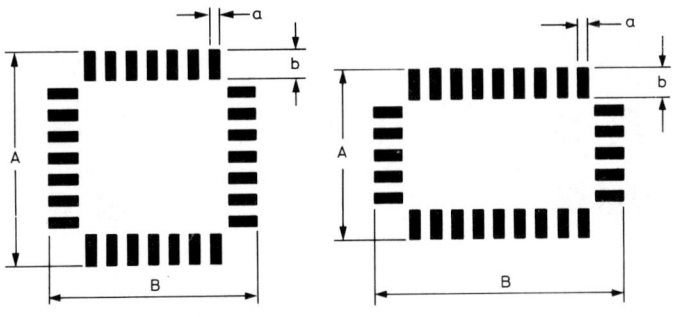

Pad pitch, p = 1.27mm

Fig. 4.22 Design guidelines for PLCC footprints (Table 4.3).

Table 4.3
Design of Footprints for 1·27 mm Pitch PLCCs (mm)

Device Package	Pad Width [a]	Pad Length [b]	Dimensions of Footprint [A × B]
20 pin (5 × 5)	0·6	1·8	10·2
24 pin (6 × 6)	0·6	1·8	11·4
28 pin (7 × 7)	0·6	1·8	12·7
44 pin (11 × 11)	0·6	1·8	17·8
52 pin (13 × 13)	0·6	1·8	20·3
68 pin (17 × 17)	0·6	1·8	25·4
84 pin (21 × 21)	0·6	1·8	30·5
18 pin (5 × 4)	0·6	1·8	8·9 × 10·2
28 pin (9 × 5)	0·6	1·8	10·2 × 15·2
32 pin (9 × 7)	0·6	1·8	12·7 × 15·2

It is for the small discrete components that the difference between the optimum pad sizes and footprints is most marked depending on their use for wave soldering or for reflow soldering. Figure 4.23 shows typical recommended footprints for the common discrete active components, SOT-23, SOT-143, SOT-89 and SOD-80. With reflow soldering, because the solder is placed exactly where it is required, the pads are smaller and placed more under the components. With wave soldering the pads have to capture molten solder and are therefore larger and more visible outside the component.

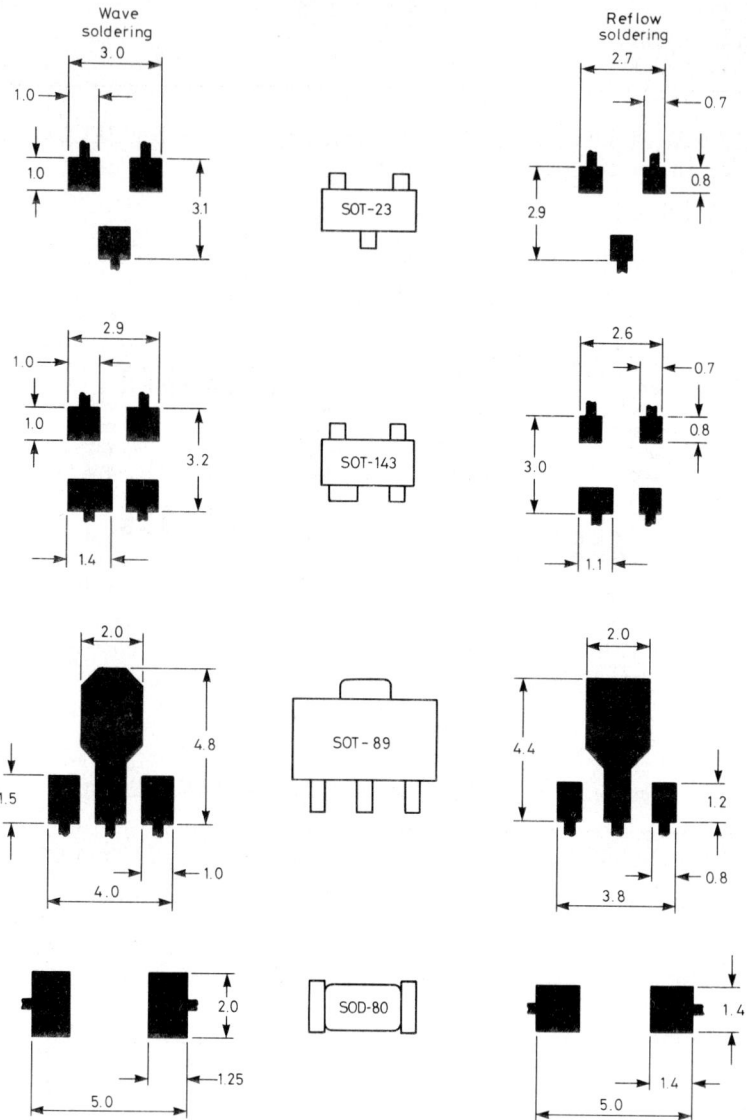

Fig. 4.23 Design guidelines for common discrete active components.

A considerable variation in recommended footprint designs for the passive components can be found in the technical literature with no apparent justification made for the choices of pad sizes and configurations. The dimensions given in

Table 4.4, taken in conjunction with Figure 4.24, result from the following guidelines. For resistor and capacitor chip components, the pad width for both wave and reflow soldering is the nominal width of the component, plus 0·2 mm. The length of the pad, and its best position relative to the component, depends partially on the height of the component which, particularly in wave soldering, governs the shape of the solder fillet. The height of a chip component depends on its value. All resistors R 0805 and R 1206 have a nominal 0·6 mm height. Capacitors C 0805 have heights in the range 0·5-1·3 mm, C 1206 in the range 0·5-1·6 mm and C 1210 and larger have heights in the range 0·5-1·9 mm. Thus, for wave soldering, better shaped solder fillets can be achieved if the pad extension is partially matched to the height of the solderable surface on the component. For wave soldering the pad extension, r, is taken in the range 0·8-1·0 mm and the under-component pad, s, as 0·4-0·6 mm. For reflow soldering the pad extension is smaller, about 0·3 mm, and less variable because of the lesser influence of component height. The under-component pad is larger, increasing from 0·5 mm for the 0805 size to 0·9 mm for the 2220 size.

For the aluminium electrolytic capacitors the electrodes are about 2·5 mm square under the component body. For reflow soldering the pad suggested is this area plus 0·5 mm pad extension. For wave soldering, the solder will not readily flow 2·5 mm under the component and so a pad 1 mm under plus 1 mm extension is recommended.

As mentioned in Section 2.7.1.4, tantalum capacitors are available in a variety of shapes and sizes. Table 4.4 gives typical footprint dimensions for the cubic chip styles a-h, whose dimensions were given in Table 2.7.

Table 4.4

Design of Footprints for Passive Components (mm)

		Wave Soldering			Reflow Soldering		
		Pad Width [a]	Pad Length [b]	Footprint Length [B]	Pad Width [a]	Pad Length [b]	Footprint Length [B]
Chip resistors and capacitors	0805	1·45	1·2	3·65	1·45	0·8	2·65
	1206	1·7	1·4	4·85	1·7	1·0	3·65
	1210	2·75	1·4	4·85	2·75	1·0	3·65
	1808	2·25	1·5	6·45	2·25	1·1	5·2
	1812	3·25	1·5	6·45	3·25	1·1	5·2
	2220	5·3	1·6	7·6	5·3	1·2	6·2
Aluminium electrolytic capacitors	1a	2·5	2·0	10·0	2·5	3·0	9·0
	1	2·5	2·0	14·0	2·5	3·0	13·0
Tantalum capacitors	a	1·5	2·0	5·0	1·5	1·1	3·2
	b	1·5	2·0	6·3	1·5	1·1	4·5
	c	1·5	2·0	7·55	1·5	1·1	5·75
	d	2·75	2·0	6·3	2·75	1·1	4·5
	e	2·75	2·0	7·55	2·75	1·1	5·75
	f	3·65	2·2	8·45	3·65	1·3	6·65
	g	3·0	2·5	9·15	3·0	1·6	7·35
	h	4·0	2·5	9·65	4·0	1·6	7·85

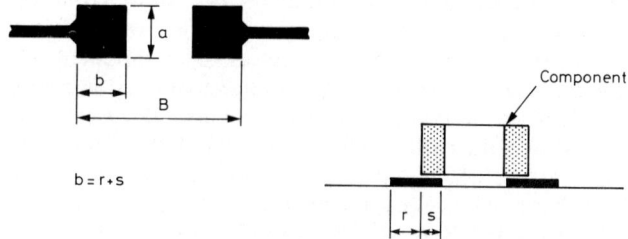

Fig. 4.24 Design guidelines for passive components (Table 4.4).

4.3.3 Solder Masks

If the assembly is to be wave soldered, the use of a solder mask[118] is recommended with a clear annulus around the solderable pads, as shown in Figure 4.25.

Two types of solder mask window, a 'gang' type and a 'pocket' type, can be used. The gang type should only be used when there are no conductors routed between lands.

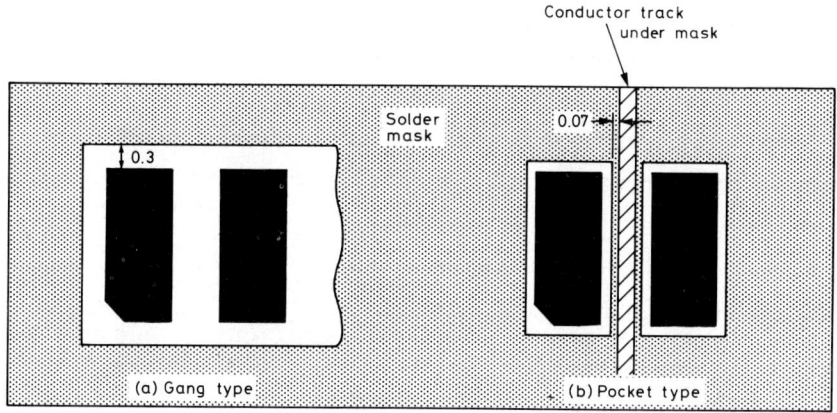

Fig. 4.25 Illustrating the gang and pocket types of solder mask application.

The purpose of a solder mask is to restrict the solder to the places where it is supposed to be. It becomes particularly important therefore as the tracking density increases on boards intended to be wave soldered. For low- and some medium-density designs, a screen printed liquid solder mask can be a suitable material; it generally can be considered to have a design accuracy, when cured, of 20-25 μm. However, with designs that have conductor tracks routed between solder pads, liquid solder mask cannot be applied with sufficient accuracy to assure complete coverage of the conductors without any contamination of the pads. In these cases dry film solder mask or photo-imageable solder mask are recommended. The additional cost of these materials is easily recouped from higher board and assembly yields.

4.3.4 Conductor Fan-outs

As mentioned already, care must be taken that there do not exist solder drain paths away from the pads. This is a problem of reflow soldering causing poor joint quality and encouraging components to move while the solder is molten. It can be overcome by the use of solder resist or by thin conductor track fan-outs from the footprint. Some illustrative examples[119] of how this should be achieved are given in Figure 4.26. They complement the use of fan-out test pads as shown in Figure 4.10. Besides large unused land areas, another drain on solder is a plated-through via hole sited nearby. Figure 4.27 shows preferred and unacceptable conditions for plated-through hole attachment to lands.[120] If solder masking is not used, the hole and the pad must be joined by, for example, a 250 μm wide track at least 500 μm long.

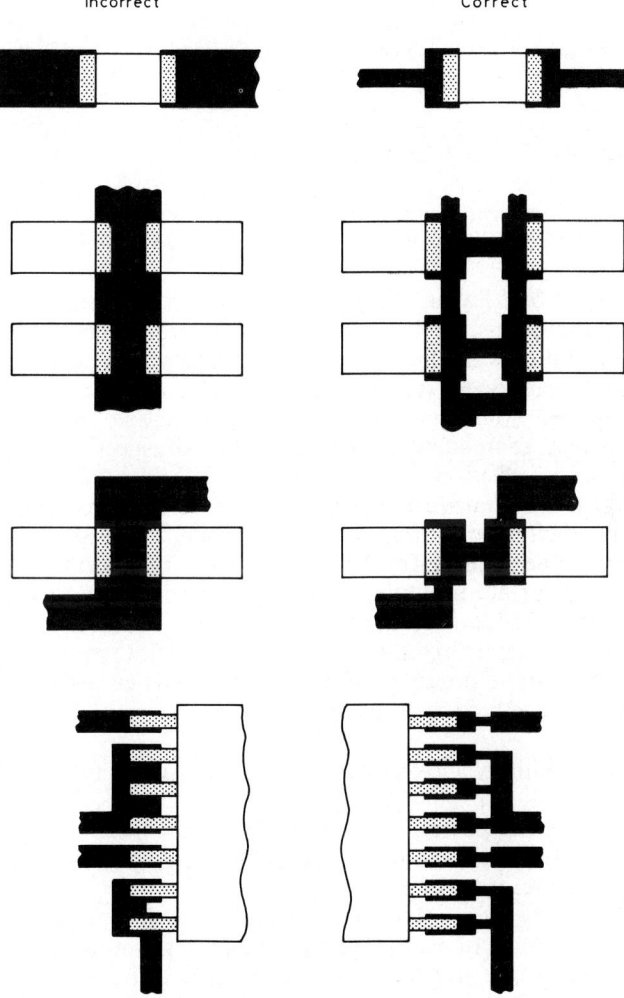

Fig. 4.26 Some examples of the use of thin conductor fan-outs to avoid solder flow away from the pads.

4.3.5 Thermal Mounting

The thermal management of individual components can be incorporated into the PCB footprint design by using thermal mounting pads of copper beneath the component with heavily plated vias through to a thermally conducting plane of the board. An alternative is to use a heat sink glued to the underside of the component and mounted through the dielectric layer of the board to contact a thermally conducting plane. This was discussed in Section 3.5.4.2.

Acceptable   Unacceptable

Fig. 4.27 Plated-through holes must be joined to lands by fine tracks to avoid loss of solder from the land.

4.4 COMPUTER-AIDED DESIGN

For all but the simplest printed circuits, computer aided design (CAD) of surface mounting assemblies is essential. CAD has two functions:

(i) To provide the best layout design of components and tracks, given certain boundary conditions and their order of precedence. Non-violatable boundary conditions might, for example, be footprint sizes, tracks widths and spacings, component spacings, number of signal layers in the board, etc. Other requested boundary conditions might include short track lengths between certain components, board size, board aspect ratio, positions of mounting holes, position of connectors etc. Within these constraints the computer is programmed to provide one or several layout options.
(ii) To produce digital co-ordinates for component placement machines and generate analogue artwork for the manufacture of the printed circuit board. A main aim of surface mounting assembly is almost always miniaturisation. With smaller pad sizes and narrower conductor widths and spacings, the demand for accurate artwork is increasing. A benefit to accuracy is available if the artwork can be produced digitally from CAD co-ordinates instead of by photographic reduction of master artwork.

CAD hardware is simply a dedicated computer consisting of five facilities:

(i) adequate computing power, memory and access;
(ii) a combination of various input devices such as an alpha-numeric keyboard, light pen, joystick, mouse, etc.;
(iii) one, or preferably two, colour video screens, the first to show the current position of the design of each conductor plane singly or in combination, and the other for listing commands, connections still to be made, etc.;
(iv) an output device for generation of hard copy layouts;
(v) adequate applications software.

It is completely impractical for the software to be written on a do-it-yourself basis and many elegant software packages are available.[121]

An advantage of surface mounting is the freedom offered the designer to place a component on either side of the substrate. A CAD system must have the potential for displaying a superimposed image of both sides of the board, as well as any internal conducting planes, on a video screen. A colour screen is therefore imperative, using different colours for each conductor plane and building up a 3-dimensional picture of the board. Furthermore, the system must have an unambiguous method of defining vias and through-holes as well as recognise the possibility of same-sized components being placed directly opposite each other.[122] Many CAD systems were designed for use with a 2·54 mm grid that is inappropriate for surface mounting assemblies. The more modern systems are gridless with no constraints on co-ordinate positions.[123]

4.5 COMPONENT PLACEMENT

4.5.1 Hand Assembly

Surface mounting assembly not only offers significant advantages, through efficiency of packaging, in assembly size and weight, but also offers the potential for fully automatic fast robotic assembly. Thus, hand assembly and surface mounting are incongruous: the components are too small to be handled with ease and the components are not in general identifiable once removed from their packing. Unless considerable organisation and concentration are instituted, hand assembly will give rise to errors if anything other than a few small prototype boards are attempted.

Components can be handled and placed with tweezers; fine pointed stainless steel tweezers are best, but to achieve a rotational movement is much simpler using a pencil shaped vacuum pipette. The vacuum can be switched on and off easily from the handle. Nevertheless, the potential for component placement error is high and the time taken to hand assemble more than a few components soon surpasses that required to program an automatic machine. When automatic placement is used, the first board can be checked and corrected and then all the following assemblies will be identical. Hand assembly involves much more rigorous checking of every board.

4.5.2 Sequential and Simultaneous Placement

Automatic placement can be roughly divided into two main modes: sequential and simultaneous.[124] Sequential placement means the components are picked up one after another and placed on to the printed circuit board. Machines of this type are often referred to as pick-and-place machines. With simultaneous placement some or all components are placed on the board in one operation, which obviously gives the potential for far higher speeds. The two principles can be combined, and many variants are available involving multiple placement heads, moving substrates on a belt or within an x-y format, and so on.[125]

The early machines were aimed at the hybrid assembly industry, handling relatively small substrates. These pick-and-place machines usually had a single moving head that sequentially picked one component from a store, orientated it, tested it electrically and placed it on a stationary substrate. These machines can operate at 1-2 thousand components an hour.

Surface mounting assembly has imposed three basic changes to this type of equipment. The size of the substrate to be handled has increased from 50×50 mm to 500×500 mm, so that the distances to be travelled are an order of magnitude larger. The variety of components to be mounted on each board and the range of component package sizes and shapes has increased considerably. Thirdly, production runs have escalated, demanding the requirement of very high throughput. Thus the new machines must be able to handle large substrates, take components of a variety of sizes and shapes from a variety of packing formats and perform at ever higher rates.

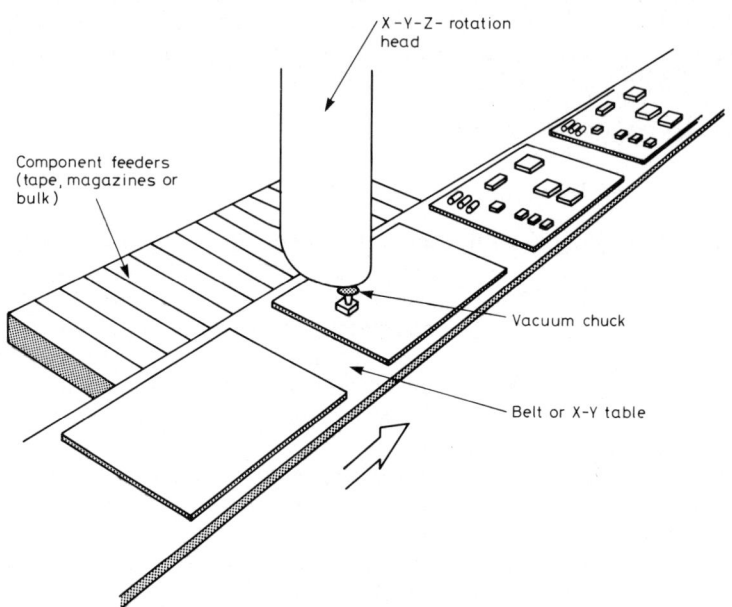

Fig. 4.28 The principle of a sequential pick-and-place single-headed machine.

The sequential pick-and-place principle of operation is shown schematically in Figure 4.28. Machines of this type have a single head which can place a large number of different components at different points on a substrate under the control of a small computer. To handle the movement times on large substrates, in some machines the substrate as well as the head freely moves on an x-y table and the programming is such that the mutual travel of the head and the substrate is minimised. Placement speeds in the range 5-15 thousand components per hour can be achieved by using more than one placement head. Each head picks and places only one component at a time. In some machines the several heads are all free to move anywhere within the substrate bounds and are programmed not to collide with each other, whereas in other machines the substrates move sequentially along a conveyor, stopping under each of a series of heads which have the responsibility of placing one or a few components, as shown schematically in Figure 4.29.

The control of a pick-and-place machine can be done either through hardware or through software. The programming of the software can be carried out either off-line or on-line.

Hardware control is the fastest but least flexible system. The pick-up placement heads move into positions defined by a programmed plate that can be generated from the PCB computer design software. This type of pick-and-place programming is obviously expensive and time consuming to edit and change. Hardware controlled machines are therefore used for high volume production where a large number of a relatively few different types of assembly are anticipated.

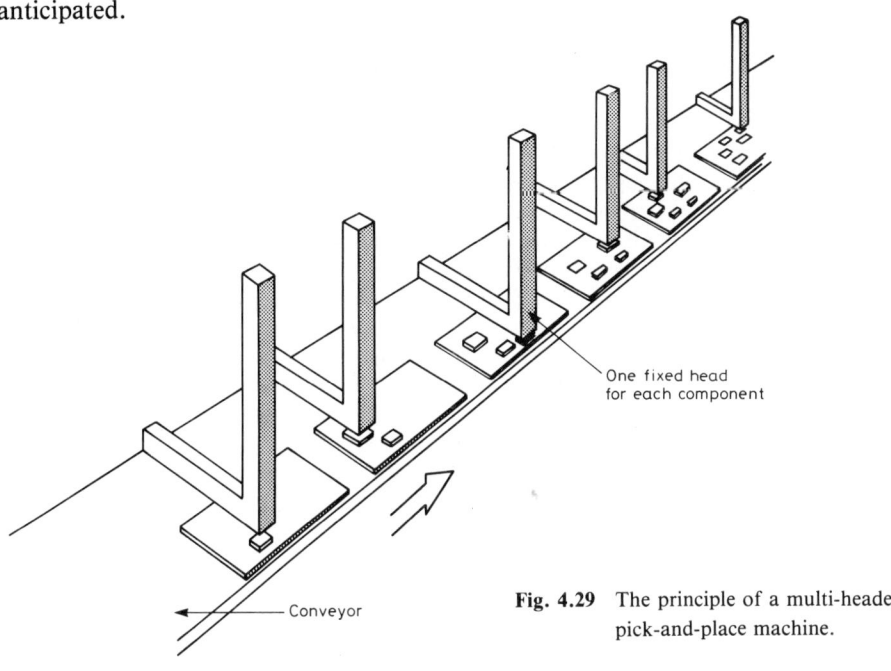

Fig. 4.29 The principle of a multi-headed pick-and-place machine.

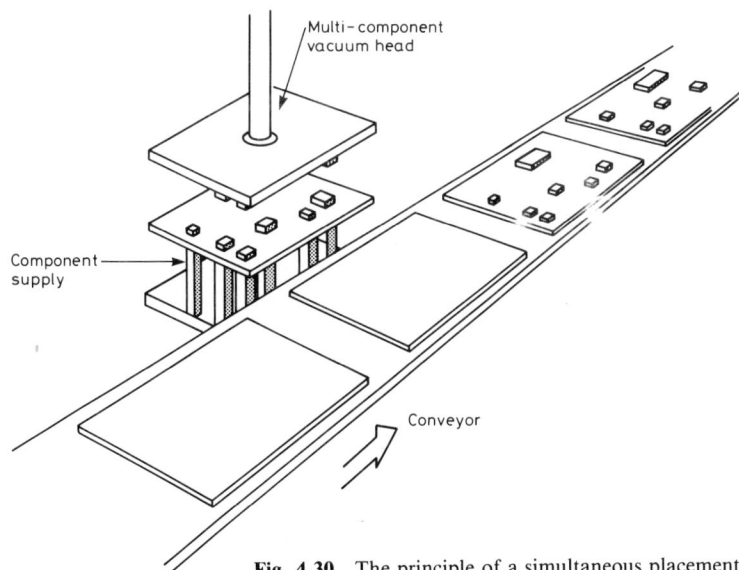

Fig. 4.30 The principle of a simultaneous placement machine.

Most pick-and-place machines are software controlled through a small dedicated computer. The computer can be programmed on-line or off-line. In the former method, usually associated with the smaller machines, the required head movements are stepped through under manual control and the machine learns the positions. Obviously, the machine cannot be used for production of one board type while programming is being carried out for another.

The most versatile machines have off-line programming, in which the placement co-ordinates and the component types are fed into the program and the machine optimises the travel distances and hence the ordering of the pick-up feeder stations. The co-ordinates can be entered in an analogue form from a screen, in digital form on a keyboard, or by a direct link to a PCB layout CAD system. This type of programming is easy and quick to edit or reprogram.

Placement machines described as simultaneous are often of the multiheaded pick-and-place variety, placing single components from each head sequentially. True simultaneous machines pick up a large number of components and place them all in one cycle, as shown schematically in Figure 4.30. This type of machine is usually hardware-controlled, is very fast, but difficult to reprogram.

4.5.3 Selection Criteria for Placement Machines

Although simultaneous placement machines feature high performance at very low cost per component, normally the sequential pick-and-place facilities offer the most suitability for small and medium production runs. A number of criteria must be considered when selecting the placement machine:

—PCB size to be assembled
—number of different PCB designs and batch sizes
—ease of programming the machine
—type and number of components
—supply packing of components
—layout restrictions on PCB
—testing of components for identity, polarity and faults
—assembly performance speed
—placement accuracy and reliability
—assembly cost
—adhesive dispensing
—flexibility.

The smallest machines basically consist of an x-y platen driven by stepper motors controlled by a small computer.[126] Components are picked up by a vacuum pipette with an automatically selected nozzle on a turret. Normally, for wave soldering assembly, adhesive is placed from a dot dispenser by the same machine prior to placement. Typically, small machines of this type can handle up to about 20 component feeders supplied in sticks. A typical placement speed is in the range of 1-2 thousand per hour.

With the requirement for larger PCBs to be populated, machines grow in physical size, allowing more component feeds. The larger machines are more flexible in their ability to handle different component packing types: vibrator bulk, stick, tape, etc. The large machines include electrical testing of each component before placement and automatic sensing of the component after placement. With a single head, it is unlikely that an average assembly rate greater than 4000 per hour can be achieved.[127] Using multiple heads the average rate specified is around 3000 per hour per head, and several machines are available that operate in the 10,000-30,000 components per hour range. The actual placement time for a component is normally between 0·5 and 1 second, and it is difficult to engineer a system that will operate faster than this.

The hardware controlled machines populate boards simultaneously. The adhesive is usually placed by a pin-transfer method and then all the components are picked up and placed together. The rate of component placement depends on the number of components per board, but machines with maximum rates in excess of 500,000 components per hour are available.

4.5.4 Placement Machine Functions

A placement machine has up to seven functions, each of which needs to be addressed when the machine is specified for a particular purpose:

(i) Substrate feed and positioning

The positional repeatability of the substrate with respect to some machine fixed point is normally better than ±0·1 mm at all points. The machine must have some means of feeding the substrate and positioning it on a platen which may be fixed or may form part of a conveyor belt or part of an x-y table.

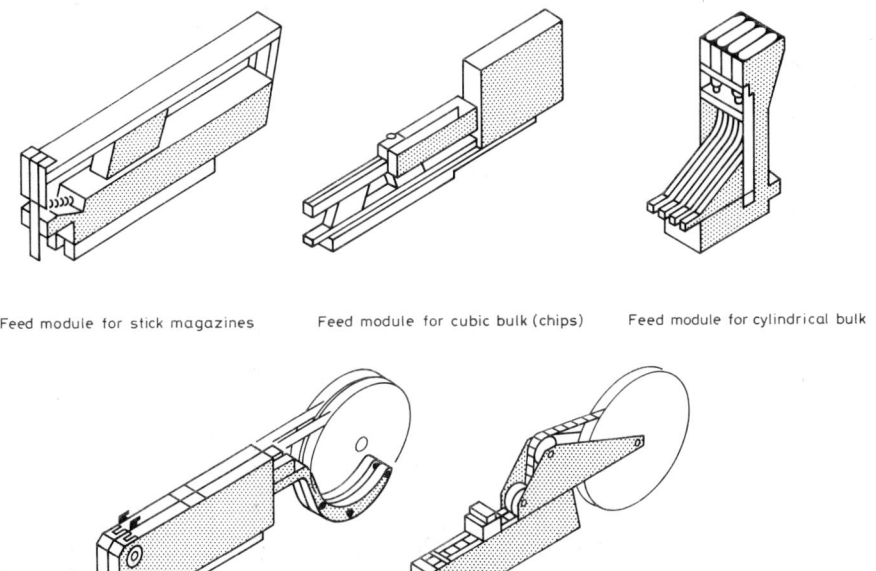

Fig. 4.31 Some typical component feeding stations designed for use with various component packaging formats.

(ii) Adhesive or solder paste application

Wave soldering is dealt with in detail in Chapter 5. Since the populated PCB is inverted, the components need to be glued to the board, and it is the job of the placement machine to place the adhesive where it is needed. The methods of achieving this are described fully in Chapter 5, but the most popular are sequentially from a pressure dispense or simultaneously by pin transfer.

For boards intended to be reflow soldered, adhesive is not required but solder paste is screen printed to the solder pad areas. This is normally done prior to feeding the PCB into the placement machine.

(iii) Components feed

The three main methods of supplying surface mounting components are in bulk, in tape and in preloaded cartridges, sticks or magazines. These are described fully in Section 2.9. The loose bulk option requires a vibratory feed to align the components which are then fed under gravity or spring pressure to their respective pick-up positions. The tape option requires a feed module that splits open the tape to expose the components one at a time. The cartridge system requires a feed mechanism that is specific to the cartridge type used, be it stick or waffle for example. Some typical component feeding stations are drawn diagrammatically in Figure 4.31. With the high speed machines it is necessary to be able to maintain a constant supply of components, and renew component tapes or cartridges without interrupting the machine operation.

(iv) Pick-up heads

The head or heads are required to pick up, identify, orientate and centre a variety of component types from a variety of feeding stations at a number of positions, and place them at any predetermined point within the placement area. Most machines use a vacuum pipette or a series of different pipettes on a turret which can be automatically chosen to accommodate special components such as cylindricals or TAB. The component identification measures and checks the resistance, capacitance, inductance, polarity, etc., normally only of two terminal chip components. The centring is necessary to eliminate positional error when a component has movement within its feeder, for example as in a blister tape feed. The centring station is usually of the form of tweezer arms which move simultaneously to centralise the component on the vacuum pipette.

(v) Computer control

All placement machines have a dedicated computer that monitors all the other functions and supplies the co-ordinates of the pick-up and placement positions.[128] As mentioned already, programming is either via the hardware or software, the latter being eligible for on-line or off-line programming.

(vi) Curing system

For boards intended to be wave soldered, the adhesive should be cured as soon as possible to avoid component movement during subsequent handling. An oven fed directly from the placement machine would be ideal, although the curing is not strictly part of an expected function of the placement machine.

(vii) Reliability features

In order to determine whether or not a component has been picked up and then placed, the vacuum is usually monitored as a function of time and the pressure change sequence compared with the stored norm. If a pick-up head is empty it can be programmed to try again several times before signalling a malfunction. It is important to monitor not only the pick-up time but the placement time in case a component has fallen off in transit and landed in the wrong place. The tip of the pick-up vacuum pipette could be accidentally dipped into the adhesive dot if such precautions were not taken.

The placement pressure must also be adjustable with care in order to ensure that it matches the requirements of the adhesive. If the component is placed with too much pressure, either it could be damaged or the adhesive could spread and contaminate the component terminations and solder pads. To avoid this, either a placement pressure or a board-to-component height needs to be specified.[129]

4.6 CHIP-ON-BOARD ASSEMBLY

Chip-on-board (COB) is the assembly technique in which bare semiconductor chips are attached directly to printed circuit boards made of organic substrates such as fibre glass reinforced epoxy and polyimide.[130-133] Regardless of the substrate used, there are three major considerations in bare chip attachment: the die bonding, the outer lead bonding and the encapsulation.

4.6.1 Die Bonding

Silver loaded epoxy adhesive is the most popular way of attaching the chip dies to the substrate. When the epoxy resin cures, the silver provides electrical connectivity and thermal conduction. The adhesives are formulated with a minimum of volatile solvents which has two beneficial effects: they do not outgas significantly and so bond without forming voids, plus there is little volatile material evolved during the cure which could contaminate the circuit pattern. The silver loaded adhesive is screen printed or spot injected on to the die-attachment areas. After the chips have been placed, the assembly is cured at around 150°C for 1-2 hours.

High-power devices are usually soldered to the board. In such cases, a copper die attachment area is designed into the circuitry. This is plated with a 5 μm nickel barrier layer and a 0·5 μm gold layer. A gold film is evaporated on to the back of the chip and a solder paste attachment technique used.

4.6.2 Wire Bonding

Wire bonding is the most commonly used method of electrically connecting the chip to the substrate.[134] The techniques used to wire bond the interconnections between the bare chip and the substrate are exactly as used by the semiconductor manufacturing industry for packaging chips, and the hybrid microelectronics industry for bonding chips to ceramic substrates, bearing in mind the inapplicability of high temperature processes for organic substrates.[135]

The bond itself is either of a ball type, as shown in Figure 4.32, or of a wedge type,[136] as in Figure 4.33. There are three methods of ball bonding, using combinations of heating, pressure and ultrasonic vibration. *Thermocompression* bonding requires pressure, plus a heated substrate and a heated tool. This is inapplicable to COB because the heating temperature required is between 300 and 400°C. The wire is normally gold. The technique is suitable for ceramic packages and substrates, offering very high bonding rates. The wire cut-off is achieved by an open flame or a spark discharge, to form a ball automatically on the end of the wire, ready for the next bond. *Ultrasonic* bonding[137, 138] is a combination of ultrasonic vibrations plus pressure. It is able to create bonds between a wide variety of dissimilar materials and is an extremely flexible process. It is more commonly used for wedge bonding with aluminium than for ball bonding. Wedge bonding requires two separate X and Y alignments of the head and so is considerably slower than ball bonding (0·7-1·0 second/wire compared with 0·2-0·3 second/wire respectively). The direction of wire feed must coincide with the lengthwise direction of the pads. Wedge bonding is also restricted due to the tool size and the consequent required access. *Thermosonic* bonding relies on a combination of ultrasonic energy and temperature.[139] Since a suitable temperature is in the range 120-150°C, thermosonic bonding can be performed on organic substrates, although high temperature epoxy or polyimide resin materials are recommended. Most thin calculators, ID cards and cameras use polyimide film for substrates.

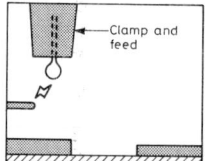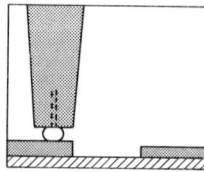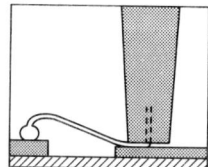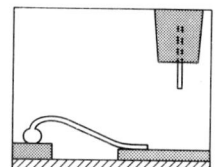

Fig. 4.32 The use of ball bonding to make a chip-to-pad wire connection.

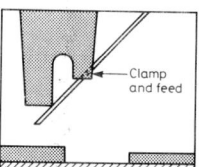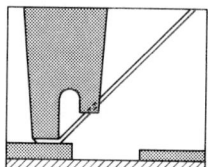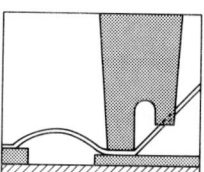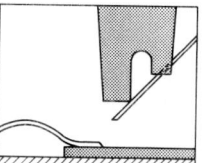

Fig. 4.33 The use of wedge bonding to make a chip-to-pad wire connection.

Normally, the bonding pads on the substrate are copper, on to which 5 μm of nickel and then a minimum of 0·5 μm, preferably 1 μm, of soft gold is plated.[140] For thermosonic bonding it is desirable to increase the nickel thickness to 10 μm to compensate the softening of the resin at the high temperature. Polymer thick film circuits are now being made that are wire bondable, by the application of a plated-metal surface over the printed polymer conductor.

The cost of gold has driven the move to aluminium wire,[141] but the aluminium-gold bonded interface is not free from problems, because of solid-state diffusion. The diffusivity of gold in aluminium is higher than that of aluminium in gold, creating voiding at the interface known as Kirkendall voids, which weaken the bond strength and increase the electrical resistance.

Another new process evolving is copper ball bonding.[142, 143] Making a good ball on the end of copper wire is very difficult, but the key to success has been found to be in controlling the thickness of the oxide film, by using an argon-hydrogen atmosphere for the bonding process. Copper has good mechanical strength and a lower electrical resistance than gold or aluminium. The mechanical properties provide excellent looping characteristics which enable higher speed bonding, but it has yet to become a commonly accepted technique.

Gold wire 18-25 μm diameter is usually used for thermocompression bonds (but not on organic substrates). Gold wire 18-50 μm diameter can be used for thermosonic ball or wedge bonds. The ultrasonic wedge method must be used for aluminium (usually 25 μm diameter). The gold is typically alloyed with a small amount of beryllium and copper to control grain growth during bonding; the aluminium is typically alloyed with 1% silicon. The resistivities of these alloys is such that 25 μm gold wire has a resistance of $0 \cdot 46$ $\Omega.\text{cm}^{-1}$ and 25 μm aluminium wire $0 \cdot 58$ $\Omega.\text{cm}^{-1}$. The 18 μm wire has twice these values of resistance and the 50 μm wire a quarter of these values. A good gold wire bond should have a pull strength of 8-10 g for 25 μm wire with failure occurring in the wire rather than at the bond. With aluminium, it is reasonable to expect a pull strength of 3-5 g.

4.6.3 Tape-automated Bonding

With the TAB system, the bare chips arrive at the assembly stage mounted by their inner lead connections on a dielectric tape in movie-film format. The conductors are etched from a metal film adherent to the film.[144, 145]

The assembly process is entirely automatic. The film is cut and the outer leads are cropped from the film. The chip is then glued into position and soldered to the circuit board, normally using a heated collet to melt and flow tin-lead plating that has been electrodeposited as part of the TAB film structure.

There are three basic options for the mounting of TAB chips, as illustrated in Figure 4.34. The standard mounting option is that in which the tape leads are flat and bonded at their inner ends to the chip and at their outer ends to the board. The lowest profile configuration, the flat mount, consists of the chip being protected within a cavity in the substrate. In the third option the tape leads are formed as shown, before assembly to the substrate. The chip is then face-up on the board, and the leads accommodate the height difference.

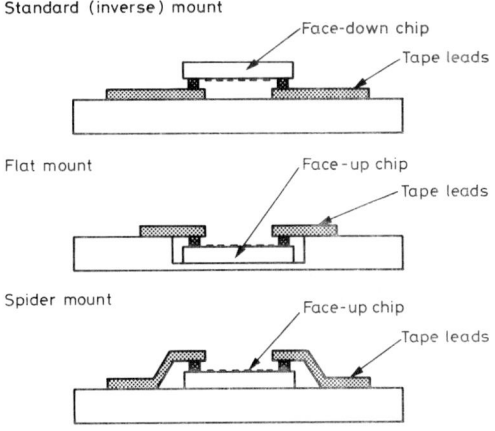

Fig. 4.34 The three basic options for mounting TAB chips.

4.6.4 Flip-chips

The flip chip is another major bonding technique for bare chips.[146] The original flip-chip concept used small solder-covered copper balls sandwiched between the chip termination lands and the appropriate lands on the circuit substrate. The chip was face-down, or flipped.[147] The solder joints were made by reflow at an elevated temperature, but the handling and placement of the minute balls were extremely difficult and costly. Modern flip-chips are produced with raised metallic bumps, usually of solder, on all the lands of the chips while they are still in a large wafer form. The individual chip is placed face-down, aligned to the circuitry and the solder bumps are reflowed. An alternative COB flip-chip method is to place the solder bumps on the substrate board rather than as part of the chip.

A schematic of the flip-chip concept is shown in Figure 4.35.

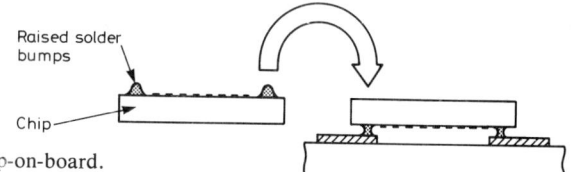

Fig. 4.35 Flip-chip assembly of chip-on-board.

4.6.5 Beam Leads

Another upside-down bare chip mounting technique is known as beam-lead assembly, shown in Figure 4.36. In this technology the leads are produced by plating during the chip fabrication processing on the undiced wafer. After dicing, excess silicon is then etched away, leaving cantilevered beam leads. The chip is then turned over and bonded directly to the substrate using the beam leads.

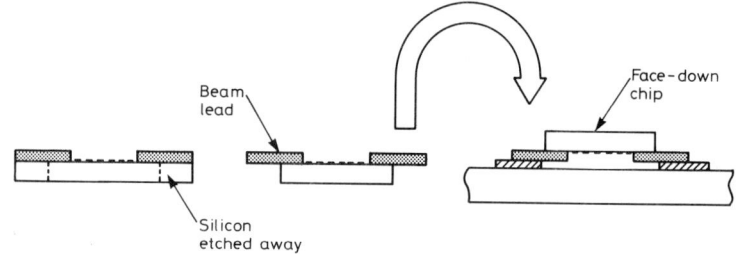

Fig. 4.36 The formation and mounting of beam-leaded chips.

4.6.6 Encapsulation

The bare chip and its fine interconnections to the PCB must be protected by encapsulation.[148] Silicon or epoxy resin is normally used to seal the chip immediately after the outer leads are bonded to the board. In many cases, a predetermined quantity of resin is automatically dispensed on to the surface of the chip. If the assembly is critical, the resin can be restrained around the chip by a barrier; a simple blob-type encapsulation is not adequate for high reliability circuits.

The sealing resin is cured at 150°C for about 2-4 hours.

Chapter 5

WAVE SOLDERING

5.1 INTRODUCTION

In Chapter 4 were shown the various assembly and soldering routes available for electronics incorporating surface mounting components. If these are to be combined with conventional leaded through-board components, the only economically feasible soldering method for automated mass production is wave soldering.

Wave soldering has been used for many years as the standard method of soldering insertion mounted printed circuit boards.[149] The equipment and the processes are well developed and the in-house expertise can be readily utilised for the wave soldering of surface mounted assemblies. This soldering method has one significant advantage over assembly via the route involving the application and reflow of solder paste, in that it can be used successfully for assembly of 'mixed-technology' boards containing both insertion mounted and surface mounted components.[150] This feature enables the electronic assembly company to make a more gradual transition towards full implementation of surface mounting technology during the period while some IC components remain available only in dual-in-line packages and many others retain an option of dual-in-line or chip-carrier packaging.

Wave soldering machines[151] comprise a conveyor which transports the populated board either continuously or in a stepwise fashion from a loading position to, in turn, a fluxing station, a preheating stage, the solder wave and a cooling station, before removal from the conveyor at the unloading point. Soldering of surface mounting assemblies is carried out at a temperature in the range 235-260°C with a contact time between 1 and 4 seconds.[152] The surface mounting components, which move through the molten solder, must be able to withstand this treatment, and remain unaffected either by the high temperature or the temperature gradients involved. Additionally, the solderable surface on the components must not unduly dissolve or leach into the molten solder.

Before the wave soldering process, however, the surface mounting components have to be temporarily bonded to the underside of the board, and this will be considered first.

5.2 TEMPORARY ADHESIVE BONDING

If wave soldering is to be used for the attachment of the surface mounting devices to the board they must pass through the wave on the underside of the board. The surface mounting components are therefore first individually bonded temporarily to the printed circuit board using an adhesive.

5.2.1 Application of Adhesive

There are three methods by which a small amount of adhesive can be applied at each site where a surface mounting device is to be placed. These are shown schematically in Figure 5.1 and are screen printing, dispensing with a syringe needle and pin transfer.

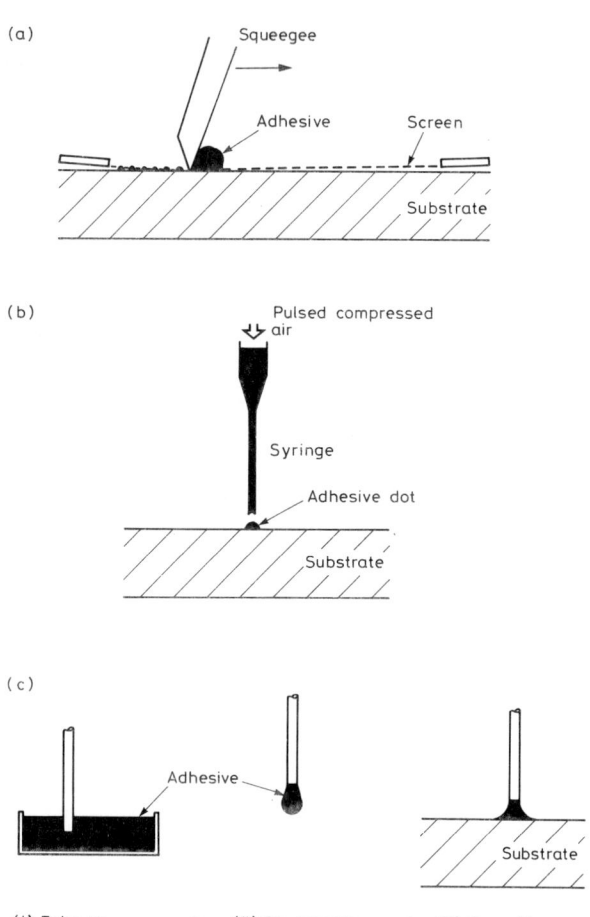

Fig. 5.1 Adhesive for temporary bonding of the surface mounting components to the circuit board can be applied by (a) screen printing, (b) syringe or (c) pin transfer.

Screen printing[153] has the advantage of speed, through multiple simultaneous application, but, as the substrate needs to be flat, screen printing cannot be used after the conventional leaded components have been inserted through the board. However, whilst placing the surface components first permits the easy application of adhesive by screen printing, it makes the mechanical insertion of through-hole components much more difficult. Moreover, this requires a larger area on the solder side of the board because the clinching or bending device of the insertion machine needs some free space around the protruding leads.

Dispensing the adhesive from a syringe is feasible for small boards with an assembly rate of less than about 5000 components per hour. The syringe can either form part of the programmed pick-and-place machine or can be hand held for prototype manual assembly or repair work. Placement machines handling great throughput of components may have several adhesive dispensers side by side but this can result in control problems and non-uniform droplets.

With the pin transfer method, either single pins can be used sequentially around the board or, for larger boards, an array of pins can be used for simultaneous application. With either configuration, equal amounts of adhesive are the result, provided the locations on the take-up reservoir from which the pins pick up the adhesive are strictly uniform and do not become locally depleted. To ensure that this is so the disturbed layer is smoothed flat with a squeegee after each pick-up. After application by the pin transfer method, the adhesive droplet takes on the form of a flattened cap of a sphere whose profile is defined by the forces of gravity and surface tension. Because the adhesive is thixotropic, however, the area of spread of the adhesive depends not only on the amount deposited but on the movement of the pin in the adhesive during pick-up and during application on to the board. Thus the downward speed and depth of immersion of the pin into the take-up reservoir plus the downward speed and the separation distance of the pin from the board during the application affect the rheological behaviour of the adhesive. Figure 5.2 shows typical adhesive surface shapes during the pin transfer process.[154]

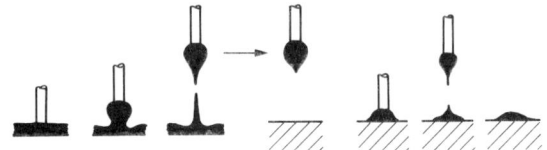

Fig. 5.2 Profiles of an adhesive dot being produced by the pin transfer method.[154]

The amount of adhesive applied, by whatever method, must be sufficient to produce a bond whose height bridges the gap between the surface mounting component and the substrate board, but whose area of spread when the component is placed does not encroach upon any conductors. Contamination by adhesive on metallised surfaces on components or on the board impairs the solderability whose restoration is then impossible by any acceptable cleaning method.

The degree of care required in defining the quantity and the rheology of the adhesive applied is illustrated in Figure 5.3. For bonded chip resistors and capacitors the stand-off height, equal to the sum of the conductors on the board and the component metallisation, is generally around 100 μm, for a reflowed through-hole-plated board. In contrast, for a leaded SOT-23 component, for example, whose leg height may be as much as 200 μm, compared with only about 20 μm for the metallisation on passive components, the resulting stand-off height is around 250-300 μm. For mass production it is obviously desirable to apply equal amounts of adhesive for each component bond and it is therefore advantageous to obtain SOT-23 components with a minimum or a reduced leg height. An alternative is to design a conductor track or a dummy track beneath the high-standoff components, as described in Section 4.2.3.

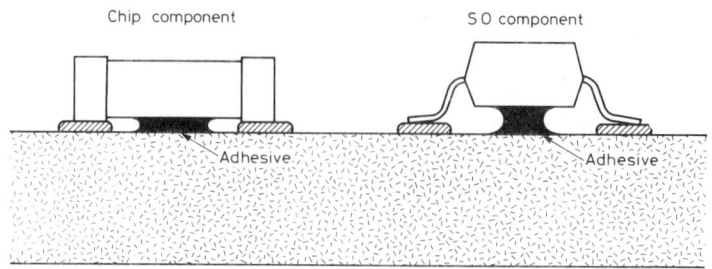

Fig. 5.3 A degree of care is required to apply exactly the correct amount of adhesive since components stand off the board by different amounts.

The three common adhesive application methods mentioned above, screen printing, syringe and pin transfer, make use of products which are polymerised (cured) either by heat or radiation to achieve their stable adhesive properties. Other methods of adhesive bonding have been used to attach the surface mounting components temporarily to the board. Two-part systems have been tried by applying one part to the component and the other to the board. This has a potential advantage in that the board can in principle be screen printed with its adhesive part but remain handleable for the subsequent insertion of conventional components, and similarly the components can have their adhesive part applied during manufacture. Another alternative has been to use adhesives which are pressure sensitive rather than those requiring a curing treatment and, indeed, even double-sided adhesive tape capable of withstanding soldering temperatures has been developed. Notwithstanding these systems, the greatest success has been found by using single part adhesives that can be readily cured by either heat or ultra-violet radiation.

5.2.2 Curing the Adhesive

A spot of adhesive is applied to the board followed by the placement of a component upon it. At this stage the adhesion must be adequate to hold the component in place while the board is being subjected to the accelerations and decelerations of the pick-and-place assembly stage, which can be quite violent, and to allow the fully populated board to be mechanically transferred to the next production stage without dislodging any component. The adhesive is then hardened. The adhesion after this curing process must be sufficient to withstand vibrational and flexural forces at room temperature caused by insertion of leaded components, if this is performed after the surface mounting components have been placed, and also to withstand the thermal shock and the temperature while entering and passing through the solder wave.[155]

The two adhesive systems which suitably meet these requirements are of the thermosetting type which are cured by heating, and of the U-V sensitive type which are cured by ultra-violet radiation.[156] The rheology of each type is different and they exhibit different viscosity-time functions during their curing. The precise behaviour is specific to the product and depends on the chemistry involved and the fillers which are used to control the rheology. However, the general form of the viscosity-cure time curves is as shown in Figure 5.4. In thermosetting, the viscosity first decreases as a result of the temperature rising. This could potentially cause a slumping of the adhesive dot, a situation which is controlled by solid filler additives. The viscosity does not increase until polymerisation is well under way. For ultra-violet curing systems, the curing is essentially linear with time.

The curing mechanism also has implications on the usable life of the adhesive. Thermosetting adhesives, usually epoxy resins, begin to cure slowly at room temperature and consequently should be refrigerated during storage. However, once incorporated into a pick-and-place machine, the room-temperature pot-life, i.e., the time during which the adhesive can be readily worked, needs to be considered. An adhesive with a long pot-life usually requires prolonged curing at a relatively high temperature, and this can cause a problem with some components incapable of withstanding such heating cycles. For a given adhesive, the curing temperature has a very significant effect on the time required to elicit a high degree of cure, as shown in Figure 5.5. Curing at 80-100°C might typically take about half an hour for an epoxy resin adhesive with a reasonable pot-life of one working day. The degree of cure dictates the firmness of the bond but, contrary to many statements, the degree of cure does not significantly affect the uptake of moisture[157] which could potentially be a source of corrosion on the finished board.

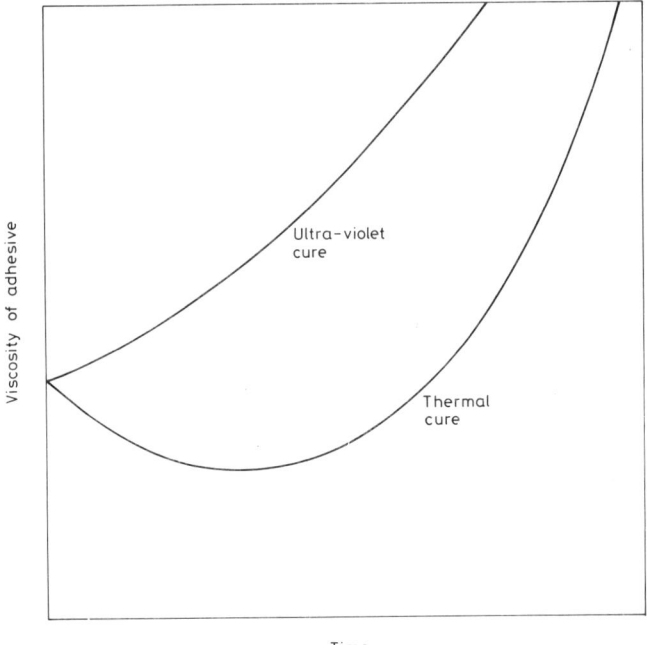

Fig. 5.4 Typical viscosity-cure profiles of adhesives cured by ultra-violet radiation and by heat.[154]

The adhesives that cure by ultra-violet radiation are usually acrylates with an added photo-initiator.[158] The radiation used is at a wavelength of around 350 nm, in the ultra-violet, and during irradiation the initiator molecules dissociate to form radicals which start polymerisation. The problem associated with adhesives curing in this manner is the accessibility of the adhesive to the radiant energy. For good curing, sufficient adhesive must protrude from beneath the component in order to initiate the polymerisation which can, to some extent, chain react to achieve cure in areas under the component, not in direct line-of-sight with the radiation source. This effect is aided by the heat generated in the process. With production equipment using a conveyorised radiation tunnel, curing can be completed in about ten seconds.

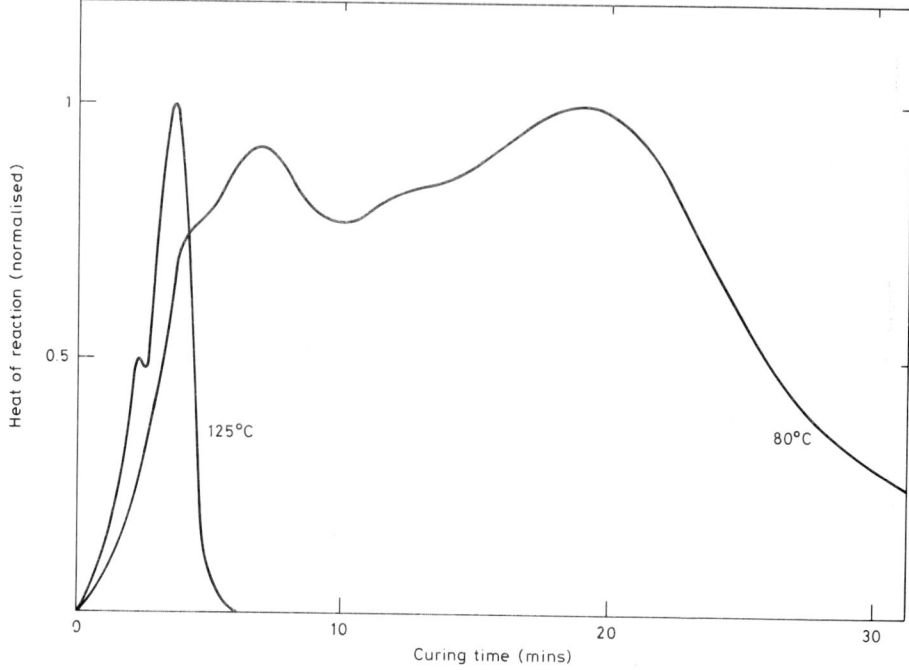

Fig. 5.5 The curing time of a thermally sensitive adhesive is strongly dependent on the temperature of cure.

5.3 FLUXING THE BOARD

In wave soldering, the flux is always applied in a liquid form in order to cover all the solder areas quickly and evenly. There is an optimum range for the amount of flux that is retained on the board to ensure satisfactory soldering and the attainment of this optimum depends on two factors:

(i) the method of application and hence the quantity of liquid applied;
(ii) the fraction of solvent to flux solute and hence its viscosity and its evaporation behaviour between application and soldering.

5.3.1 Application of the Flux

Five methods of flux application have been successfully employed in wave soldering machines: foam, wave, spray, dipping, and brush, of which the first three are most suitable for surface mounted assemblies with plated-through holes and will be discussed here. The application method should produce a continuous film of flux on the underside of the board and hence promote capillary rise up into the plated holes. Depending on the type and activity of the flux used, between 0·7 and 3 g.m^{-2} of solid flux is required to be evenly distributed over the board parts to be soldered.[4] For application, these solids are dissolved in a solvent carrier and these figures correspond to a wet flux layer thickness in the range 3-20 μm.[159] During soldering this layer is washed forward by the molten metal and helps to remove the oxide film in the back-wash region of the wave (see Section 5.5), thus reducing the occurrence of excess solder drag-out from the wave, to form solder bridges and icicles. However, an excess of flux is not only wasteful, but adversely affects the operation of the soldering machine by contamination and would require a change in the machine parameters to effect good quality soldering.

5.3.1.1 FOAM FLUXING

The liquid flux is usually applied from a large tank and one method of application to the board is by means of an aerator to produce a very turbulent bubbling surface through which the underside of the populated board passes, as shown in Figure 5.6. To produce this aeration, low pressure air is blown through the pores of a tubular porous stone and the generated fine bubbles are guided to the surface by baffle plates. The bursting of the bubbles at the surface assists in the coating of the walls of the through-board holes.

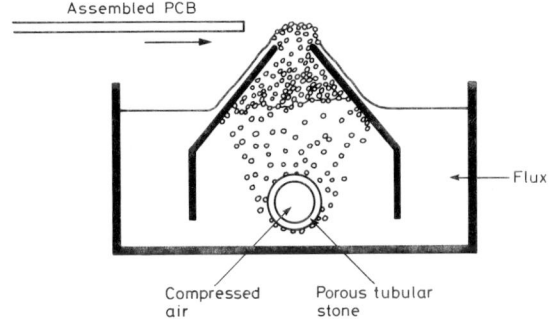

Fig. 5.6 The principle of foam fluxing of circuit boards: a foam mixture of flux and air contacts the underside of the board.

If correctly adjusted, the wetting of the board by the foam method is very quick and the amount of liquid flux applied is independent of the conveyor speed. Thus in order to control the dosage of flux to the board the solute-to-solvent ratio (i.e., the solids content) of the flux must be well defined and controlled. This is usually done by monitoring the density of the liquid and maintaining close limits, and by the replenishment of solvent to replace that lost by evaporation. This replenishment can be incorporated as an automatic facility on the wave soldering machine.

The flux density is also an indirect measure of its viscosity. If the viscosity is too high the bubbles will not burst correctly and the foam may rise uncontrollably high, and overflow. If the viscosity is too low, difficulty in achieving any foaming action may result. When within specification, a foaming flux should produce bubbles with diameters between one and two millimetres. The nature of the diffusion stone determines the uniformity of the size of the bubbles and is therefore the most important parameter in controlling the quality of the foam. The air must be dried, be free of oil, and run at as low a pressure as possible to maintain the head of foam.

5.3.1.2 WAVE FLUXING

The second method suitable for the application of flux on mixed-technology boards for wave soldering is to pass the board across the crest of a standing wave of the flux, as shown in Figure 5.7. In contrast to foam fluxing, the height of the wave is quite critical to the amount of flux deposited, since the liquid flux can easily be forced up by the hydrostatic pressure of the wave into the holes in the board and penetrate to the upper surface of the board—a situation that is undesirable, especially if the boards are not to be cleaned after soldering. In order to maintain careful wave height control, not only must the impeller be controlled but concomitantly so must the liquid level in the reservoir tank. Following passage over the wave, excess flux is wiped away with a soft brush.

Any liquid flux with solids content up to about 60 wt% can be applied in this manner whereas for foam fluxing the flux must have suitable rheological properties which usually limit the solids content to around 35 wt%.

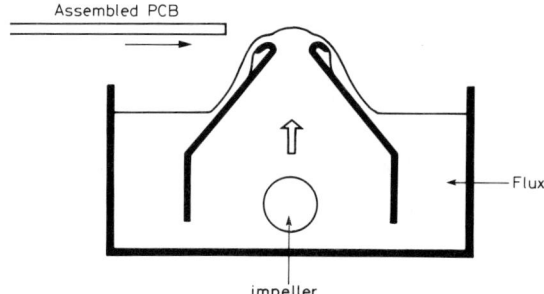

Fig. 5.7 The principle of wave fluxing: a standing wave of flux is formed by pumping up through a baffle.

5.3.1.3 SPRAY FLUXING

Several methods exist for the production of a directional spray of liquid flux on to the underside of the populated printed circuit board. A common system consists of a fine stainless steel mesh drum, rotating in the flux reservoir, while air is blown into the drum, generating a fine spray from the top surface of the drum. Another method, shown schematically in Figure 5.8, consists of a drum of closely spaced radial spring leaves which are loaded with flux by rotating through the reservoir and then the flux is flicked off each leaf in turn, in the direction of the circuit board.

5.3.2 Monitoring Flux Density

The dosage of flux to the board depends on the amount of solid flux dissolved in the solvent carrier, and this parameter is usually monitored and maintained

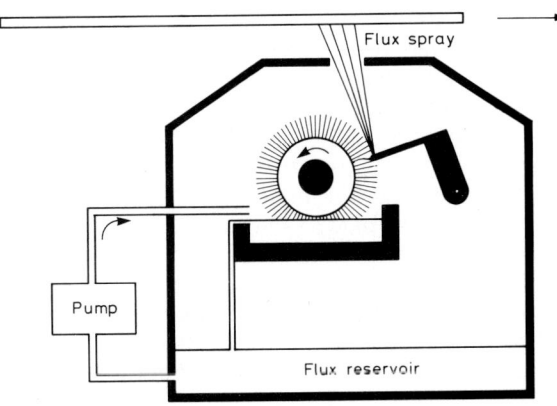

Fig. 5.8 One type of spray fluxer: the flux is flicked off successive spring leaves as the drum rotates.

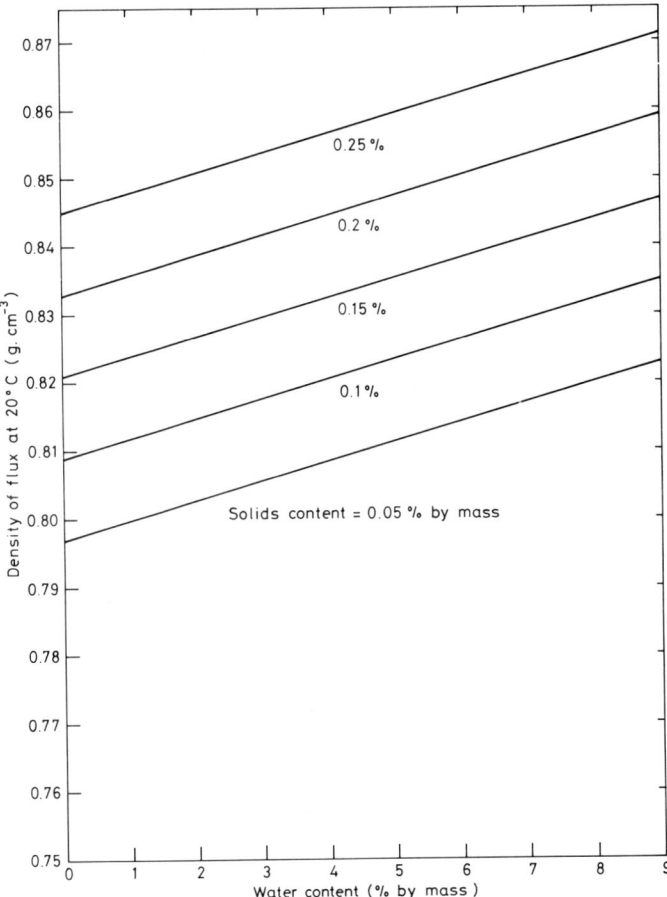

Fig. 5.9 The effect of absorbed water on the density of a flux of rosin dissolved in isopropyl alcohol.[4]

through the liquid density, according to the manufacturer's specification. Some care must be taken here, however, because the flux may absorb water directly from the air, or from that condensing from the air because of temperature differentials in the machine. The water content can have a significant effect on the density and thus give rise to a misleading value for the solids content.[4] This is shown in the data in Figure 5.9 for a rosin flux dissolved in isopropanol as the carrier. Clearly, either the water must be removed from the isopropyl alcohol before the density measurement is made—but this requires an inconvenient distillation—or an estimate of the water content must be made and then the solids content adjusted using the data of the type given in Figure 5.9. A quick test for estimating the amount of water present in fluxes using isopropyl alcohol as the solvent has been given,[4] as follows: Mix 10 ml of the flux with 10 ml of heptane. Then add deionised water a drop at a time from a small burette with a resolution better than 0·1 ml, thoroughly mixing after each drop. When the water content of the mixture is 1·5 ml the liquid becomes turbid and begins to separate. Thus the difference from 1·5 ml of the amount added before this situation pertains is the amount of water in the original flux under test. Generally speaking, if the flux contains more than about 5% water, replacement is recommended. Solderability tests on leaded components show that a water content below this level has no discernible effect on the efficacy of a rosin flux.

5.4 PREHEATING THE BOARD

Between the application of the flux and the wave soldering of the board, a preheating stage is always incorporated. This serves several purposes, the most important of which is to volatilise the solvent flux carrier. To remove this solvent requires an amount of heat to be supplied, determined by the latent heat of the solvent. If this heat has to be supplied from the solder wave, the solder bath temperature is significantly affected. Also the volatilising vapour of the solvent between the board and the wave, besides causing solder splatter, precludes good and uniform thermal contact between the board and the wave.

The preheating stage additionally increases the heat content in the board and reduces, to some extent, the thermal requirement of the solder wave to raise the board to the soldering temperature. However, depending on the design of the wave soldering machine, this beneficial effect can be very small. The preheating does, however, reduce the thermal shock experienced by the components, especially the board itself and the surface mounting devices that are obliged to traverse, submerged, through the solder wave. Thermal shock to the board can lead to its bending during the soldering.

The amount of heat required during the preheat stage depends on the board design and its component materials, but more importantly on the flux properties. Foam fluxes, for example, generally use solvents with high boiling points and hence require a longer dwell time at the preheat station.

If the preheating is inadequate, the viscosity of the flux will be too low and, at the solder wave, it may be prematurely washed from the board, giving rise to poor wetting of the solderable surfaces. Additionally, in this case, the flux will not be in evidence at the exit meniscus, as the board leaves the wave, increasing the likelihood of solder bridges and icicles. The heating of the flux, besides drying it, brings it to the point at which its activity is initiated and it begins to break up the oxide surface layer on the conductors to be soldered.

If the preheating is excessive the viscosity of the flux will increase and, if it is a rosin flux, the colophony will oxidise and begin to polymerise. Subsequently, in the solder wave the colophony will not melt adequately and hence will not be readily swept aside by the molten solder.

The preheating in wave soldering machines is achieved either by convection of circulating hot air or by radiation from infra-red lamps or hot-plate panels, or a combination of both. The radiant heat is sometimes applied from above in addition to the underneath, solder side of the board. The conveyor can either be of the continuously moving type, traversing the preheat zone, in which case the total heat applied is controlled by the temperature; or it can be of the stepping type in which case the heat is usually controlled by the dwell time at the preheat station. The increased ability to remove the unwanted vapours when using the hot air blowing type of preheater makes this type more effective than the radiant type. Also, with the radiant heating, any areas with reflecting surfaces, such as copper, have a marked effect on the preheat temperature attained. The time and temperature of the preheating stage should be adjusted by monitoring the temperature of a test board equipped with thermocouples. Figure 5.10 is a typical

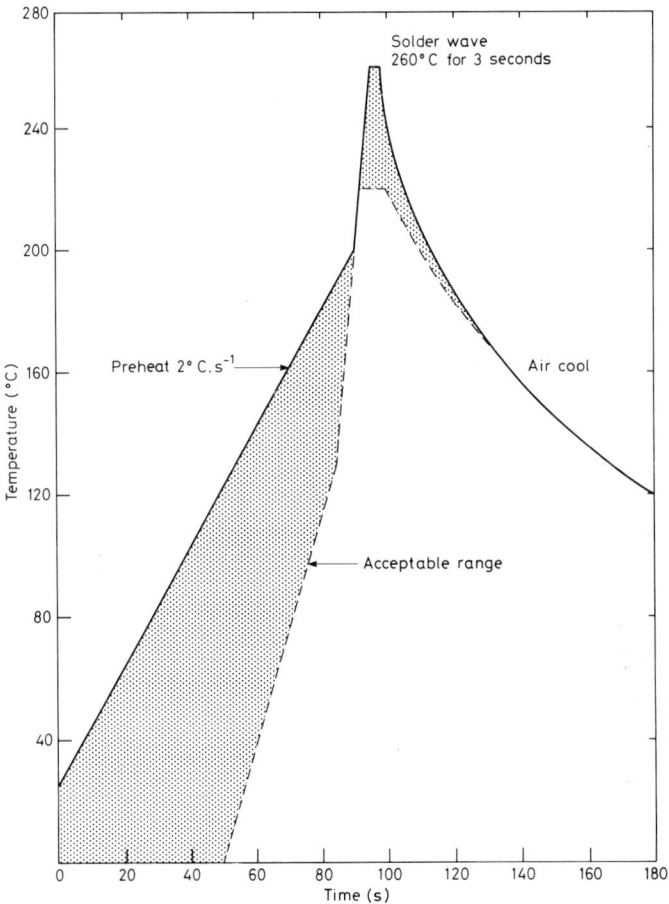

Fig. 5.10 A typical temperature-time profile of a circuit board traversing a wave soldering machine.

temperature-time profile of a circuit board traversing a wave soldering machine. The hot air heaters are less subject to the effects of dripping flux since they do not have to be positioned directly under the board.

The preheating stage of a wave soldering machine raises the temperature of the board to somewhere in the range 80-120°C. The boiling point of isopropyl alcohol, the most common flux carrier, is 82·4°C, so that evaporation and eventual volatilisation are very rapid while the board is being pre-heated.

The amount of heat Q(joules) required to be transferred to the PCB from the preheater is related to the temperature rise required ΔT (°C) and the heat capacity C (J.K^{-1}) of the PCB:

$$Q = C\Delta T$$

The heat capacity is a product of the density ϱ (kg.m^{-3}), the specific heat c (J.kg^{-1}.K^{-1}) and the volume V (m^3) of the PCB:

$$C = \varrho c V$$

Thus, if the wave soldering machine conveyor speed and the length of the preheater are such that each part of the fluxed board spends a time t over the preheater, the power required of the heater is

$$W = \frac{\varrho c V \Delta T}{t} \text{ watts}$$

For a PCB that has a heat capacity ϱc of about 2.10^6 J.m^{-3}.K^{-1} requiring to be heated from 20°C to around 100°C, the necessary power is typically 2-5 kW. Taking into account the inefficiency of heat transfer from the heater to the board, a heater of 8-10 kW is normal. The power of the preheater must be such that the specified board temperature can be realised at the highest conveyor speed.

5.5 THE SOLDER WAVE

In wave soldering, a continuously replenished wave of molten solder is generated by pumping upwards from a sump, while the printed circuit boards to be soldered traverse in one direction across the crest of the wave,[160] as shown schematically in Figure 5.11. On single-sided boards the inserted component leads are soldered to lands on the board underside while, on double-sided and multilayer plated-through-hole boards, the solder rises through the hole around

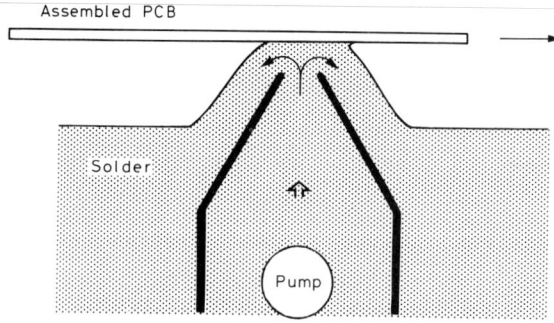

Fig. 5.11 Simple schematic of the principle of wave soldering: the circuit board passes through the crest of a standing wave of molten solder.

Wave Soldering

the component lead, by a combination of hydrostatic pressure and capillary forces to fill the hole and flow over the solderable land on the top surface of the board. The component leads are either slightly splayed or are crimped over after insertion, in order to avoid the buoyancy effect which can cause the components to float up from the board, as they pass over the wave.[161]

5.5.1 Wave Generation

The solder wave is generated either by use of a mechanical pump submerged in the molten solder or by applying Lorentz forces generated by external electric and magnetic fields. The latter type has no moving parts and is shown schematically in Figure 5.12. The molten solder can be made to travel at quite high speeds by this method and the solder is forced out through an orifice to produce a jet and a hollow wave, in contrast to the 'solid' wave usually produced by a pump.

When a pump is used, molten solder is forced upwards from a sump into an ejection chamber fitted with baffles to divert the flow, and out through a nozzle. The molten solder forms a stable standing wave, linearly across the machine, perpendicular to the direction of traverse of the conveyor carrying the printed circuit board. The wave falls away along either one or both sides of the ejection chamber and back into the solder sump. The soldering time is determined by the speed of the conveyor across the wave and the width of the contact area of the board with the crest of the wave.

The hydrodynamic behaviour of the solder wave has been described phenomenologically.[4] The surface of the standing wave is covered with an oxide skin which remains static, except at the extreme edges where some turbulence usually occurs. Thus the flow velocity of the solder at the outer surface of the wave is zero and the molten solder streams through between the nozzle and the skin. When an approaching board contacts the surface of the wave, the oxide skin is broken and the part in front of the board is pushed forward without being rumpled. The whole oxide skin moves at the same speed as the board and in this

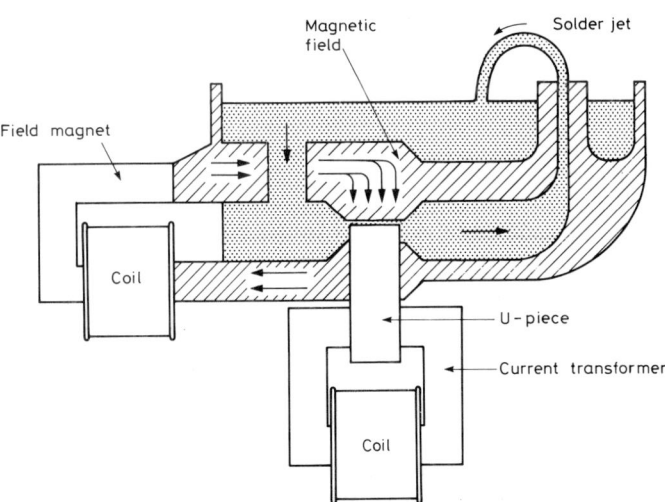

Fig. 5.12 As an alternative to mechanical pumping to produce the solder wave, crossed electrical and magnetic fields can be used to generate the pumping action.

way the machine works with minimum dross formation. When the board is on the wave, the solder in contact with it moves at the same speed as the board, irrespective of the nominal direction and speed of the wave.

5.5.2 The Soldering Process

In considering the effectiveness of the wave in soldering the circuit board, the solder wave can be divided conveniently into three distinct regions having differing physical mechanisms:[162] the wave entrance, the main contact or heat transfer zone, and the exit wave, commonly called the peelback or backwash region as shown in Figure 5.13.

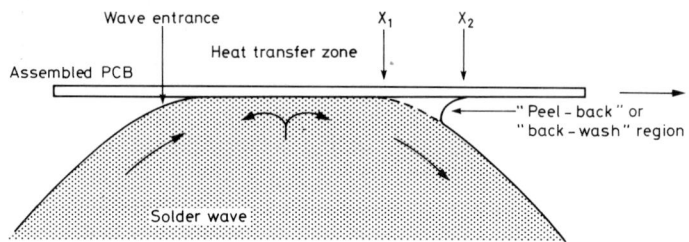

Fig. 5.13 Definition of the three zones of a solder wave: the entrance zone, the heat transfer zone and the peel-back zone.

The characteristic phenomena associated with the wave entrance are those involving the physical and chemical qualities of the flux; any remaining solvent is immediately evaporated and flux activation begins. Because of the hydrodynamic pressure of the molten solder on the board surface as it passes over the wave, hot chemically reactive flux accumulates at the wave entrance, thus enhancing the surface preparation of the solderable surfaces prior to entry into the wave. The precise mechanism describing the removal of oxide films and contamination from the solderable surfaces is contentious, but it seems safe to assume that the resultant reaction products are either dissolved in the liquid flux or lifted from the surface and washed away by the solder wave. The wave entrance is therefore primarily responsible for the surface contamination removal.

It is in the heat transfer zone that the wetting and spreading of the solder is accomplished. The contact time for soldering is simply the length of the heat transfer zone divided by the conveyor speed. The length of the heat transfer zone is easily measured and monitored for its uniformity from side to side of the wave by using a sheet of glass as a dummy board.

During the soldering of a printed circuit board two mechanisms occur within the heat transfer zones. First, there is a nucleation and growth of a new metallurgical phase at the interfaces between the liquid solder and any solderable surface. Because of the temperatures involved and the fact that the solder is molten, this growth is extremely rapid, as will be discussed in Chapter 10. The reaction product formed at the interface is an intermetallic compound or compounds, such as those formed between copper and tin on the PCB pads and between silver and tin, or gold and tin, at the solderable pads or terminations of components. The second physical process occurring is that of dissolution of the solid surfaces into the molten solder. Although governed by the same diffusion rates as are controlling the growth of intermetallic compounds, dissolution can occur even in the absence of any chemical reaction and the dissolving species is swept away by the wave.

The rate of bonding between the component surface and the solder has been studied extensively by means, for example, of wetting balances and high speed filming. The mechanism of the solder spreading across a surface has also been studied in great detail and, although several quite different theories exist regarding the rôle of intermetallic formation and growth in the spreading phenomenon, they are common in their need for the intermetallic layer to promote good bonding.

Within the heat transfer zone a certain amount of flux survives the wave pressure where it continues to perform its cleaning action. However, it can also have a detrimental effect by becoming trapped at pockets between components or leads and hence prevent intimate contact between the molten solder and the solderable surfaces.

Finally, the exit, or back-wash region: this is where the joints are made and where the associated soldering defects are generated. No solder joints are made in the bulk of the wave; that is where wetting occurs. The individual joints are made during the dynamic phase separation as the board parts company with the wave. The substrate, solder, flux and air must all be considered at the backwash region. Tin oxide, which forms much more readily than lead oxide, can produce a skin on the solder and consequently a dramatic increase in the surface tension. This prevents an efficient separation of the board from the wave, and hence results in an excessive amount of solder drag-out from the wave. The form of the back-wash is dependent on the amounts and configurations of the solderable to non-solderable parts. On un-wetted regions the liquid solder will break away from the surface at point X_1, while on wetted regions the break-away point X_2 may well be 25 mm further on. The actual position of break-away occurs in an unpredictable manner because of the complex configurations of the component leads and copper tracks. The critical situation occurs if, owing to the board layout, the solder meniscus has to retract abruptly in the direction X_2 to X_1. The amount of solder in the back-wash region must then decrease suddenly, increasing the likelihood of the formation of solder bridges between component leads or pads. The vertical distance of point X_2 above the wave is determined by the wave shape and the inclination of the conveyor as well as the surface tension of the solder. This distance governs the vertical force of the solder to pull away from the board. If too high, then there is not enough time for solder bridges to open and the excess molten solder to flow away. Since the surface tension plays a part here, the rôle of flux still on the board at this point is important, and oil if a solder oil inter-mix is used (Section 5.7).

5.5.3 Wave Shape

The conventional solder wave, depicted schematically in Figure 5.11, falls back to the sump on either side of the nozzle. This type of wave is usually provided with extension plates, as shown in Figure 5.14, in order to define better the wave profile.[163] The soldering quality is found to be higher if drainage of excess solder is encouraged by inclining the board transporter upwards at a small angle of 5-10° to the horizontal.[164] This has the overall effect of producing joints with less solder and this reduces the possibility of bridging or icicling faults, because there

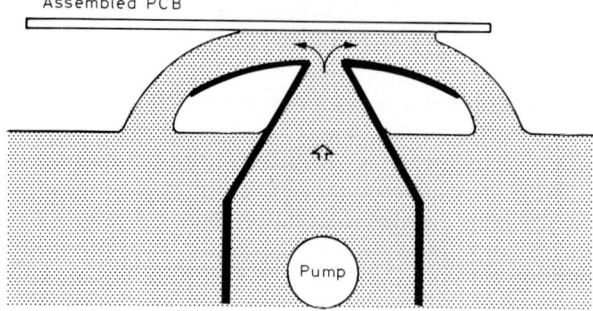

Fig. 5.14 Extension plates are added to the solder wave baffles to control the shape and contact time of the wave.

is then an additional shearing force to aid break-away of the soldered joint from the wave, as shown in Figure 5.15.

Drainage of the solder can also be assisted by positioning the break-away point of the board from the solder at the point where the relative speed of the solder and the board is zero. This can be done, whilst retaining a sufficient heat transfer zone, by using a very wide nozzle,[165] but better control is obtained by using wide extension plates to produce the same effect, as shown in Figure 5.16. This type of extended wave also has the advantages of first providing a longer heat transfer zone and hence enabling higher transporter speeds and greater production rates to be obtained, and secondly allowing the wave shape and relative speeds of the solder in the different zones to be optimised. Several wave shapes are available;[166] that shown in Figure 5.16 is one type in which the backward flow rate of solder over the front plate is high, which additionally provides an effective washing action to promote wetting.

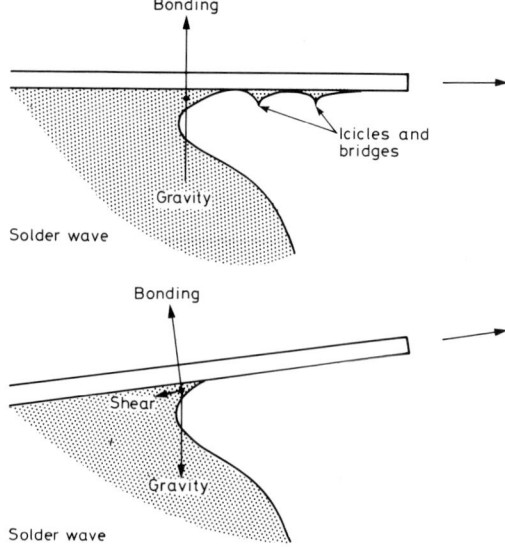

Fig. 5.15 The effect of inclining the circuit board conveyor to the solder wave is to induce a shearing force at the exit point, and hence reduce the likelihood of bridges and icicles.

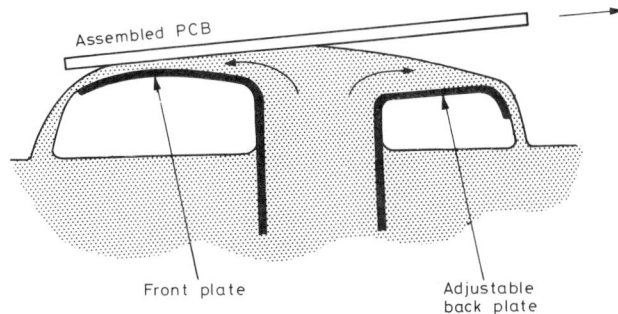

Fig. 5.16 One type of modified wave shape giving (i) a long heat transfer zone, (ii) an exit point where the speed of the solder relative to the board is zero and (iii) a high solder speed over the front plate which aids wetting.

5.5.4 Shadowing

Conventional double-sided waves have been used effectively to solder arrays of simple discrete passive surface mounting chip components,[167,168] but the technique is very restricting in terms of component orientation and layout and can be used only for small chip resistors and capacitors. The problem is that, when passing through the solder wave on the underside of the board, the body of the component is not wetted by the solder and creates a depression in the wave which, in combination with the effect of the surface tension of the solder, can cause a shadowing so that the solder makes no contact with the lands on the trailing side and hence produces a solder skip fault. This effect is shown schematically for a surface mounting leaded component in Figure 5.17. Clearly, the higher the component body and the smaller or shorter the terminations, the more severe is the effect. In the early years of surface mounting, trapped flux and flux fumes were often mentioned as also contributing to this effect with the result that gas-releasing holes were incorporated in the board,[169] but such measures are not now thought to be at all beneficial.

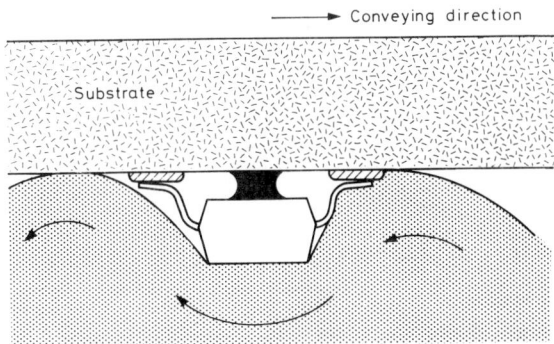

Fig. 5.17 The shadowing effect of a non-wettable component body as it passes through the solder wave.

The reason that the shadowing effect is not a problem with chip resistors and capacitors is that these types of components have three- or five-face metallisation around each end and, provided the solderability of these surfaces is adequate, the molten solder is pulled around into the recessed corners as shown in Figure 5.18.

Thus, although the conventional wave machine can be used for certain components and circuit board configurations, its effectiveness is very limited.

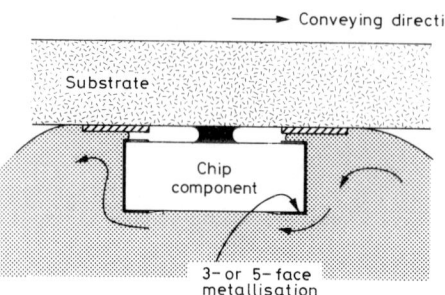

Fig. 5.18 The shadowing effect is not a problem for chip resistors and capacitors because of their five-face metallisation which allows the surface tension of the molten solder to wet the trailing-edge pad adequately.

5.5.5 Jet Wave

A completely different type of solder wave can be generated by the use of a backwardly inclined nozzle ejecting the molten solder at a higher speed than for the conventional standing wave, as shown in Figure 5.19. Such a wave has the characteristics of being fast moving, hollow and turbulent.[170] It was originally designed as an alternative to the conventional wave for soldering through-hole boards but was soon recognised as having distinct advantages for the soldering of surface mounting assemblies.[171]

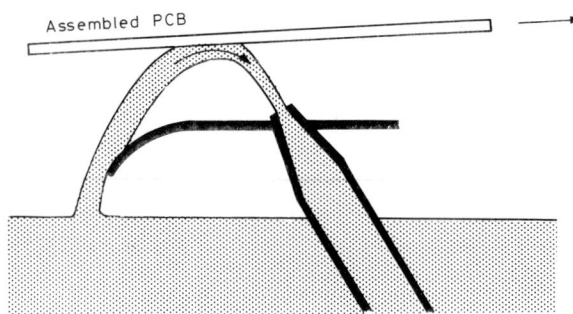

Fig. 5.19 Schematic of a hollow jet wave which can be used successfully for wave soldering surface mount assemblies.

The solder in the jet moves very quickly at about 2 m.s^{-1} and this is particularly well suited for surface mounting assembly because the Bernoulli effect associated with fast-moving fluids gives rise to a reduced pressure on the convex, outer surface of the wave, causing the solder to pull completely around each component body. In addition, the scrubbing action of the fast, turbulent wave helps remove excess flux and eases the escape of any gases generated.

Bernoulli's law states that, in a flow of incompressible fluid, the sum of the static pressure and the hydrodynamic pressure along a streamline is constant if gravity and frictional effects are disregarded. The law arises directly from energy conservation considerations. From Bernoulli's law it follows that where there is a velocity increase in a fluid flow there must be a corresponding pressure decrease.

The hollow form of the wave also has a degree of flexibility, allowing it to adapt to non-planar boards. The wave height adjustment is not so critical, because the downward force to depress the wave is not so great as for a conventional standing wave. Furthermore, this type of wave requires only a short heat transfer zone of 2-3 cm and this minimises possible thermal damage to the components which have to pass, board underside, through the wave.

Jet wave soldering machines have been shown to be capable of high quality cost-effective soldering of surface mounted assemblies and of mixed-technology boards comprising both surface mounting components and through-hole leaded devices.

5.5.6 Dual Wave

In recent years, with the growth of surface mounting assembly, a new generation of wave soldering machines has been designed to meet the flexible requirements of full-scale production of surface mounting and mixed technology boards.[172]

Such machines combine a first wave that is turbulent and a second wave that is smooth. The turbulent wave may be of the jet type, producing a double wave configuration as shown in Figure 5.20. Alternatively the turbulence may be promulgated either by an array of small baffles in a standing wave or by incorporating an oscillatory motion to the wave.

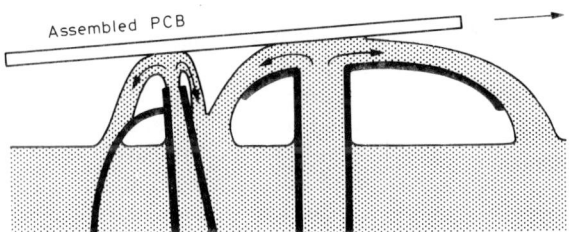

Fig. 5.20 A dual wave soldering system for surface mount assemblies: a turbulent first wave and a smooth second wave.

The first, turbulent wave is required to enable the molten solder to drive between the components and ensure that all the component terminations are heated and have achieved full wetting, before contact with the smooth second wave which then controls the meniscus of the molten solder at each joint, as the board leaves the calmer wave at near zero relative velocity.[173]

The beneficial effect of turbulence on the surface of a solder wave is illustrated in Figure 5.21: data taken on one specific type of surface mounting panel using a single solder wave whose turbulence amplitude, at 60 Hz, could be varied. Increasing the turbulence ensured the elimination of solder skips on pads that were shadowed by component bodies, but at the same time caused an increasing trend of solder bridging as the turbulence increased.

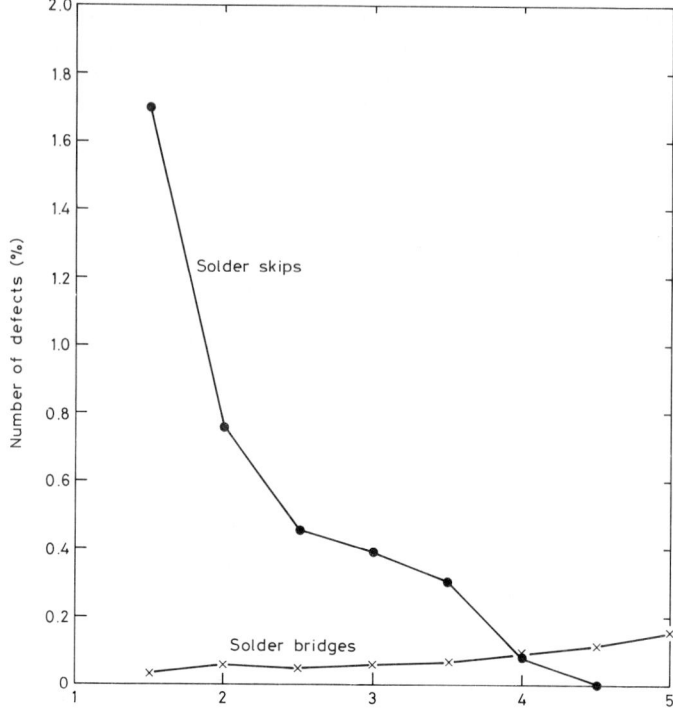

Fig. 5.21 The effect of the degree of turbulence in a solder wave on the occurrence of solder skips and solder bridges.

Some simple machines drive both the waves from the same solder sump with a single impeller and produce the two waves using a system of baffles. Most machines, however, use separate pumps from a single sump so that the heights of each wave can be independently controlled. For maximum versatility two completely separate sumps can be used so that the temperature, shape, velocity and even flux conditions can be controlled independently.

For satisfactory operation of a dual wave system for surface mounting assembly it is important that the turbulence of the first wave is as uniform as possible over its full width, and that the second wave is as smooth as possible, preferably with a small amount of a suitable oil applied. The dual wave concept consistently achieves better soldering quality than the single jet wave when soldering plastic leaded chip carriers[174] which are prone to shadowing effects because of their height. This is thought to be because of the strongly unidirectional nature of the single jet when compared with the more symmetrical first turbulent wave incorporated in dual-wave machines. The latter can to some extent reach into the passing shadow of the PLCC body.

5.6 THE SOLDER ALLOY

5.6.1 The Choice of Alloy

A very wide variety of low melting point alloys are classed as solders, being combinations of tin, lead, antimony, silver, bismuth, indium, copper, and so on.

Of these alloys, only a few have found popular applicability for the assembly of electronic components to PCBs, and the one alloy of tin and lead, at a composition near its eutectic, with the tin in the range 59-63 wt.%, is the major one. Only when special properties are required of the solder are other alloys used.

For surface mounting assembly there is, in fact, frequently a special requirement: because of the manufacturing process for many leadless components, their solderable surfaces often consist of a silver-based fired-on deposit. Silver has a high dissolution rate in molten tin-lead solder and consequently the solderable surface rapidly disappears and the resultant solder bond can be weak or non-existent. The best solution to this problem is to use components with a barrier layer, but a solder that is pre-saturated with silver (customarily 62 Sn:36 Pb:2 Ag) will significantly reduce pad dissolution. The cost precludes its widespread use in wave soldering but the alloy is common for solder pastes.

The almost exclusive use of solder alloys based on around 60 wt.% tin and the remainder almost all lead, for electronics assembly, arises because of their uniquely satisfying properties and applicability for the purpose.

The melting point of these alloys (near 180°C) lies well above the maximum service temperature of electronic equipment but is low enough to permit the use of commonly available materials for the construction of components and substrates; materials that, in the case of surface mounting assembly, must pass submerged through molten solder for 2-3 seconds. Tin-lead solders offer adequate mechanical properties, in particular being ductile with no brittle phases. The soldering process is accomplished in air and the surface of the liquid oxidises, but the tin-oxide film poses relatively few problems, being easily removed and not forming a wasteful dross as do many other low temperature alloys. In addition to these favourable properties, the major benefit of molten tin, that is attributable to its alloys also, is its ability to wet many other metals and spread with the aid of only mildly acid fluxes.

The range and the choice of suitable solder alloys for assembling surface mounting boards is discussed in Section 6.4, as well as the magnitude of the problem of leaching of silver from pad terminations, and the levels of impurities that are tolerable.

5.6.2 Impurities in the Solder

The problem of a build-up of impurities in the solder of a wave soldering machine is very different from the tolerable levels of impurities in solder pastes, because of metallic dissolution from the PCB, its components, auxiliary tools and the fabric of the machine itself, into the recirculating molten solder.

After fresh solder is loaded into a cleaned machine, the impurity levels will gradually rise as metals dissolve, mainly off the assembled PCB, until a dynamic equilibrium is reached. The solder bath is being continuously topped up with fresh solder to replace that carried out by the PCBs.

Analysis of a sample of the bath is carried out either at regular intervals of between two weeks and three months depending on the usage made of the machine, or when the soldering quality begins to deteriorate. It is found that the tin-lead ratio changes with time due to dissolution of different composition solder coatings from components passing through the wave and due to the preferential oxidation of tin. Under normal operation it is the copper from PCB conductor tracks that gives rise to the most common need for replacement of the solder in a wave soldering machine. The copper level in fresh solder is normally specified at less than 0·08%. The maximum tolerance level is about 0·3%, although an effect may be noticeable at the 0·2% level; the solder begins to exhibit a stickiness that manifests itself as bridging between component leads or pads, and the retention of a solder web on insulating parts of the board. Maximum tolerable concentrations of impurities in the solder of a wave soldering machine[175] are shown in Table 5.1. These values can be compared with the maximum levels specified for fresh solder, given in Table 6.3.

Table 5.1

Maximum Tolerable Levels of Impurities in Solder Baths
(% by Mass)

Copper	0·3	Antimony	0·5
Arsenic	0·03	Bismuth	0·25
Iron	0·02		
Nickel	0·01		
Aluminium	0·006		
Cadmium	0·005		
Zinc	0·005		

Antimony and bismuth are sometimes added to the solder at levels higher than stated in Table 5.1 in order to produce specialised properties. Antimony at levels in excess of 0·5% gives a deterioration in wetting properties, but can be added up to about 5% as a cheaper replacement for some of the tin or because it is argued that it is necessary if the solder is designed to operate at below the allotropic transformation temperature of tin at about −30°C. The transformation begins to occur at +13°C but at −30°C the rate is highest, causing metallic β-tin to disintegrate into the grey powder α-tin. About 0·1–0·2% antimony will completely suppress the transformation but the evidence suggests that the lead in tin-lead solders has the same effect and the antimony is therefore unnecessary. Bismuth is sometimes used to reduce the working temperature; the Sn-Bi eutectic has a melting point of 139°C and the Sn-Pb-Bi eutectic a melting point of only 96°C. Additionally lower bismuth levels are used to impart a uniformly matt finish to the solder joints and this renders visual inspection much less fatiguing.

5.6.3 Oxidation of the Solder

In wave soldering a new clean solder surface is being generated continuously at the crest of the wave and so any oxide found there is not a problem. However, a fresh molten metal surface is very reactive with the oxygen in the air and a build-up of oxide and dross on the solder surface away from the crest is inevitable. The layer of dross formed does not, unfortunately, inhibit further dross formation but, because of turbulence of the molten metal caused by the dross, it actually enhances its further formation. If the dross is not removed at regular intervals it may build up to the extent where particles are carried around with the circulating solder, impeding the solder flow and sticking to the board. The most effective way of significantly reducing dross formation is to use oil on the wave as described in the next section.

The oxidation rate of molten, near-eutectic tin-lead solder in a static situation can be quantified in terms of a weight increase as oxygen reacts with the surface. This weight increase occurs with a parabolic growth rate, i.e., proportional to $\sqrt{\text{time}}$. Thus if w (kg.m^{-2}) is the weight increase per unit area of static solder in time t (seconds), then

$$w = \text{const.} \sqrt{t} \exp\left(-\frac{E_A}{kT}\right) \qquad (5.1)$$

where T is the temperature is kelvin, E_A (joule) an activation energy for the reaction and k is Boltzmann's constant ($1\cdot38\ 10^{-23}$ J.K^{-1}). For near-eutectic tin-lead solders at soldering temperatures the activation energy $E_A \approx 6\cdot8\ 10^{-20}$ J and the constant about 16 g.m^{-2}.s$^{-½}$. This oxidation rate is about half that of pure tin.

Equation (5.1) is not entirely applicable to the solder wave because of the formation of dross in addition to the oxide when the solder is agitated. Dross is a much more severe problem than an oxide film, consisting of solder metal enveloped in the oxide. The oxygen content of dross is only a few per cent. The dross is generated in the turbulence because the oxide skin is being continuously folded over, trapping small quantities of solder within oxide enclosures.

5.7 OIL IN THE SOLDER WAVE

The use of oil as an additive to the molten solder has been found beneficial for the effective wave soldering of devices such as SOTs, SOICs and quad packages, helping to eliminate solder bridging and heavy solder build-up by controlling the formation of an oxide skin especially in the back-wash region of the second wave in a dual-wave machine.[176]

5.7.1 Reduction of Solder Bridges

To attain the desired effect the oil must cover the wave as a very thin film. The oxide skin which forms on the surface on an unprotected wave is very coherent and difficult to displace by an incoming board. As the board leaves the wave, this skin drags solder from the wave, increasing the break-away distance X_1-X_2 in Figure 5.13, and thus greatly increasing the likelihood of icicles and bridges. The oil on the surface of the wave provides a barrier against the oxide formation on the surface of the molten solder and helps to break up and carry away any oxide skin that is formed. It is said[177] that the oil, intermixed with the molten solder, reduces its surface tension and its benefit is through this mechanism. In fact the surface tension of the mixture is greater than that of the pure solder[4] and the beneficial mechanism is purely through the reduction in oxide formation and the break-up and easier removal of any existing oxide film on the wave.

The surface tension of the oil film itself can be reduced by the addition of wetting agents, enabling the oil to spread much more evenly and thinly on the wave surface. This has the additional benefit that less oil is left on the board after soldering. Mildly acidic additives can also be blended into the oil, which scavenge any dross formed, to produce harmless soaps. These additives apparently have no harmful effect on the solder.

The oil layer on the surface of the solder in the sump inhibits dross formation. Dross is an expensive consumer of solder and its abrasive action can also wear out moving parts such as the solder pump. Since dross, namely the oxides of tin, lead, and solder impurities enveloping solder, is formed when the molten solder comes into contact with air there is an obvious advantage in establishing a barrier between the air and the solder. This also eliminates the health hazard encountered when the dross is removed from the surface of the solder sump, during which procedure tin and lead particles can become airborne dust.

The disadvantages of using oil are the resultant contamination of the machine with thickened and decomposed oil and a build-up of solder impurities normally removed with the dross. Additionally, the board itself may require extra cleaning in organic solvents to remove all the oil residues.

5.7.2 Types of Oil

The oil used in wave soldering can be either a purely vegetable peanut oil or a mixture of mineral oils and fatty acids known as tinning oil. The tinning oils have a high flash point and long service life at soldering temperatures, but this is not always important if the oil is in contact with the air and the wave for only a short time. The oils generally exhibit a mild fluxing action. They react with tin and lead oxides to form soaps which disperse in the oil. Whichever oil is used it must be virtually chemically inert at 250°C and suffer little decomposition at that temperature.

5.7.3 Application of the Oil

Three methods have been successfully employed for getting the oil on to the surface of the wave and maintaining it there at the desired level. These make use of an oil-solder intermix system, an oil injection into the wave, and oil drawn or dripped directly on to the wave. For the intermix system the oil is dispersed in the solder at the pump. For the injection system the oil is pumped through a fine nozzle into the solder wave, some centimetres below the surface. These two methods are shown diagrammatically in Figure 5.22. For the direct application methods the oil can be sprayed or dripped on to the wave, where it immediately spreads out over the solder, or in some situations the oil can be drawn up by capillary flow from a reservoir.

In the best systems the oil reservoir is submerged within the molten solder bath so that the oil is supplied at the solder temperature.

5.7.4 Colophony Addition

Very effective avoidance of solder bridging by the addition of colophony to the second wave of a dual wave soldering system has been reported.[178] The colophony replaces the oil used in conventional systems. The test work was conducted using both solid and liquid colophony, with and without halide additives. The best results were obtained using colophony dissolved in a high boiling point solvent activated with a bromine-containing addition.

The liquid is introduced on to the second wave in very small quantities and it spreads very easily over the entire wave, even against the solder flow. The use of colophony obviates the problem of cleaning relatively large quantities of oil from the board after soldering, and colophony residues can even be left on the board if the liquid is unactivated and if the solidified colophony is not going to give rise to test probe contact problems.

(a) Intermix

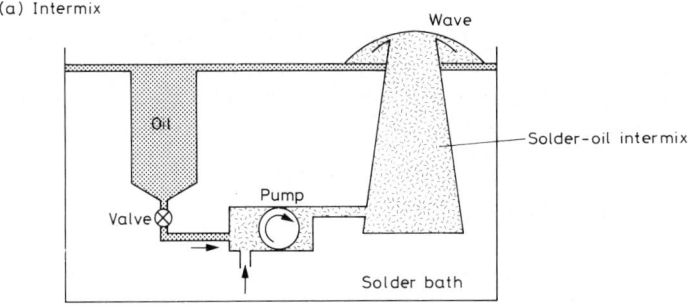

(b) Injection

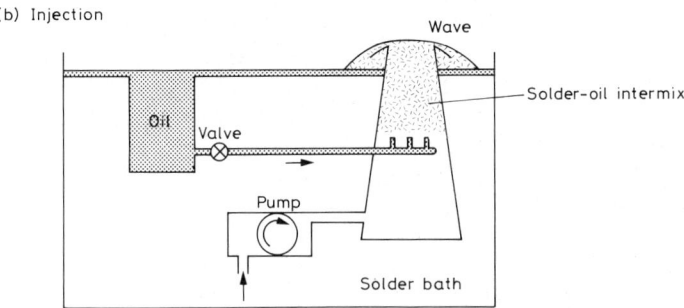

Fig. 5.22 A representation of (a) the intermix method and (b) the injection method applying oil to a solder wave.

5.8 AIR KNIFE

The occurrence of the most prevalent problem during the wave soldering of surface mounting devices, namely that of shorting bridges of solder between terminals, is associated with the interaction of the board with the solder wave as the two part company, that is in the back-wash region. The most significant changes made to the wave soldering process since its innovation have all been aimed at controlling this region: the inclined conveyor, the addition of oil to the wave and the reshaping of the basic wave. A further development, the hot air knife,[179] also addresses this problem and has been found to be specifically beneficial when wave soldering surface mounting devices.

The hot air knife consists of a fine jet stream of high velocity hot air directed across the underside of the soldered board as it leaves the solder wave, as shown in Figure 5.23. The hot air simply blows away the excess solder between joints and reshapes the solder fillets before solidification.

It has always been considered a mistake to disturb a solder joint during solidification, since this may greatly reduce joint strength. However, the hot air knife disturbs the molten solder and removes it readily from non-wetted surfaces, but leaves it still molten with enough heat content to self-anneal and not leave any excessive residual strain within the joints.

The hot air knife can in principle be fitted retrospectively to an existing wave soldering machine but more commonly comes as an intrinsic part after the smooth wave of a dual wave machine. The distance of the air jet from the wave and from the underside of the circuit board, the angle of the jet to the board, the air pressure and temperature are all potentially adjustable as well as the synchronisation of the air jet with the conveyor.[180] Usually these parameters are controlled through a microprocessor. Figure 5.23 shows a preferred configuration of a hot air knife together with the typical conditions for the parameter values.

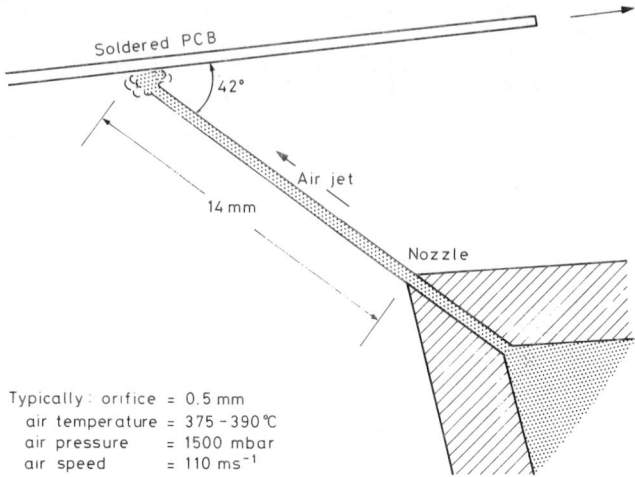

Fig. 5.23 The parameters of a hot air knife applied as the soldered circuit board leaves the wave.[181]

The performance of the hot air knife is directly dependent upon the proximity of the nozzle to the board and the wave. The closer the air knife can be brought to the workpiece while the solder is at its hottest, the greater will be its benefit.[181]

If the air pressure is increased indiscriminately in order to gain more benefit and reduce the occurrence of bridging, too much solder can be removed from some joints resulting in weak, very lean joints, even completely stripping through-hole joints of solder. This is especially so if problems of poor solderability occur.[182] The strength of the adhesive bonding the components to the board is very much reduced at the soldering temperature and there is the possibility that components can be blown away by excessive air knife pressures.

5.9 CIRCUIT BOARD FINISH

There are numerous and varied routes to produce, from copper-clad laminate, a circuit board with plated-through holes and solder mask, and the characteristics of the finished board reflect the production processes chosen.

The most common technique is to use tin-lead electroplate as an etch-resist for the underlying copper and then apply solder mask to eliminate the possibility of solder shorts between tracks, during wave soldering. However, applying the solder resist over the electroplated tin-lead means that during soldering this electroplate melts under the mask and, when solidified again, causes the mask to have a wrinkled appearance. This technique of solder mask over tin-lead does not guarantee total elimination of solder bridges, especially on narrow circuit tracks when the electroplate is thick. For through-hole mounted boards the use of solder mask over tin-lead is not a great problem, but for surface mounting boards it is definitely not recommended. This is because, if the adhesive mounting for a surface mounting component is placed on solder mask over tin-lead, it may move or flow away when the soldering is taking place and the tin-lead melts and reflows. The mounting adhesive for surface mounting components should only be placed on solder mask on bare laminate or solder mask on bare copper.

This is illustrated in Figure 5.24. In (a) the adhesive is on solder mask on the board, but in (b) it is on solder mask on fused tin-lead electroplate, which will lead to movement of the component during the soldering stage. In (c) the board has been manufactured by the hot air levelling route and the adhesive is on solder mask over bare copper.

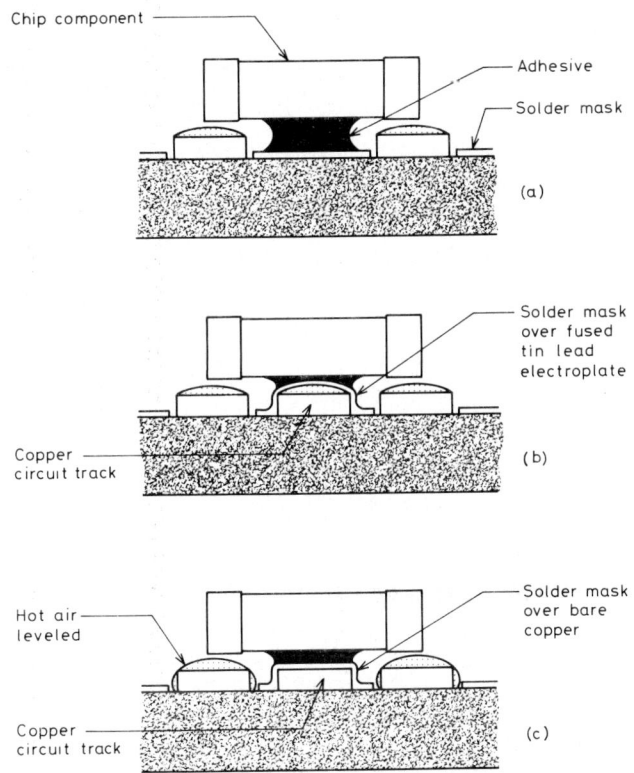

Fig. 5.24 Adhesive bonding of surface mounting components on solder mask is acceptable if the mask is over (a) laminate or (c) bare copper, but not if it is over (b) reflowed electroplated Sn-Pb.

The hot air levelling process has been described in detail in Section 3.5.1.8 and specific problems encountered have been discussed. In brief, a bare copper circuit board is produced, the solder mask is then applied and the board dipped in molten solder to apply a solder coating to the exposed copper lands to retain their solderability during storage. As the board is being withdrawn from the solder, the plated-through holes are cleared of excess solder and the overall level of solder is controlled by a stream of hot air directed perpendicularly at the board.

Hot air levelled boards with an adequate solder-coating thickness have excellent solderability with mild fluxes and a very long shelf life. It is a board finish that is recommended for surface mounting assemblies.[183]

There are several types of solder mask available—printable inks, imageable inks and dry film, for example, whose relative merits have been discussed at length.[184] On boards for surface mounting applications, a new problem has arisen, with the use of the thicker solder masks. Because the solderable lands on a surface mount board are so small, there is a distinct possibility that some can be skipped in the hot air levelling solder bath, if the solder mask is thicker than the bare copper, which causes the copper to be recessed from the surface.[185] This is illustrated in Figure 5.25. The 1 oz.ft^{-2} copper thickness is about 36 μm while a thick solder mask may be as much as 75 μm.

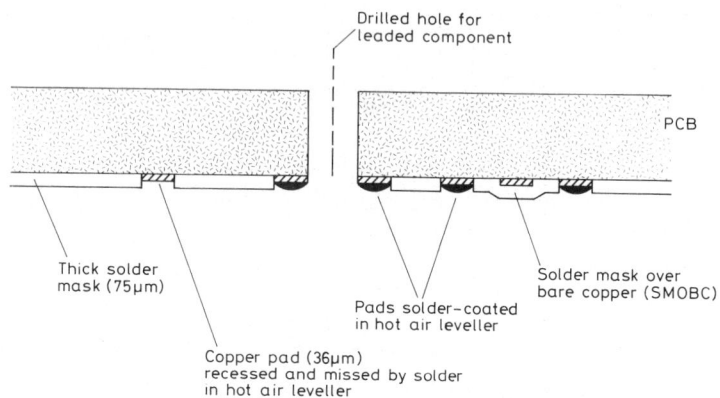

Fig. 5.25 If the solder mask is thicker than the copper tracks, some copper lands may be skipped during hot air levelling.

5.10 WAVE SOLDERING DEFECTS

Inspection of wave soldered boards with surface mounted and insertion mounted components reveals several endemic faults for which there may be a readily realisable solution, or which may point to a necessary change in board design or assembly method. Alternatively they may have occurred because the whole assembly process is operating within a very small acceptable volume in multi-dimensional space and the faults observed are simply the trailing edges of Gaussian distributions of unquantifiable parameters.

5.10.1 Inserted Components

The major advantage of the wave soldering route for making the component-to-board joints is that surface mounting devices can be used in conjunction with conventional leaded through-hole devices on the same board. This technology is not the subject of interest here, but nonetheless it forms part of the integrated interconnection technology used for surface mounting. Numerous articles have been produced concerning the optimisation of the process for leaded components in plated-through holes[186] and the identification of the causes of soldering faults,[187-192] to which reference may be made.

There are three families of endemic faults associated with soldering leaded components in plated-through-holes.

(i) Outgassing faults which are due to the volatilisation and evolution of moisture absorbed in the epoxy of the board laminate, arising from the thermal spike generated by the molten solder in the plated-through hole. Outgassing faults are manifested as visible blowholes in the surface of the solder fillets, internal hidden voids within the fillets and considerable or even total blow-out of the solder from the holes. These problems have been discussed in great detail with regard to their harmfulness in service[193, 194] and to the procedures available for their elimination.[195-199]

(ii) Solder bridges and icicles are caused in similar ways to those associated with surface mounting joints and their occurrence can be reduced by changing the mechanisms associated with the back-wash region of the wave[200] through the use of an oil intermix with the solder or a change in the conveyor speed.

(iii) Poor solder pull-through in the plated-through hole is evidenced by the solder not flowing to the top of the hole and over the top-surface solderable land. This may be a problem of an incorrectly adjusted wave height for the conveyor speed used, but more likely it is a solderability problem of the through-hole plating due, for instance, to the formation of the copper-tin intermetallic layer being exposed at the knee of the plated barrel.

Notwithstanding the problems of solderability, the soldering parameters can be adjusted to minimise these problems, although these may well be mutually exclusive, requiring a compromise solution to be found. Figure 5.26 shows how the three types of fault described above, outgassing, excess solder and poor solder pull-through, are related to the temperature of the solder and the speed of the conveyor over the wave. In some situations, of a particular board type and a particular wave soldering machine, the working area in the middle may be non-existent or very hard to locate.

5.10.2 Surface Mounted Components

The main problems arising in the wave soldering of surface mounting components are solder skips and solder bridges.[201] These problems have already been discussed because they are intrinsic to the structure of the solder wave and the design of the printed circuit board. With the use of a dual wave, solder skips can be virtually eliminated provided the layout of components is sensible and the conveyor speed is correct for the soldering conditions of temperature and wave height.[202] Solder skips which do occur are now invariably associated with wetting problems of the solderable surfaces of components.

Several methods of minimising solder bridges have been discussed already, including the use of additives such as oil or colophony to the wave to reduce the drag-out of solder by the board in the wash-back region of the wave, and the incorporation of a hot air jet to remove excess molten solder from the board. A further method of minimising the problem of solder bridging lies in the modification of the solder pad footprints and the board layout.[203] This will be dealt with in Section 5.10.2.1.

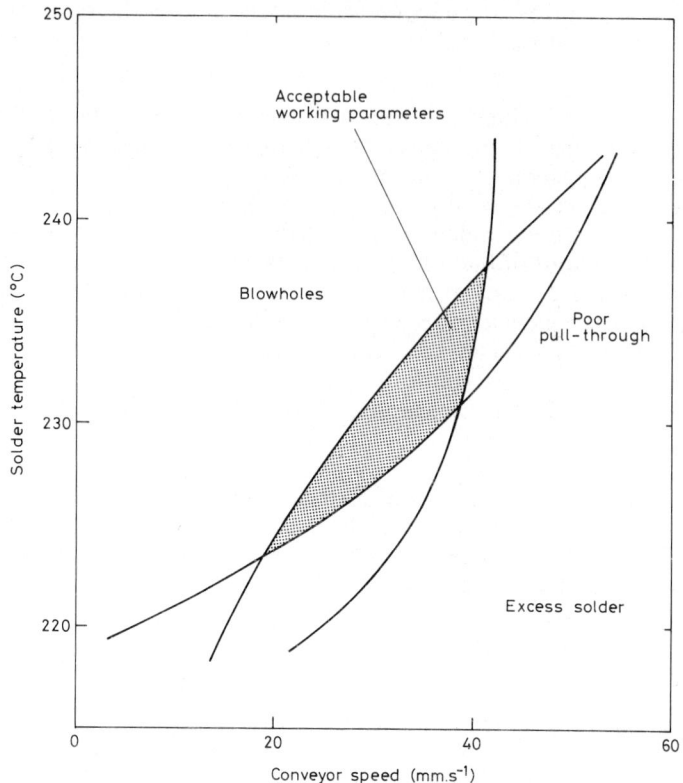

Fig. 5.26 Representative curves defining acceptable limits of the three generic types of fault encountered in solder fillets on leaded components in plated-through holes. These curves define a working area on the axes of solder temperature and conveyor speed.

The shifting of a component, or even its complete removal, is another observed defect. The shear strength of the adhesive, if properly used and cured, with a typical contact area between a component and an epoxy-glassfibre board, of 1 mm^2, is about 25 newtons. The hydrostatic force of the molten solder acting on the component as it drives through the wave is only a few millinewtons, so the shifting of the component is not a direct result of the force of the wave upon it. There are two causes of the problem. The cured adhesive is susceptible to thermal shock cracking when it first contacts the solder wave. It is recommended that the preheated temperature of the assembly is at least 100°C in order to alleviate such problems. The second cause of component shifting is the incorrect use of the adhesive in conjunction with solder resist. This was dealt with in Section 5.9.

With a suitable choice of materials, process conditions and pattern design, defects can mostly be avoided provided solderability is adequate. The resultant joints will have a different shape profile, a different metallurgical microstructure and therefore different metallurgical properties from those produced by reflow of a solder paste. The joints produced by wave soldering are capable of surviving all the mechanical and environmental conditions that are likely to be met in service.

5.10.2.1 AVOIDANCE OF BRIDGES BY PATTERN DESIGN

The hydrostatic pressure in a smooth and laminar solder wave is not adequate to give the solder surface a sufficiently sharp curvature to reach the lands and solder terminations which are shadowed in a corner between a non-wettable component body and non-wettable printed board, as was shown in Figure 5.17. As the thrust and turbulence of the wave are increased, fewer and fewer of these solder skips will occur. However, it is known that, in order to avoid solder bridging between neighbouring conductors and terminations, the solder wave should be as quiet and smooth as possible. Thus these two demands of upward thrust and smoothness are not compatible and have led to the evolution of the new generation of dual-wave soldering machines.

With the incorporation of a turbulent first wave, the problem of skipped joints is avoided and the remaining difficulty in wave soldering surface mounting assemblies is to prevent solder bridging, especially in multi-leaded devices. Bridges are most common within the trailing rows of terminations of transversely placed SOICs, between the trailing terminations of SOICs placed parallel to the transport direction and on the corresponding terminations of quad flat packs. For devices such as the VSO-40 which have a smaller leg-to-leg pitch than the standard SOs (0·762 mm compared with 1·27 mm) and also longer feet, bridge-free wave soldering can only be achieved if the device is placed parallel to the transporter direction.

A reduction in the area of the solder lands gives no improvement in the bridging problem, and in fact increases the incidence of solder skips.

Some empirical recommendations, however, can be made[178] for the reducing of solder bridges between SOIC terminations. All soldering machines behave differently and the degree of bridging is dependent upon the adjustment of the many parameters, both those available to the operator for adjustment and those defined by the manufacturer of the machine. However, some universal suggestions can be made as follows. For SOICs positioned transversely to the transport direction, (a) a lesser protrusion of the solder land beyond the trailing end of the lead is beneficial and (b) a reduction in the conveyor speed is beneficial. For SOICs parallel to the transport direction, (c) the addition of solder robbers to the footprint pattern is beneficial. The beneficial use of non-central alignment of the device on its footprint, and the use of solder robbers are now discussed.

Bridges occur only on the downstream side of SOICs that are positioned transversely to the transport direction. Assuming that the component is accurately placed in the centre of the footprint, the length of pad protruding from

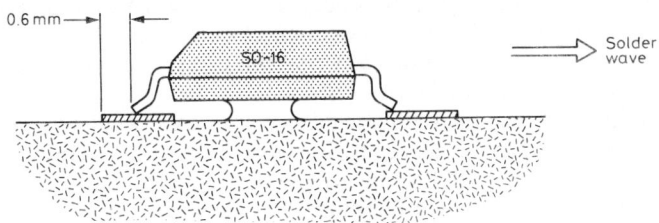

Fig. 5.27 Showing the displacement of a transverse SOIC with respect to its footprint, to reduce solder bridges on the downstream side of the device.

the flying end of each leg of the device is about 1·0 to 1·3 mm. For example, for an SO-16 component, the overall width of the footprint is about 7·2 mm (Table 4.2) and, whilst the overall width of the component can be a maximum of 6·2 mm (Table 2.1), more usually it is closer to 5·0 mm. A reduction of the protruding length on the trailing side of the device is a very efficient way of eliminating bridges.[178] This can easily be realised by programming the placement co-ordinates of such devices with a sufficiently large offset from the mid-position. For SOIC footprints of the small type, for example an SO-16, a protruding length of no more than 0·6 mm is ideal, as shown in Figure 5.27, necessitating a linear offset depending on the exact packaging of the device. For SOs using the larger footprint such as the SO-28, which are inherently easier to solder than those using the smaller, a protruding length of 1 mm is small enough to eliminate bridges effectively.

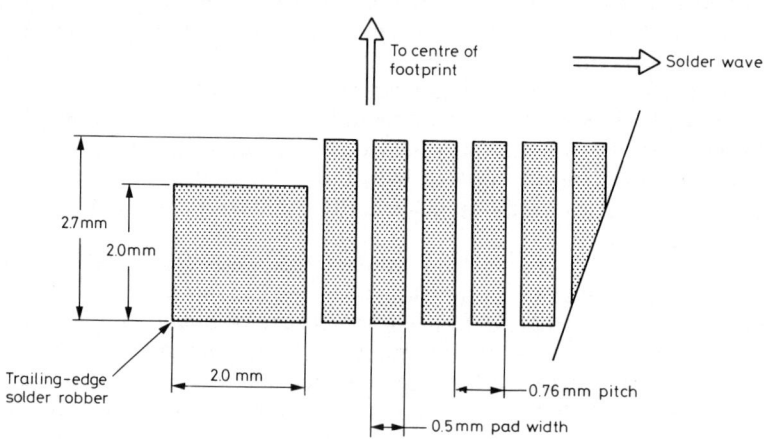

Fig. 5.28 The incorporation of solder robbers after the trailing lands of an SOIC mounted parallel to the transport direction helps to avoid solder bridges.[178]

Additionally the ends of the solder leads should be limited by solder resist, since open tracks at the end of the lands negate the benefit of offsetting the component to reduce the protruding solderable land.

The benefit to be gained by the offset is also limited by the sideways accuracy with which the component legs are placed on the pads. Obviously the likelihood of bridging is reduced if the legs are placed exactly centrally (sideways) on the footprint. It has been pointed out that the optimum value of the protruding length of the land is quite critical and a few tenths of a millimetre make a significant difference.[178] If the consequences of such close tolerances are unrealistic in regard to the automated placement system used, other ways of avoiding bridges must be invoked.

For the quad flat packs it is obviously not possible to offset the component towards the trailing edge of the footprint without also incurring a detrimental displacement on the sides of the device parallel to the transport direction. In addition, the quad flat pack has finer pitch (0·65-1·0 mm) gull wing shaped leads which make it invariably prone to solder bridging. The most successful wave soldering of quad flat packs is achieved by placing them at 45° to the transport direction and also incorporating solder robbers in the corners of the footprints.

The use of solder robbers is also beneficial for SOICs mounted parallel to the transport direction. If solder bridges occur, they almost always do so between the last and the next-to-last legs, at the trailing end of the device. The solution to this problem is to add a fake solder land, downstream from the last pad in the footprint, or to enlarge the last solder pad. In this way, if a bridge were to form it would not be in a position to cause an electrical shortcircuit. Suitable robbers in the footprints are illustrated by a VSO example in Figure 5.28.

Chapter 6

SOLDER PASTES

6.1 INTRODUCTION

Solder paste is a homogeneous, stable suspension of solder powder particles in a flux binder. The shape and size of the metal particles in combination with the rheological properties of the binder are tailored to match the method used to apply the solder paste to the PCB pads and to match the design of the PCB.[204-206]

The binder not only contains the flux but also other substances that determine its viscosity, its thixotropy and its flow properties that enable the solder to be applied to the places where it is needed, to remain in those places, to retain and support the components placed on it, and to enable the solder to melt, set and spread to form a reliably strong joint. The flux binder must have the property of keeping the heavy solder particles evenly dispersed through the paste at room temperature. The paste must maintain its consistency during application and the binder must not inhibit the action of the flux during soldering.[207]

The flux binder has four constituents: (i) the flux dissolved in (ii) the solvent, with (iii) an activator and mixed with (iv) thickeners and lubricants that determine the rheological properties of the paste. The properties of the solder powder will be discussed first.

6.2 THE SOLDER POWDER

6.2.1 Solder Particle Shape

The control of solder particle shape within the desired size range is extremely important. For two reasons the attainment of particles as nearly spherical as possible is deemed to produce the most satisfactory results. First, a sphere is the physical object with the lowest surface area-to-volume ratio and hence, for a given oxide thickness, will contain the smallest amount of oxide. Secondly, non-spherical particles can jam the screen through which the paste is printed or clog the needles if the application is via a syringe. As well as poor quality solder application, undue wear is caused to the screening or syringing equipment.

Some manufacturers have, in the past, advocated the use of non-spherical particles which interlock to help prevent slump and outflow of the paste as the flux melts. It has been shown[208] that using spherically shaped particles alone can induce a degree of slump and flow which can upset good fine-line soldering. The spherical shapes permit the solder to sink unduly just before melting occurs and it is possible for adjacent tracks to have solder touching before then splitting unequally back on to their respective pads. Also, if the solder particles spread apart with the melting flux they are left stranded at the edges of the fluxed area and then remain as undesirable discrete solder balls, not drawn in to the main solder fillet. By using non-spherical particles, a reduction in the problems of slump and solder balling can be achieved since fewer solder particles are carried outwards with the spreading flux.

For some applications, a proportion of non-spherical solder particles may therefore be considered to be important to retain a profile that is as good as possible without movement during the reflow.[209] When a solder paste is specified, the relative proportions of spherical to non-spherical particles is sometimes taken as a measure of the reproducibility from batch to batch. However, modern solder pastes contain proper rheological modifiers to minimise deleterious outflow of the particle even when they are perfectly spherical.[210] For surface mounting applications the sphericity of the solder particle is now taken as one measure of the quality of the paste,[211] and the degree of sphericity achieved is demonstrated by the micrograph in Figure 6.1.

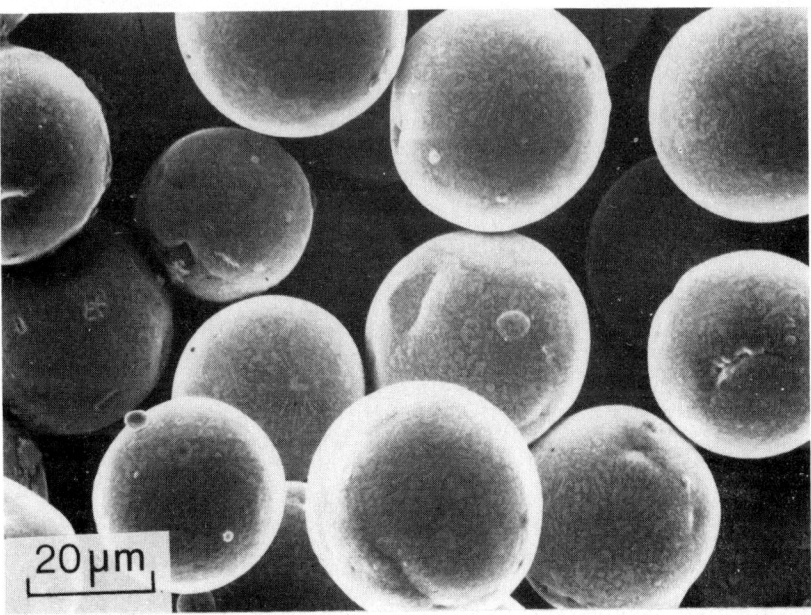

Fig. 6.1 A scanning electron micrograph of atomised solder particles, illustrating the sphericity obtained.

The shape of the solder particles depends on the powder production process and on proprietary skills. The powder is produced by atomising the liquid solder alloy in an inert atmosphere. The elimination of non-spherical particles and agglomerated clumps must be made at this atomising stage since it is not possible during the separating of the powder into different sizes by sieving.[212] Irregularly shaped particles can pass a sieve with one dimension being much longer than the mesh size.

6.2.2 Solder Particle Size

The solder particles are sorted into various size ranges by sieving through meshes. A solder paste can in principle be manufactured using any size range or combination of size ranges specified by the customer, but solder paste manufacturers have, in the main, standardised on three or four particle size distributions,[213] as given in Table 6.1.

Table 6.1

Common Mesh Sizes Used for Sieving Solder Powders

Mesh Size	Particle Sizes
−325	less than 45 μm diameter
−270	less than 55 μm diameter
−200	less than 75 μm diameter
−200, +325	between 45 and 75 μm diameter

The relative dimensions of the meshes used are shown schematically in Figure 6.2. The mesh size is in units of 'lines-per-inch', with the minus sign meaning the particle fraction that has passed through the mesh is used and the plus sign meaning the fraction that has not passed through is used.

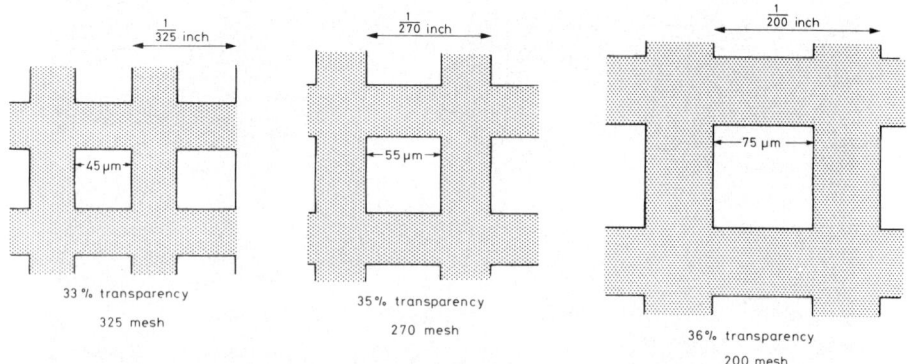

Fig. 6.2 The relative dimensions of common meshes used for sieving atomised solder powders.

The choice of size of the solder particles is a compromise between ease of application and definition of the applied paste pattern versus the problems of surface oxide and of solder balling.[214] On the one hand, the finer the particles the easier their passage through the printing screen or syringe and hence the finer the screen that can be used and the greater the definition of the screen-printed paste. On the other hand, the use of a smaller-sized particle leads to two problems. First, the surface area-to-volume ratio of a sphere increases as the particle size reduces and hence, for a given oxide thickness (corresponding to a given exposure of the particles to the air), more oxide will be present in the paste. The degree of this effect is shown in Figure 6.3. In general, the behaviour of the solder paste improves as the oxide content is reduced, as discussed in Section 6.2.3. Secondly, fine particles of solder (known as 'fines') are the cause of solder balling for two

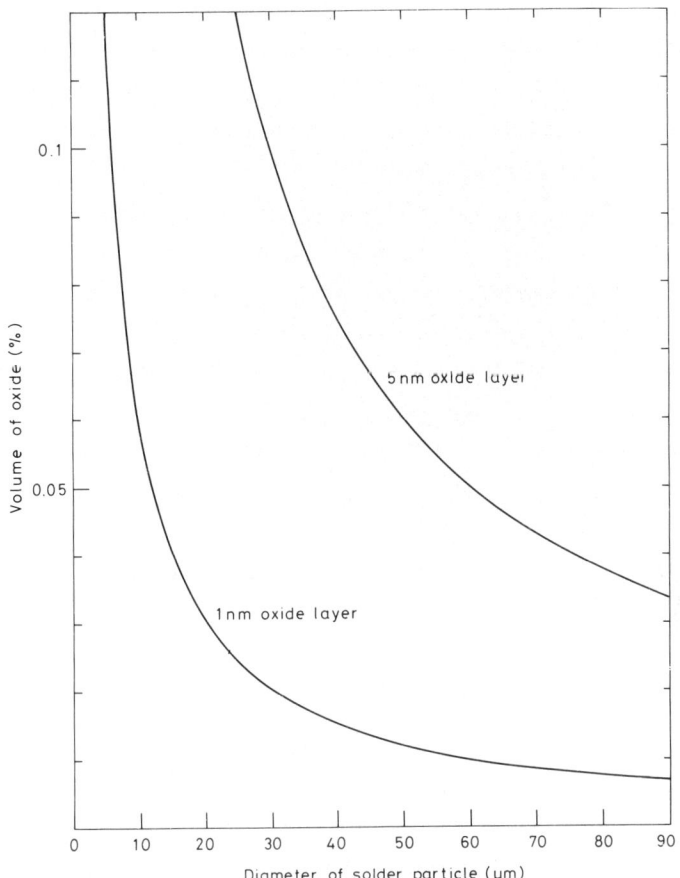

Fig. 6.3 The relative volume of oxide increases rapidly as the size of a solder particle decreases.

reasons. First, the smallest particles are those flushed out of the paste deposit as the flux melts and spreads, and are carried to the extremities of the flux. Secondly, as the temperature rises the solder melts and the main body of the paste coalesces to form a solder fillet, but the smaller particles are left behind and, having more oxide content, they do not coalesce as rapidly. Thus the smallness of the fines means they are carried furthest from the main body of solder and also have relatively more oxide and so are less likely to coalesce readily. This effect is shown in Figure 6.4.

The relation between solder balling and particle size distribution is illustrated in Figure 6.5. Using the materials and parameters shown in the figure, solder balling occurs significantly only when the minimum particle size is below about 40 μm. This size varies very little if the parameters are changed. Thus the −200, +325 pastes, whilst more grainy in consistency and not wholly suitable for printing pads of less than 250 μm width, exhibit virtually none of the solder balls experienced by the finer pastes and they are thus growing in popularity. For higher resolution work, a −270 powder is preferred to the −200 grade, enabling a finer printing screen to be used.[215]

158   *Solder Pastes*

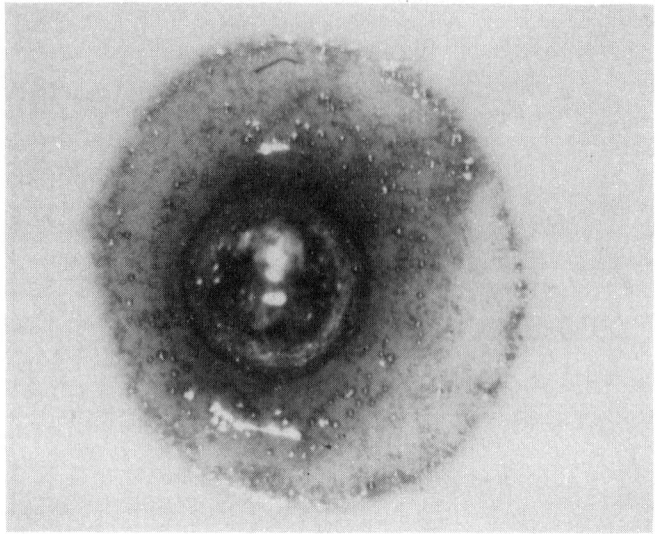

Fig. 6.4 An illustration of how solder balls are formed as the 'fines' are swept out by the spreading liquid flux and are then not able to coalesce as the solder melts.

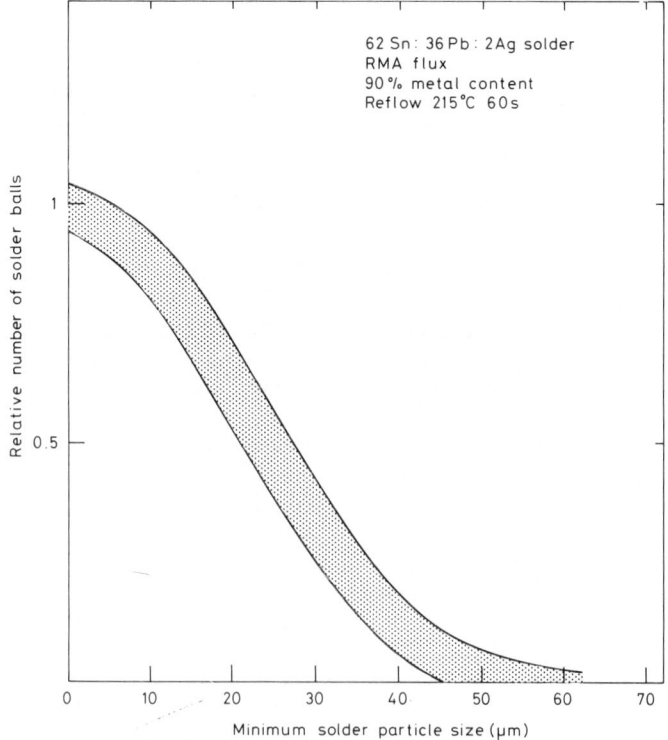

Fig. 6.5 The degree of solder balling is very dependent on the minimum particle size of the solder powder in the paste. Solder balls are due to solder 'fines'.

6.2.3 Oxide Content

As mentioned previously, oxide in the solder powder must be at a minimum to obtain rapid coalescence of the individual particles upon melting. During soldering, oxides should be removed from the metal by the activity of the flux in the paste, but if the oxide level is excessive the flux may not be active enough to accomplish sufficient oxide removal. Excessive oxides inhibit solder coalescence during reflow which results in the formation of solder balls and incomplete wetting of the substrate solderable pads.[216]

After atomising the molten solder into powder, some manufacturers are now able to separate the particles into their various size ranges and introduce the powder to the flux binder, all without the solder seeing anything other than an inert atmosphere,[217] thus virtually eliminating the oxide layer. Such high quality powder is sometimes referred to as 'oxide free'.

Solder pastes can now be manufactured with an oxide content below 0·03 wt% of the metal content. When the oxide content rises above 0·15 wt%, solder balling is almost certain to be a problem particularly with low activity flux in the paste. Since the density of solder is about 8 g.cm^{-3} and the density of the oxide is about 6·5 g.cm^{-3}, 0·15 mass% oxide-to-metal content corresponds to an oxide thickness of 6 nm on a 20 μm diameter solder sphere and 25 nm on an 80 μm diameter solder sphere. The full data are given in Figure 6.6.

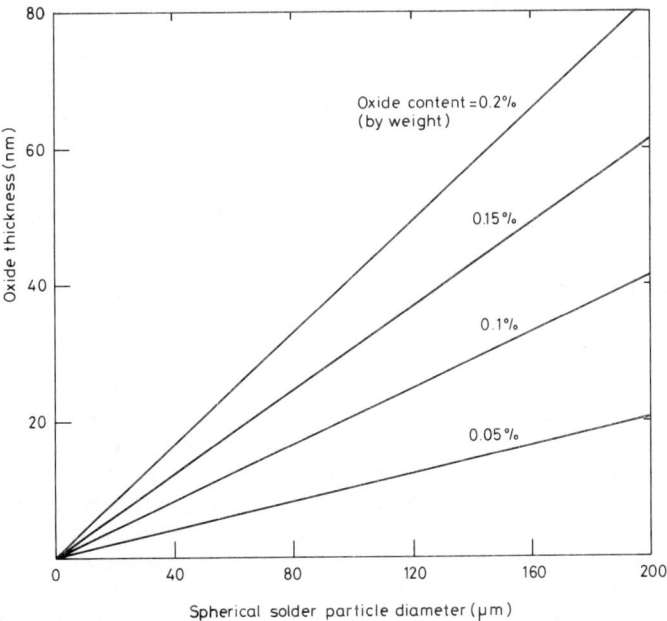

Fig. 6.6 The mass % of oxide on spherical solder particles as a function of their size.

An inexpensive and easily implemented procedure has been developed to test qualitatively the oxide content of a solder paste.[3] The test involves the use of peanut oil to extract the flux materials from a sample of the paste and then reflowing the resultant flux-free paste. Since the oxides on the solder particles are much less dense than the metal after melting, the oxide floats to the surface. Because the flux has been removed, the oxide is left on the surface of the solder as it cools and solidifies. The percentage of oxide covering the metal indicates the relative amount of oxide in the paste. The details of the peanut oil test are:

1. Place about 10 g of solder paste in a small porcelain crucible.
2. Add enough peanut oil to about three-quarters fill the crucible.
3. Heat the crucible in an oven at about 215°C for 30 minutes.
4. Remove the crucible from the oven and allow to cool to room temperature.
5. Pour off the peanut oil and add a flux cleaner to remove any oil residue.
6. Remove the metal button from the crucible and inspect under a low-power microscope.

Oxide-free solder has a smooth shiny appearance, but oxides appear as rough dull growths on the top surface of the button. If more than about 15% of the button's surface is covered by the oxide growths, the solder is likely to have an excessive oxide content.

A solder paste that initially has low oxide content and performs well may have an increase in oxide content due to improper handling or storage. Therefore it is recommended that solder pastes be evaluated periodically to check such oxidation. A quick method to test a paste's usefulness is simply to reflow a small amount on a clean substrate. The solder should form a single shiny ball. If many balls form, the paste is no longer usable. This test does not measure the oxide content because the flux is still active when the solder reflows. It does, however, give an indication of any deterioration in the oxide content during storage. A standard procedure for a solder balling test is given in Section 6.10.1.

6.2.4 Solids Content

Solder paste is a suspension of solder powder particles in flux dissolved in a solvent with additional rheology-modifying agents. The metal powder content, as a percentage of the mass of the wet solder paste, is called the solids content. For surface mounting applications the solids content lies in the range 80–95% by mass, which is about 30–70% by volume. Typical volume-weight ratios are shown in Figure 6.7.

In general, the lower the solids content, the easier the paste is to print through a screen to give fine dimensions, or the easier it is to apply by syringe, but the more it slumps during reflow. The optimum solids content is therefore a compromise.

Solder pastes with about 85% and 90% solids content are most commonly used for screen printed surface mounting applications. The 85% paste can be screen printed to very fine dimensions of less than 250 μm pad width and spaces, without bridging or wicking. A 90% solids-content paste can be printed to equally fine dimensions with less slumping but a stencil rather than a screen is often preferred to prevent jamming the pattern, and care must be taken to restrict the printed thickness in order to avoid bridge formation during reflow. The 90% solids-content paste is recommended if a thick deposit is required at the expense of fine dimensions. For paste application by syringe, a metals content in the range of 80–85% is usually required.

The high solids content ensures that the paste height is maintained and results in a taller solder fillet after reflow, standing the component off the board to a

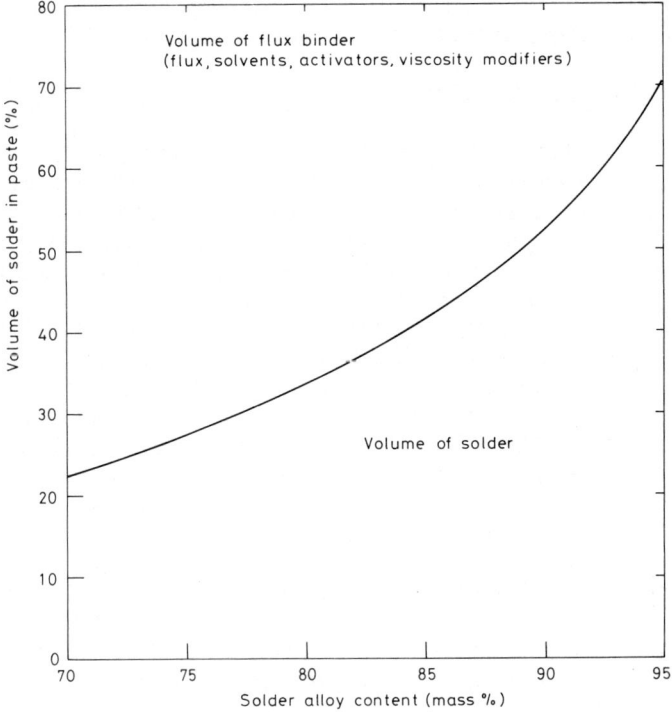

Fig. 6.7 The relationship between the solids content of a paste measured by mass and measured by volume.

greater height. This has several advantages:

(i) The higher the component is off the board, the easier it is to clean away flux residues from beneath it;
(ii) increasing the fillet height improves the joint fatigue strength by enabling a degree of flexibility;
(iii) the more solder there is in the joint the less detrimental the effect on the mechanical strength of the joint are the dissolved impurities leached from the solderable surfaces of the component and the board.

The stand-off height can be increased by one of several methods. For example, solder bumps of an alloy with a higher melting point can be applied to chip components. An easier, one-step process is to use a solder paste in which inert particles such as glass spheres have been incorporated. These glass beads are of some 150 μm diameter and act as spacers during the solder reflow to give the necessary stand-off. They apparently do not have any deleterious effects on joint lifetimes.

The lower the solids content the longer the paste will remain tacky and accept components, all other things being equal. This is simply due to the presence of more flux binder. At one time solder paste had to be populated and then reflowed as soon as it was deposited but, due to high volume requirements the available times between deposition and component population and then between component population and reflow have increased because of the development of quite complex rheology modifiers added to the paste. The times can be increased further if the boards are stored in a freezer after application of the paste, or stored in nitrogen after placement of the components.

6.3 THE FLUX IN THE PASTE

When solder is used to join two metallic pads the inherent strengths of the pads and of the solder are known constants and relatively high. The weakest links of the joint usually are at the two interfaces whose strengths depend on how well the solder wets the metallic pads. The ability of the molten solder to wet the solid surfaces is largely dependent on how clean the surfaces are. A clean surface has a high surface energy and hence is wet by a molten metal with a low surface energy, since this thermodynamically reduces the total energy of the system.

There are many different types of dirt on the solderable surfaces. Inorganic soils and greases can be removed easily by washing with solvents and detergents, but the biggest barrier to wetting usually consists of metal oxide and, to a smaller degree, certain metal sulphides, both of which must be removed if the solder is to wet the surface. Fluxes provide chemical cleaning. They contain polar and non-polar solvents that will remove inorganic grease and oils and also contain active ingredients that reduce metallic oxides to the metal or form a salt that can be dissolved or flushed away by the flux or the flux residue cleaner.

In principle, the same fluxes are used in solder pastes as are used for hand, dip or wave soldering, of which much has been written on their chemistry, their corrosivity and their classification. In the industry, several ways of categorising fluxes have evolved. At present there are two main groups used in the electronics industry:

1 Fluxes soluble in organic liquids
2 Water-soluble fluxes.

6.3.1 Fluxes Soluble in Organic Liquids

The fluxes in this category are resins dissolved in a combination of aliphatic alcohols. Most of the resin used is based on colophony (commonly called rosin). Colophony is a natural product, obtained from the sap tapped from various species of pine tree, which is then steam distilled to produce liquid turpentine and solid colophony. Each source of the product has its own chemical fingerprint.

Colophony has been used universally as a flux throughout the electronics assembly industry because of a combination of favourable properties. As a solid it is not reactive but at soldering temperature, as a liquid, it wets tarnished metal surfaces and has a sufficiently low viscosity to remove reaction products. At these soldering temperatures it is reasonably stable and, on cooling, it can remain on the board as a layer having good insulating properties and no corrosivity.

When pure colophony is dissolved in alcohol or similar solvent, a very mild flux is produced whose efficacy is too low to be used for most soldering operations. Activators are therefore added. These are mainly organic acids or organic salts that are not active at room temperature but are chemically active at the soldering temperature. Despite use of the widely recognised term 'activator', these added chemicals do not activate the resin, but simply act directly in addition to the colophony.

Two salts are commonly used as activators, not only in colophony fluxes but also in water soluble fluxes. These are dimethyl-ammonium chloride (DMA-HCl), $(CH_3)_2NH \cdot HCl$ and diethyl-ammonium chloride (DEA-HCl), $(CH_3 \cdot CH_2)_2NH \cdot HCl$. These salts are readily soluble in alcohol but do not dissolve in the colophony. At slightly below the soldering temperature these activators decompose, yielding hydrochloric acid. Both the activity and the corrosivity depend on the amount of activator added to the flux.

The activity of a colophony flux activated in this way is expressed as a percentage mass ratio of chloride ions Cl^- to the colophony content. For example, a mildly activated flux contains some tenths of a percentage of activator. Figure 6.8 illustrates clearly the effect of increasing the Cl^- ion content in colophony flux on the wetting time for clean copper, soldered at 235°C.

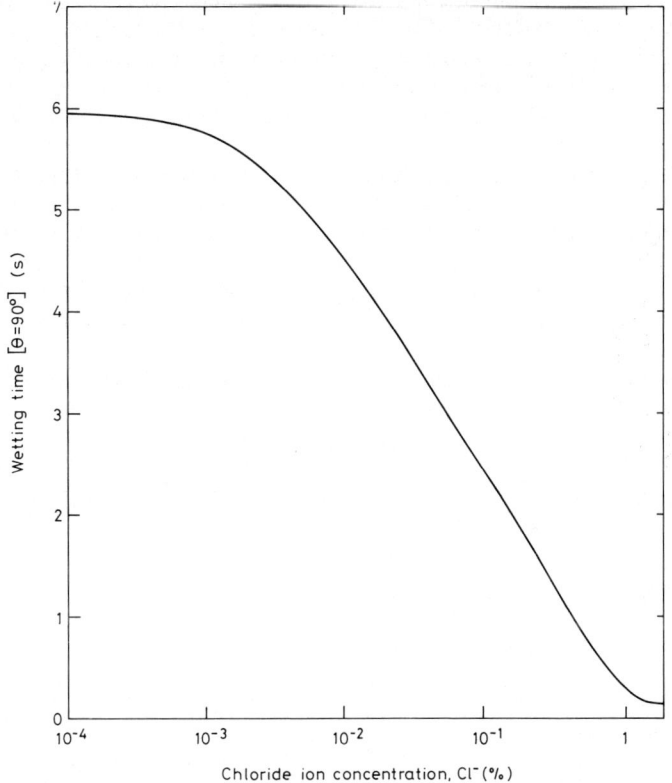

Fig. 6.8 The effect of increasing the chloride content of the flux, on the wetting of clean copper by solder at 235°C.[4]

Besides DMA-HCl and DEA-HCl, a number of organic acids are also used as activators in colophony fluxes. The choice is restricted by considerations of corrosivity, volatility and health. To attain the same efficacy as the DMA-HCl and DEA-HCl activated fluxes a much larger addition of suitable organic acid is required, and hence they are generally used only for mildly activated fluxes. These fluxes are generally termed 'halide-free'.

A rough classification of the activity of colophony fluxes has come into common usage, deriving from a USA Federal Specification:

R—Rosin, non-activated colophony. This flux depends on the very weak organic acids in natural rosin. It is a very mild flux and finds use only with metals whose oxides can be removed very easily.

RMA—Rosin, mildly activated. This flux contains rosin plus a very mild activator to supplement the cleaning action of the rosin. It is used for metals whose oxides are moderately bound to the surface of the base metal. It is the most widely used flux in the assembly of electronics. Generally, the flux is neither corrosive nor conductive before or after soldering, since the activator remains inert at room temperature, and is released and finally dissipated during the soldering operation.

RA—Rosin, activated. This flux contains rosin plus a strong ionic activator to strengthen the cleaning action of rosin. It is used with metals whose oxides are difficult to remove. It is slightly corrosive and conductive before soldering, as is the flux residue after soldering. Thus, because of the difficulty in completely washing off all ionic residues and the potential for future corrosion and possible failure of parts, this flux is only rarely used in electronics. RA fluxes can be used for joining structural, non-porous metal members if the flux residue can be cleaned off thoroughly.

Colophony, or rosin, is a natural product and hence subject to the vagaries of its source and its processing. The possibility of a synthetic substitute is therefore attractive, and one that is commercially available is based on ester pentaerythritol tetrabenzoate which fumes less and spatters less than colophony fluxes during the soldering operation. The flux residues are even less corrosive than those of colophony, and so need not be removed after soldering, unless further activators have been added.

Thus, in summary, fluxes soluble in organic liquids are based on resins, and can be divided into classes as follows:

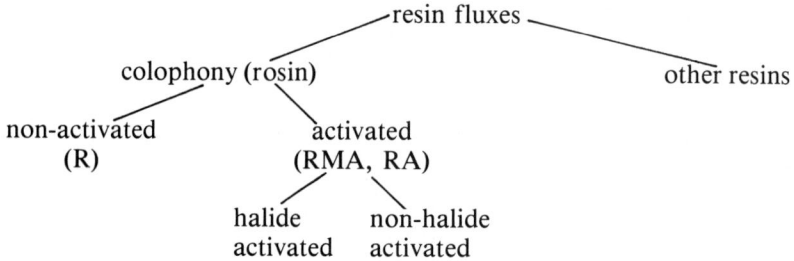

6.3.2 Water Soluble Fluxes

Because of the possible safety problems and restrictions in the use of solvent systems, there is a strong interest in the use of fluxes whose residues can be removed with water (aqueous cleaning) or water plus detergent ('semi-aqueous' cleaning). Many RMA fluxes can be converted to water soluble forms by 'saponifiers'. Also, detergents and semi-aqueous cleaning apparatus are available that effectively clean most RMA fluxes.

Water soluble fluxes usually have high fluxing activity and the residues are more corrosive than those of the resin fluxes, and must be removed after soldering. The residues are soluble in water, and it is usual to wash first in hot water, at 60°C, then cold water, followed by a final rinse in deionised water.

The range of constituents used in water-soluble fluxes is very wide, but they usually consist of four necessary components: a solvent such as water or alcohol to enable the viscosity to be modified and enable an even distribution of the active ingredients on the surfaces; a carrier such as a glycol or polymer that is water soluble and with a low evaporation to keep the active ingredients on the surface in semi-liquid form; a wetting agent to promote spreading; and finally the chemically active ingredient that removes oxides and greases.

Although the solvent can be water since all the chemicals are soluble in water, more usually it is an organic solvent. This is because water has a tendency to spatter and solvents with higher boiling points are preferable.

Unlike rosin fluxes which frequently remain on the soldered board and are not washed, water-soluble fluxes are always washed off. This means that a higher activity is acceptable in water-soluble fluxes, even though the cleaning may be relatively more difficult. The activator used may be either an organic salt, acid or amine, or an inorganic salt or acid. The organic salts are halide containing chemicals such as DMA HCl used as the activator in rosin-based fluxes. The organic acids used are weak, such as lactic acid or citric acid, and care must be taken that they do not decompose at too high temperatures. The third type of organic activators are amines and amides such as triethanol-amine or urea. The inorganic activators are based on salts such as zinc chloride and ammonium chloride or acids such as hydrochloric.

6.4 THE SOLDER ALLOY

Depending upon the metallisation and the solder alloy, different materials have varying solderabilities even when the surfaces are clear of any oxide or contaminants. The choice of solder alloy therefore requires a knowledge of the metallisation involved and of the temperature at which the joint can be made without giving rise to deleterious effects on the components.

Solderable metallisations are classified as fusible, soluble or insoluble depending upon the solder alloy and the soldering conditions. Fusible coatings are molten at the soldering temperature. Soluble coatings, although not molten, will react or dissolve rapidly in molten solder. This process is often termed leaching or scavenging and can result in the destructive stripping of plating on leads or thick film metallisations.

To counter leaching, the solder alloy can be partly saturated with the soluble metal to inhibit further dissolution from the metallisation. For example, 95 Sn:5 Ag and 62 Sn:36 Pb:2 Ag solders are used on silver-containing metallisations and 80 Au:20 Sn is used on gold coatings. A 50 Pb:50 In solder alloy is also used on gold coatings since both lead and indium are poor solvents for gold. The rate of dissolution of silver into molten 62 Sn:36 Pb:2 Ag solder is markedly less than that into 60 Sn:40 Pb, as shown in Figure 6.9.[218] The rate of dissolution dD/dt varies exponentially with the inverse temperature:

$$\frac{dD}{dt} = \text{const.} \exp\left(\frac{-E_D}{kT}\right) \quad (6.1)$$

where E_D (joules) is the activation energy for dissolution, k (J.K^{-1}) is Boltzmann's constant and T is the temperature in kelvin, and hence in Figure 6.9 the data are plotted in the form of an Arrhenius plot to give a straight line dependence.[219, 220]

It is also possible to make components with solderable metallisations that are more resistant to dissolution by tin, although low dissolution often goes hand-in-

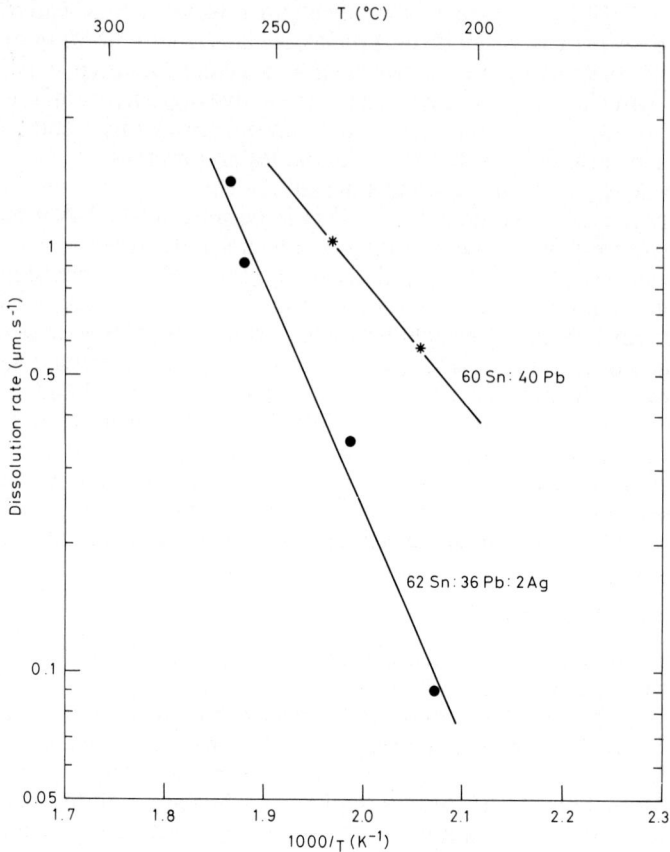

Fig. 6.9 The dissolution rate of silver in tin-lead solder is markedly reduced by the addition of silver to the alloy.[218]

hand with poor solderability and so the choice of metallisation becomes a compromise. Dissolution-resistant coatings used commonly are palladium-silver instead of silver and platinum-gold instead of gold. Figure 6.10 gives the dissolution rates of the relevant elements and shows that the dissolution rates of palladium and platinum in molten 60 Sn:40 Pb solder are some hundred times less than those for silver and gold.

Copper and nickel are used as coatings and are 'insoluble'. They are not, however, readily solderable using mildly activated fluxes and so are often used as intermediate barrier layers to prevent leaching from a palladium-silver metallisation to a tin-lead hot dipped or electroplated solderable coating.

6.4.1 Reflow Temperature

The temperature at which a solder alloy is liquid is a prime consideration when choosing the alloy composition. Processing restrictions, such as heat sensitivity of components and the differences of expansion between the parts, may set an upper

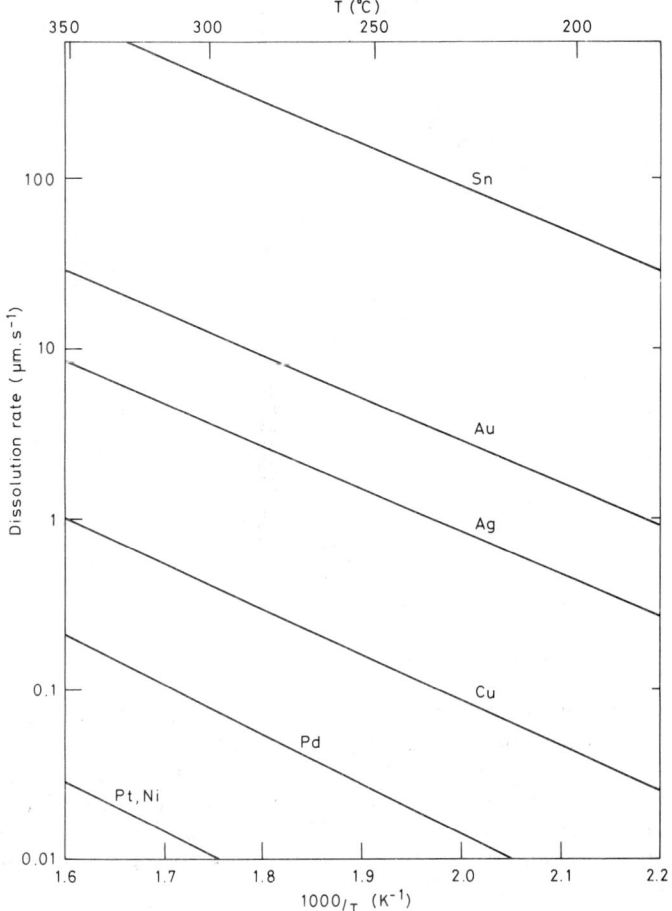

Fig. 6.10 The dissolution rates of various metals in liquid 60 Sn:40 Pb solder.[4]

limit on the temperature. The requirement for sequential soldering, that is to solder subsequently with a second alloy at a lower temperature, would set a lower limit on the temperature of the first solder used. In operation, electronic equipment often develops considerable heat and, if an alloy is used with too low a melting point, the solder joints may soften and break loose.

Solders composed of one metal melt at a single well defined temperature. In general, multi-component alloys melt over a range of temperatures. As the temperature is raised, they start melting at the solidus temperature (S) and complete melting at the liquidus temperature (L). In between these two temperatures the solder is in a 'pasty' state, being composed of a mixture of solid crystals of one composition and a liquid of another composition. This is illustrated by the phase diagram of the tin-lead binary alloy[221] shown in Figure 6.11. It can be seen that there is one composition (and this is true for all solder alloys) at which the solidus and the liquidus temperatures are the same. This point is called the eutectic. For the tin-lead system, the eutectic composition is 61·9% Sn by mass (74·0 Sn in atom%) and the eutectic temperature is 183°C.

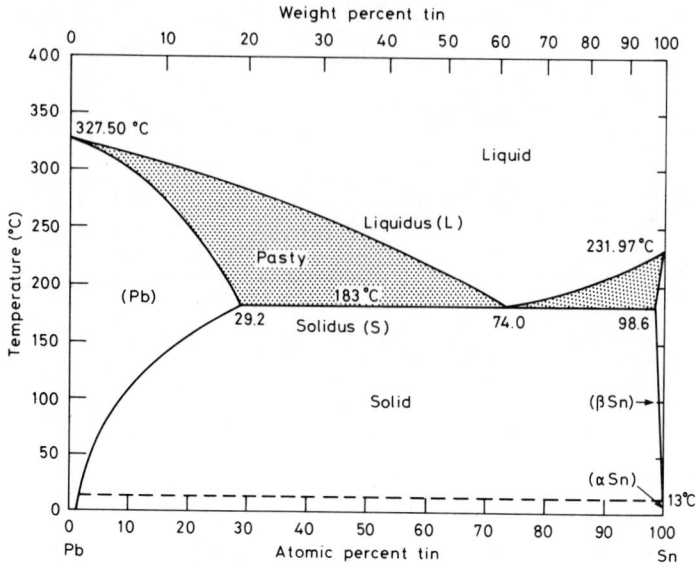

Fig. 6.11 The phase diagram of the tin-lead system.[221]

6.4.2 Physical Properties

Since the solder alloy becomes the filling material between the components and the board, its physical properties may be important. This aspect will be considered in Chapter 11, but in brief, two aspects of joint strength affect the choice of alloy to be used. These are the intrinsic strength of the solder itself and the effect on that strength of the intermetallic compound layer that exists at the joint interfaces as a result of dissolution and solid-state diffusion between the solder and the metallisation.

6.4.3 Metallurgical Systems

6.4.3.1 TIN-LEAD SYSTEMS

This group of pastes is used extensively because of user experience with 60 Sn:40 Pb and 63 Sn:37 Pb alloys in assembly methods prior to surface mounting. The lower tin systems, 50 Sn:50 Pb for example, have poorer wetting and spreading properties but the 10 Sn:90 Pb and 5 Sn:95 Pb alloys are used because they are inexpensive, for such high temperature compatible applications as solderable coatings on chip capacitors and solder stand-off bumps on components.

Tin-lead solders can crack under repeated thermal cycling when expansion mismatched components and substrates are used.

6.4.3.2 TIN-LEAD-SILVER SYSTEMS

These systems were primarily designed to be used with silver bearing materials such as palladium-silver thick film conductors or chip capacitor end terminations. In particular 62 Sn:36 Pb:2 Ag paste is now used extensively. These alloys exhibit good tensile and shear strength and the silver content improves the spreading properties of the solder. Some users are evaluating cheaper high-lead systems such as 5 Sn:93·5 Pb:1·5 Ag and 10 Sn:88 Pb:2 Ag with higher melting points suitable for sequential soldering. They do, however, tend to have short shelf lives.

6.4.3.3 TIN-SILVER SYSTEMS

Typically 95 Sn:5 Ag or the eutectic composition 96·5 Sn:3·5 Ag is used not only for higher melting temperature but for superior wetting and higher joint strength, compared with the tin-lead systems. Tin-silver and tin-antimony alloys are the most resistant to thermal fatigue.

6.4.3.4 TIN-ANTIMONY AND TIN-LEAD-ANTIMONY SYSTEMS

Antimony improves the tensile and shear strengths of solders and can improve the creep strength by as much as 30%. Antimony in small amounts is beneficial in tying up detrimental aluminium impurities as an intermetallic compound. It is sometimes incorporated into Sn:Pb solder at the rate of 1% Sb replacing 2% Sn, up to 5% Sb addition as a cost saving measure, without the wetting properties becoming unmanageable.

6.4.3.5 LEAD-INDIUM AND TIN-INDIUM SYSTEMS

The lead-indium systems are soft and ductile, and exhibit lower crack propagation than more common alloys, but their main use is in soldering gold plated parts since the dissolution of gold in both lead and indium is low. The tin-indium systems are chosen when a lower melting point is required (the 48 Sn:52 In alloy has a 118°C melting point) and they exhibit excellent wetting properties.

6.4.3.6 TIN-BISMUTH AND TIN-LEAD-BISMUTH SYSTEMS

Bismuth is incorporated into solder alloys for low temperature soldering of heat sensitive components.[222] Tin-bismuth alloys are able to combine high strength with low melting point.

Some examples of solder pastes suitable for surface mounting applications are as follows:

60 Sn:40 Pb 63 Sn:37 Pb	For standard attachment of components to printed circuits
10 Sn:90 Pb 5 Sn:95 Pb	For applying solder bumps under chip carriers to elevate them from the board. For attaching surface mounting components to the bottom of a board that is then wave soldered
62 Sn:36 Pb:2 Ag	For components with thick film silver based metallisations

| 10 Sn:88 Pb:2 Ag | High melting points makes them suitable for sequential |
| 5 Sn:93·5 Pb:1·5 Ag | soldering with 62 Sn:36 Pb:2 Ag paste |

| 60 In:40 Pb | |
| 50 In:50 Pb | To avoid dissolution of gold plating |

| 43 Sn:57 Bi | |
| 43 Sn:43 Pb:14 Bi | Soldering of heat sensitive components |

Solder pastes can be manufactured from almost any solder alloy composition. Because the metal in solder paste is not maintained in a molten state for more than a few seconds during reflow, and because of the intimate contact between the powder and the flux, the tendency for dross formation within solder pastes is minimal. This means that alloys which would not otherwise be considered for soldering applications can be readily used in this way. In fact, the variety of alloys that may be specified is an outstanding feature of solder pastes. However, for soldering surface mount assemblies, three compositions are used in quantities vastly greater than any others. These are 62 Sn:36 Pb:2 Ag, 63 Sn:37 Pb and 60 Sn:40 Pb, of which the first has, in recent years, been growing fastest in its popularity because of the widespread use of silver-based metallisations on surface mounting components.

In Table 6.2 are listed the melting points and relative measurements of shear strength of solder pastes used in surface mounting. The mechanical properties of a solder joint are very dependent on the materials being joined, the shape of the joint, the type of test and the speed of testing. These representative values of shear strength are all from one source,[223] between copper surfaces and tested at a strain rate of 1 mm.min^{-1}.

6.4.4 Metal Purity

When a solder paste is manufactured the main constituents must be within specified limits and any impurities must be below allowable levels.[224] Concerning the control of impurities in the solder, there are two advantages of soldering using a paste. First, the periodic composition checks are not required, as they are for the solder used in a wave soldering machine which is continually dissolving impurities, notably copper, from the boards. Secondly, the amount of solder available at a joint is small and static, virtually eliminating the capacity for the formation of dross.

Precise purity requirements should be ascertained from the relevant specification and compared with those of the supplier. Most national and international specifications are in close agreement concerning those elements that are included and the levels that are set. Typical levels are given in Table 6.3 for the three common solder paste alloys used in SMT.

The effects of low level traces of Al, As, Bi, Cd, Cu, P, S, Sb and Zn have been established[225] and elsewhere at slightly higher concentrations of Al, As, Bi, Cd, In and Zn.[226] Whilst Cu and Fe contaminations have marked effects on the viscosity, solderability and appearance of the solder, silver is beneficial in these respects. Aluminium is an oxide-promoting element and the effects of aluminium oxidation are seen as poor wetting and a gritty solder appearance.

Solder Pastes

Table 6.2

Alloy Systems for Solder Pastes

Alloy System (mass %)						Code	Melting Temperature (°C)		Shear Strength 1 mm.min^{-1} (N.mm^{-2})	
Sn	Pb	Ag	Sb	In	Bi		sol.	liq.	20°C	100°C
100						Sn	232		22·1	19·0
63	37					Sn63	183 *	183	-	-
60	40					Sn60	183	188	33·6	21·6
50	50					Sn50	183	216	30·0	24·0
40	60					Sn40	183	234	34·3	13·7
10	90						275	302	28·9	14·7
5	95						310	314		
62	36	2				Sn62	179 *	179	43·0	18·6
10	88	2					268	299	-	-
5	93½	1½					296	301	23·8	15·7
96½		3½				Ag3½	221 *	221	37·7	22·5
95			5			Sb5	236	243	37·2	21·1
	40			60		In60	174	185	-	-
	50			50		In50	180	209	-	-
37	37			25		In25	138 *	138	-	-
42					58	Bi58	139 *	139	50·0	19·5
15	33				52	Bi52	96 *	96	-	-
34	42				24	Bi24	100	146	34·3	17·5
43	43				14	Bi14	143	163	-	-

*nominally eutectic composition

Table 6.3

Solder Composition Specification (% by mass)

		Sn60	Sn63	Sn62
Tin (Sn)		59-61·5	62·5-63·5	61·5-62·5
Copper (Cu)		<0·08	<0·08	<0·08
Antimony (Sb)		<0·1	<0·1	<0·1
	*	0·2-0·5	0·2-0·5	0·2-0·5
Bismuth (Bi)		<0·25	<0·25	<0·25
Arsenic (As)		<0·03	<0·03	<0·03
Aluminium (Al)		<0·005	<0·005	<0·005
Zinc (Zn)		<0·005	<0·005	<0·005
Iron (Fe)		<0·02	<0·02	<0·02
Silver (Ag)		<0·015	<0·015	<1·75-2·25
Cadmium (Cd)		<0·001	<0·001	<0·001
Lead (Pb)		remainder	remainder	remainder
Phosphorus (P)	**	<0·005	<0·005	<0·005
Sulphur (S)	**	<0·0005	<0·0005	<0·0005

*Antimony containing alloys   **Not usually specified

High levels of antimony can also cause poor wetting, but small amounts are beneficial because antimony helps to overcome the detrimental effects of aluminium, by tying it up as an intermetallic compound. Alloys for solder paste can be specified with or without small (<0·5%) additions of antimony, which is required by some specifications but barred by others. The use of antimony additions is contentious, and the arguments for and against have recently been reviewed,[227] and outlined in Section 5.6.2.

Bismuth and arsenic[228] degrade wetting, although the effect on solder properties is insignificant at the levels encountered. Cadmium and zinc, like aluminium, are oxide formers and these are generally the most deleterious to the performance of the solder paste. Other elements, such as sulphur and phosphorus, not generally specified or tested for, are quite detrimental. Sulphur will impart to the solder a gritty appearance through formation of SnS and PbS when present at a level of only a few parts per million. The sulphur content is especially important if the metallisation is silver, as sulphur reacts with silver producing dendritic 'whiskers' that can lead to shorts between conductors. Phosphorus induces dewetting of solder at very low concentrations (0·01%). Carbon content in solder paste has also been shown to correlate with poor solder fillet quality.

If periodic checks are to be made on the purity of the metal content of a solder paste, the atomic absorption method is recommended.

6.5 APPLICATION OF THE PASTE

Application of the solder paste in the proper amount and in the right place is extremely important to produce high yields in surface mount assemblies. Solder pastes used in surface mounting can be applied in one of several ways. The most common method for volume production is screen printing but other methods such as stencilling, dispensing from a syringe, roller coating and dotting are used as the situation warrants.[229]

6.5.1 Screen Printing

Screen printers are available in many sizes to accommodate single printed circuit boards or large arrays.[230] They are available to be operated manually for small batches or entirely automatically for large runs. Screen printers are commonly available for single-sided boards, but modifications can be made to allow the screening of paste on to both sides of double-sided boards. During screen printing the solder paste is forced through the exposed mesh of a patterned printing screen. The screen consists of a frame holding a mesh on which is deposited an emulsion. The emulsion is the part that provides the actual printing pattern to the screen. The pattern is made by exposing the emulsion to ultra-violet light through a film artwork which blocks the light from the emulsion and prevents it from curing. The uncured emulsion is then rinsed off to produce the required screen pattern. Polyvinyl alcohol, polyvinyl acetate or other suitable materials are used for the emulsion which has been deposited by the supplier to a required thickness, appropriate to the thickness of paste required.

The screen mesh which supports the emulsion is a structure of woven fibres, of a size and spacing to allow the solder paste to readily pass through. The most commonly used screen mesh materials are monofilament nylon, monofilament polyester and stainless steel. For fine-line deposition the mesh is aligned with its axes at 45° to the PCB axes so that the shadowing effect of the mesh filaments is minimised and maximum line definition is available. The screen mesh count refers to the number of openings or lines per linear inch (l.p.i.); for screen printing solder pastes the mesh count is usually in the range 60-200. If the solder particles are larger than the openings in the mesh they will obviously not pass through. As a general rule-of-thumb, for the free passage of the paste through the screen, the maximum particle size should be no larger than one third the mesh opening to ensure that the screen does not become jammed. For example, an 80 mesh screen has openings of about 224 μm. Therefore, the particle size should not exceed 75 μm, which corresponds to a -200 mesh powder size (Table 6.1). The parameters of some commonly used screens are given in Table 6.4 and defined in Figure 6.12.

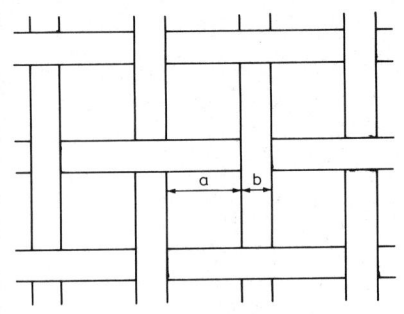

Mesh thickness $h_m \approx 2b$

Mesh opening = a
Filament diameter = b
Open area $A_o = a^2/(a+b)^2$

Fig. 6.12 Definition of parameters of a screen for printing solder paste.

Table 6.4

Mesh Count (l.p.i.)	Nominal Wire Diameter (b) (μm)	Mesh Opening (a) (μm)	Open Area (%)	Nominal Mesh Thickness (μm)
60	114	310	53·5	235
80	94	224	49·5	215
105	76	165	47	175
165	51	104	45	115
180	46	97	46	100
200	41	86	46	90

The thickness of the solder deposit that is obtained on the printed circuit board is largely determined by the emulsion thickness and the screen mesh thickness, although it also depends to some extent on the pattern dimensions, the printer parameters and the solids content and rheology of the paste. Some typical empirical deposition data for a series of mesh sizes are given in Figure 6.13. In general, finer mesh screens are used for thinner deposits. A 180 mesh and a fine powder paste could be used to produce a deposit with a thickness of 100-150 μm, but the most commonly applied thickness is 150-250 μm, for which the usual combination is an 80 mesh, −325, +200 particle size paste with 90% solids content.

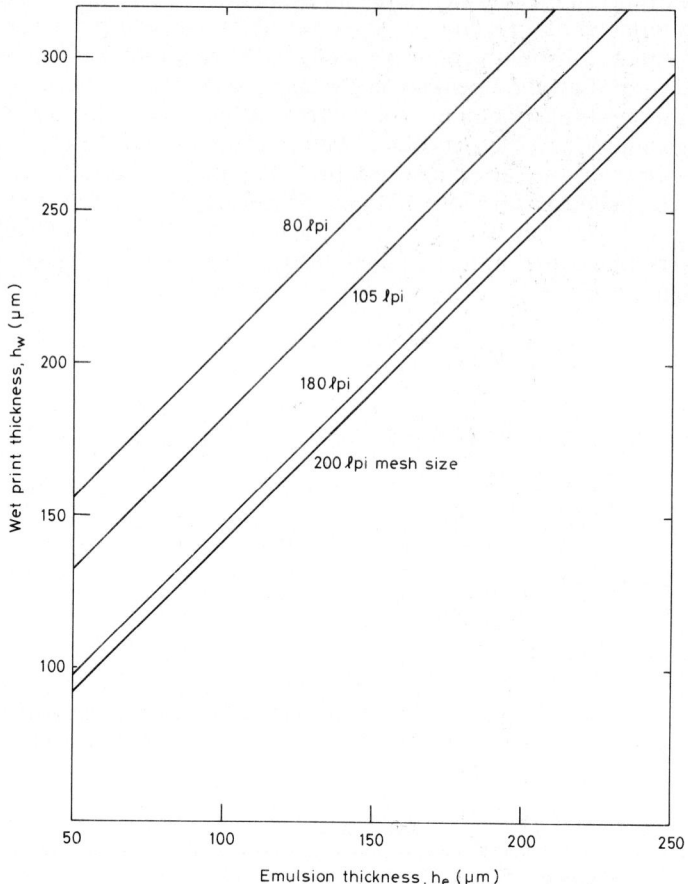

Fig. 6.13 Empirical data on solder paste deposits printed using various mesh sizes.

As an approximation, the wet print thickness h_w is given by

$$h_w = (h_m \times A_o) + h_e \qquad (6.2)$$

where h_m is the mesh thickness, h_e is the emulsion thickness and A_o is the fractional open area.

In some applications, the areas of the open pattern in the emulsion may require an enlarging adjustment to allow for the mesh's blocking the flow of paste to the board. However, increasing the open area also increases the potential for the paste to bridge as it dries. The occurrence of bridging also increases as the emulsion thickness is increased. It is therefore not usual to exceed an emulsion thickness of 250 μm, and an emulsion thickness in the range 100-250 μm is customary for most surface mount situations. With the development of new emulsions it is now possible to produce emulsion screens up to 1000 μm thick, suitable for printing solder pastes, if the edge definition is not critical.

The difference between the thickness of the screened wet layer and the dried layer is not large because, as the solvent evaporates, the metal powder particles soon make contact with each other, preventing significant slump. However, the metal content of the paste may be as low as 30% by volume and so during melting and reflow there is a substantial change in the shape and size of the deposit. If a solder paste deposit is screen printed as a cylindrical dot on to a circular pad, upon reflow it would form a spherical cap whose volume V is given by

$$V = \frac{1}{6}\pi h_r (h_r^2 + 3r^2) \qquad (6.2)$$

where h_r is its height and r the radius of the cap. Provided r is small, h_r is almost independent of its value. A guide to the relationship between the reflowed solder thickness, h_r, and the applied wet print thickness, h_w, is given in Figure 6.14 for a range of solids contents. Also shown is the relationship between h_r and h_w when a track of solder paste, rather than a circular pad, is reflowed. It should be noted that, even though there is always a volume reduction upon reflow, in some

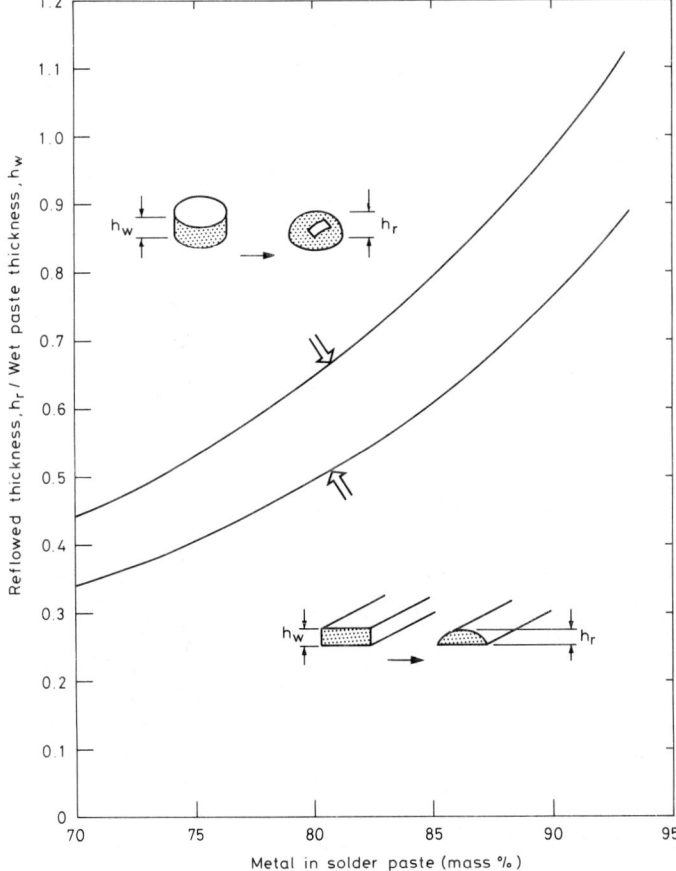

Fig. 6.14 The calculated relationship between the thickness of printed wet paste, h_w, and the thickness of the reflowed deposit, h_r.

instances the maximum height of the reflowed deposit, h_r, is greater than the printed thickness, h_w.

For some applications the screen can be prepared by not only removing the pattern emulsion, but also etching away the mesh. This 'etched mesh' screen then acts like a stencil, but with the flexibility of a screen. An etched mesh is used in two extreme situations:

(i) for very wide conductor lines or large solder pads where more than 250 μm thickness is required;
(ii) for very fine lines, less than 250 μm wide, for which the mesh wires represent a significant proportion of the print area.

The two vital parts of a screen printer are the screen and the squeegee. The remainder of the machine is simply to hold and control these two parts. The squeegee is a straight rubber bar, used to wipe the solder paste along the surface of the screen and force it through the screen's open pattern area. The basic mechanism of the screen printer is demonstrated in Figure 6.15. The board is aligned beneath the screen, solder paste is spread on the screen either by hand or automatically, and the squeegee is pushed across the top of the screen. During this operation, the screen must stretch and then snap back to its equilibrium position. The distance between the board surface and the screen when mounted in the printer is called the 'snap-off'. If this setting is too small or too large, the print may well smear. The proper snap-off setting depends on the size of the board and is nominally specified by the printer manufacturer, but can be optimised by small trial and error adjustments to suit the particular mesh and paste combination used.

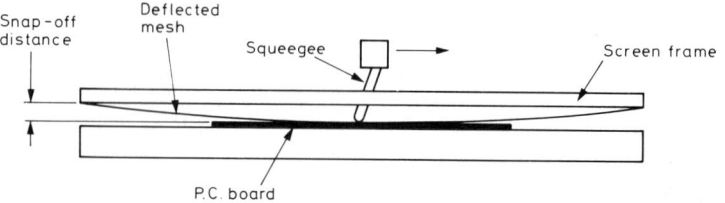

Fig. 6.15 The basic components of a screen printer.

The hardness of the squeegee and its angle of attack at the screen also affect the results of the paste printing. If the squeegee is too hard it will not take up the shape of the paste-covered screen, as it stretches and deforms. Squeegees of several hardness values are available and the choice depends mainly on the size of the printer and should have been optimised by the manufacturer. Likewise the angle of attack is adjustable, and depends on the size of the printer but also on the pressure exerted by the squeegee on to the screen, which must be optimised for a particular combination of solder powder size, screen mesh size, paste viscosity and print thickness. The edge of the squeegee is subject to wear, specially when using steel screens; the contacting edge should be restored as necessary.

At the end of a printing run, excess solder paste must be thoroughly cleaned from the screen and the squeegee. The best cleaning method is in an ultrasonic bath, using a chloro-ethyl solvent. Whichever solvent is used it must in no way affect the emulsion on the screen.

6.5.2 Stencil Application

For thicker paste deposits, metal masks (stencils) instead of mesh screens are used. This was the method used originally to deposit a pattern of solder paste. The stencil is usually made from a sheet of beryllium-copper, brass, nickel or stainless steel, and the pattern is chemically milled or etched with the desired solder pattern.[231] The etched stencil is then placed over the substrate board and solder paste forced through the openings using a wiper, in a similar manner to the squeegee used in screen printing.

After deposition, the stencil has to be removed very carefully to prevent smearing at the edges of the paste deposits. For more uniform printing the metal sheet can be mounted on a mesh screen which provides some 'snap-off' as used in screen printing. However, in screen printing when the emulsion is pressed against the printed circuit board, a seal is formed around the open area into which the paste is deposited. Metal stencils lack the ability to deflect and seal in this manner. For this reason, pastes made from irregularly shaped solder particles are preferred for use with stencils, to increase the viscosity and reduce the pull-away of the paste when the mask is removed. Because of these difficulties, the use of metal stencils is normally relegated to deposition of non-critical, heavy concentrations of solder paste.

Metal stencils tend to have a longer useful life than mesh screens, which can lose their resilience and shape after much use. However, mesh screens are less expensive and more quickly made than stencils.

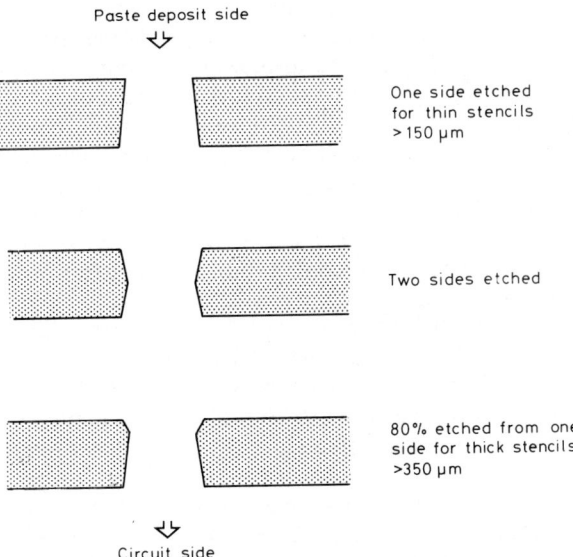

Fig. 6.16 When producing a metal stencil there is some undercutting of the pattern and, for thicker stencils, it is necessary to etch from both sides.

During etching of the stencils, which is performed through an etch resist printed on to the metal mask, some undercutting of the sheet occurs as shown in Figure 6.16. For thicknesses greater than about 150 μm, etching is usually done from both sides, which can give rise to pattern alignment problems. In order to obtain a satisfactory etched profile of the holes through a thick mask, it is necessary to etch more from the underside of the mask than from the top surface of the mask, as shown in Figure 6.16. For stencils greater than about 350 μm, an 80:20 ratio is used for the bottom:top etching depths.

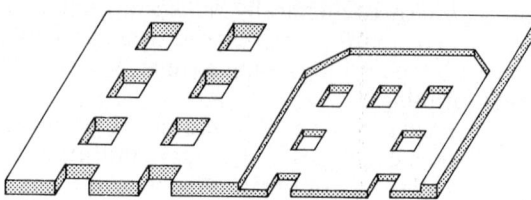

Fig. 6.17 A metal stencil can be differentially etched to enable different thicknesses of paste to be printed simultaneously.

In order to print different thicknesses of paste for different types of component, a metal stencil can be differentially etched as shown in Figure 6.17. In this way each solder pad can be customised for the specific component, but the solder still applied with a single pass of the squeegee. Normally only three levels of recessing of the mask are adequate to meet requirements for different deposited thicknesses.

6.5.3 Syringe Dispensing

For certain applications, particularly where the substrate surface is uneven or populated very sparsely with components, the solder paste is best deposited at the required locations by dispensing through a syringe. These are cartridges of solder paste fitted with sliding pistons. The dispensing is controlled either manually or pneumatically and is usually metered so that the amount of paste can be precisely specified. Hand syringes are normally used only for circuit repair. It is possible to gang many dispensing nozzles so that an array of solder paste dots is deposited.[232] Such a system usually forms an integral part of an automatic component pick-and-place assembler. In that way the components are mounted on the board immediately after deposition of the paste by the same machine, so that any difficulties of computer software compatibility or registration of different X-Y movement tables are avoided.

The solder pastes formulated for syringe dispensing have different thixotropic agents, to ensure that flow occurs only when pressure is applied and that the deposited dot has a sharp cut-off with no stringing of the paste from point to point. The solder particle size needs to be more tightly controlled and of a finer size distribution than is normally necessary for screen printing, in order to avoid clogging of the syringe tube.

It is usual for automatic dispensing machines to require syringes that are pre-loaded by the solder paste manufacturer. These are usually made of rigid polyethylene or polypropylene and will not bow or bulge, so that, at normal working pressures, the piston is both air-tight and paste-tight. Pre-loaded syringes are most satisfactory, but, if self-loading is required, great care must be taken to exclude any trapped air which will cause misses to occur in the dispensing and also cause the paste to exude after the pneumatic pulse is off. Some machines are designed to exert a slight negative pressure, or pull-back, on the piston after each pulse, to give a clean break to the flow of paste.

Besides adjustment of the air pressure (usually in the range 10-40 psi, 70-300 kPa) and the length of the pneumatic pressure pulse, the size of the paste deposit is determined by the diameter of the orifice of the needle that is attached to the syringe. Using pastes suitable for syringe application, the deposited dot diameter will be about 1·5 times the needle internal diameter. Suitable needles are in the range 250 μm to 1·6 mm inside diameter, the most popular standard sizes being 510 μm and 840 μm inside diameter (1/50 and 1/30 inch respectively). To ensure repeatability, the selection of the needle should be such that it is as short as possible (about 5 mm) and that there are no sudden changes in the internal cross-section between the syringe and the needle. This latter criterion is to ensure that there are no entrapments of the paste which will gradually separate leaving compacted solids.

6.5.4 Pin Transfer

Systems have been developed for solder dot placement using multiple pin transfer. The pin array is first immersed to a controlled depth in a level trough of solder paste and then touched down on the pads of the circuit board. The size and design of the pins and the properties of the solder paste determine the quantity of paste transferred. Different quantities can be placed on different pads by using different size pins.

Usually the pins are spring-loaded and are often tulip shaped to hold the paste. The paste selected for this placement method must be sufficiently tacky to stick to pins during transfer, but also be fluid enough to rapidly refill the gaps in the trough left by the pins. The trough can be designed to move slightly between each pin pick-up, in order to reduce this latter problem.

6.6 PASTE RHEOLOGY

6.6.1 Viscosity

Solder pastes are thixotropic; the viscosity decreases as the shear stress increases but slowly recovers its original value when the shear is removed. Such thixotropic consistency is important for pastes so as to prevent them from slumping or moving after application.

The viscosity of the solder paste depends on the solids content of the flux, on the quantity, size and shape of the solder particles and on the thickeners that are added to make the paste thixotropic.[233] The selection of viscosity depends on the method of application of the paste, and recommended values are given in Table 6.5.

Viscosity is a very temperature-sensitive property and at higher temperatures it decreases markedly, as shown typically in Figure 6.18. It is therefore recommended to apply the paste in a temperature controlled room. The viscosity of a paste can increase substantially during storage use. It should therefore be measured and appropriate thinner added during the mixing, just prior to application.

Table 6.5
Recommended 25°C Viscosities of Solder Pastes

Application Method	Metal Content (% mass)	Viscosity (Pa.s)*
Stencil print	90	600-1000
Screen print	85	400-700
Syringe dispenser	80	200-450
Pin transfer	75	50-250

*1 centipoise = 0·001 Pa.s

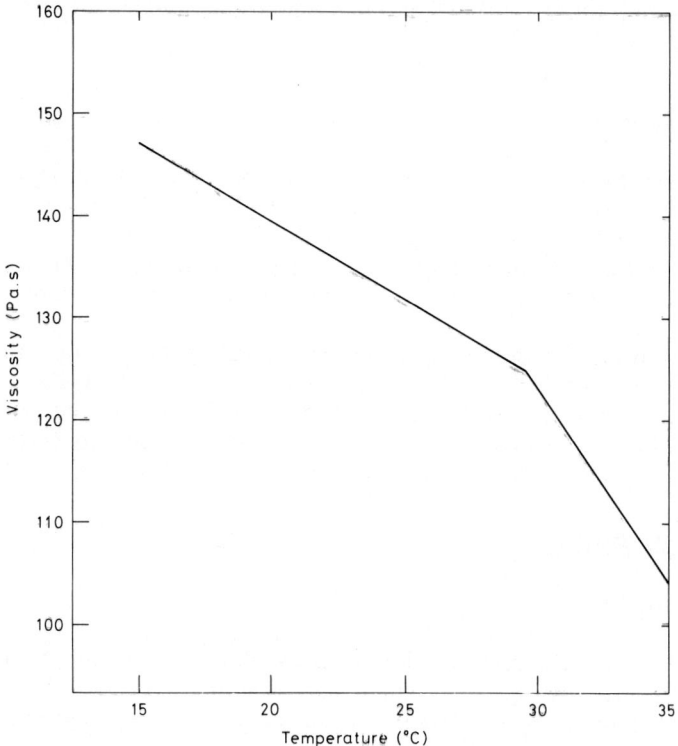

Fig. 6.18 The temperature-viscosity relationship for a typical paste. The break in the curve is caused by a phase change in the flux binder.

The lower the metal content, the more dependent the solder paste is on the thickening agents for its viscosity control, particularly when relatively high viscosities are required.[234] However, the ability of such thickeners to control viscosity rapidly decreases as the temperature rises. Thus, the lower the metal content, then the greater the rate of spread of the solder paste during the drying out heat cycle. For high definition, therefore, both high viscosity and high metal content are required.[235]

6.6.2 Measurement of Paste Viscosity

Viscosity is a very important parameter of the solder paste as it is essential that the consistency is correct for the method of application. The best available method of measurement for the appropriate range of viscosities is believed to be the rotating concentric cylinder method.

Newton's definition of viscosity, that the ratio of tangential stress to velocity gradient is a constant, holds for liquids that are homogeneous in the sense that the liquid is present only in one phase. Such liquids are called Newtonian liquids. Solder pastes are representative of non-Newtonian liquids, i.e., liquids containing discrete particles of another phase dispersed through the continuous liquid phase. During the flow of a non-Newtonian liquid, its viscosity decreases as the velocity gradient increases. In such cases the viscosity coefficient obtained from measurements made assuming that the viscosity coefficient is constant do not give the true viscosity but an apparent viscosity. This is equal to the viscosity of a Newtonian liquid which would give the same flow rate under identical conditions in the particular apparatus used.

Viscosity is the resistance to deformation of a liquid. The force F required to maintain a constant velocity difference dv between two layers in the liquid of area A and distance dx apart is given by

$$F = \eta A \frac{dv}{dx} \qquad (6.4)$$

In this equation η (Pa.s) is called the coefficient of dynamic viscosity. Its reciprocal is called the fluidity.

To measure the apparent coefficient of viscosity of a solder paste, by the concentric cylinder method, the paste is placed in the annular gap between two coaxial cylinders, one rotating at a constant speed and the other stationary. In this configuration the liquid, here the solder paste, assumes a rotary motion which it tries to communicate to the stationary cylinder. A torque is therefore required to keep this cylinder at rest. From the measured torque, the geometrical dimensions and the rotational velocity, the shear rate existing in the paste can be calculated.

With reference to Figure 6.19(a), if p is the radius of the inner cylinder, q the radius of the outer cylinder and ℓ the length of the inner cylinder, then the torque developed is[236]

$$M = 4\pi\ell \frac{p^2 q^2}{q^2 - p^2} \eta\Omega \quad \text{N.m} \qquad (6.5)$$

where the dimensions of η are Pa.s and Ω is the rotational velocity in radians.s^{-1}.

Equation (6.5) holds strictly only if ℓ is infinitely large. In practice the lower end of the moving cylinder is exposed to various forces and an end correction is necessary. This can either be eliminated physically by protecting the end with a guard ring or by using different lengths of the inner cylinder on which the torque is measured.

Commercial viscometers of this type are available that have been calibrated so that η may be found directly from a reading of the torque M. A range of angular velocities, from 0·1 to 100 radians.s^{-1} is convenient to improve the precision of measurement and also to cover the full range of solder paste viscosities encountered. The viscometer should, in any case, be calibrated from time to time by measuring the viscosity of standard (Newtonian) oils of well-defined viscosities. If a range of angular velocities is available the viscosity of the solder paste should be measured over as wide a range as possible, both as the angular velocity Ω is increased and as it is decreased. Because the solder paste is not a Newtonian liquid, η varies with changing Ω and it is therefore necessary to present the whole curve or specify the angular velocity at which the viscosity value pertains.

When filling the viscometer, it is imperative that no air bubbles are included in the paste, and that the sample be kept at a temperature that is constant to at least 0·2°C. The normally specified temperature for the measurement is 23 or 25°C.

A convenient commercial instrument is available for which the solder paste can be kept in its own container and a spindle immersed to a known depth shown systematically in Figure 6.19(b). The spindle, with a paddle attached, executes a number of axial rotations on a helical path, so that the immersed depth (ℓ) is changing, either increasing or decreasing. In this way, the indeterminacy of the outer cylinder size, q, can be compensated for. Two spindle types rotating at 5 rpm ($\Omega = 0·52$ radians.s^{-1}) cover the viscosity range 50 to 1400 Pa.s.

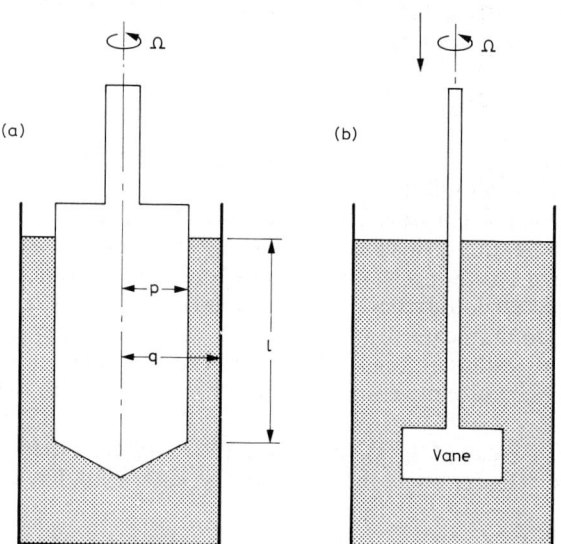

Fig. 6.19 Schematic configurations of viscometers suitable for solder paste: (a) concentric rotating cylinder and (b) rotating helipath paddle.

An entirely different approach for comparing viscosities of solder pastes is the nozzle flow method in which the paste is dispensed at different defined pressures on to a glass slide. The mass of paste deposited is an arbitrary but relevant measure of viscosity. This measurement is appropriate to pastes intended for syringe application rather than screen printing. A typical set-up is illustrated in Figure 6.20. The glass slide is weighed. The dosing needle length and internal diameter are defined (15 mm and 840 μm or 510 μm respectively being standard) and pressure is exerted on the piston for 10 seconds at 1, 2, 3, 4 ... bar. The slide is then reweighed.

6.6.3 Screenability

A paste that is designed to be screen printed can be said to have good screenability if the proper wet height, coverage and edge resolution are maintained on the printed substrate. Screenability of a paste includes such subjective evaluations as how easy it is to screen print and how well the solder paste spreads.

Although in many ways a qualitative assessment of the solder paste, attempts have been made to define the properties that ensure good screenability. The shearing rate of the paste can be calculated from the squeegee speed and the properties of the screen and its emulsion. Knowing the shear rate, the shear stress can be determined by using a viscometer. For the paste to screen well, the shear stress needs to be less than 3 N.cm^{-2}.

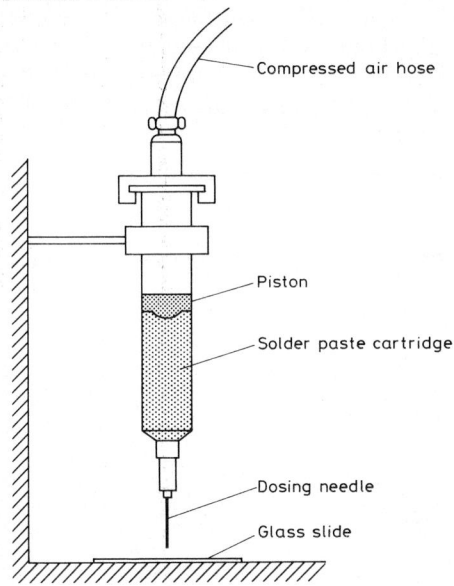

Fig. 6.20 The viscosities of syringeable pastes can be compared using a well defined and controlled syringe, measuring the weight of paste dispensed as a function of pressure and temperature.

6.6.4 Slump

After the solder paste has been screen printed or syringe dispensed, because it is a fluid, it tends to slump and spread over a period of time. The degree of slump depends on the viscosity of the paste but also on the volatility of the binders which governs the speed with which the paste begins to dry. The height of the paste deposit and the temperature are critical parameters.

Whilst there is no quantifiable measure of slump, it is useful to have a comparative standard test that can be used to rank different pastes. A suitable test involves the determination of spread of a well defined deposit of the paste on a flat ceramic substrate, under specified conditions.

The test requires the application of a dot of solder paste, about 300 μm thick and 5·5 mm diameter. The choice of this volume of paste dispensed arises from the requirement of simplicity in carrying out the test. Most standard self adhesive paper labels have a thickness of 100±10 μm, and a standard paper-punch produces holes that are 5·5±0·1 mm diameter. Therefore, three labels can be stuck one on top of another and several holes punched through using a paper-punch. This is then stuck to a ceramic substrate to provide a series of circular wells, 5·5 mm diameter and about 300 μm deep. An excess of paste is applied and wiped with a flat metallic squeegee before carefully removing the paper mask.

The dot diameters are monitored during the first few minutes using a travelling microscope, and then during storage at 23±2°C, 60±20% RH at 15 minute intervals. It is also informative to carry out the test at elevated temperatures, say 80°C and 150°C, the former temperature to test for slump during the paste drying procedure and the latter to test for slump as the workpiece is heated towards the reflow temperature.

By changing the number of sticky labels, the height of the applied dot can be adjusted in 100 μm increments. The significance of the height of the deposit on the slump is illustrated in Figure 6.21; measurements for one particular paste on alumina during heating at 150°C for 2 minutes. The 500 μm height deposit slumps to an extent that results in an increase of one third in its diameter—almost twice the printed area.

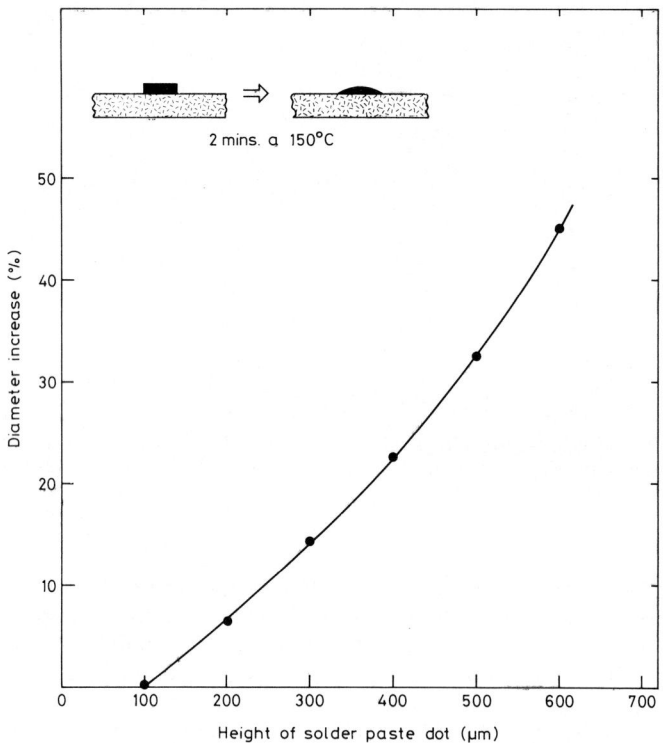

Fig. 6.21 The significance of the height of a printed solder paste on its slumping characteristics.

6.6.5 Tackiness

After screening or deposition of the solder paste on a PCB, the board is transferred to a placement machine where it is loaded with components. The paste, at that time, must be adequately tacky to ensure that the components are retained in their placement positions both while the remainder of the components are placed and during the handling of the board before reflow. On some placement machines the placed components are subjected to considerable acceleration and deceleration forces as the board is aligned under the placement head for subsequent components. The tackiness and the retention of tackiness after deposition are therefore important characteristics of a paste.

As already mentioned, the lower the solids content the longer the paste will remain tacky and accept components. Thus the retention of tackiness must be a compromise with the limit of slump specified.

Several tests have been suggested for ranking the tackiness of solder pastes after deposition, in the context of the adhesion and retention of placed components.[237] These tests involve the measurement of the force necessary to break the adhesive bond between the paste and a solid test piece. The way in which this measurement is carried out has not been standardised. One method is to use a 1·6 mm diameter stainless steel cylindrical probe, the bottom end of which must be square to its axis and polished flat. The paste is applied, possibly in the same manner as for the slump test, 200 μm thick, on to a copper substrate, and then stored at 25°C, 40% RH for periods 0-48 hours. The end of the probe is then brought into contact with the centre of the deposit with a pressure of 15·0±0·5 g.mm^{-2}, similar to that experienced by components when being placed. After five seconds the probe is withdrawn from the deposit at a well defined speed, and the maximum force recorded.

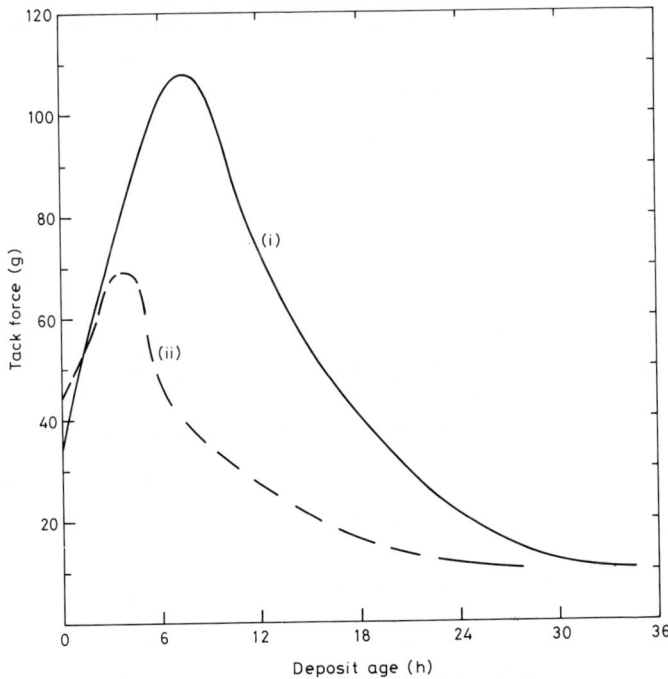

Fig. 6.22 The change in tackiness of two solder paste deposits during the time after deposition.

The measured tack strength increases over the period of some hours after the paste has been deposited as shown for some common solder pastes in Figure 6.22. It is clear from these data that components should be placed on the same working day that the paste is deposited, if at all possible.

6.7 DRYING THE SOLDER PASTE

The solder paste is applied to the board in a condition that is sufficiently wet to be printable or syringeable with good definition and then to allow the components to stick to it. This condition of wetness is required to be maintained for as long a time as possible in order to accommodate production schedules. This is achieved by selecting solvents with a sufficiently low vapour pressure at room temperature, which implies a solvent with a relatively high boiling point.

Pastes with high boiling point solvents, by definition, take a long time to dry out at room temperature and very careful accelerated drying out of the paste is required after placing the components and before reflow. If drying out is omitted there is a great danger of three types of problems occurring:

(i) components will be blown away or moved from their rightful positions by too rapid an evaporation of the solvent during the reflow process;
(ii) solder voids will form within the fillet, arising from volatile constituents of the paste not escaping from the liquid solder during reflow, before solidification occurs;
(iii) if the paste is heated rapidly without predrying, some individual solder particles or clumps of particles may be carried away from the main mass of paste and not then coalesce into the fillet when melting occurs, giving rise to the solder balling problem.

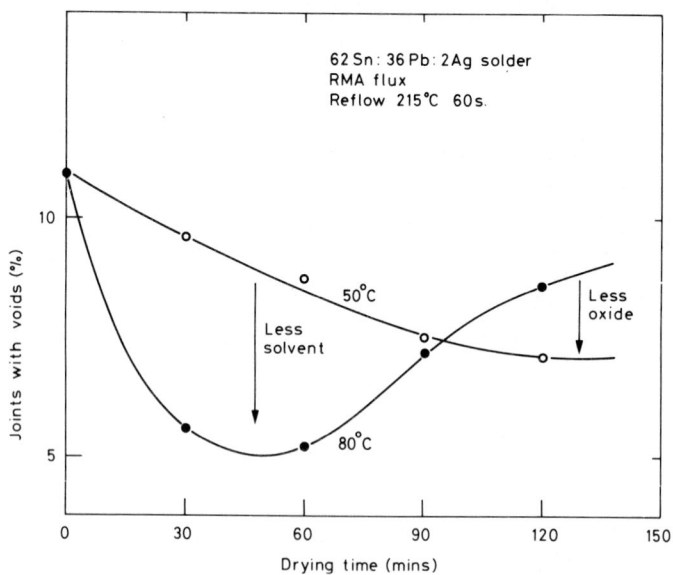

Fig. 6.23 The variation in the occurrence of solder voids as a function of the pre-bake time and temperature. The curves result from a combination of the kinetics of volatilistion of the solvent and the kinetics of oxidation of the solder particles.

The dependence of voids in the solder fillet upon the drying procedure is demonstrated in Figure 6.23 in which the degree of voiding during reflow is plotted as a function of the drying time of the paste at both 50°C and 80°C, for one particular solder paste. It can be seen that, for this paste, the solvent is lost by evaporation after about 45 minutes at 80°C, but has not been fully lost after 2 hours at 50°C. However, since the drying takes place in air, too long a time is as detrimental as too short a time, since the solder particles begin to oxidise to a significant degree and then, during reflow, the performance of the solder paste is severely hindered, giving rise to both voids and solder balls.

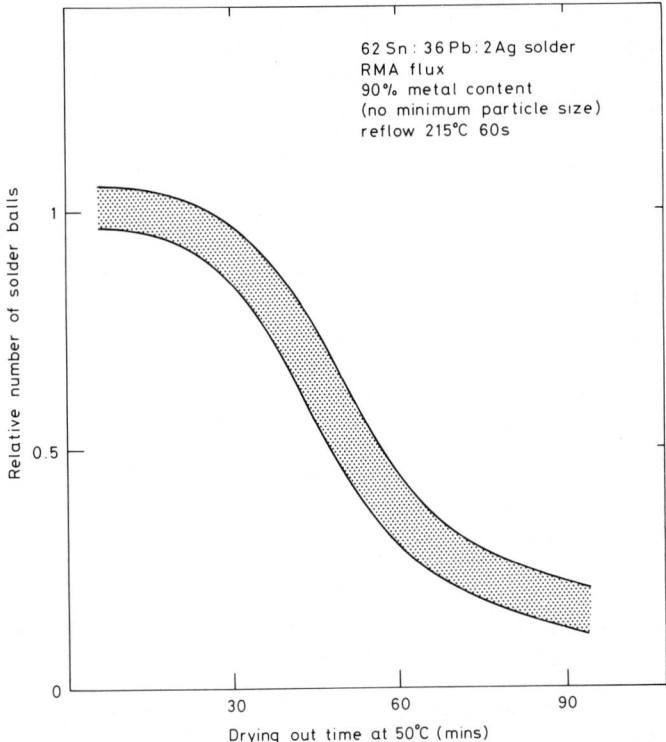

Fig. 6.24 The solder balling problem is exacerbated if the applied paste is not adequately dried by pre-baking.

The effect on the solder ball problem of drying the paste is shown in Figure 6.24 in which one particular paste is dried at 50°C for a range of times.

The necessary amount of drying depends on the paste used, the shape of the joint and the parameters of the soldering procedure that follows, such as the method of heat input and the rate of heating. It is clear that for laser soldering and vapour phase soldering, for example, the heating rate is extremely high and the presence of any volatiles could be catastrophic, whereas, using infra-red heating for the reflow, the drying out could be incorporated within the initial stages of the heating profile.

The drying out of the solder paste is almost always performed in air, at a temperature from 50°C up to 170°C. Manufacturers' guideline times are in the ranges 1-2 hours at 50°C, 30-60 minutes at 70°C, 5-20 minutes at 90°C down to 10 seconds at 170°C.

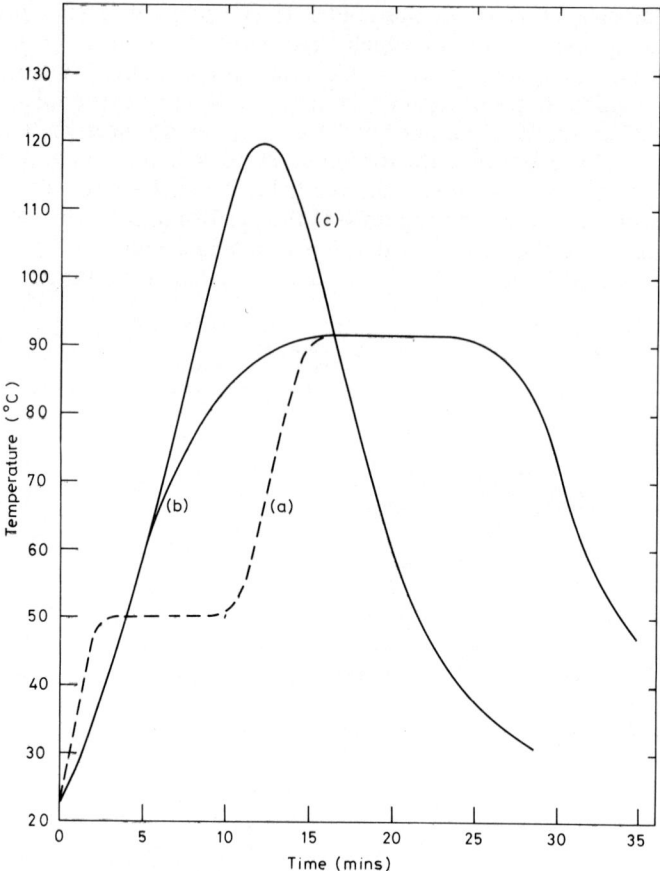

Fig. 6.25 Curves illustrating typical drying cycles for normal solder pastes. (a) step heating (b) single temperature heating and (c) flash heating.

The best retention of definition of the deposited solder paste is achieved by using a step heating cycle as shown in Figure 6.25 (a). Normally, however, a straight heating-up temperature profile is used, as in (b). If some flux spread and solder powder spread is acceptable, a flash drying profile can be used as in (c), but the time saving is not normally significant.

6.7.1 Storage after Drying

The drying of the paste usually takes place directly after populating with components, but reflow soldering can then be left for a period of several days. This time can be extended to periods in excess of a year, but an application of a liquid or paste flux on the board may then be necessary to ensure satisfactory reflow of the solder.

If a period of more than a few hours is envisaged between the drying of the solder paste and its reflow, some care should be taken to consider the storage environment,[238] in particular the relative humidity. Figure 6.26 demonstrates a marked deterioration in the reflow properties of the solder paste, manifested as an increase in the solder ball problem, if the dried solder pastes are kept in a humid environment. Parts kept in a relatively dry atmosphere showed only minimal solder balling whilst parts kept in a humid atmosphere deteriorated rapidly and continued to do so.[239]

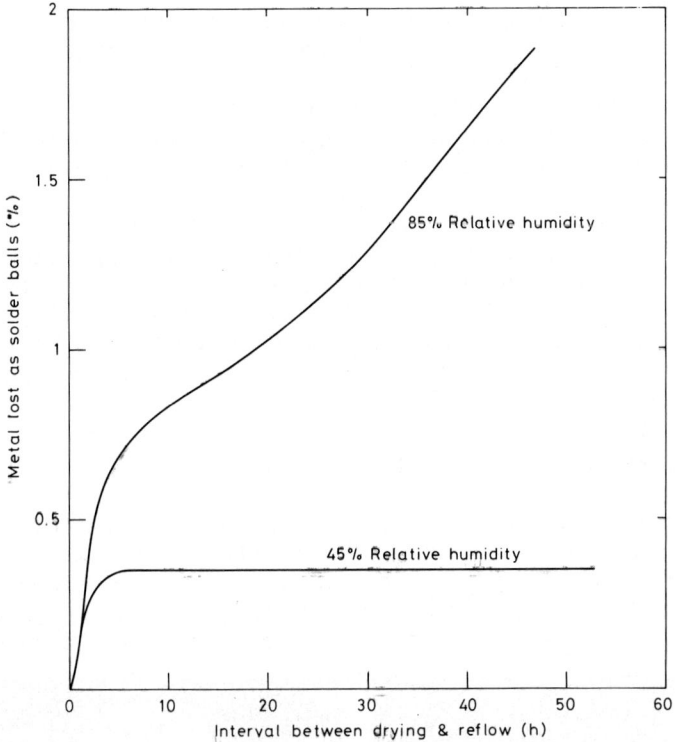

Fig. 6.26 The effect of storing dried solder paste in dry and humid environments.[240]

6.8 HANDLING AND STORAGE OF SOLDER PASTES

The use of pre-mixed solder pastes is both easy and convenient. However, the metal powder and the flux binder can be purchased separately and mixed in-house. The powder is shipped in air-tight argon filled packages to protect it from oxidation and contamination. The advantage of mixing solder paste in-house is that just enough paste for immediate use need be produced, although for consistent results it is really necessary to use all the powder from one package in one mix. For both pre-mixed and in-house mixed pastes, any unused paste must be stored in small tightly sealed containers.

Solder paste reacts with the atmosphere, the degree of reaction increasing more or less linearly with time and exponentially with temperature. The flux binder absorbs moisture and reacts with oxygen. The individual solder particles, although submerged in the flux binder, increase their oxide surface layer. To minimise oxidation, only that paste to be used forthwith should be opened and the proper recommended storage condition followed for unopened containers. The shelf life is increased considerably by storing at 4°C in a refrigerator. Before refrigerated solder paste is opened it must be allowed to warm to room temperature. If the container is opened too soon, atmospheric moisture condenses on the paste and this severely degrades its performance.

Besides oxidation during storage, the solder particles and the flux binder have a tendency to separate to a greater or lesser extent depending on the type and source of the paste. A small degree of separation is normal and can be remedied by stirring the paste in its container. A large amount of separation is indicative of a paste that is poorly dispersed during manufacture.

Some pastes may lose solvent from the flux binder and dry out during storage or handling. This can be minimised by storing at a cool temperature and avoiding contact with the atmosphere. If paste has begun to dry out it may be possible to add solvent to reduce the viscosity, but great care must be taken to ensure that the solvent is compatible with the overall flux binder system. Most solder paste manufacturers offer fluxing composition thinners that are suitable for reducing the viscosity of their pastes. Such thinners are always used very sparingly, at the rate of 1 drop to 20 g of paste at maximum. In particular, thinning cannot be used to convert a printable paste into a syringeable paste. If oxidation is so severe that the powder particles have clumped together or the paste has become crusty it is unsalvageable and should be discarded.

6.9 VOIDS IN THE SOLDER FILLETS

One of the problems associated with the use of solder paste to form a reflowed solder fillet is the formation of voids, as shown in Figure 6.27. Although invisible and not detrimental to the cosmetic appearance of the board, such voids affect the integrity of the circuit, structurally weakening the joint in fatigue[241] and producing spot overheating.[242] The voids are caused by volatile components of

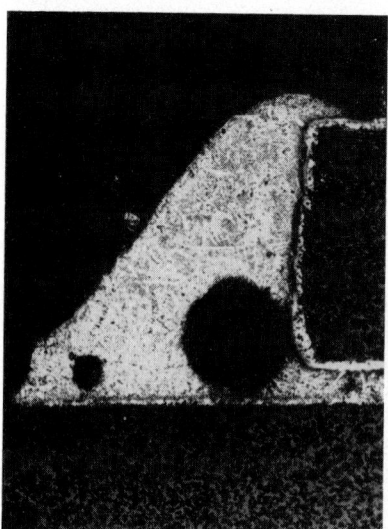
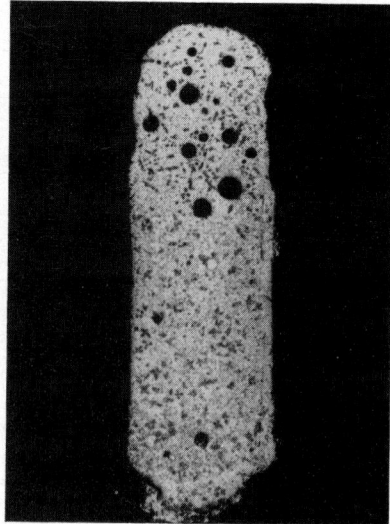

Fig. 6.27 Porosity demonstrated in vertical and horizontal microsections through reflowed solder paste joints on leadless chip carriers. The pores are both under the component and in the solder constellation. Note that in the vertical section, the solder contains coarse $AuSn_4$ crystals because of dissolution from the Au/Pt metallisation during an excessive solder reflow time. (Photographs courtesy of Dr H. A. H. Steen, Swedish Institute for Metals Research, Stockholm).

Solder Pastes

the paste vaporising during soldering and being unable to escape to the surface of the molten metal before solidification occurs.

Some correlation is found between the occurrence of voids in the solder fillets and the solderability of the thick film metallisation used on surface mounting components; as the solderability declines, void formation increases. This is probably due to an organic component of the metallisation ink not being fully fired out during manufacture of the component. Notwithstanding this observation, the degree of voiding is usually a direct result of the formulation of the paste and how it is dried and handled before reflow.[243] For this reason the problem of voids is treated in this chapter on pastes. We are concerned here only with solder fillets whose voids have occurred during the reflow process. Voids and cracks that occur after solidification, usually due to thermal cycling stresses, are a separate problem that will be taken up in Chapter 11.

6.9.1 Flux Activation

For a given set of circumstances, voiding in the solder fillets increases as the flux activation of the paste increases. Some illustrative data are shown in Figure 6.28 for identical samples of 62 Sn:36 Pb:2 Ag paste with a 90% metal content reflowed at either 215°C, 235°C or 255°C on a hotplate for 60 seconds. Prior to the reflow the samples were left to stand at room temperature for 20 hours with no further drying-out preheat.

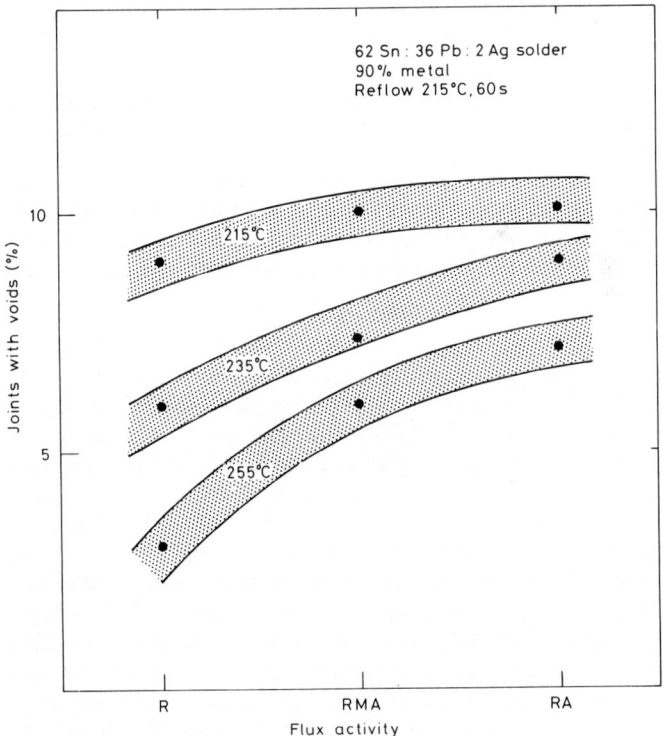

Fig. 6.28 The effect of changing the flux activity and the reflow temperature on the occurrence of voids in surface mount solder fillets.

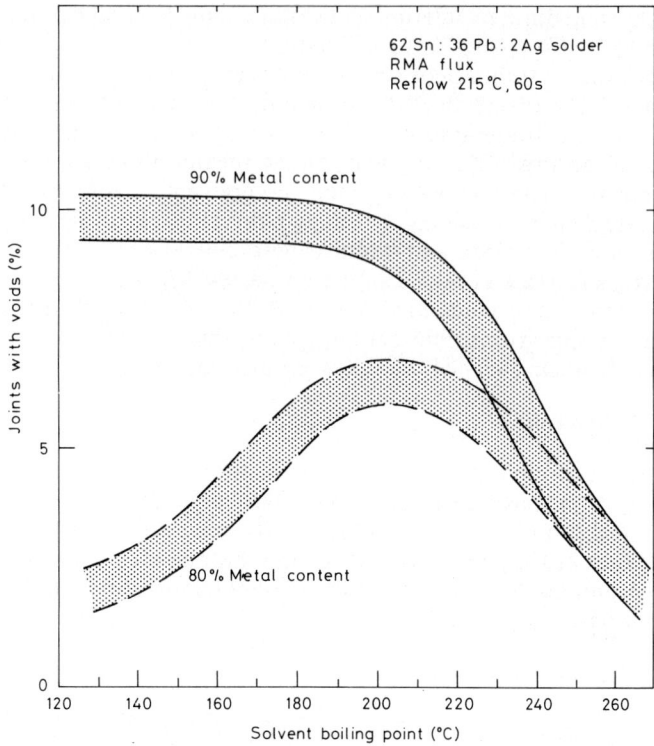

Fig. 6.29 The effect of changing the solvent boiling point on the outgassing of the paste during reflow. With low metal content the paste slumps and allows the solvent to evaporate more readily.

It has been suggested that it is not only the flux activity that is important but the relative amount of activator present.[244] As the amount of activator or the activity level increases, more oxide is reduced, consequently generating a larger amount of gas and resulting in a greater propensity for void formation.

6.9.2 Solvent

It is difficult to isolate the effect of the type and quantity of solvent present since its rôle in producing voids in the solder fillet depends strongly on the print thickness, the pre-bake and the reflow temperatures. It is generally believed that the solvent either should be completely evaporated before solidification of the reflowing solder begins, or should maintain an insignificant vapour pressure during the reflow process.

Figure 6.29 gives some results for a series of test solder pastes manufactured all at the same time, from one batch of solder powder and one batch of rosin flux, but with solvents of different boiling points. The alloy used was 62 Sn:36 Pb:2 Ag with a 90% metal content, screen printed with a wet print thickness $h_w \approx 150$ μm, and left to stand at room temperature for about 20 hours before reflow. The samples were reflowed at a hotplate temperature of 215°C for 60 seconds contact time. These observations demonstrate the importance of an adequate drying out of the paste, which was not achieved in this case by a room temperature stand for 20 hours. A large number of voids were created, reduced only in the paste with a high boiling point solvent for which the vapour pressure at the reflow temperature was low enough to reduce the voiding.

If, however, a paste is chosen with a low metal content, 80% by mass, the results are different. This is because after printing, during the drying out period and the initial stage of reflow, the paste slumps so that the heat absorption is maximised. The solvent volatilises earlier in the reflow cycle and is able to escape. Thus, even though there is more solvent in the paste, the voiding is less.

6.9.3 Pre-heating

Two effects occur during the pre-heating or drying of the deposited solder paste.

(i) The solvent is removed, the success of which depends upon the temperature-time profile and also the vapour pressure of the solvent.
(ii) The solder powder increases its oxide content, because the drying is invariably carried out in an air oven.

The first process reduces the occurrence of voids by removing the source of the gas whilst the second process increases the occurrence of voids by causing more reaction products to be formed between the solder particles and the flux. The effect of these concomitant phenomena was discussed in Section 6.7.

6.10 SOLDER BALLS

The problem of solder balls is one directly related to the formulation and reflow procedures of the solder paste. Several factors need to be considered to ensure the elimination of what can be a catastrophic problem in service.

(i) The powder particle size distribution: The exclusion of solder 'fines' from the powder is very important to the reduction of solder balls. This was discussed in Section 6.2.2 and demonstrated in Figure 6.5.
(ii) Oxide content of the paste: Solder balls will be a problem if the oxide content is greater than $0 \cdot 15$ mass% of the metal content, especially in a paste with a low flux activity. This was discussed in Section 6.2.3 and a procedure given for making a quick assessment of the oxide content of a paste.
(iii) Pre-heating of the paste after component populating: The solder ball problem is exacerbated if the applied paste is not adequately dried out before reflow. This effect was discussed in Section 6.7 and illustrated by the data in Figure 6.24.
(iv) Environmental conditions before reflow: The solder paste reflow performance deteriorates as the time between drying and reflow increases unless care is taken with control of the storage environment. This deterioration is manifested as a contribution to the solder ball problem, as discussed in Section 6.7.1 and illustrated in Figure 6.26.

6.10.1 Solder Balling Test

Whilst many characteristics of a solder paste should be considered when making an assessment of a suitable choice of supplier and paste type, it is generally impractical to carry out regularly a full appraisal of new batches of paste. The occurrence of solder balling is, however, a good indicator of many of the important reflow properties of a solder paste. In particular, it is a measure of the solder particle size and shape distributions and of the amount of oxide present on the solder particles. The solder ball situation deteriorates significantly if either of these properties falls below specification. It is therefore useful to delineate a simple standard procedure for solder ball testing. Such a test procedure has been proposed.[4]

The test simply considers the behaviour of a standard deposit of solder paste on a non-wettable substrate, either ceramic or oxidised metal. The standard deposit is about 200 μm thick and 5·5 mm diameter applied into a hole punched through two self-adhesive paper labels stuck one on top of the other in the manner described for the slump test (Section 6.6.4) and the tackiness test (Section 6.6.5).

After deposition of the paste, the substrate is heated to 250±2°C for 3 seconds. It is pertinent to conduct the test both on wet solder paste and on paste that has been dried according to recommended procedure, as well as paste that has been left in the air for a period that might be encountered in production, between component population and reflow.

During heating, the solder powder particles should melt and coalesce into one single solder ball, leaving no separate particles in the liquid flux.

The substrate should have a thickness of up to 1 mm and, to ensure good heating, should be floated or partially immersed on the cleaned surface of a molten solder bath. A modified wetting balance solderability tester is ideal for the job, since an exact immersion depth and a contact time can be chosen and used with good repeatability.

6.11 SOLDER PASTE DESIGN APPRAISAL

Solder pastes which are nominally of the same alloy composition, the same flux activity and designed for the same mode of application can vary substantially from one manufacturer to another, from batch to batch and with age after manufacture.[244]

It is recommended that a batch of paste be carefully evaluated to ensure that it meets all of the requirements of the application. It is suggested that the following properties should be evaluated:

(i)   metal purity;
(ii)   oxide content;
(iii) particle shape and size distribution;
(iv) flux activity;
(v)   screenability/syringeability;
(vi) shelf life;
(vii) working life;
(viii) solder ball formation;
(ix) solder wetting;
(x)   ease of cleaning;
(xi) general appearance;
(xii) vendor expertise and support;
(xiii) vendor control over his product.

The last of these points arises because some suppliers of pastes mix components that have been produced elsewhere.

In designing or making an assessment of a solder paste the processing conditions must also be considered:

(i) type of metallisation;
(ii) amount of oxide and contaminants on the metallisation;
(iii) pad dimensions, separation and required print thickness;
(iv) application method;
(v) amount of slump that can be tolerated;
(vi) time between application of the paste and positioning of components;
(vii) time prior to reflow;
(viii) desired drying conditions;
(ix) ambient temperature during application and prior to reflow;
(x) reflow method and temperature profile;
(xi) procedure for cleaning reflowed parts.

First, the proper solder alloy must be selected. Next, one must balance the cleaning and the solderability requirements of the flux. The viscosity and metal content are specified by the fillet height required, the application method and the pad spacing on the circuit. Pad spacing, working life and reflow profiles impose potentially conflicting demands on the solvent and viscosity modifier selection. There are many competing conditions and conflicts that can be resolved by one of many compromises, and it is therefore important to assess carefully the type and the suppliers of solder paste.

Chapter 7

REFLOW SOLDERING USING RADIANT HEATING

7.1 INTRODUCTION

For the making of a solder joint the applied solder paste must be melted and allowed to wet and flow over the solderable surfaces of the component and the substrate. The three basic modes of heat transfer, namely conduction, convection and radiation, may be used separately or in combination to elevate the temperature to that required for reflow. This chapter deals with the principles of heat transfer using radiation[245] for the successful application of radiant heating to the soldering of heat sensitive components and circuit materials.

The rate of heat transfer by the conduction and convection modes is proportional to the difference of the temperature between the heat source and the heat receptor. Thus the amount of heat transferred is independent of the absolute magnitude of the temperature as long as the difference is the same. This is not the case for thermal radiation. The quantity of heat exchanged by radiation is proportional to the difference between the fourth power of the absolute temperatures of the two bodies. Thus, for a given temperature difference, the heat transferred is much greater at high temperatures than at low temperatures.

The inherent advantages of radiant heating for the reflow soldering of SMDs are as follows.

(i) There is no contact with the workpiece, thus reducing the possibility of contamination and the need for extra cleaning. This also enables a relaxation in the mechanical constraints imposed on the substrate during soldering as, for example, necessary in wave soldering.

(ii) The heat source can be remote, enabling heat transfer into chambers under vacuum or other protective environments.

(iii) There is a fast response and good control of the radiant flux, allowing feedback process control to be incorporated.

(iv) The equipment can be manufactured and installed at significantly lower cost than that required for other reflow techniques.

The full realisation of these advantages is usually restricted by the materials properties of the components to be soldered.

7.2 RADIATION HEAT TRANSFER

7.2.1 The Electromagnetic Spectrum

All bodies with a temperature higher than absolute zero continuously emit energy in the form of electromagnetic waves. The electromagnetic spectrum extends from the very short wavelength cosmic rays to the very long wavelength radio waves, as shown in Figure 7.1. The radiation throughout the spectrum propagates rectilinearly and obeys the same laws of reflection and refraction. Our human senses are able to detect radiation only if its wavelength falls within the spectrum region between 0·1 and 100 μm. Radiation in this wavelength range causes appreciable heating, and within the narrow band 0·38 to 0·76 μm can also be detected by the human eye. Energy from any part of the electromagnetic spectrum will result in the heating of a body in which it is absorbed, but specifically the radiant heating region is that consisting of energy emitted by virtue of the temperature of the source. For practical purposes this radiation is limited to wavelengths in the range 0·1 to 100 μm. The radiant energy other than that included in this infra-red zone can be useful and thus the heating method is properly called radiant heating rather than infra-red heating.

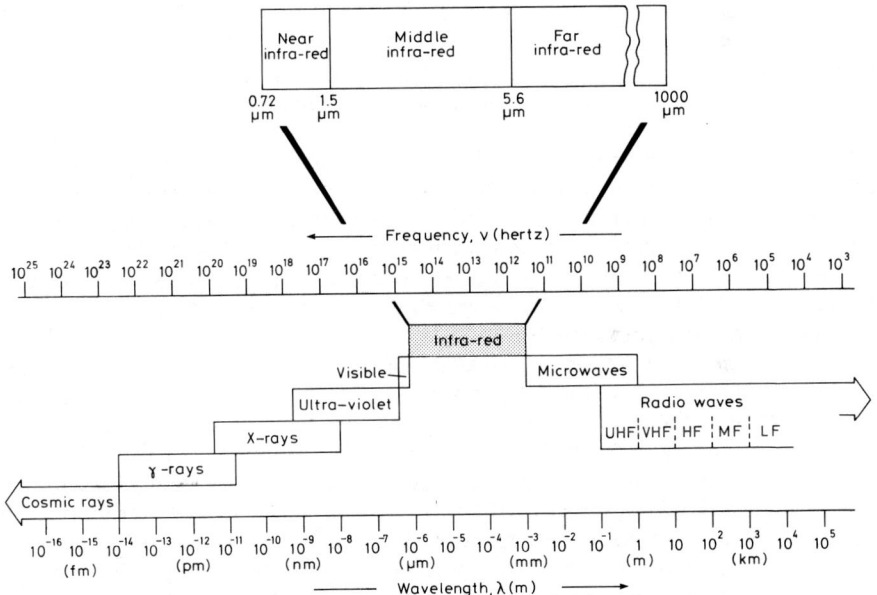

Fig. 7.1 The electromagnetic spectrum.

7.2.2 Basic Definitions

When radiant energy falls on a body, part may be absorbed, part reflected and the remainder transmitted through the body, as shown in Figure 7.2. The part of the radiant energy which is absorbed is transformed again into thermal energy. That which is reflected or transmitted falls on surrounding bodies and is absorbed by them. Thus, after a series of absorptions, the radiant energy is completely distributed amongst the surrounding bodies. Consequently, each body not only continuously emits but also continuously absorbs the radiant energy[246].

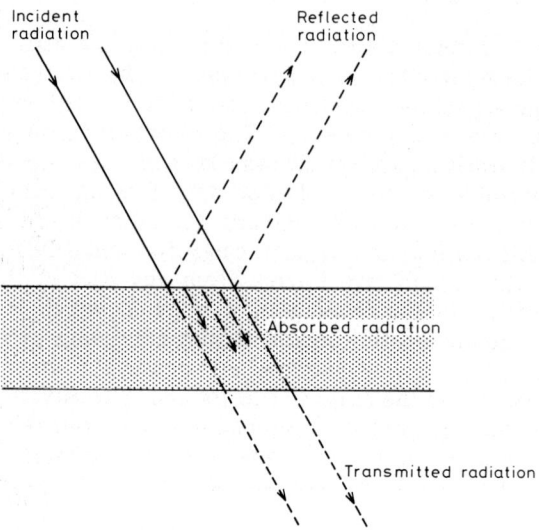

Fig. 7.2 When radiation is incident upon a body, part is reflected from the surface, part is absorbed in the material and part is transmitted.

As a result of these phenomena of mutual transformation of energy (thermal-radiant-thermal), the process of heat transfer by radiation takes place. The emissive or absorptive power of a body is determined by the difference between the amounts of emissive and absorptive energy associated with the body. This difference is not zero if the temperatures of the participating bodies are different from each other. When these bodies are all at the same temperature, the system is in dynamic thermal equilibrium.

The *total emissive power*, W (watts), of a body denotes the total *emitted* thermal radiation energy (joules) per unit time. If A is the emission source area, the term W/A is then the total emissive power per unit area, or the *radiant flux density* (watts m^{-2}). The total emissive power is defined to consist only of the original emission from a surface and does not include any energy leaving the surface that may be the result of the reflection of some incident radiation. The emissive power is found to be dependent upon the temperature of the emitting surface, the substance of which the surface is composed and its structure and roughness. The emissive power from a given elemental surface may have both a spectral and directional dependence. The spectral emissive power, i.e., that at a particular radiant wavelength, λ, is denoted W_λ.

The *total radiosity*, J (watts), is the term used to denote the total radiant energy leaving the surface per unit time. This quantity differs from W in that it includes reflected energy as well as the original emission. As in the case of the emissive power, the total radiosity of a surface element consists of all the radiation leaving a surface, regardless of any directional or spectral dependence.

The *total irradiance*, G (watts), is the term which denotes the total radiant energy incident upon the surface per unit time. This incident radiation is the result of emissions and reflections from other surfaces and as such may be directionally or spectrally preferential. The total irradiance, however, is again the total energy incident upon the surface, regardless of these other factors.

When radiation falls on a surface, part of it may be absorbed by the body, part may be reflected away from the surface and part may be transmitted through the body.

The absorptivity α, reflectivity ϱ and transmissivity τ are defined as the fractional amounts of the total radiant energy falling on a body, that are respectively reflected, absorbed and transmitted. Thus, using these definitions:

$$\alpha + \varrho + \tau = 1 \qquad (7.1)$$

If all the radiant energy is absorbed by the body, $\alpha = 1$, and it is called a black body. Similarly, if $\varrho = 1$ the body is white and if $\tau = 1$ the body is transparent. In nature, there are no absolutely black, white or transparent bodies except over limited wavelength ranges. For absorption and reflection in the infra-red the rugosity of the surface has a greater effect on α and ϱ than does its colour.

In general, the absorptivity, reflectivity and transmissivity of a surface are dependent upon the direction of the incident radiation and its spectral distribution as well as the composition, structure and temperature of the irradiated surface. We are generally concerned here with thermally opaque materials, so that

$$\alpha + \varrho \doteq 1$$

The definitions of total emissive power W, radiosity J, and irradiance G now lead to the following equations

$$\begin{aligned} J &= W + \varrho G \\ &= W + (1 - \alpha)G \end{aligned} \qquad (7.2)$$

7.2.3 Planck's Law

The emissive power W of a body is the energy emitted by the body from area A per unit time at all wavelengths. The emissive power W within the wavelength interval λ to $\lambda + d\lambda$ was derived theoretically by Planck for a perfectly black body at temperature T:

$$\frac{W_\lambda}{A} = \frac{d}{d\lambda}\left(\frac{W}{A}\right) = k_1 \lambda^{-5} \{\exp(k_2/\lambda T) - 1\}^{-1} \qquad (7.3)$$

where k_1 and k_2 are the radiation constants which can be expressed in terms of the fundamental constants:

c	speed of electromagnetic radiation	$= 2 \cdot 998 \ 10^8$ m.s^{-1}
h	Planck's constant	$= 6 \cdot 626 \ 10^{-34}$ J.s
k	Boltzmann's constant	$= 1 \cdot 381 \ 10^{-23}$ J.K^{-1}
k_1	first radiation constant $= 2\pi hc^2$	$= 3 \cdot 742 \ 10^{-16}$ W.m^2
k_2	second radiation constant $= hc/k$	$= 1 \cdot 438 \ 10^{-2}$ m.K

Planck's theory requires that radiant energy be considered to be emitted in discrete amounts called quanta, thus differing from the classical theory of continuous emission. The energy quanta are pulses of electromagnetic waves called photons with each photon propagating at the speed of light, c (m.s^{-1}), and containing electromagnetic energy E_{ph} (J), equal to $h\nu$, where ν(Hz) is the frequency of the radiation and h is the universal Planck's constant. Thus:

$$\lambda \nu = c \tag{7.4}$$

Figure 7.3 shows the spectral energy distribution, given by Equation (7.3) for a black body, as a function of wavelength and temperature. It is seen that, at a temperature of, for example, 3000 K, the energy in the visible emission ($\lambda = 0\cdot 38$ to $0\cdot 76$ µm) is extremely low compared with that in the infra-red ($\lambda = 0\cdot 8$ to 40 µm). Although the black body emissive power is composed of radiation of *all* wavelengths, the principal contribution is made between the wavelengths of $0\cdot 1$ and 100 µm.

The calculation of the radiation emissive power is facilitated by dividing Equation (7.3) by T^5, to yield a function of only the product λT. This function is

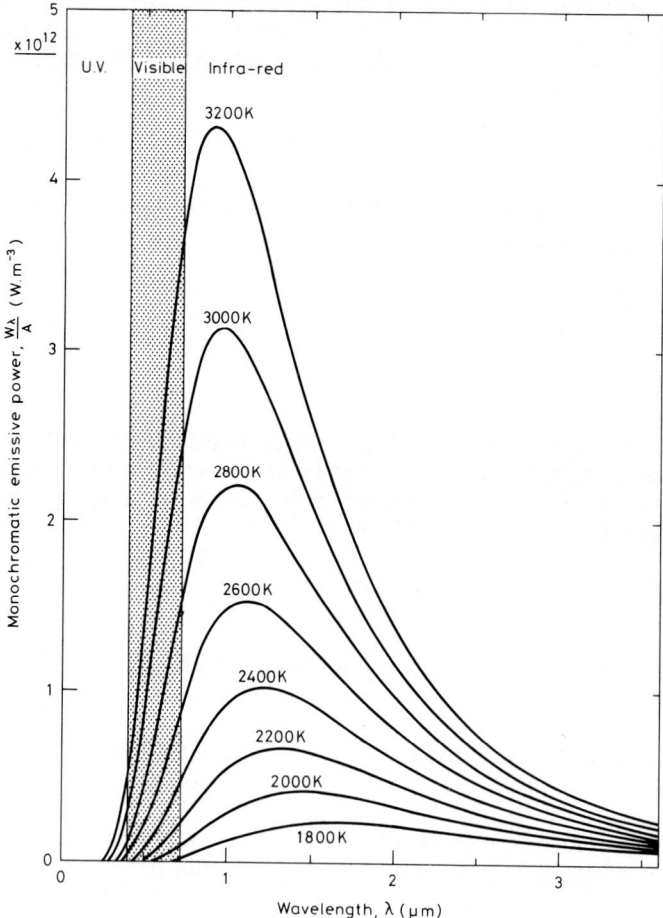

Fig. 7.3 The spectral emissive power of a black body as a function of temperature.

given in Figure 7.4, and can be used directly for all temperatures by simply calculating λT and T^5.

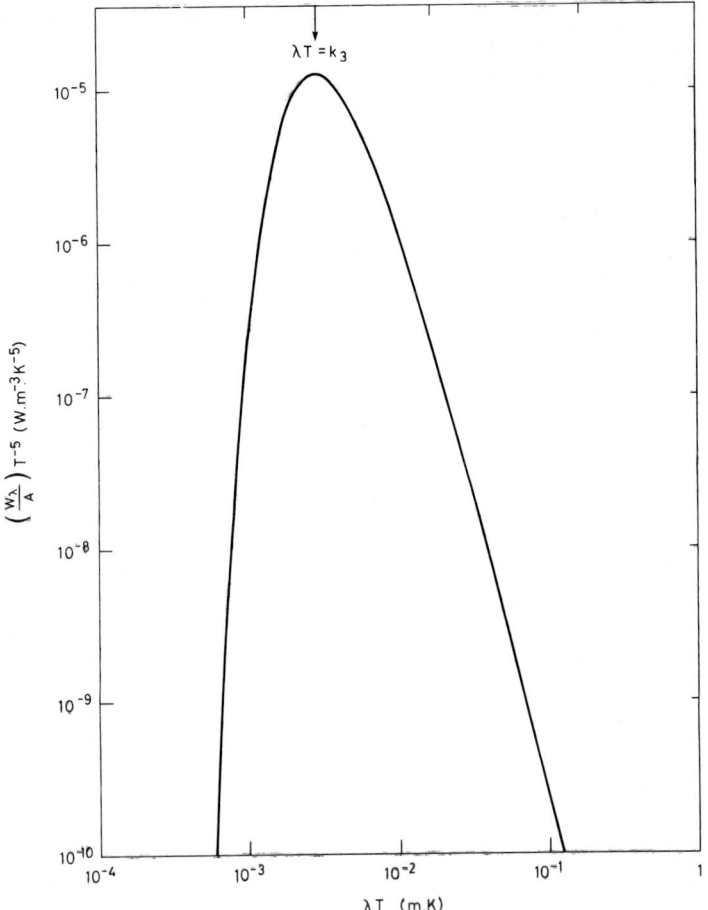

Fig. 7.4 The curves in Figure 7.3 can be represented by a universal curve by changing the variable to λT.

7.2.4 Wien's Law

It can be seen in Figure 7.3 that as the source temperature is increased the maximum emission moves to lower wavelength, i.e., higher energies. The wavelength at which the maximum energy is emitted at a particular temperature is shown in Figure 7.5. A quantitative inspection of the curves demonstrates that

$$\lambda_{max}(T) = k_3/T \tag{7.5}$$

where $\lambda_{max}(T)$ is the wavelength at which W_λ has its maximum value for a particular temperature T. This is in agreement with everyday experience that, as the temperature of a radiant source increases, its colour shifts from red to white as increasingly more radiant energy is emitted in the shorter wavelength region.

By differentiating Equation (7.3) and equating to zero, to find the maximum, we derive $k_3 = 2\cdot 90$ mm.K. Equation (7.5) is known as Wien's displacement law of spectral distribution, whose general form was derived theoretically before Planck's quantum theory, by considering a cavity filled with radiation and taking it through a thermodynamic cycle of expansion and contraction.

Substituting λ_{max} from Equation (7.5) into Equation (7.3) yields the following

$$\left(\frac{W_\lambda}{A}\right)_{max} = 1\cdot 29 \ 10^{-5} \ T^5 \ (W.m^{-3}) \tag{7.6}$$

for the power emitted at the maximum of the spectral distribution curve. This function is also plotted in Figure 7.5. Thus the maxima of the curves in Figure 7.3 are proportional to the fifth power of the absolute temperature.

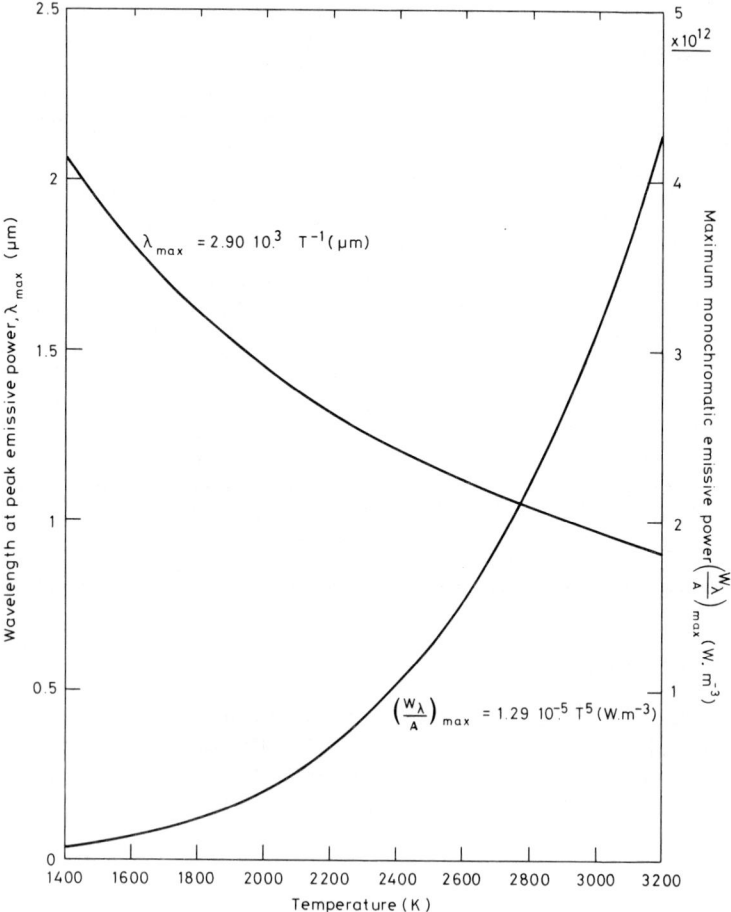

Fig. 7.5 Left: the wavelength at which the maximum emissive power occurs in Figure 7.3. Right: the value of that maximum emissive power.

7.2.5 Stefan-Boltzmann Law

Equation (7.3) was used to produce the series of curves in Figure 7.3. The integral of, or the area under, each curve is the total amount of energy W emitted from area A, per unit time over the entire wavelength range. Thus, for a black body, $W = W_b$ where

$$\frac{W_b}{A} = \int_0^\infty \frac{W_\lambda}{A} d\lambda \qquad (7.7)$$

whence, by integrating Equation (7.3), $\frac{W_b}{A} = \sigma T^4$ (7.8)

where $\sigma = \frac{k_1}{15}\left(\frac{\pi}{k_2}\right)^4$ (7.9)

which is the universal Stefan-Boltzmann constant, $5 \cdot 67 \ 10^{-8} \ W.m^{-2}.K^{-4}$. Thus the emission energy of a body is proportional to the fourth power of its absolute temperature, as plotted in Figure 7.6.

Strictly speaking, the Stefan-Boltzmann law holds only for an absolutely black body. However, the experimental data show that the law may be applied to grey bodies as well by introducing an emissivity ε which ranges from 0 to 1 depending on the nature, surface conditions and temperature of the emitting surface. Thus, for a grey surface:

$$\frac{W_b}{A} = \sigma \varepsilon T^4 \qquad (7.10)$$

The emissivity is the ratio of the power emitted by the grey body to that emitted by a black body at the same temperature. A grey body is therefore a special type of non-black body, the assumption being that the emissivity ε is independent of the wavelength λ, such that the spectral distribution curves of grey and black bodies at the same temperature are affine to one another, as shown in Figure 7.7, there being no shift in the peak of the curve. Some real surfaces do approximate to greyness, but also shown in Figure 7.7 is the emissive power of a more typical non-black, non-grey surface for which the emissivity ε_λ is strongly dependent on wavelength.

For making calculations of the emissive power of real, non-black surfaces it is useful to have available numerical values of the emissive power between specific wavelengths, for which the following integral is required

$$\frac{W_b(o-\lambda)}{A} = \int_0^\lambda \frac{W_\lambda}{A} d\lambda \qquad (7.11)$$

Then, dividing Equation (7.11) by Equation (7.7) in combination with Equation (7.3) and changing the variable to (λT):

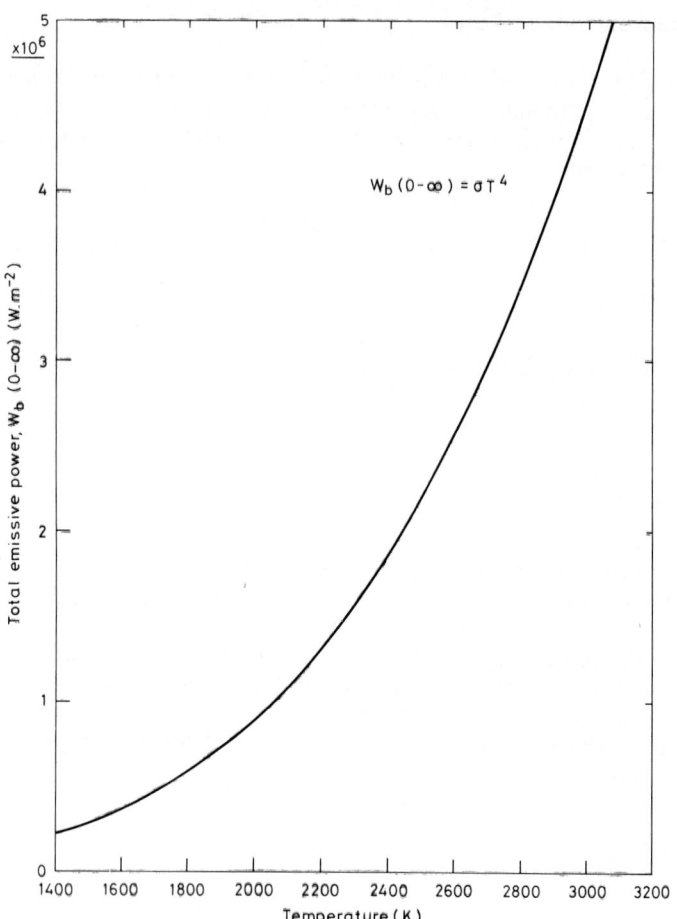

Fig. 7.6 The total emissive power over the entire wavelength range, 0-∞, of a black body (area under each curve in Figure 7.3) as a function of temperature.

$$\frac{W_b(o-\lambda)}{W_b(o-\infty)} = \frac{W_b(o-\lambda)}{\sigma T^4} = \int_0^{\lambda T} \frac{k_1}{(\lambda T)^5(e^{k_2/\lambda T}-1)} \, d(\lambda T) \qquad (7.12)$$

Figure 7.8 is a plot of $W(o-\lambda)/\sigma T^4$ as a function of (λT). The curve gives the area under the emittance curves of Figure 7.3 from zero wavelength up to a wavelength λ, as a proportion of the area of the total curve between zero and infinity. For the 3000 K curve, for example, the radiation at wavelengths up to 1 μm is given by the point where $\lambda T = 3.10^{-3}$m.K, namely 2·2% of the total radiation. At wavelengths from zero up to 2 μm, the emittance is given by the point where $\lambda T = 6.10^{-3}$m.K, namely 35% of the total radiation. Thus, within the window 1-2 μm about one third of the total radiation is emitted, at 3000 K.

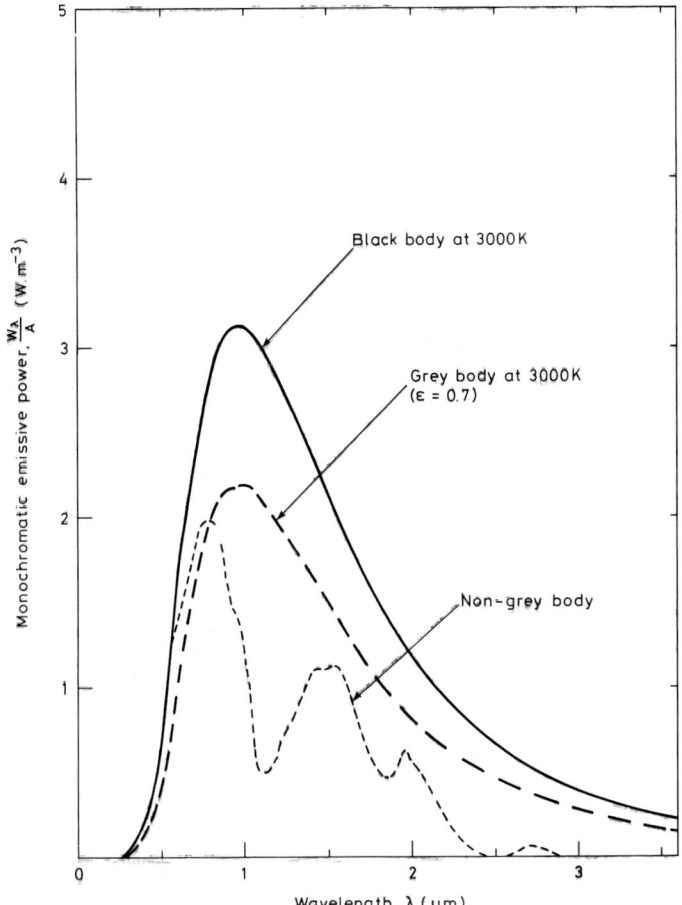

Fig. 7.7 A black body has an emissivity, $\varepsilon = 1$; a grey body has a constant emissivity, $\varepsilon < 1$; a non-grey body has a wavelength dependent emissivity, ε_λ.

7.2.6 Kirchoff's Law

There is an interesting relation between the efficiency of a surface as an emitter of radiation as defined by the emissivity ε, and its efficiency as an absorber as defined by the absorptivity α. This relation is due to Kirchoff and is simply

$$\varepsilon = \alpha \tag{7.13}$$

This was proved in thermodynamic argument, independent of any detailed assumptions about the emission and absorption processes, by considering the thermal equilibrium of surfaces of different nature which are exchanging thermal energy by emission and absorption. It has also been verified experimentally. The law also holds for non-equilibrium conditions since α and ε are surface properties.

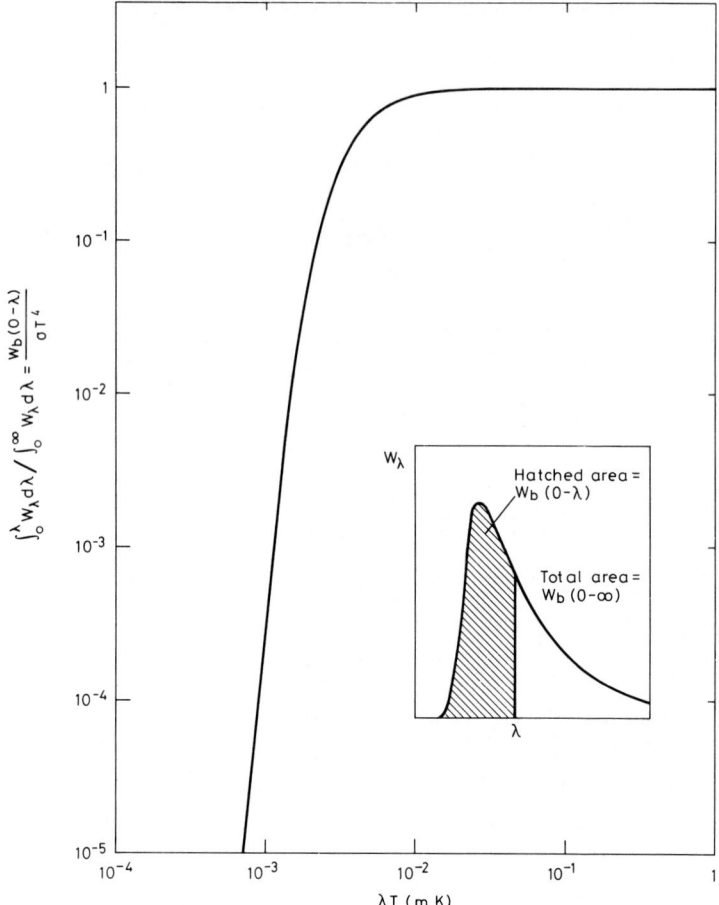

Fig. 7.8 The emissive power of a black body in the wavelength range o-λ as a proportion of the total emissive power.

7.2.7 Lambert's Law—Diffuse Emission

The Stefan-Boltzmann law determines the total emissive power of a body in all directions. The changes of the emissive power along specific paths is determined by Lambert's law. Figure 7.9 defines the radiation from a source emitted at an angle θ from the normal to the plane of the source and at an angle φ from a line in the plane of the source. Lambert's law states that the radiant energy incident on elemental area B from source A is proportional to the cosine of angle θ and inversely proportional to the square of the distance between A and B, assuming the source to be a diffuse emitter. By the term diffuse emitter is meant that the emittance ε does not vary with angular direction φ.

7.2.8 Heat Transfer by Diffuse Radiation

An approximate average value for the radiant heat transfer from a diffuse source can be calculated by considering two grey plane surfaces at different

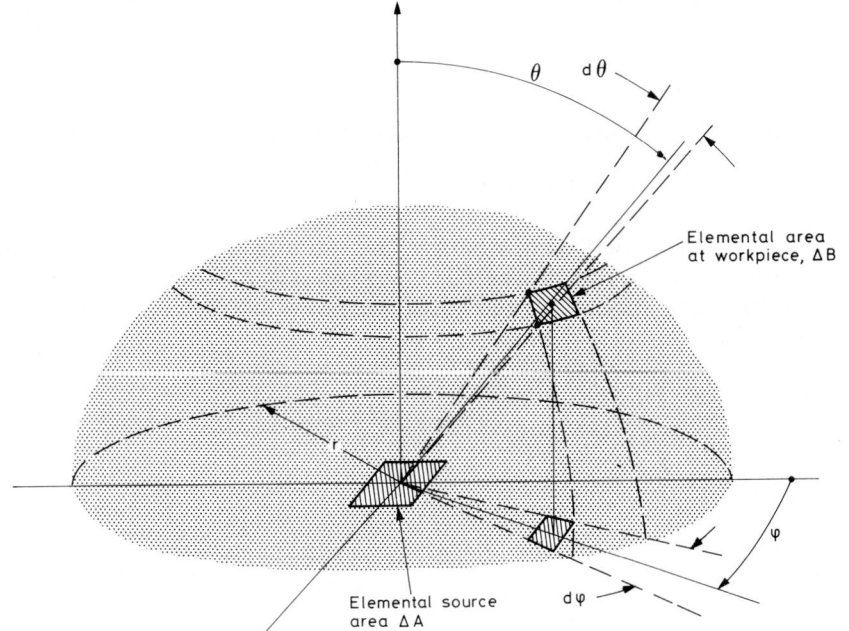

Fig. 7.9 Lambert's law states that the radiation received at B varies as cosine θ. If the source is diffuse, there is no dependence on angle φ.

temperatures, mutually irradiating one another. Assuming that all the energy leaving each plane strikes the other, then the net radiant heat exchange is found by evaluating the net power given up or absorbed by one of the surfaces. Since the planes are grey, the relation between emittance radiosity and irradiance is given by Equation (7.2) for each plane:

$$J_1 = W_1 + (1 - \alpha_1)G_1$$
$$J_2 = W_2 + (1 - \alpha_2)G_2 \qquad (7.14)$$

but, since all the energy leaving one plane is deemed to strike the other,

$$J_1 = G_2 \qquad (7.15)$$
$$\text{and } J_2 = G_1$$

The four equations above, (7.14) and (7.15), can be solved for J_1, J_2, G_1 and G_2 in terms of W_1 and W_2. Then, from Equation (7.10):

$$W_1 = \sigma A \varepsilon_1 T_1^4$$
$$\text{and } W_2 = \sigma A \varepsilon_2 T_2^4$$

and from Equation (7.13):

$$\alpha_1 = \varepsilon_1, \ \alpha_2 = \varepsilon_2$$

one finally arrives at the following equation for the net exchange of radiant power between two grey bodies at temperatures T_1 and T_2:

$$W_{12} = \sigma A \frac{(T_1^4 - T_2^4)}{\frac{1}{\varepsilon_1} + \frac{1}{\varepsilon_2} - 1} \tag{7.16}$$

The denominator reduces to 1 for the black body case with $\varepsilon_1 = \varepsilon_2 = 1$.

7.2.9 Heat Transfer Coefficient

The heat transfer coefficient h $(W.m^{-2}.K^{-1})$ of any heat transfer process is defined in terms of the rate of transfer of heat energy, dQ/dt across an interface of area A, between a hot source and a cold receptor with a temperature difference $(T_1 - T_2)$. Thus:

$$\frac{dQ}{dt} = hA(T_1 - T_2) \tag{7.17}$$

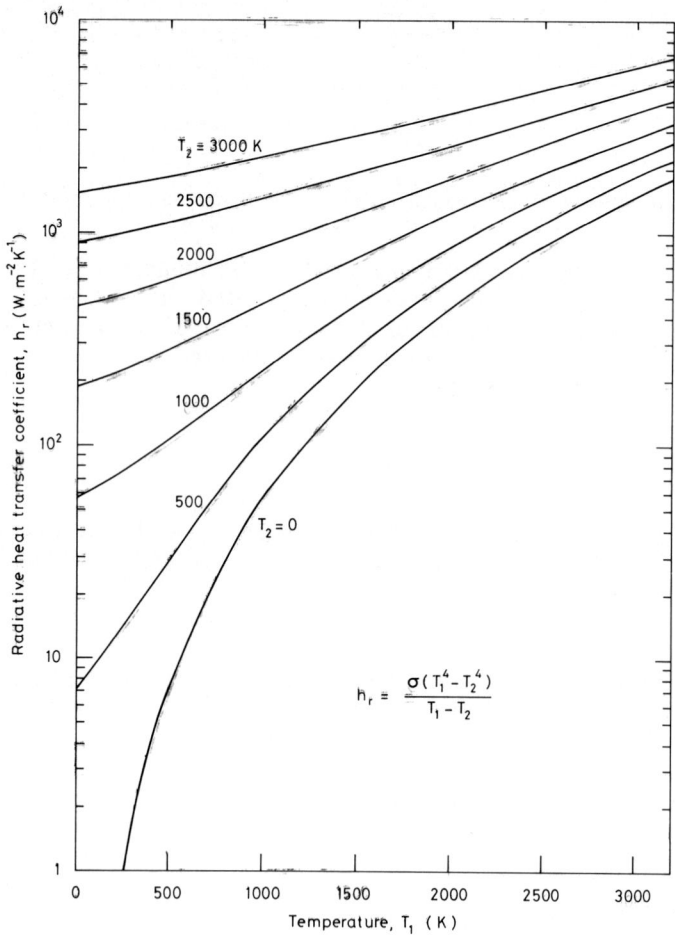

Fig. 7.10 Radiative heat transfer coefficient between two bodies at temperatures T_1 and T_2.

For conduction and convection the heat transfer is driven by the temperature differential and so Equation (7.17) can be readily solved. For radiant heat flow the net exchange of energy is driven by the difference in the fourth power of the two temperatures, given in Equation (7.16). Thus, for a black body,

$$\frac{dQ}{dt} = A\sigma(T_1^4 - T_2^4) \qquad (7.18)$$

For systems in which the heating results from a combination of radiant, conductive and convective transfer it is convenient to combine Equations (7.17) and (7.18) to define a radiation heat transfer coefficient:

$$h_r = \frac{\sigma(T_1^4 - T_2^4)}{T_1 - T_2} \qquad (7.19)$$

so that the overall heat transfer coefficient at a given temperature can then be taken as the sum of the radiant, the conductive and the convective heat transfer coefficients. Whereas the conductive and convective heat transfer coefficients are not generally temperature dependent but only temperature-difference dependent, h_r varies with T_1 and T_2 even if $T_1 - T_2$ remains constant. It is therefore convenient to view the form of Equation (7.19), and this is presented in Figure 7.10, where h_r is plotted as a function of both T_1 and T_2.

7.2.10 Penetration of Radiation

The ability of radiant energy to be absorbed within the bulk of the workpiece and hence deliver heat below the surface is significant in the heating for reflow of solder paste. The proportion of the incident radiant energy Q_i that is absorbed is dependent on the absorptivity α_λ of the workpiece. Assuming that the intensity of the radiation $Q(\lambda,z)$ at a given wavelength λ and a depth z from the irradiated surface decays exponentially away from that surface, it is possible to write

$$Q(\lambda,z) = Q_i \alpha_\lambda e^{-\mu_\lambda z} \qquad (7.20)$$

where μ_λ is the linear absorption coefficient or extinction coefficient for that wavelength of radiation. Thus

$$Q_i \alpha = \int_0^\infty Q(\lambda,z) dz \qquad (7.21)$$

and the change of energy absorption at depth z is proportional to the derivative of the intensity with respect to z:

$$\frac{dQ}{dz} = \frac{d}{dz}Q(\lambda, z)$$
$$= -\mu_\lambda Q(\lambda, 0) e^{-\mu_\lambda z} \tag{7.22}$$

During preheat of the solder paste the volatiles must be removed without mechanical disturbance of the solder print. Methods that heat the surface exclusively by conduction and convection impose limits on the rate of heating of the wet paste to ensure the removal of volatiles before the development of a surface structure that could cause eruptions. Infra-red reflow equipment, however, emits in the wavelength region for which the organics are largely transparent, allowing greater absorption of the available energy into the solder particles throughout the depth of the paste. This penetration allows the organics to be rapidly removed without the disturbance of the solder print normally associated with fast temperature rises. When an incident infra-red ray strikes the paste, it experiences multiple reflections from the reflective particles and is largely captured by the flux film, although it is virtually transparent, as shown in Figure 7.11. Once the solder is melted and reflowed the surface becomes highly reflective because no transparent path exists to carry energy into the film.

The penetration of the infra-red radiation allows the stress induced on substrates to be substantially reduced during rapid heating. If the surface alone is heated, a greater temperature differential is developed, creating an expansion stress between the different regions of the workpiece.

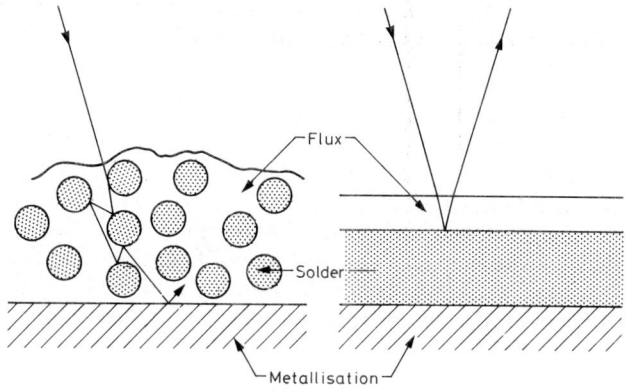

Fig. 7.11 Infra-red radiation incident on solder paste experiences multiple reflections and is largely captured by the flux film despite its being virtually transparent.[247]

7.3 INFRA-RED ENERGY SOURCES

Modern infra-red radiation sources are constructed using either a primary emitter, i.e., a hot resistive element, or a secondary emitter in which the primary source is embedded in an appropriate thermally conductive ceramic-based material. The properties and applicability for surface mounting reflow are first listed[248] in Table 7.1 and then the reasons for the advantages and disadvantages of each will be discussed.

Reflow Soldering Using Radiant Heating

Table 7.1

Characteristics and Suitability for SMT, of Infra-red Sources[248]

Emitter Type	Emission	Wattage	Suitability
Focused tungsten tube filament lamp	near I-R	300 W.cm^{-1}	Shadowing by components. Thermal degradation: board delamination board warping charring Colour selectivity
Diffuse array of tungsten tube filament lamps	near I-R	50-100 W.cm^{-1}	Colour selectivity
Diffuse array of nichrome tube filament lamps	near to middle I-R	15-50 W.cm^{-1}	Greater component densities are possible Little colour selectivity problem
Area source secondary emitter	middle to far I-R	1-4 W.cm^{-2}	No shadowing No colour selectivity

7.3.1 Tungsten Filament Sources

A typical tungsten filament lamp consists of a sealed tube of quartz or high-silica glass evacuated or argon-filled, enclosing a helically-wound tungsten filament, resistively heated, supported along its length by small tantalum discs, as shown in Figure 7.12. The end seals of the lamp are the most critical part of the design due to the different expansion coefficients of the envelope and metal lead wires. This necessitates adequate cooling at these terminations. Such lamps are usually rated in the range 50-250 watts per centimetre of tube. The lower wattage lamps are designed to operate at about 2500 K for a lifetime of around 5000 hours. The variations in the operating temperature and the expected life as the operating power is changed are shown typically, as in Figure 7.13.[249] Lamps of this type can be operated for short times at up to twice the design voltage with a trade-off of a drastic shortening of life.

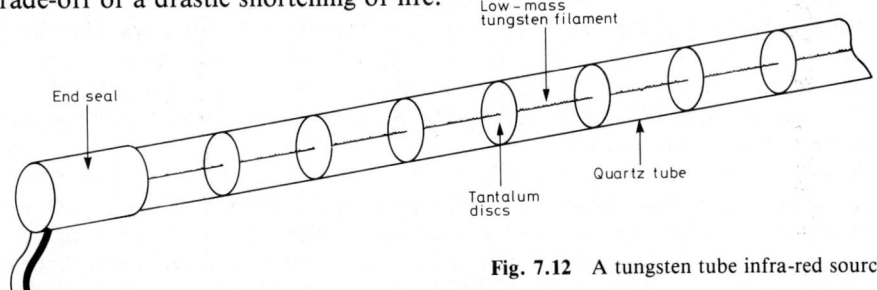

Fig. 7.12 A tungsten tube infra-red source.

Tungsten filament lamps intended to operate at higher wattages can have design operating temperatures of up to 3000 K. During the life of these lamps tungsten slowly deposits on the inside of the quartz tube envelope, reducing the transmission of radiant energy.

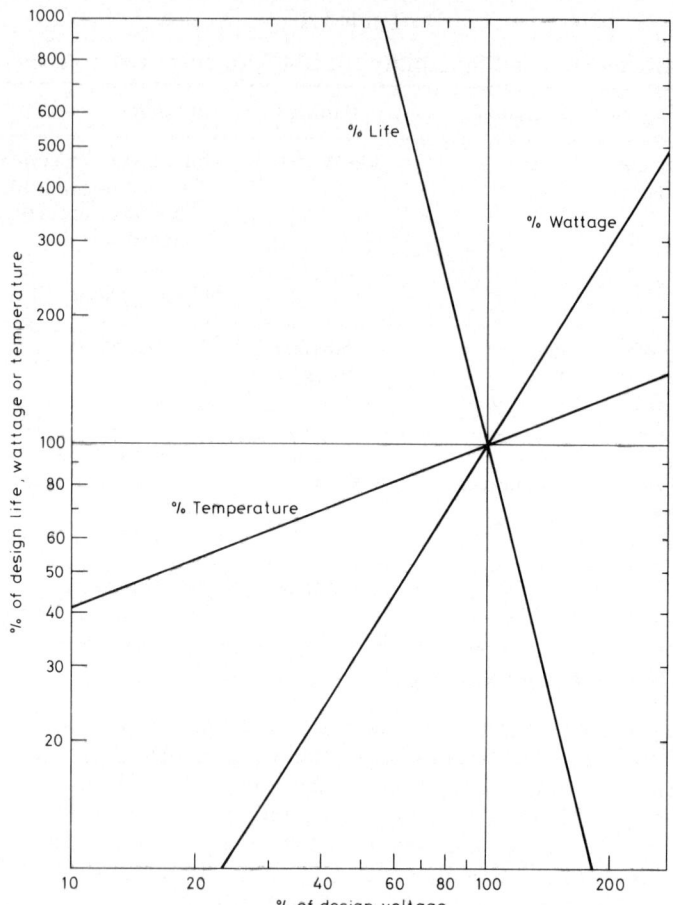

Fig. 7.13 Typical wattage/temperature-life characteristics of a tungsten tube infra-red source as a function of input voltage.[249]

In order to prevent tungsten evaporation, and enable the operating range to be increased to 3400 K the evacuated quartz or high-silica glass tube can be filled with a halogen vapour, either iodine or bromine. The tungsten chemically combines with the vapour and as the tungsten-halide compound strikes the filament the tungsten is redeposited.[250] This cycle provides a long lamp life at high filament temperatures and permits a constant energy transmission throughout the operating life.

The measured emissive power ($W.m^{-2}$) of tungsten is shown in Figure 7.14. This family of curves agrees with that theoretically derived using Figure 7.8, which gives the emissive power of a black body over the wavelength range 0-λ, except for a small adjustment to account for the change in the total emittance of tungsten, which rises from 0·26 at 1500 K to 0·37 at 3400 K. The curves give the emissive power of tungsten at all wavelengths up to that marked on each curve, i.e., at all energies greater than hc/λ.

The tungsten wire in the lamp is enclosed within a quartz or high-silica glass envelope. The transmittivity of these materials is shown in Figure 7.15. It can be

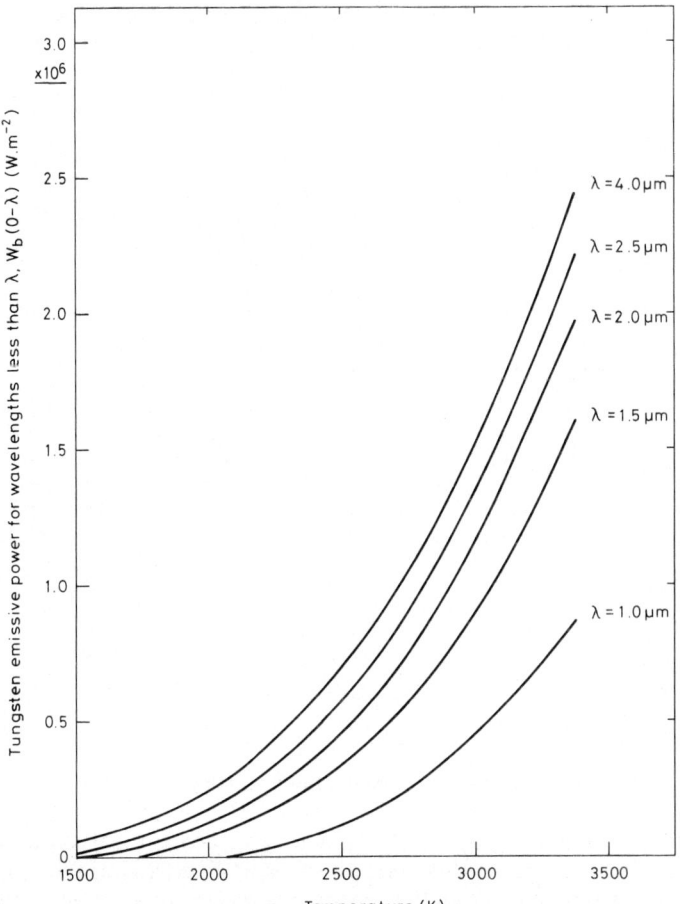

Fig. 7.14 The emissive power of tungsten in the wavelength range 0-λ.[251]

seen that there is a sharp cut-off in transmission at about 4 μm wavelength, and no energy emitted at wavelengths greater than 5 μm is transmitted. Thus, to a good approximation, the top curve (4 μm) in Figure 7.14 is a representation of the emissive power of the lamp.

The temperature range over which tungsten filament lamps operate is such that the peak emission, as given in Equation (7.5) is at wavelengths less than 1·5 μm, and occurring in the near infra-red (high energy infra-red) region of the spectrum, as defined in Figure 7.1.

Thus tungsten filament lamps are a source of intense, short wavelength radiant energy. The filaments have very little mass and thus heat and cool quickly in response to the applied voltage. Typically some 85% of the equilibrium temperature is achieved within 3 seconds of switching on, making the source suitable for feed-back process control situations. The quick response removes the power losses associated with long heating and cooling periods experienced with more massive sources.

In order to concentrate the radiant energy into a smaller well defined work zone the source can be placed at the focal point of one of many refractive and

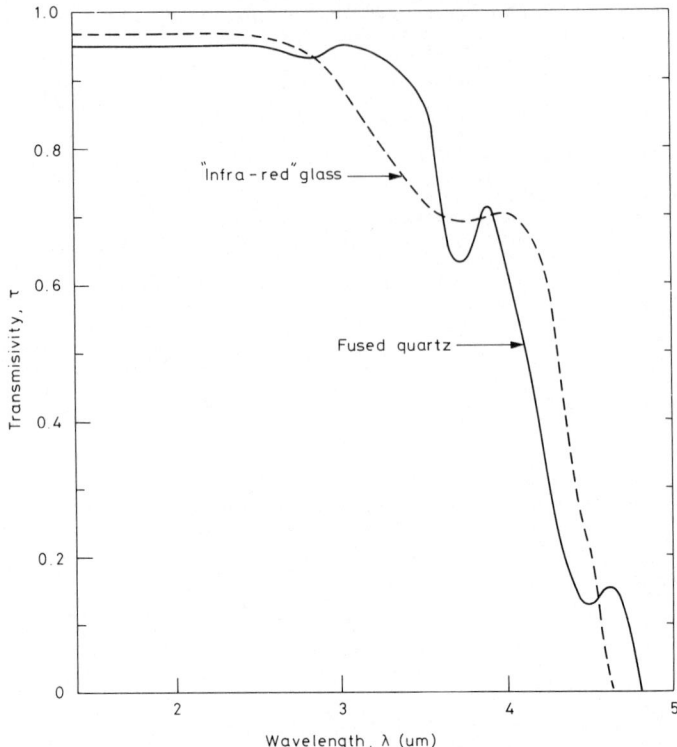

Fig. 7.15 The transmissivity of quartz and high silica infra-red glass.

reflective optical systems. The most efficient and economical systems use an elliptical reflecting surface. In this case, all the emitted energy at the source focal point theoretically passes through the conjugate focal point after one reflection at the reflector. The tubular lamps described above are used with cylindroidal reflectors, i.e., a cylinder having an elliptic right section, to produce linear heating zones. Aluminium surfaces are satisfactory provided consistent polishing procedures are used, but gold plated, electro-formed nickel has been found to be the best, low cost reflector material. The image forming characteristic of a cylindroidal reflector is illustrated in Figure 7.16. The lamp used in this study dissipates 40 watts per centimetre over a working length of 60 centimetres, providing a peak heat flux of 33 $W.cm^{-2}$ at the lamp rated voltage of 480 V. The curves shown in Figure 7.16 include a reduction in this heat flux of some 10% due to the quartz window placed across the minor axis plane. This window serves to keep contamination from processing operations, such as flux fumes, from degrading the reflector surface.

With a lamp of this type, over 20% of the available radiation can be focused into a strip of width only 1 cm wide, if the workpiece is at the image focal plane. By increasing the distance between the lamp and the workpiece, defocusing occurs, producing a relatively uniform radiation flux distribution. This effect can be used either to reduce the incident heat flux or when the workpiece is large.

A better method for large area heating, suitable for reflow of surface mounting assemblies, is to use one of the many types of available diffuse reflectors[251] such

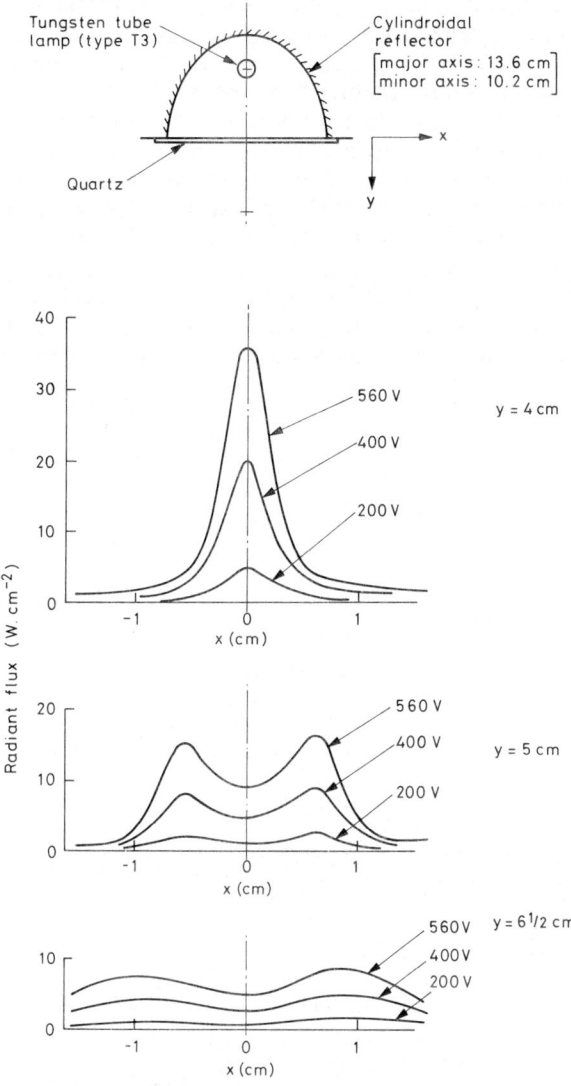

Fig. 7.16 Typical distribution of radiant flux from a tungsten tube emission source positioned at the focus of a cylindroidal reflector.[251]

as shown in Figure 7.17. Here the reflector is of a silica foam type and, with the same power dissipation of 40 watts per centimetre at 480 V rated voltage, the heat flux is about 17 $W.cm^{-2}$, i.e., half the peak flux obtained when using a focused reflector. More than 75% of the radiation arrives at the workpiece directly from the lamp so that fast changes of radiant flux remain possible. This type of reflector helps to avoid shadowing effects and also is very insensitive to the distance between the lamp and the workpiece.

The focused type of tube lamp, i.e., with a reflector, is normally used for high emission density, small area requirements. These sources are not very suitable for the reflow of solder pastes in surface mounting assembly because of shadowing effects and the fact that generally the power of the sources is too great, commonly causing board delamination, board warping and board discoloration or charring. Delamination occurs because moisture trapped between laminations is heated too rapidly above its boiling point. Prior baking of the board can minimise this problem. Charring and discoloration of the laminate is another problem associated with the inherent rapid energy transfer and short wavelength emission, characteristic of near infra-red emitters. Additionally, temperature-induced component failure is more likely to occur at these radiation wavelengths since the absorption of this radiation is quite colour sensitive, so that some components absorb significantly more heat than others. Colour selectivity is the differential heating of two identical materials of different visible colour. It occurs when the infra-red emission extends into the visible region of the spectrum, and is thus a characteristic of near infra-red sources. This effect also causes non-uniform

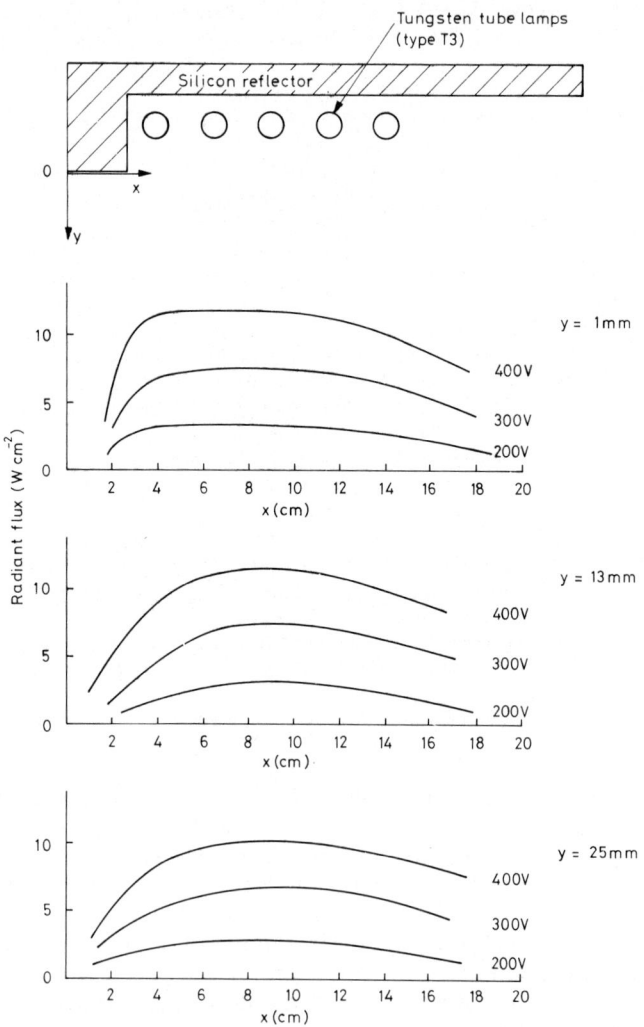

Fig. 7.17 Typical distribution of radiant flux from an array of tubes with a diffuse reflector.[251]

heating of the substrate which, together with the shadowing effect of nearby components, can cause poorly reflowed fillets to occur and board warping associated with temperature gradients.

A more satisfactory mode of use of tubular tungsten filament lamps is at the lower power end of the operating range, without focusing reflectors. Satisfactory solder paste reflow can be achieved but again, because of the wavelength range of the radiation, colour selectivity is a problem.

7.3.2 Nichrome Filament Sources

An alternative tube filament source is the nichrome alloy quartz lamp, similar in construction to the tungsten lamp except that the filament is contained in a non-evacuated quartz tube. The end terminals do not need cooling in this design. These lamps operate most effectively at lower temperatures, namely 750-1400 K, at power ratings of around 15 watts per centimetre and corresponding emission maxima at wavelengths in the range 2·1 to 3·9 μm, in the middle infra-red region of the spectrum. The colour selectivity is not usually a problem provided the lamp is operated at the lower end of the temperature range, i.e., away from the visible spectrum.

The nichrome quartz tube sources operate at lower temperatures, they emit radiation at longer wavelengths which is beneficial, but have a lower overall energy flux.

7.3.3 Area Emission Sources

Infra-red sources which emit from an area, in contrast to a tube, operate on the secondary emission principle. This is accomplished by embedding resistive-element primary emitters in appropriate thermally conductive ceramic based materials, as shown in Figure 7.18. These elements are placed in close proximity to the emitting side of the panel whilst the other side is backed with refractory insulation to ensure efficient emission in one direction only. A thin, low mass, electrically insulating material with a high emissivity is then attached to the front, emitting side of the panel. The heating elements are well protected from oxidation and hence the panels generally have a relatively long life, some 50% increase over filament-tube sources at the same emission densities.

Sources of this type are designed to emit in the temperature range 190-700°C corresponding to the wavelength region 3-6 μm, in the middle-to-far infra-red, as shown in Figure 7.19, compared with the emission range of the primary sources.

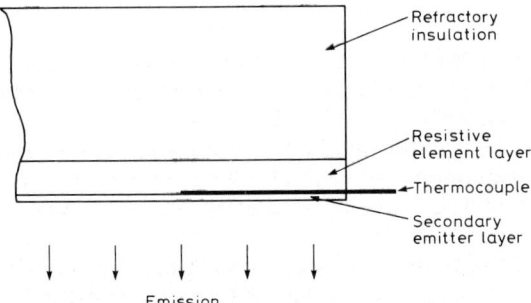

Fig. 7.18 Construction of an area source infra-red emitter.

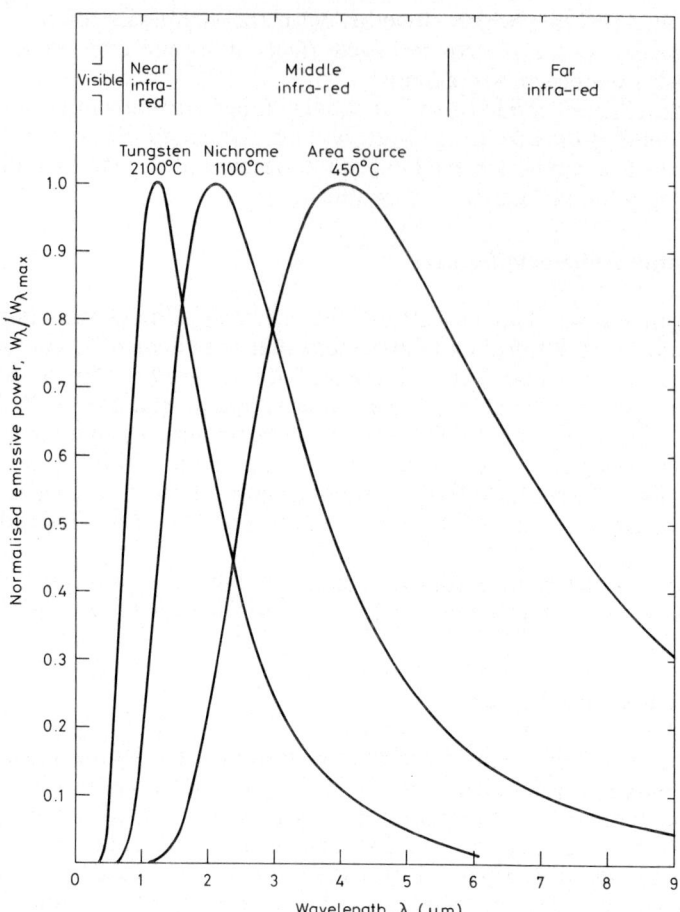

Fig. 7.19 Normalised emissive power of a tungsten tube source, a nichrome tube source and an area source.[252]

Being well away from the visible region of the spectrum there is virtually no colour selectivity problem with area-source panels.

An area emission source has several distinct advantages over the tubular type of source.[251]

(i) Because of the large diffuse area of the source, compared with a series of line sources, shadowing effects are eliminated and the incident radiation is uniform over the entire working area.

(ii) The peak emission of radiation is in the middle-to-far infra-red rather than the near infra-red, hence overcoming substantially the problem of colour selectivity. This has been achieved by using secondary sources which are efficient emitters at low temperatures.

(iii) Because of the secondary nature of the emission, the wavelength is tuneable to that most suitable for the absorption characteristics of the substrate.

(iv) Because of both the much larger area of emission and because of the different energy of the emitted radiation, the area source is much more efficient at heating the air within the furnace. This effect not only increases the rate of heat transfer, but also heats solder where it is not in line-of-sight with the emitter.

Additionally the conversion from primary to secondary emission is very efficient at 80-90% so that very little increase in running costs is incurred.

7.4 WORKPIECE CHARACTERISTICS

In addition to the spectral emittance of the radiation source, an understanding of the spectral absorptivity of the workpiece is required for the successful operation of radiant heating for solder reflow.

7.4.1 Absorptivity of Materials

For metals, as shown in Figure 7.20, the monochromatic reflectivity ϱ_λ usually increases with increasing wavelength such that the most effective heat transfer into a metal occurs with energy of wavelength below 1 μm. It can be seen from the figure that smooth metals all behave very similarly in the infra-red spectrum but appear different from each other to the eye because of the wide spread of reflectivity in the visible region (0·38-0·76 μm). Referring to Figure 7.3, it is seen that from a source at 2500 K about 17% of the total radiated emission is at wavelengths less than 1 μm, and this proportion doubles if the temperature is raised to 3200 K. The higher filament temperatures are thus much more effective in heating metallic surfaces. However, as discussed previously, such radiation,

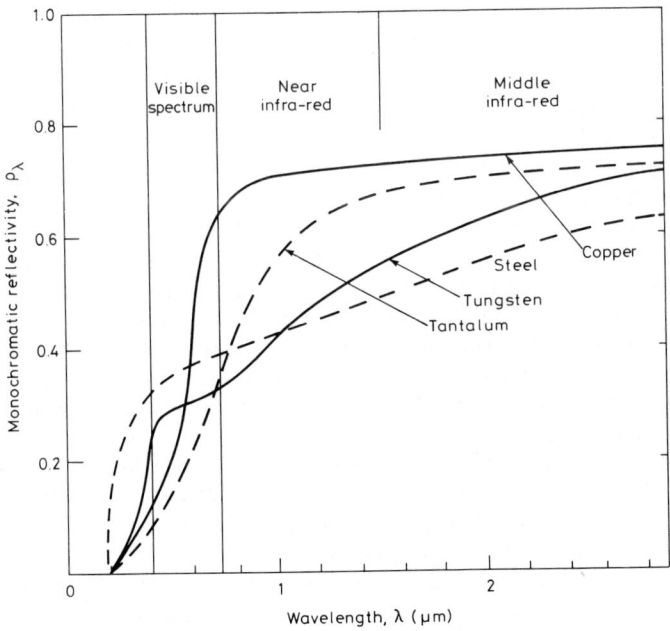

Fig. 7.20 Reflectivity of some common metals.

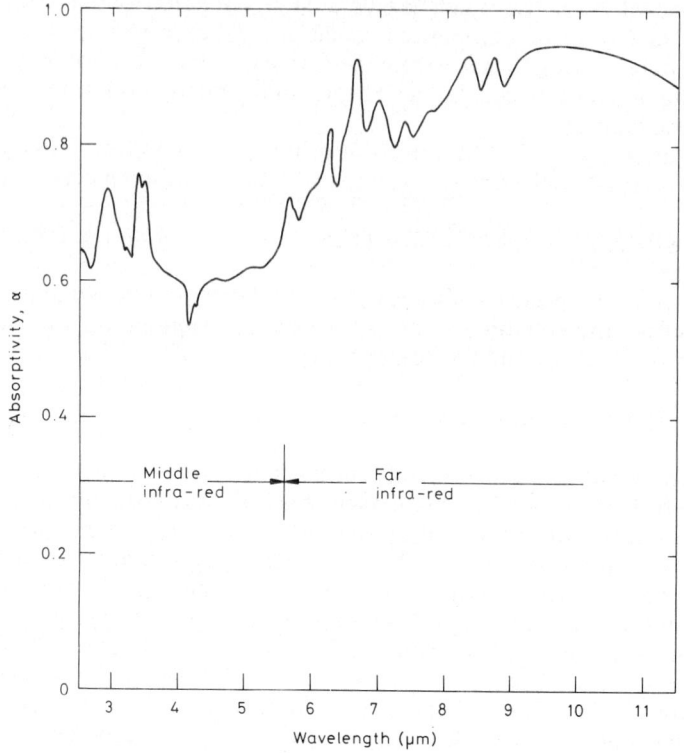

Fig. 7.21 An infra-red absorption spectrum of FR-4 epoxy-fibreglass circuit board laminate.[248]

being near the visible, is colour selective and hence is not conducive to uniform heating of the workpiece.[251]

Insulating materials usually have absorptance characteristics opposite to those of metals, exhibiting low reflectivity and a high absorptivity in the infra-red. Their wavelength dependence is generally not well defined and many materials have distinct absorption bands. As with metals, the reflectivity properties of insulators may vary significantly according to the surface properties and chemical compounding. Epoxy-fibreglass circuit board laminates vary substantially according to the supplier, with absorptivity very dependent on wavelength. A typical infra-red absorption spectrum is shown in Figure 7.21, but because of the wide range of optical properties of materials, both metals and non-metals, it is not useful to generalise.

Surface contamination, including oxides and adsorbed water or oils, results in a wide range of absorption for most metals. Solder coatings vary over a very wide range from 0·1 to 0·8 in their absorptivity depending on their surface roughness if electroplated, the extent of oxidation and the possible use of chemical brighteners during electroplating. In reflow soldering of solder paste, the absorption variation is complex because of the phase change from an essentially dull absorptive surface to a reflective molten metal. In practice, the absorptive properties are also considerably altered by the presence of the flux.

7.4.2 Surface Roughness and Directional Emissivity

Radiative heat transfer is very dependent on the surface rugosity of the workpiece. If the surface roughness or any imperfections are smaller than the wavelength of the radiation, the surface is considered smooth and exhibits a high reflectance.

In addition to its variation with wavelength, the emissivity of many bodies also has directional properties which do not conform to Lambert's cosine law. This is illustrated in Figure 7.22 where the directional emissivity ε_θ is plotted on a polar diagram.[253] For surfaces whose radiation intensity follows Lambert's cosine law and depends only on the projected area, the emissivity curves would be arcs of circles. Measurements, however, show that for non-conductors such as glass or oxide films the emissivity decreases at large values of the emission angle θ, whereas for polished metals the opposite trend is observed. For example, the emissivity of polished chromium, which is widely used as an infra-red radiation shield, is as low as 0·06 for infra-red radiation in the normal direction but increases to 0·14 when viewed near grazing incidence. The importance of the local angle of incidence of the radiation on to the workpiece becomes less important as the radiation wavelength is reduced below 1 μm.

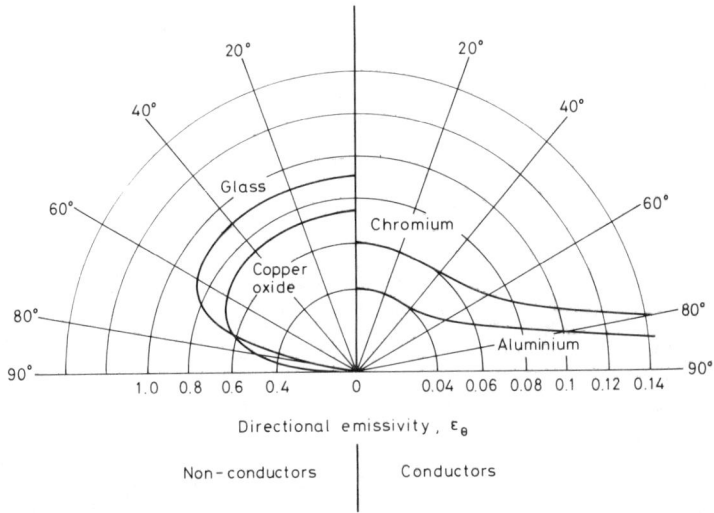

Fig. 7.22 Directional emissivities, ε_θ for polished metals, illustrated by chromium and aluminium, and for insulators, illustrated by glass and copper oxide.[253]

7.4.3 Thermal Degradation

An important consideration when using radiant heat transfer on printed circuit assemblies is that of heat sensitivity and thermal degradation. Many materials commonly used for microelectronic components and their interconnection are heat sensitive either because of a temperature effect or a heat flux effect. Polyester materials used for flexible circuitry are an example of the former kind since they distort at 150°C, some way below the melting point of solder. Epoxy-fibreglass circuit boards, on the other hand, are an example of the latter kind since they can apparently withstand soldering temperatures of up to 260°C for 20 seconds without exhibiting evidence of thermal degradation, but extended times may give rise to charring, discoloration, blistering or measling. Measling is seen as discrete spots in the laminate where glass fibres have internally debonded from the epoxy matrix. The rate of formation of these thermal degradation faults is dependent upon the heating rate, time and temperature. Figure 7.23 illustrates the dependence of thermal degradation on the heat flux. Samples of multilayer circuit board were exposed to predetermined levels of radiant energy flux on one surface whilst supported by the other side such that conductive losses were negligible. The samples were heated until thermal degradation was visibly observed, at which time the incident energy was removed, and the samples were cooled in air to room temperature.

In Figure 7.23 are plotted both the temperature and the time at which thermal degradation was first seen, for a given incident heat flux. Apparently, thermal degradation effects occur at both high and at low incident heat fluxes, at a

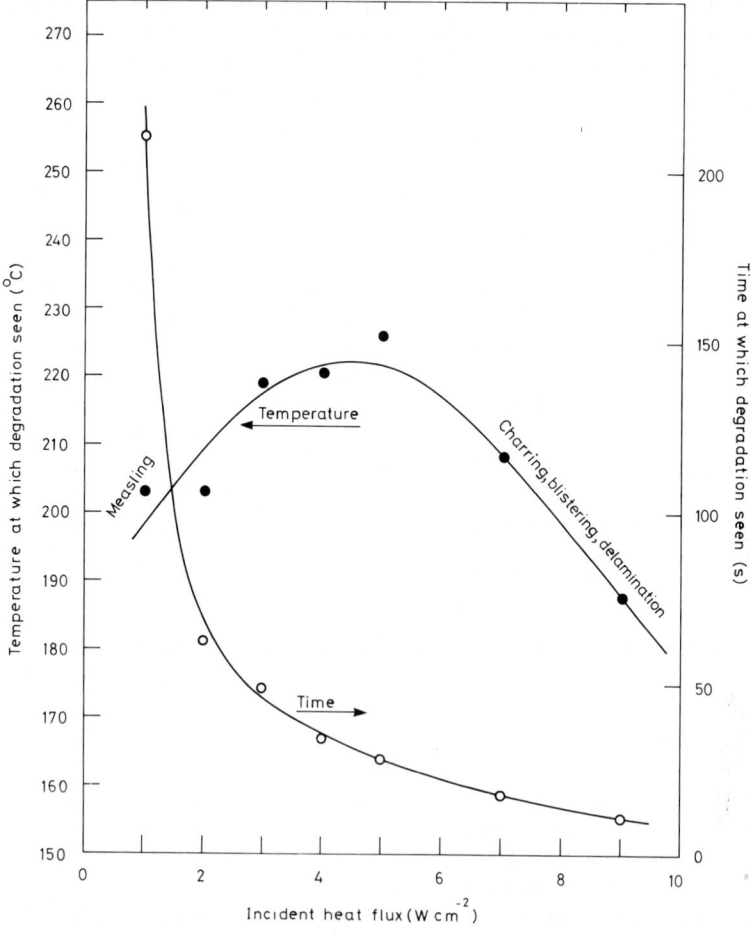

Fig. 7.23 The time and temperature at which thermal degradation was first observed using test samples of laminate subjected to radiation at various heat fluxes.

temperature-time combination less than that required for solder reflow. At high incident radiation flux levels, greater than about 7 W.cm^{-2}, charring, blistering and delamination occur while at low incident radiation flux levels, less than about 2 W.cm^{-2}, measling occurs before solder reflow. An intermediate radiation flux level is therefore indicated in this case.

Thus there is an optimum heat flux to avoid thermal degradation and this will differ for different board materials and, indeed, different component packages. Figure 7.23 serves only to illustrate the problem that can exist.

7.4.4 Geometrical Effects

The rapid heating available when using infra-red radiation induces temperature differentials across the workpiece, as a function of the heating rate and its thermal complexity. The thermal complexity is increased by intricate geometries, low or variable thermal conductivity, variable specific heat, variable mass and variable absorptivity. Absorptivity is usually the least important of these factors except for large, highly reflective surfaces being heated by systems with high infra-red character.

The geometry and mass effects are usually the most predominant and are greatest with materials of low thermal conductivity.[247] This is confirmed by the fact that the first area of a printed circuit board assembly to overheat if processed incorrectly is a corner and not the most absorptive surface or the one with the lowest specific heat. The origin of the geometric effect can be understood with reference to Figure 7.24, considering the surface area-to-volume ratio of elements of the workpiece located at a corner and at the centre. The element at the centre has only one surface whilst the element at the corner has three. Since the heat transfer delivers more energy to the surface than anywhere else, elemental analysis shows that temperature differentials would develop as a function of the thermal conductivity of the workpiece.

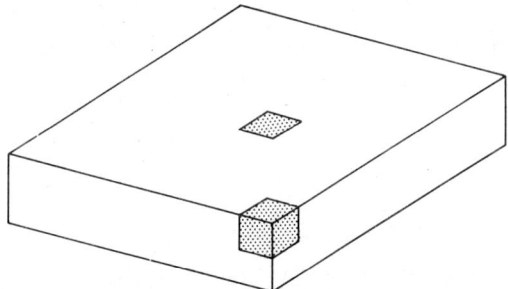

Fig. 7.24 Demonstration of why the corners of boards reach the highest temperatures during infra-red heating.[247]

Two points should be considered for the understanding of the creation of heating processes for the reflow of a complex and heat-sensitive printed circuit board.[254] First, the temperature differentials are almost solely a function of heating rate for any given assembly; the slower the heating rate, the less will be the differential. Secondly, these differentials disappear to within a couple of degrees for any system when it is in thermal equilibrium. However, when using infra-red radiation as the thermal energy source the reflow process is arrested before equilibrium is achieved since it is an excess heating, rather than an equilibrium heating, method. This means that a well defined temperature-time cycle must be decided upon and the process control must be reliably capable of maintaining that cycle.

7.5 THE GASEOUS ENVIRONMENT

The efficiency and effectiveness of the radiant heat transfer can be modified by changing the nature of the environment within the infra-red furnace, since the gas absorbs a proportion of the radiation. Figure 7.25 shows the absorption, by air, of radiation in the infra-red region. It can be seen that sources emitting in the middle to far infra-red (5-8 μm wavelengths), such as the area emission panels, heat air more effectively. By heating moving air in this way, any shadowing effects due to the geometry of the workpiece are greatly reduced. Such shadowing effects are often evident when large components are placed on substrates such as epoxy-fibreglass which are poor thermal conductors.

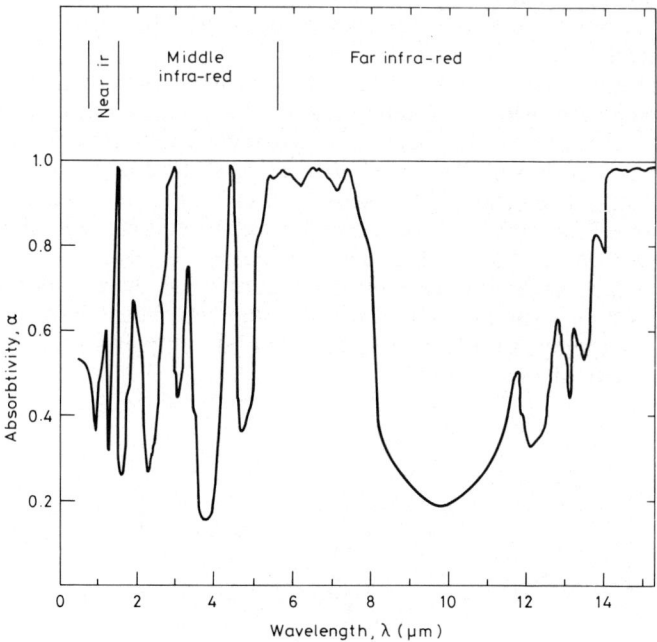

Fig. 7.25 Infra-red absorptivity of air.

7.5.1 Air

Advanced infra-red equipment provides the ability to control the atmosphere very precisely, maintaining contamination levels as low as 1 ppm. Air is the most commonly used gas in a reflow furnace. It is useful for the removal of volatiles and maintaining a directed exhaust flow. However, the oxygen level degrades both the organic and the metallic components when they are at elevated temperatures. In particular, the problem of combustion of the flux in air gives rise to only a small available process-temperature window for the reflow, and the problem of oxidation of metallic parts gives rise to difficulties in subsequent or secondary soldering processes.

7.5.2 Nitrogen

The most common oxygen-free environment for the reflow process is nitrogen, in which the oxygen can readily be maintained at below 5 ppm. This level of oxygen is sufficiently low to eliminate significant oxidation of the metallic parts and to extend the available process-temperature window for printed circuit boards and fluxes. In such an atmosphere, epoxy-glassfibre boards can be exposed to 300°C temperatures without discoloration and degradation, which enables the process time to be reduced typically by 30-40%. Besides the obvious economic and productivity benefits, possible damage to components is lessened because the shorter time reduces temperature excursions experienced by the devices inside their packages. Furthermore, cleaning of the finished board is rendered easier since the flux is not oxidised and burnt.

7.5.3 Hydrogen-nitrogen

The most expensive atmosphere normally considered for the reflow furnace is a hydrogen-nitrogen mixture which is slightly reducing at the reflow temperatures.[255] The benefit of such an atmosphere results to some extent from the reducing effect on metals, although generally not sufficient to clean up oxidised surfaces, but mainly from the effect of the hydrogen on the surface energy of fluxes. As little as 5% hydrogen added to a nitrogen environment increases the contact angle of the liquid flux and consequently reduces the spreading of the flux on polymers and ceramics. This reduction of the spreading of flux on laminate, soldermask and ceramics in surface mounting assembly affects three important areas of the process, as follows.

The reduction of the spreading of the flux reduces the radius of the liquid solder mass formed just before reflow, reducing the solder balling problem. Solder balls are created when solder paste particles spread with the flux but then, on melting, fail to pull back and coalesce on the pad. Secondly, the reduction of the spreading of flux reduces component movement during soldering. Components swim and slide on a film of flux, much less so on molten solder. When a device is supported by a continuous liquid film of flux, external vibrations and other disturbances can cause its location to be translated or rotated. If the device ends up displaced by more than half a pad pitch, the reflow of the solder may seize it into a wrong position. Finally, the increase in the flux contact angle decreases the amount of flux flow under the components. The removal of flux from under low-lying packages is very difficult and can leave damaging ionic residues if not achieved completely. The limitation of the flux to the pad area greatly eases the subsequent cleaning operation.

7.6 INFRA-RED PROCESSING

The temperature cycles that infra-red equipment currently uses to reflow surface mounting solder joints differ from competing heat transfer technologies in several distinct ways. Most importantly, infra-red heating provides controlled pre-heat rates in the critical initial stages during which the volatile organic constituents are removed from the solder paste.

7.6.1 Infra-red Character

There are a number of different types of infra-red emission sources, of which the tungsten tube, the nichrome tube and the panel secondary emitter types are generally used in the electronics assembly industry. Each type radiates in a different portion of the infra-red spectrum and therefore has differing heating effects on the materials of which the furnace is constructed and on the gas that forms the environment. Direct radiation from the source is therefore not the only heating mechanism, but also radiation from the furnace walls, conduction and convection from the gas. The concept of infra-red character[247] of a system can be used to examine the parameters useful in the study of radiative heating systems. Infra-red character, R, is defined as

$$R = Q_r/Q_{cc} \qquad (7.23)$$

where Q_r is the net radiative heat transferred from the furnace to the product and Q_{cc} is the net conductive and convective heat transferred from the furnace to the product. Thus, infra-red character is a measure of how significant radiative transfer is in the heating of an object within a particular environment, and it allows different types of equipment to be placed on a scale, from which the nature of other parameters follows. A chart listing equipment types and the interrelation of parameters through the infra-red character is shown in Figure 7.26.

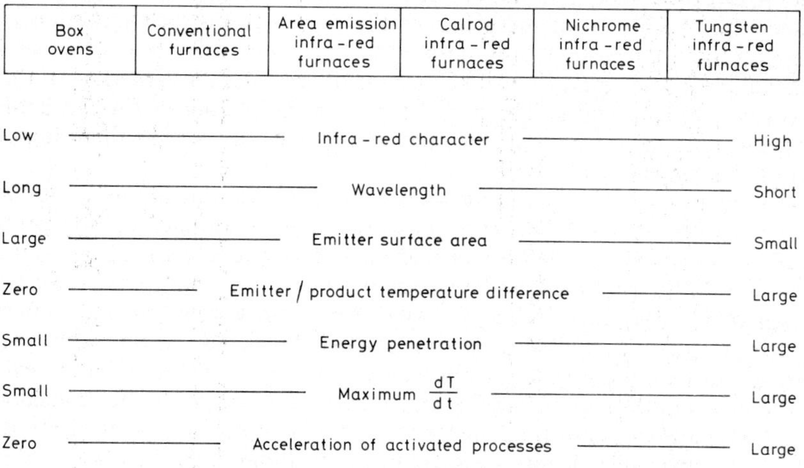

Box ovens	Conventional furnaces	Area emission infra-red furnaces	Calrod infra-red furnaces	Nichrome infra-red furnaces	Tungsten infra-red furnaces

Low	Infra-red character		High
Long	Wavelength		Short
Large	Emitter surface area		Small
Zero	Emitter / product temperature difference		Large
Small	Energy penetration		Large
Small	Maximum $\frac{dT}{dt}$		Large
Zero	Acceleration of activated processes		Large

Fig. 7.26 Infra-red processing equipment can have an infra-red character that is either high or low. Other parameters are then related to this property.[247]

The choosing of a radiative wavelength that is particularly absorbed or transmitted by a substance can only be done after it is established that the environment in which the wavelength dominates is one in which significant heat is transferred by radiation.[256]

7.6.2 Furnace Design

Figure 7.27 shows the basic design of a tube lamp furnace in end view. Furnaces designed for surface mounting assembly are typically constructed with alumina/silica back-up insulation around a firebrick inner shell, the whole having an outer steel case. The conveyor belt rides on quartz rods or wear strips as it moves through the tunnel. Accurate and repeatable conveyor speed is best achieved with a light-sensing closed-loop feedback system directly controlling the motor. The degree of control of the temperature can be designed to requirements, but a ±3°C switching hysteresis is typical.

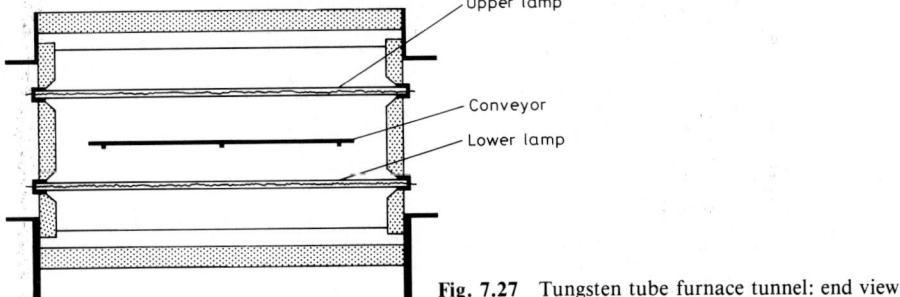

Fig. 7.27 Tungsten tube furnace tunnel: end view.

The nichrome tube furnace is constructed similarly, the main difference between it and the tungsten tube furnace being the lower emission temperature and the consequent lower radiant flux. With both types of tube it is not very satisfactory to run them at temperatures much below their rated values with a view to reducing the exposure of the workpiece to the radiation. This is because, at low-temperature lamp operation, the emission is not black-body and also the longer wavelength radiation is absorbed in the quartz or glass of the tube. Consequently, if a reduction of exposure is required, the speed of the conveyor must be increased, which is not generally a positive feature when processing materials with poor thermal conductivity such as epoxy-glass laminate.

Figure 7.28 is a schematic end view of an area source furnace suitable for soldering surface mounting assemblies. The process area is constructed with the emitter panels as the internal tunnel-wall medium. If inert atmospheres are to be used, the panels can be tightly sealed together.

The inert gas atmosphere can be introduced by way of the support rods carrying the conveyor or directly through the inner wall porous insulation. This

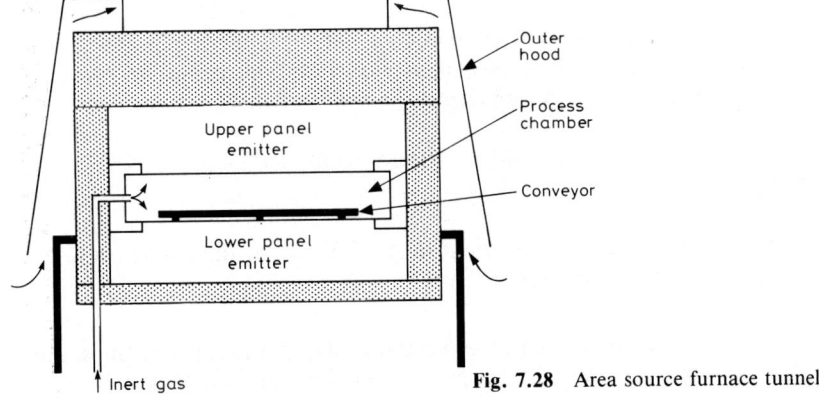

Fig. 7.28 Area source furnace tunnel: end view.

latter method of gas distribution, as shown in Figure 7.29, results in a very uniform and low gas velocity throughout the chamber, and eliminates variations in flow rate across the tunnel that arise when the gas is introduced in a laminar flow manner.

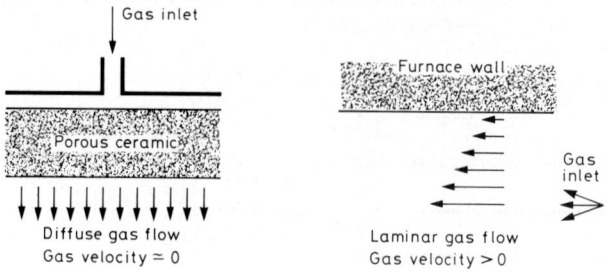

Fig. 7.29 Infra-red reflow can be carried out in a controlled atmosphere. The gas can be diffused into the workspace through the porous walls or flowed in from a series of orifices.

7.6.3 Estimate of Heating Requirements

Before purchasing or using any given infra-red reflow system, the user must confirm that adequate radiant power is available and can be delivered to the workpiece, to keep the process temperature within specified limits during the reflow cycle. In an ideal case the heating requirements can be calculated very precisely but for surface mounting assembly we do not have sufficient knowledge concerning the thermal properties of the materials or the boundary conditions of the reflow process. In this case it is futile to attempt such a rigorous calculation, and it is possible only to make an estimate which can then be confirmed by practical experiment.

The total heating energy available, Q (joules), for the reflow process is defined as being equal to the product of the net heat input, W_n (watts), and the heating time, t_h (seconds),

$$Q = W_n t_h$$

and $W_n = w A\alpha - W_\ell$ \hfill (7.24)

where w (watts.m^{-2}) is the incident radiant flux, A(m^2) is the irradiated surface area to be reflowed and α the total absorptivity of the solder paste. W_ℓ (watts) is the heat loss. The heating energy required is the sum of three parts:

Q_1, that required to raise the solder particles from room temperature to their melting point;

Q_2, that required to supply the latent heat of fusion of the solder and hence melt it while not increasing its temperature;

Q_3, the energy required to raise the molten solder to the reflow temperature at which the molten solder wets and spreads over the component pads.

In principle Q_1, Q_2 and Q_3 can be calculated for each joint on the board and the time and energy required for reflow then estimated. For a single joint, Q_1 is the sum of heat energy required to raise the temperature of each component of the joint through a temperature excursion ΔT_1, from room temperature to the melting point of the solder:

$$Q_1 = \sum_i V_i \varrho_i c_i \Delta T_1 \qquad (7.25)$$

where V_i (m³) is the volume, ϱ_i (kg.m⁻³) the density and c_i (J.kg⁻¹.K⁻¹) the specific heat of each component part, i.e., the solder paste, the metallisations, etc. Q_2 is the heat energy needed for the latent heat of fusion of the solder

$$Q_2 = V_s \varrho_s \lambda_s \qquad (7.26)$$

where V_s (m³), ϱ_s (kg.m⁻³) and λ_s (J.kg⁻¹) are, respectively, the volume, density and latent heat of the solder component part.

Also, similar to Q_1:

$$Q_3 = \sum_i V_i \varrho_i c_i \Delta T_2 \qquad (7.27)$$

where ΔT_2 is the temperature excursion from the solder melting point to the reflow temperature. Now, however, in Equation (7.27) the density and specific heat of the solder take values appropriate to the liquid metal, whereas in Equations (7.25) and (7.26) the values for solid solder are used.

The above equations relate to the thermal requirements. The efficiency of supplying this heat from a radiant source depends on the absorptivity of the component parts of the joint and also on the heat loss, occurring mainly by conduction away from the joint through the copper and the laminate.

7.6.4 Heating Cycles

A basic tube-emitter infra-red furnace comprises two temperature zones, and can be used to reflow solder paste on epoxy-glassfibre laminate. Typically, the first zone runs at about 1200°C with peak emission at a wavelength of 2 μm, to preheat the laminate and the solder uniformly. The second, reflow zone is set at around 2100°C with a peak wavelength at 1·2 μm. This shorter wavelength energy is absorbed by the solder but transmitted by the laminate, with the effect that the solder temperature rises while maintaining a lower laminate temperature. The resultant temperature profiles of solder on epoxy-glassfibre laminate when using this preheat-spike system are shown in Figure 7.30. The need for the preheat is to ensure that the temperature differentials between the various parts of the board are kept to a minimum. Only when the soldering temperature is approached does the thermal spike raise the solder temperature significantly above that of the laminate and components.

The furnace type that has proved most useful for solder attachment of surface mounting devices has four zones. The first zone is the same preheat used in a two-zone furnace. The second is of high infra-red character, i.e., heating is chiefly by radiation, in which heating is accomplished very rapidly. The heating is at wavelengths which allow safe, rapid removal of the volatiles from the paste. Heating rates in this zone are usually in the range 2-7°C.s^{-1}, which are high enough to afford fast processing but not so high as to stress ceramic components. Because the heating is rapid in this zone, temperature differentials are created

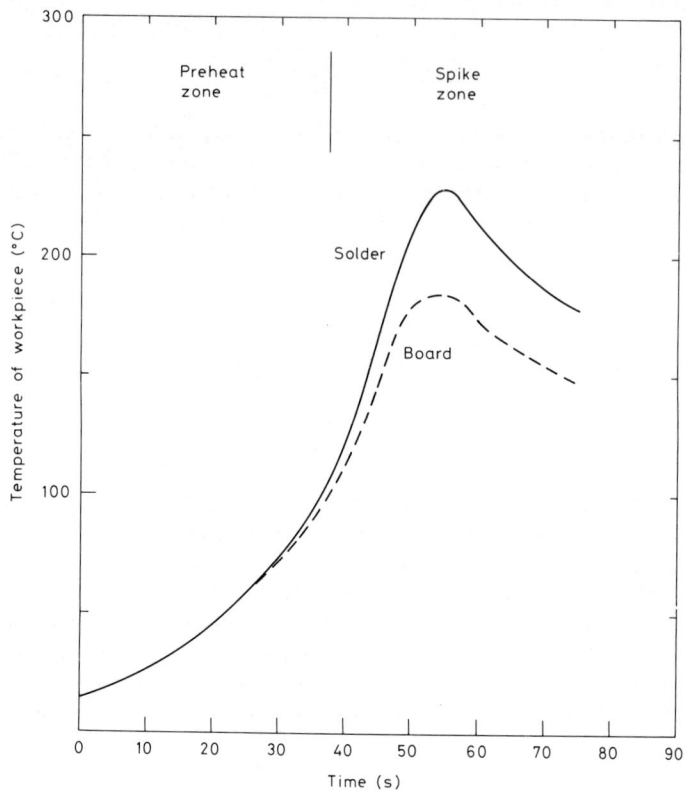

Fig. 7.30 Typical temperature-time profile of a workpiece passing through a two-zone radiant emission furnace.

within the board. If maintained, this heating rate would ensure that these temperature differentials were unacceptably high during the solder reflow. In order to avoid the thermal damage caused by such differentials, a third zone is used to allow the temperatures to equilibrate. This zone operates with its source at a lower temperature, near the equilibrium temperature, and is a zone of low infra-red character. Within the zone the heating rate is around 0·5°C.s^{-1}, allowing the temperature differentials created in the fast heating zone to disappear. The board equilibrates at a temperature just below the melting point of solder, i.e., about 170°C. The final zone returns to a regime of high infra-red character. The heating rate is high and the reflow time is tightly controlled and minimised. This affords the lowest possible exposure of components to high temperature and at the same time, because of the equilibrating stage, keeps the temperature differentials to a minimum. A typical temperature-time profile of a four-stage furnace is shown in Figure 7.31.

The heating is usually applied both from below and from above the board, but the two need not necessarily be at the same temperature. In the fast heating ramp zone the temperature of the heaters beneath the workpiece is higher, such that the shorter wavelength radiation penetrates into the board and results in more uniform heating of the laminate. The radiation incident on the top surface is of longer wavelength so as not to heat the devices unduly inside their packages. In the reflow spike zone, most of the heating is applied from the top, to heat the solder only, and is of short wavelength (1·2 μm) to be better absorbed by the solder.

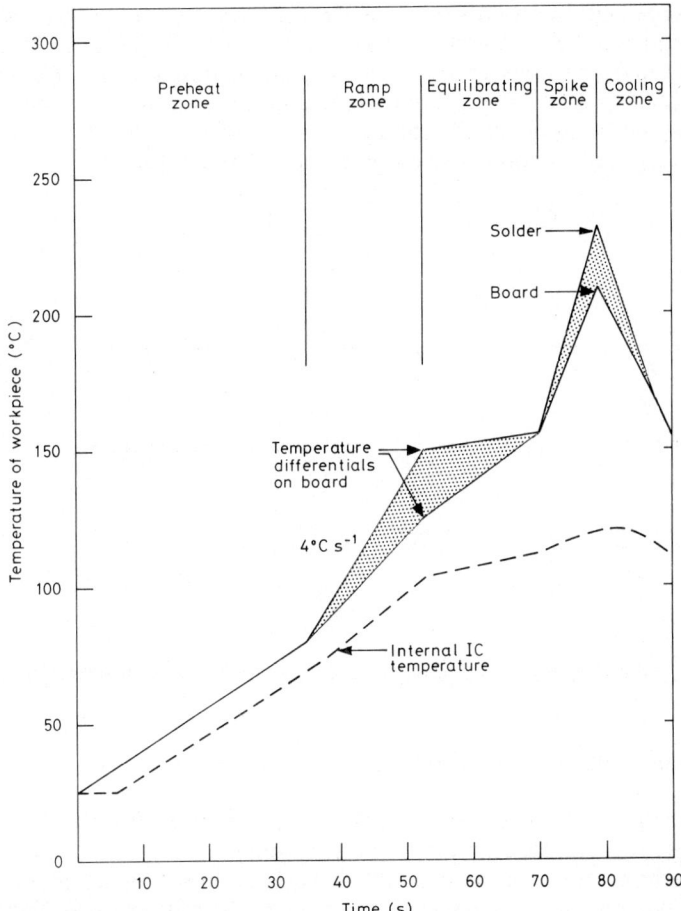

Fig. 7.31 Typical temperature-time profile of a workpiece passing through a four-zone radiant emission furnace.

The temperature profile of an assembly passing through an area emitter furnace is very similar to that associated with a tube furnace. The temperatures of the emitters themselves, however, are much lower and the process is more closely operated in an equilibrium heating regime than an excess heating regime. The system, then, has a lower infra-red character. A commercial arrangement of area emitters is as shown in Figure 7.32. The panel temperatures are monitored with thermocouples directly behind the secondary emitting surface and typical values are shown in the figure. In this particular design, the first preheating is mainly on the top surface to remove volatiles from the paste, which are allowed to escape before further rapid heating from panels 2 and 3. The equilibrating zone is defined by panels 4 and 5, the reflow spike zone by panels 6 and 7.

For the efficient use of infra-red reflow furnaces it is invariably necessary to conduct a thermal-profiling procedure to adjust the conveyor speed and the temperatures of the zones until satisfactory reflow is achieved for the particular assembly. To reduce the number of variables, it is usual that all of the temperatures are kept in proportion to each other. One method of monitoring the profile is to attach a thermocouple to the top surface of a fully loaded board. The thermocouple must be shielded from direct incident radiation by, for example, burying it in the top layer of board or under a J-lead of a component and in either case attaching it with a small quantity of porcelain cement. If only one thermocouple is used, it should be close to the centre of the board.

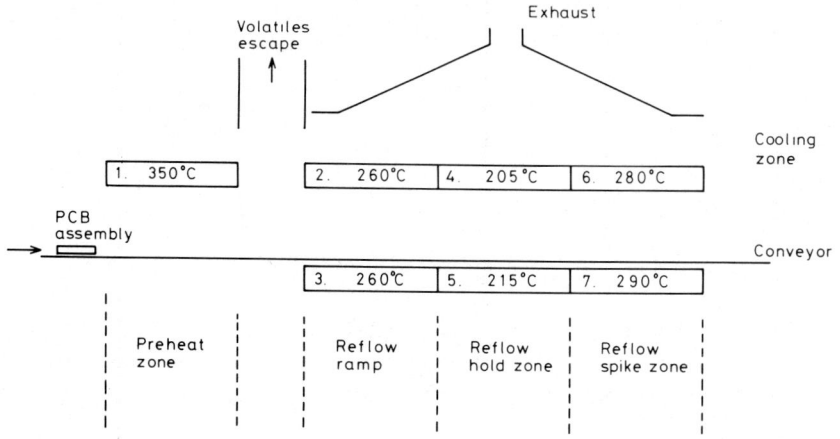

Fig. 7.32 Schematic of a typical infra-red reflow furnace using area panel emitters.[252]

7.6.5 Instrumentation and Control

Radiant heating reflow systems require three main control functions: the radiant power, the exposure time and the atmosphere.

Several methods are used to provide a process control signal to the power controller. The transducer used measures either the radiant heat flux or the temperature in the work zone, and provides a suitable voltage. Either a thermocouple or a radiometer can be used, but care must be taken when the simpler and cheaper thermocouple is chosen. The thermocouple reading depends on its size, shape, emittance and thermal response. It must be small enough for a steady state temperature to be reached in a time period significantly less than the time required for the process being controlled. It is also important that the emittance of the thermocouple remains unchanged and therefore not become contaminated by flux fumes. In general a thermocouple will provide only relative measurements of radiant flux, and must be calibrated using a radiometer. A Gardon thin circular foil radiometer[257] is suitable, which delivers a voltage that is directly proportional to the intensity of radiation over a range of about 20-1500 $W.cm^{-2}$, with a time constant typically 50 milliseconds.

The most accurate methods of process control involve the direct measurement of the workpiece temperature. Non-contacting methods using one of the many types of optical pyrometer introduce errors through the workpiece emittance, and through the reflectance of the incident radiation. Thus the best method requires the placing of a thermocouple directly on the workpiece in a non-destructive manner. The thermocouple must read the temperature of the workpiece and not be influenced by the incident radiant energy.

Chapter 8

SOLDER REFLOW BY FLUID HEAT TRANSFER

8.1 INTRODUCTION

The heat required to reflow the solder paste in a surface mount assembly can be supplied by fully immersing the workpiece in a hot fluid, be it a gas, a condensing vapour or a liquid. In these reflow methods, heating is applied from all directions. The fluid is maintained at a constant temperature and the workpiece is heated until its temperature has asymptotically approached that of the fluid. In this respect the principles and practice of the heating are quite different from heat transfer by radiant or by conducted heating, in which the temperature of the heat source is much higher than the desired soldering temperature such that excess heat is available and the soldering of the workpiece is controlled by the time. Thus hot fluid reflow is an equilibrium process and therefore intrinsically a more reproducible process.

By definition, a fluid may exist in three distinct states—gas, vapour or liquid. Each of these physical states has a unique set of properties and so has led to three branches of the technology of heat transfer fluids. The science, however, is essentially the same.

The use of a hot gas, namely air, as the heat transfer fluid has been used successfully for reflow soldering, for example for the soldering of pins into boards[259] and also for the fusing of solder coatings on boards,[260] but the prolonged heating time in hot air furnaces makes the method unsuitable for use with solder paste. Generally, a time of about five minutes is required, during which the flux is consumed and its activity insufficient by the time the solder melts. The most popular fluid in use for heat transfer to surface mount assemblies is a condensing vapour of an inert fluorocarbon fluid, a process known commonly as vapour phase soldering or condensation soldering. With the development of new fluorocarbon fluids that remain liquid at soldering temperatures, the process of liquid phase soldering has also become a recent practical possibility.[261]

This chapter will consider first the principles of heat transfer by a fluid before discussing the mechanisms involved in the use of a condensing vapour and a liquid. Finally, the practical and engineering aspects of these reflow methods will be covered.

8.2 HEAT TRANSFER FROM A FLUID

8.2.1 Heat Transfer Coefficient

The primary relation between the temperature T_s of a solid surface submerged in a fluid at temperature T_f and the rate of heat transfer dQ/dt ($J.s^{-1}$, namely watts) through the surface was formulated by Newton:

$$\frac{dQ}{dt} = hA\,(T_f - T_s) \tag{8.1}$$

where A is the total surface area exposed to the fluid and h, the constant of proportionality, is called the heat transfer coefficient ($W.m^{-2}.K^{-1}$).

The rate of heat transfer is related to the rate of temperature rise of the submerged workpiece by its heat capacity $C(J.K^{-1})$:

$$\frac{dQ}{dt} = C\,\frac{dT}{dt} \tag{8.2}$$

and, in turn, the heat capacity is a product of the density $\varrho(kg.m^{-3})$, the specific heat $c(J.kg^{-1}.K^{-1})$ and the volume $V(m^3)$ of the submerged object:

$$C = \varrho cV \tag{8.3}$$

The rate of heat transfer is the property of concern in determining the necessary time of immersion in a fluid to accomplish soldering. Since the definitions of A, T_s and T_f in Equation (8.1) are the same whether the fluid is a gas, vapour or liquid, the controlling parameter is the heat transfer coefficient, h.

The heat transfer coefficient is an extremely complex function of the physical properties and state of the fluid as well as the geometry of the system and the fluid velocity across the surface. Four basic equations for the heat transfer coefficients in the situations of natural fluid convection, forced fluid convection, condensing vapour and boiling liquid, have been derived.[262] It is therefore possible, in principle, to calculate the appropriate value of h but, in general practice, values have to be measured experimentally. Since h depends upon the specific geometry used, quoted values are taken only as relative or average for a typical geometry, of a populated printed circuit board, for example. The properties of the fluid upon which h depends are its viscosity, its density, its thermal conductivity, and its specific heat.

Combining Equations (8.1)–(8.3):

$$\varrho cV\,\frac{dT}{dt} = hA(T_f - T_s) \tag{8.4}$$

and hence the temperature rise $T_s(t)$ of the workpiece at time t after immersion is given by:

$$(T_s - T_o) = (T_f - T_o)(1 - \exp^{-t/t_o}) \tag{8.5}$$

in which T_o is the starting temperature of the workpiece and t_o is the characteristic time of the exponential temperature rise from T_o towards the equilibrium temperature T_f. The characteristic time of the exponential temperature curve is such that, when $t = t_o$, the temperature T_s has risen 63% of the way to the equilibrium temperature T_f. From Equations (8.4) and (8.5):

$$t_o = \frac{\varrho c V}{hA} \tag{8.6}$$

and hence

$$\ln\left(\frac{T_f - T_s}{T_f - T_o}\right) = -\frac{hA}{\varrho c V} t \tag{8.7}$$

which is the equation of practical usefulness for comparing experimental measurements in regard to the heat transfer capabilities of fluids. The validity of Equation (8.7) for use in soldering has been discussed at length.[263]

In Equation (8.7), ϱc is the specific heat of the workpiece which, for almost all materials, lies within a relatively narrow range, $1-5$ MJ.m^{-3}K^{-1}, so that for a given workpiece it is h which significantly controls the soldering time, since its value may range over three orders of magnitude. The choice of fluid is therefore crucial. Not only is a high speed of soldering important for economy of production, but also it reduces the possibility of thermal damage to components and minimises the amount of intermetallic compound formed in the solder joint.

8.2.2 Heat Flow in the Workpiece

When the object to be heated is immersed in the fluid (gas, vapour or liquid) the time taken for its temperature to reach that of the fluid is not governed only by the heat transfer through the surface (defined by the heat transfer coefficient, h, of the fluid) but also by the heat conductivity of the object itself (defined by the thermal conductivity coefficient, K, of the solid). The thermal conductivity of the workpiece, whilst not affecting the heat transfer rate when in the fluid, does control the rate of rise of T_s towards the value of T_f. Thus a perfect insulator attains thermal equilibrium at the boundary layer immediately whereas a good thermal conductor with a thermal path out of the fluid never attains equilibrium. The real situation lies between these two extremes.

The temperature profile of the object at time t after immersion can be calculated for the particular materials and geometry involved. Consider the case of a plate of thickness 2ℓ immersed in the hot fluid at temperature T_f, for example a printed circuit board or an alumina substrate. The temperature of the surface atom plane of the solid must be that of the adjacent fluid atoms and we assume that the surface of the workpiece is at temperature T_f. The temperature $T(z)$ at a distance $-\ell < z < \ell$ from the centre of the plate is given by [40]

$$\frac{T(z) - T_o}{T_f - T_o} = 1 - \frac{4}{\pi} \sum_{n=0}^{\infty} \frac{(-1)^n}{2n+1} \exp\left\{\frac{-(2n+1)^2 \pi^2 \beta}{4}\right\} \cos\left\{\frac{(2n+1)\pi\xi}{2}\right\} \tag{8.8}$$

where $\xi = z/\ell$ and $\beta = \varkappa t/\ell^2$, both dimensionless parameters. Here $\varkappa$ is the thermal diffusivity (m^2.s^{-1}) of the material of the plate which itself is a function of the thermal conductivity K(W.m^{-1}.K^{-1}), the specific heat c(J.kg^{-1}.K^{-1}) and the density ϱ(kg.m^{-3}):

$$\varkappa = \frac{K}{\varrho c} \tag{8.9}$$

The functional form of T(z) given by Equation (8.8) is shown in Figure 8.1 for a range of β values. The temperature profile at time t is clearly strongly dependent upon $\varkappa$ for a given plate thickness. Values of $\varkappa$ are given in Table 8.1 for some electronic materials.

For copper, the thermal diffusivity is very high and, after immersion of a copper plate in a hot fluid, the thermal front has progressed 1 mm into the metal in about ten milliseconds. In the same time, the thermal front would have progressed only about 30 μm into epoxy. Consider, by way of illustration, plates of each material given in Table 8.1, 1·6 mm thick ($\ell = 0·8$ mm). From Figure 8.1, the centre plane of the plate (z = 0) reaches 90% of the surface temperature when $\beta \approx 1$. The time for this situation to be achieved is therefore $6·4 \, 10^{-7} \varkappa^{-1}$ and these times are given in the last column of Table 8.1. The rate of increase of the

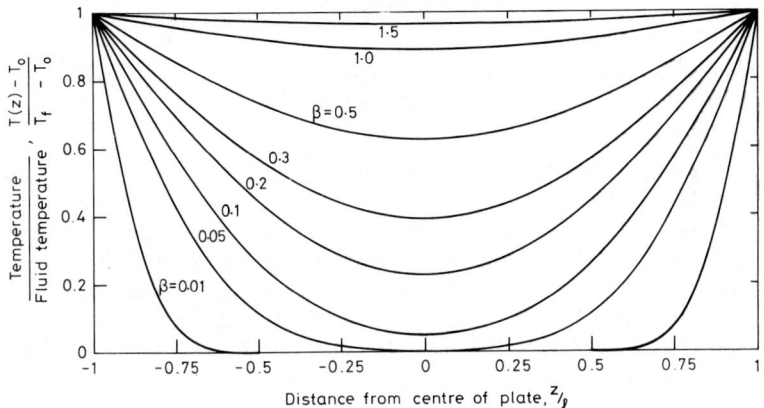

Fig. 8.1 The temperature T(z) at a distance z from the centre plane of a flat plate of thickness 2ℓ, for a range of thermal parameters $\beta = \varkappa t/\ell^2$, immersed in a fluid at temperature T_f.[40]

Table 8.1

Thermal Diffusivity Data of Some Electronic Materials

	Thermal Conductivity	Specific Heat	Density	Thermal Diffusivity	Time for Centre of 1·6 mm Thick Plate to Reach $0·9 \, T_f$
	K	c	ϱ	$\varkappa = \dfrac{K}{\varrho c}$	
	$W.m^{-1}.K^{-1}$	$J.kg^{-1}.K^{-1}$	$kg.m^{-3}$	$m^2.s^{-1}$	milliseconds
Copper	370	390	8920	$1·1 \, 10^{-4}$	6
Solder	50	180	8155	$3·4 \, 10^{-5}$	19
Alumina	35	1000	4020	$8·7 \, 10^{-6}$	74
Glass	0·8	630	2250	$5·6 \, 10^{-7}$	1130
Epoxy	0·15	1500	1350	$7·4 \, 10^{-8}$	8640

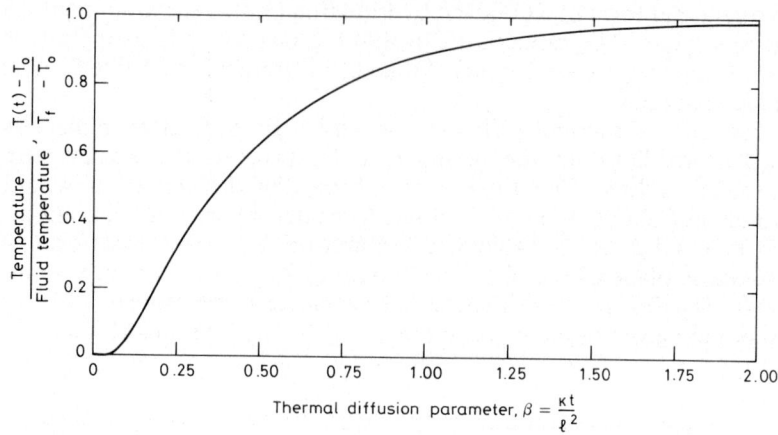

Fig. 8.2 The rate of temperature increase at the centre of a flat plate with thermal diffusivity κ, immersed in a fluid at temperature T_f.

centre-plane temperature is shown in Figure 8.2 which is simply the function given in Equation (8.8) at $z = 0$.

The significance of these data concerns both the experimental measurement of heat transfer coefficients and the practicalities of soldering by fluid immersion. In most cases of reflow soldering by total immersion in a heat transfer fluid, however, the thermal gradient across the assembly is quite small. Consequently the necessary total energy, Q(joules), for the workpiece to reach the fluid temperature is simply

$$Q = \varrho c V (T_f - T_s) \tag{8.10}$$

8.2.3 Evaluation of Heat Transfer

Heat transfer coefficients of fluids operating under different conditions vary over several orders of magnitude and some values are given in Figure 8.3 for oils and glycerol used in the past for liquid phase soldering, together with fluorocarbon fluid commonly used for vapour phase soldering. The values are compared with those for water and air. It is immediately clear that forced convection of a liquid significantly improves its heat transfer properties. What is less certain is the possibility of any improvement to be gained by using a condensing vapour rather than a liquid. It has often been suggested that,[264] because a vapour can release its latent heat very quickly upon condensing on to a cold surface, the heat transfer rate could be substantially greater than that derived from a convecting liquid. Condensing steam has a heat transfer coefficient twice that of boiling water and ten times that of agitated hot water. However, for the fluorocarbon liquid the latent heat of vaporisation is very much less and the change of phase apparently contributes little to the heat transfer. Mean values for h are given[265] as 115, 1830 and 750 W.m^{-2}.K^{-1} respectively for the fluorocarbon condensing vapour, the boiling liquid and the agitated liquid, i.e., the condensing vapour phase is a less efficient heat transfer medium than the liquid. Elsewhere[266] a value of 570 W.m^{-2}.K^{-1} for condensing fluorocarbon

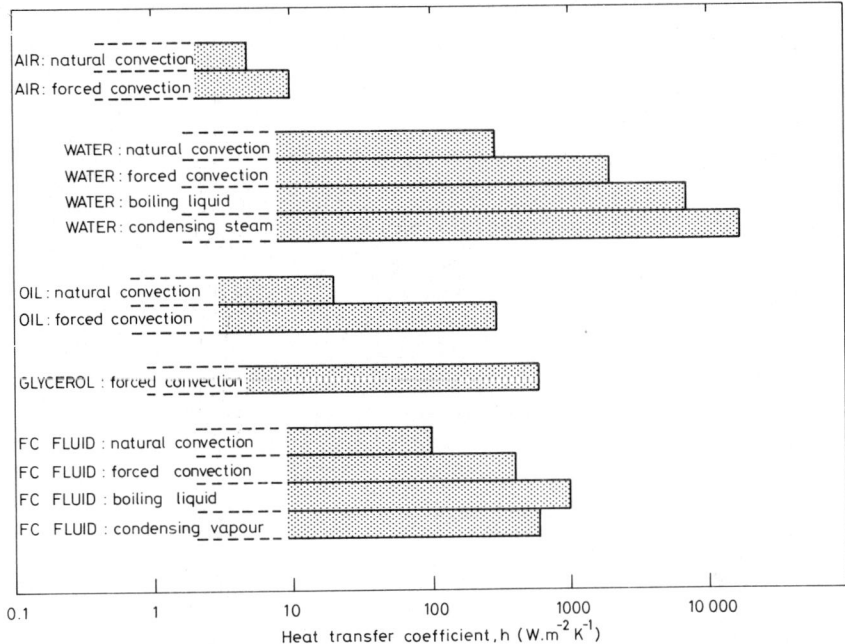

Fig. 8.3 Some measured heat transfer coefficients taken from published values.[261]

vapour is quoted but this is the same value given for liquid glycerol, again indicating no heat transfer advantage from the vapour of fluorocarbons. Also, in the original work on vapour phase soldering[267] using fluorocarbons, the workers obtain a heat transfer coefficient whose 'values are about the same as for immersion in a heated liquid'.

Experimentally, the heat transfer coefficient can be calculated from the measured thermal response of a transient plate calorimeter when immersed in the fluid. Comparative measurements using the same calorimeters and measurement chain have been made for fluorocarbon liquids and condensing vapours[261], some of which are shown in Figures 8.4 and 8.5. From Equation (8.7) a plot of $ln(T_f - T_s)/(T_f - T_o)$ against time t should yield a straight line, from the slope of which the heat transfer coefficient h can be calculated. Figure 8.4 presents measurements made in a fluorocarbon liquid at three temperatures and shows the effect that stirring the liquid has on the measured heat transfer. If the liquid is not stirred, the movement of liquid across the sample surface is controlled only by natural convection currents. This increases with increasing temperature and hence the measurements give a temperature dependent heat transfer coefficient as shown.

At specimen temperatures T_s in excess of about 90°C an approximately constant heat transfer coefficient of 400 $W.m^{-2}.K^{-1}$ has been measured. Thus, above 90°C the temperature of the immersed object rises in an exponential fashion to approach asymptotically the temperature of the liquid. Initially, however, at low specimen temperatures, a much higher heat transfer coefficient is found. This arises prior to the attainment of a thermal gradient in the liquid. Since the calorimeter is capable of conducting heat away from the solid-liquid

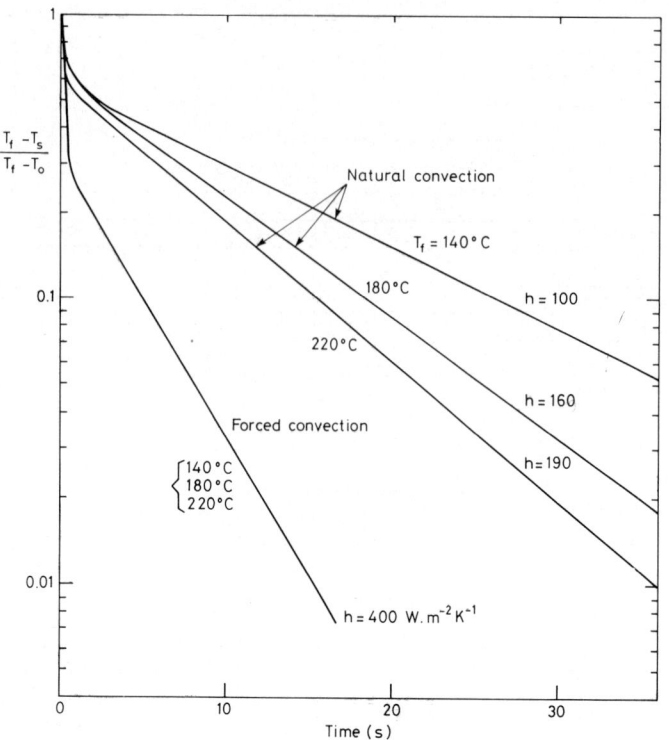

Fig. 8.4 Heat transfer measurements made in a fluorocarbon liquid at three temperatures, showing the effect of stirring the liquid.

interface faster than it can be supplied by the liquid, the temperature of the liquid at that interface must be reduced. Equation (8.1) assumes the thermal gradient in the liquid close to the interface to be of constant width; the heat transfer coefficient measured is only constant when this situation pertains. The measurements of sample temperature against time in Figure 8.4 must therefore show an instantaneously high heat transfer at time zero, until the temperature drops in the adjacent liquid.

Some typical data for heat transfer from a condensing vapour of a fluorocarbon, measured using the same calorimeter, are given in Figure 8.5. In common with the liquid heat transfer medium, at temperatures in excess of about 120°C the heat transfer coefficient is constant, with the specimen temperature rising exponentially towards that of the vapour. On vertical surfaces the measured coefficient is about 400 $W.m^{-2}.K^{-1}$, closely comparable to that of the liquid. On horizontal surfaces it is somewhat less, about 300 $W.m^{-2}.K^{-1}$. This difference will be discussed in the next section. However, in contrast to the liquid, immediately after immersion and during the initial stages of heating, the measured heat transfer coefficient is lower. This is because there are two phase interfaces present: the solid-condensate and the condensate-vapour, and at time $t=0$ there is no requirement for a step function in the temperature profile at either interface.

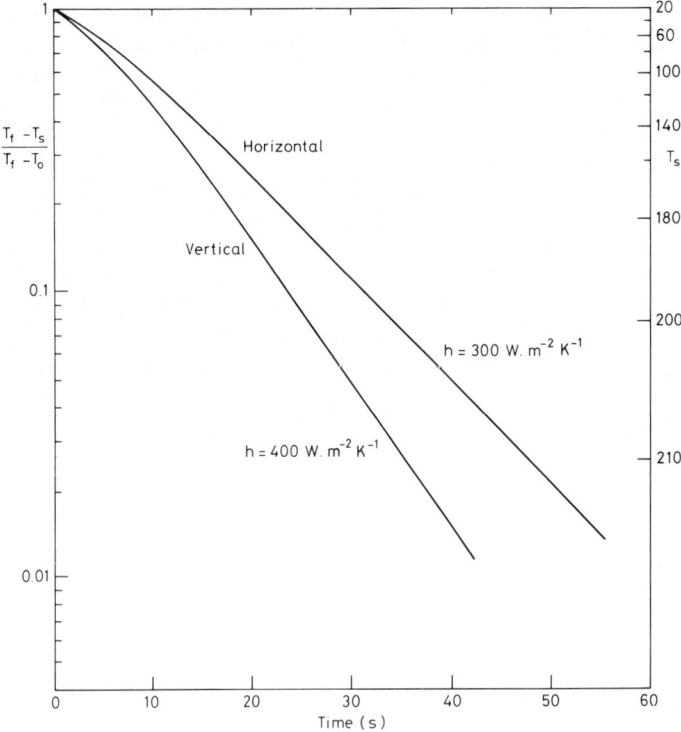

Fig. 8.5 Heat transfer measurements made in a condensing fluorocarbon vapour using the same calorimeter as used in Figure 8.4, showing a difference between horizontal and vertical surfaces.

These experimental data demonstrate that there is no significant difference in the heat transfer coefficients of typical fluorocarbon condensing vapours and liquids. A greater effect exists, depending upon the orientation of the workpiece surface to the vertical. An explanation of this effect will now be given.

8.2.4 Condensation Heat Transfer

Vapour phase or condensation soldering is a transient condensation heat transfer process. However, the transient is slow compared with the time required to attain steady film condensation, and the heating process can be considered essentially as being quasi-steady state, and the steady state heat transfer equations used.[268] Physically this means, as is evident from Figure 8.5, that the film establishes itself at a rate faster than that of the change of temperature of the workpiece. The error in making this assumption decreases rapidly as the temperature of the workpiece increases.

When a vapour comes into contact with a solid surface below its condensation point (at the pressure concerned), the vapour will normally release its latent heat of vaporisation and condense either to form discrete droplets on the surface and run off the surface as rivulets or, if the liquid wets the surface, to form a continuous thin liquid film. The former is termed dropwise condensation, the latter film condensation. Dropwise condensation is heavily dependent on the surface condition and the heat transfer coefficient changes as the surface ages. On the other hand, film condensation tends to be more consistent as long as a continuous film forms. The fluorocarbon fluids used for vapour phase soldering have low surface tension and wet the surfaces well. Therefore only filmwise condensation need be considered in detail.

An equilibrium thickness of the liquid film is reached when the rate of growth by condensation is balanced by the rate of removal of fluid by hydrodynamical flow. When the film attains a state of constant thickness it also attains a state of constant thermal resistance. Assuming the heat is delivered to the film solely as latent heat, the problem of heat flow becomes essentially a problem of transport of matter and can in principle be calculated as follows.

The temperature of the submerged surface T_s is related to the mass flow rate of condensing fluid $M(kg.m^{-2}.s^{-1})$ by[269]

$$T_c - T_s = \frac{M \delta \lambda_f}{K_L} \tag{8.11}$$

where T_c is the temperature of condensation, K_L $(W.m^{-1}.K^{-1})$ is the thermal conductivity of the liquid film and δ is its thickness. $\lambda_f (J.kg^{-1})$ is the latent heat of condensation of the fluid. Furthermore, the heat transfer rate dQ/dt $(J.s^{-1})$ from the film to the solid surface, across an area A (m^2), is related to the mass flow:

$$\frac{dQ}{dt} = AM\lambda_f \tag{8.12}$$

Combining Equations (8.11) and (8.12):

$$\frac{dQ}{dt} = \frac{AK_L}{\delta}(T_c - T_s) \tag{8.13}$$

and therefore, for a given heat transfer fluid (for which K_L and T_c are fixed naturally) the rate of heat flow is a function only of the thickness of the liquid boundary layer. Inspection and comparison of Equation (8.13) and the general equation (8.1) shows, since the fluid temperature T_f is, in this case, the condensation temperature T_c, that for a condensing vapour:

$$h = K_L/\delta \tag{8.14}$$

Because the flow of the condensed film is obviously higher on vertical surfaces its thickness δ is less and the heat transfer rate is therefore greater than on horizontal surfaces.

8.2.4.1 VERTICAL AND INCLINED SURFACES

Analytical expressions for the boundary layer thickness have been derived[270] and depend on the viscosity and the density of the fluid. Whilst not being rigorous, the treatment does contain all the significant physical mechanisms and the results agree reasonably with experiments. Analysis was made for steady state condensation on vertical and inclined surfaces where the condensate liquid film

Solder Reflow by Fluid Heat Transfer

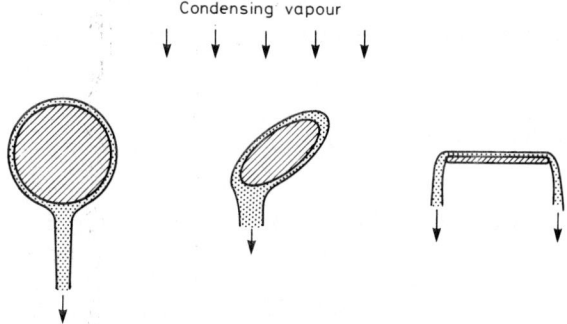

Fig. 8.6 Situations in which a condensing fluid film flows downwards along the surface, under gravity.

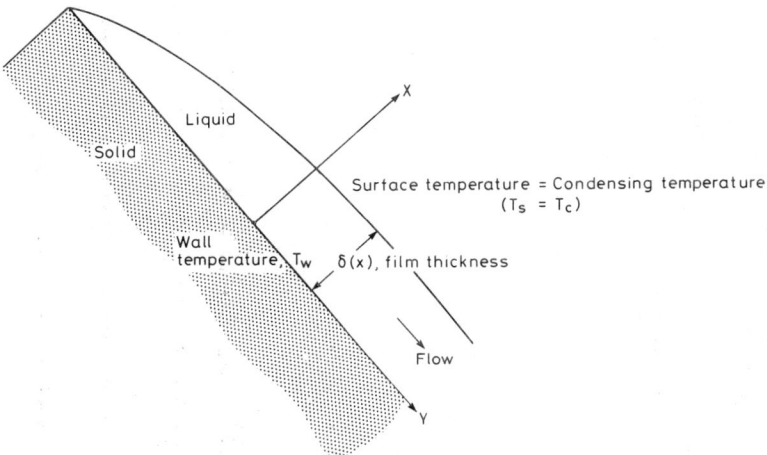

Fig. 8.7 The form of a liquid film used for the analysis of heat transfer through a liquid on an inclined plate.[270]

flows downwards along the surface under the action of gravity, as shown schematically in Figure 8.6. Three assumptions are made to simplify the problem of determining the effect of the film thickness:

(i) The film flows down by gravity which, for a thin film, is balanced by the viscous force in the liquid.
(ii) The heat flow is determined by the thermal resistance of the film. Since the film is thin, convective effects are negligible and the rate of condensation heat transfer is determined by the rate of conduction through the film.
(iii) The condensate at its vapour interface is at the saturation temperature T_c of the vapour and the temperature gradient in the film is linear from T_s to T_c.

In the analysis for a flat plate, the film thickness at the top edge of the surface is taken as zero as shown in Figure 8.7, and increases in the downward direction as the film accumulates the condensate from upstream. At any location x from that top surface the local film thickness is

$$\delta(x) = \left(\frac{4K_L(T_c - T_s)\mu_L x}{\varrho_L(\varrho_L - \varrho_V) g \lambda_f \sin\theta} \right)^{1/4} \tag{8.15}$$

where μ_L (Pa.s) is the viscosity of the liquid, ϱ_L and ϱ_V (kg.m^{-3}) are the densities of the liquid and the vapour respectively, g(m.s^{-2}) is the acceleration due to gravity and θ is the angle of inclination from the horizontal.

From Equation (8.14) with Equation (8.15) the local heat transfer coefficient is readily derived and hence, by integration over the full length b of the plate, the average heat transfer coefficient is found to be:

$$h = 0 \cdot 94 \left(\frac{K_L^3 \varrho_L (\varrho_L - \varrho_V) g \lambda_f \sin\theta}{(T_c - T_s) \mu_L b} \right)^{1/4} \tag{8.16}$$

Modifications to this theory[271] include the effect of surface tension when the film is very thin. Another important refinement is to include the heat capacity of the condensate. In the original approach, the heat flux available is simply the latent heat released during the phase transition from vapour to liquid. There is, however, additional energy released when the condensate heats up the liquid film which is at a temperature less than T_c. This is usually called the sub-cooling effect and can be accounted for by the use of a modified latent heat value:

$$\lambda_f' = \lambda_f \left(1 + \frac{0 \cdot 68(T_c - T_s) c_L}{\lambda_f} \right) \tag{8.17}$$

where c_L (J.kg^{-1}.K^{-1}) is the specific heat of the condensate. For a common fluorocarbon vapour phase fluid, Equation (8.17) increases the modified latent heat by a factor 3.2 if the workpiece is at room temperature, thus increasing the heat transfer coefficient, given in Equation (8.16), by one third.

8.2.4.2 HORIZONTAL SURFACES

For understanding measurements of heat transfer at vertical and inclined surfaces, one major assumption is the draining of the condensed liquid film due to gravity. For horizontal surfaces this mechanism of liquid film removal is no longer operable and furthermore the two cases of upward facing and downward facing surfaces must be treated differently.[272]

The upward facing surface has been treated analytically and experimentally[273] by making the simplification of allowing liquid flow off only one of the four edges of a rectangular plate. The flow is driven by the variation in hydrostatic pressure due to the changing film thickness, it being thinnest at the draining edge. The work shows that if this minimum film thickness is less than 40% of the maximum equilibrium thickness, remote from the draining edge, then the heat transfer coefficient can be expressed:

$$h = 0 \cdot 16 \left(\frac{K_L^3 \varrho_L^2 g \lambda_f}{(T_c - T_s) \mu_L a} \right)^{1/4} \qquad (8.18)$$

This equation is akin to Equation (8.16) except that a, the dimension of the draining edge of the workpiece, is now important rather than b, its length parallel to the direction of liquid flow.

Equation (8.18) is found to agree quite closely with observation. However, in practical applications the entire perimeter, rather than one edge, is usually available for drainage of the condensate. A rigorous analysis is not available but an approximation can be made by modifying Equation (8.18) by replacing the dimension a with a characteristic length a' where

$$a' = \frac{\text{area of condensing surface}}{\text{total drainage length}}$$

Several practical considerations affect the accuracy of Equation (8.18), in particular the edge profile of the specimen being heated and the surface tension of the condensed fluid. A sharp corner on the draining edge restrains the liquid film, thus lowering the heat transfer coefficient. The difference between a sharp and a rounded edge can be as much as a factor of two.[273]

The above analysis refers to an upward facing horizontal surface. When vapour condenses on the underside of a workpiece, instead of the liquid draining towards the edge, it falls off in an apparently random manner of droplets. The critical size of these droplets and their separation distance depends on the surface tension of the liquid. The heat transfer coefficient is now independent of the physical size of the surface but is defined by a characteristic length d between the droplets.[274]

$$h = \frac{K_L}{d\gamma} \left(\frac{0 \cdot 9 \, R_a^{1/6}}{1 + 1 \cdot 1 \, R_a^{-1/6}} \right) \qquad (8.19)$$

where $R_a = \dfrac{(\varrho_L - \varrho_V) g \lambda_f' \, d\gamma^3}{K_L \mu_L (T_c - T_s)}$

is a dimensionless quantity called the Rayleigh number, and $\gamma (\text{N.m}^{-1})$ is the surface tension of the liquid. In vapour phase soldering R_a is initially around 10^5 but, as the workpiece surface temperature T_s increases, so does R_a. The greater the Rayleigh number the more realistic are the approximations leading to Equation (8.19).

The above equations can be used to calculate the heat transfer coefficient, h, as a function of the surface temperature of an immersed workpiece. This has been done using typical parameters for a vapour phase soldering condensing fluid, and the calculated data are given in Figure 8.8. Note that, in general, as the workpiece temperature increases, the differential $T_c - T_s$ decreases, and so the heat transfer coefficient, calculated from the Equations (8.16)–(8.19), rises. This is not in direct agreement with the experimental data in Figure 8.5, where a constant value of h is reached when $T_s \geq 100°C$.

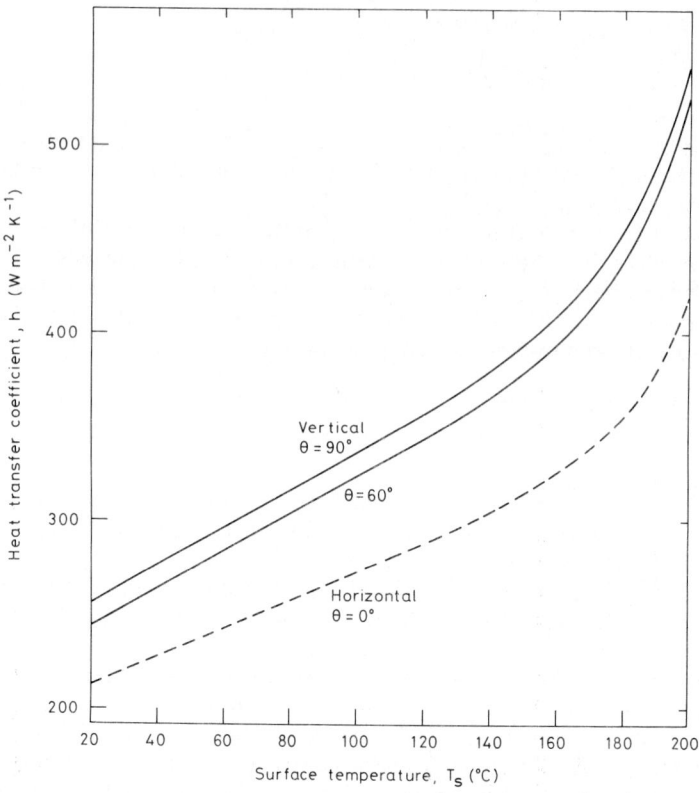

Fig. 8.8 Calculated values of heat transfer coefficients during heat transfer from a condensing vapour to an immersed plate, using Equation (8.16) for vertical and inclined surfaces and Equations (8.18) and (8.19) for horizontal surfaces.

8.2.4.3 COOLING TUBES

The condensation of the working vapour and the consequent heat transfer has been considered above in relation to the total immersion of the surface mount assembly in the vapour. Consideration must also be given to the condensation of the vapour on the cooling coils which guard against the escape of vapour from the heat transfer system. The cold condensing tubes are mounted close to the horizontal and the flow of condensed liquid from them is as shown schematically in Figure 8.6. There is no inhibition of the liquid flow from the surface and so the problem can be treated in a manner similar to that of the inclined flat surface. For a horizontal tube of diameter 2r, the average heat transfer coefficient is given by an expression very similar to Equations (8.16) and (8.18):

$$h = 0 \cdot 61 \left(\frac{K_L^3 \varrho_L (\varrho_L - \varrho_V) \, g \lambda_f'}{(T_c - T_s) \, \mu_L r} \right)^{1/4} \tag{8.20}$$

This expression is required for the design of the cooling coils in production vapour phase soldering facilities, as discussed in Section 8.4.4.2.

8.2.4.4 WORKPIECE SURFACE CONDITION

The rate of heat transfer predicted by Equation (8.14) and the calculations of the boundary layer thickness δ often differ significantly both from laboratory measurements and industrial practice. This is because the analyses above assume a laminar flow of the liquid boundary layer whereas in the practical situation some turbulence generally occurs; if the boundary layer is unstable, surface waves are generated resulting in higher heat transfer. Even for laminar flow the thickness of the boundary layer can be strongly dependent on the roughness; for example, a rough oxide surface usually creates a thicker boundary layer, reducing the heat flow.

The effect of surface roughness on the heat transfer process is not only relevant in determining the boundary layer thickness of the condensing fluid but also in the heat transfer from heaters submerged in the boiling liquid in the sump of the vapour phase system. This will be discussed more fully in Section 8.4.2.4.

8.3 PROPERTIES OF THE HEAT TRANSFER FLUID

In order to use a heat transfer fluid effectively, be it a liquid or a condensing vapour, all the physical, chemical and physiological properties that determine and limit the application must be considered. These properties can usefully be taken in three groups:

1 *Thermal properties* of the fluid, which determine its heat transfer characteristics
   — Density ϱ, and its variation with temperature, i.e., the expansion coefficient
   — Heat capacity (specific heat) c_L of the liquid boundary layer, which determines the amount of heat within that layer
   — Latent heat λ_f of the fluid, if there is a phase change during the heat transfer, as in vapour phase soldering
   — Thermal conductivity K_L which determines the rate of heat flow through the boundary layer
   — Viscosity μ_L which, if the fluid is a condensing vapour, controls the thickness of the liquid boundary layer

2 *Limiting properties*, which fix the range of application of the fluid
   — Physical stability
   — Chemical stability
   — Thermal stability
   — Cost

3 *Handling properties*, which define the compatibility of the fluid and its environment
   — Surface tension
   — Corrosion
   — Flammability
   — Toxicity
   — Cleaning and removal.

All these properties must be borne in mind when considering any available choice between either a liquid and a vapour heat transfer fluid or between different chemicals suitable for using as a heat transfer fluid.

8.4 VAPOUR PHASE SOLDERING

Reflow soldering in a saturated vapour, commonly known as vapour phase soldering or condensation soldering, utilises the latent heat of a condensing, saturated vapour to heat the workpiece. The temperature of the saturated vapour is the same as the boiling point of the liquid used, and the heat transfer due to the phase change from vapour to liquid is very rapid. As soon as the workpiece reaches the temperature of the vapour, condensation stops and there is no further heat flow. Thus the system does not need temperature control. Furthermore, condensation takes place simultaneously on all surfaces, resulting in very uniform heating. Vapour phase heating is thus an elegant way of applying controlled heat rapidly and uniformly to a complex surface in the absence of air.

An ideal fluid for vapour phase soldering must have a boiling point sufficiently high to produce high quality joints consistently. However, the temperature must be sufficiently low to be compatible with the thermal properties of components. Also, at the operating temperature, the vapour must be non-flammable, inert, chemically and thermally stable and not a danger to health. The vapour density at the boiling point should be significantly greater than air and the fluid must have a well defined boiling point for the advantages of vapour phase soldering to be fully implemented.

Vapour phase soldering was first shown to be a practical and advantageous proposition for electronics assembly in 1974 using a fluorinated polyoxypropylene liquid with a molecular weight of 950·2 and a boiling point at atmospheric pressure of 223·9°C.[275] When vapour phase soldering developed as an industrial production process,[267] the highest boiling point fluorinated liquid consistently available was the top of the 3M Fluorinert* range of perfluorinated hydrocarbons, a perfluoro-triamylamine with a trade name FC70 and a boiling point of 215°C. Thus 215°C became, by default rather than design, the accepted standard for the vapour phase soldering process.[276]

The commercial success of vapour phase soldering, particularly in recent years with its rôle in soldering surface mounting devices, has encouraged other chemical companies to introduce new fluids suitable for the process. Because of their different chemistry, these fluids offer different combinations of properties and they have also extended the temperature range available for solder reflow.

8.4.1 The Soldering System

An equipment as simple as that shown in Figure 8.9 can be used for producing and maintaining a saturated vapour of the heat transfer fluid into which the surface mounting assembly can be immersed. Although automated in-line systems are now available, the majority of vapour phase soldering facilities remain of the batch type. Small laboratory systems are available with external heaters, but for production soldering, heaters immersed in the liquid are necessary to deliver the required power to maintain the vapour. To prevent the vapour from escaping, water cooled condensing coils are mounted around the top rim of the vessel.

---

*Fluorinert is a proprietary name of the 3M company.

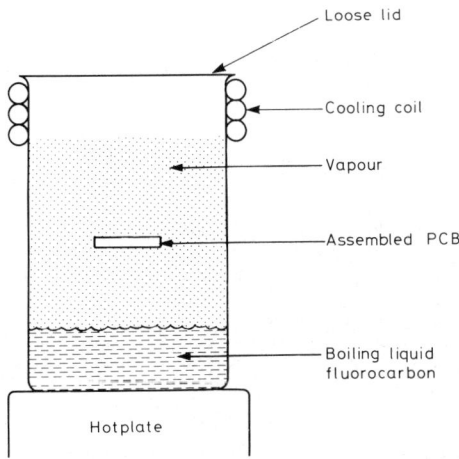

Fig. 8.9 The simplest system for vapour phase soldering.

The vessel is open to the atmosphere (apart possibly from a loose fitting lid) so that the vapour temperature is that of the atmospheric boiling point of the chosen working fluid. The density of the vapour of all of the available fluids is significantly greater than that of air, resulting in a sharp interface between the vapour and the air. Despite this, the loss of working fluid from such a system can be large[277] because of diffusional and aerosol losses into the air as will be discussed fully in Section 8.4.2.8. The running costs of such a system are greatly reduced by the vapour blanket technique[277] in which a layer of lighter and less expensive secondary vapour floats on top of the hot primary soldering vapour to act as a shield against the loss of the primary fluid. The secondary fluid must be thermally and chemically stable when in contact with the hot primary vapour and remain inert with respect to the primary vapour and the workpiece. The fluid that is used almost universally for providing the secondary vapour in soldering systems is 112-trichloro-trifluoro-ethane (the refrigerant R113) whose cost is only a few per cent of that of the primary fluids.

A batch loading soldering system with a secondary vapour blanket is shown schematically in Figure 8.10. The primary vapour is generated continuously in the boiling sump and condensed at the primary coils, returning under gravity to the sump to be reboiled. The secondary vapour, also originating in the sump, boils before the primary fluid, but once the boiling point of the primary fluid is reached, the secondary fluid is maintained above the primary coils and is vaporised by contact with the hot primary vapour. The secondary vapour lost to the atmosphere is replenished from a reservoir.

In Figure 8.11 are plotted the temperature, the density and the pressure profiles along a vertical axis of the vapour phase soldering system. The two interfaces between the primary and secondary vapours and between the secondary vapour and the air can be maintained sharply as demonstrated by the temperature and density gradients measured. The convective mass transfer across the interfaces is quite small. However, although the boiling point of the secondary fluid R113 is 47·6°C, in practice the vapour blanket temperature is in the range 70-90°C, which is due to the mixing and diffusion at the interface. The vapour blanket is a

250 Solder Reflow by Fluid Heat Transfer

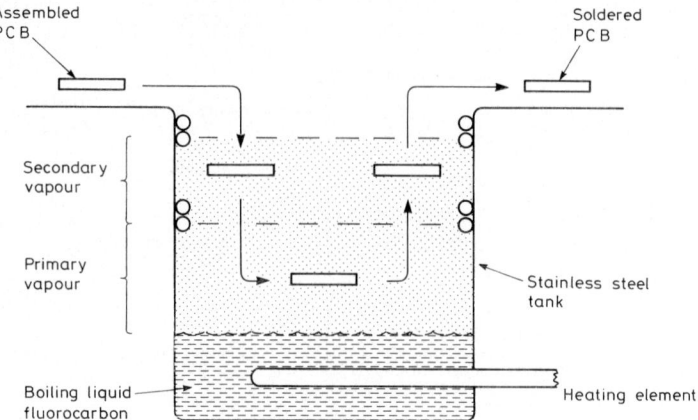

Fig. 8.10 Schematic of a production batch vapour phase soldering machine.

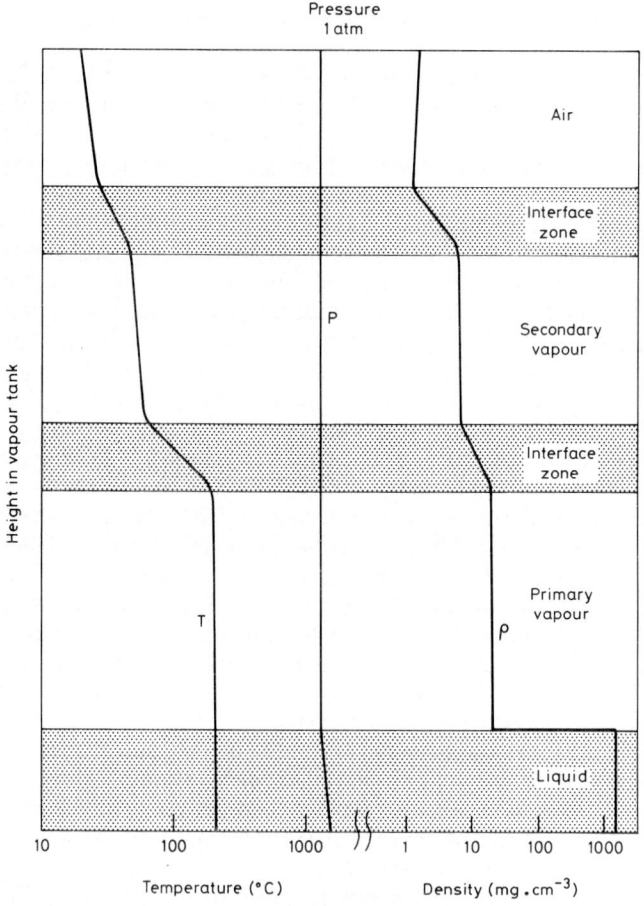

Fig. 8.11 Typical variation of the pressure, temperature and density within a vapour phase soldering machine.

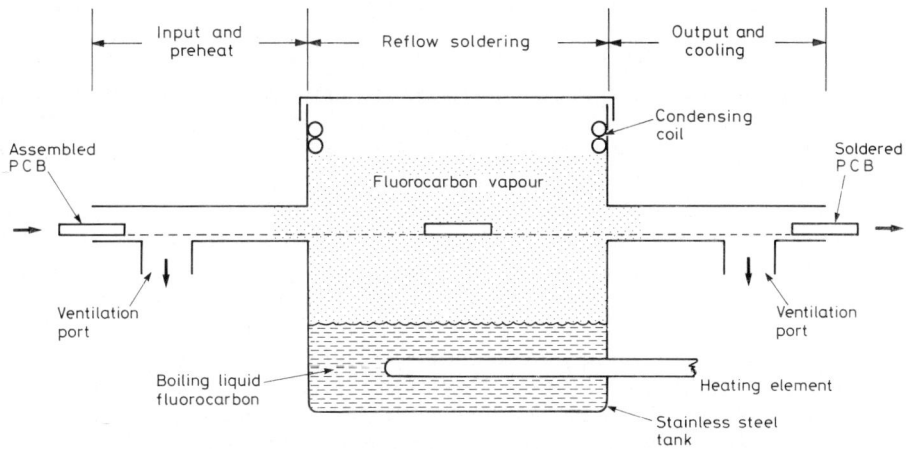

Fig. 8.12 Schematic of an in-line vapour phase soldering machine.

mixture of R113 with a small amount of the primary vapour, whose precise composition is determined by the rate of generation of the secondary vapour and the geometry of the system. In the interest of reducing vapour losses, it is desirable to operate the blanket at as low a temperature as possible.

For in-line, rather than batch loading, vapour phase facilities, the secondary vapour can be dispensed with, because the loss of primary fluid by evaporation is negligible. An in-line system is shown in Figure 8.12.

8.4.2 The Primary Fluid

The primary fluid for vapour phase soldering must have a boiling point sufficiently high to produce high quality metallurgical bonds consistently between solder and metals. Experience has dictated that an excess temperature of about 30°C over the solder melting point is the minimum needed to fulfil this aim. The fluid temperature must, however, be sufficiently low to minimise thermal damage to any part of the product. The vapour and the liquid must both be non-flammable, inert, chemically and thermally stable and with low toxicity. Regarding thermal stability, the fluid is generally in contact with heaters that are at a temperature substantially higher than its boiling point. In order to take full advantage of the heat transfer concept of vapour phase soldering, it is desirable for the fluid to have a sharp boiling point, although a range of 10°C is of little consequence. For ease of operation the fluid should have a pour-point below room temperature. The vapour density of the fluid should be significantly greater than that of air in order to aid retention of the fluid within the chamber. Finally, the fluid must have excellent dielectric characteristics to avoid the possible degradation of the electrical properties of the soldered assemblies.

8.4.2.1 FLUID CHEMISTRY

A number of suitable fluids are currently available commercially for vapour phase soldering. Several such examples are now described and compared in order to illustrate the necessary properties outlined above. The three fluids described are not the only ones available and doubtless new ones will appear on the market.

However, the same scrutiny should be given to them as has been to these selected examples.

The commercial quality liquids specifically marketed for vapour phase soldering are all derived from common organic compounds by replacement of all the carbon-bound hydrogen atoms with fluorine atoms. All are colourless, odourless, non-flammable, thermally stable while boiling, exceptionally inert chemically, entirely non-toxic, and expensive.

Fluorinert FC-70 has been the most widely used fluid during the first ten years of vapour phase soldering and has been the focus of a substantial volume of research. This has considered its heat transfer properties,[267] boiling characteristics[278] and its thermal degradation. FC-70 is a perfluoro-triamylamine, being three fully fluorinated amyl groups linked by a nitrogen: $(C_5F_{11})_3N$, with a molecular weight of 821 and a boiling point of 215°C, that has set the standard. The fluid has a room temperature density of 1·93 g.cm^{-3} and the vapour density at the boiling point is 20·3 mg.cm^{-3}, which is about 17 times that of air at room temperature. The latent heat of condensation of FC-70 is 67 J.g^{-1}.

Whilst FC-70 has been the main heat transfer fluid used for vapour phase soldering since its inception, it has now, for reasons of lowering the toxic thermal degradation by-products, been replaced by a fluid FC-5311, in the USA and seems likely to be so in Europe. This fluid was first marketed under the tradename *Flutec PP11, but has recently been incorporated into the Fluorinert range.

phenanthrene → perhydro-phenanthrene → perfluoro-perhydro-phenanthrene
$C_{14}H_{10}$ $C_{14}H_{24}$ $C_{14}F_{24}$

FC-5311 is perfluoro-perhydro-phenanthrene which has a molecular weight of 624 and a boiling point of 215°C. Its room temperature density of 2·03 g.cm^{-3} is slightly greater than that of FC-70 and its latent heat of condensation is equivalent.

The third fluid commonly available is LS230 which is one boiling fraction with a boiling point of 230°C, of the **Galden range of perfluoropolyethers. These compounds have a linear chain structure of low molecular weight polymers represented by the formula

$$F-\underset{F}{\overset{F}{C}}-\left[O-\underset{CF_3}{\overset{F}{C}}-\underset{F}{\overset{F}{C}}\right]_m-\left[O-\underset{F}{\overset{F}{C}}\right]_n-O-\underset{F}{\overset{F}{C}}-F$$

---

*Flutec is a proprietary name of ISC Ltd.   **Galden is a proprietary name of Montedison.

with, typically, molecular weights in the range $10^3 - 10^4$ and the ratio m/n in the range 20-40. Several restricted molecular weight fractions have been isolated by fractional distillation, of which Galden LS230 is just one. A 260°C boiling fraction is also available for soldering with alloys of higher temperature, extending the technique of vapour phase soldering to the more specialist higher tin alloys such as 96½Sn 3½Ag (melting point 221°C), 95Sn 5Sb (melting range 236-243°C) and 100Sn (melting point 232°C).

Some physical properties of the three common fluids FC-70, FC-5311 and LS230 are given in Table 8.2.

Table 8.2
Some Physical Properties of Representative Fluids for Vapour Phase Soldering

Property	Units	R113	FC-70	FC-5311	LS230
Boiling point or range	°C	47·6	215	215	230±5
Molecular weight	-	187	821	624	~650
Pour point	°C		−25	−20	−80
Density of liquid at 25°C	g.cm^{-3}	1·57	1·93	2·03	1·82
Density of saturated vapour at BP	mg.cm^{-3}	7·38	20·3	15·6	19·5
Viscosity of liquid at 25°C	cP	0·7	27	16	8
Surface tension of liquid at 25°C	mN.m^{-1}	19	18	19	18
Specific heat of liquid at 25°C	J.g^{-1}.K^{-1}	0·95	1·05	1·07	1·00
Thermal conductivity at 25°C	mW.m^{-1}.K^{-1}	74	70	53	70
Electrical resistivity	Ω.cm		2.10^{15}	$>10^{15}$	10^{15}
Heat of vaporisation, at BP	J.g^{-1}		67	68	63

8.4.2.2 HEAT TRANSFER PROPERTIES

As already demonstrated, the heat transfer from a condensing vapour to a solid surface is very dependent upon the nature and the orientation of the surface and is not an absolute property of the heat transfer fluid. The angle of the solid surface to the horizontal, the ability of the surface topography to induce waves in the flowing condensed liquid, the ability of the microroughness of the surface to inhibit hydrodynamic flow, whether or not the surface is wet by the condensed liquid, and finally the thermal conductivity of the solid all determine the measured rate of its temperature rise. This helps to explain why available measured or observed rates of heat transfer of vapour phase soldering fluids range over at least a factor of ten.

Measurements of heat transfer coefficients of the fluids have been made under identical experimental conditions by measuring the thermal response of a transient plate calorimeter when immersed in the vapour. Measurements have been made with the surfaces horizontal and vertical. The values of heat transfer coefficient at 200°C are given in Table 8.3. It is clear that the values for the different fluids are similar, ranging ±9% about the mean for the horizontal surfaces and only ±5% about the mean for the vertical surfaces. The differences between the horizontal and vertical measurements are greater than the differences between the fluids.

Table 8.3

Measured Values of Heat Transfer Coefficients of Condensing Fluids, Made Using a 2 mm Plate Copper Calorimeter

	h at 200°C ($W.m^{-2}K^{-1}$)	
	Horizontal Surfaces	Vertical Surfaces
FC-70	300	400
FC-5311	250	360
LS230	280	380

The reason that the values of heat transfer coefficients are given for a temperature of 200°C is because the workpiece rises to a temperature near the melting point of the solder in just a few seconds and the initial low value of h does not significantly affect the required immersion time. In practical soldering the greater proportion of the time required is not that to raise the temperature of the solder to its melting point but to supply its latent heat of fusion. Because this is the case, even though the heat transfer coefficients of the liquids are very similar, the immersion times required for reflow soldering are quite different, depending upon the condensing temperatures T_c of the fluids, as will now be discussed.

8.4.2.3 SOLDER REFLOW PROPERTIES

The time to raise solder from its initial room or preheated temperature T_o to its melting point ($T_s \approx 185°C$) by immersion in a condensing vapour at temperature T_f is given by Equation (8.7). To that time must be added the time t_L at which the solder remains at its melting temperature while it changes phase from solid to liquid. This time depends on the latent heat of fusion of the solder $\lambda_s(J.kg^{-1})$ and is given by:

$$\lambda_s \varrho V = hAt_L \,(T_f - T_s) \qquad (8.21)$$

where the other symbols are as before.

Thus the time t_L is inversely proportional to the excess temperature $T_f - T_s$ of the condensing temperature over the melting point of the solder. The significance of this, in respect of reflow properties, is shown in Figure 8.13. Here the rate of temperature rise of a 1 mm^3 spherical cap of 62:36:2 SnPbAg solder has been calculated, when immersed in condensing fluids with boiling point either 215°C or 230°C, both with a heat transfer coefficient h of 400 $W.m^{-2}.K^{-1}$. The sample reaches its melting point in one second in both fluids but the 215°C fluid requires 50% longer time to melt the solder. Thus the sample reaches a temperature of 210°C in 7·1 seconds in the 215°C fluid but in only 4·8 seconds in the 230°C fluid.

Such calculations have been readily confirmed by experiment, either by recording the temperature-time curve of a solder paste sample during reflow by vapour phase or by using a glass vapour phase system and simply observing the reflow timescale using different fluids. The temperature and time of the reflow soldering operation also affect the solder fillet shape, its metallurgical structure

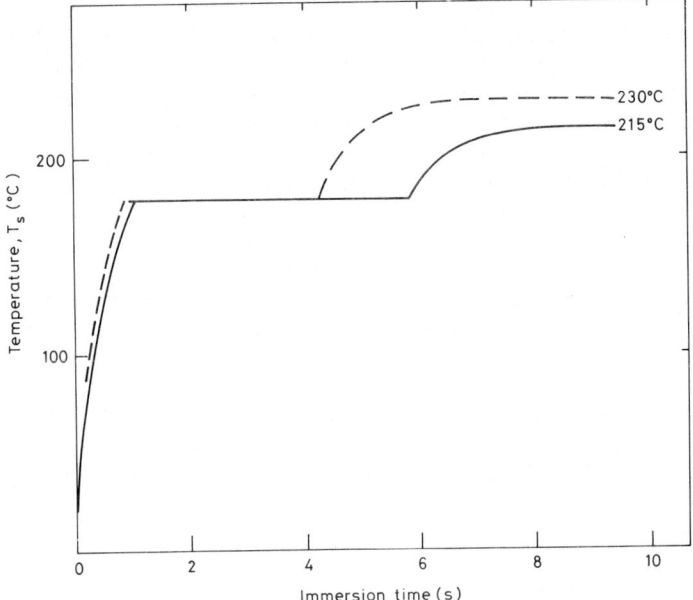

Fig. 8.13 Calculated temperature-rise curves during reflow of a 1 mm^3 circular cap of 62:36:2 SnPbAg solder by condensing vapour of fluids with boiling points of 215°C and 230°C. A heat transfer coefficient of $h = 400$ W.m^{-2}.K^{-1} is assumed throughout, for both fluids.

and the amount of intermetallic compound formed at the solder-substrate interface.

8.4.2.4 OPERATING POWER REQUIREMENTS

The heating method most frequently used in commercial vapour phase soldering machines is electric heaters immersed in the primary fluid sump. Such a case of a heating surface submerged in a boiling liquid without any external agitation of the liquid is known as pool boiling, and is a situation that pertains in both batch and in-line vapour phase soldering machines, with or without a secondary vapour.

As long as the surface temperature of the submerged heater does not exceed the boiling point of the liquid by more than a few degrees, heat is transferred to liquid near the heating surface by free convection. Convection currents circulate this superheated liquid and evaporation takes place at the free surface. As the temperature of the heating surface is increased a point is reached where, in certain places, the energy level of the liquid adjacent to the surface becomes so high that vapour nuclei spontaneously develop at favourable sites which eventually grow to form vapour bubbles. This regime is called nucleate boiling. The vapour bubbles are at first small and condense before reaching the surface but, as the heater temperature is raised further, they become more numerous and larger until they finally are able to rise to the free surface. As the rate of bubble emission from a site increases, bubbles collide and coalesce with their predecessors, eventually merging into essentially continuous columns. As the number of columns increases a limit is reached when the space between them is no longer sufficient to accommodate the stream of liquid which must move towards the heater surface to replace the liquid evaporated to form the vapour columns. Eventually a vapour film blankets the entire heater surface and a stable film-boiling regime is entered.

If the surface temperature of the heater is monitored as the heating energy input is increased, a curve of the type shown in Figure 8.14 is obtained. This is known as a boiling curve, and the various regimes described above are clearly delineated. At point (a) the surface temperature of the heater can increase as the energy input decreases because part of the surface is insulated with a vapour film. The maximum heat flux at this point is called the critical heat flux $(Q/A)_{crit}$, corresponding to a critical temperature excess ΔT_{crit}. At point (b) this vapour film is continuous over the heater surface and heat is transferred to the liquid by conduction and radiation. The shape and relative positions of the features of a boiling curve depend on the different physical properties of the liquid but the general form will be as shown in the figure.

In practice, the heat flux, i.e., input energy per area, is controlled and hence, if at a level greater than $(Q/A)_{crit}$, a transition from nucleate to film boiling takes place with a rapid rise in the surface temperature of the heater during a jump from point (a) to point (c). The heater in a vapour phase soldering machine must be kept well below $(Q/A)_{crit}$ to avoid the transition from nucleate to film boiling which would incur a large spontaneous increase in heater surface temperature. Such a situation could cause degradation of the fluids especially if flux residues are adherent to the heater.

Boiling curves can be obtained for the fluorocarbon liquids using a method of monitoring the power and temperature of a submerged platinum wire heater.[278] A schematic diagram of a suitable boiling cell is shown in Figure 8.15. A 0·5 mm

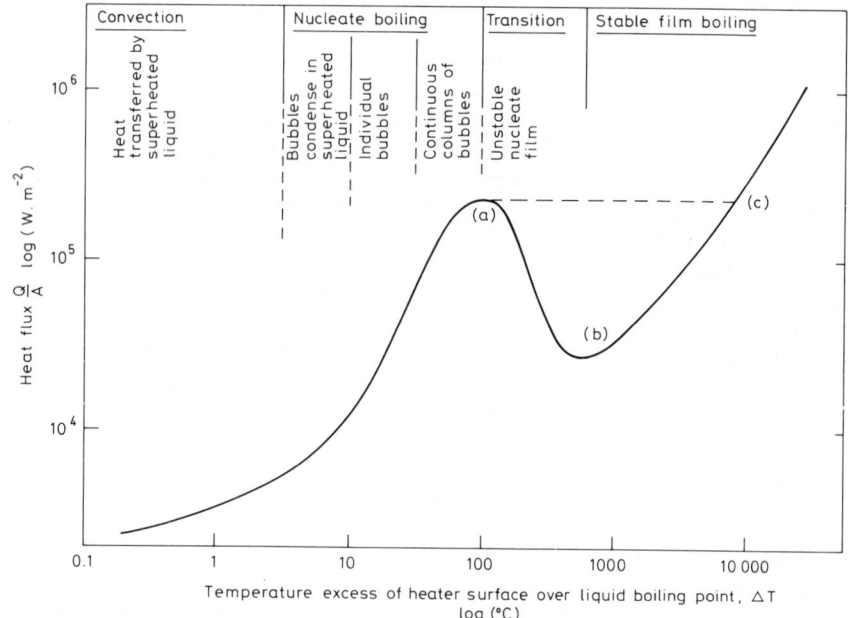

Fig. 8.14 The general form of a boiling curve for a non-decomposing liquid.

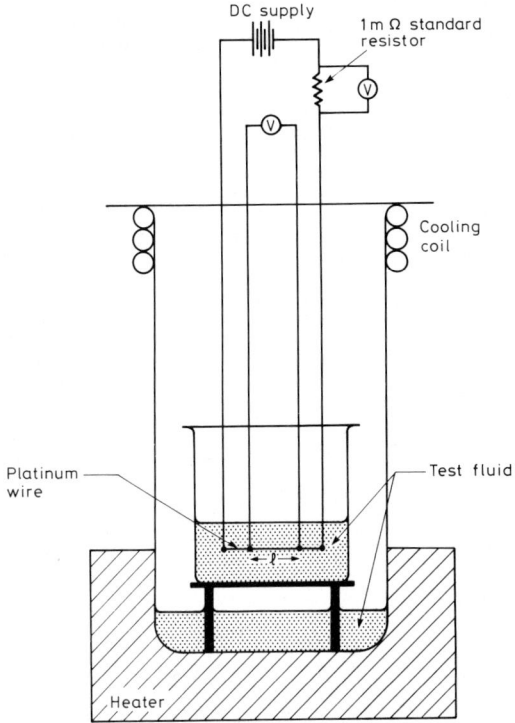

Fig. 8.15 A simple boiling cell for production of boiling curves. The temperature is determined from the resistance of the platinum wire heater.

diameter platinum wire acts as the heater in the inner cell. The current through, and the voltage across the central section of the wire are monitored and the resistivity of the platinum used to determine its temperature from platinum resistance thermometry tables. The curves are obtained by incrementally increasing the power applied to the wire until it just begins to glow red-hot and then incrementally decreasing the power, recording the voltage-current relationship. The ordinate of the boiling curves is plotted as a heating flux, i.e., watts per surface area of the wire, as shown in Figure 8.16. Boiling curves of this type have been published for FC-70[278] and for FC-5311.[279]

Vapour phase soldering machines operate in the nucleate boiling regime. Two features of the boiling curves are therefore important: the values of $(Q/A)_{crit}$ and ΔT_{crit} plus the slope of the nucleate boiling section of the curve. The higher the critical excess temperature, the higher can be the heater temperature excess over the fluid boiling point and therefore, all else being equal, the faster the initial heat-up time and the recovery time after condensation of the vapour. The nucleate boiling part of the curve is always approximately straight and may therefore be described thus:

$$\frac{Q}{A} = k\Delta T^m \tag{8.22}$$

The steeper the nucleate boiling part of the curve (i.e., the higher the value of m),

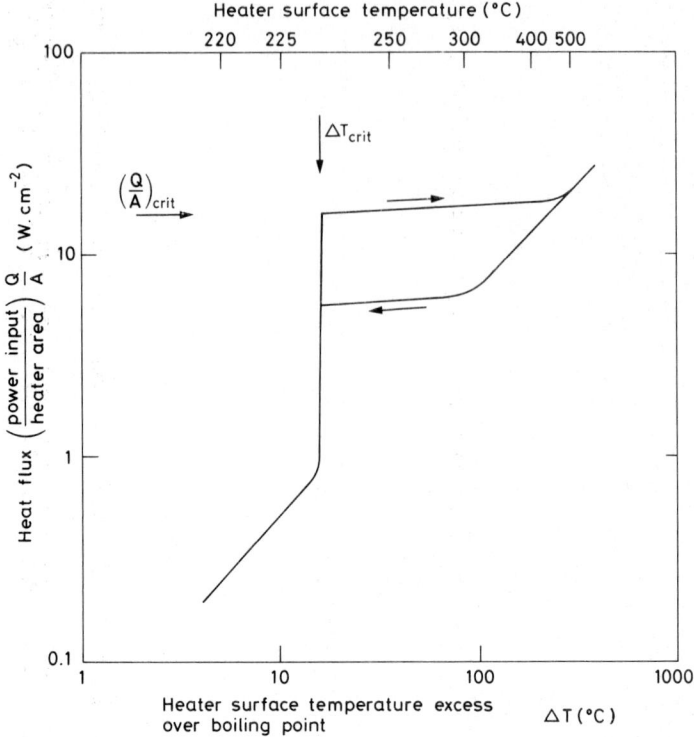

Fig. 8.16 An experimentally determined boiling curve for FC-5311 fluid in air at one atmosphere pressure.

the less important is the control of the power input to maintain a constant heater surface temperature excess. A compilation of these parameters as measured for the test fluids is given in Table 8.4.

Table 8.4

Power Requirement Parameters

	ΔT_{crit} °C	$\left(\dfrac{Q}{A}\right)_{crit}$ W.cm^{-2}	Nucleate boiling slope $\dfrac{Q}{A} = k\Delta T^m$	T_c °C	ΔT at 5 W.cm^{-2} °C
FC-70	20 ± 3	15·5	m = 4·5	215	16
FC-5311	16 ± 2	15·7	m = 50	215	15·5
LS230	14 ± 5	13·9	m = 4	230	11

8.4.2.5 ROSIN FLUX SOLUBILITY

At room temperature the solubility of rosin, in the form of a mildly activated rosin flux, in all the condensation soldering fluids under test, is about 10 ppm. As they are heated to their boiling points the solubility rises about a hundred times to around 0·1%. This is made use of in production equipment since, as the fluid is circulated and cooled, the rosin precipitates from solution and is captured by the filtering system. An estimate of the solubility is therefore a prerequisite for designing the type and circulation frequency of the filtering system.

It has been pointed out[279] that the quality of soldering can be affected by the loss of active flux from the workpiece due to its dissolution in the heat transfer liquid, an effect that is enhanced as the required soldering time in the condensing vapour increases. Such an increased immersion time may arise because of the heat capacity of the workpiece or because of the type and activity of the flux used. In general, therefore, the lower the flux solubility in the condensing fluid, the greater the proportion of flux will remain on the workpiece where it is intended to be. Additionally, the more flux that is dissolved and removed from the workpiece, the greater is the demand placed on the filtering system (perhaps partially compensated by a benefit gained for the board cleaning process).

The flux solubility and wash-off is measured by taking a sample of, typically, either a mildly activated rosin flux or a water soluble flux and applying it, using an eye-dropper, to a clean glass microscope slide, allowing it to spread unrestricted. This is then held horizontally in a freely ventilated chamber at 25°C for about 20 hours, as might be typical of the delay between the loading of a board with solder paste and components and its soldering. After immersion of the microscope slide vertically into the condensing vapour for various times the percentage weight loss is found, as shown in Figure 8.17.

As well as reducing the effectiveness of the flux by removing it from the workpiece, the dissolution of flux in the fluid has a marked effect on the boiling characteristics.

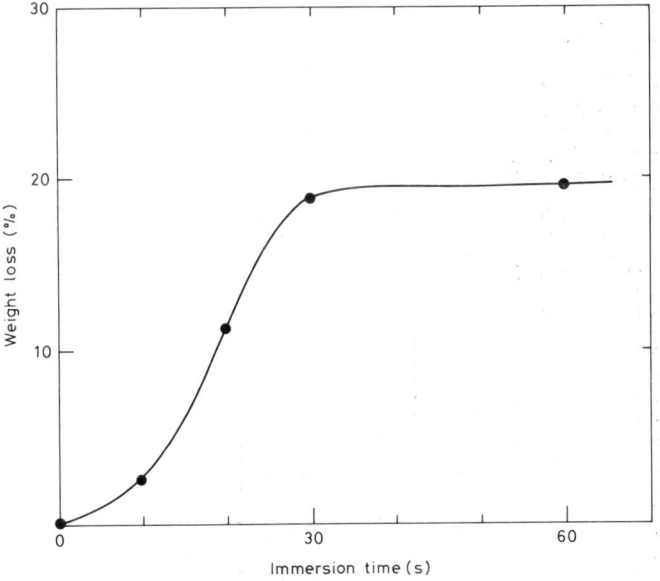

Fig. 8.17 Percentage of flux (RMA type) washed off a glass slide during vertical immersion in condensing vapour of Galden LS230 fluid.

It is found that the slope of the boiling curve in the nucleate boiling regime always decreases with rosin build-up and hence the control of the surface temperature of the heater becomes more difficult. This effect is shown in Figure 8.18 as a function of time and in Figure 8.19 as a function of cumulative concentration of rosin. In the first case 0·1 wt% solid rosin has been added to the heat transfer fluid and boiled using a 5 W.cm^{-2} power flux. After one hour and subsequent hours the critical part of the boiling curve is traversed to produce the data given.

In practical soldering the rosin concentration in the boiling fluid increases with the throughput of fluxed boards to be soldered. In Figure 8.19 the rosin concentration has been increased at the rate of 0·03% per hour, while boiling with a power flux of 3 W.cm^{-2}, and the boiling curve determined at the end of each hour of operation. The slope m of the boiling curve, in Equation (8.22), falls from its clean value of about 4·5 to a value of 1·0 as the fluid (in this case FC-70) becomes saturated with rosin.

8.4.2.6 THERMAL DEGRADATION—TOXICITY

The thermal stability of fluorocarbons has been the subject of considerable study.[280]

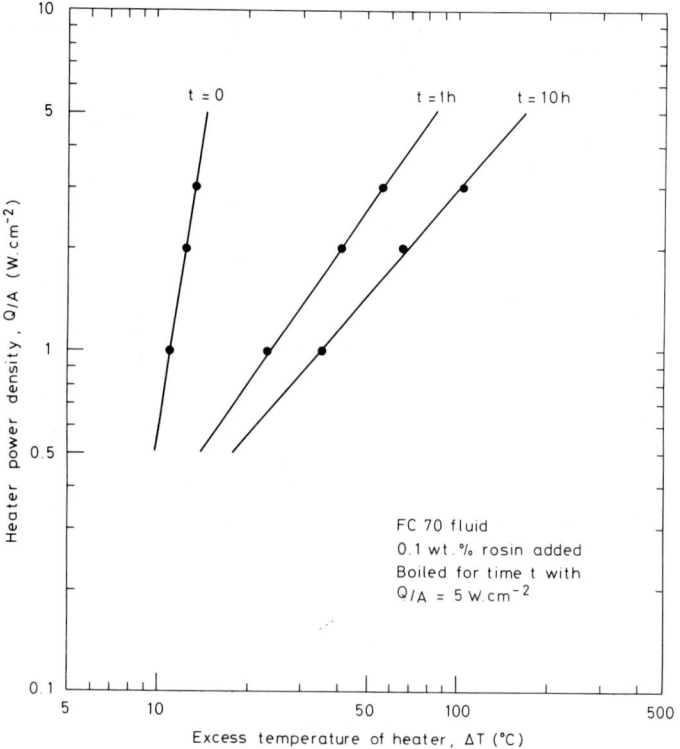

Fig. 8.18 Measurements showing the effect on the slope of the nucleate boiling curve of FC-70 when contaminated with 0·1 wt.% rosin.

The chemical properties of all fluorocarbon compounds are dominated by the short interatomic distance between carbon and fluorine and the strength of the bond joining the two, resulting in very high stability. The fluorocarbon fluids used for vapour phase heat transfer are non-toxic and biologically compatible both in the vapour and the liquid phases.[281] However, they do have an inherent potential to produce toxic thermal degradation products if subjected to superheating or thermal stress. In particular, fluorinated alkenes arise which are highly toxic, the most dangerous being perfluoro-isobutylene (PFIB), $(CF_3)_2-C=CF_2$, a pulmonary irritant some ten times more toxic than phosgene. It is a colourless, odourless vapour and difficult to detect. The concern over PFIB generation was a contributory factor that prevented the widespread acceptance in Japan of vapour phase soldering as an alternative to infra-red reflow. PFIB has an acute inhalation toxicity with a reported accumulated lethal concentration (ALC) of 0·5 ppm over 6 hours,[281] giving rise to an acceptable time-weighted

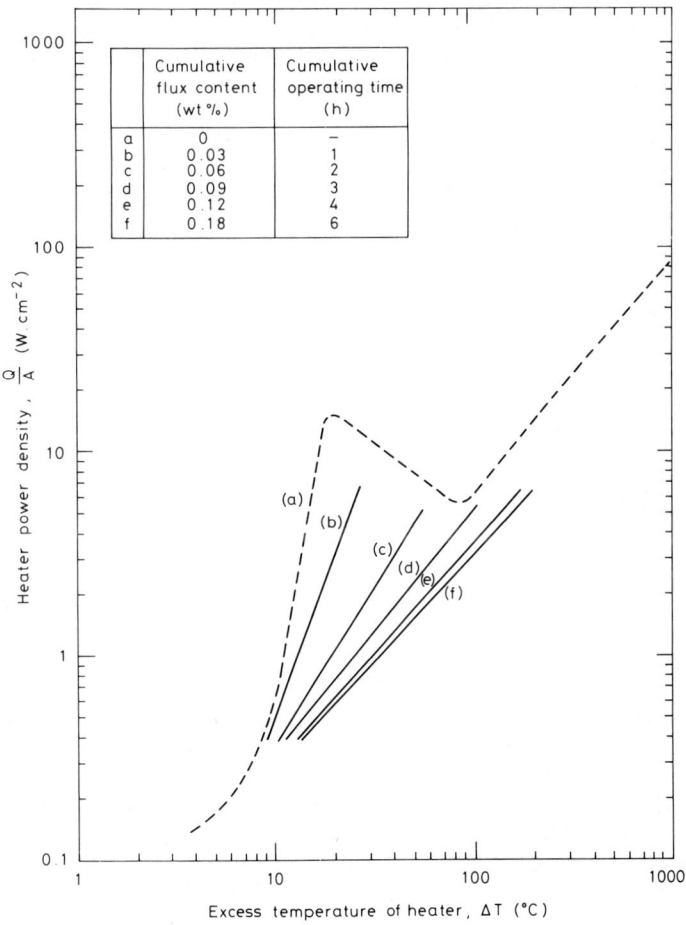

Fig. 8.19 The effect on the heater performance submerged in FC-70 fluid, as a function of cumulative rosin concentration and time of boiling.[278]

average inhalation, over an 8-hour day, set at 1 part in 10^8. Short-term exposure hazards are set in terms of an inhalation concentration which gives rise to 50% mortality of the exposed people (LC_{50}) which for PFIB is 180 ppm.min^{-1}.

An analysis of PFIB in the generated vapour has been made for FC-70[282] and quoted as 2 ppm.h^{-1}, equivalent to $2.10^{-4} LC_{50}$.

For Flutec PP11 (FC-5311) no PFIB could be detected using equipment with a detection limit of 1 ppb,[283] i.e., over 1000 times less than FC-70. The technique employed for these measurements was based on a method using a gas chromatograph with an electron capture detector.[284] Trade literature for Galden fluids quotes a PFIB generation rate of 0.5 ppb.h^{-1}, comparable with FC-5311.

8.4.2.7 THERMAL DEGRADATION—CORROSIVITY

Some degree of corrosion of the stainless steel secondary condensing coils in vapour phase soldering machines is not uncommon. The corrosion residues have been identified as metal fluorides and chlorides resulting from condensate at these coils, containing hydrofluoric and hydrochloric acids. The latter arises from the decomposition, under thermal stress or high temperature, of the 112 trichloro-trifluoro-ethane secondary fluid used to produce the secondary vapour blanket.

Because this solvent vapour is not in contact with its own boiling medium, some of the secondary blanket can become superheated up to the temperature of the primary vapour even though its own boiling point is only 47.6°C. Concentrations of the secondary fluid of about 3% have been measured within the primary vapour,[285] but the amount of available oxygen satisfactorily restricts the production of phosgene. Trichloro-trifluoro-ethane is thermally and chemically stable only up to 120°C whence, in the presence of air, it begins to decompose producing carbonyl halides such as phosgene:

$$F_2ClC\text{-}CFCl_2 + O_2 \rightarrow COF_2 + COCl_2$$

Also, at the superheated temperatures, in the presence of hydrocarbons such as the soldering flux, it can undergo hydrogen exchange reactions to give hydrochloric acid:

$$F_2ClC\text{-}CFCl_2 + CFCl_2 + RH \rightarrow F_2ClC\text{-}CHFCl + RCl$$
$$\downarrow$$
$$F_2C = CFCl + HCl$$

The higher the temperature of the primary vapour the greater will be the superheat available to the secondary vapour and hence the greater the potential of these decomposition reactions.

The corrosivity problem associated directly with the primary fluids is the production of hydrofluoric acid. This arises by the reaction of the generated perfluoro-alkenes such as PFIB with water. (The water arises in the vapour phase machine during the cool-down cycle, condensing from the air on the cooling coils.) Thus, for example,

$$(CF_3)_2C = CF_2 + H_2O \rightarrow (CF_3)_2CH\text{-}OOH + 2HF$$

Hydrofluoric acid will therefore build up within the machine unless either chemical measures are undertaken to neutralise it on a continuous basis or the production of PFIB is eliminated or negligible.

8.4.2.8 CONSUMPTION OF FLUID

Notwithstanding any variation between the fluids in their heat transfer efficiency or their toxicity, it is often the relative cost of replacing consumed liquid that contributes most to the practical choice of fluid. Unfortunately this issue is the most contentious and the most difficult to quantify because measured losses depend very much on the specific soldering machine.[286] All the vapour phase soldering fluorocarbon heat transfer fluids are expensive and the periodic replenishment of the fluid sump becomes a major factor in determining the operating costs of the facility.

The fluid losses incurred depend on:

(i) the type of machine (batch or in-line);
(ii) the efficiency of the secondary vapour blanket;
(iii) the type of product and the consequent drag out;
(iv) the rates of entry and exit, and dwell times of the workpiece;
(v) the control and housekeeping of the machine (maintenance, fluid cleanliness, flux filter upkeep, ventilation exhaust rates, condensing water temperature, cleanliness of heaters, and control of the heater power).

The ratio of loss rates from one fluid to another can be very different from one machine to another. Reports on fluid usage[279, 287] have considered losses specific to certain vapour phase machines, but little work has been done to measure the relative importance of the various physical mechanisms which contribute to loss of the primary fluid.

Five main sources of primary fluid loss have been identified:[279]

1   *Diffusion and convection:* Because of the large difference in densities between the vapour and air, and the constant maintenance of vapour at the condensing coils, there exists a very sharp temperature and concentration gradient across the interface. A natural convection current of the air therefore exists above the vapour carrying out a convection loss of vapour with it, whose degree is dependent upon the inter-diffusion between the vapour and the air.

At all temperatures there will be some loss by evaporation from any liquid with a free surface. At any instant some molecules are moving with a velocity component perpendicular to the surface that is sufficient for them to escape into the air. By the loss of its faster molecules, the liquid loses both matter and heat. The vapour pressure of a liquid at a temperature below its boiling point gives an indication of the evaporative loss of the liquid at that temperature, whilst the latent heat of vaporisation gives an indication of the cooling effect induced by evaporation when the heating is removed.

In vapour phase soldering machines, evaporative losses are greatest during close-down because, in order to avoid condensation, the water in the cooling coils is turned off before the fluid is cold.

Evaporation losses can be compared using a method based on the ASTM standard D2595. A weighed sample of the liquid is placed in an evaporation cell and maintained at a given temperature ±1°C for 22±0·1 hours during which pre-heated, filtered air is passed over the surface at a flow rate of 2±0·02 litres.min^{-1}. The preheating is adjusted such that the exit air temperature is at the test temperature ±1°C. This is a harsh test compared with the conditions encountered in a vapour phase machine and generally, in practice, the diffusional losses are small compared with fluid loss by aerosol formation and by drag-out. The losses by the evaporation mechanism are highest for the LS230 fluid over an initial period of about 100 hours. This is consistent with a rise in the working temperature of about 6-7°C over that period, followed by a constant boiling temperature and minimal further losses by this mechanism.

2 *Aerosol or mist* During operation a small portion of the hot vapour condenses into small droplets upon encountering the colder air at the interface. This mist is carried upwards by natural convection currents. In contrast to the diffusional losses, the problem is worse for in-line machines rather than batch machines, since the moving conveyor results in a forced convection of the droplets.

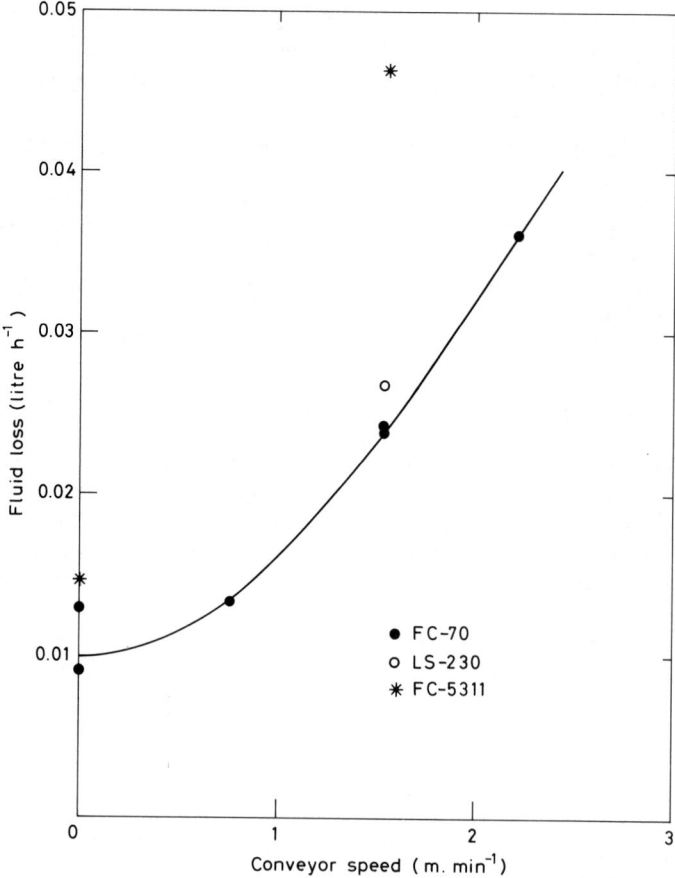

Fig. 8.20 Fluid loss from a commercial 30 cm belt in-line vapour phase soldering machine, as a function of the conveyor speed, demonstrating aerosol losses (data from reference 279).

The significance of fluid loss by aerosol transport on in-line facilities is demonstrated in Figure 8.20. The total fluid loss is measured from a commercial 30 cm belt in-line machine as a function of the conveyor speed. The aerosol loss using FC-5311 is similar to FC-70 when the belt is stationary but is twice as high when the conveyor is moving at $1 \cdot 5$ m.min^{-1}. The effect is due to the difference in aerosol droplet size, that of the FC-5311 fluid being considerably smaller than that of FC-70, enabling the droplets to be more readily swept out.

3   *Ventilation*   An over-zealous ventilation system on batch vapour phase machines can give rise to serious losses of the fluids, due to an excessive air speed across the surface of the vapour.

The critical effect of excess ventilation is demonstrated in Figure 8.21, giving the total fluid loss from a commercial 30 cm belt in-line machine charged with FC-70, running at a fixed conveyor speed of $1 \cdot 5$ m.min^{-1}. An abrupt increase in the fluid loss occurs at a particular ventilation air speed which is presumably related to the aerosol droplet size.

4   *Drag-out*   During soldering operations, losses can be caused by drag-out of condensed fluid on the soldered assembly and by disturbance of the air-vapour interface due to vapour entrainment into the air. Such disturbances

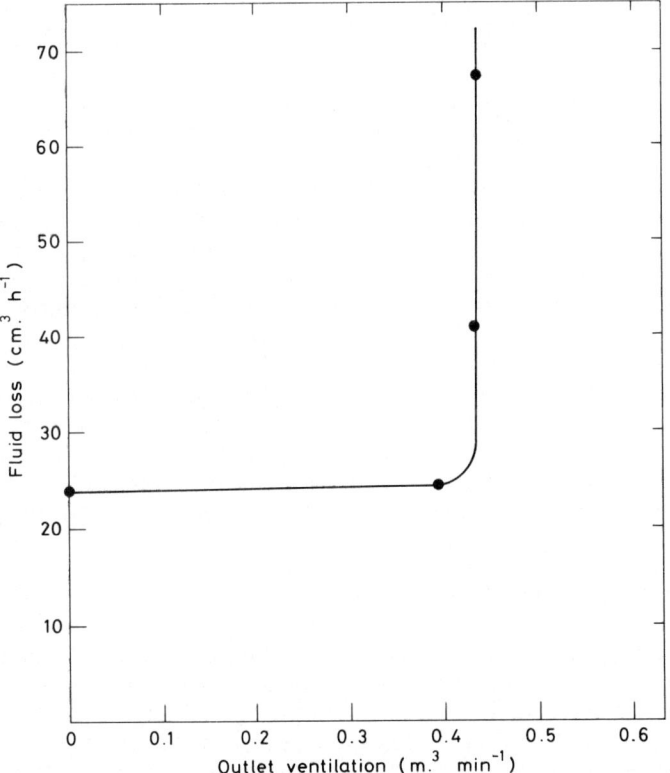

Fig. 8.21 A demonstration of the critical effect on fluid loss of increasing the ventilation air speed. These data, from reference 279, are from a commercial 30 cm belt in-line machine charged with FC-70 fluid, running at $1 \cdot 5$ m.min^{-1}.

are induced as the workpieces are introduced into and removed from the soldering machine. The amount of drag-out is not only a function of the workpiece but also of its temperature profile during exit.

5 *Filtration* The working fluid is required to be filtered in order to remove flux and particulate debris as well as to be acid scrubbed to remove corrosive thermal degradation products. Both the flux filtration plant and the acid neutralisation system with their associated pumping requirements result in fluid loss.

8.4.3 The Secondary Fluid

The secondary fluid acts as a stable vapour barrier between the expensive primary vapour and the air. The secondary vapour must therefore have a density between those of the primary vapour and air. The secondary liquid and vapour must be thermally and chemically stable and inert with respect to the primary liquid and vapour, and the workpiece. Additionally, because of the continuous boiling process of vapour production used in the vapour phase machine, the secondary fluid must have a boiling point that is lower than that of the primary fluid. The two fluids should preferably be miscible. Finally, the cost of the secondary fluid must be low, since it is used sacrificially to protect the primary fluid.

The fluid used to produce the secondary protective vapour blanket in production vapour phase soldering systems is almost exclusively 112 trichloro-trifluoro-ethane (commonly referred to as refrigerant R113 under the nomenclature scheme invoked for fluorinated hydrocarbons). Some of its relevant physical properties are given in Table 8.2.

As mentioned in Section 8.4.1, the vapour blanket is actually a binary mixture of the two working fluids. For an evaluation of the fluid losses incurred and the protective performance of the secondary blanket, a knowledge of the precise mixture of the two constituents is desirable, and its dependence on temperature. These inter-relations have been determined for the liquid state and so, from a measure of the density of the liquid and the temperature, the level of contamination of one liquid by the other can be determined directly. For the primary fluid FC-70, which has a density of $1\cdot 93$ g.cm^{-3} at room temperature (20°C), the reduction in its density as the concentration of R113 increases is shown in Figure 8.22. For the effect of low concentrations of the primary fluid in the R113, the equivalent data are given in Figure 8.23. These curves can be used for determining the composition of fluid in the primary sump and the purity of the fluid in the secondary fluid holding tank.

8.4.4 Production Vapour Phase Systems

A typical batch production vapour phase soldering system consists of a hooded vapour tank with a fully programmable batch conveyor defining the workpiece preheat through the secondary vapour, the time of solder reflow in the primary vapour and the rate of cooling through the secondary vapour into the air. The rate of withdrawal also controls the amount of drag-out of the primary fluid by the workpiece and the conveyor. The vapour tank has three zones, each defined by a set of condensing coils: the primary vapour zone, the secondary vapour zone and the air freeboard zone at the top. The freeboard zone is chilled by its cooling

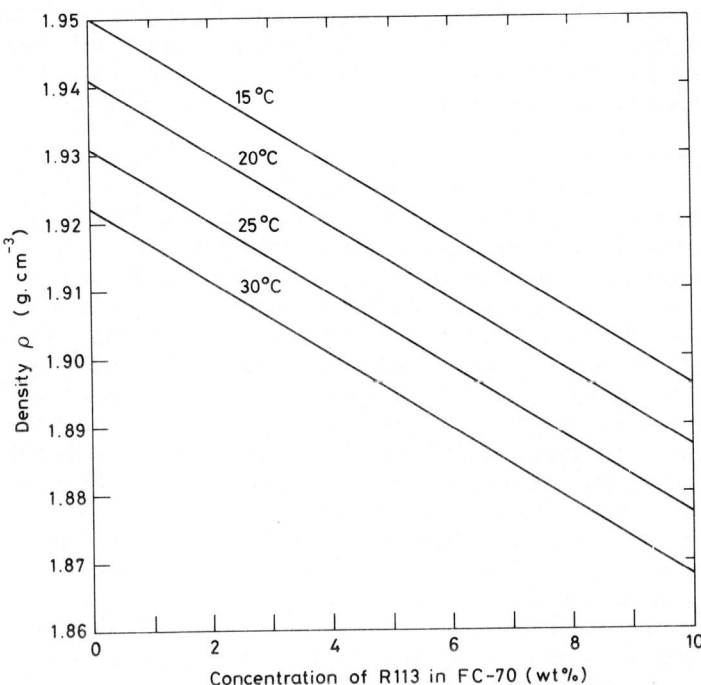

Fig. 8.22 The density of liquid FC-70 primary fluid as a function of the temperature and the degree of contamination by the secondary fluid, R113.

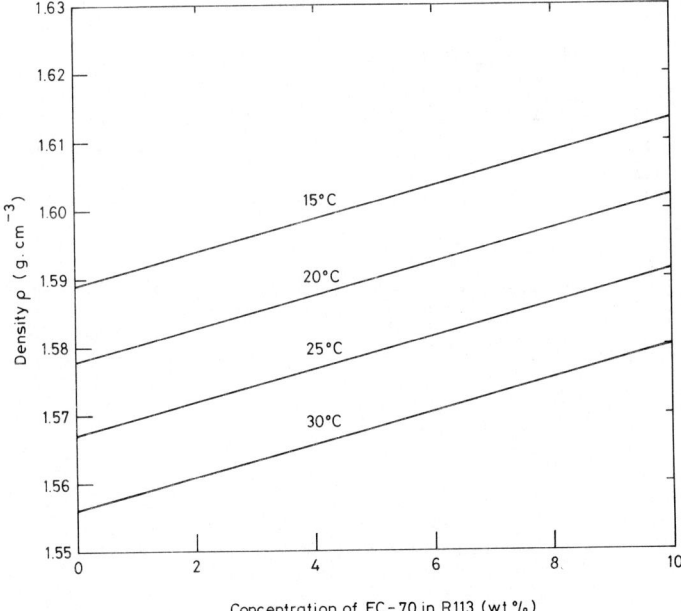

Fig. 8.23 The density of the liquid secondary fluid, R113, as a function of the temperature and the degree of contamination by the primary fluid FC-70.

coil in order to create a cool denser layer of air above the vapours and reduce the influx of moisture from the room air. The heating energy is delivered to the working fluid by means of electrical immersion heaters. A filtration system is needed for removing flux and particulate debris from the fluids.

8.4.4.1 HEATING

Low power density (3 W.cm^{-2}) immersion heaters are normally used to boil the fluids in order to avoid thermal degradation problems. It is usual for the input power to be carefully controlled and the heaters run at power densities below their rated value in order to keep their surface temperature low. For small systems up to about 20 kW, a constant power can be used but this becomes wasteful for larger systems and an idling state is used to maintain the two stable vapour zones while the reserve power is automatically controlled by thermostats situated in the vapour zones. As a workpiece to be soldered enters the primary vapour zone, condensation on it causes a drop in the vapour level which is detected by the thermostat sensor and the reserve heaters are energised. This thermostat is set some 5°C below the boiling point of the primary fluid.

Vapour phase systems have approximately 4 kW of immersion heater power per kilogram of product to be soldered in each batch.

The effect of surface finish of the heaters is of significance,[278] as shown in Figure 8.24, in which the two heaters are identical except one is smooth and the other roughened to about 0·5 mm. The open structure of the roughened surface promotes bubble nucleation, making it a more efficient nucleate boiler. This produces a lower surface temperature difference ΔT and therefore a higher thermal transfer coefficient for a given power density.

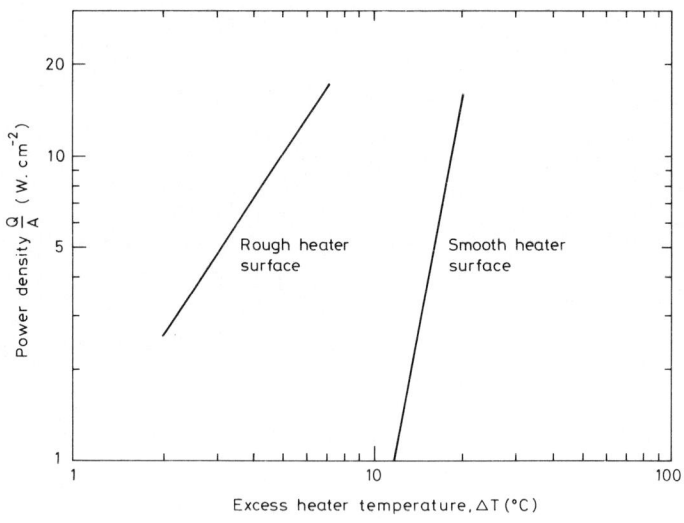

Fig. 8.24 The effect of surface roughness on heat transfer of submerged heaters in a boiling liquid. The heaters are identical apart from a surface roughening of about 0·5 mm.[278]

8.4.4.2 CONDENSING

The condensing coils are usually stainless steel tube with diameters of between one and three centimetres. The cooling rate should be such that during idling each vapour zone condenses on the lowest turn of the respective coil. Heat transfer rates for tubes of different sizes and varying surface temperature can be calculated directly from Equation (8.20). For actual design of the cooling coils, however, the water temperature rather than the outer tube surface temperature is stipulated and therefore the following calculation is required.

If h_c (W.m^{-2}.K^{-1}) is the heat transfer coefficient given by Equation (8.20) for the vapour condensing on a cooled tube, h_w is the heat transfer coefficient between the cooling water and the tube and K_t(W.m^{-1}.K^{-1}) is the thermal conductivity of the tube material (stainless steel), then the effective overall heat transfer coefficient can be shown by simple algebra to be:

$$h_e = \left(\frac{r_o}{r_i h_w} + \frac{r_o}{K_t} \ln \frac{r_o}{r_i} + \frac{1}{h_c} \right)^{-1} \quad (8.23)$$

where r_i and r_o are the inner and outer radii of the cooling tube. The total heat transfer is then as given in Equation (8.1):

$$\frac{dQ}{dt} = 2\pi r p h_e (T_c - T_w)$$

where p is the tube length, r is the mean of r_i and r_o, T_c the condensation temperature of the fluid and T_w the bulk temperature of the cooling water.

The surface temperature of the outer wall of the cooling curve, T_s, is given by

$$h_c (T_c - T_s) = h_e (T_c - T_w) \quad (8.24)$$

Since h_c depends on T_s, an iteration is required between Equations (8.23) and (8.24) if exact values are required. However, the dependence of h_c on T_s is not strong, and a quick approximation can be made.

The heat transfer coefficient for the cooling tubes is generally very large and the relevant data are given in Figure 8.25 for primary fluid FC-70 and in Figure 8.26 for secondary fluid R113 as a function of the cooling tube size, the rate of water flow and the bulk water temperature. Similar curves can be calculated readily for any of the fluids used.

The primary cooling coil must operate at a temperature higher than 50°C to avoid condensing out the R113 secondary vapour blanket. During the soldering cycle the primary coil is subject to varying loads and consequently a recirculating surge tank is used to supply it, as shown in Figure 8.27. This tank contains heaters to ensure the water temperature, especially at start-up, is above 50°C. Once operation has commenced the water in the surge tank is cooled by use of a heat exchanger containing chilled water.

8.4.4.3 SECONDARY INJECTION

The secondary vapour blanket is controlled by a continuously recirculating injection process. A monitored amount of secondary fluid is pumped into the soldering tank from a reservoir through injection ports in the tank wall at the

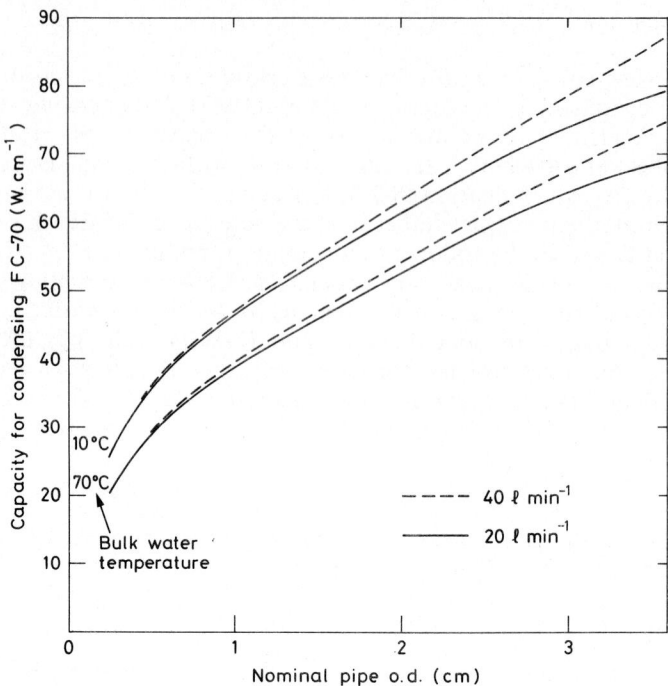

Fig. 8.25 The condensing capacity for Fluorinert FC-70 vapour phase soldering fluid, of stainless steel pipe, as a function of the pipe size, the bulk water temperature and the water flow rate.[276]

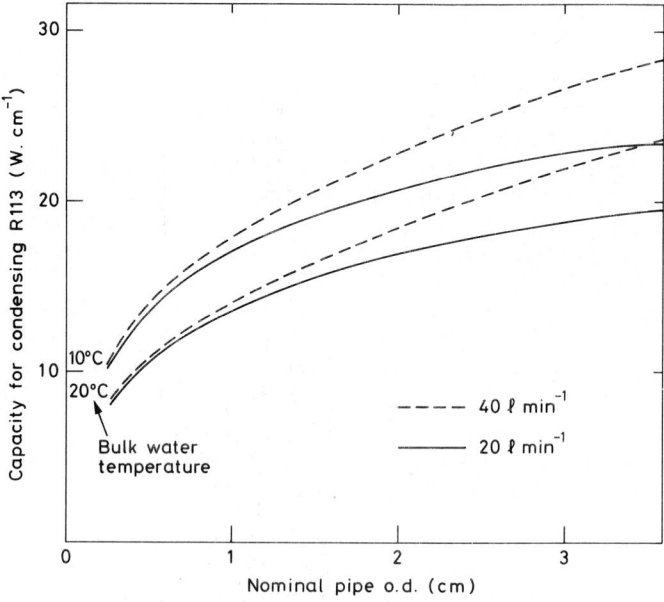

Fig. 8.26 The condensing capacity for R113 vapour (112 trichloro-trifluoro-ethane), of stainless steel pipe, as a function of the pipe size, the bulk water temperature and the water flow rate.[276]

height of the secondary blanket. The liquid is vaporised as it flows down the wall towards the primary zone. The secondary fluid injection system is included schematically in Figure 8.27. A large portion of the secondary fluid condensing at the coil is collected and recirculated back to the reservoir through a water remover.

Thermal decomposition of the R113 fluid is responsible for the formation of hydrochloric and hydrofluoric acids, as mentioned in Section 8.4.2.7, which can initiate the corrosion of the cooling coils. The acidity of the R113 fluid is neutralised by the incorporation of a chemical filter, for example granular soda lime. A mechanical filter is also used to remove particulate debris and insoluble chemicals and so prevent clogging of the injection nozzles. Another method used to remove the acid is water extraction, whereby the recirculating R113 is turbulently mixed with clean water which dissolves the acid before the mixture reaches the water separator.

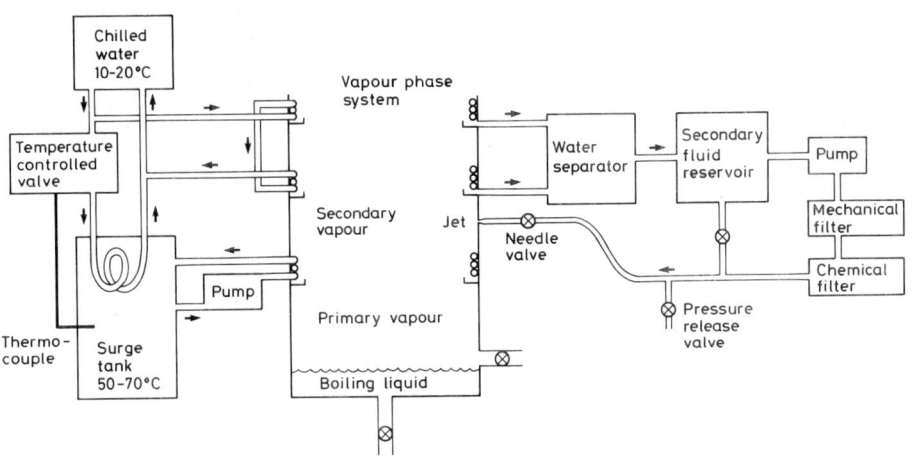

Fig. 8.27 Diagram of systems for (i) the recycling of cooling water for the primary coil and the secondary coils and (ii) the secondary injection and recycling.

8.4.4.4 FLUID FILTRATION

During soldering, as much as 40% of the flux applied to the workpiece can be washed off into the fluid sump,[288] together, possibly, with other chemicals and debris. The flux degrades at the sump temperature and changes the performance of the primary fluid. Furthermore, when there are excessive amounts of flux in the sump, it begins to adhere to the heaters, eventually carbonising and causing local hot spots on the heaters which can result in the thermal degradation of the primary fluid.[278] It is therefore necessary to remove the flux from the working fluid and a number of methods have been used such as continuous filtration, batch filtration, distillation, and chemical treatment,[288] of which batch filtration is the most common. This involves the cooling of the primary fluid to precipitate the flux and its removal by filtration. A typical system offering both continuous and batch filtration is shown in Figure 8.28.

For batch filtration the working fluid is drawn off from the sump, through a strainer to remove particular matter, to the pump. It then passes through a heat exchanger to be cooled and cause the rosin flux to precipitate. The solubility of rosin in the boiling primary fluid is about 0·1% by weight whereas at 10°C it is a hundred times lower. The precipitated rosin is then filtered out using a cartridge filter system.

If the amount of rosin washed into the sump during soldering approaches or reaches the solubility limit, continuous filtration can be used. In this case additional heat must be supplied to the cooled filtered fluid before returning it to the tank. Additionally, a counter flow heat exchanger can be used to conserve some of the thermal energy.

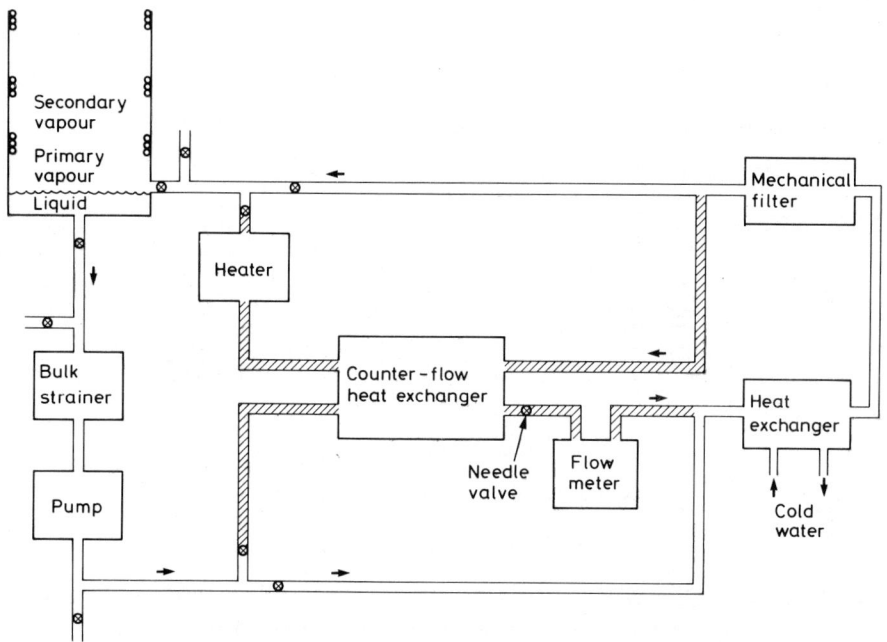

Fig. 8.28 Filtration system to remove rosin from the heat transfer fluid. The system can be used either in a continuous mode (shaded pipework) or in a batch mode.

8.4.4.5 CYCLING PROCESS CONTROL

Most vapour phase soldering machines operate on a pre-determined timed cycle of operations. At start-up all of the coils are cooled and the secondary injection system is activated. Both the idle and reserve heating power are used until the vapour levels are established stably. This takes about 30 minutes. During this initial period the secondary vapour condenses on both the primary and secondary coils until the boiling point of the primary fluid is reached. The temperature of the water in the surge tank then rises until eventually the secondary fluid no longer condenses on the primary coil.

During the soldering cycle, as the workpiece travels through the vapour blanket, secondary vapour condenses on it, dripping to the primary-secondary interface where it rapidly boils. As the conveyor enters the primary vapour zone it may condense enough vapour to drop the primary vapour level and thus turn on the reserve power momentarily. The conveyor then dwells in the high temperature zone for a preset time before traversing back to the vapour blanket where it is held for the preset secondary dwell time.

There is no advantage in stopping in the vapour blanket during entry to preheat the workpiece, provided the volatiles have been removed from the solder paste by a pre-solder bake. A dwell in the secondary blanket may well wash off some of the flux. The time that the workpiece remains in the primary vapour is best determined by an initial trial run, but it should be at least ten seconds after it reaches 30°C above the melting point of the solder. To minimise primary fluid drag-out this dwell time should be long enough for all the surfaces of the workpiece to reach the primary vapour temperature.

As the conveyor is removed from the primary vapour, the dwell time in the secondary vapour must be adequate to allow for draining and evaporation of the primary fluid, and the assembly should be left stationary long enough for the solder to freeze.

8.5 LIQUID PHASE SOLDERING

Liquid phase soldering with hot oil as the heat transfer fluid has been used for many decades. However, because of the unwanted decomposition products of available oils operating at soldering temperatures, the technique has been restricted traditionally to situations where the surfaces can be scrubbed, e.g., fusing of electroplated solder coatings.[289] However, in 1974 an experimental evaluation of eighteen liquids was reported[290] specifically tested for soldering multilayer glass-epoxy PCBs. The operating temperature range considered was 200-250°C. The most satisfactory liquids were found to be glycerol (which could be run 'safely' up to 230°C) and fluorinated polyoxypropylenes (two liquids were tried with boiling points at 224·2 and 290·0°C). Glycerol has a high heat transfer coefficient despite a relatively high viscosity. Also it is cheap and non-toxic, but its low flash point and high vapour pressure require careful equipment design for safety. Such equipment was not built for production soldering and the other, more viable techniques of infra-red and vapour phase heating cornered the market for reflow soldering of electronic assemblies.

Ten years on, however, the high temperature fully fluorinated hydrocarbons,[291] which have boiling points at atmospheric pressure that are above their decomposition temperatures in air, in excess of 300°C, apparently offer all the properties, both physical and technological, for safe and effective liquid phase heating. These are thermal stability, chemical inertness, low vapour pressure, adequate heat transfer properties, absence of flash or fire point, lack of carbonisation and total biological inertness. The surface tension of the liquids is very low, which leads to good wetting and flow, increasing the heat transfer coefficient and also ensuring good drainage of the workpiece on removal from the liquid. The workpiece is readily cleaned using inexpensive fluorinated solvents such as 112 trichloro-trifluoro-ethane and the heat transfer fluid then easily distilled from the washings and returned to the soldering tank. These properties indicate technical acceptability as an improved substitute for conventional oils in liquid phase soldering and have opened new technological areas for liquid phase soldering to become a viable process, in particular for the attachment of surface mounting components by reflowing screen-printed solder paste.

Production systems for liquid phase soldering of surface mount assemblies are at present under development and should be available shortly. The most suitable liquid readily available is the highest molecular weight fraction of the Galden perfluorinated polyether range, which has a very low evaporative loss at 300°C in air.

Many of the points discussed in Section 8.4 concerning the performance and handling of the heat transfer fluid as well as the design and operation of the soldering facility apply in a similar manner to the concept of liquid phase soldering. There will also be additional considerations to be made such as the lifting and floating of components as they enter the liquid. Such problems are equated against advantages to be gained over vapour phase soldering, for instance in the choice of soldering temperature. Consider first the heat transfer and the speed of the soldering process.

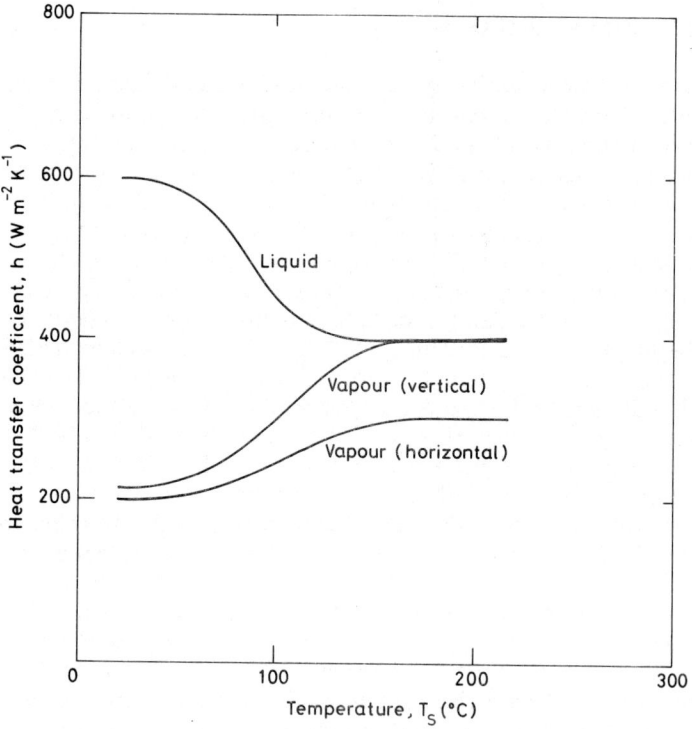

Fig. 8.29 Heat transfer coefficients as a function of the rising temperature of the immersed workpiece, for liquid phase and vapour phase heat transfer, both with a fluid temperature of 215°C.

8.5.1 Comparison of Heat Transfer in Liquid and Vapour

The measurements presented in Figures 8.4 and 8.5, made using the same procedures and measurement chain, showed no significant difference in the heat transfer coefficients of a liquid and a condensing vapour of a fluorocarbon, except at the start of the heat input into the workpiece.

The heat transfer coefficient h is proportional to the slope of the curves and, by differentiating them, h may be computed directly. In Figure 8.29, h is plotted as a function of specimen temperature. These curves are for the fluids FC-70 vapour at $T_f = 215°C$ and Galden LP liquid at the same temperature $T_f = 215°C$. The liquid offers a faster rate of heat transfer during the initial stages of the temperature rise and this is expressed in more practical terms in Figure 8.30 which presents heat-up curves for a plate calorimeter, 2 mm thick, representative of a PCB. In a vapour phase soldering machine with a 215°C condensing vapour, the centre plane of the plate reaches the melting point of solder, say 185°C, in about 20 seconds if held vertically and in about 25 seconds if horizontal. Using a liquid phase as the heat transfer medium, at the same temperature of 215°C, the time is cut to about 12 seconds. If the components on the board are such that a higher temperature, say 235°C, can be tolerated the corresponding time to reach 185°C is only 9 seconds. Thus, even though the measurements presented in Figures 8.4

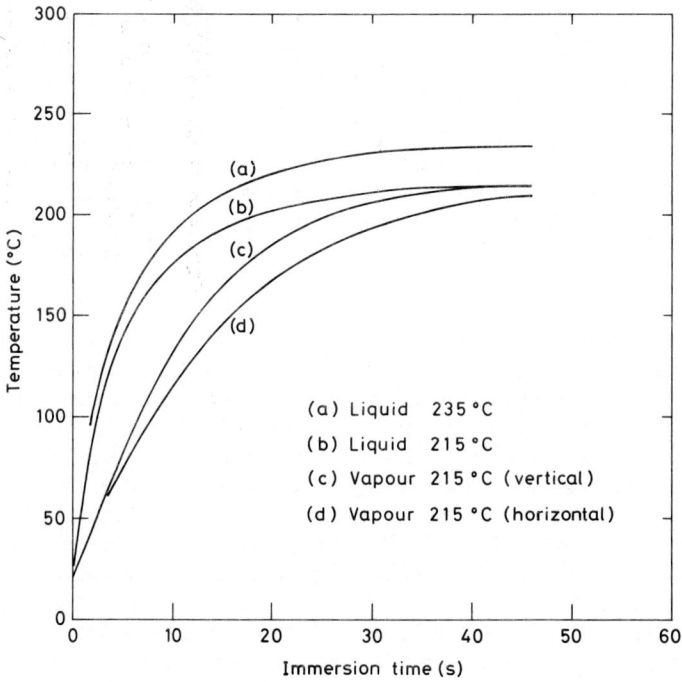

Fig. 8.30 Heat-up curves measured with a plate calorimeter and an identical measurement chain in each case.

and 8.5 seem to indicate a similar heat transfer coefficient of 400 $W.m^{-2}.K^{-1}$ for liquid and condensing vapour media, the immersion time for the workpiece to reach the melting temperature of solder is significantly less in the liquid. For practical soldering, however, this time difference is not a major consideration because the greater part of the time required is not usually for the heating of the workpiece and the solder to its melting point, but for supplying the latent heat of fusion to the solder during melting, as was shown in Figure 8.13.

8.5.2 Practical Tests of Liquid Phase Soldering

Some tests have been carried out to observe the practical efficiency of liquid phase heat transfer, in particular for the wetting and spreading of solder and for joint integrity.

As an example, some data of reflow using a liquid fluorocarbon as the heat transfer fluid are shown in Figure 8.31. The specimen is as shown in the inset of the figure: a 63Sn:37Pb preform annulus stamped from 0·5 mm thick sheet, with an outside diameter of 1·9 mm and an inside diameter of 1·0 mm. The preform was coated with mildly activated flux, which, when removed, left a solder volume of 1·0 mm³. This ring specimen, supported by threading on a 0·6 mm diameter solder coated copper wire, was immersed in the liquid with the wire horizontal. The solder was deemed to have fully melted when it wet and spread along the wire. Two parameters were varied, the temperature T_f of the heat transfer liquid and the temperature T_o of the specimen just prior to its entry into the liquid. The pre-heating was done using a second bath of the same liquid held at a temperature below the solder melting point, in which the specimen was held for ten seconds before transfer to the hotter bath at temperature T_f.

In Figure 8.31 are given reflow times as a function of T_f obtained with the preheat bath at 140°C giving, after fast transfer through the air to the adjacent reflow bath, an estimated entry temperature T_o of 100-120°C. In fact the preheat temperature does not significantly alter the time for reflow and dispensing with all preheating for samples of this type gives indistinguishable quality of reflow in only marginally longer times. This is because, as discussed in Section 8.4.2.3, the major portion of heat required to reflow a small solder volume is not used to raise its temperature to the solder melting point but to supply its latent heat of fusion.

The time to raise the solder specimen from its temperature T_o on entry into the heat transfer liquid is given by Equation (8.5). To that time must be added the time t_L given by Equation (8.21) spent at the melting point while the solder melts. The latent heat of pure tin is 7·2 $kJ.mol^{-1}$ and that of pure lead is 4·9 $kJ.mol^{-1}$. Using simple mixing rules the latent heat of eutectic-composition solder is 47 $kJ.kg^{-1}$. Taking the heat transfer coefficient h as constant at 400 $W.m^{-2}.K^{-1}$, the calculated reflow times are shown in Figure 8.31 for the case of no preheating ($T_o = 20°C$) and a high-temperature preheating ($T_o = 150°C$). Clearly the predicted effect of preheating is minimal, in agreement with the measurements.

The heating rate of an item in a liquid heat transfer fluid reduces as the ratio A/V, surface area to volume, increases. For a solder paste comprising many small solder spheres, this ratio is large and the time to reach the melting temperature is therefore short. Hence the difference initially in the heat transfer coefficient for liquid and for vapour phase soldering is not important. The process of reflow is controlled more by t_L, which is inversely proportional to the

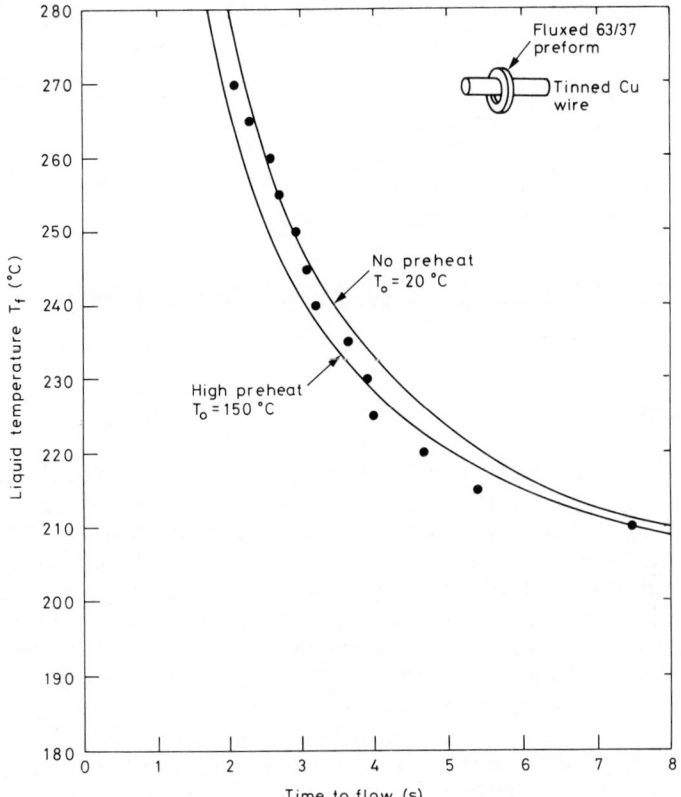

Fig. 8.31 Reflow times for a 1 mm³ annular solder preform on a 0·6 mm diameter solder-coated wire, immersed in Galden LP fluid at various temperatures. The lines are theoretical predictions using Equations (8.5) and (8.21).

excess temperature $T_f - T_s$ of the fluid over the melting point of the solder. Thus, raising the fluid temperature reduces significantly the necessary immersion time as was demonstrated in Figure 8.13. The main advantage therefore in liquid phase soldering is the opportunity to use any reflow temperature and to choose the highest possible that is compatible with the materials being soldered.

Chapter 9

OTHER ATTACHMENT METHODS

9.1 INTRODUCTION

There are three main approaches to meet the requirement of attaching surface mounting components to their interconnection substrate, as discussed at length in Chapters 5, 7 and 8. They are, respectively, wave soldering, infra-red reflow of a solder paste and vapour phase reflow of a solder paste. These have their own advantages and disadvantages which give each of them an appeal for use by a specific manufacturer or for a specific product line. These three options meet the needs of the electronics assembly industry for standard products, but there is always a place for new ideas to develop in conjunction with new technology, and new mass assembly methods cannot be discounted. Additionally there are attachment methods required for particular components or particular substrates.

This chapter deals with some of the alternative attachment methods available for the implementation of surface mounting technology[292]. First, the use of a hot conveyorised plate is considered, this being the traditional reflow method for hybrid microelectronics assembly. Next are considered some methods employing localised heating that require sequential soldering joint-by-joint or component-by-component. Such techniques often find more application for replacement or rework of components than for primary assembly. Finally, as a total replacement for the soldering process, the use of conductive adhesives in surface mounting technology is briefly reviewed.

9.2 HOT-PLATE SOLDER REFLOW

The simplest and cheapest method of reflowing deposits of solder paste to attach surface mounting components is to conduct the heat through the substrate, using a hot plate as the heat source. In order to control reliably the quantity of heat supplied to the workpiece, a conveyorised belt is usually employed, and there are numerous commercially available machines of this type.

The heat energy is transferred directly by conduction at the bottom surface of the printed circuit substrate. The heat for reflow of the solder paste must therefore be conducted through the substrate, for which good thermal conductivity and good thermal stability are obviously required. In addition, to ensure thermal contact with the heated belt, the bottom of the substrate must be very flat. These requirements generally restrict the use of hot-plate or hot-belt reflow soldering to single-sided ceramic substrates of modest proportions. Thus the technique is used extensively for the assembly of hybrid microelectronics.

Even when used to reflow small ceramic substrates, this method of solder reflow is restricted by two prevalent problems: the thermal resistance of the system and the lack of uniformity of the heat received at the top surface of the substrate. The main cause of both these problems is the magnitude and variation of the air gap between the hot-plate surface and the substrate. This is exacerbated by the different thermal demands of components. To some extent the problems can be alleviated by using a conveyorised belt rather than a stationary hot-plate, and traversing the belt across a series of hot plates, providing a heating and cooling cycle.[293]

9.2.1 Heat Conduction through the Workpiece

When reflowing solder paste on the top surface of a substrate, it is the time dependence of that top-surface temperature, $T_s(t)$, that is required to be assessed. This temperature depends on two parameters: the heat transfer from the hot-plate to the bottom surface of the substrate and the conduction of the heat through the substrate. For a substrate material with relatively high thermal conductivity, such as alumina, T_s is controlled by the heat transfer parameter, whereas for a substrate material with low thermal conductivity, such as reinforced organic laminate, the heat transfer is subordinate to the substrate thermal conductivity, in controlling T_s.

This effect is shown in the calculated data in Figure 9.1 which give the time taken for solder to melt ($T_s \approx 183\,°C$) on the top surface of a substrate either 1 mm or 0.1 mm thick, when placed on a hot-plate at 240°C. With alumina, its thermal conductivity is such that its top and bottom temperature are always close and the time taken is strongly dependent on the heat transfer coefficient for the heat conduction into the substrate. With the organic substrate, the attainment of a high heat transfer is almost irrelevant, especially when the substrate is thick. It is the low thermal diffusivity through the substrate that is important.

9.3 LOCAL CONDUCTIVE HEATING

There exist a number of techniques in which heat is supplied to the joint from a tool that is either heated resistively itself or causes the joint components to heat resistively. The most readily available of such methods is the soldering iron, but its use for anything more than small-scale prototyping or limited rework of surface mounted boards is both tedious and fraught with danger for both components and substrate. If local heating is to be used, it has to be supplied to the joint in a more controlled manner than is possible with a hand-held soldering iron.

9.3.1 Single-lead Soldering

Single point conductive heating simply involves a semi-automated application of a well-controlled soldering iron. It has the advantages, however, that the angle which the hot contacting surface makes with the component lead is well defined and the pressure exerted by the tool on to the lead is carefully controlled. The technique can be used only for leaded devices of the 'gull-wing' type such as SO and quad packages. Figure 9.2 illustrates the technique; the diameter of the tip

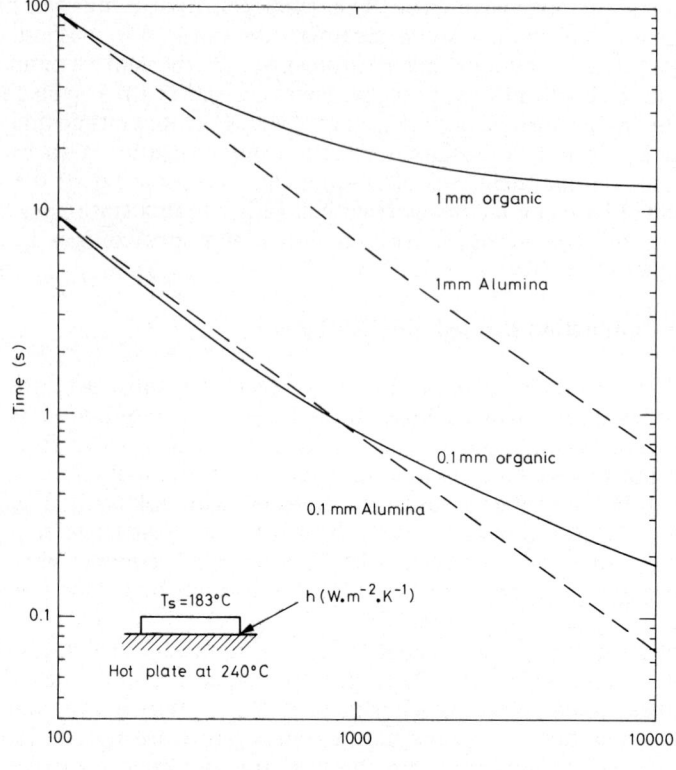

Fig. 9.1 The time taken for solder to reach its melting point on the top surface of a substrate lying on a hot-plate at 240°C. Alumina and an organic laminate are considered, with both 1 mm and 0·1 mm thickness. The heat transfer coefficient between the hot-plate and the bottom surface of the substrate is taken over the range 100 to 10,000 $W.m^{-2}.K^{-1}$.

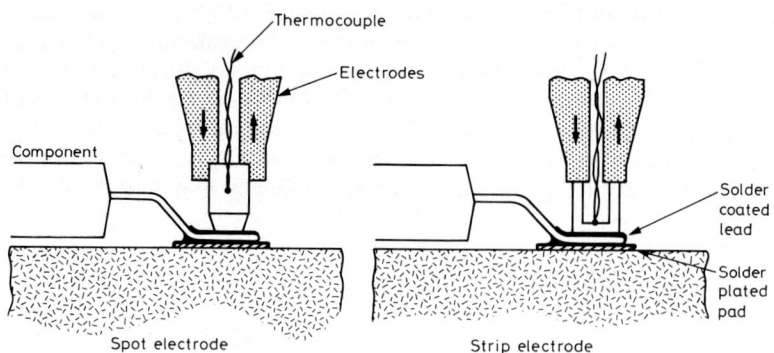

Fig. 9.2 Showing resistively heated devices for sequential single-lead soldering of 'gull-wing' leaded components.[4]

contact area can be as small as 0.1 mm but more suitably is the width of the component lead, in order to maximise the heat transfer. The electrode is normally molybdenum through which the electric current is passed.

The solder can be available at the joint in the form of a printed paste as for normal reflow, or as a solder deposit on the substrate board, although a fluxing stage is then required. In both cases the component preferably needs to be solder dipped in order to aid solder flow over the top surface of the lead. The electroplated solder layer which has served as an etch resist during the manufacture of the printed circuit board will probably not exhibit adequate solderability, and a hot-dipped, hot-air levelled board is preferred. The soldering is achieved more readily if the solder-dipped deposit is thick, ideally in the range 15–30 μm. Above 30 μm the positioning of the component leads with accuracy becomes quite difficult.[4]

The technique of single-lead conductive heating can achieve very high speeds such that, if the low capital equipment costs are also considered, it is very attractive economically, for boards of suitable design. Since the heat transfer takes place by conduction, the solderable lands and component leads should not have a high thermal demand, to ensure success.

9.3.2 Multiple-lead Soldering

Multiple-lead soldering is an extension of the conductive single-lead soldering, especially developed for fast attachment of SO and quad-flatpacks in surface mounting technology. Figure 9.3 illustrates how the heated soldering tip is extended so that all of the leads along one straight row of 'gull-wing' leads can be soldered simultaneously. It is, of course, not hard to imagine an array of such soldering heads that would solder both sides of an SO package or all four sides of a quad-flatpack simultaneously.

The heat is supplied to the head either continuously, maintaining a constant temperature, or by a short pulse, triggered by a given pressure on the leads by the head. If the constant temperature mode is used, the heater element will attain an

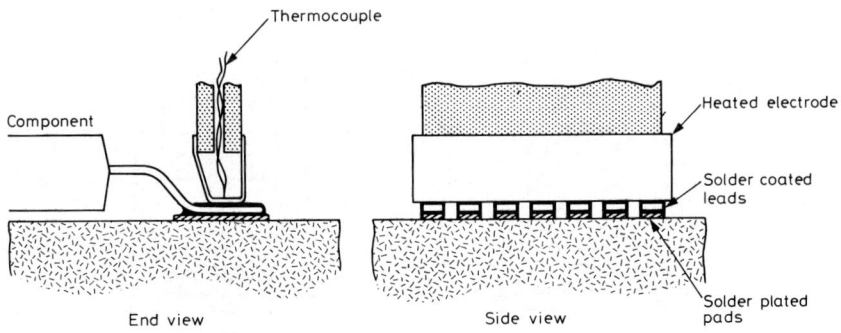

Fig. 9.3 Multiple-lead reflow soldering of a row of 'gull-wing' leads.[4]

equilibrium that is nearly homogeneous along its length, but if the impulse mode is used the temperature distribution along the length of the heater element may need adjusting by changing its cross-section locally. The impulse heating mode gives better results because the solder joints can be allowed to cool to solidification before the pressure of the tool is relaxed. It thus overcomes any problems of coplanarity that frequently arise with large gull-wing packages. If the head is hot, and the solder therefore molten, when the pressure is removed, one or more of the leads may spring up off its pad. For soldering a row of leads on an SO packaged device, a pulse of about two seconds, giving rise to a maximum element temperature of 400°C, is required, with a pressure exerted on the leads of some 50 N.[4]

9.3.3 Collet Soldering

The multiple-lead soldering concept can be additionally applied to PLCCs by using a heated collet. Because the PLCC has J leads, pressure cannot be applied directly on to the leads and has to be applied to the component body as shown in Figure 9.4. The component is held within the collet by friction against its leads. The figure shows the sequence of placement and heating by the collet.

A collet of this type for PLCCs or a ganged heater element for SOs and quad-flatpacks form an essential part of a repair station used to replace components after inspection and test. The technique is described in Section 12.4.2.1.

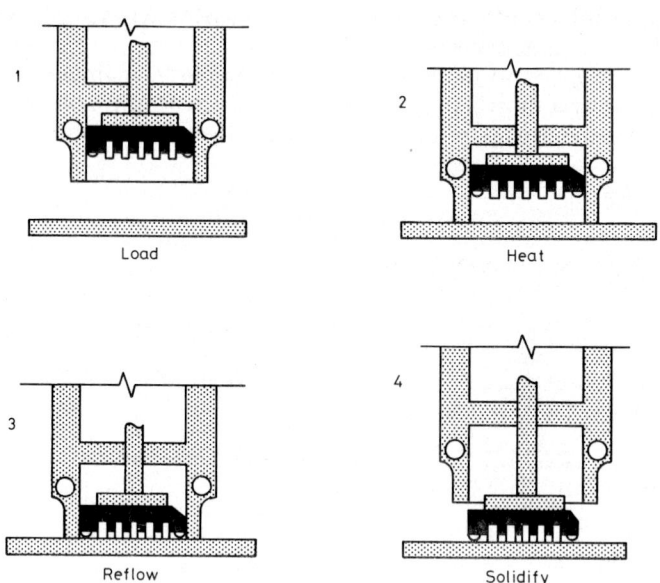

Fig. 9.4 The sequence of operations when using a heated collet to place and solder a PLCC.

9.3.4 Resistance Soldering

A variation on conductive heating is to use the component lead as part of the current-carrying resistive path, as illustrated in Figure 9.5. Two adjacent electrodes are pressed on to a lead and a pulse of current is passed to melt the solder and make the joint. The figure shows an ideal situation in which the substrate, the component lead and the ends of the electrodes are all perfectly flat and parallel. In general, this is not so and it is therefore usual to spring-load the electrodes to compensate for any irregularities. The temperature reached for a given current pulse depends on the thickness and width of the lead as well as its material identity and the separation of the two electrodes. This last parameter should be adjustable so that it can be matched to the electrical resistance of the particular component lead.

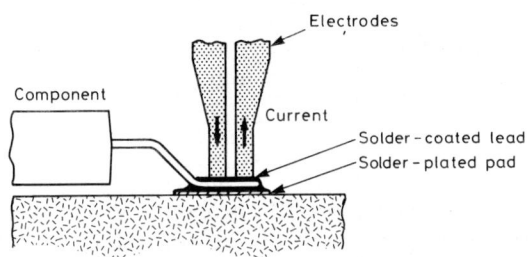

Fig. 9.5 Resistive heating; passing a heating current directly through the part of the lead to be soldered.[4]

9.3.5 Local Heat Conduction in a Workpiece

During single-component or single-terminal soldering, heat is supplied very locally, by some means such as a hot iron or a jet of hot gas. When the specific area on the top of the interconnection substrate is heated, the heat diffuses into the board both transversely and sideways, heating a larger area than intended. If the substrate has a relatively high thermal conductivity, such as alumina, this heat sinking is rapid and the rate of supply of heat flux to the substrate surface where it is needed must be fast. If however the substrate is, for example, epoxy-fibreglass with a relatively low thermal conductivity, soldering on its surface can be achieved much more readily since the heat remains localised.

If the board is relatively insulating, thermally, and the time of heating relatively short, the sideways heat flow can be ignored when estimating the relationship between heat flux supplied and surface temperature achieved. Quantitatively, this assumption can be expressed as:

$$b \gtrsim \sqrt{\frac{10Kt}{\varrho c}} \tag{9.1}$$

where b is the radius of the heated area on the surface of the substrate, and t is the time that the heating is applied. The thermal properties of the substrate are defined by its thermal conductivity K (J.m^{-1}.s^{-1}.K^{-1}), its density ϱ (kg.m^{-3}) and its specific heat c (J.kg^{-1}.K^{-1}). With an epoxy-fibreglass substrate the glass is considerably more thermally conductive than the epoxy (see Table 8.1) but since the glass fibres run in the plane of the substrate board, whereas the heat flow is through the board, the glass cannot act as a thermal short-circuit, and it is the epoxy that controls the heat flow. Thus for a 5 mm diameter heating spot on the surface of an epoxy-fibreglass board, Equation (9.1) means that lateral heat flow can be essentially ignored for heating times of less than about 8 seconds. Although this limiting time is dependent upon the size of the heating spot, for times less than the limit the surface temperature rise is, of course, independent of the heating spot size, since there is no lateral flow of heat. In such cases, the solution to the thermal diffusion equation is[40]

$$\Delta T = \Phi \sqrt{\frac{4t}{\pi K \varrho c}} \qquad (9.2)$$

where ΔT is rise in the surface temperature as a result of a heat flux Φ (W.m^{-2}) applied for time t(s). Whence, for epoxy-fibreglass, using the data in Table 8.1,

$$\frac{\Delta T}{\Phi} \approx 0 \cdot 002 \sqrt{t}$$

To melt solder, a temperature rise ΔT of about 160°C is necessary, typically within 2 seconds. The required heat flux is therefore about 6 W.cm^{-2}. This is the product of the temperature (K) of the heating tool or the hot gas jet, and the heat transfer coefficient h (W.m^{-2}.K^{-1}).

If the requirement of Equation (9.1) is not fulfilled, because of long heating times or because the substrate, such as alumina, has a high thermal conductivity, then lateral flow of heat cannot be ignored, and therefore the size, b, of the heating spot becomes relevant. In these cases the solution to the thermal diffusion equation is[40]

$$\Delta T = \Phi \sqrt{\frac{4t}{\pi K \varrho c}} \left\{ 1 - \sqrt{\pi} \text{ ierfc} \left(\frac{b}{2 \sqrt{\frac{Kt}{\varrho c}}} \right) \right\} \qquad *(9.3)$$

---

*Many problems of mathematical physics, notably in the theory of heat conduction, diffusion and hydrodynamics, can be reduced to the solution of an ordinary second order differential equation of the type

$$\frac{d^2y}{dx^2} \pm ax \frac{dy}{dx} \pm by = 0$$

whose solution involves the integral error function ierfc(x). The nomenclature used is:

$$\text{erf}(x) = \frac{2}{\sqrt{\pi}} \int_0^x e^{-\zeta^2} d\zeta$$

$$\text{erfc}(x) = \frac{2}{\sqrt{\pi}} \int_x^\infty e^{-\zeta^2} d\zeta = 1 - \text{erf}(x)$$

$$\text{ierfc}(x) = \int_x^\infty \text{erfc } \zeta \, d\zeta = e^{-x^2} - x\sqrt{\pi} \text{ erfc}(x)$$

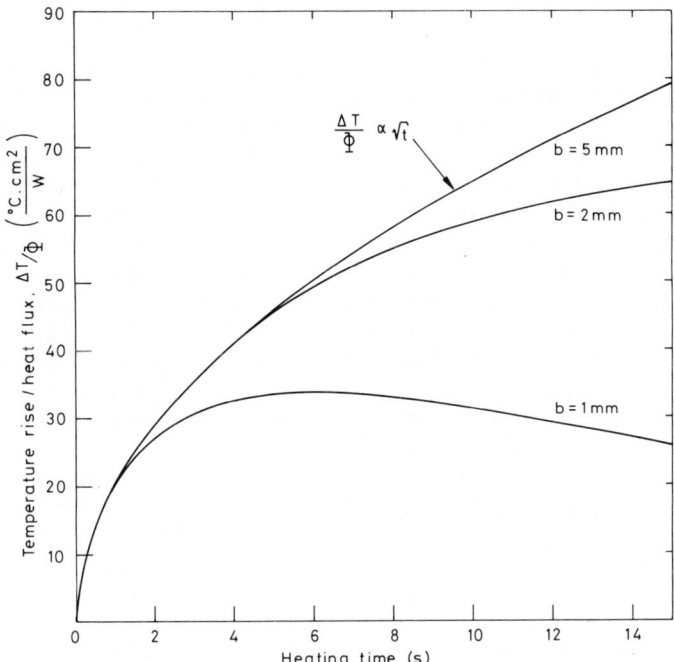

Fig. 9.6 Local heating on an epoxy-fibreglass substrate. Using a heat spot of radius b and flux Φ, the surface temperature rise is ΔT. If the spot is small, lateral heat flow is important in restricting the temperature achieved. For short times or large heating spots, lateral heat flow is negligible.

Some representative data are plotted in Figure (9.6) for heating epoxy-fibreglass substrates with local heating diameters of 2, 4 and 10 mm. When using large heating spots, lateral heat flow is negligible and the temperature rise is parabolic as in Equation (9.2). When using small heating spots or long heating times, loss of heat laterally becomes influential.

9.4 LOCAL HOT GAS SOLDERING

As mentioned in Section 8.1, the use of a hot gas as a heat transfer fluid for simultaneous reflow soldering of a whole assembly is not successful because the heat transfer is slow and the consequent temperature-time profile is not conducive to reliable soldering since the flux activity peaks too soon before the solder melts. Hot gas reflow soldering can, however, be used very effectively if localised to one component at a time, such that a jet of gas can be employed. The technique has found an ideal application in rework stations for the removal and replacement of individual surface mounted components. For prototyping boards or for small batch assembly such a rework station can be used very efficiently for the primary assembling process. The use of hot gas in a rework station is described in Section 12.4.2.2.

9.5 LASER SOLDERING

A laser is easily capable of generating the energy required to microsolder joints between electronic components. The paramount advantage of laser soldering lies in the fact that heat is applied only to the joint to be soldered, and not the entire circuit board and its components. A corollary of this is that both sides of the printed circuit board are available for mounting surface components. A laser soldering system applies a tightly focused beam of energy to one joint at a time, stepping rapidly from joint to joint under control of a computer-commanded x-y table. Because only the joint is heated, damage to the package or its internal semiconductor chip is averted. The danger of unsoldering a hybrid device's internal connections is eliminated too. Also, due to the very short duration of the soldering process (as brief as 5 ms), formation of the brittle intermetallic compounds within the solder bond is virtually eliminated. Because the joints are heated sequentially, the method also minimises the formation of short-circuit solder bridges between leads and pads.

The use of lasers for soldering is a comparatively recent development. Two ancillary technological developments have led to its use as a heat source for soldering systems. One is the advent of the microprocessor, which allows workpiece positioning, via a numerically controlled x-y table, and the output energy of the laser to be readily programmed. Since laser soldering is essentially a sequential rather than a simultaneous process, such control is essential, but once accomplished is also a virtue of the process. The second development is that of dispensable and screenable solder pastes such that the application of solder can be restricted to very localised areas.

The first laser soldering system with computer numerical control (CNC) was available commercially in 1976[294]. This was a 50 watt, continuous wave, carbon dioxide, flowing gas laser. Although pulsed yttrium-aluminium-garnet (YAG) lasers had been available since around 1971, their application was directed at cutting, drilling and welding rather than soldering. Only in 1982 did the first continuous wave YAG laser soldering system become commercially available. In recent years with the advent of surface mounting assembly methods and the development of VLSI component packages with as many as 250 leads at 300 μm spacings, the technique of laser soldering has been coming more and more to the forefront[295-297]. Although not a panacea for all the needs of soldering SMDs, the laser solves many common problems in state-of-the-art circuitry manufacturing.

9.5.1 The Laser

Before discussing the application of the laser in surface mount technology, a very brief synopsis of the physics involved is given.

9.5.1.1 THE PHYSICAL MECHANISM

The acronym LASER stands for Light Amplification by Stimulated Emission of Radiation. The manifestation of the atomic phenomenon is a unique source of radiation capable of delivering intense coherent electromagnetic fields in the spectral range between the ultra-violet and the far infra-red. The laser beam coherence means that the beam is highly monochromatic and is very directional.

The device consists of an active lasering medium which may be a solid, liquid or gas. Power is pumped into the medium, exciting atoms, ions or molecules, which then relax to a lower atomic energy state, rendering possible the amplification of a particular wavelength energy. In the laser, oscillation of the amplified light signal is achieved by the use of mirrors to form a closed optical resonant cavity. Thus stimulated emission in the medium results in the required amplification, while the mirrors supply the feedback required for regeneration and oscillation.

An atom may exist only in energy states of very well defined energy. When moving between two energy levels E_m and E_n, electromagnetic radiation is absorbed or emitted by the atom, at a characteristic frequency ν_{nm} given by

$$h\nu_{nm} = E_n - E_m \qquad (9.4)$$

where h is Planck's constant, equal to $6 \cdot 626 \, 10^{-34}$ J.s. There are numerous and quite varied methods by which an atom may be raised from a lower to a higher energy level, but when an atom finds itself in an excited state there are only two independent processes through which its excess energy may be radiatively emitted. The atom may decay spontaneously, emitting an electromagnetic wave in a random manner shown in Figure 9.7(a) or it may be induced to emit radiation by an already existing electromagnetic field, as in Figure 9.7(b). In the latter case, the radiation wave is in the same direction as the inducing or stimulating field, has the same phase, and possesses identical polarisation.

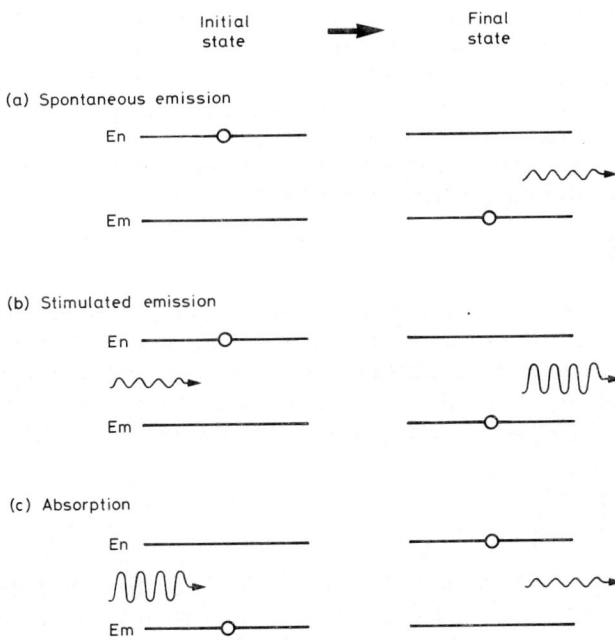

Fig. 9.7 The interaction of a two-energy-level system with an electromagnetic field.

The rate of stimulated emission is proportional to the energy density of the inducing field $\Lambda(\nu_{nm})$ (J.m^{-3}.Hz^{-1}) at the characteristic frequency, as well as the population N_n, of the excited state. The total power radiated through the process is therefore

$$P_{\text{stimulated}} = B_{nm} h\nu_{nm} N_n \Lambda(\nu_{nm}) \tag{9.5}$$

where B_{nm} is the proportionality constant representing the strength of the induced transition between the two energy levels.

The spontaneous emission occurs equally well in the presence or absence of external radiation and hence the rate of the process is independent of any external field. Thus

$$P_{\text{spontaneous}} = A_{nm} h\nu_{nm} N_n \tag{9.6}$$

where A_{nm} is a constant representing the rate of spontaneous transition between levels E_n and E_m. The proportionality constants A_{nm} and B_{nm}, which are called the Einstein coefficients, are related using fundamental quantum mechanics:[298]

$$A_{nm} = \frac{8\pi h \nu_{nm}^3}{c^3} B_{nm} \tag{9.7}$$

where $c = 2 \cdot 998 \, 10^8$ m.s^{-1} is the velocity of light. Thus, once the excited energy state is populated, both stimulated and spontaneous emission will occur simultaneously but the ratio of the power radiated spontaneously to that radiated by stimulation increases as the cube of the frequency. This makes the attainment of laser action in the ultra-violet and higher frequencies exceedingly difficult.

The inducing field, $\Lambda(\nu_{nm})$, as well as stimulating radiation, will, if there are atoms in the lower energy state E_m, give rise to energy absorption and stimulate transitions from the lower to the upper state as depicted in Figure 9.7(c). The power absorbed from the electromagnetic field is then

$$P_{\text{absorbed}} = B_{mn} h\nu_{nm} N_m \Lambda(\nu_{nm}) \tag{9.8}$$

where B_{mn} is the proportionality constant representing the strength of the upward energy transition. It can be shown that $B_{nm} = B_{mn}$ and therefore the ratio of the power absorbed to the power radiated through stimulated emission is given by the ratio of the population of the lower and upper energy states. At thermal equilibrium, these populations are related by the Boltzmann distribution:

$$\frac{N_n}{N_m} = \exp-\left(\frac{E_n - E_m}{kT}\right) \tag{9.9}$$

where k is the Boltzmann constant, $1 \cdot 38 \, 10^{-23}$ J.K^{-1} and T is the temperature in kelvin. At room temperature $kT \approx \frac{1}{40}$ eV (4.10^{-21} J) so that for transitions in the visible or infra-red range, where $E_n - E_m \gtrsim 1$ eV, the population of the upper level N_n is vanishingly small in comparison with that of the lower energy level. Thus it requires a special set of circumstances to succeed in establishing a state of population inversion where $N_n > N_m$ such that the stimulated emission exceeds the absorption, and a phase coherent amplification of the incident beam can occur.

A simple atomic model of a system which allows population inversion to be established is shown in Figure 9.8. This four-level model is the basis of both the Nd:YAG laser and the CO_2 laser which are the two most appropriate for use in a surface mount soldering station.

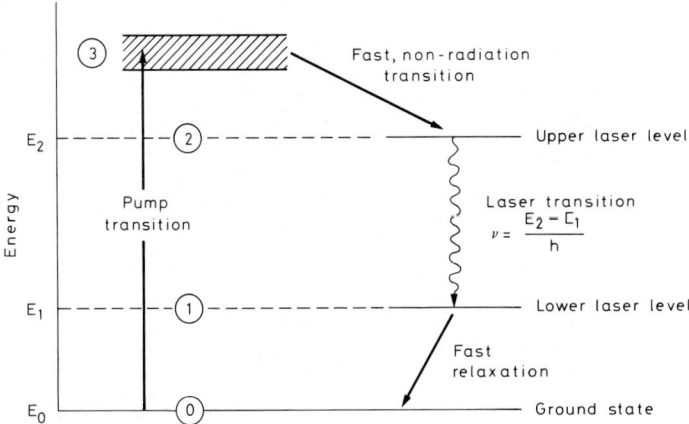

Fig. 9.8 A model of a four-energy-level system that can achieve population inversion and consequent lasering.

Atoms are initially raised from the ground state into a broad energy level 3 either by absorbing electromagnetic radiation from an optical pump such as a flash lamp or by electron impact. A fast non-radiative relaxation from the excited state ensues, primarily into level 2 which serves as the upper laser level and is usually a metastable state. By virtue of the long relaxation time of level 2, an accumulation of atoms in that state occurs. Level 1, which serves as the lower laser level, possesses a fast relaxation time and is usually appreciably above the ground state ($E_1 - E_0 \gg kT$). These two characteristics will ensure the existence of population inversion, since the first condition guarantees the absence of a build-up in the population of level 1, and the second implies the virtual emptiness of that level in thermal equilibrium. With such a four-level system, only a small percentage of the total atomic population has to be lifted out of the ground state in order to create the desired population inversion between level 1 and 2.

Lasering will occur whenever the amplification through stimulated emission is sufficiently high to balance the attenuation suffered by the field due to various loss mechanisms which are invariably present in the system, including, of course, the power extracted from the laser as a useful output.

A schematic representation of a single laser cavity is shown in Figure 9.9 to explain the onset of laser oscillations. Initially, no electromagnetic radiation is present in the cavity. Therefore, once population inversion is achieved through some pumping scheme, atoms can only emit spontaneously. Since spontaneous emission is randomly distributed in all directions, a great part propagates away from the cavity and is lost. However, the small part that has a direction of propagation perpendicular to the mirrors may stay in the cavity, and this part

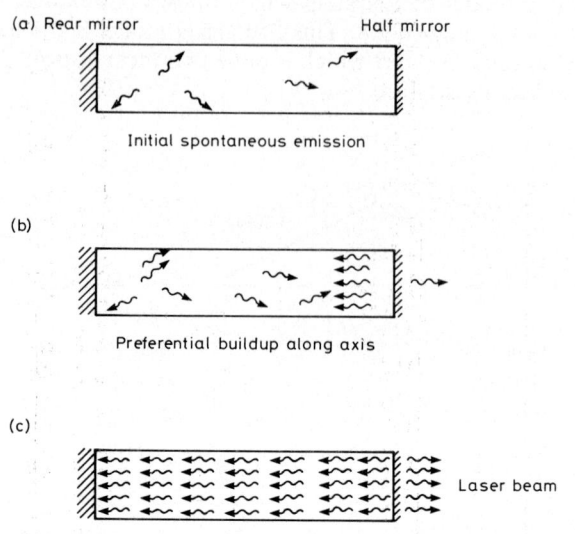

Fig. 9.9 A demonstration of the onset and build-up of laser oscillations.

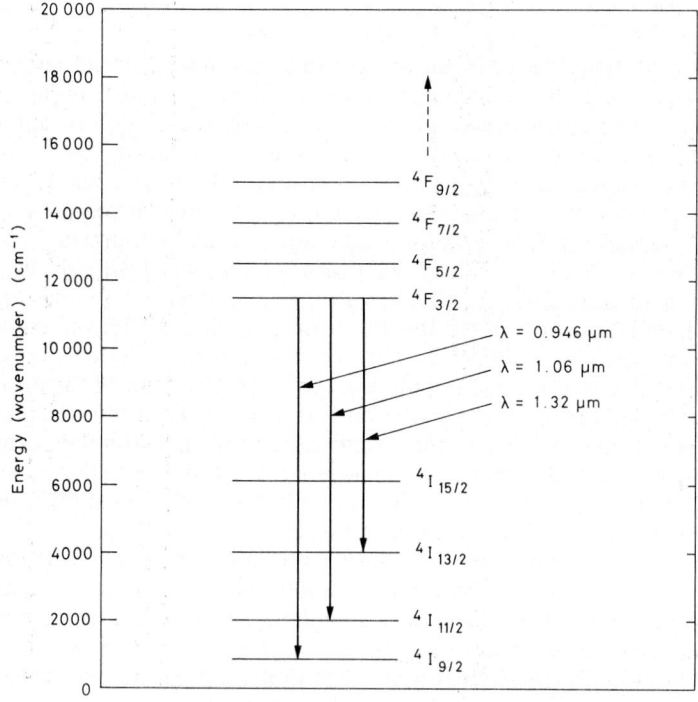

Fig. 9.10 Energy levels and laser transitions of neodymium ions in YAG.

provides the source for stimulated emission. With successive reflections by the mirrors, light travelling through the laser medium is further amplified coherently. If the gain of the lasering medium is sufficiently high, eventually the gain adequately compensates for losses sustained in the cavity, and oscillation occurs.

9.5.1.2 THE Nd:YAG LASER

The trivalent neodymium ion, Nd^{3+} will operate as a four-level laser system, within a suitable host material such as yttrium-aluminium-garnet (YAG), to form a practical laser that has found numerous industrial uses. To attain population inversion in the Nd:YAG system, ions are optically pumped by a broadband, incoherent source such as a flash lamp, from the ground state $^4I_{9/2}$ to any or all of the energy levels above the $^4F_{3/2}$. The appropriate energy levels of the Nd ion are shown in Figure 9.10. For atomic energy levels it is customary and convenient to express energy levels in terms of a wavenumber, w, being the number of wavelengths in one centimetre. Thus

$$w = 1/\lambda \text{ cm}^{-1}$$

and the energy of a photon, E_{ph}, is then given by

$$E_{ph} = h\nu = \frac{hc}{\lambda} = hcw = 1 \cdot 99 \; 10^{-23} \; w \text{ joules}$$

or $1 \cdot 24 \; 10^{-4} \; w$ electron volts

Ions excited into these higher energy states decay non-radiatively and with near-unity quantum efficiency into the $^4F_{3/2}$ level. This level has a lifetime of 250 μs and is therefore metastable and serves as the upper laser level. The main laser transition is obtained between the $^4F_{3/2}$ and $^4I_{11/2}$ levels, producing a wavelength of $1 \cdot 06$ μm. The population of the $^4I_{11/2}$ level at normal operating temperatures is negligible. This permits attainment of population inversion and oscillation, at a typical threshold of a few hundred watts pumping power, with a laser output power of the order of several watts.

9.5.1.3 THE CARBON DIOXIDE GAS LASER

The most important gas laser is that based on carbon dioxide, in which transitions between vibrational:rotational levels can provide a lot of power at relatively high efficiency (up to 30%) at a wavelength of $10 \cdot 6$ μm. In comparison with atoms and ions, the energy-level structure of molecules is complicated and originates from three sources—electronic, vibrational, and rotational motions. The energy separation between vibrational levels of the same electronic energy state generally corresponds to frequencies in the middle infra-red range. Figure 9.11 shows three possible modes of vibration within the CO_2 molecule in its lowest electronic energy state. The asymmetric longitudinal vibrational mode is the highest energy state, having a wavenumber $w = 2331$ cm^{-1}, and acts as the upper laser level. The symmetric longitudinal vibrational mode has an energy $w = 1388$ cm^{-1} and acts in the lower laser level, leading to radiation of wavelength $(2331-1388)^{-1}$ cm or $10 \cdot 6$ μm. Atoms then decay radiatively to the

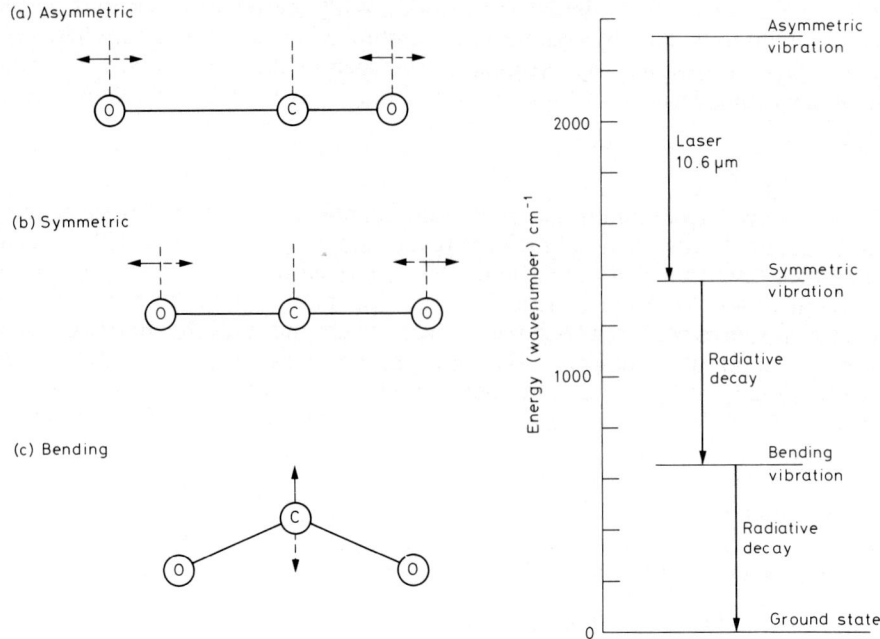

Fig. 9.11 The symmetric and asymmetric vibrational motions of the CO_2 molecule.

bending mode ($w = 667$ cm^{-1}) and the ground state. When a gas discharge is established, CO_2 molecules are excited to the vibrational states and favourable relaxation rates permit the establishment of a population inversion between the upper (asymmetric) level and the lower (symmetric) level.

9.5.2 Selecting the Laser for SMT

Of the commercially available lasers, the two types that have emerged as the most suitable for microsoldering are the carbon dioxide laser and the neodymium:YAG laser. Both can be operated in the continuous mode and can be interfaced with computer positioning systems. Both types make solder joints that are microscopically very similar. However, the two have very different properties, and a cost-performance study is required before selection of one type in preference to the other.

9.5.2.1 EFFICIENCY OF HEATING

The Nd:YAG laser emits light at a wavelength of $1 \cdot 06$ μm, in the near infrared. This wavelength is readily transmitted by glass and most plastics but is reasonably well absorbed, rather than reflected, by most metals. Some typical reflectivity data[299] are shown in Figure 9.12 for solder and PCB laminate material irradiated with laser light of $1 \cdot 06$ μm wavelength. Because of the relatively high absorption of light of this energy by solder it is an efficient method of bringing

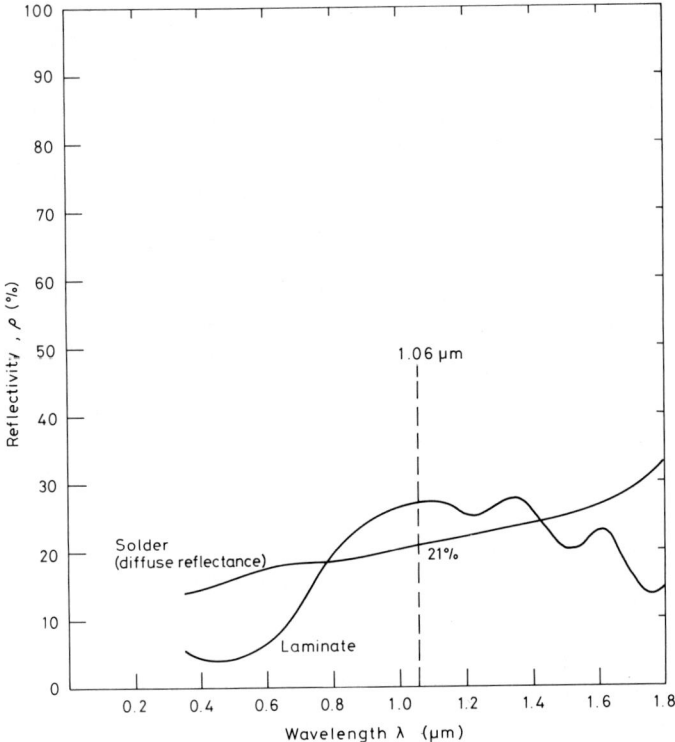

Fig. 9.12 Reflectivity measurements from non-reflowed solder and from a polyimide printed circuit board material in the region of the 1·06 μm wavelength light emitted by the Nd:YAG laser.[297]

the solder to its melting point and making the joint. Because of the transparency of the laminate board material and plastic components to light of this energy, it is not necessary to shut off the laser beam when moving from one joint to the next.

The CO_2 laser operates at a wavelength of 10·6 μm in the far infra-red. Electromagnetic radiation at this wavelength is reflected by most metals but absorbed by glass and plastics. The difference in the operating wavelength gives rise to certain advantages and certain disadvantages compared with the Nd:YAG laser. From Figure 9.13 it can be seen that the reflectance of light at 10·6 μm wavelength, from the same sample of unfused electroplated solder as used to obtain the data in Figure 9.12, is about 74%. It might seem therefore that it takes about three times as much energy to make a joint with the CO_2 laser, compared with the Nd:YAG laser, since the bulk of the light at 10·6 μm wavelength is reflected whilst at 1·06 μm wavelength it is absorbed. However, in practice this is not altogether the case, because of the presence of flux on the solder at the joint. Flux absorbs about 95% of the 10·6 μm wavelength light and transfers this energy conductively to the solder.[300] Once the solder is molten, the reflectance drops dramatically. Under normal circumstances, therefore, the energy required for CO_2 laser soldering is very similar to, and may be somewhat less than, that required using a Nd:YAG laser.

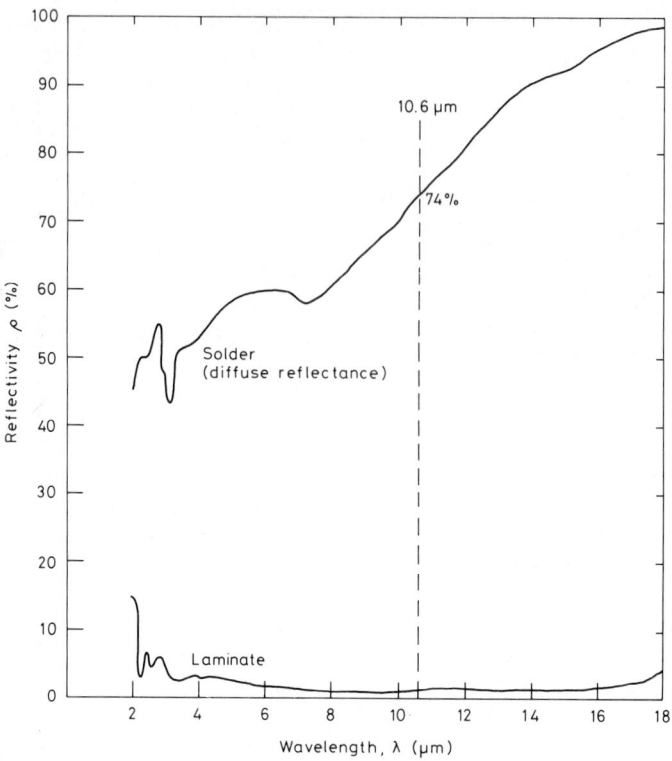

Fig. 9.13 Reflectivity measurements from non-reflowed solder and from a polyimide printed circuit board material in the region of the 10·6 μm wavelength light emitted by the CO_2 laser.[297]

Additionally, the energy efficiency of a CO_2 laser is, at 10-15%, much higher than a similarly powered Nd:YAG laser which is only about 1%. This leads to higher operating and maintenance costs for the Nd:YAG laser in addition to an initial capital cost that is twice as high as the CO_2 laser. The CO_2 laser is therefore the more cost-effective alternative for the great majority of standard soldering situations encountered in surface mounting assembly, but the Nd:YAG laser may be more convenient in certain situations.

9.5.2.2 POWER LEVELS AND PULSE DURATION

Laser soldering, unlike laser cutting and welding operations, is more time dependent than total-energy dependent. Time is needed for the highly concentrated energy to diffuse through the solder, melt the solder and allow it to wet and flow to form the joint. For example, a 100 watt continuous wave pulse applied for a duration of 10 milli-seconds contains an energy of 1 joule. This pulse would probably burn a hole through the solderable pad. The same energy is contained in an 8 watt pulse of 120 milli-seconds duration. This would make an excellent joint. With a 10 watt beam focused into a 100 μm spot, the target is receiving energy at the rate of over 100 $kW.cm^{-2}$.

The energy required to make the solder joint is

$$Q = \Delta T \sum_i \varrho_i c_i V_i + \lambda_s \varrho_s V_s \tag{9.7}$$

where ϱ_i, c_i and V_i are the density, the specific heat and the volume of each part of the joint, namely the solder, the flux, the component lead and the solderable pad; ΔT is the temperature rise required, say from 20°C to 220°C. The second term is the latent energy of fusion of the solder, i.e., the energy required to melt the solder without increasing its temperature. λ_s, ϱ_s and V_s are the latent heat, the density and the volume of the solder. The total energy per joint is usually about 1 joule for surface mounting components.

Using a CO_2 laser, surface mounting components have been successfully soldered to PCBs at rates exceeding 125 joints per minute[301], with the solder applied as solder paste. The power setting was 33 W and the heating times varied according to the joint type, from 200 to 400 ms.

Using a Nd:YAG laser the heating time has been reduced to as little as 4 ms before damage to the lead and the pad occurs.[4] Allowing time for board loading and indexing of the joints, a throughput of some 3000 joints per hour can be achieved using a single beam laser system. This rate is comparable with that of batch vapour-phase soldering when the complete processing time is included.

9.5.2.3 DAMAGE TO THE BOARD

The light emitted by the Nd:YAG laser, at 1·06 μm wavelength, is transmitted through glass and plastics. This has the advantage that the printed circuit board material is not significantly heated if the laser beam or part of the laser beam falls upon it. The disadvantage is that the operator cannot be effectively shielded from the laser light by a transparent screen, and the beam can certainly harm the retina of the eye.

The light emitted by the CO_2 laser, at 10·6 μm, is also harmful to the eye, but is completely absorbed by ordinary glass and perspex. Thus, using the CO_2 laser, the system operator can be very effectively shielded from the laser beam by a screen or eye glasses, yet still maintain unimpeded visual contact with the workpiece. The concomitant disadvantage is that the light energy is absorbed by the board material and can damage it. CO_2 laser soldering systems are limited to spot sizes of 50-100 μm while comparable Nd:YAG systems can easily produce spot sizes in the range of 10-20 μm. By accepted definition, the energy contained within the spot size is only 86% of that available. The remaining 14%, around the periphery of the spot, may, for the CO_2 laser, be sufficient to damage the board material because of the greater absorption by the board compared with the solder. It is suggested that, to minimise this effect, the spot diameter should be, at most, one-third the minimum dimension of the solder pad.

Another manifestation of this problem appears as damage to the board as a result of reflected energy. The Nd:YAG laser radiation, although not 100% absorbed by the solder, generally tends to have minimal reflected beam energies, and this, coupled with shorter exposure times, minimises reflected beam damage compared with the CO_2 laser. However, the incident light at 10·6 μm wavelength from the CO_2 laser is reflected from the metal parts it is intended to heat, and can damage the side walls of components which, being plastic or ceramic, readily absorb the incident energy. The damage is usually evidenced by some surface carbonisation. As mentioned in Section 9.5.2.1, these problems with the CO_2

laser are overcome very adequately by ensuring that flux is present at the solderable surfaces. This absorbs the light efficiently and minimises reflection, thus coupling the heat energy into the metallic parts that make up the joint.

9.5.2.4 FLUXLESS SOLDERING

Unlike the CO_2 laser, whose energy absorption is through the flux, the Nd:YAG laser output is efficiently coupled directly into the metallic pads and leads. It is therefore an ideal choice for fluxless soldering applications. With this method, clean, oxide-free surfaces are a prerequisite since the laser is used solely as a thermal energy source and does nothing to remove any of the existing oxides that inhibit solder wetting. It is therefore usual to carry out such soldering in a reducing atmosphere. This type of soldering is primarily used where corrosive action of the flux cannot be tolerated and subsequent cleaning operations are not possible or difficult, as is often the case in surface mounting assemblies.

9.5.2.5 MULTIPLE-BEAM LASER SOLDERING

The Nd:YAG laser can be used with a multiple optical fibre delivery system such that many joints can be made simultaneously. This technique is particularly suitable for surface mounting assemblies containing a large number of similarly packaged components, enabling a whole component to be soldered by one pulse. Optical fibres or other flexible delivery systems are not currently available for CO_2 lasers although there is no physical principle forbidding their development.

A second method of multiple joint laser soldering that is being developed is the splitting of the beam with a hologram of the board, so that, again, a large number of specific places on the circuit board can be heated simultaneously. The problems with this technique arise from the requirement to dissipate a high energy beam by a hologram.

9.5.3 The Laser-soldering System

Figures 9.14 and 9.15 show typical system configurations for laser soldering using a Nd:YAG laser and a CO_2 laser. Whichever laser is used, the basic requirements are an associated power supply and water cooling system, an objective lens to focus the beam to a suitable spot diameter, and some means of positioning the part to be processed under the beam. Since both types of laser energy are invisible, some means of beam positioning and focal point indication are necessary.

Most Nd:YAG systems use a coaxial viewing and focusing arrangement. The laser beam is aligned with a reticle located within a closed circuit television or binocular, and the visual image is set to be coincident with the laser focal point. Thus, when the part is in focus visually, it is also in focus to the laser.

Although co-axial viewing with the CO_2 laser is possible, it requires expensive far-infra-red transmissive optics, and this is done only in the most high-quality systems.[302] More commonly, an off-axis binocular and/or closed circuit television viewing set-up or a helium-neon laser spotting beam is used for CO_2 laser alignment. Although effective for larger parts at fixed working distances, these viewing systems lack versatility and accuracy, and generally require more set-up time than coaxial viewing systems.

In common with all laser systems the x-y-z parts motion interface is the key area. All successful production systems use computer or micro-computer controlled positioning. These systems, when interfaced with the laser, not only position the parts sequentially at the focus of the laser beam, but sequence the timing shutter and control the laser power as well.

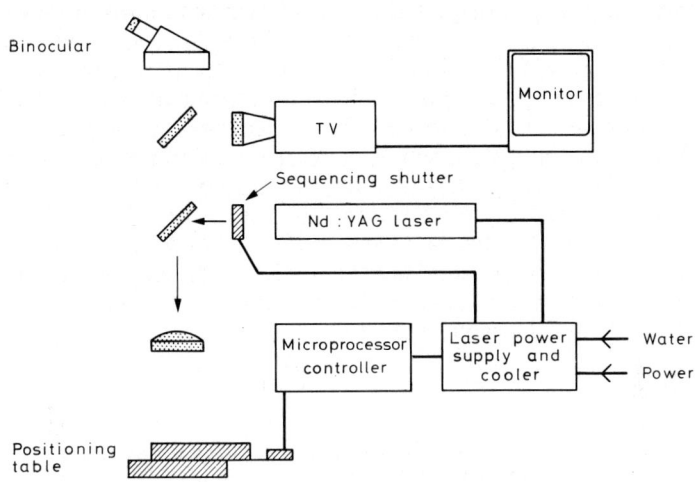

Fig. 9.14 Schematic diagram of a typical Nd:YAG laser soldering system.

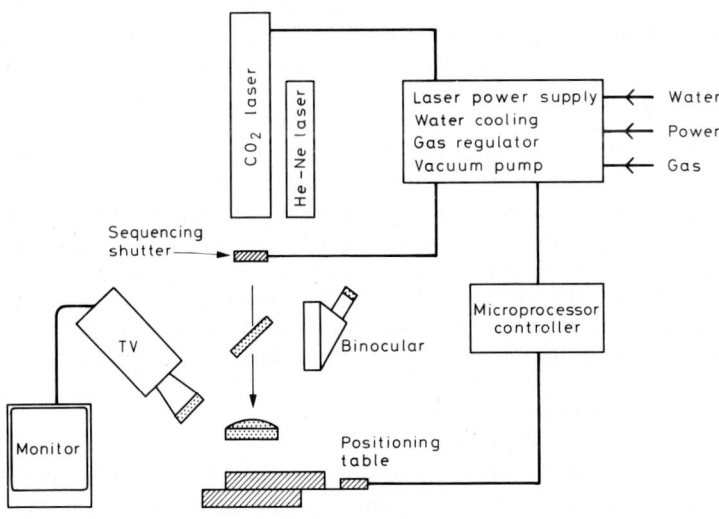

Fig. 9.15 Schematic diagram of a typical CO_2 laser soldering system.

9.5.4 Practical Soldering with a Laser

9.5.4.1 APPLICATION OF SOLDER

For practical surface mounting the board can be prepared in the usual manner using solder paste.[297] The solder paste can be applied either by dispenser or by screen printing. The laser parameters are not critical for obtaining complete reflow, except that, when using a CO_2 laser the laser spot must impinge upon the mound of paste rather than any bare area of the pad, in order that the heating energy can be coupled in by the flux. The laser focus is quite critical in the controlling of spatter of the paste and the consequent formation of solder balls.[300] It is often necessary to defocus the incident beam or place the joint below the focal point of the optical system or increase the heating times in order to heat the mound of solder paste uniformly and minimise spatter.

Great difficulty can arise in preventing the formation of solder balls, and in such cases the use of thickly plated tin-lead pads is advocated[303]. For this, all components are solder dipped, and the solder pads on the board are some 30-50 μm thick of unfused Sn-Pb plated deposit. The flux is then applied universally to all the solderable surfaces.

9.5.4.2 LASER-BEAM ANGLE

If solder paste is used to make the joint, some degree of component lead or pad height can be accommodated while still retaining a continuous thermal path between the component and the board. In this case the laser power can be applied for any direction into the vicinity of the joint. If the dipped and plated solder approach is used considerably more care is required, as shown in Figure 9.16. The device to be soldered is normally dipped in solder. The solder coated pads will then be at slightly different heights and, when placed on the board, the component will have, in general, only three-point contact. If the laser beam is applied at normal incidence, although the solder melts, surface tension of the liquid prevents its flow and a gap will probably remain between the component and the substrate, as shown in Figure 9.16 (a). Three modifications should be made to alleviate the problem. First, the laser beam is applied at an angle, as shown in Figure 9.16 (b), so that both the dipped solder on the component and the plated solder on the substrate are heated and melt. Secondly, the hot-dipped solder deposits on the component are flattened so that the device sits in more intimate contact with the board, as shown in Figure 9.16 (c). This can be done by placing the device, after solder dipping, on black anodised aluminium on a hot-plate and reflowing the solder[299]. The third requirement is to thicken the plated Sn-Pb deposit on the board, so that when it melts and reflows its central area rises sufficiently to ensure liquid contact with the solder on the device.

9.5.5 Joint Metallurgy

Laser soldering forms shiny ductile solder joints because the solder is melted and solidified so rapidly that appreciable intermetallic compound cannot form. The intermetallic compound layer, whilst strengthening the joint, dramatically reduces its ductility. For surface mounted assemblies it is the ductility that is of the essence because of stresses arising from thermal expansion mismatches. Thus

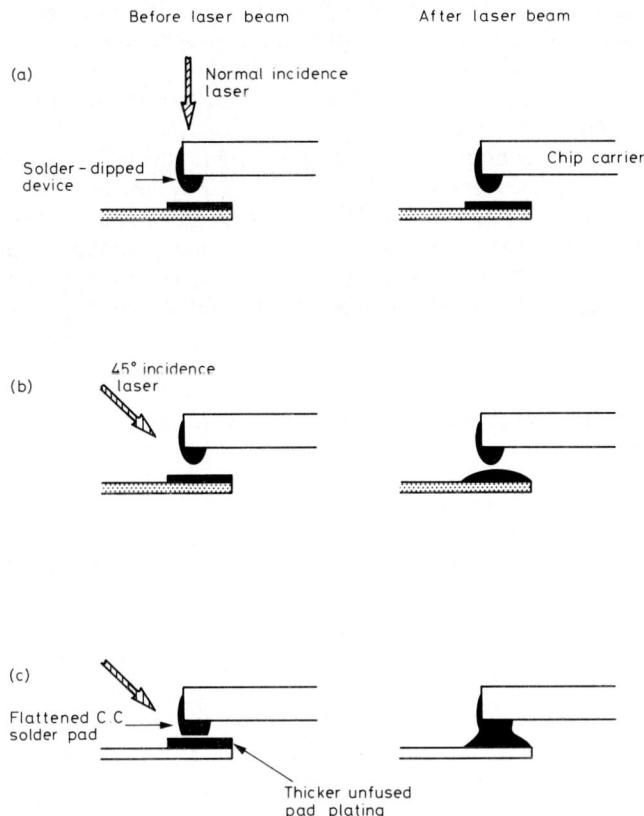

Fig. 9.16 Components can be laser soldered without the use of solder paste. If this is done, the pads should have a thick unfused Sn-Pb deposit, and the devices should be solder dipped and then the solder flattened to reduce possible gaps. The laser must be directed at an angle.[297]

the thickness of the intermetallic compound layer needs to be minimal, and in this respect laser soldering has a distinct advantage over other soldering techniques because of the very short time of heating employed.

In wave soldering the component interface is at a temperature of about 230°C for 3 seconds, resulting in an intermetallic compound thickness of about 0·5 μm. In laser soldering the time at temperature is reduced by a factor of 10-20, such that the layer thickness should be reduced by a factor of 4-5. In most joints made by laser soldering, however, the intermetallic compound layer is indiscernible by standard microsectioning techniques.

9.5.6 Intelligent Laser Soldering

An alternative to the 'dumb' laser soldering machines described above, i.e., those that apply a pre-determined amount of energy to each joint, lies in a newly introduced concept that uses an infra-red detector to monitor the soldering process as it is occurring[304]. Heat, via the laser, is applied until the solder melts and flows. The point at which melting occurs is determined by the detection of infra-red radiation emitted by the solder. The infra-red detector used to monitor the emitted radiation shares the laser's optics but is of a design that is blind to the wavelength emitted by the laser. Accordingly, the response of the detector is based on the infra-red energy radiated from the joint throughout the sequence of heating and melting.

The infra-red sensor monitors the heating process, and develops a command signal to shut off the laser energy shortly after melting has been detected. The monitoring process is accomplished by digitising the infra-red detector's electrical output and processing the signal to create a thermal 'signature' during the formation of the solder joint. In Figure 9.17 are given two such thermal signatures from nominally identical joints with identical laser exposure times.

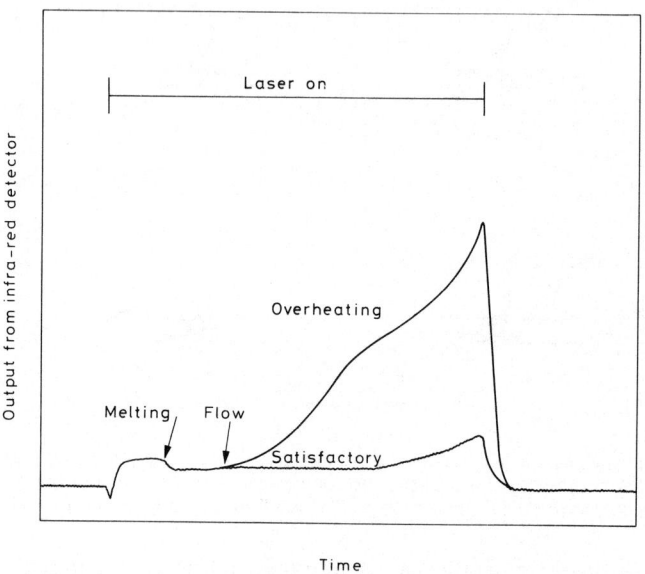

Fig. 9.17 Two 'thermal signatures' from nominally identical joint configurations. The overheating was due to the amount of solder being too small.[304]

Because of a difference in the amount of solder present, to be melted, the signatures are very different and one shows considerable overheating to have occurred. With an intelligent soldering system this situation is obviated by using some point in the thermal signature to trigger the cutting off of the laser power. One such point might be that at which the solder melts. As the solder melts and begins to flow, the solder surface becomes very shiny which causes a sudden and quite drastic reduction in the surface emissivity. The output signal of the infra-red detector therefore drops proportionately and digital control circuitry translates the changing pattern of output into a command to shut off the laser energy.

A schematic of a typical processing system is shown in Figure 9.18.

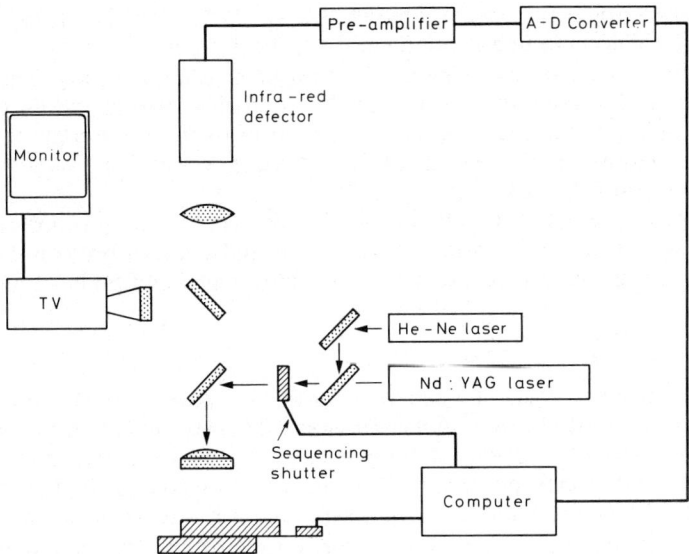

Fig. 9.18 Schematic of an intelligent laser soldering system.[304]

9.5.7 Desoldering and Repair by Laser

Many surface mounted assemblies are so densely populated with components that desoldering and repair are a major problem. The ability of the laser to direct the necessary heat just to the joint where it is needed makes it suitable for a soldering repair station. It is also ideal for post-assembly operations such as the attachment of heat pipes to circuit boards. Heat pipes are used for the removal of heat from high-power devices on the board and they cannot be soldered by conventional techniques because they conduct heat so rapidly and so non-uniformly away from the circuit board. A laser soldering system, however, can apply sufficient heat rapidly enough directly to the required joint.

9.5.8 Summary of Advantages of Laser Soldering

1. Laser soldering is an ideal process for attaching heat sensitive devices since the heat is applied only to the area of the joint.
2. Laser soldering is ideal for dense packaging of devices on a board since it can reach closely spaced components without disturbing adjacent parts.
3. Because each component is soldered without disturbing any others, both sides of a circuit board can be populated.
4. Since the laser solders each joint sequentially, solder on adjacent pads is not molten simultaneously, and hence the potential for solder bridging from one pad to another is minimised.
5. The amount of brittle intermetallic compound formed in the laser soldered joint is minimal. The joints are shiny and ductile.
6. Laser soldering forms low-stress joints because the whole PCB is at room temperature, except for the momentary solder fusion that forms each successive joint.

7 Laser soldering is the only technique that will solder devices already attached to heat sinks, including those attached to heat pipes.
8 The performance of laser soldering has unsurpassed uniformity. The amount of energy delivered to a joint can be controlled within a few per cent. Repeatability is limited only by the tolerances of the joint components.
9 Precise positioning is possible because there is no physical contact between the laser and the workpiece.
10 Computer compatibility allows precise control of all parameters in an automated soldering station. Power and pulse duration can be varied independently and automatically by a programmed computer sequence.

9.5.9 Laser Safety

Radiation from a laser is coherent both in space and in time. These characteristics give rise to two safety problems: the laser beam can traverse great distances with very little change in the power density and extreme intensities can be obtained. This second point is readily illustrated by comparing the effects on the eye of a 100 watt incandescent light bulb and a 1 watt laser beam.

To calculate the power density at the retina of the eye we require two parameters, the power density at the cornea and the image size on the retina. For a source, size a, that the eye can resolve, at a distance r from the eye, the size of the retinal image is

$$d_r = \frac{af}{r}$$

where f (≈ 17 mm) is the focal length of the eye. The retinal power density P_r is then given by

$$P_r = P_c \left(\frac{d_c}{d_r}\right)^2 = \left(\begin{array}{c}\text{corneal power} \\ \text{density}\end{array}\right) \frac{\text{(corneal area)}}{\text{(retinal area)}}$$

where d_c is the pupil diameter. Assuming the light bulb is 8 cm diameter, radiating isotropically and located 50 cm from the eye, we obtain

$$P_r \approx 200 \text{ W.m}^{-2} \quad \text{[light bulb]}$$

For a laser with a Gaussian energy distribution, the beam image on the retina is limited by the diffraction of the lens of the eye. This is given by

$$d_r = 1 \cdot 27 \frac{\lambda f}{d_c}$$

where λ is the wavelength of the laser light. For 1 μm wavelength radiation, d_r is some 3 μm, so that the power density is

$$P_r \approx 100 \text{ GW.m}^{-2} \quad \text{[laser]}$$

which is some 500 million times greater than the case of the light bulb.

The near infra-red radiation of the Nd:YAG laser is partially absorbed by the ocular medium, but a significant proportion is transmitted, and focused on to the retina. The far infra-red radiation of the CO_2 laser is fully absorbed at the cornea and does not reach the retina.[305]

The effect of laser radiation on the retinal tissue may be temporary or may leave permanent damage with impairment of vision. Radiation absorbed in the cornea can cause painful damage and result in conjunctivitis.[305]

As mentioned previously, the 10·6 μm wavelength radiation of the CO_2 laser is absorbed by glass and plastics and so visually transparent shielding of the laser station can be implemented. Such a screen must of course be fail safe and removal of any part of it must automatically render the laser inoperable. The 1·06 μm wavelength radiation of the Nd:YAG laser is not absorbed by glass or plastics, and metallic shielding is required. The main beam must be closely confined so that specular and diffuse reflections will not damage the enclosure or associated equipment. All commercially available laser soldering stations meet industrial safety requirements, but nevertheless the key to successful implementation of safety is the training of the personnel involved. It is essential that all the people involved with the use of lasers are well informed concerning the hazards and the control measures required. Neither 10·6 μm nor 1·06 μm radiation can be seen and, as invisible beams, they are quite susceptible to being accidentally intercepted if they are in the open. The probability of damaging the eye in such instances is very high.[305]

9.6 ASSEMBLY WITH ADHESIVES

The assembly of surface mounting components on printed circuit boards using conductive adhesives, as an alternative to solder, is a technology that is growing in popularity.[306] For adhesive assembly the leaded or leadless surface mounting components are placed on to spots of uncured conductive adhesive just as components to be reflow soldered are placed on to spots of solder paste. For adhesive assembly the board is bare copper. The populated boards are then subjected to a thermal treatment to cure the adhesive, during which it sets to form a permanent conductive board.

At that point the assembly of the board is complete. Unlike soldering, cleaning is not needed after the curing. This assembly method, using conductive adhesive, has both advantages and disadvantages when compared with the more common method of soldering,[307] and it is not an ideal alternative in every case. It does, however, offer the designer another choice for assembly processing.

9.6.1 Conductive Adhesives

The adhesives used for surface mounting assembly are primarily epoxies although some polyimides are also suitable. The polyimide adhesives have a higher glass transition temperature (T_g) which means they retain their physical properties to higher temperatures. This, however, is of no particular advantage in component attachment since the substrates used in surface mounting are generally epoxy-based laminates. In fact, this high glass transition temperature is detrimental during repair and component replacement, because of the correspondingly high softening temperature. Additionally, the polyimide adhesives are solvent based and so are prone to uncontrolled drying during processing as the solvent evaporates. Polyimide adhesives need to be carefully dried to remove the volatiles, before the curing cycle. There are therefore several reasons for choosing epoxy adhesives which have no solvent component.

Many epoxy adhesives are suitable for use in surface mounting.[308] These compounds can be divided into two major categories, one-part and two-part. The one-part epoxies are easier to handle and often lead to more consistent results because they are mixed by the supplier. They do have a much shorter shelf life and must be refrigerated during storage.

Conductive adhesives, be they epoxies or polyimides, achieve their conductivity using a dispersion of powder of a conductive material, most commonly pure silver. The metallic content is usually about 20% by volume (85% by weight) which typically gives a volume resistivity for the cured adhesive, in the range 1-10 μohm.m, compared with a value of $0 \cdot 17$ μohm.m for 60:40 SnPb solder at 25°C.

If making a direct choice between solder or conductive glue, the latter technology has several positive points and several negative ones.

(i) Using conductive adhesives eliminates the problem of leaching of component metallisations with its concomitant uncertain effect on joint strength.

(ii) The manufacturing process steps are considerably fewer when using conductive adhesives, these being a screen print of the adhesive on a bare copper board, component placement, curing of the glue, test and rework. Notable is the absence of any requirement to clean the board after assembly.

(iii) It is usually accepted that repair, replacement and rework are easier when the assembly route is with conductive adhesives[309]. A component can be readily removed once the bond strength of the adhesive has been reduced by heating. The temperature to which the joint must be heated for component replacement is the glass transition temperature of the glue. This is a critical temperature at which it starts to change from a crystalline to an amorphous state. The transition temperature can be tailored by choice of the type of resin used in the glue and the curing cycle as illustrated in Figure 9.19. A glass transition temperature in the range of 80-100°C is usually ideal, high enough for normal service but low enough for easy reworking.

(iv) Joints made with epoxy adhesives can withstand, with much greater reliability than solder joints, prolonged temperature cycling or mechanical flexing, because of the greater flexibility of epoxy resin.

(v) A potential problem of epoxy resins is their characteristic of absorbing moisture from the atmosphere, which may give rise to a reduction in bond strength after prolonged service.

(vi) Another potential problem for silver-loaded adhesives is that, in the presence of moisture and an electrical field, silver is prone to grow fine whiskers than can eventually cause short circuits between closely spaced conductive areas. This growth may be a problem when the conductors are spaced at less than 500 μm, which is the case for almost all multileaded surface mounting components. To avoid this problem, other materials such as gold, copper or silver-palladium can be used in place of silver. The cost of using gold for surface mounting applications is usually prohibitive, considering that about 20% by volume of each joint comprises the metal.

(viii) Finally, there is a consideration of the degree of spreading the adhesive. A molten solder will wet and spread only on a metallic pad. An adhesive will wet almost any material and so is not constrained, as a solder is, to the pad. A consequence of this is that the accuracy of both the screen printing alignment and the thickness of the deposit are more critical for adhesive assembly than for solder assembly.

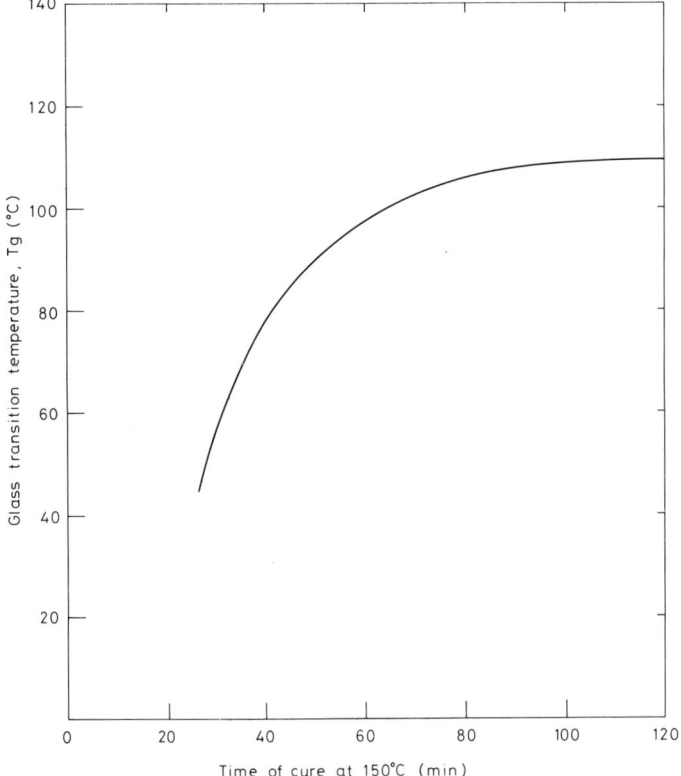

Fig. 9.19 Glass transition temperature, T_g, as a function of the cure time at 150°C for a typical epoxy adhesive.[308]

When considering conductive adhesives for the joints on surface mounting assemblies, a range of properties should be assessed:

—Screenability: the quality of screen printing achievable
—Pot life: usable period once pot is opened
—Slump: the maintenance of print height over a period after printing
—Wet strength: the ability to retain components before cure
—Cure time: to be as short as possible compatible with the desired properties
—Outgassing during the cure: use solventless glue
—Strength after cure: shear strength can be equivalent to that of solder
—Fatigue life: fatigue properties should be much superior to those of solder
—Electrical conductivity
—Thermal conductivity
—T_g: should be low (100°C) to facilitate rework
—Corrosivity
—Moisture resistance.

9.6.2 Application of the Adhesive

The uncured adhesive may be applied to either the substrate or the component by one of several methods. Substrate application is more common.

For application of the adhesive to the substrate, the methods include, in order of popularity, screen printing, syringe, and pin transfer using techniques exactly analogous to those for solder paste. The method chosen depends on several factors which include the previous operations, for example whether or not the board is already partially mounted with components which would preclude screen printing.

The techniques of screen printing, syringe application and pin transfer have already been described in detail for the application of adhesive for temporary bonding during wave soldering and for the application of solder paste. In screen printing the adhesive, a finer mesh is used than those used for the larger particulate solder pastes. The mesh is normally of stainless steel and of mesh size 165 or 200, which are compatible with the size of the metal loading powder used.

Epoxies are easier to screen print than are the polyimides which can encounter problems with certain screens because the solvent in the polyimide may cause the emulsion to swell. Another advantage of the absence of solvent in the epoxies is that they do not jam the screens as the solvent dries out during application.

If a measured amount of adhesive at each joint is preferred, a dispensing technique can be used, employing a syringe, akin to those used for applying solder paste. This dotting of the adhesive can be accomplished with a hand-held pneumatic syringe or with a computer programmed automatic machine with either one or multiple dispensing heads. The dispensing machine can be either a stand-alone unit or, more usually, it can form part of the component placement machine, relying on common software for placement of the adhesive dots and placement of the components. A stand-alone dispenser is, however, quite adequate since most modern adhesives used for surface mounting allow component placement hours or even days after application.

The dispensing syringes are controlled by compressed air and the rheological properties of the adhesive must be carefully defined to prevent stringing from dot to dot, and to prevent clogging of the fine dispensing needle.

9.6.3 Adhesive Curing

The curing schedules for the various conductive adhesives used to make the joints in surface mounting assemblies vary widely. The polyimides require a slow evolution of the solvent and a typical curing treatment would be 150°C for 30 minutes. In contrast, one of the most popular epoxy adhesives used in surface mounting can be fully cured at 175°C for only 45 seconds.[310] The same epoxy can also be cured at 120°C for a time of 90 minutes.

9.6.4 Strength of Conductive Adhesive Joints

Conductive adhesive joints have a relatively high shear strength at operating temperatures below the glass transition temperature, as illustrated in Figure 9.20 for epoxy and polyimide types. The shear strength of an interconnection is very dependent on its configuration, but typically for 60:40 SnPb solder the shear strength of a similar joint would be about 20 MPa at 20°C and 10 MPa at 100°C. It is therefore clear that, at normal service temperatures, the joint strength using conductive adhesives is comparable to that of solder.

In low-cycle thermal or mechanical fatigue, the adhesive joints, notably those based on epoxy, can exhibit significantly better reliability than equivalent solder joints. This is because of the intrinsic elasticity of epoxy.

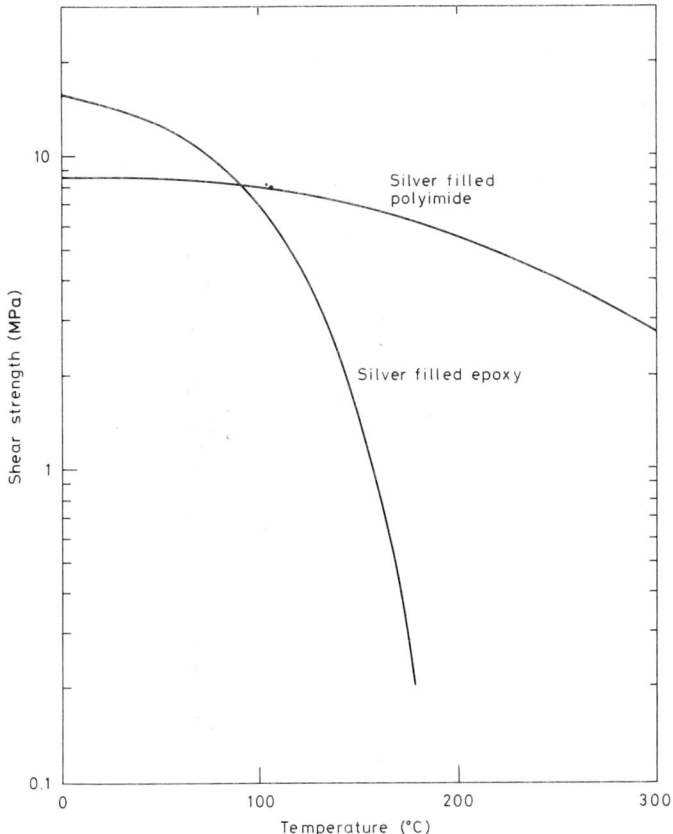

Fig. 9.20 Shear strength as a function of temperature of electronic component joints made with conductive epoxy and conductive polyimide.[309]

Chapter 10

SOLDERABILITY

10.1 INTRODUCTION

The effectiveness of automated soldering processes is dependent upon the solderability of the two surfaces being joined. Problems of poorly solderable components can, to some extent, be overcome by the use of more active fluxes, but the trend of surface mounting assembly, towards denser component packing, leads to difficulties in cleaning the flux after assembly; less active fluxes are therefore preferred. The problems of inspection, testing and component replacement or solder fillet touch-up on SMT boards all put pressure towards no-defect soldering. Solderability is therefore an increasingly important parameter.

The solderability of a component is a complex parameter, as yet undefined in terms of traceable quantifiable properties of the solderable surfaces. The property of solderability defines the total suitability of a component to be soldered by the desired method. There are three aspects to solderability:[4] (i) thermal demand, (ii) wettability and (iii) resistance to soldering heat. The thermal characteristics of the component must enable the solder joint areas to be heated to the desired temperature for soldering within the time available for the soldering operation. The solderable surfaces must allow the molten solder to wet and spread during the available time without subsequent de-wetting. The soldering heat and the induced thermal stresses associated with it must not affect the functioning of the components beyond a specified limit. Each of these three aspects can be engineered to fit the particular application by a suitable choice of component materials and attention to quality control. The most restraining of the three parameters in regard to design for performance during soldering is the component wettability, and this will be considered first.

10.2 WETTABILITY

10.2.1 Speed and Degree of Wetting

When discussing the wetting characteristics of a component by molten solder, there are two aspects which are both important.[311] The *degree of wetting*, i.e., how far the solder spreads, is an equilibrium situation governed by the laws of thermodynamics and dependent on the surface and interfacial tensions involved at the liquid and solid front. The *speed of wetting*, i.e., how fast the solder wets and spreads, is governed by the thermal demand of the system, the ability of the heat source to supply that heat, the efficacy of the flux and the chemical reactions occurring at the interfaces.

10.2.2 Surface Tension

The surface tension of a liquid is equal to the amount of energy needed to enlarge the liquid surface area. It has dimensions of $J.m^{-2}$.

Inside the liquid the intermolecular attraction is uniform in all directions. At the surface, due to the lesser number of neighbouring atoms in the surrounding environment, the outside attraction is smaller and there exists a resultant force acting inwards perpendicularly to the surface and consequently a pressure across the surface. This pressure draws the surface atoms into the interior of the liquid more rapidly than their replacement by other molecules moving outwards so that the surface contracts to a minimum area for the particular volume of liquid. Thermodynamically speaking, the system strives to obtain a minimum value of its free energy, and this it does by minimising its surface area. A floating droplet therefore assumes the shape of a sphere since a sphere has the minimum surface-to-volume ratio.

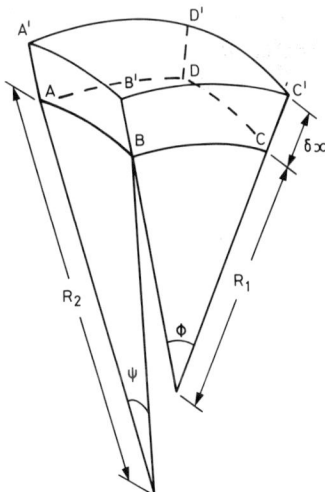

Fig. 10.1 A liquid surface ABCD having principal radii of curvature R_1 and R_2 expanding a distance δx radially.

Figure 10.1 is part of a curved liquid surface, having principal radii of curvature R_1 and R_2. If γ is the surface tension, the energy of surface ABCD is γ times the area, i.e., $\gamma.R_1\delta\phi.R_2\delta\psi$. If the area of the surface is increased to $A'B'C'D'$, the energy is increased to $\gamma.(R_1+\delta x)\delta\phi.(R_2+\delta x)\delta\psi$. If the pressure difference across the curved surface, arising from the resultant forces acting inwards is δp, then the work done to cause this expansion is (pressure) × (area) × (expansion distance), i.e., $\delta p.R_1\delta\phi.R_2\delta\psi.\delta x$. Thus

$$\gamma\delta\phi\delta\psi[(R_1+\delta x)(R_2+\delta x)-R_1R_2]=\delta p\delta\phi\delta\psi R_1R_2\delta x$$

whence
$$\delta p=\gamma\left[\frac{1}{R_1}+\frac{1}{R_2}\right] \qquad (10.1)$$

This is known as the Laplace equation.

In Equation (10.1) a radius is positive when it is measured inside the liquid and negative if measured outside the liquid. For a floating droplet, a sphere of radius R, therefore

$$\delta p=\frac{2\gamma}{R} \qquad (10.2)$$

The sphericity of a drop of liquid standing on a flat plate which is not wet by the liquid is only an approximation, valid in the limit as R approaches zero. As the volume of the sessile drop increases, so its top surface becomes flattened by the effect of gravity, as shown in Figure 10.2(a). The difference in pressure across the surface, at any point on the liquid surface is then

$$\delta p = g\varrho z + \text{constant} \qquad (10.3)$$

where g is the acceleration due to gravity, ϱ the density of the liquid and z the vertical co-ordinate measured from the apex of the drop as shown in Figure 10.2(b). In the case of a sessile drop on a flat plate, the surface has cylindrical symmetry about the vertical axis, and the vertical sections in Figure 10.2 define the shape of the drops.

The normal to the curve at any point (x,z) makes an angle φ to the vertical, and the radius of curvature of the surface in the plane subtended at angle φ is $R_2=x/\sin\varphi$. The radius of curvature in the plane of the section is R_1 as shown. At the apex of the drop, $z=0$, $R_1=R_2=r$, the maximum radius of curvature, and hence the constant in Equation (10.3) is $2\gamma/r$.

Thus the equation for the shape of the cylindrically symmetric sessile drop is obtained by substituting into Equation (10.1):

$$\frac{r}{R_1}+\left[\frac{r}{x}\right]\sin\varphi=2+\frac{r^2 g\varrho}{\gamma}\left[\frac{z}{r}\right] \qquad (10.4)$$

The form of this equation is useful because the linear dimensions, R_1, x and z, appear as ratios to the maximum radius of curvature, r. Also, the parameter:

$$\zeta=\frac{r^2 g\varrho}{\gamma}$$

is a dimensionless constant for the particular liquid and its quantity. For 60:40 tin-lead solder

$$\zeta=0\cdot 20\, r^2$$

if r is expressed in millimetres. In turn, r depends on the volume of liquid solder present. Thus, a given value for the constant ζ gives the same drop shape regardless of drop size, and changing r influences only the size of the drop without altering its shape. Of course, changing the size of the drop while holding the surface tension and density constant causes both ζ and r to change simultaneously.

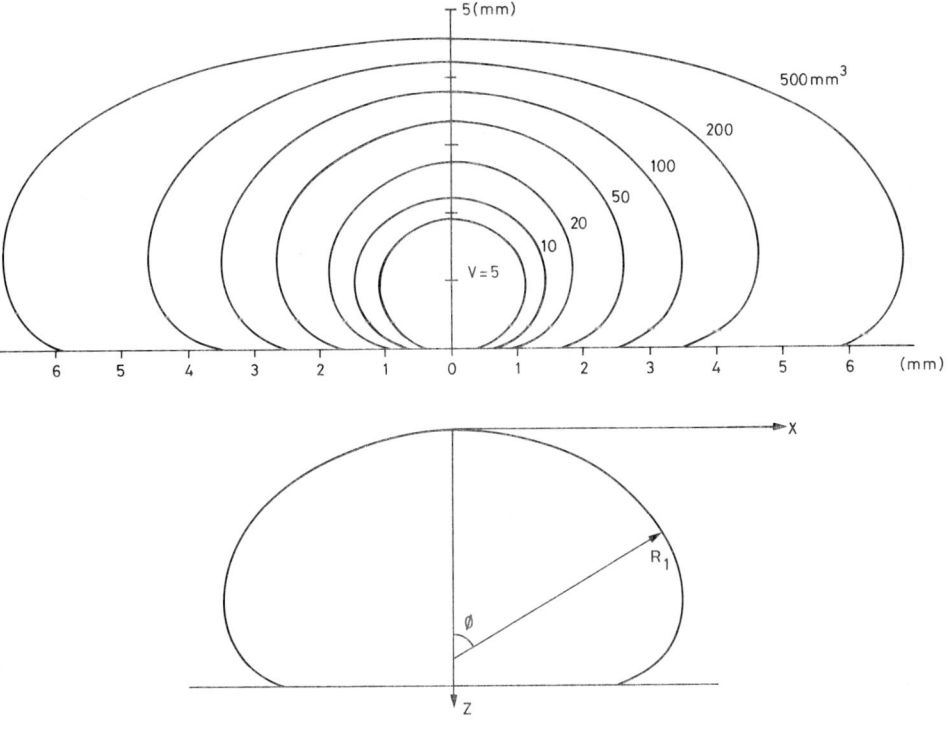

Fig. 10.2 Vertical sections through sessile drops of liquid on a flat surface, defined by Equation (10.5): (a) calculated for solder with $\frac{g\varrho}{\gamma}=0\cdot 2$ mm^{-2} and (b) defining the x, z axes and the radius of curvature R_1 at angle φ.

Both R_1 and φ involve derivatives of z with respect to x, to give:

$$r\frac{d^2z}{dx^2}+\frac{r}{x}\left\{1+\left[\frac{dz}{dx}\right]^2\right\}\frac{dz}{dx} = \left[2+\frac{\zeta z}{r}\right]\left\{1+\left[\frac{dz}{dx}\right]^2\right\}^{3/2} \quad (10.5)$$

This equation is that for the shape of a liquid sessile drop, illustrated in Figure 10.2.

An analytical solution is not available but tables of numerical solutions were produced as long ago as 1883.[312] Modern curve-filling routines have eased the task of parametrically characterising the profiles of the liquid drops.[313, 314]

10.2.3 Thermodynamics of Wetting

When a drop of liquid is brought into contact with a flat solid surface which it wets, the final shape taken up by the drop depends on the relative magnitudes of the molecular forces that exist within the liquid (cohesive forces) and between the liquid and the solid (adhesive forces).[315] The index of this effect is the contact angle, θ, which the liquid subtends with the solid.

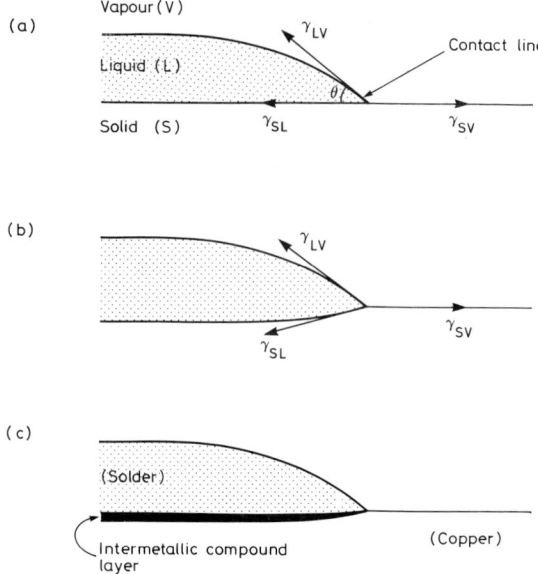

Fig. 10.3 The thermodynamic model of a wetting front of a liquid spreading on a solid surface, (a) defining a contact angle θ in the model case, (b) where some distortion of the solid surface occurs and (c) where the liquid is solder which reacts with a metal substrate to give rise to an interfacial intermetallic compound layer.

The situation is illustrated in Figure 10.3. Young originally considered the equilibrium qualitatively in 1805 and Depré, in 1869, put it in mathematical terms. Every system tends towards minimum total free energy and hence the areas of the interfaces and surfaces strive to become as small as possible.[316] In so doing they must counteract each other: one becomes smaller at the expense of another surface becoming larger. Thus an equilibrium condition is reached when the total free energy, F, given by

$$F = \sum_i \gamma_i \text{(surface area)}_i \qquad (10.6)$$

is a minimum.

If the contact line in Figure 10.3(a) moves so that an additional area of solid is wet, then there is a surface free energy increase at the solid-liquid interface, a decrease at the solid-vapour interface and an increase at the liquid surface. Thus, at equilibrium,

$$\gamma_{SV} = \gamma_{SL} + \gamma_{LV}\cos\theta \qquad (10.7)$$

where the subscripts SV, SL and LV represent the solid-vapour, the solid-liquid and the liquid-vapour interfaces respectively.

This is known as the Young-Dupré equation. It may be argued that Equation (10.7) is not rigorously correct since it is algebraically equivalent to resolving the horizontal components of the surface forces, and no account is taken of the vertical component of γ_{LV}. There is evidence for some distortion of the solid surface by this component of force and therefore, strictly, the stresses in the solid which in part balance the vertical force, as shown in Figure 10.3(b), should be included in the derivation of the contact angle equilibrium. However, where the solid is thick and the temperature is such that there is little elastic or plastic deformation, the neglect of these effects is usually justified. In soldering, the solid material dissolves in the liquid and the final equilibrium shape tends to that shown in Figure 10.3(c). Additionally, the interdiffusion of the substrate and the solder by dissolution usually results in a separate solid intermetallic phase forming at the interface. Thus, in the practical case of soldering on a soluble substrate, Equation (10.7) is only an approximation.

The term 'wetting' is often used loosely. For practical purposes it is usually said that if the contact angle $\theta > 90°$ the solder has not wet the substrate, but, strictly speaking, only if $\theta = 0°$, has wetting truly occurred. A small contact angle θ is promoted by small values of γ_{LV} and γ_{SL} in combination with a relatively large value of γ_{SV}. This usually means that solder will not wet and spread on an oxide-covered substrate since the surface tensions of oxides (γ_{SV}) are distinctly lower than the values of their corresponding un-oxidised metals.[317] The major function of the flux in soldering is to remove the oxide on the substrate and, in so doing, increase its surface energy, making it thermodynamically advantageous for the liquid solder to wet and spread over it, thereby reducing the total surface energy of the system.

As the contact angle θ approaches zero, the imbalance of surface free energies is defined by the spreading energy σ_{SLV}:

$$\sigma_{SLV} = \gamma_{SV} - (\gamma_{SL} + \gamma_{LV}) \tag{10.8}$$

If σ_{SLV} is positive, spreading is accompanied by a decrease in the energy of the system, and is therefore spontaneous. The spreading energy is the difference between the work of adhesion of the liquid to the solid ($\gamma_{LV} + \gamma_{SV} - \gamma_{SL}$) and the work of cohesion of the liquid ($2\gamma_{LV}$), which parameters are defined schematically in Figure 10.4. Whether or not spreading occurs therefore is a manifestation of the relative magnitudes of the molecular forces.[318]

Depending on the magnitudes of the surface tensions it is possible to attain equilibrium values of the contact angle θ between 0° and 180° for, respectively, complete ideal wetting and complete non-wetting. In soldering, even with an active flux, a contact angle $\theta = 0$ is not a normal situation because of the magnitude of the interfacial tension γ_{SL}, which is changing with time as the thickness and the morphology of the intermetallic compound layer change.

10.2.4 Liquid Meniscus Shapes

The surface profiles of sessile liquid drops in contact with a surface that is not wet by the liquid, as shown in Figure 10.2, are defined by the Laplace equation (10.1) plus gravity in the form of a hydrostatic pressure $\varrho g \Delta z$, where Δz is a difference in height between two points in the liquid.

The same mathematical expressions apply also in the case when the liquid wets the solid surface. In a stable situation and with constant volume, the equilibrium contour of the liquid surface represents the state of minimum energy with respect to small changes of profile. For the two-dimensional solution, relevant to an infinite plate dipped into a liquid, a mathematical representation of the profile of the meniscus was derived by Rayleigh and the curve is known as the elastica. This

314 *Solderability*

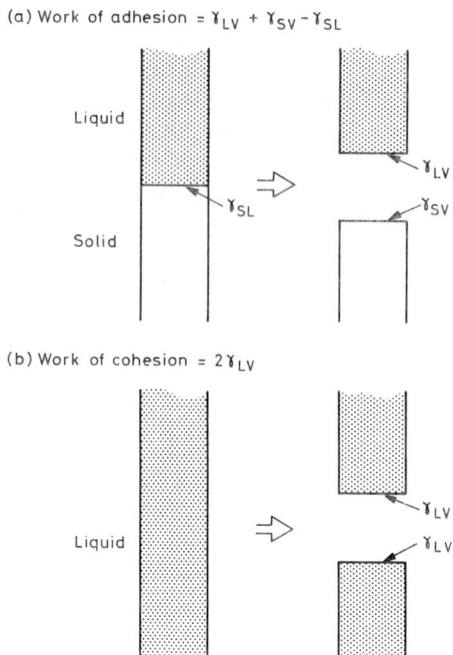

Fig. 10.4 Defining the work of adhesion between a liquid and a solid, and the work of cohesion of a liquid, in terms of the three surface energies γ_{SL}, γ_{SV} and γ_{LV} between solid, liquid and vapour interfaces.

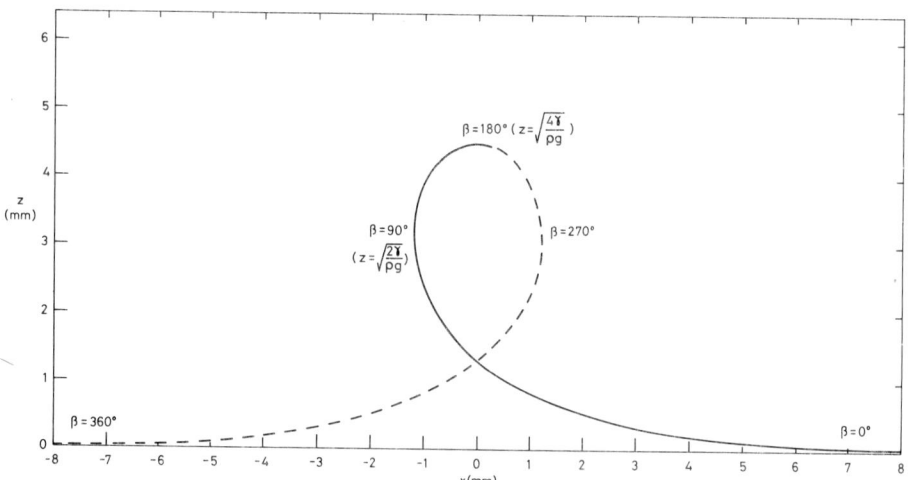

Fig. 10.5 The elastica curve for liquid solder, defined by its density ϱ and its surface tension γ, using Equation (10.9).

curve is shown for solder in Figure 10.5. At any point on the curve, the x and z coordinates can be expressed as functions of the angle β that the curve at that point subtends to the x axis:

$$x = \sqrt{\frac{\gamma}{\rho g}} \left[\ell n \cdot \cotan \frac{\beta}{4} - 2 \cos \frac{\beta}{2} \right] \tag{10.9}$$

$$z = 2 \sqrt{\frac{\gamma}{\rho g}} \sin \frac{\beta}{2}$$

The curve has been drawn using values for the parameters γ and ρ appropriate to 60:40 SnPb solder at 235°C under a rosin-based flux: the surface tension γ takes a value of $0\cdot 4$ J.m^{-2} and the density ρ a value of 8 g.cm^{-3}. Also, the acceleration due to gravity, g, is $9\cdot 81$ m.s^{-2}, so that the parameter $\sqrt{\frac{\gamma}{\rho g}} = 2\cdot 25$ mm, as appearing in Equation (10.4). If an infinitely long flat plate is immersed in solder at an angle α to the liquid surface, and the contact angle defined by the wetting is θ, then the angle that the meniscus, at its contact line, makes with the horizontal is

$$\beta = \alpha - \theta$$

and the height of the meniscus is therefore given by

$$m = 2 \sqrt{\frac{\gamma}{\rho g}} \sin \left(\frac{\alpha - \theta}{2} \right) \tag{10.10}$$

The profile of the meniscus is defined by the angle β on the elastica as shown in Figure 10.6 for four illustrative cases. In the case of perfect wetting ($\theta = 0$) with the plate normal to the solder surface, the meniscus rise m is 3.2 mm.

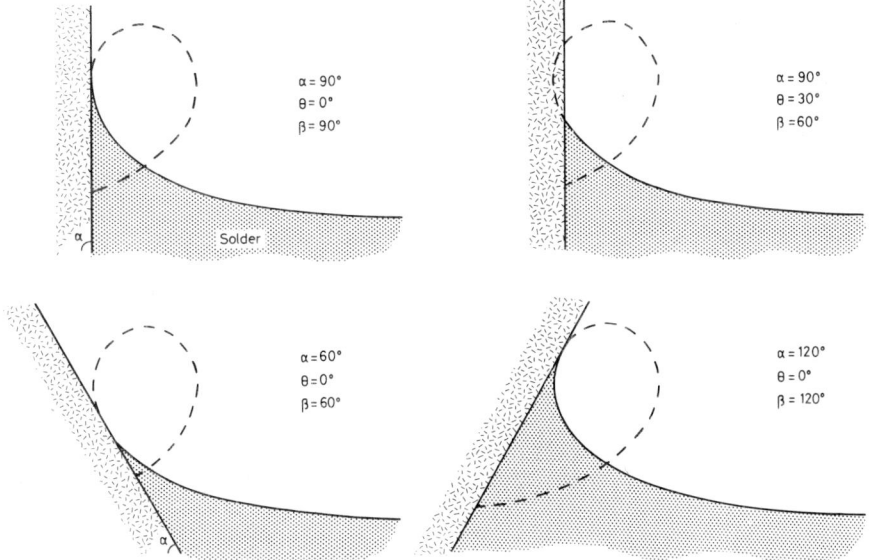

Fig. 10.6 Illustrative examples of how the elastica curve defines the shape and the height of a solder meniscus at a plate surface. α is the angle of the plate to the horizontal, β is the angle of the elastica curve to the horizontal, and hence the contact angle θ is defined as $(\alpha - \beta)$.

10.3 WETTING BY SOLDER

10.3.1 Effect of Solder Alloy

The surface tension of the molten solder γ_{LV} depends on the alloy composition and the flux. (γ without a subscript is used as a shorthand notation for the surface tension of the liquid solder where there is no ambiguity. Where subscripts are required, the subscript LV, liquid-vapour, continues to be used for the solder-flux interface, for simplicity. Young's equation applies equally, whether the 'vapour' in Figure 10.3(a) is a gas, a vacuum or, as in the case of practical soldering, a liquid flux.)

Most surface tension measurements of solders have been made in an inert gas atmosphere rather than under a flux.[319] Besides removing the oxide from the substrate and thereby increasing γ_{SV}, another advantageous effect of the flux is to reduce the surface tension γ_{LV} of the solder. Figure 10.7 gives measured surface tension measurements of tin-lead alloys across the composition range, measured under an electronic-grade flux, compared with measurements made in an inert reducing atmosphere of hydrogen.[320]

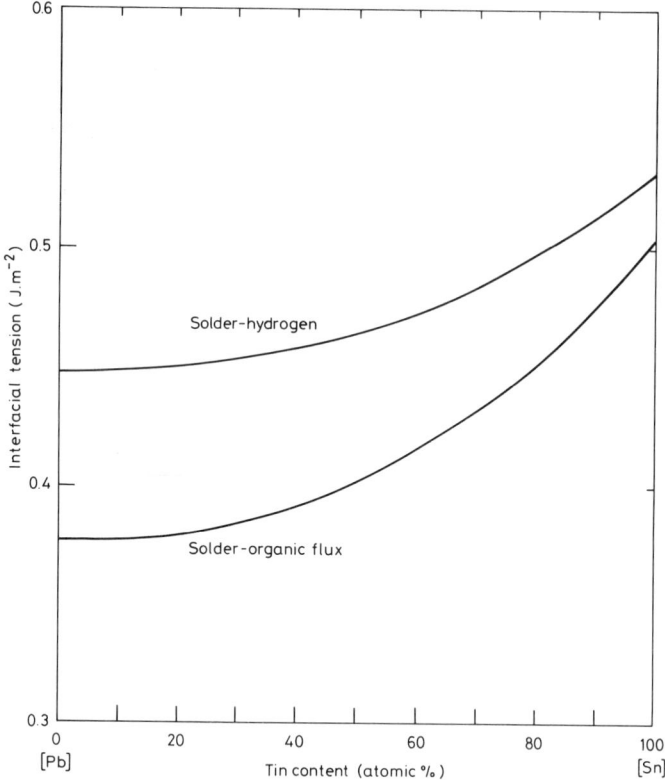

Fig. 10.7 Surface tension measurements of tin-lead alloys, in hydrogen and under an electronic-grade flux.[320]

A weighted mean value of the measurements in the literature for the surface tension of 60:40 SnPb solder at 235 °C under non-activated colophony flux is 0·41 J.m^{-2} reducing to 0·35 J.m^{-2} under activated flux. Measured values of the temperature dependence of surface tension of solder alloys show considerable spread but the weighted mean value is about -10^{-4} J.m^{-2}.K^{-1}, near soldering temperatures. For most calculations a value of the surface tension of molten solder of 0·4 J.m^{-2} can be used with sufficient precision.

The maximum area of spread and hence the lowest equilibrium value of θ for a given volume of SnPb solder alloy is found[317] to occur for a composition of about 50:50 SnPb. However, as was seen in Figure 10.7, increasing the lead content reduces the surface tension of the solder. These two observations must mean that either the interfacial tension γ_{SL} increases or the interfacial tension γ_{SV} decreases with increasing lead content. Probably both of these possibilities occur. Firstly, γ_{SL} increases as the lead content is increased because, with a higher tin content in the molten alloy, there is a faster growth of the copper-tin intermetallic compounds at the interface. The observation that any exposed intermetallic compound on a solderable surface renders that surface hard to solder is verification of a surface tension lower than that of a surface without any intermetallic exposed. Secondly, the interfacial tension of the solid, γ_{SV}, can decrease as the lead content of the solder is increased because there is evidence that lead diffuses ahead of the spreading front to form an atomically thick halo on the substrate surface. This is discussed in the following Section.

10.3.2 The Rôle of Surface Composition

Halos of one of the solder alloy species, diffusing preferentially ahead of the spreading liquid, have been measured for Sn-In and Sn-Pb alloys on copper, using surface analytical techniques. The halo referred to here is not the macroscopic, visible ring sometimes swept out by the flux, but a layer only one or two atoms thick, produced by an atomic surface diffusion mechanism.[321] The surface analytical observations of this mechanism have been made in vacuum or hydrogen because of instrumental restrictions of the technique, but similar transport of species may well exist under or in a liquid flux.

Figure 10.8 shows a surface compositional distribution across a solder front spreading on copper,[322] measured at a temperature above the liquidus, using Auger electron spectroscopy.[323] This is the composition of only the outermost few atomic layers on the surface, and clearly demonstrates the presence of lead on the solid surface, decreasing the surface energy and inhibiting wetting as the lead content of the alloy increases.

On the left side of Figure 10.8 the surface composition of the bulk 60:40 SnPb alloy shows an enhancement in lead. This is because lead, having a lower surface tension than tin, segregates at the liquid surface, being thermodynamically driven to reduce the total surface free energy of the system. At the spreading solder interface, the tin component of the spectra begins to decline and a contribution from the copper in the intermetallic compound layer is seen. However, a partial atomic monolayer of lead continues to extend over the surface and, since the Auger electron spectroscopic technique analyses only a few atoms in depth, the lead signal remains dominant. The lead halo on the 'bare' copper substrate is some 50% monolayer coverage and, in this experiment, extends about 80 μm in front of the spreading front.

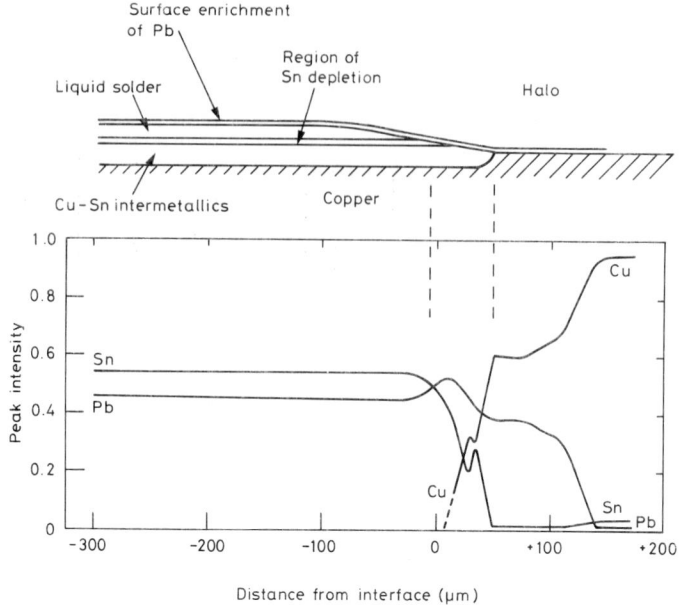

Fig. 10.8 A surface composition of liquid tin-lead alloys, measured by Auger electron spectroscopy, demonstrating significant enrichment of lead at the surface atomic layer.[322]

The compositions shown in Figure 10.8 are typical of those after spreading has ceased and a quasi-equilibrium has been reached. It shows that the spreading stops when the copper diffusion from the substrate into the molten solder is such that the tin-copper intermetallic compound is exposed at the liquid surface.

When the measurements are made well before this quasi-equilibrium has been reached, the data support the equilibrium model in that, if the spreading liquid front is monitored as it passes a given point, the lead is observed prior to the tin. Furthermore, the onset of the lead signal is very sharp whereas the tin signal increases less rapidly.

The surface segregation of lead on liquid tin-lead alloys has been measured directly, again using Auger electron spectroscopy.[324] Some data across the Sn:Pb compositional range are shown in Figure 10.9. The measurements demonstrate a significant surface enrichment of lead across the whole range of alloys, and the ratio of the surface atomic fraction of lead X_{Pb}^s to the bulk atomic fraction of lead X_{Pb}^b shows a close fit to the theoretical curves obtained from a quasi-chemical bond energy model[325] and a semi-empirical predictive relationship.[326] In the case of a liquid binary alloy, the contribution to the free energy of segregation from the strain energy is nil since a liquid cannot support a lattice strain, and as a result comparison between theory and measurement is more precise than for the corresponding solid state case.

The important point to note from Figure 10.9 is that, at the eutectic composition, although the bulk composition X_{Pb}^b is 26·0 at% (38·1 wt%), the surface atom layer composition X_{Pb}^s is about 67 at%. This surface enhancement of lead explains why, in Figure 10.7, the surface tension of the solder alloy remains close to that of lead over a wide decline in lead bulk content. The

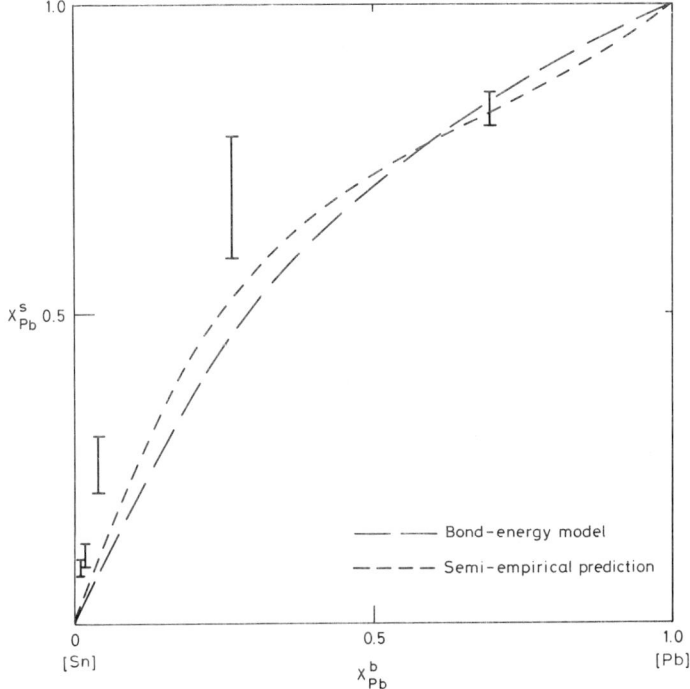

Fig. 10.9 The surface composition of liquid tin-lead alloys, measured by Auger electron spectroscopy, demonstrating significant enrichment of lead at the surface atomic layer. The bulk composition is plotted as the abscissa and the surface composition as the ordinate. The bars are experimenal measurements, the lines from a bond-energy model[325] and a semi-empirical model.[326]

relationship between the surface energy of a binary alloy and the surface energies of its components has been the subject of much investigation.[327] For an ideal solution (with complete mixing, no interaction, no segregation) the surface energy of the alloy is simply equal to the sum of the surface energies of the components added in proportion to their *bulk* concentrations. In the case of a regular solution, such as liquid SnPb alloys, the surface energy is, to a good approximation, equal to the sum of the surface energies of the components added in proportion to their *surface* concentrations. For SnPb solders this holds to within a few percent across the whole compositional range. The surface tension of pure lead under flux, from Figure 10.7, is $0 \cdot 375$ J.m^{-2} and that of pure tin is $0 \cdot 504$ J.m^{-2}. Near the eutectic SnPb composition the surface concentration of lead is about $0 \cdot 67$ at% and that of tin about $0 \cdot 33$ at%. Thus, summing the surface tensions in proportion to their surface compositions gives a value of $0 \cdot 41$ J.m^{-2}, exactly as measured.

Relatively low concentrations of impurities or alloy additions in tin-lead solders can have a marked effect on their wetting and spreading characteristics. This can arise in several ways. As seen above, the surface tension of a liquid is determined by its surface composition rather than its bulk composition and impurities with low surface energy will segregate at the free liquid surface, reducing the surface energy γ_{LV}. Being a liquid, this diffusion process occurs in a very short time. The other two interfacial energies, γ_{LS} and γ_{SV}, can also be affected by relatively small changes to the bulk alloy composition, affecting the growth rate of the intermetallic compound at the solder-substrate interface and affecting diffusion processes across the solid surface ahead of the spreading liquid front.

Some solder impurities arise from the basic constituents, it being uneconomical to remove them or reduce them to lower levels. Other impurities arise during the production and the use of the solder, for example during casting or during the assembly of electronic components, by dissolution of metallic surfaces. In addition, some solder constituents are added at low levels to improve particular properties such as strength or fatigue resistance or for economic reasons and these additions may also have an effect on the wetting and spreading characteristics.

Low-level alloy additions and impurities can affect the wetting and spreading characteristics of solder in several ways, for example by resulting in second-phase crystals within the liquid which increase the viscosity or by affecting the surface composition and hence the surface tension. The specific effect of low-level additions on the surface tension has been studied experimentally and theoretically.[324] The effect of 1% addition of either silver or copper is negligible because these two elements, having relatively high surface energies, do not segregate to the free surface and compete with the lead atoms. Antimony on the other hand has a lower surface energy than lead and therefore, in the liquid alloy, has an enhanced concentration at the surface, reducing the surface tension of the alloy as shown in Figure 10.10. The curve is a theoretical prediction based on a calculation of the synergistic bond energy effects between the surface atoms, to predict the surface composition and hence the surface tension.

10.3.3 Effect of Surface Roughness

Surface roughening and texturing can significantly modify and control the wetting of the solid by a spreading liquid. Using a thermodynamic theoretical treatment,[328] the additional surface area produced by roughening the substrate can be regarded as causing effectively an increase in its surface energy, leading to the prediction

$$\cos\theta_R = W_R \cos\theta$$

where W_R is the roughness area ratio, i.e., the true area/nominal area, θ_R is the contact angle observed on the rough surface and θ, taken as an inherent material parameter, is the contact angle when the substrate is smooth. This assumes implicitly that the surface features of the substrate are insignificantly small compared with the dimensions of the liquid drop and that their geometry is of no consequence except insofar as it affects the surface area.

In a contrasting approach,[329] the roughness can be considered to pose barriers to the flow of the liquid as it attempts to take up the minimum contact angle, so that

$$\theta_R = \theta + \alpha_m$$

where α_m is the maximum slope of the surface feature at the spreading front. While, mathematically, α_m can be positive or negative, and hence θ_R greater or less than θ, considerations of the minimisation of the surface energy lead to the conclusion that an advancing liquid front comes to rest on a descending slope

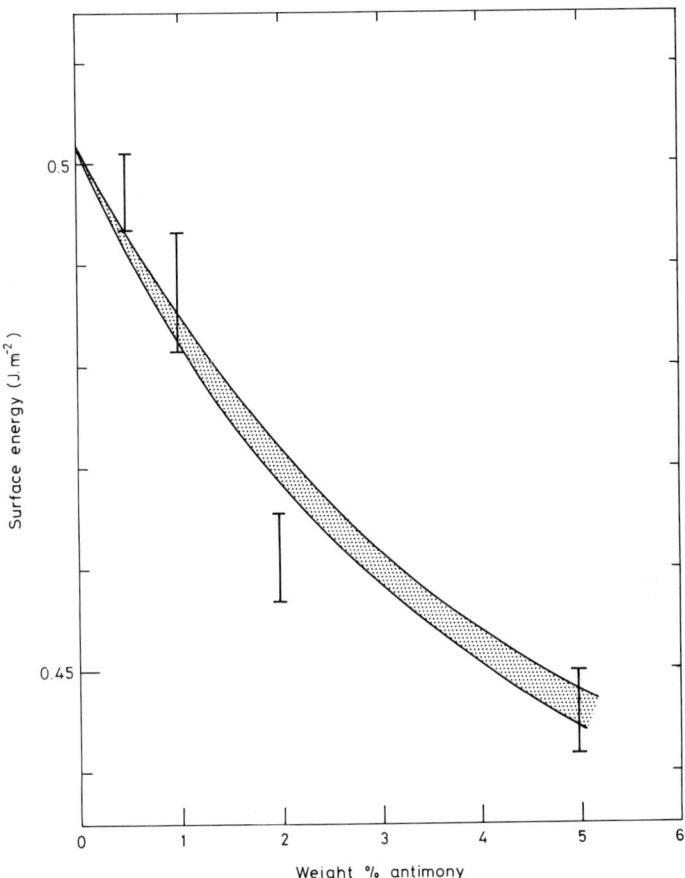

Fig. 10.10 Antimony in a molten tin-lead solder is enhanced at the surface, reducing the surface tension of the alloy. The points are experimental and the curve is a calculation based on the synergistic bond energy effects between the surface atoms.[324]

with α_m positive, but a receding front comes to rest on an ascending slope with a negative α_m value.

These two approaches unfortunately lead to conflicting predictions about the effect of surface roughening. The first predicts that the contact angles of non-wetting liquids will be increased and those of wetting liquids decreased by roughening the substrate, whereas the second leads to the expectation that the advancing contact angles of both non-wetting and wetting liquids will be increased.

The two models apply in different régimes of roughness and inherent wettability. In recent work a whole range of liquid-solid combinations with well-defined roughnesses have been carefully studied experimentally.[330, 331] The broad conclusions are that, usually, roughening a substrate causes its wettability by both wetting and non-wetting liquids to decrease. Exceptions to this generalisation occur when the surface texture is very rough and when the liquid is inherently very well wetting. The contact angles in both wetting and non-wetting situations often increase linearly with the substrate surface texture parameter

R_a/λ_a, where $\dot{R}_a$ is the average amplitude and λ_a the average wavelength of the surface features.

If the roughening is not of a random texture, but comprising grooves parallel to the spreading direction, the liquid spreading can be considerably enhanced by the grooves acting as capillaries.

10.3.4 Hysteresis of Wetting

The measurement of contact angle, or some other parameter directly related to it, is an important input towards the understanding of the mechanisms of wetting at the solid-liquid interface. There is, however, always a hysteresis of the wetting, such that the contact angle measured as equilibrium is reached with an advancing liquid front is not the same as that measured when the liquid front is receding.[332] The liquid front is somehow 'pinned' when the liquid is static and requires an extra activation energy to start its movement again, in the same way that in solid-solid sliding movement static friction is greater than friction measured once sliding has commenced. This 'pinning' of the liquid occurs even between virtually non-reactive materials such as glass and water.

Surface roughness and surface heterogeneity have been identified as reasons for the hysteresis. In soldering, where the nature of the solid-liquid interface is time-variant, the intermetallic layer pins the spreading liquid in two ways, first by increasing the effective roughness of the solid surface over which the liquid has to flow when receding and secondly by freezing at the spreading front.

10.3.5 Degrees of Wetting

When a solid is immersed in molten solder and when withdrawn and allowed to cool, the degree of wetting of the solidified solder on the solid substrate can be assessed visually, often by comparison with sets of photographs of model surfaces. The success of the soldering is described in terms of the fractional areas that can be respectively described as 'wetted', 'partially wetted', 'non-wetted' or 'de-wetted', as illustrated in Figure 10.11.

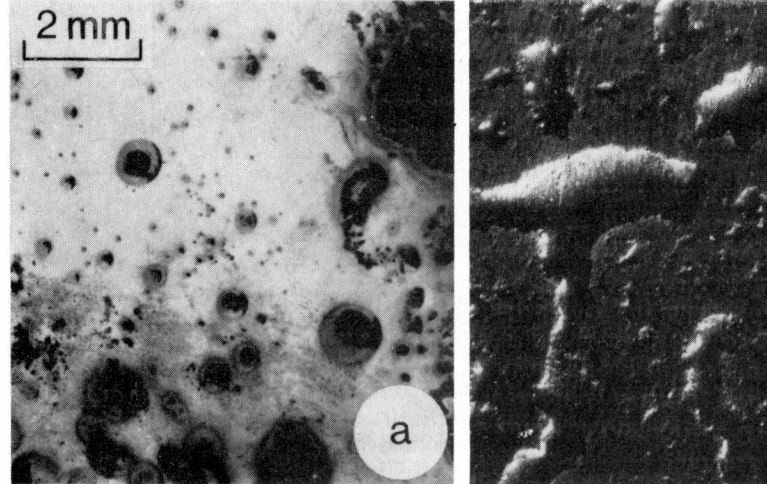

Fig. 10.11 (a) A partially wetted surface and (b) a de-wetted surface of copper with tin-lead solder.

Wetted: As the substrate is removed from the solder bath, the molten solder drains off but a thin layer remains, ideally uniform, smooth, unbroken and adherent.

Partially wetted: The surface of the substrate is, in some areas, well wetted with a smooth adherent coating of solder, but in other areas, often in an island array, the substrate remains non-wetted.

Non-wetted: Upon withdrawal of the substrate from the molten solder, the solder runs off the surface leaving it essentially unchanged by the immersion. This is usually a result of the surface oxide on the substrate being too thick to be removed by the chosen flux within the time available in the solder.

De-wetted: As the substrate is being withdrawn, it appears initially to be fully wet, but before the solder coating solidifies it recedes from some parts of the surface to leave a very thin layer of solder on de-wetted areas and a thickened coating on other areas.

10.3.6 The Phenomenon of De-wetting

De-wetting of a solder film has the appearance of water on a greasy surface, as shown in Figure 10.11(b). It arises when a surface is wetted initially and the solder adheres to the solid metal surface but retracts after a time due to an increase in the contact angle causing the solder to collect up into discrete globules and ridges. Although the remainder of the base metal surface retains the colour of solder, this solder layer is thin and its surface has poor solderability. De-wetting is a practical problem occurring on a variety of substrates and affecting the quality of solder joints by reducing the size of the solder fillets.

De-wetting arises from some time-variant mechanism that changes the surface tension balance and hence changes the contact angle locally so that it becomes energetically favourable for the solder to ball up rather than remain as a flat layer. As the balling up occurs, bare substrate metal is of course not re-exposed because a reaction has occurred at the solder-substrate interface and the exposed surface is mainly an intermetallic compound.

Several time-variant mechanisms can occur to give rise to de-wetting. There may be contamination on a base metal under a coating of tin, tin-lead, silver or gold. During soldering the coating dissolves and the contamination is exposed. Alternatively the growth of the intermetallic compound at the interface might be such as to change the spreading energy significantly, since generally intermetallics rapidly become unsolderable when exposed to the air. In both these mechanisms, small non-wettable areas exist. A simple model has been developed[333] to demonstrate the manner in which such small non-wettable areas give rise to de-wetting. In Figure 10.12(a) a wettable surface with non-wettable spots is completely covered with a thick layer of solder. In Figure 10.12(b) the same base surface is partly covered with solder droplets, and partly with a very thin film of solder on the wettable surface and no solder on the non-wettable spots. If the fractional area of spots is f, then, in the first case the free surface of solder per unit substrate area is simply

$$SS(a) = 1 + f \qquad (10.11)$$

The thickness of the solder coating is d and hence the volume per unit area is

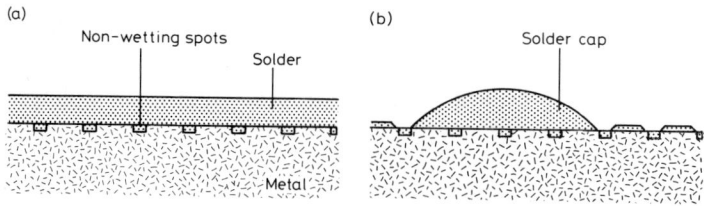

Fig. 10.12 The simple model to demonstrate mathematically the occurrence of de-wetting.[333] The small non-wettable spots cover a fractional area f. After de-wetting the solder droplets have a base area A_1 and a cap area A_2.

also d. In the second case the surface is partially de-wetted. The solder volume has been taken up into N droplets assumed here to be of equal size, with base area A_1 and cap area A_2. The free surface of solder, per unit substrate area, is therefore

$$SS(b) = (NA_2 + fNA_1) + (1 - f)(1 - NA_1) \qquad (10.12)$$

The difference between these two solder surface areas $\Delta SS = SS(a) - SS(b)$ is a measure of the change in the total surface energy of the system because the wettable part of the substrate surface is completely wetted in both cases and so contributes nothing to the energy change. When ΔSS is positive, de-wetting can occur favourably. Its value is a function of the solder thickness d, the fractional area of non-wettable spots f, the number of droplets N and the contact angle θ of the droplets which defines the ratio $A_1:A_2$.

In Figure 10.13 are plotted some values of the area difference, ΔSS as a function of the contact angle θ, taking a solder coating thickness $d = 10$ μm and a non-wettable area of 1%. It can be seen that as N decreases the solder surface area of the de-wetted substrate also decreases. This is possible because the droplets on a de-wetted surface are still in mutual contact via the thin layer of liquid solder on the de-wetted zone. Hence the smaller droplets, which have the higher liquid pressure, can disappear into the larger droplets via the thin layer of solder. In principle, minimum solder surface energy is attained when all the solder is taken up into one single drop. With thinner coatings (lower d) and larger non-wettable fractions (higher f) the more likely de-wetting becomes.

10.3.7 The Need for a Flux

A liquid will only flow over a surface if, in so doing, the total surface free energy of the system is thereby reduced. A clean, non-oxidised surface has a higher surface energy than a dirty one and hence a liquid will more readily wet and flow over a clean surface. In the specific case of soldering it is possible to clean the solder surface and promote wetting *in situ* by the use of a suitable flux. The word 'flux' comes from the Latin meaning 'flow', and indeed the main rôle of the flux is to promote flow of the solder. The mechanisms of this phenomenon are however quite complex.[334] The flux affects both the rate and the degree of solder spreading but does not enter into the bond formation.

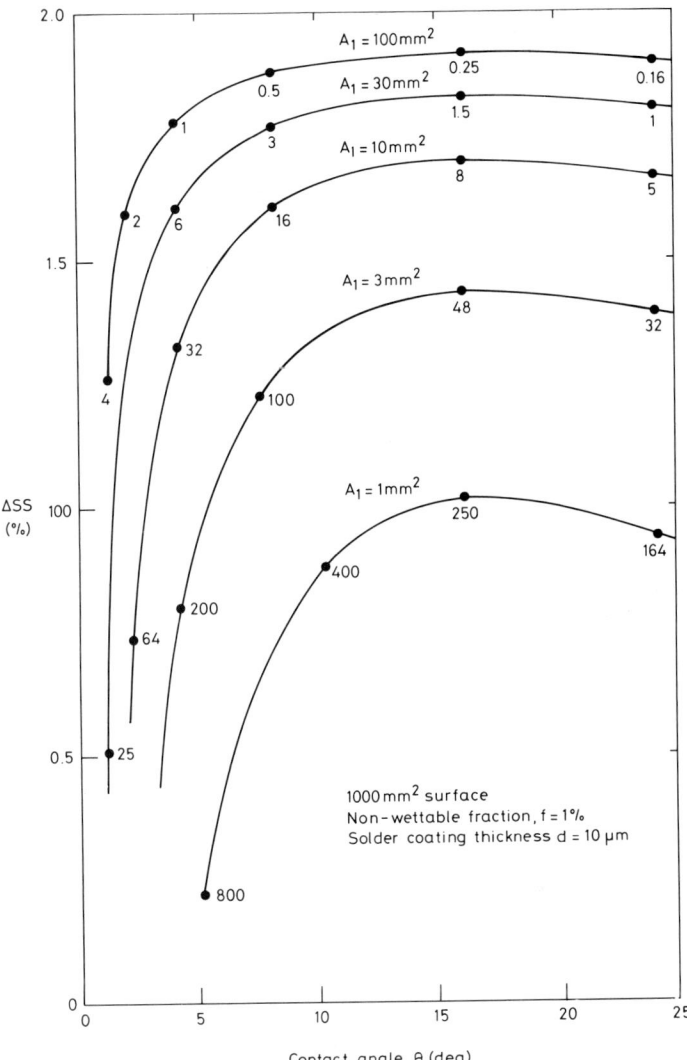

Fig. 10.13 Some values of the area difference of solder surface before and after de-wetting of a 10 μm thick solder coating over non-wettable spots covering 1% of the substrate. The curves are calculated, each assuming a constant base area A_1 for the de-wetted droplets. The figures beside the lines are the values of N, the number of droplets per 1000 mm² plate area.[333]

The major functions of the flux are:

(i) to provide clean oxide-free surfaces of the solid substrate by dissolving or breaking up and sweeping away the surface layer;
(ii) to retain the oxide-free nature of the hot substrate ahead of the wetting front;
(iii) to influence the surface tension equilibrium such that the dihedral contact angle is reduced;
(iv) to retard oxidation of the molten solder surface during flow and subsequent cooling.

Most fluxes are virtually inert at room temperature but become acidic and corrosive as the soldering temperature is reached. Thus there must be an initiation period between the application of heat to the joint and the commencement of solder spreading, first because the activation temperature of the flux must be reached and secondly because a finite time is required for the dissolution of the oxide film. The process of spreading is shown schematically in Figure 10.14.

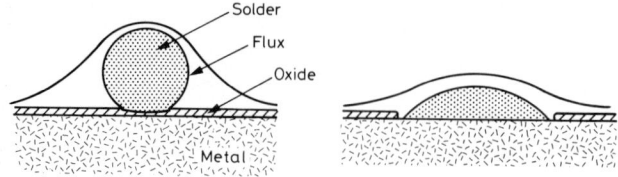

Fig. 10.14 The stages of spreading of a solder droplet on an oxidised base metal. Spreading proceeds upon removal of the oxide layer by the action of the flux.

10.4 TIME-VARIANT CHANGES IN WETTABILITY

In the context of electronic assembly, the deterioration in solderability which occurs as a result of ageing during storage is usually identified by specifying a solderability test parameter, such as a wetting time from immersion in solder to the condition $\theta = 90°$, under standard conditions. A plot of wetting time against ageing time has the general form[335] shown in Figure 10.15 for components which have a protective solder coating.

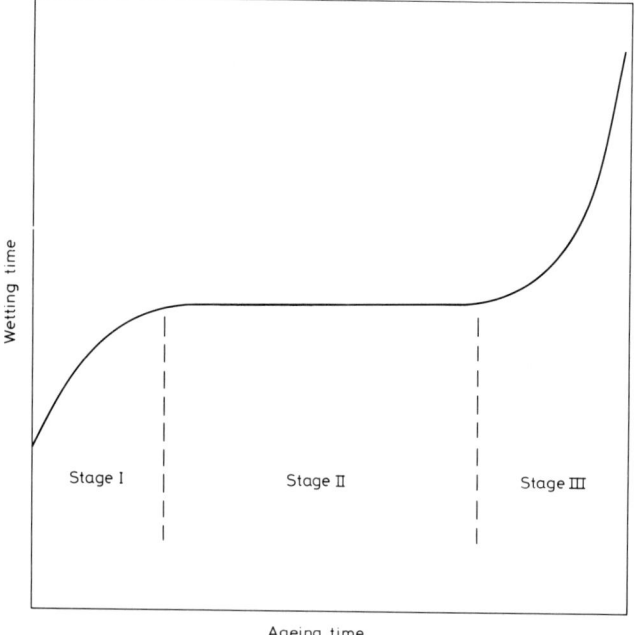

Fig. 10.15 Schematic diagram of the wetting time of a solderable surface as it ages.[335]

During the first stage of ageing the solderability decreases slightly due to oxidation or corrosion of the solder coating. During the next stage the surface layer of oxide is often essentially protective against further deterioration. This may be either a chemical passivation of the surface or a limitation placed by the reduced diffusivity of species through the oxide layer. After a usually extended period, the intermetallic compound layer at the coating-substrate interface begins to penetrate the surface and during this third stage the solderability deteriorates rapidly. Under corrosive conditions, no passive layer is produced and the second stage is eliminated. If the coating is so thin that the intermetallic compound reaction consumes a substantial part of it during the coating process (for example, by excessive reflow of a thin electrodeposit), neither the first or second stage exists and the solderability deteriorates drastically as soon as ageing begins.

In order to predict the ageing behaviour of solderable surfaces during storage, samples can be subjected to artificial accelerated ageing treatments. The aim of artificial ageing is to produce, in a short time, the same deterioration in solderability as would be produced by natural ageing over a typical storage time of up to several years and thus distinguish between components that will maintain good solderability during storage and those whose solderability will deteriorate to below an acceptable level. Ideally, an artificial ageing procedure should be readily usable in production quality control and take only a matter of hours or days to predict natural ageing behaviour over one or two years.

10.4.1 The Rôle of Solderable Coatings

In order to be readily solderable, the surfaces of both the component terminations and the PCB lands must have a relatively high surface energy, and this implies that any oxide or other chemical layers which decrease the surface energy must be easily removable by a flux that is sufficiently mild to be acceptable for electronic assembly. Unprotected surfaces of copper, and to a lesser extent iron-nickel alloys and fired-on solderable inks, become oxidised quite rapidly in air at room temperature. It is therefore usual to apply an impervious protective coating to the terminations.[336] This coating is almost always tin or tin-lead but may be gold, silver, nickel, etc.[337] During the manufacture of PCBs, an electroplated tin-lead layer is frequently used as an etch resist for the underlying copper, and is subsequently reflowed to produce an impervious layer. Boards with a hot-air solder levelled finish likewise have an impervious layer of tin-lead, in this case applied by hot dipping.

In order to improve the solderability of surface mounting components, dipping into molten solder, to provide a surface of recognisably good solderability, is becoming popular and is strongly recommended. Automated machines are now available to coat terminations on surface mounting devices at very high throughput. Pre-tinning is important for surface mounting components for the following reasons:

(i) It allows the surfaces of the leads and pads to be tested and inspected for wetting before assembly.
(ii) It increases the strength of the joint.[3]
(iii) The additional solder slightly elevates the component from the board, which aids in cleaning beneath the body of the component after soldering.
(iv) It removes gold from the pads of LCCCs. The gold is there because of the

manufacturing route used and also to provide a solderable surface with very long term retention of solderability. However, it is preferable to replace this with a solder coating prior to assembly, which is readily achieved by hot dipping.

10.4.2 The Ageing Process

Ageing in the context of electronic assembly refers to the deterioration in solderability of the surfaces to be joined, as a function of the time interval between component manufacture and assembly, as a result of the tendency towards thermodynamic equilibrium and a decrease in free energy of the surfaces.

During the storage of components there may be some corrosion of the solderable surfaces if flux (used during reflow of solder coatings for example) or other process and handling contaminants are retained on the surfaces after manufacture. However, the main source of change in the surface energy is due to reaction with the gaseous environment of storage: oxygen, nitrogen, carbon dioxide, sulphur dioxide, moisture, etc. For a given chemical composition of the storage environment, the rate of reaction of the surface of the component depends on the temperature. Thus, the outer surface of the solderable terminations is changing with time during storage, resulting in a deterioration in solderability that can be correlated directly with the surface composition.

If the component terminations have a protective coating, it is usually of a metal that will either melt during the soldering process or dissolve and react sufficiently fast that, in either case, there is a large composition gradient within the molten solder at the liquid-solid interface. Because of these processes it is not sufficient to consider the effect on the solderability only of the outer surface of a solderable area but also the composition and morphology of the coating-substrate interface. During storage at room temperature the composition of that interface is changing by solid-state diffusion of the substrate species into the coating, and *vice versa*. At that interface the main time-dependent phenomenon occurring is the growth of a layer of intermetallic compound, or compounds between one species of the coating, usually the tin, and one or more of the substrate elements.[338] Additionally, metallic or organic impurities may segregate to that interface (or indeed to the free surface), either during deposition of the coating, or subsequently during burn-in treatments or room-temperature storage. The segregation takes place by diffusion-controlled mechanisms of impurity species within either the substrate or the coating. The contamination by organic species usually arises through their co-deposition with the coating species during electroplating because of excessive organic brighteners present in the plating bath. A further source of change at the coating-substrate interface may arise by diffusion of species from the environment through the coating if it is porous and therefore not fully protective.

Thus, in summary, when assessing the deterioration of the solderability of a coating during storage, consideration must be given to:

(i) the oxidation or corrosion of the outer surface, the possible changes in the nature of the surface due to segregation of one or more species, the way in which these changes increase or decrease the susceptibility to oxidation or corrosion and the changes in resistance to removal of the surface contamination by fluxes;

(ii) the nature of the coating-substrate interface and its changes during storage as a result of impurities, inter-diffusion temperature and the thickness and porosity of the overlying coating.

10.4.3 Intermetallic Compound Growth

When two metals or alloys are in intimate contact with one another there will be an interdiffusion of one or more of the species of each alloy into the other. In some situations, if there is a strong interaction between species, an intermetallic compound, i.e., a material of well-defined stoichiometry, may form. The relevant phase diagram can be used to determine whether this will be so, and to determine the final equilibrium state of the system. When both the solder and the substrate are in the solid state, however, as during storage, the equilibrium situation as predicted by the phase diagram is reached only after very long times. When the solder is liquid, during the component manufacture and during the soldering, the approach to equilibrium is much more rapid, since the rate determining step is the interface dissolution rather than solid state diffusion.[339] Since the protective, solderable coating of, for example, tin-lead is applied either by electroplating with or without fusing, or by hot dipping in the molten alloy, the growth of intermetallic compounds in contact with both liquid and solid solder must be considered.

Intermetallic compound layers can have a detrimental effect not only on the solderability but also on the mechanical properties of the soldered joints because these layers are brittle compared with solder. This problem and the control of mechanical properties of the solder fillets associated with surface mounted components are considered in Chapter 11. The discussion concerning the rate of growth of the intermetallic layer is equally relevant to the deterioration of both the component solderability and the joint strength.

In the general case of a solid component substrate with a fusible coating (usually tin or tin-lead) in intimate contact, the metal substrate S, or one species of a metal alloy, is able, thermodynamically, to form an intermetallic compound SC with one species C (usually tin) of the coating. Figure 10.16 shows the situation schematically.[340] The thickness of the intermetallic compound layer

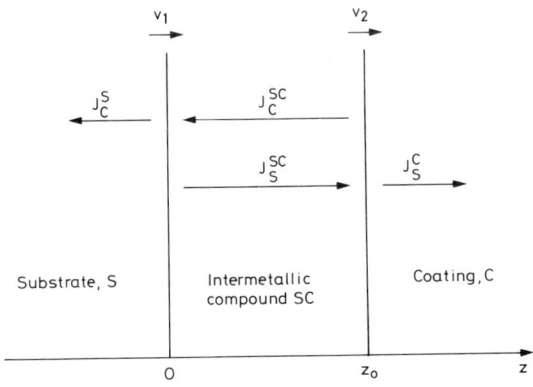

Fig. 10.16 Schematic diagram illustrating the growth of an intermetallic layer between a substrate S and a coating C. J_C^S is the flux of element C through phase S, etc. and v_1 and v_2 are the velocities of movement of the respective interfaces.[340]

after a given time depends on the solubilities of S in C and C in S, and the rate of diffusion of S and C in the substrate, the coating and the intermetallic compounds.

Defining:
- v_1 = velocity of the S-SC interface in the z direction
- v_2 = velocity of the SC-C interface in the z direction
- J_i^j = flux of element i through phase j
- D_i^j = diffusivity of element i in phase j
- $\dfrac{dC_i}{dz}$ = concentration gradient of element i in the z direction
- z_0 = thickness of intermetallic compound layer

Then from Fick's law

$$J_i^j = D_i^j \frac{dC_i}{dz} \tag{10.13}$$

and the rate of thickening of the intermetallic phase SC is approximately

$$\frac{dz_0}{dt} = v_1 - v_2 \propto (J_C^S - J_C^{SC}) - (J_S^{SC} - J_S^C) \tag{10.14}$$

In practice, the rate of diffusion of the coating species C is much greater in the intermetallic than in the substrate (for example, the solubility of tin in all the commonly used substrates is very low), and hence v_1 is always negative. This means that the intermetallic grows into the substrate.

At the coating interface, however, the situation depends on whether the coating is in the liquid or the solid state. In solid state diffusion, the solubility of the species S in C is generally negligible so that the flux of S atoms through the intermetallic is much greater than the flux away from the SC-C interface into the coating. Thus $J_S^{SC} > J_S^C$, v_2 is positive and the intermetallic grows into the coating.

If, on the other hand, the coating is liquid, there is an appreciable solubility of S in C and, in general, $J_S^C > J_S^{SC}$. Since J_S^{SC} depends on the composition gradient in the intermetallic layer and hence decreases as the thickness z_0 of the intermetallic increases, a steady state condition can be attained where the rate of dissolution of intermetallic at the SC-C interface is equal to its rate of growth at the S-SC interface. If the volume of liquid is limited (as it is for a solder coating during fusing or a solder paste fillet during reflow), the concentration of element S in the liquid C rises, and hence J_S^C decreases, leading to an increase in v_2 and an increase in the steady state thickness of the intermetallic.

In summary, when the coating is solid, $J_C^{SC} \gg J_C^S$ and $J_S^{SC} \gg J_S^C$, so that the growth rate of the intermetallic layer is

$$\frac{dz_0}{dt} \propto (J_C^{SC} + J_S^{SC}) \tag{10.15}$$

The rate-controlling process is therefore the diffusion through the intermetallic layer, giving a parabolic dependence of thickness on the time, t:

$$z_0^2 = Dt \tag{10.16}$$

where D is the overall diffusivity for growth of the intermetallic layer, which varies with temperature, according to an Arrhenius equation:

Solderability

$$D = D_o \exp(-Q/RT) \tag{10.17}$$

where D_o is the diffusion coefficient, and Q the activation energy for intermetallic growth, R the Gas Constant and T the temperature in kelvin.

When the coating is liquid, $J_C^{SC} \gg J_C^S$, so that the growth rate is

$$\frac{dz_o}{dt} \propto (J_C^{SC} + J_S^{SC} - J_S^C) \tag{10.18}$$

As time progresses, $\frac{dz_o}{dt}$ reduces to zero and the thickness z_o is maintained at a constant value.

10.4.3.1 COOLING AND SOLIDIFICATION

The treatment above, of the growth of the intermetallic compound, pertains to the isothermal situation, either in the solid-solid or the solid-liquid case. In the latter case, where the temperature is higher than the melting point of the coating, the thickness and the structure of the intermetallic compound layer also depend on processes occurring during cooling and solidification.

In the isothermal situation, the continuous dissolution of intermetallic results in a layer of liquid C (coating) at the interface, which is enriched in species S (substrate). The exact composition of the liquid at that SC-C interface is that which is in equilibrium with the solid intermetallic at that temperature in the phase diagram. The concentration of the enrichment falls off exponentially, away from the interface. When the sample is cooled, the liquid close to the interface becomes supersaturated with species S and the intermetallic grows further into the liquid as it cools and freezes. This effect is shown in Figure 10.17, (a) part of the S-C phase diagram drawn schematically and (b) the concentration of species S in the region near the intermetallic-liquid interface. A zone of liquid, width z_s, is supersaturated as it cools and, on freezing, results in an additional thickness of intermetallic compound, as shown in Figure 10.17.

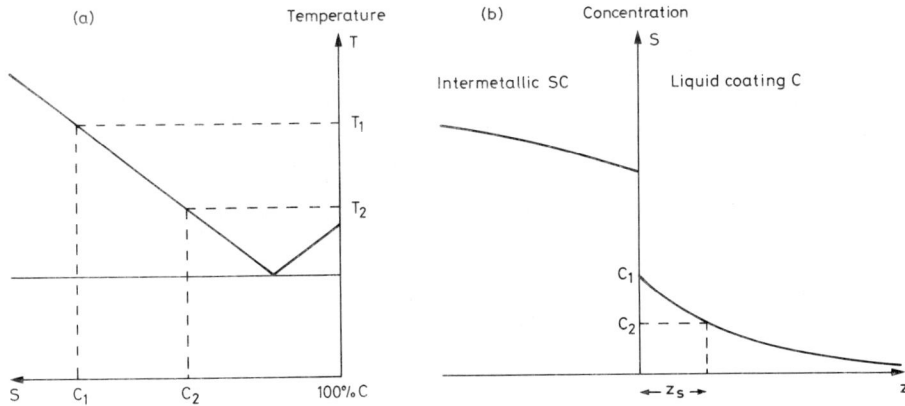

Fig. 10.17 (a) Part of the S-C (substrate-coating) phase diagram. Cooling from T_1 and T_2 results in a supersaturated zone up to a distance z_s ahead of the interface; (b) the concentration profile of S as a function of the distance from the SC-C front.[340]

Under the steady-state conditions where the coating is liquid, the intermetallic-liquid interface is relatively planar as drawn in Figure 10.18(a) because any projection of intermetallic into the liquid will find itself in a region of lower solute content and hence be preferentially dissolved. However, during cooling and freezing, the sweeping away of other species (such as the lead and the impurities of the coating in a tin-lead solder) causes a nodular or a dendritic structure to crystallise. The shape and the size of the morphological features are governed by the rate of cooling, but also the ability of the liquid to minimise the concentration gradient at the interface.[340] Thus, using copper/tin-lead as an example, when hot-dipping in a solder bath or when wave soldering, the surface is swept by the liquid, and an intermetallic with a smooth 'cobblestone' appearance results, as shown in Figure 10.18(b), whereas in solder paste reflow, where the amount of solder is small and convection currents are very restricted, a more fragile dendritic structure can arise, as shown in Figure 10.18(c).[341] It is clear that these different types of interfacial structure can have very different effects on the wetting and spreading of molten solder over them, even though their chemical composition is identical.

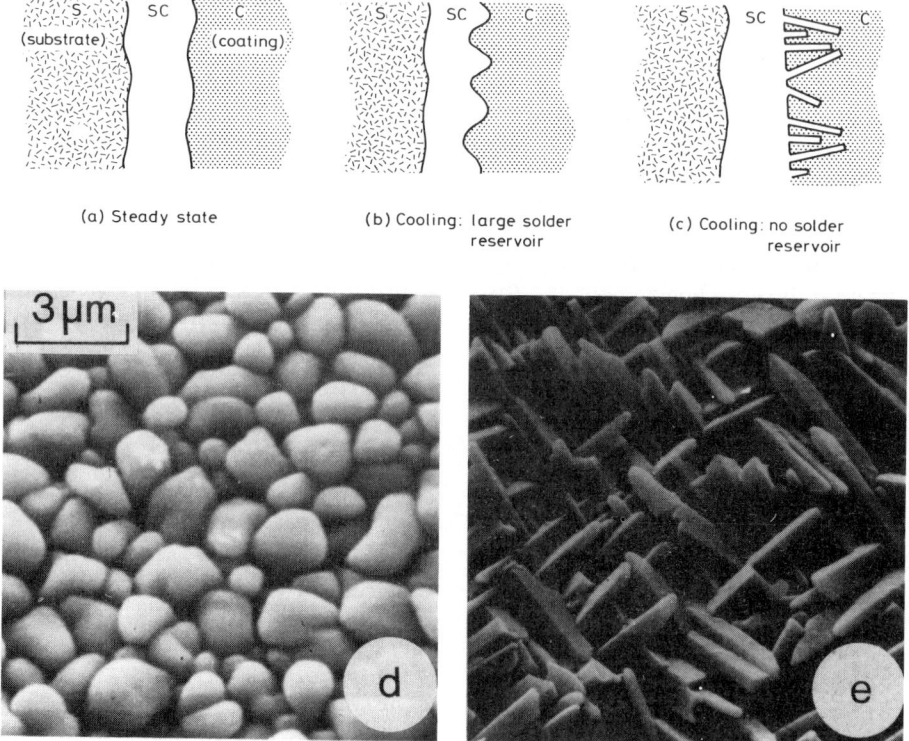

Fig. 10.18 Schematic diagrams illustrating the growth of an intermetallic layer in contact with a liquid coating.[340] (a) In the steady state the interface is relatively smooth. During cooling and solidification the intermetallic layer develops either (b) in a cellular fashion if the amount of liquid is unrestricted, but (c) in a dendritic fashion if the liquid is limited. (d) and (e) are scanning electron micrographs of intermetallics illustrating these effects; the outer coating has been removed chemically.[341] (Photographs courtesy of Professor Kh. Schmitt-Thomas, Technical University of Munich)

10.4.3.2 INTERMETALLIC PHASES

The coating used almost exclusively to maintain the solderability of terminations and lands on electronic components is tin, or a solder in which tin is the major constituent. Consequently, all of the intermetallic phases that are important, in their effects on solderability and on joint strength, comprise tin plus one or more constituents of the substrate. For surface mounting components and their interconnections, the following are used mainly: copper or copper-rich alloys, kovar (FeNiCo), iron-nickel alloys, nickel barrier layers, gold coatings, silver-palladium fired inks and gold-platinum fire inks.

The relevant intermetallic phases formed in the solid state have been studied extensively and are summarised in Table 10.1.

Table 10.1

The Main Intermetallic Compounds Present in Electronic Assemblies

System	Intermetallic Phases		
Cu-Sn		Cu_3Sn (ε)	Cu_6Sn_5 (η)
Ni-Sn	Ni_3Sn_2	Ni_3Sn_4	Ni_3Sn_7
Au-Sn	AuSn	$AuSn_2$	$AuSn_4$
Fe-Sn	FeSn	$FeSn_2$	
Ag-Sn		Ag_3Sn	

10.4.4 Growth of Intermetallic Phases in Contact with Solid Sn or Sn-Pb

The rates of solid-state diffusion[342] and of the growth of intermetallic compound layers at the interface between tin or solder and various substrates have been studied comprehensively.[343] In all cases a parabolic growth rate of thickness z_o increasing with $\sqrt{t}$ has been found, but often only after a somewhat faster growth rate over an initial period. Typical results (here for the combined thickness of both the copper-tin intermetallics) are shown in Figure 10.19, covering the temperature range from room-temperature to 170°C. For each substrate-solder system, the reported results vary within a factor of three depending on whether the coating is tin or solder, whether fused or unfused, matt or bright plating, on the tin-lead ratio, the condition of the base substrate and how exactly the intermetallic layer thickness is measured. That the growth rate is higher initially than would be predicted from the parabolic relationship is because, only when the interface is completely covered by the intermetallic layer, does diffusion through the layer become the rate-determining step, and a parabolic growth law is established.

Unfortunately, it is the growth rate from zero up to one or two micrometres in thickness that is of interest in its effect on solderability, and several factors limit the practical use of measurements of the type shown in Figure 10.19:

(i) The initial growth rate is apparently faster than the final steady-stage parabolic growth rate and so the mean measurements cannot be used to predict reliably the thickness formed after the short ageing times which are practical for predictive accelerated ageing tests.
(ii) For natural ageing during storage, the thickness of the intermetallic layer is

generally less than 1 μm which makes measurement of natural ageing effects and comparable accelerated ageing tests very difficult.
(iii) The variations in the layer thickness from point to point of the interface can often result in a large spread of the measurements.
(iv) The solderability of the coating is most affected when the intermetallic compound penetrates the free surface, so that a very irregular layer is more deleterious than a smooth layer of the same mean thickness.

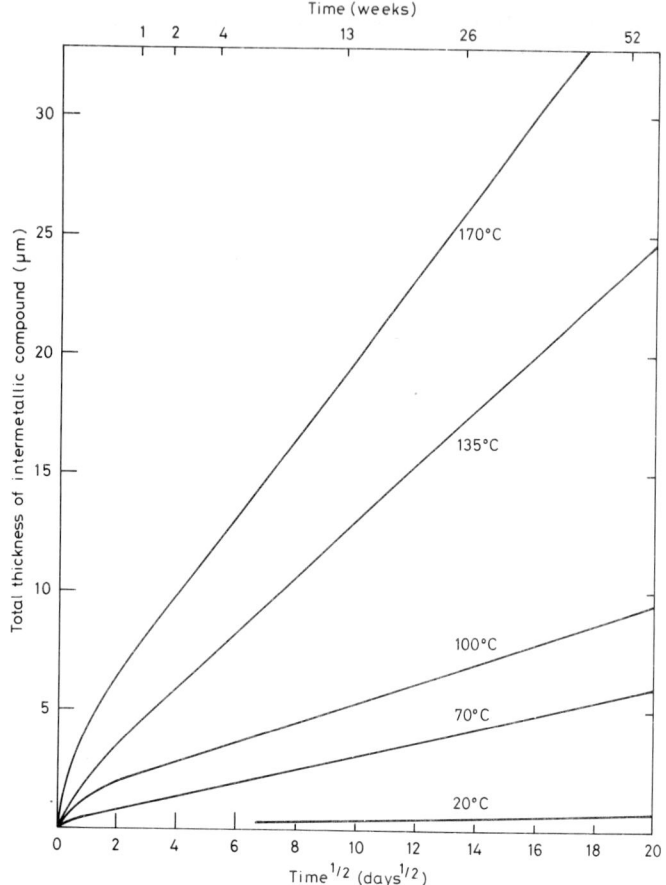

Fig. 10.19 The intermetallic compound thickness when 60:40 SnPb solder coatings on annealed copper are stored for various times and temperatures.[343]

10.4.4.1 COPPER SUBSTRATES

The copper-tin intermetallic system has been given most attention because of its relevance to PCBs and insertion mounted components with tinned copper leads. In the solid state, two phases grow at the interface.[344, 345] The Cu_6Sn_5 (η) phase is formed at all temperatures; it is relatively coarse grained and is that observed in the scanning electron micrographs in Figures 10.18(c) and (d). At temperatures in excess of about 60°C, the Cu_3Sn (ε) phase begins to grow at the copper-η interface.

Solderability

At normal storage temperatures the growth rate of the total intermetallic layer reduces as the tin-lead ratio reduces, as shown in Figure 10.20, but at higher temperatures there is very little difference.[346] This is expected from atomic mobility considerations alone, that the maximum growth rate would occur for the highest tin content alloys when the growth rate is diffusion controlled. Also, some differences in the growth rates have been observed with differences in the metallurgical condition of the copper. Further, by considering all the data on the diffusion coefficients of copper in tin and *vice versa* it has been asserted that the transport mechanism operating in the growth of the intermetallic layer is dominated by bulk diffusion above about 120°C, but dominated by grain boundary diffusion below 120°C[347] and therefore the growth rate would be additionally modified.

The standard deviation of the measurements of intermetallic compound thickness is relatively large (±30%). In the earliest stages of growth, the compound is discontinuous and nodular, and this leads to an uneven coating when it becomes continuous, even if great care is taken to prepare the copper surface to be homogeneous both chemically and physically.

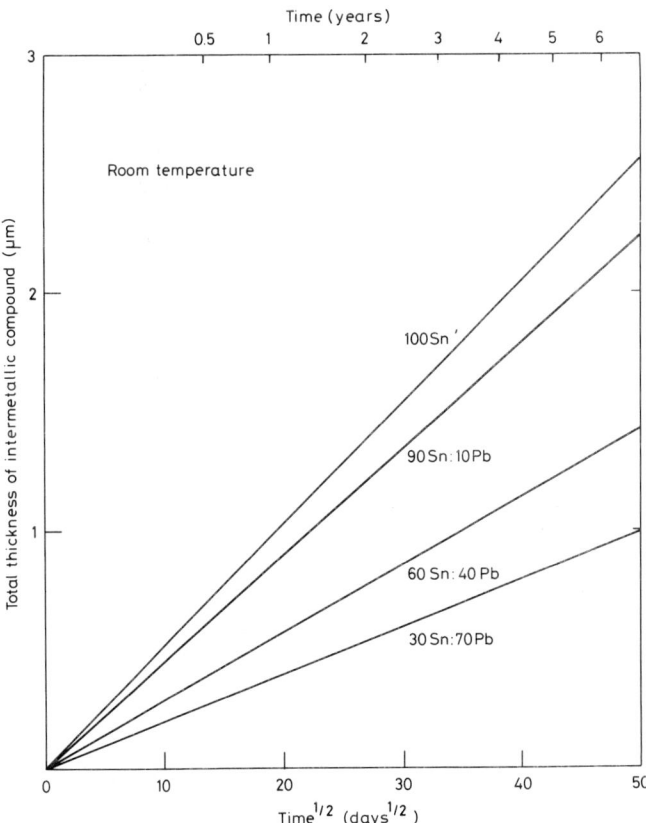

Fig. 10.20 The intermetallic compound thickness obtained at room temperature at a copper-solder interface, as a function of the coating alloy composition.[346]

Notwithstanding all these variations in measurements, Equations (10.16) and (10.17) can be given mean numerical values[348] to act as a rough guide for estimating the total intermetallic layer thickness z_o in the solid-state Cu-solder system, as a function of the time t and the temperature T:

$$z_o^2 = Dt = D_o t \exp(-Q/RT) \qquad (10.19)$$

where the diffusion coefficient $D_o \approx 10^{-6} m^2.s^{-1}$ and the activation energy for intermetallic growth, $Q \approx 80$ kJ.mol^{-1}. Thus, for example, a calculated z_o of $2\mu m$ is found after about 8 hours at 150°C but after 20 years at 20°C.

10.4.4.2 OTHER SUBSTRATES

The growth of the tin intermetallics on gold, silver, nickel and iron substrates has been investigated,[343, 349, 350, 351] and weighted mean values for the respective diffusion constants are given in Table 10.2. The points made about the precision of such measurements are valid here too, and they should be used only as a guide. With these data, values of the intermetallic layer thickness can be calculated for a particular time and temperature. The values of D calculated are probably good only to a factor of ten, and hence z_o to a factor of about three.

The growth of the intermetallic layer between tin and gold is extremely rapid, while the growth rates on iron, silver and nickel are about the same as for copper, except that at high temperatures the growth rate of the Ni-Sn intermetallic is significantly lower and nickel is often used as a barrier layer between copper and solder layers.

Table 10.2

Diffusion Constants for the Solid-state Growth Rate of Intermetallic Layers

$(z_o^2 = Dt, \quad D = D_o \exp(-Q/RT), \quad R = 8 \cdot 314 \text{ J.mol}^{-1}.K^{-1})$

System	Diffusion Coefficient, D_o $(m^2.s^{-1})$	Activation Energy, $Q(J.mol^{-1})$
Cu-Sn	1. 10^{-6}	80000
Ni-Sn	2. 10^{-7}	68000
Au-Sn	3. 10^{-4}	73000
Fe-Sn	2. 10^{-9}	62000
Ag-Sn	8. 10^{-9}	64000

The intermetallic growth rate in some situations can vary quite significantly from these predicted values; for example, the use of a tin-lead-indium alloy for soldering gold plated terminations reduces the growth rate by a factor of five. Also, with surface mounting components, the solderable surface is often a fired thick-film conductor, rather than a metal. The solid state interdiffusion between the metallisation and a solder coating has been investigated.[352, 353] A copper thick-film conductor produces the same intermetallic compounds with tin-lead solder at much the same rate as solid copper base material, but the solder, during reflow, penetrates the porous conductor and so the overall amount of intermetallic compound is considerably greater than at a smooth metal surface. Likewise, with gold-platinum thick-film metallisations, the gold behaves in a manner similar to electroplated gold, forming mainly $AuSn_4$ at a comparable rate. Plated nickel coatings on surface mounting components contain phosphorus, which concentrates ahead of the advancing intermetallic front, producing, in addition to any tin-nickel compound, a nickel-phosphorus compound. Ultimately, this phosphorus diffusion contributes to the ageing of the conductor since the nickel-phosphorus compounds are brittle and the reaction is accompanied by a volume change.

10.4.5 Growth of Intermetallic Phases in Contact with Liquid Sn or Sn-Pb

If the solid substrate is in contact with molten solder, an intermetallic layer is formed at the interface but, at the same time, as discussed in Section 10.4.3, significant amounts can be dissolved by the molten solder as long as the solder is not saturated with the substrate species.[354-356] These combined processes of growth and dissolution ultimately result in a constant layer thickness. In this case, therefore, as well as the time and temperature, also important are the amount of solder available per unit area of substrate and the degree of mixing in the liquid.

It has been pointed out,[340] in a review of intermetallic growth, that almost all of the work concerning the growth rates at a liquid solder-solid substrate interface has neglected to consider either the quantity of solder present (which controls the dissolution rate) as a variable or the amount and form of the intermetallic which forms during the solidification process. Some data, however, do show[357] that, during dipping of copper in a large solder bath, over a range of temperatures, the intermetallic layer thickness attains its equilibrium thickness very rapidly, in no more than a few seconds, the thickness ranging from about $0.5\ \mu m$ at $240°C$ to $0.9\ \mu m$ at $345°C$. When the dissolution rate of the intermetallic layer into the molten solder is increased, for example by ultrasonic vibrations, the equilibrium thickness is significantly decreased.

The interface structure of the intermetallic, observed in solidified solder joints, almost certainly forms during the cooling and solidication process, unless the adjacent solder is saturated in the molten state. Variations in the structure are due to variations in the amount of supersaturation and the thickness of the super-saturated layer formed during cooling.

10.4.6 Effect of the Intermetallic Layer on Solderability

The effect on the solderability, of intermetallic compounds at a coating-substrate interface of a component has been the subject of considerable work,[358-361] especially on the Cu-Sn system. If a tin coating on copper is heated, in vacuum so that no oxide forms, gradually the tin is consumed until the outer surface consists of Cu_6Sn_5, the η phase. This phase is then slowly transformed to the copper rich Cu_3Sn, ε, phase as the heating continues. Both phases, on the outer surface, are wettable by molten solder using a strongly activated flux. However, any oxide on the intermetallic, even as thin as 1.5 nm (only some 5 atomic layers) renders it unsolderable by normal standards of electronic assembly.

It is therefore important, as far as retention of solderability is concerned, to be able to estimate the time taken for the tin in a solder coating to be consumed by the growth of intermetallic. It is axiomatic that the thicker the initial solder coating the less deterioration in the solderability will occur as the result of intermetallic compound formation.

Assuming an extreme situation, where all the tin in a coating becomes combined to form intermetallics, it is possible, knowing the densities and the compositions, to calculate the resultant thickness of the intermetallic layer. In the case of a tin coating on a copper substrate for example, Cu_6Sn_5 contains 60 wt%Sn and Cu_3Sn contains 38 wt%Sn. Assuming that the growth rates of both phases and the densities of both phases are approximately equal, the thickness of the intermetallic when all the tin is consumed is about twice that of the original coating. If, however, the coating is 60:40 SnPb solder, the lead is not consumed, so obviously the thickness of the intermetallic layer per initial thickness of coating is not so great, but also there develops an enrichment of lead in the remaining coating. This by itself can lead to a deterioration in the solderability since lead-rich solders do not wet and spread as well as the near-eutectic tin-lead compositions, and the lead oxidation and corrosion products are significantly harder to remove from the free surface when using colophony fluxes.

Thus, a number of factors have to be considered when estimating the storage lifetime of solderable terminations on components and in the application of any accelerated method of ageing aimed to predict the storage lifetime.

(i) The coating composition determines the intermetallic growth rate and the ratio of layer growth to coating consumed.
(ii) The coating thickness determines the ageing time needed to consume the coating.
(iii) The morphology of the intermetallic layer is important because an irregular growth will lead to premature exposure of the intermetallic at the surface or an inhomogeneous lead-rich coating.
(iv) Local variations in the coating thickness can also give rise to premature loss of solderability, even though the mean thickness appears to be sufficient.

10.4.7 Dissolution of Terminations in Molten Solder

During soldering, when a component termination or a PCB land is in contact with molten solder, its constituents will dissolve in the liquid until the solubility limit is reached locally. As has been seen, this dissolution occurs when an intermetallic layer is being formed, but it also occurs when there are no intermetallic phases. Besides the compositions of the termination and the solder, and the temperature, the dissolution rate also depends on the rate of flow of the solder across the surface since the solute may or may not be carried away from the dissolving interface. In soldering processes where there is an abundance of solder, such as wave soldering, the dissolution rate remains constant during the time at temperature, but where there is a limited amount of solder, such as in reflow soldering, the dissolution rate decreases continuously with time.

Many of the metals used in the manufacture of electronic components have relatively high dissolution rates in tin-lead solders. Some values of dissolution rates are shown in Figure 6.10.

10.4.8 The Oxidation and Corrosion of Solderable Surfaces

The solderability of a solder-coated component termination degrades with the passage of time because of the interfacial reactions between the solder coating and the metallic substrate. It also deteriorates because of reactions between the outer surface and the atmosphere.

The oxidation of solid tin has been investigated extensively.[362, 363] The oxide film that forms is protective and comprises SnO, attaining a limiting thickness of 2-6 nm at room temperature in dry air.[364] At higher temperatures and in moist air the oxide growth rate is faster and the equilibrium film thickness is somewhat higher with significant cavity decohesion of the oxide from the metal. The oxide growth rate and the oxide structure are also affected by impurities in the tin, depending on the affinity of the impurity species for oxygen and moisture.

Solid lead also forms a protective oxide, PbO, which is ductile and non-porous. Its growth is diffusion-limited and therefore the thickness has a parabolic relationship with time. Whilst solid solder coatings in general contain both tin-rich and lead-rich phases, the tin has a greater affinity for oxygen and, since tin is soluble in lead to several %, SnO is to some extent formed preferentially on both phases.[365] Any PbO that is formed on the lead-rich areas during the initial stages of oxidation tends to be reduced to SnO as the tin diffuses to the surface.

In pure oxygen, both tin and lead form protective oxides which inhibit further surface reactions. There is the possibility that other constituents of air might be more reactive than oxygen and produce surface films that are either non-protective or have a greater detrimental effect on the solderability than do the oxides. Besides oxygen and nitrogen, the gases SO_2, NO_2, H_2S, Cl, NH_3 and CO_2 have all been used for tests.[366] The formation of sulphides is normally insignificant since the sulphur content of air is very low and because the reaction of tin-lead at these sulphur levels, even with 100% humidity, is negligible. On solderable terminations containing silver, however, a black film of sulphide forms even at very low concentrations. Artificial sulphurisation of copper using aqueous sodium sulphide has been suggested as a means of reproducing a scale of standard solderability levels (Section 10.6).

Nitrous oxide (NO_2) at low concentrations (10-100 ppm) corrodes SnPb coatings by reacting with the lead to form a film of the lead nitrate $Pb(NO_3)_2$ which is not protective, can grow to thicknesses of greater than 1 μm and renders the surface difficult to solder. The reaction rate is sensitive to humidity as shown in Figure 10.21,[366] causing marked deterioration of solderability when the relative humidity is in the range 15-75%.

The presence of low concentrations of chlorine (<10 ppm) in a storage atmosphere, in conjunction with a relative humidity of more than 25%, corrodes SnPb coatings. Both tin chloride ($SnCl_2$) and lead chloride ($PbCl_2$) are formed by reduction of the respective oxides, thickening the corrosion products linearly with time, indicating a film of a non-protective nature. Generally the chlorine acts only as a catalyst, since the lead chloride reacts with carbon dioxide in the air to form the basic lead carbonate ($PbCO_3$), that is virtually unsolderable with normal electronic-grade fluxes.

10.4.9 Accelerated Ageing Treatments

The solderability of components decreases with time due to a variety of mechanisms jointly termed 'natural ageing'. Since components are very often stored by the manufacturer or the user for periods of months or even years before use, the solderability of a sample of each batch should be tested at intervals during storage or shortly before assembly. Once the solderability level has fallen below that specified, the components must be discarded.

When components are purchased, the assessment of solderability is invariably

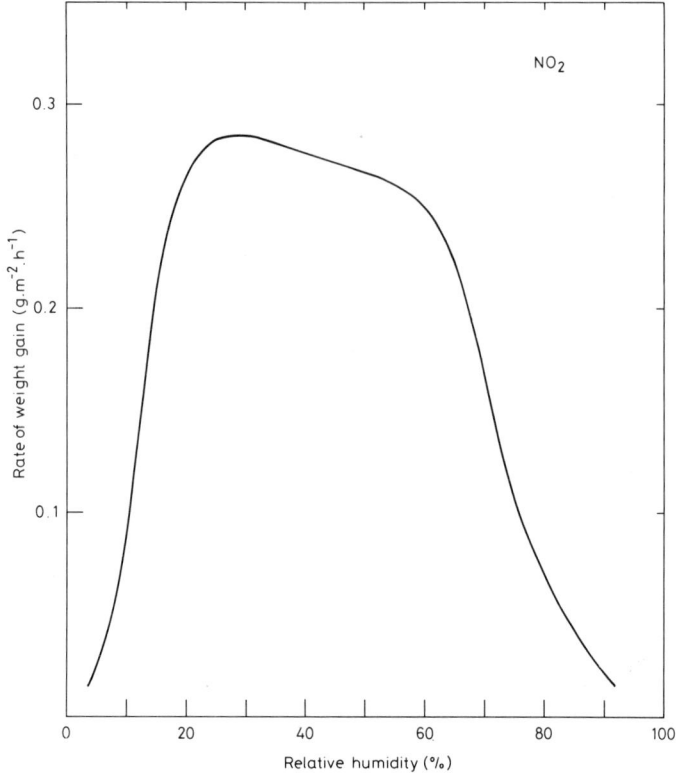

Fig. 10.21 The rate of weight gain for a 50:50 SnPb solder surface exposed to an atmosphere with 42 ppm NO$_2$, at various humidities.[366]

incorporated as part of the goods-inward inspection procedure. It is also pertinent to make a predictive assessment of how rapidly and in what manner the solderability will decline during storage. In order to accomplish this, a sample of the components is subjected to a treatment that attempts to simulate natural ageing in an accelerated manner.

In practice, the environmental conditions of storage are never precisely known and do not remain constant. Among other parameters the effects of storage depend on the temperature, the humidity, the pollution species and levels, and the packaging and storage facilities used. Natural ageing is therefore quite unreproducible and unpredictable and an accelerated ageing treatment cannot simply reproduce natural ageing at a faster rate by, for example, increasing the temperature. Accelerated ageing treatments do not in general reproduce either the composition or the microstructure of solderable surfaces that have been aged naturally.[367] An accelerated ageing treatment is designed to reduce the solderability of a component at a rate that has a known relationship to the reduction of solderability by natural ageing, at least over a limited time range.

The lack of reliable knowledge concerning the processes of natural ageing has led to the development of a number of different accelerated ageing tests, which have been incorporated in national and international solderability specifications.[368, 369] Some of these were originally intended to artificial age the component itself and have subsequently been taken, for convenience, into the régime of solderability testing. The most popular accelerated ageing conditions appearing in specifications are:

1 *Dry Heat*: in air at 155°C for 2, 16, 72 or 96 hours
2 *Damp Heat*: in moist air at 40°C, 93% RH for 4, 10, 21 or 56 days
3 *Steam*: in steam with or without extra gas (air, oxygen, sulphur dioxide)
4 *Cyclic Damp Heat*: in damp air at 95% RH, cycling between 25°C and 55°C with 3 hours for the temperature ramps and 9 hours at temperature (24 h cycle).

The rate of ageing can vary greatly even using just one of these tests. For example, if natural air is used, it may contain a variety of industrial pollutants that will affect the resultant surface composition of the sample. However, if an oxygen-nitrogen gas mixture is used, apparently to place greater reproducibility

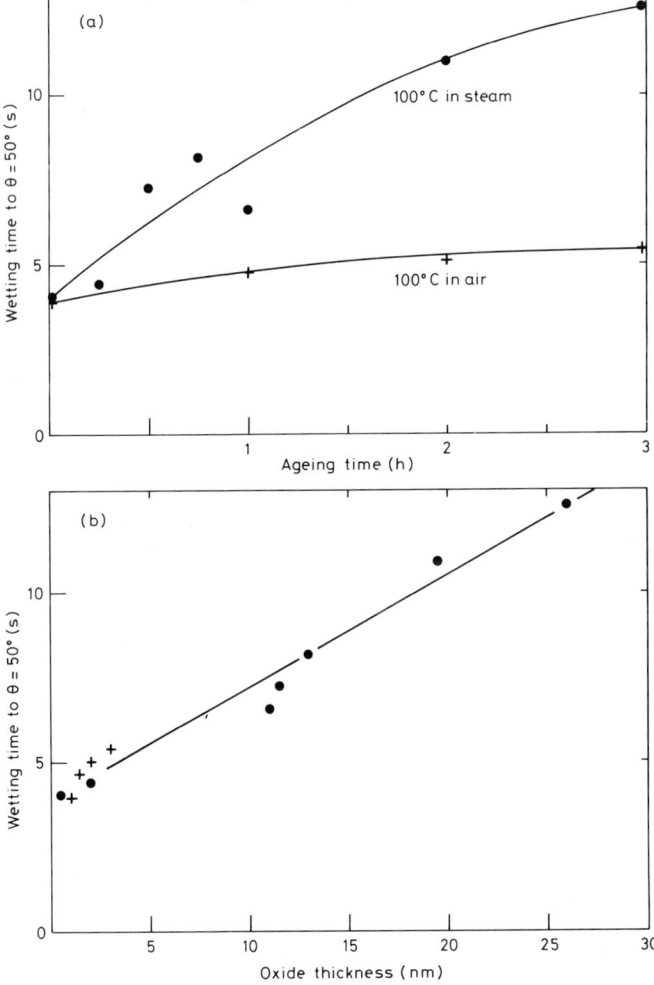

Fig. 10.22 Wetting times of solder-coated copper coupons aged either in steam at 100°C or in an air oven at 100°C. The times, measured using the wetting balance, are for the contact angle θ to equal 50°. When the oxide thicknesses are plotted (measured by Auger electron spectroscopy) the wetting times give a linear function.

on the test, the absence of carbon dioxide has an effect out of proportion to its concentration in natural air, since the unsolderable basic lead carbonate cannot then be produced on the sample.

The accelerated ageing procedures modify mainly either the outer surface or the intermetallic layer between a solderable coating and the base metal. The tests that rely on the humidity or the corrosivity of the environment affect mainly the outer surface, whilst those that rely on higher temperatures increase the diffusivity of the species and so modify internal interfaces such as the intermetallic compound layer.[370]

An example of the first case is shown in Figure 10.22 for copper samples with a thick solder coating such that the intermetallic growth at the copper-solder interface is not as important as the changes occurring at the free surface of the solder. The time of wetting by solder, measured using time t_6 in the wetting balance (Section 10.5.4 and Figure 10.33), is plotted as a function of the oxide thickness. The samples have been aged using both dry and wet accelerated ageing procedures: in air at 100°C or in steam. Both the composition and the thickness of the oxide were measured using Auger electron spectroscopy combined with inert gas ion sputtering. The figure shows that the wetting time does not correlate directly with the accelerated ageing time but with the oxide thickness in a linear fashion over this range. These oxide films, grown at 100°C, are very thin and the wettability can be fully restored by removing the oxide using a 15% hydrochloric acid dip.

The spatial resolution of Auger electron spectroscopy can be as good as 20 nm. When used in combination with ion sputtering to determine the thickness of the oxide, it must be remembered that solder has a two-phase structure and the tin-rich surface areas and the lead-rich surface areas will not be oxidised in a homogeneous fashion. Therefore, either both separate phases must be identified and analysed, or the spectroscopic probe be defocused such that the analysed area is large enough to give a representative average of the oxide nature.

The combined effect on wetting, of changes in both the surface oxide and the intermetallic layer, is illustrated[335] in Figure 10.23. In this case specimens of tin coated copper were prepared with the coating thickness varying from 0·5 μm to 8 μm, and suspended in an enclosure over boiling water. The solderability of the thin coatings deteriorated rapidly during ageing, and in fact gave a reduced solderability even without any ageing, presumably due to a degree of porosity in the coating. The 2 μm coating shows a significantly decreased solderability only after ageing for more than 20 hours and demonstrates clearly the three-stage ageing behaviour of Figure 10.15. The thicker coatings do not reach stage III of the solderability curve.

Thus short-term ageing will expose coatings that already have deficient solderability or which will deteriorate after only a few months storage. With such short term accelerated ageing, however, the mechanisms are not the same as for natural ageing. This is because intermetallic growth and diffusion mechanisms change at temperatures over 100°C, and corrosion in many cases is no longer self-limiting.

In the case of medium term accelerated ageing, for example 40°C, 93% relative humidity, tests appear to simulate natural ageing on tin and tin-lead coatings and, theoretically, the mechanisms would seem to be the same. Simulation of natural ageing for more than one year requires at least one day of steam ageing to produce a corresponding amount of diffusion and intermetallic growth.

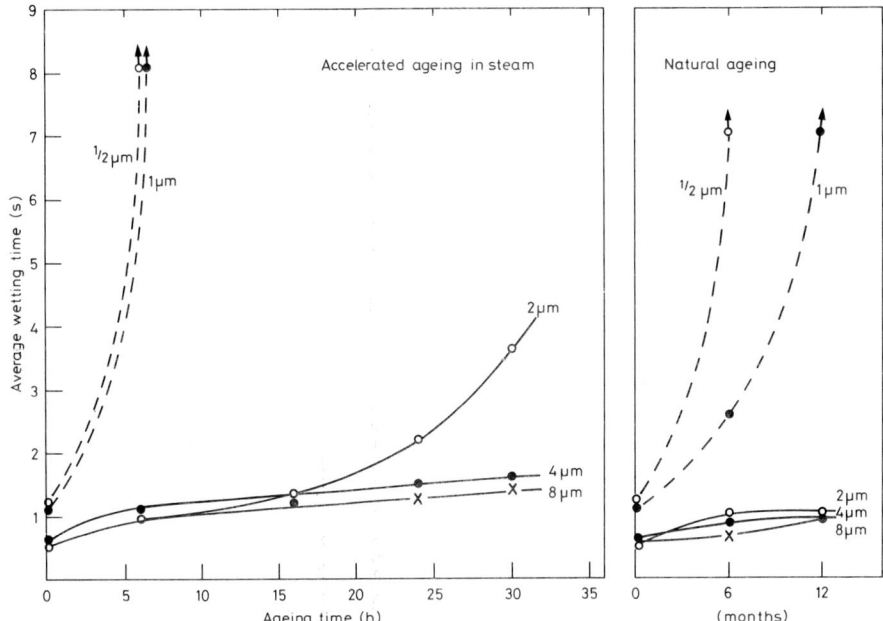

Fig. 10.23 The wetting time for tin-coated copper wire as it ages naturally and in an accelerated steam-ageing test, showing the effect of different coating thicknesses.[335]

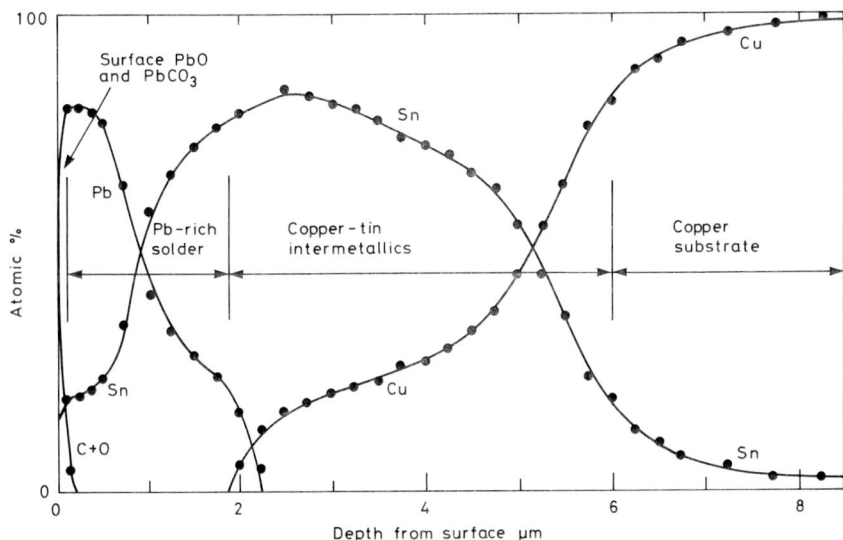

Fig. 10.24 A composition depth profile through a solder coating which exhibited very poor solderability because of ageing; the intermetallic layer is very thick and the surface is consequently enriched in lead, forming the oxide and the carbonate. Auger electron spectroscopy was used to obtain the data.

Surface analytical techniques such as Auger electron spectroscopy and X-ray photoelectron spectroscopy, combined with ion sputtering, are now allowing depth-composition profiles to be constructed through solderable coatings several micrometres thick, as shown illustratively in Figure 10.24 for a tin-lead coating on copper which exhibits poor solderability because of a thick intermetallic compound layer and a consequent lead enrichment at the surface. The surface has reacted with oxygen and carbon dioxide in the air to produce a very thin film of lead oxide and lead carbonate.

10.4.9.1 RELATIVE HUMIDITY

The moisture content of an environment in which components are stored is a major factor in determining the degree of retention of solderability of their terminations. In addition to the solderability aspect, moisture also diffuses into the plastic packages of components and devices giving rise to electrical deterioration. Thus the humidity of storage space should be given some consideration.

Air at any given temperature cannot contain more than a definite amount of water vapour. This amount, called the saturated vapour density (SVD), is shown in Figure 10.25, in $g.m^{-3}$. When the air contains this maximum amount it is saturated. A parametric fit to the tabulated experimental data allows interpolation and extrapolation from the Arrhenius equation:

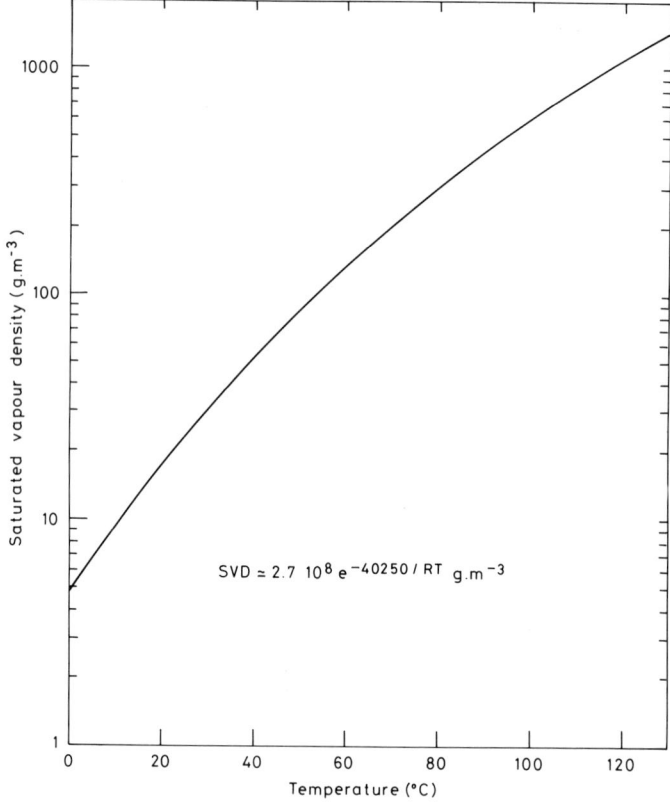

$$SVD = 2.7 \; 10^8 \; e^{-40250/RT} \; g.m^{-3}$$

Fig. 10.25 The saturated vapour density of air measured experimentally. The line is a parametric fit to the data.

$$SVD = 2\cdot 7\ 10^8 \exp\left(-\frac{4840}{T}\right) \text{g.m}^{-3}$$

where T is the temperature in kelvin.

The term humidity describes the water content of the air; the absolute humidity is the mass of water vapour in a unit volume of air and the relative humidity (RH) is the ratio of the actual amount of water present in the air to that required to saturate it at the same temperature (i.e., the SVD).

The degree of corrosion of a solderable surface occurring as a result of moisture is dependent upon the chemical potential of the moisture at the surface and it is the relative humidity that is a measure of this. It is therefore the relative humidity that is always stipulated in ageing tests and storage environments, rather than the absolute moisture content.

10.4.9.2 THE RELEVANCE OF ACCELERATED AGEING TESTS

The aim of artificial ageing is to produce in a short time the same deterioration in solderability that would occur by natural ageing during storage over a period of up to two years, and then predict which components will maintain satisfactory solderability if such a period of storage is a necessity. Ideally an accelerated ageing would achieve this by simply accelerating the mechanisms involved in natural ageing. Since all the relevant reactions and processes are speeded up by increasing the temperature it might be thought possible to choose an elevated temperature and a short time that would be equivalent to ageing at room temperature for a long time. Unfortunately there are a number of factors which prevent this.

(i) To accelerate the diffusion and the intermetallic compound growth processes such that one or two years of natural ageing are condensed into a few hours would require a temperature close to or higher than the melting point of the solder.
(ii) The intermetallic compound layer grows at an anomalous rate during the first few hours at high temperatures, so that thicknesses cannot be predicted reliably by extrapolation of measurements made after longer times.
(iii) The mechanism of diffusion changes with temperature; at room temperature, diffusion is via grain boundaries whereas at T>120°C bulk diffusion predominates.
(iv) Some of the corrosion mechanisms require the presence of water films, resulting from atmospheric humidity. Such films cannot survive at temperatures greater than 100°C.

For these reasons it must be borne in mind that, in general, accelerated ageing procedures are designed for convenience only: to give a change in solderability that can be related to the change in solderability during natural ageing. No assumption of physical or chemical equivalence can be made.

10.5 THE ASSESSMENT OF SOLDERABILITY

There exist, for the solderable surfaces of printed circuit boards and for the leads on insertion mounted leaded components, a number of methods for the measurement of wetting parameters which act as a guide to the solderability in production assembly.[371, 372] The development of a variety of tests has arisen because of the complex character of the property of solderability and the lack of reference materials of standard solderability values. Not only is there no common test used throughout the industry, there is often no standardised procedure for carrying out the tests[373, 374] and no consensus as to how the output data be used to physically describe the property of solderability in terms of the requirements of the electronic assembly industry. These requirements of solderability, as mentioned in the introduction to this chapter, consist of characteristics of thermal demand, wettability and heat resistance.

With regard to wetting, two properties of a given solder/flux combination, the onset of wetting and then the spreading on a substrate, can be assessed in a solderability test, either separately or in combination. The non-equilibrium speed of wetting is dependent upon the efficacy of the flux and the thermal demand of the subjects under test, whereas the degree of wetting obtained in the equilibrium situation is a thermodynamic property of the flux-solder-substrate combination. Since, in soldering practice, the times during which the solder is molten on the PCB are usually shorter than required to reach equilibrium, the non-equilibrium tests are more akin to the real situation.

In the solderability testing of leaded insertion components, qualitative visual inspection has always been the most commonly used method and only in recent years have objective quantitative evaluations become more acceptable. Both qualitative and quantitative tests used on insertion components have been adapted and miniaturised to offer viable tests for the solderability of both leaded and leadless surface mounting components.

10.5.1 Solder Dip Method

In this test method a small bath of solder of given composition and purity is maintained at a constant temperature. The test component is then fluxed and preheated before dipping into the molten solder at a known rate of immersion to a given depth, or floating on the solder surface for a specified time and then withdrawing it, again at a known rate. It is the simplest of all solderability tests to perform. The difficulty of the test lies in the subjectivity of the subsequent visual assessment.

The surface mounting component is best held in a stainless steel clip or tweezers and the whole completely immersed in the flux to be used in the manufacturing process. Any drops of excess flux are then removed by contact with absorbent paper. After the oxide film has been skimmed off the solder bath, the specimen is immersed in the liquid solder. The immersion can be performed by hand but is best achieved by a controlled mechanical apparatus providing a constant immersion speed between 20 and 25 $mm.s^{-1}$. The recommended dwell time at full immersion is given in Table 10.3. For most components, the areas to be examined should be immersed at least 2 mm below the solder meniscus, with the component plane vertical as shown in Figure 10.26. When considering the resistance to soldering heat, certain large flat components such as ceramic chip carriers, if immersed in this manner, will not experience the thermal gradient across their thickness that they would in practical soldering. In such cases, the component should preferably be floated on the solder surface.

The solder used in the test is normally 60Sn:40Pb, 63Sn:37Pb or 62Sn:36Pb:2Ag as used in manufacturing assemblies. If components have

Solderability

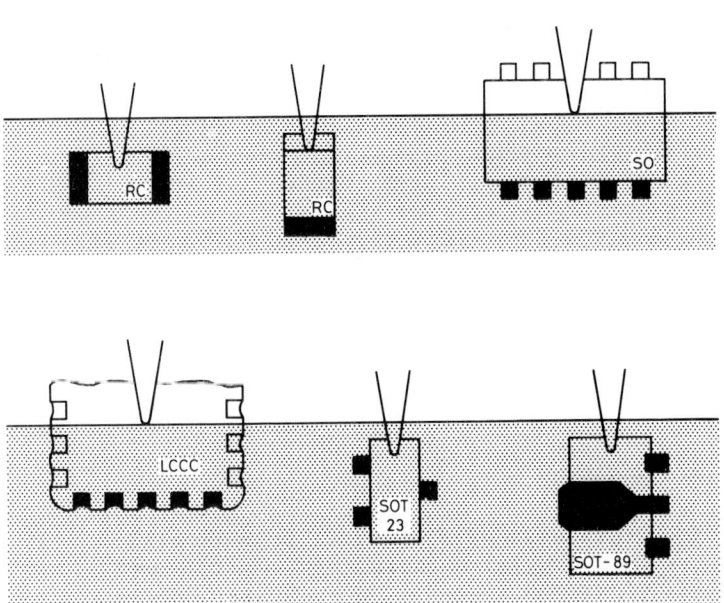

Fig. 10.26 The dip and inspect solderability test for surface mounting components.

terminations that are plated with pure tin, there might be a discrepancy between the results of the dip test and those of manufacturing practice, using methods, such as 215°C vapour phase, which operate below the melting point of tin.

Table 10.3

Recommended Immersion Conditions for the Solder Dip Test

	(a)	(b)	(c)	(d)
Dwell Time	3±0·3 s	2±0·2 s	5±0·5 s	30±1 s
Solder Temperature	215±3°C	235±5°C	260±5°C	260±5°C
Test of Wettability	✓	✓		
Test Resistance to De-wetting			✓	
Test Resistance to Soldering Heat and Dissolution of Metallisation				✓

Condition (a) of 3 seconds at 215°C allows testing at the relatively low temperature normally used for vapour phase soldering. Condition (b) of 2 seconds at 235°C is the normal solderability assessment for wave soldering assembly, while conditions (c) and (d) are designed to test the resilience of the components to de-wetting and to immersion in the solder wave.

The dip test is a single-condition, quasi-equilibrium test method and gives no indication of the speed of wetting; it does, however, show whether adequate wetting can be achieved within the specified time. After removal from the solder bath, the flux residues are removed with a suitable solvent and the solderable surfaces examined and assessed visually. The dipped surfaces should be covered with a smooth and bright solder coating with no more than small amounts of scattered imperfections such as pinholes, unwetted areas or de-wetted areas. Any imperfections that do exist should not be concentrated in one area. In visual assessments of this type, specifications usually refer to a 95% coverage requirement, in which it is unacceptable for there to be more than 5% of the area of a solderable surface non-wetted or de-wetted.

A variant of the single-component dipping test is to glue an array of the components under test on to a solderable plate or a PCB with appropriate solder pads and then, after fluxing, slowly immerse into molten solder at 45°. Once immersion is complete, withdrawal is commenced immediately. Each row of the array has therefore experienced a different immersion time, and the minimum time needed to give adequate soldering can be assessed by visual inspection. Any variation in the solderability of the batch of components can be assessed by examining the differences of soldering quality along a single row of the components.

10.5.2 Area-of-spread Test

In this test, the solderability of a surface is assessed by measuring the area over which a given mass of fluxed solder will spread on the sample substrate, at a given temperature in excess of the melting point of the solder and in a given time. Although, in principle, this can be used in a non-equilibrium manner to assess the speed of spreading, in practice, because of the necessarily slow thermal response of the system, the area-of-spread test is invariably used to measure equilibrium spreading. The heat source is removed only when spreading has effectively finished. This quick and simple test is frequently used to assess the efficacy of a range of fluxes for a particular substrate/solder alloy combination. In this context, a standard sized pellet of the solder is placed on the surface under test and fluxed, and the whole is lowered on to a hot plate or preferably a molten solder bath. The area of spread after a suitable time is then assessed either qualitatively or quantitatively. The test is obviously not suitable for solder-coated substrates.

A miniaturised version of the area-of-spread test has been proposed[375], suitable for the assessment of the solderability of leads of surface mounting components. The final area wetted by the solder must be smaller than the lead width, limiting the mass of solder used in the test to some tens of micrograms. Pellets of such size are difficult to prepare, handle and weigh, which negates this test as a quick production quality control procedure. Nevertheless, it is an area-specific test and so has its place as a laboratory quality assurance procedure for the assessment of differences in solderability of the various parts of a lead on a leaded surface mounting device. It is also suitable for footprint pads on PCBs. It is often the only realistic technique available for a quantitative assessment of solderability of specific parts of a component or a non-coated substrate.

If the substrate is homogeneous compositionally and structurally, and if the combination of substrate, flux and solder is such that wetting occurs, then, upon melting, the solder will form a spherical cap, whose dimensions are defined in Figure 10.27. The contour of the liquid is not truly part of a sphere because its top surface is flattened by the gravitational force downwards. This effect may be significant when interpreting the standard large-scale area-of-spread test but, for the micro-test, the surface tension is sufficiently large compared with the weight of liquid that making the assumption of sphericity results in an error of less than

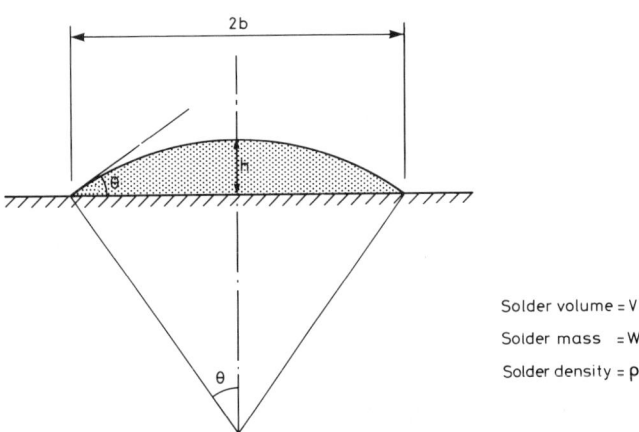

Fig. 10.27 Dimensions of a molten solder cap.

1% in the diameter of the wetted area.

If the mass of the solder pellet is W and its density is ϱ, the volume of the spherical cap of solder is V, given by:

$$V = \frac{W}{\varrho} = \frac{1}{6}\pi h (h^2 + 3b^2) \tag{10.20}$$

where h is the height of the cap and b the radius of the measured, circular area-of-spread. The dimensions of h and b are related by the contact angle θ:

$$\frac{h}{b} = \frac{1 - \cos\theta}{\sin\theta} \tag{10.21}$$

so it follows, combining Equations (10.20) and (10.21):

$$\frac{W}{\varrho} = \frac{\pi b^3}{6}\left[\left(\frac{1-\cos\theta}{\sin\theta}\right)\left\{\left(\frac{1-\cos\theta}{\sin\theta}\right) + 3\right\}\right] \tag{10.22}$$

from which, for a known solder pellet size, the contact angle can be determined from the diameter, 2b, of the circular area-of-spread. Some calculated values from Equation (10.22) are shown in Figure 10.28, relating the degree of spreading to the contact angle, over a range suitable for testing surface mounting component leads.

In area-of-spread tests, a non-dimensional spread factor is often expressed, for example the ratio of the diameter of the spread area of the solder to the diameter D of a sphere of the same solder volume. In this case the spread factor $S = 2b/D$ where $V = \frac{1}{6}\pi D^3$ whence, equating this volume with that given in Equation (10.20) for a spherical cap, the spread factor S can be calculated as a function of the contact angle θ, as shown in Figure 10.29. Another spread factor used is defined as $S' = (D - h)/h$ where D and h remain as defined above.

In the practical application of the test, the value 2b can be measured on the solidified solder assuming that the periphery has not changed during solidification. A measurement of h cannot be made after solidification since the

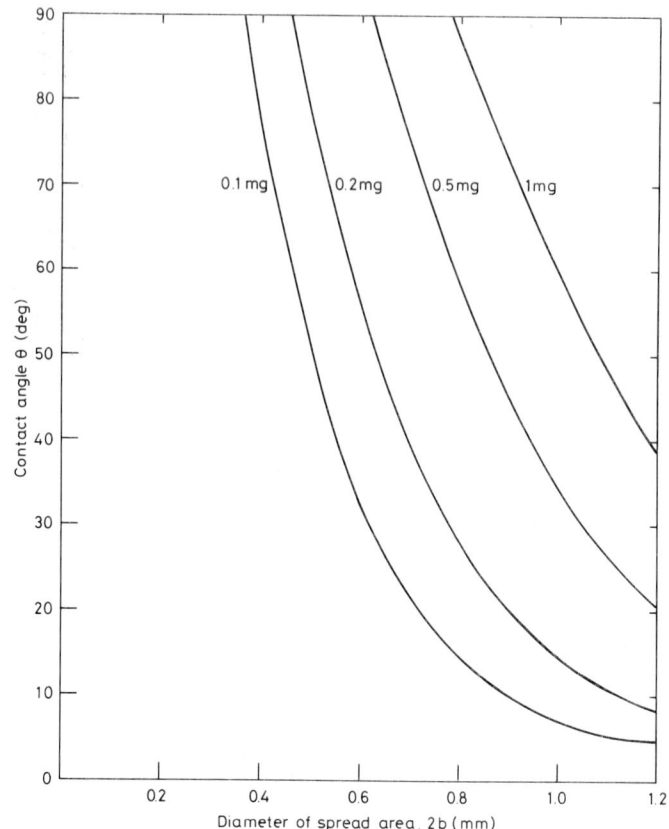

Fig. 10.28 The determination of an average contact angle from an area-of-spread test using four sizes of solder pellets.

height will always change in a fairly unpredictable way during solidification as the volume of the solder shrinks. The volume change of solder during solidification is about 4% followed by a further 1·2% solid-state shrinkage during cooling down to room temperature.

Generally, on electronic components, the area-of-spread will not be circular because of inhomogeneities of the surface composition on an atomic scale. In such cases, either several measurements of the diameter 2b must be made along pre-specified axes or the area must be measured and an average diameter of the contact area calculated from $2b = (4 \times \text{area}/\pi)^{1/2}$. The area can be measured either automatically using an optical imaging system with an area measuring computer or by use of area comparators which are available for use with stereo microscopes.

A relationship between contact angle and wetting quality has been proposed, and is used in some specifications. It is given in Table 10.4.

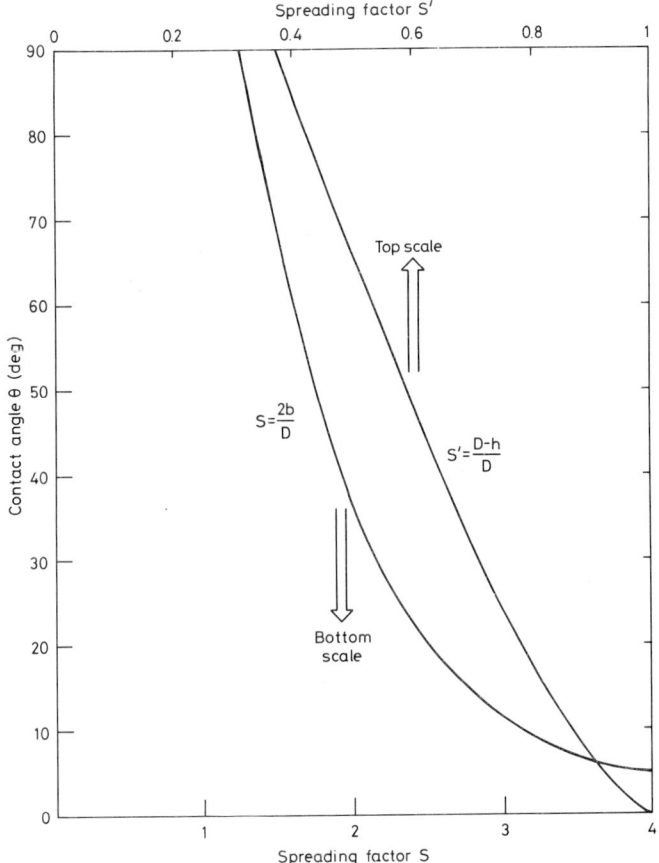

Fig. 10.29 Two spreading factors that can be defined from the area-of-spread test.

Table 10.4

Classification of Wetting Quality in terms of Contact Angle

Contact Angle	Wetting Quality
$0° < \theta < 10°$	perfect wetting
$10° < \theta < 20°$	excellent wetting
$20° < \theta < 30°$	very good wetting
$30° < \theta < 40°$	good wetting
$40° < \theta < 55°$	adequate wetting
$55° < \theta < 70°$	poor wetting
$70° < \theta < 90°$	very poor wetting
$90° < \theta$	non wetting

10.5.3 Meniscus Shape Method

In this test method, the solder meniscus is rapidly scanned with a near-vertical laser beam as the solder rises around a surface mounting component in contact with the molten metal.[376] The reflected laser light from the solder surface is detected by a number of photocells arranged in a vertical arc to form a coarse goniometer. The photocell outputs are used to compute the meniscus shape as a function of time. With sufficiently rapid scanning of the meniscus by the laser beam, the changing shape of the meniscus during the soldering process can be followed, once the contact angle has fallen below 90°.

At present, the instrument is large and complex, and difficulties are envisaged in the use of this concept in a production model for routine quality assessment.

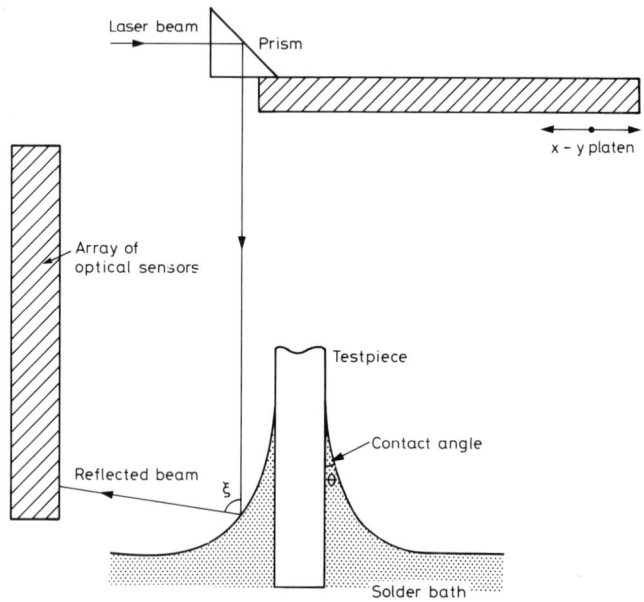

Fig. 10.30 Schematic of an optical reflection technique that can be used for determining the shape of a rising meniscus, and hence a contact angle.[376]

The principle of the test instrument is shown in Figure 10.30. A laser beam with a diameter of about 100 μm impinges vertically on the liquid solder, but, because it is positioned to strike the meniscus of the immersed sample, it is reflected at an angle ξ which depends on the radius of curvature of the meniscus at the point of reflection. As soon as the sample is dipped into the solder bath, a motorised precision table is used to enable the laser beam to scan the meniscus rapidly, starting from the periphery of the sample. The measurements stored are the reflection angle ξ as a function of the x-coordinate. In order to calculate the meniscus height m at the contact line, the shape of the surface is assumed to be exponential in form. Thus

$$z = \text{const.} \; e^{kx}$$

and $\xi(x=0) = 180° - 2\theta$

where θ is the contact angle between the molten solder and the sample, i.e., at $x=0$, $z=m$.

10.5.4 The Wetting Balance

In the wetting balance test method, the specimen is suspended from a sensitive balance and immersed edgewise, at a controlled rate and to a set depth, into a bath of molten solder at a controlled temperature.[377–379] The specimen is subjected to time-variant vertical forces of buoyancy upwards and surface tension downwards, which are detected by a transducer and converted to a signal that is usually continuously recorded on a high-speed chart recorder.[380, 381]

This apparatus is the most versatile solderability test method and since its development has been used extensively not only for assessing the solderability of leaded insertion components but also the efficacy of fluxes. For use with surface mounting components, some modifications are required, but in order to understand the reasons for such modifications the wetting balance method for general use is first described and discussed.

The specimen is first fluxed and hung on the balance over the solder bath for a predetermined time such that it is effectively pre-heated by convection from the solder. The specimen is then immersed, usually at a speed of 20±5 mm.s^{-1}, to a predetermined depth and held in that position for a specified time, usually about 10 seconds, before withdrawal. The force experienced by the specimen as a function of time is recorded.

In Figure 10.31 the six stages of the testing of a specimen that is readily wet by the solder are shown:

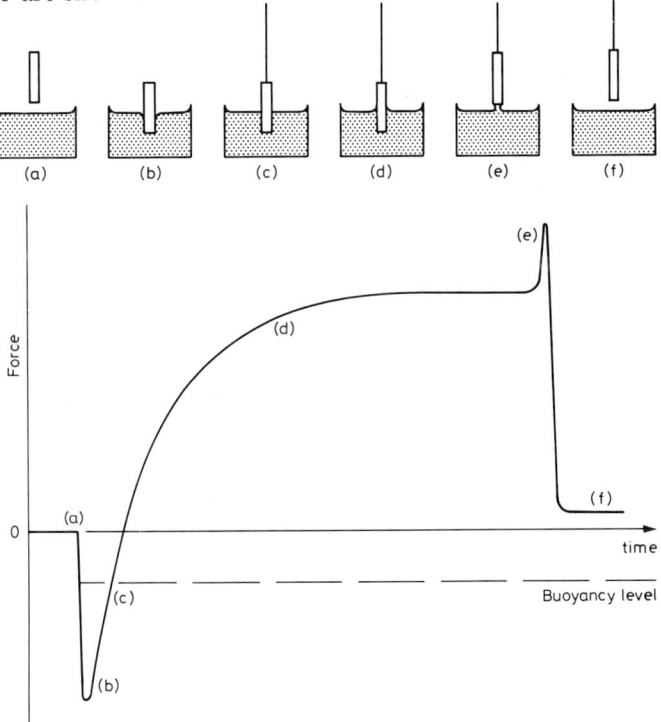

Fig. 10.31 A wetting balance curve.

(a) just prior to the moment of immersion;
(b) immediately after the moment of immersion, when wetting has not begun and there is an upward summation of buoyancy and surface tension forces;
(c) after wetting has begun and the meniscus has risen up the specimen to the point where the vertical force from surface tension is zero and the net force acting in the specimen is that due to its buoyancy;
(d) when the meniscus is curved upward and the surface tension force is acting downwards;
(e) as the specimen is being withdrawn and surface tension and a possible oxide film on the solder are causing a drag out of solder;
(f) when the specimen has been withdrawn and is heavier than at the start of the test because of the adherent solder coating.

Some representative curves are shown in Figure 10.32. In each case the full horizontal line represents the force condition at the start of the test cycle and the dotted horizontal line the buoyancy level at which the wetting force is zero, i.e., the contact angle is 90°. The buoyancy of the specimen is simply the product of the immersed volume (measured from the mean liquid level, not from the meniscus level) and the density of the molten solder which it displaces. At the usual test temperature of 235°C the density of 60Sn:40Pb solder can be taken as 8000 kg.m^{-3}.

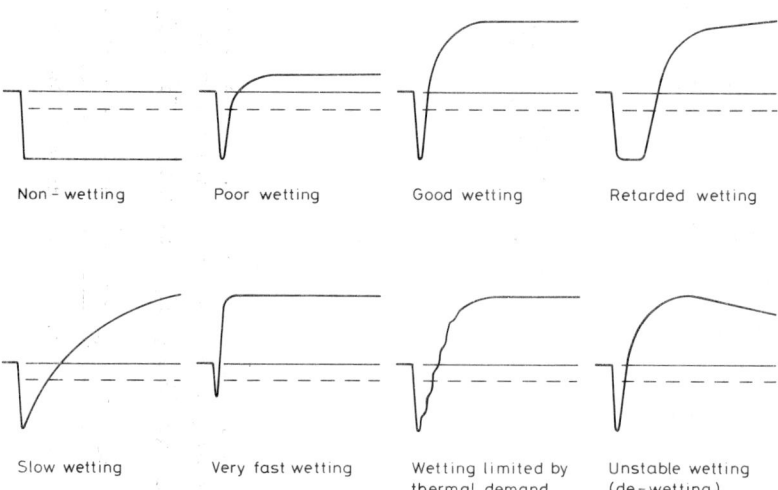

Fig. 10.32 Some representative shapes of wetting balance curves.

For samples that are readily solderable, the measured forces change very rapidly and a chart recorder with a fast response time is essential. Unless care is taken, the recorder response time can influence test results.

10.5.4.1 THEORETICAL WETTING FORCE

The wetting force measured when equilibrium wetting has been achieved can be compared with a calculated value. If the specimen has a cross-section in the vicinity of the meniscus that is constant through the length of the specimen, has a perimeter or circumference c and an immersed volume v, then the net force causing the balance to record heavier is:

$$\text{Force} = \gamma c \cos\theta - \varrho v \qquad (10.23)$$

where γ is the surface tension of the solder in contact with the flux, θ is the wetting contact angle and ϱ the density of the solder (8000 kg.m^{-3}). If c is measured in mm, v measured in mm^3, γ taken as 0·4 J.m^{-2} at 235°C, and perfect wetting assumed so that $\theta = 0$, then the theoretical maximum force is

$$\text{Force}_{(th)} = 0 \cdot 4c - 0 \cdot 08v \text{ mN}.$$

10.5.4.2 INTERPRETATION OF WETTING BALANCE CURVES

The wetting balance test equipment is the most useful tool for investigating the soldering properties of combinations of solder, fluxes and surfaces. However, even though the test was suggested as early as 1949, no generally accepted method for evaluating the wetting force curves exists. Much work has been published, but almost as many different points of significance on the curves have been chosen as indices of solderability. There is a need for an evaluation method that can be applied to all types of specimen in different fields of soldering, which would make the data comparable.

The reason for the ambiguity in the interpretation of the wetting force-time curves lies in the fact that no rigorous theoretical interpretation of the rate of wetting of the meniscus up the specimen has been applied to the experimental technique. Also, the wetting balance test gives information about both the speed and the degree of wetting and it is arguably not possible to present both these two independent facets of soldering within a single solderability index.

The wetting balance curve, in practice, exhibits many different forms, some smooth and some with perturbations depending upon the interplay of the various controlling parameters such as thermal conductivity, oxide thickness, flux activity, etc. Broadly speaking, the wetting part of the curve can often be approximated to by an exponential, although even in the ideal theoretical case of a non-reactive system the wetting curve is not a true exponential. The theoretical wetting curve is closest to an exponential for an infinite flat plate being wetted non-reactively. The deviation from an exponential may be quite large for a small coupon or a circular wire in solder. Figure 10.33 is a schematic of an experimental curve in which all the important events are represented. These events may or may not be relevant. With the interfacing of wetting balances to desk-top computers the curve can be digitised easily and the defining points readily accumulated and processed.

In the published literature, curves have been analysed in terms of the time taken to achieve a specific event on the curve or in terms of a wetting rate. The time t_3 for the meniscus to pass through a contact angle $\theta = 90°$ or the time t_4 for the wetting force to return to zero are common, and indeed these values correlate well with the wetting times obtained for wires in the globule test. The globule test is the most commonly used solderability test for wire-leaded components, in which the fluxed wire is made to bisect a defined small volume of molten solder and the time measured for the solder to wet the wire and coalesce on the top side of the wire. The time t_4 on the wetting balance is the easiest to measure and for wires up to about 1 mm in diameter this measurement corresponds 1:1 with globule test measurements. For larger wires the globule test is affected more

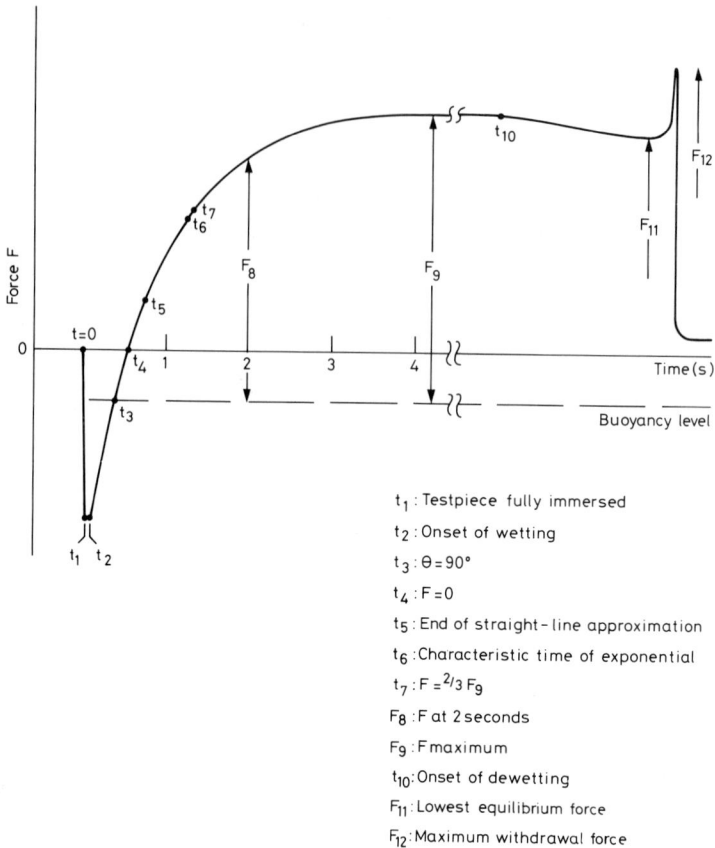

Fig. 10.33 A range of time and force parameters that can be defined on a wetting balance curve.

adversely by the thermal drain and the direct equivalence between the two tests is lost.

In the wetting balance curve the initial rate of wetting is determined by the slope of the straight line approximation between times t_2 and t_5, but such an interpretation takes no account of any incubation period prior to the onset of solder spreading.

A very common solderability index is the time t_7, being that time passing until the force has reached two thirds of its maximum, measured from the buoyancy line. No specification is however placed upon what that maximum force should be and the use of t_7 as an index implies a universal form to the wetting curve. The commonly chosen time t_7 at ⅔ maximum force is of interest because it closely approximates to t_6 which defines the complete exponential curve. This is the time taken for the force to reach 63·2% of its maximum, assuming that the wetting part of the curve is a true exponential. In practice, the curve between points 2 and 9, where the maximum force is achieved, is parametrically fitted to an exponential

$$F = F_9(1 - e^{-t/t_6})$$

where t_6 is the characteristic time of the exponential and, when $t = t_6$, $F = 0 \cdot 63\, F_9$.

Perhaps of most practical significance is F_8, the wetting force achieved after a specific time, e.g., 2 seconds, which is comparable to the soldering times applicable to automated soldering. In some standards which call upon the wetting balance, it is required to express this force F_8, measured after a given time, in terms of a contact angle, using Equation (10.23). By using θ rather than F, the size and cross-sectional shape of the test specimen need not be specified. The quality of the soldering can be classified according to Table 10.4.

To complete the characterisation of the wetting force curve, t_{10} is the time of the apparent onset of de-wetting and F_{11} is the lowest wetting force of the de-wetted surface. F_{12} is the maximum withdrawal force.

In attempts to incorporate both rate and force information into a single solderability index, several workers have suggested specific combinations of parameters or geometrical constructions within the curves, as shown in Figure 10.34. In (a) the wetting force curve has been integrated to a given point, t', of intersection with the time axis.[382] An advantage of this appraisal of the curves is that both high wetting force and short wetting time make their effect felt in the same direction, producing an increase in area A. Similarly, the immersion force and the incubation time are included with a measurement of area B. A solderability index is therefore achievable as, for example, area A/area B.

In Figure 10.34(b) the tangent to the curve from the point defined by $t = 0$ and the buoyancy line has been constructed.[383] The slower the wetting time and the higher the wetting force, the greater will be $\tan \chi$ ($= F^*/t^*$). Now F^* depends upon the circumference c wetted and the surface tension γ of the solder. For a given set of test conditions, γ is constant and the wetting angle at time $t = t^*$ is given by $\theta = \cos^{-1} F^*/c$, providing a measure of the degree of wetting and hence the solderability.

A solderability index is then defined:

$$S = F^*/ct^*t_3$$

which has units $kg.m^{-1}.s^{-2}$.

Another construction involves measuring the wetting force (from the buoyancy line) achieved either as a maximum or within a given time, say 2 seconds.[384] This dimension F' is then transposed along the buoyancy line from the zero time axis, as shown in Figure 10.34(c). The end point, i.e., time $t = F'$ is taken as the centre of a constructed circle of radius H, where H is the largest possible value such that the circle remains completely beneath the wetting curve. The wetting force and the wetting time are taken into account to an identical extent since any increase in force will cause the centre of the circle to be displaced to the right and any reduction in the wetting time will cause the curve to be displaced to the left, which both result in an increase in the size of the circle. The area of the inscribed circle is thus taken as a measure of the solderability.

A modified circle method combines the tangent and the circle constructions and is claimed to have the 'advantages' of both. The tangent to the curve re-defines the position of the centre of the circle along the buoyancy line.

These constructions have little or no physical significance and, although useful in ranking wetting curves in terms of a solderability index, they cannot be interpreted in physical terms and so cannot be extended to the general case.

Using a purely pragmatic approach, the wetting time is best defined by t_3, t_6 and t_9, the times to reach a 90° contact angle, two-thirds of the equilibrium force and the maximum wetting force respectively, whilst the wetting quality is best

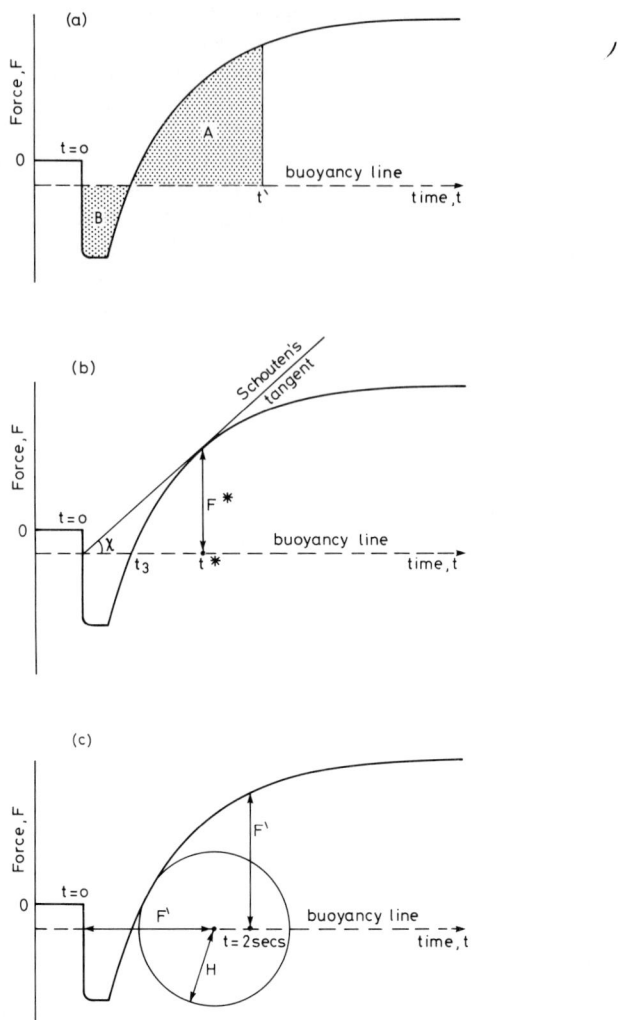

Fig. 10.34 Several solderability indexes have been suggested to incorporate both the speed and the degree of wetting given by a wetting balance curve. (a) Evaluation in terms of the integrated area up to a given time.[382] (b) Construction of a tangent to the curve.[383] (c) Construction of an inscribed circle whose area is a measure of the solderability.[384]

defined by F_8 and F_9, the wetting force after a given time identifiable with industrial soldering practice (i.e., 2 seconds), and the maximum wetting force respectively.

The wetting balance is capable of supplying considerable information but, in practice, for quality control of solderability, the requirement is to know whether or not a specific wetting force is achieved within the time available during automatic soldering, namely about 2 seconds. Thus if a point A is defined at time $t=2$ seconds and at an acceptable wetting force F_o, then the measured wetting curve must pass over that point A as shown in the inset of Figure 10.35, i.e., $F_8 > F_o$. In this context, whether or not the curves approximate to exponentials, or

Solderability 359

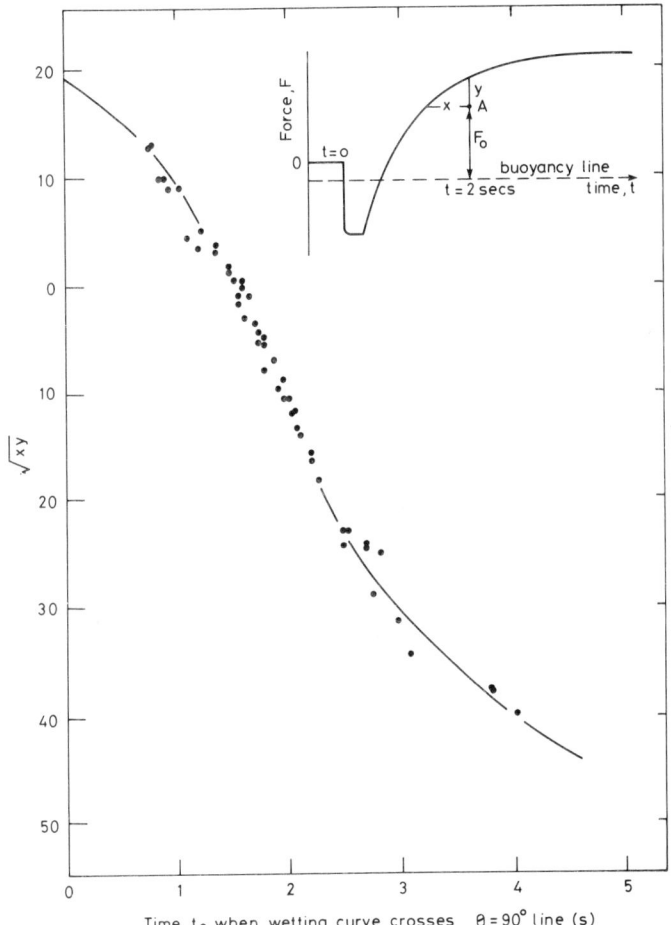

Fig. 10.35 A defined acceptance point for a wetting balance curve gives rise to the function $\sqrt{xy}$. This function is plotted against the time t_3 (for $\theta = 90°$) for copper specimens passivated in BTA to varying extents. The line is theoretical assuming the wetting curve is a true exponential.

whether or not the equilibrium wetting force is very high, or whether or not there is an incubation period, are all irrelevant. An index of solderability can be defined in terms of x and y, the displacements of the curve from point A. If the curve is a true exponential, then a plot of $\sqrt{xy}$ against the intersection point of the curve with a horizontal axis, e.g., the buoyancy line, takes the form of the solid line in Figure 10.35. The points in the figure are measurements taken from wetting curves of aged copper coupons and serve to show how close the measurements can align to the theoretical curve.

The line of acceptable solderability is $xy = 0$. but this of course depends upon the original choice of F_o. This value can be placed by practical experience or theoretically using Equation (10.23) from a knowledge of the surface tension of the molten solder under flux γ, its density ϱ, and the geometry.

The contact angle θ then becomes the quality assurance parameter. A value $\theta < 40°$ would be considered good and hence setting $\theta = 40°$ defines the point A.

10.5.4.3 WETTING BALANCE FOR SURFACE MOUNTING COMPONENTS

In the standard application of the wetting balance for the assessment of solderability of leaded insertion components, the immersion depth is commonly 5 mm, and not less than 2 mm. During the wetting, the solder meniscus can rise up to 3·2 mm above the mean solder level, as was shown in Section 10.2.4, so that at least 3 mm length of solderable termination is required to be in contact with the solder.

Surface mounting components do not in general have the 5 mm of termination available for such an immersion procedure, i.e., 2 mm immersion plus 3 mm meniscus rise. For some components, ceramic chip capacitors for example, the wetting balance can still be used provided the metallisation to be wetted is present around the entire periphery of the component.[385] With the length of the component vertical, the test can be carried out with an immersion depth of almost zero, i.e., relative movement between the component and the solder bath is halted as soon as contact is made. Unfortunately, the electrical conductivity of chip capacitors in particular is poor and the required contact or immersion depth is difficult to attain. Control has been achieved either by using the mechanical contact to stop the movement or by using a high frequency a.c. make-break circuit through the capacitor.[386] A force-time curve of the zero-immersion type shown in Figure 10.36 is obtained. Because the effective immersion depth is only about 0·01 mm, the buoyancy offset of the measured force is near zero. As

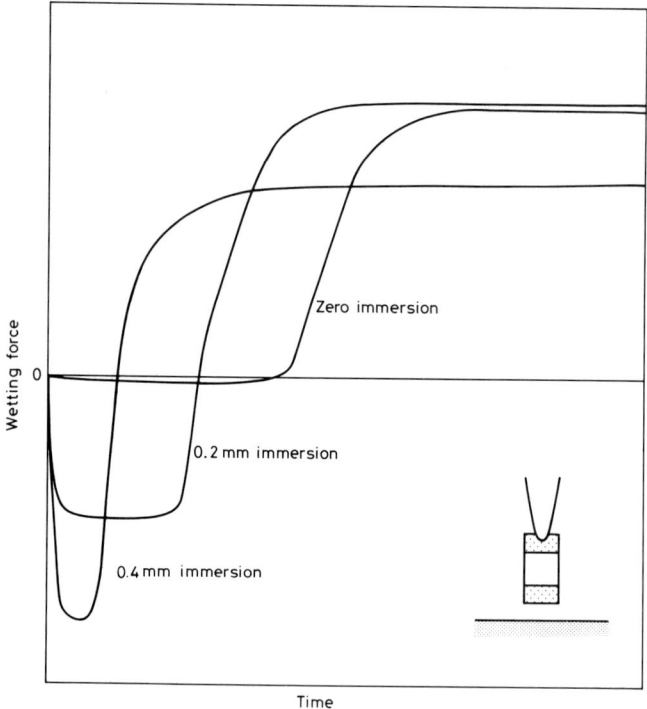

Fig. 10.36 Wetting balance curves obtained from a single terminal of a chip component, varying the immersion depth.

illustrated in Figure 10.36, increasing the immersion depth speeds up the wetting because of greater heat transfer, but reduces the wetting force because the meniscus is more limited by the height of the component metallisation. Since the shape and heat transfer conditions of the specimen are different from those of the standard mode, the force-time curve needs to be interpreted differently and the test results compared with the best solderable components of the same type. It has been found that, for ceramic chip capacitors, satisfactory soldering results if the test procedure gives a time to commence initial wetting, t_2, of less than 0·5 second, and a wetting force after 3 seconds of at least 25% that obtained with an optimal component of identical size and shape. The 3 seconds time limit is longer than the 2 seconds normally specified for insertion components because the solderability of the fired-on metallisation of chip capacitors is generally poor compared with that of solder coated leads. This poorer solderability is acceptable because it is offset by the advantageous geometry of a surface mounting joint, whose capillary spaces between components and solder lands favour wetting and spreading of the solder.

If the metallisation dissolves rapidly in the solder, this may cause some alteration to the observed force-time curve as the dissolution occurs, usually resulting in a gradual lowering of the force.

10.5.4.4 SCANNING MODE WETTING BALANCE

An adaption of the normal mode of operation of the wetting balance, which gives it potential to identify small-scale variations of solderability on a surface and consequently potential for use with surface mounting components, is the scanning mode of operation.[387-390]

In the normal mode of operation, the rate of immersion is much higher than the rate of wetting in order that at zero time the part of the specimen of interest can be assumed to be immersed in the solder, but not wetted. Wetting can then be followed as it occurs up the specimen. The immersion rate could be increased further to ensure this initial state if it were not for the consequent production of unwanted waves in the solder bath. In the scanning mode, by contrast, the rate of immersion is less than, or comparable to, the speed of wetting—usually about 1 mm.s^{-1}. During immersion the buoyancy force increases at a rate proportional to the immersion speed, as shown in Figure 10.37. A perfectly homogeneous specimen produces a straight line parallel to the buoyancy line: wetted specimens above the buoyancy line and non-wetted specimens below. The initial steep part of the curve corresponds to the height of the wetting or the non-wetting meniscus. Any spots on the immersing surface with a different solderability from that of the main area show up as perturbations on the curve as the meniscus passes across them.

This type of test can, in principle, be used to assess the solderability of the pads on a leadless surface mounting component. Imagine, for example, dipping a ceramic substrate with a series of solderable pads, as shown in Figure 10.38, slowly into a solder bath. The meniscus reaches the lower edge of the metallised area, not at depth a but at depth $m + a$, where m is the depth of the non-wetting meniscus. The meniscus then rises as the metallised area is wet, producing a reduction in the non-wetting force, until the upper edge of the metallisation is reached, after which a negative meniscus again occurs. For each metallised pad, a peak in the force-time curve therefore arises. It is narrower if the wetting is rapid

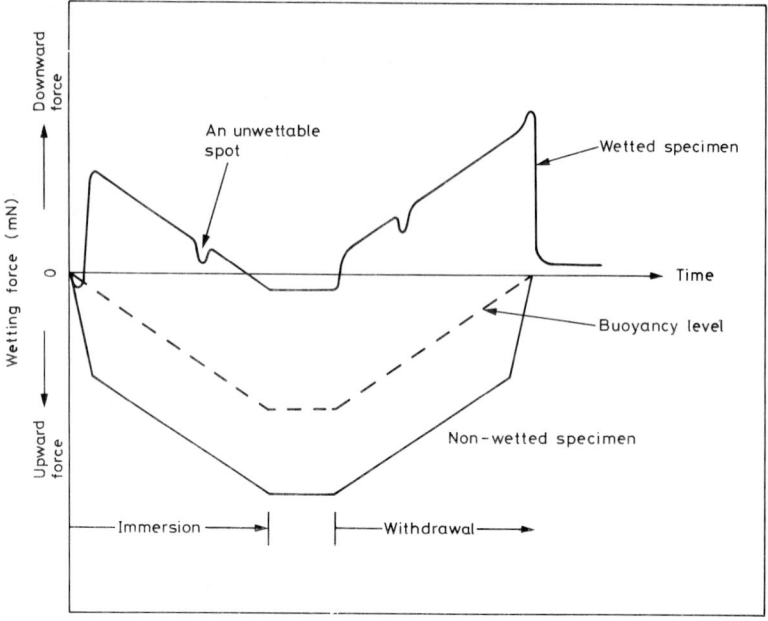

Fig. 10.37 Scanning-mode wetting balance curves, obtained when the speed of immersion is less than the speed of wetting.

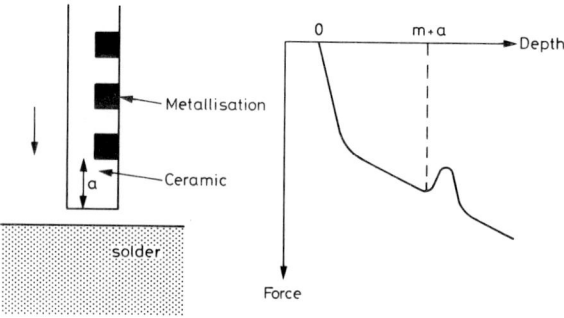

Fig. 10.38 The effect of dipping a non-wetting ceramic substrate with solderable metallisations into solder, using a wetting balance.[390]

and higher if the degree of wetting is good. Because the wetted areas are small, the signal generated is also small, and improved resolution is obtained if the signal is differentiated, such that the changes in slope are displayed.

When the test is used on real components where the solderable pad may be a three-dimensional castellation, or other than rectangular, or one of an array, or in a protruding lead, computer evaluation of the wetting force signal is necessary.

10.5.5 The Globule Balance

The globule balance is a variant of the wetting balance, in which the bath of molten solder is replaced by a small globule of molten solder.[386, 391] This has two advantages for the testing of surface mounting components; first the spatial selectivity of a small globule allows individual component leads or metallised pads to be assessed separately in turn, and secondly, by using a curved solder surface of a globule instead of a flat surface of a solder bath, there is a signal resolution benefit to be gleaned.

This measurement instrument has demonstrated the greatest potential for use as a standardised test and measurement method for assessing the solderability of surface mounting components. Existing wetting balances can normally be modified relatively easily, and commercial instruments of this type are available.

10.5.5.1 GLOBULE SIZE

Two sizes of globule and globule block cover the whole range of surface mounting components.[386]

A 200 mg solder globule on a 4 mm diameter iron pin, as used in the standard globule test for insertion components, may be used successfully for testing chip capacitors and resistors, SOTs, SODs, MELFs and any other package configuration where a single wettable termination of restricted length is to be tested.

A 25 mg solder globule on a 2 mm diameter pin in a conical-topped support block, as shown in Figure 10.39, is used for testing the individual terminations of closely spaced multi-terminal devices such as SOIC legs, PLCC leads and LCCC pads, where the termination pitch does not permit access of the larger globule to individual terminations.

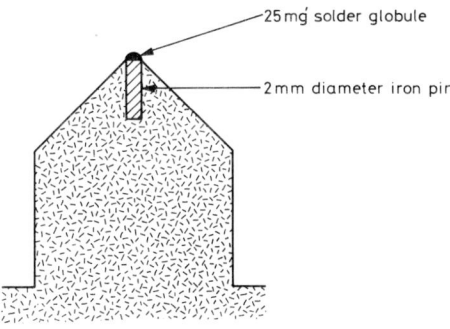

Fig. 10.39 The geometry of a 2 mm iron pin in a conical-topped support block, for the globule balance.[386]

The globule balance test is tolerant of variations of globule size, as shown in Figure 10.40: a plot of the measured wetting force after 3 seconds on one batch of chip capacitors (each point is the mean of ten capacitors) as a function of the globule size on a 4 mm diameter pin. The tolerance of ±15% on the 200 mg globule seems acceptable.

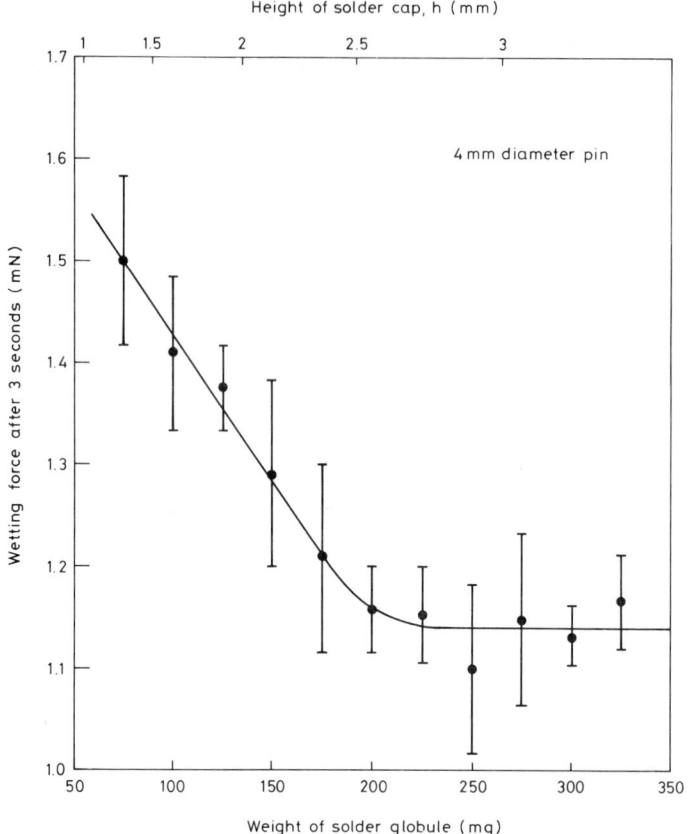

Fig. 10.40 The measured wetting force after 3 seconds for chip capacitors, using the globule balance with a range of globule sizes.

10.5.5.2 SPECIMEN-SOLDER CONTACT

The globule balance solderability test is a non-immersion method, and it is necessary to avoid immersing the specimen beneath the solder surface. Commercial instruments allow the stepper motor drive of the globule block to stop within less than 10 μm when contact between the specimen and the solder is established. The contact detection systems are therefore crucial to the test method.

The detection of contact between the specimen and the solder is best made using a high frequency a.c. circuit that is effective not only for resistors but for capacitors and some semiconductors which, for d.c., may be open circuit. The presence of a dried flux film on the specimen can lead to delays in the detection unless the speed of approach of the globule and specimen is less than 5 mm.s^{-1}. At this relatively low speed the flux has time to soften by heating as it approaches the solder but before contact is made.

The accuracy of attainment of zero immersion is much less critical when PLCC J-leads or SOIC gull-wing leads are under test, as these metallic terminations do not impose the severe restrictions on solder rise necessary with chip capacitors, chip resistors and other leadless devices. For such components it may not always be convenient to detect the point of contact with the solder using an a.c. conductance method, because of the problem of ensuring electrical continuity through the device. Far more convenient is to detect the contact by sensing the first upthrust exerted by the solder on the component. This method will not allow the high accuracy of depth control possible with the conductance method, but the requirements in this regard are far less demanding when testing metallic terminations than when testing leadless metallisations, and the sensor method is entirely satisfactory in such cases.

10.5.5.3 RESTRICTION OF SOLDER RISE

As already mentioned, the length of the wettable region on a surface mounting component during test may be less than the height to which the solder would naturally rise on a large metallic specimen. In such cases the measured force will be limited by the height restriction and not therefore be a true function of the contact angle. When using the curved solder surface of a globule instead of the flat surface of a solder bath, this unwanted effect is alleviated.

Equation (10.10) showed that, if an infinitely long plate is immersed edgewise into molten solder, the meniscus can rise to some 3·2 mm if wetting is perfect. In practice, however, specimens cannot be considered to be infinitely long. The effect when immersing a small specimen into a liquid is to introduce a curvature into the meniscus in the plane of the solder surface, which has the result of generally reducing the meniscus rise as the specimen gets smaller. The magnitude of this effect can be calculated since the surface tension force limits the mass, and therefore the volume, of solder that may rise up under the meniscus. For a given volume of solder under a meniscus per unit perimeter length of the immersed specimen, the meniscus height must obviously be less if the surface is curved. The meniscus height is less on a fine wire than on a thick wire, and reaches a maximum for an infinite flat plate. The effect can be calculated for wire specimens of circular cross-section (for which there are no anomalous corner effects) as shown in Figure 10.41 for a wire entering the solder at $\alpha = 90°$ and assuming ideal wetting with $\theta = 0°$ and also $\theta = 30°$.

The restriction in solder rise with decreasing specimen size is fortuitous, but the effect is not sufficient to permit successful solderability testing of all but a few surface mounting components using the wetting balance in its normal mode. The globule balance, however, offers a further benefit by effectively reducing the specimen entry angle α, as shown in Figure 10.42. If, for example, α is reduced from 90° to 70° in Equation (10.10), the meniscus rise in the case of ideal wetting of an infinite plate would be reduced by the ratio of sin $\alpha/2$, from 3·2 mm to 2·6 mm. Further, by curving the solder surface there is another restriction to the meniscus rise up the sample, as illustrated in Figure 10.43. Because initial contact between the component, for example a chip capacitor, and the solder is on the end surface of the component, the meniscus height m has to include also the distance m'. Since the volume of molten solder is a fixed quantity, as it spreads out across the end face termination of the hanging component, the surface profile of the globule must distort.

Thus in summary, by using the globule balance, the solder rise up the termination on a specimen is restricted, for a given wetting angle, in four ways:

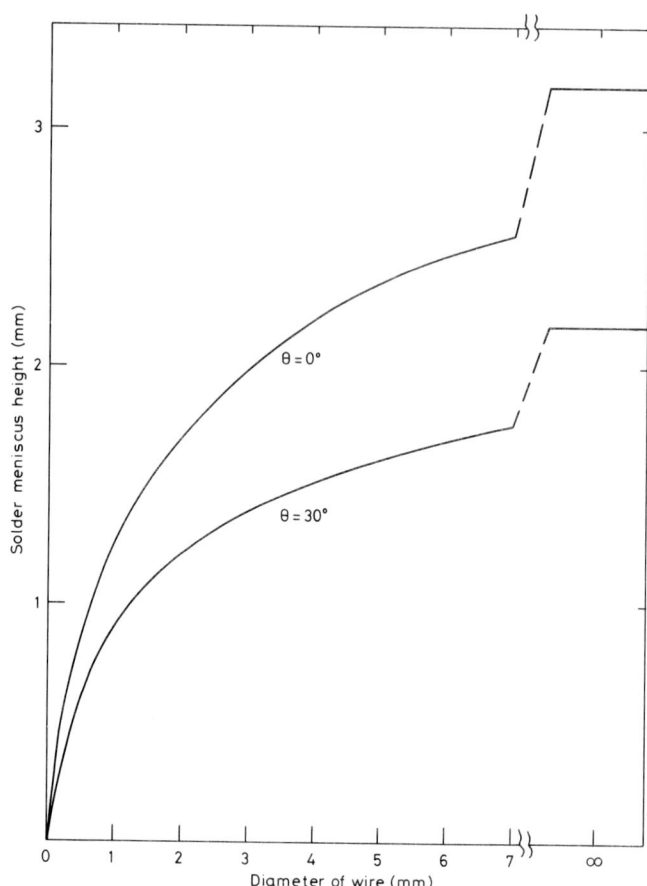

Fig. 10.41 The calculated solder meniscus rise up a circular wire as a function of its diameter and the contact angle θ.

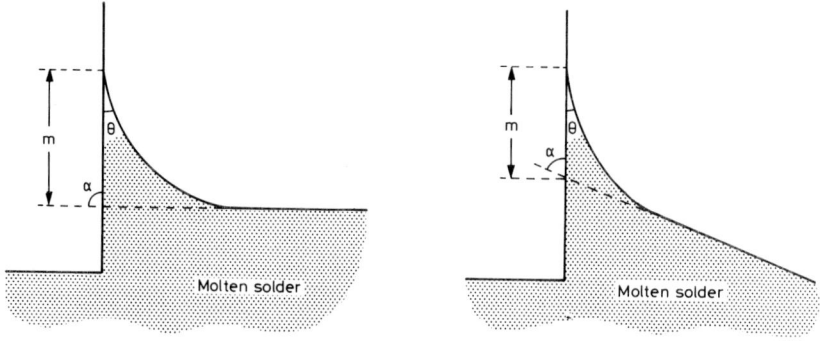

Fig. 10.42 By using a solder globule rather than a solder bath the meniscus rise m is reduced because of the change in effective entry angle α of the component at the solder surface.

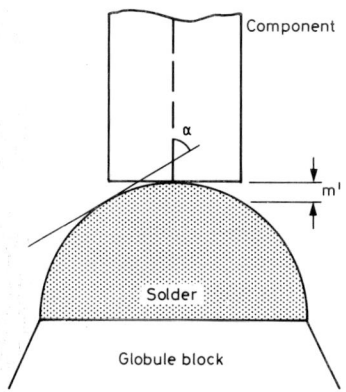

Fig. 10.43 Zero immersion of a component at the apex of a solder globule means that the meniscus is further restricted in its rise because of the gap m'.

(i) by the small specimen radius;
(ii) by the effective specimen entry angle $\alpha < 90°$;
(iii) by the additional height m';
(iv) by a possible additional factor because the solder surface is deformed by the wetting forces upon entry of the component.

An example of the magnitude of the combined effect of these factors has been determined for a 0·75 mm diameter solderable wire. Using the normal mode wetting balance the mean meniscus rise is measured as 1·05 mm, close to the theoretical value for ideal wetting shown in Figure 10.41. Using a 200 mg solder globule, producing a radius of curvature of about 2 mm, the mean meniscus rise on the wire is only 0·5 mm.

10.5.5.4 THERMAL RESPONSE OF THE GLOBULE BLOCK

In solderability tests, the thermal demand of the component may affect the test result. This is most noticeable when the amount of solder is small such as in the globule balance method. Not only is the solder temperature very important but also the quantity of heat that can be supplied during the few seconds of the test. Therefore the thermal response of the block on which the globule is sitting plays an important rôle. This is particularly true for components with metal leads, some of which may be made of a high thermal conductivity copper alloy.

The problem of the thermal response of the globule block has been addressed in some detail by considering the heat flow into a wire immersed in the molten globule. A surprisingly large supply of heat is required, so that the problem of thermal response cannot be ignored. For testing surface mounting components the degree of the problem is rather less because heat conduction away from the globule is much less, but concomitantly the amount of heat held in the molten solder is also less.

10.5.5.5 COMPARISON BETWEEN GLOBULE BALANCE AND WETTING BALANCE

The use of a globule instead of a bath of solder has several other practical advantages. The small quantity of solder used permits economic testing by, and development of, relatively expensive specialised solder alloys which are becoming more prevalent with the advent of surface mounting technology. Also, the small globule reacts to the effects of dissolution of solderable coatings and metallisations which could be swamped in a large solder bath. This better represents assembly methods used in surface mounting technology and also allows a new, uncontaminated solder sample to be used for each test if required.

Data are available of a comparison of solderability tests made under identical conditions[386] using ceramic chip capacitors (a) side dipped using the normal mode wetting balance, (b) end dipped using the wetting balance and (c) end dipped using the globule balance, as illustrated in Figure 10.44. The typical curves shown are for a zero immersion depth. The side dip into a solder bath gives maximum scope for solder rise and hence the greatest wetting force. However, usually separate results are required for each end, or the component is multileaded, or the component cannot be dipped sideways because it is cylindrical. In (b) the wetting force is much less than half that in (a) because the meniscus runs out of wettable termination. Note that, in curve (c), not only is the wetting force greater but the resolution of the features on the curve is considerably better.

The force-time wetting curves obtained from the globule balance belong to the same family as those obtained from the wetting balance and therefore offer full interpretive compatibility with the quantitative methods standardised for the wetting balance and insertion components.

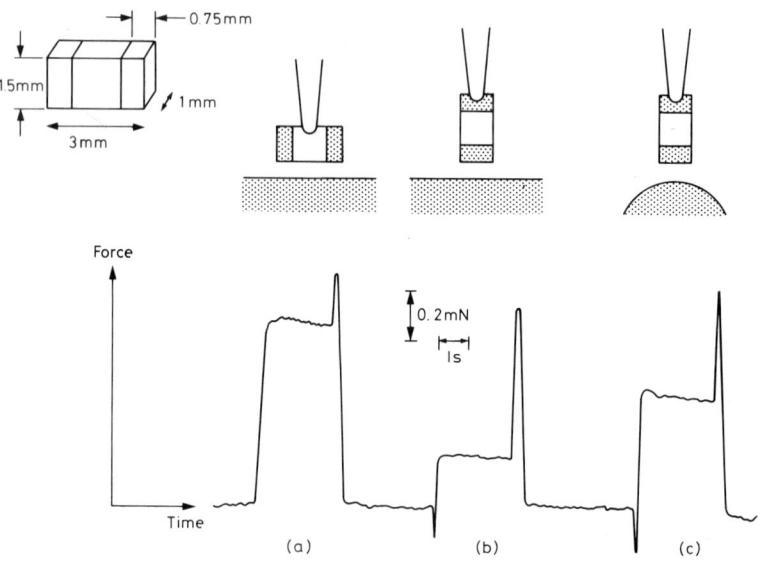

Fig. 10.44 A direct comparison of wetting balance curves obtained when a chip component is tested (a) sideways, (b) end-on and (c) with the globule balance.[386]

10.5.6 Rotary Dip Method

Every solder joint on a printed circuit assembly comprises a solder fillet between a component and a copper land on the board. Thus the solderability of the PCB, and the testing procedures used to assess it, are as crucial as that of the components. The rotary dip method[392] has been designed to test the solderability of flat surfaces such as PCBs. It is such that the specimen skims the flat surface of a bath of molten solder. The sample of board follows a circular path about a horizontal axis as shown in Figure 10.45(a). The time of contact between the specimen and the solder surface can be varied in three ways: by changing the length of the arm, by changing the speed of rotation and by interrupting the rotation while the specimen is in contact with the solder. The time of contact with the solder is measured using a needle whose tip is at the same radius on the rotating arm as the centre of the lower face of the specimen. The needle is not wetted by the solder and contacts the surface of the solder at some distance from the specimen so as not to disturb the wetting.

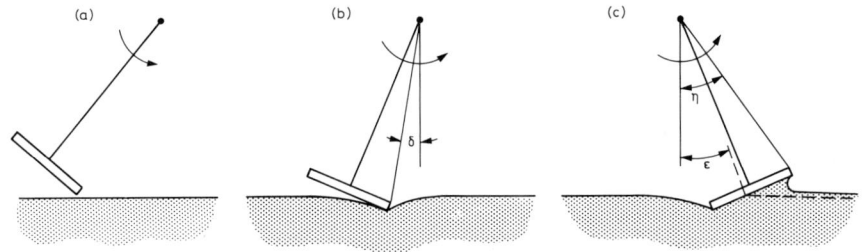

Fig. 10.45 Definition of parameters for the rotary dip method of solderability testing for PCBs.

The assessment of the solderability is by visual inspection after one rotation. The straightforward dip and inspect method is easier to execute, but the rotary dip method was developed to simulate nominally the sweeping motion of a solder wave over a board.

The significance of various parameters of the rotary dip test method has been discussed in some detail.[393] Clearly the radius of rotation of the ends of the specimen is different from that of its centre and consequently the contact times of the ends with the solder are longer. The contact time, for a given arm length and rotation speed, also depends on the particular surface nature of the board. Suppose, as is usual, that a point on the board, for example the leading edge of the board, is non-wetted at the moment of entry, but is fully wetted by the time it exits. At entry, the solder is depressed as shown in Figure 10.45(b) whilst on exit the solder is dragged out as shown in Figure 10.45(c). The total arc traversed by the leading edge is $\delta + \eta$ whilst that traversed by the firing needle is 2ε.

The ratio of the arc traversed, in contact with the molten solder, by the firing needle and that traversed by any part of the board can be calculated[393] using simple trigonometry and the elastica curve given in Equation (10.9) which defines the shape of the meniscus rise above the horizontal. It is found that, if the immersion is shallow, the deviation between the needle time and the real contact time can be considerable. As the immersion depth is increased, the discrepancy is reduced. As an example, for a non-wettable surface at an immersion depth of 1 mm, the contact time at its middle is only 0·85 times the needle time because of the re-entrant meniscus at the solder surface. Increasing the immersion depth to 2 mm, the contact time is 0·91 times the needle time. On the other hand, for a wetting specimen whose contact angle on entry is 180° but on exit is 0°, the contact time of its middle is 1·57 times the needle time when the immersion depth is 1 mm, reducing to a factor of 1·35 at 2 mm immersion. If the leading and trailing edges of the specimen are considered, the situation is even worse. The practical manifestation of this, however, is that bad boards are in contact with the solder for shorter times than good boards, resulting in an increase in the qualitative discrimination.

10.6 SURFACE OF STANDARD SOLDERABILITY

Because of the importance in quality assurance of solderability testing of components and printed circuit boards, a large number of test procedures have been developed. Each type of test provides one or more criteria upon which the solderability of a component (and hence the batch of which it is representative) can be accepted or rejected. In order to quantify the tests and enable direct comparisons to be made, standard reference surfaces of defined solderability are required. The solderability of such a surface must be tunable to fall within the range encountered in practice, such that components with a solderability worse than the standard are rejected while those better than the standard are accepted, for all test procedures.

For a given standard surface the work involves:

(i) defining the preparation procedure;
(ii) characterising it using surface analysis techniques;
(iii) correlating the surface chemistry with solderability;
(iv) assessing the reproducibility of the surface and the effect of storage;
(v) the development of a 'solderability index' as a criterion for acceptance or rejection.

A copper surface chemically passivated using benzotriazole (BTA) has been suggested[371] as providing samples of different but reproducible degrees of solderability. The importance of closely controlling the preparation, cleaning and rinsing of the copper to obtain reproducible solderability, has also been demonstrated.[394] X-ray photoelectron spectroscopy (XPS) shows that a prepared surface is covered with a very thin layer of Cu_2O. BTA is conveniently adsorbed on to this surface from acidic aqueous solutions and, for all immersion times and temperatures, peaks in the XPS spectrum corresponding to both Cu(I)-BTA and Cu(II)-BTA are observed.

For short times and low temperatures of immersion, the Cu(I)-BTA state predominates but as oxidation proceeds the ratio of the oxidation states Cu(II):Cu(I) increases, to an equilibrium value. As an example, the oxidation equilibrium of copper in 0·02M BTA at 60°C is reached in about two hours. The degree of oxidation, as determined by the immersion time in BTA, correlates quite well with the solderability of the surface,[395] as shown in Figure 10.46. Because of this correlation, BTA-passivated copper has the potential of providing a range of surfaces of standard solderability. However, great care has to be taken

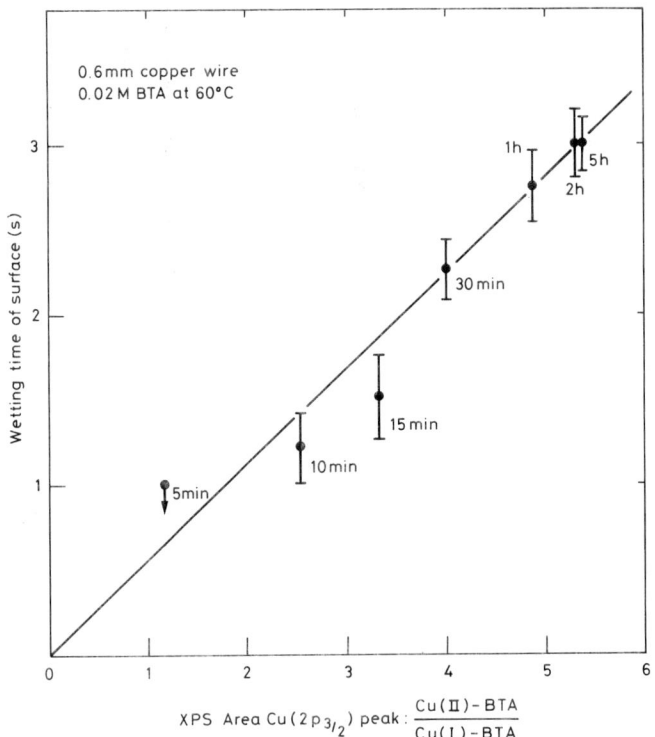

Fig. 10.46 BTA-passivated copper has the potential of providing a range of surfaces of standard solderability. The ratio of oxidation states Cu(II):Cu(I), measured by X-ray photoelectron spectroscopy, correlates very well with measured wetting times.

in the control of many parameters, notably the preparation of the basis copper, the ageing of the BTA solution and its temperatures.

Another surface that has been proposed as a standard of solderability, and indeed is already incorporated in the French standard, comprises copper that has been subjected to an oxidising treatment by immersion in a solution of sodium sulphide. The French standard stipulates an immersion time of 30(+0·5, −0·0) seconds, a temperature of 22·5±0·5°C and concentrations specified to 0·1%, so it is evident that much care needs to be taken to achieve reproducibility. Additionally, a specific source of copper is specified, comprising electroplated copper on sheet copper.

A problem of this procedure is the attainment of solutions of sodium sulphide with specified concentrations because the solid is hydrated and is very hydroscopic. The best method of preparing the solutions is probably to use anhydrous powder but this is very difficult to maintain in storage. Alternatively, solutions of approximate concentrations can be used if then carefully analysed, and the effect interpolated from known data.

10.7 MOVEMENT OF COMPONENTS DURING SOLDERING

When surface mounting components are being assembled by the solder paste and reflowing route rather than the wave soldering route, there is no physical retention of the component during the reflow process. Thus, when the solder is molten there is a possibility that the component will move with respect to the substrate. Under some circumstances the component may be preferentially pulled into better coincidence with its footprint by the liquid surface tension forces, but under other circumstances it may move away from its true position, or one or more legs may lift clear of the solder.[396] These effects are termed 'floating' or 'swimming' of the component on the molten solder, which terms refer to movement in a plane more or less parallel to the substrate. The component may even rotate about a horizontal axis to stand up on one of its edges; this is termed 'tombstoning'.

10.7.1 Floating and Swimming of Components

When the solder paste at all the terminations is molten and the component is effectively floating it will 'swim' into a position such that the total surface area of molten solder is minimised and hence the free energy of the system is minimised. In most cases this means that the component will self-align with the footprint if it has been placed askew, provided the solderability of the component and the substrate lands is good and the footprint is symmetrical with respect to the component.

For chip resistors and capacitors, with only two terminations, the effects of the solderability and land dimensions are more marked than with multi-termination components. Provided both ends have similar and correct dimensions and good solderability, the chip component will self-align from quite large skew angles and displacements in both x and y directions. The degree of 'swimming' depends on how much molten solder is present under the component terminations. If the initial component placement on the solder paste is too misaligned, a problem of solder bridging may be encountered.

Whilst the 'swimming' effect usually acts in a beneficial manner, under some circumstances, with chip components especially, the effect worsens the alignment. If the solderability is poor and inhomogeneous the component may

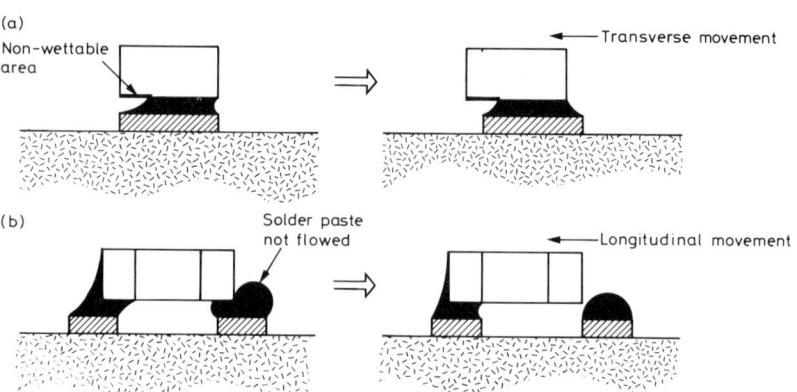

Fig. 10.47 Swimming of placed components: (a) A non-wetted area on a component termination may cause the component to swing out of alignment. (b) If wetting at both ends of a chip component is not simultaneous the component may move longitudinally and lose contact at the other end.

Solderability

swing from good alignment to poor alignment as shown, for example, in Figure 10.47(a). A non-wetted area on the component termination provides a thermodynamic advantage to the system if it is not aligned over the substrate land.

If the lands are small, as shown in Figure 10.47(b), and the solder at one end of the component wets a fraction of a second earlier than at the other end, the component may be pulled along its axis, thereby breaking contact with the solder at the other end, as shown in the figure.

10.7.2 Tombstoning of Components

When a chip capacitor or a chip resistor is reflow-soldered using a solder paste, one end sometimes lifts or flips off the substrate during the reflowing process. The component may completely stand up on its end surface, as shown in Figure 10.48, and in practice, all angles of incline are observed up to the vertical position. This phenomenon is known variously as the tombstone, Manhattan, Stonehenge or drawbridge effect, for fairly obvious reasons, and has been examined in detail.[397]

Fig. 10.48 Tombstoning of chip components.

In general, it can be said that the problem occurs most frequently when using vapour phase soldering and is more common with rectangular chip resistors than with other components. As is nearly always the case with problems encountered in soldering, tombstoning is a multi-parametric phenomenon and the degree of severity of the problem can be changed by adjustment of one or a combination of the following variables:

(i) the speed of heating;
(ii) the direction from which heat is applied;
(iii) the physical characteristics of the solder paste;

(iv) the predrying of the solder paste;
(v) the pre-heating temperature before soldering;
(vi) condensation of fluid between the component and substrate;
(vii) the wettability of component metallisations;
(viii) the dimensions of the solder lands of the footprint;
(ix) the shape and size of the metallisations on components.

Based on observations made during control tests, the explanation of the tombstone effect can be summarised as follows:

The thermal expansions during the early stages of heating up the assembly cause a decreasing, or even breaking, of contact between the component metallisation and the solder paste at one end of the component. The paste at this end does not, then, heat up as quickly as the paste at the other end because most of the heat is transported via the components. Thus a meniscus of molten solder forms at one end before the other and, under certain circumstances, the rotational moment of the surface tension force is greater than that of the weight of the component, causing the component end that is not in contact with molten solder to rise or move. This initial movement increases the thermal resistance of the free end still further, increasing the time taken for the paste there to melt.

This sequence of events explains why tombstoning occurs most frequently when using vapour phase soldering, in which a substantial part of the heat supplied to the paste is through the component. In infra-red or hot plate reflow soldering, the heating of the paste is mainly through the substrate, and the opposite ends of the component are heated more evenly.

That the surface tension is sufficient to cause lifting of a chip component can be demonstrated with the aid of a simple model.[397] Referring to Figure 10.49, it is assumed that, as a first approximation, the meniscus is a straight line at all angles of the lifting component. Three moments are acting on the component to rotate it about its line of contact with the land on the substrate:

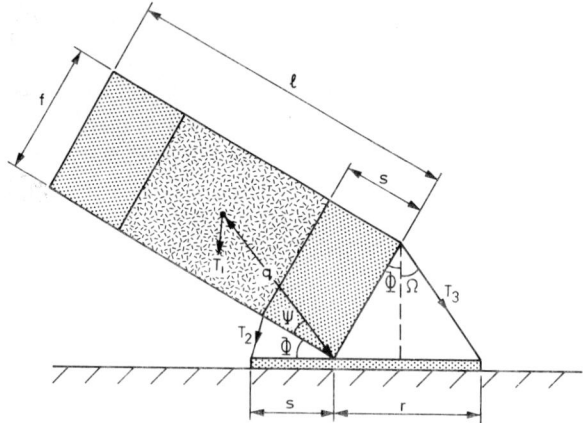

Fig. 10.49 Definition of parameters for the calculation of the tombstone effect on chip components.

(a) Gravitational moment, downwards:

$$T_1 = Mgq \cos(\Phi + \Psi)$$

where M is the mass of the component per unit width, g is the acceleration due to gravity, q is the distance between the centre of gravity of the component and the point of rotation, $q = \frac{1}{2}\sqrt{(\ell^2 + f^2)}$ where ℓ and f are, respectively, the length and the height of the component. The angles Φ and Ψ are respectively the angle of rise of the component from the horizontal and $\tan^{-1}(f/\ell)$.

(b) Downwards moment, due to the surface tension of the liquid solder under the component:

$$T_2 = \gamma s \cos \frac{\Phi}{2}$$

where γ is the surface tension of the liquid solder and s is the width of the underside metallisation of the component.

(c) Upwards moment, due to the surface tension of the liquid solder meniscus:

$$T_3 = \gamma f \sin(\Phi + \Omega)$$

where the angle Ω can be seen from Figure 10.49 to be

$$\Omega = \tan^{-1}\left(\frac{r - f \sin\Phi}{f \cos\Phi}\right)$$

and r is the protruding length of the solderable land on the substrate.

In Figure 10.50 have been calculated the downward, restoring moment $(T_1 + T_2)$ and the upward, tombstoning moment (T_3) as a function of the tilting angle Φ, for an R0805 type chip component. The downward moment has been calculated for an underside metallisation width s of 0·3 mm, 0·15 mm and 0 mm, this last case being equivalent either to metallisation on the end face only or to no substrate land extending under the component, so that $T_2 \to 0$. The upward moment has been calculated for three different land extensions, with r/f values of 1·0, 0·75 and 0·5. Similar sets of curves are produced when considering other sizes of chip components, but, because of the small mass of the R0805, the effect is most severe in this case.

It can be seen from Figure 10.50 that the balance between the downward and the upward tilting moments is quite fine. If, when $\Phi = 0$, the upward moment is greater than the downward moment, the component will begin to rise if its other end is free, i.e., has not been wet by the solder paste. Furthermore, whereas the downward moments decrease as Φ increases, the upward moments increase; hence the situation is unstable and it is preferable for Φ to increase and the component to rise. Thus, once a component begins to lift it will always tombstone, provided the solder at the pivoting end remains molten for a long enough time, and provided there is some means whereby the molten solder in the end fillet can channel away, usually by moving underneath the pivot point.

In order to overcome the problem of tombstoning, the downward moment can be increased by increasing s, the underside metallisation and under-component land size. The upward moment can be decreased by decreasing the length of the protruding solder land, r, thereby decreasing the effective arm along which the surface tension force acts. Thus, the solution to the problem lies in modifications to the component metallisation and the layout of the printed circuit board.

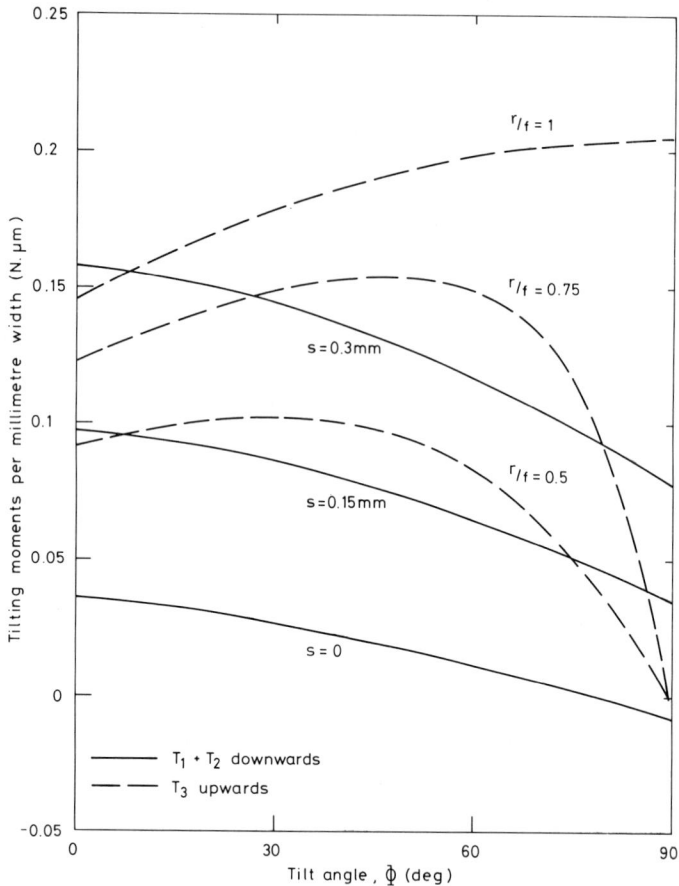

Fig. 10.50 Calculations of the downwards (T_1+T_2), restoring moment and the upwards (T_3), tombstoning moment, relating to R0805 chip components that have wet at one end only. The parameters s, r and f are defined in Figure 10.49.

These broad conclusions, drawn from the simple two-dimensional model, are in general agreement with observations made in control tests. Data are reported[397] on the degree of tombstoning of R1206 chip resistors with four configurations of substrate lands, as shown in Figure 10.51. The results show that the occurrence of tombstoning increases as the size of the solder land increases, as predicted, since the downward moments T_1 and T_2 remain constant whilst the upward moment T_3 increases as r increases. A minimum of tombstones occurs when the protruding land, r, is between 0·2 and 0·4 mm. In view of the tolerances of component length and placement, this requirement is difficult to fulfil.

Unfortunately, also, whilst tombstoning may be alleviated by using smaller lands, the faster wetted end is pulled towards the centre of the land, thereby possibly breaking contact at the other, as yet unwetted, end of the component, as shown in Figure 10.48(b).

Besides altering the land sizes, tombstoning can be alleviated by increasing the underside metallisation on the component, thereby increasing the downward

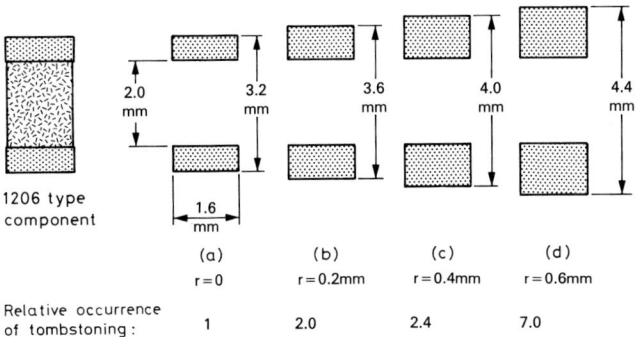

Fig. 10.51 Four configurations of a footprint for a 1206 chip component used for the tombstoning tests, showing the relative occurrence of tombstoning encountered.[397]

moment T_2. This has been demonstrated using resistors with specially made terminations and by mounting standard chip resistors upside down on the circuit board, since the metallisation on the resistor side is normally larger than on the underside. It is found that underside component metallisation and the underside land overlap, s, should be at least 0·3 mm.[397]

In addition to these controls of component and footprint geometries, some further recommendations have been made from observations made in controlled tests, for decreasing the occurrence of tombstoning:

(i) Ensure sufficient predrying of the solder paste before soldering.
(ii) Use a low oxide solder paste with an oxygen content less than 0·01% by mass.
(iii) Avoid the use of tin-lead under the solder resist.
(iv) Prevent the entrapment of any fluid under the chip component by selecting the gap between the component and the board to be as large as feasible.
(v) Specify equal solderability and equal geometry of both ends of chip components. This problem arises because of sequential, rather than simultaneous, production of the terminations at each end of some chip components.

Chapter 11

THE SOLDER FILLET

11.1 INTRODUCTION

This chapter deals with the properties of the solder fillets on a surface mounted board, once the solder has solidified. The mechanical strength of the solder fillet in service depends on its shape and size, and its metallurgical microstructure. In the solid state, diffusion processes continue to alter the microstructure but at rates very much reduced from those associated with molten solder.

Low cycle thermal fatigue is the prime cause of solder joint failure in surface mounted assemblies, the cyclic strain arising from temperature gradients induced during power-up and power-down. A substantial amount of investigatory work has been reported, aimed to predict thermal fatigue behaviour, and this represents the major part of this chapter.

11.2 METALLURGY OF THE SOLDER

11.2.1 Tin

Tin is the principal component of almost all solders used in the assembly of electronic components. Its ubiquity arises from an adventitious combination of properties, not least of which is its ability to wet and spread on a wide range of substrates using mild fluxes.

Tin does, however, have some unusual mechanical properties which tend to mitigate against its use in an unalloyed form. These characteristics arise because its crystal structure is body-centred tetragonal and therefore anisotropic. Consequently, the thermal expansion of tin is also anisotropic so that repeated heating and cooling causes plastic deformation and eventual cracking at grain boundaries. This effect has been observed in thermal cycling over a range as small as 30–75°C. Thus thermal fatigue can be induced in tin, or the tin-rich phases of solder alloys, even when no external mechanical strain is imposed on the solder by the thermal cycle.

Deformations of this type can be reduced by alloying with elements that, for instance, give rise to hard, second phase particles in the tin matrix. The strengthening effects of alloying additions, giving rise to an increase in the hardness of the metal, can be produced either by additions of soluble elements (Figure 11.1) or of elements which form particles (Figure 11.2).[398] Hardness is measured by indentation using a small hard tool (usually a pyramid), pressed into

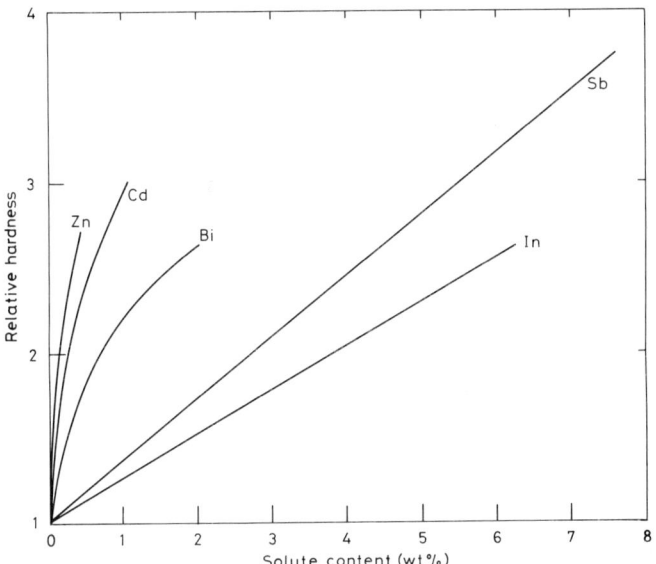

Fig. 11.1 The hardness of tin is increased by additions of soluble elements.[398]

the surface by a standard load, and the size of the indentation produced is then measured.

11.2.2 Tin-lead Alloys

Tin is normally alloyed with lead to produce solders, which find applications over almost the entire Sn:Pb compositional range. For soldering electronic assemblies, an alloy containing 60–63% by mass of tin, which is close to the eutectic composition, is most commonly used. The melting point of pure tin (232°C) is reduced by additions of lead to a minimum of 183°C at the eutectic composition of 61·9% Sn. This lower temperature is more compatible with the thermal properties of electronic component packages. In addition, the cost of tin is some 20 times higher than that of lead, and consequently the higher tin alloys find only limited application. If the tin content is too low, however, besides the melting point increasing, there is a general reduction in the wetting properties of the alloy.

Some physical properties of tin-lead alloys (density, electrical conductivity, thermal conductivity and temperature coefficient of expansion) over the compositional range are given in Table 11.1. As expected, the conductivities increase with increasing tin content whereas the densities increase with increasing lead content. Also given are the corresponding values for some alloying elements found in solders. The effect of such additions on the properties of the alloy can be judged readily.

A significant property of these low-melting point alloys is that during service at room temperature or above they are functioning at more than ⅔ melting temperature in kelvin, and consequently have a tendency to undergo structural changes. The low mechanical strengths of solders, especially creep strength, are a result of this structural instability.

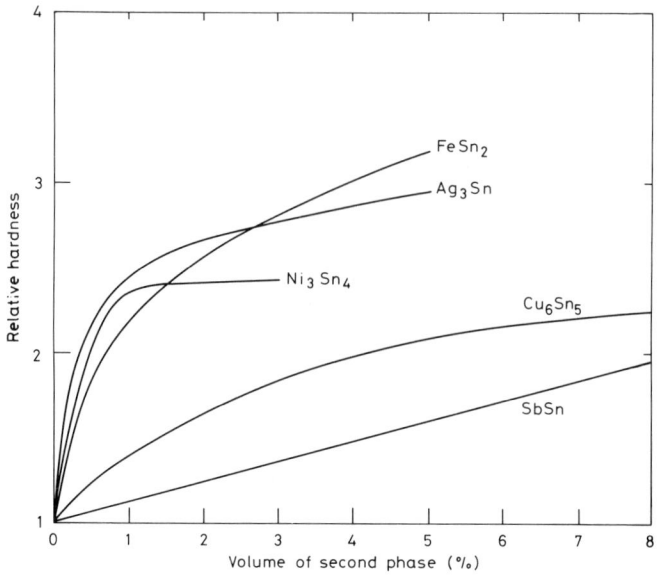

Fig. 11.2 The hardness of tin is increased by second-phase particles.[398]

Table 11.1

Some Physical Properties of Tin-lead Alloys and Solder Alloying Elements

	Density at 25°C (g.cm^{-3})	Electrical Resistivity at 25°C ($\mu\Omega.m$)	TCE ($10^{-6}.K^{-1}$)	Thermal Conductivity at 25°C ($J.m^{-1}.s^{-1}.K^{-1}$)
Tin (Sn)	7·29	0·115	20	66·8
63%Sn:37%Pb	8·46	0·148		
60%Sn:40%Pb	8·52	0·150	25	51·0
50%Sn:50%Pb	8·90	0·158		
40%Sn:60%Pb	9·28	0·171		
Lead (Pb)	11·34	0·206	29	35·3
Antimony (Sb)	6·68	0·390	9	24·4
Bismuth (Bi)	9·80	1·068	13	7·9
Copper (Cu)	8·92	0·017	16·6	401
Indium (In)	7·30	0·084		81·8
Silver (Ag)	10·52	0·016	19	429

11.2.3 Strength Properties of Solder

Solid materials can be subjected to considerable deformations and then regain their original shape when the constraints are removed. Such complete recovery is observed up to certain levels of stress, characterising the *elastic* range, over which there is a linear relationship between the deformation and the stress. Beyond the elastic range, the material acquires a permanent set and a *plastic* deformation remains when the stress is removed. Figure 11.3 shows a stress-strain curve of a ductile material such as solder. The linear elastic range terminates at a stress σ_e, the elastic limit, while gross plastic yielding occurs above the yield stress σ_y. If at a certain point P, the stress is removed, the material would behave elastically along the line PA. The total strain at P consists of the irreversible plastic component OA and the elastic contribution AB. Young's modulus E, of the material, is given by tan θ which represents the slope of the line in the elastic range, being the ratio of tensile stress to tensile strain in uniaxial tension.

When in uniaxial tension, a ductile material such as solder will elongate considerably with a consequent reduction in cross-sectional area. Poisson's ratio for a given material is the ratio of lateral contraction to longitudinal extension in uniaxial tension. The stress at any point is the load divided by the reducing area of cross-section. For solder, a considerable reduction in area occurs before failure. If the applied load, rather than the stress, is plotted against the strain, the curve passes through a definite maximum, as shown in Figure 11.4, and the material will fracture at a load less than the maximum. The maximum load divided by the cross-section area of the undeformed test piece is referred to as the ultimate tensile strength, UTS.

The mechanical behaviour of materials is generally investigated by tests each of which is characterised by the use of a specific stress system, such as tension, compression, torsion, bending, or a combination of these, as well as by the time-

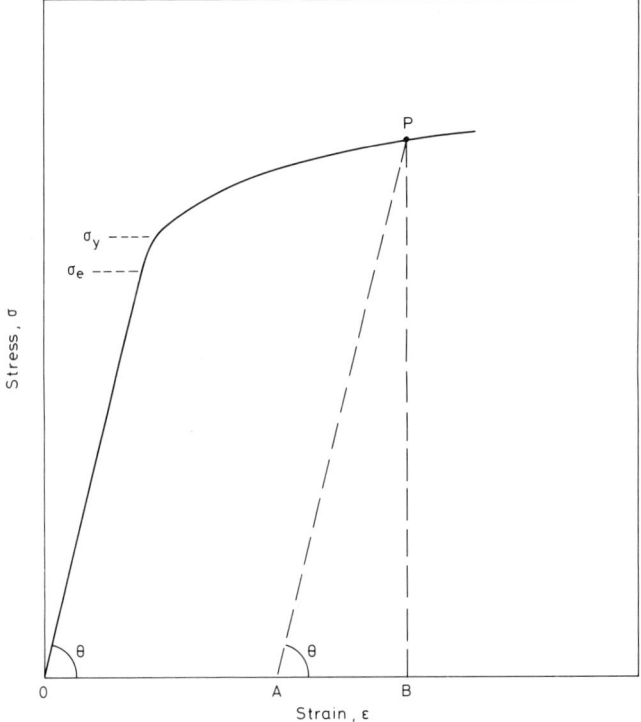

Fig. 11.3 A typical stress-strain curve for a ductile material such as solder; σ_e is the elastic limit, σ_y the yield stress and tan θ is Young's modulus.

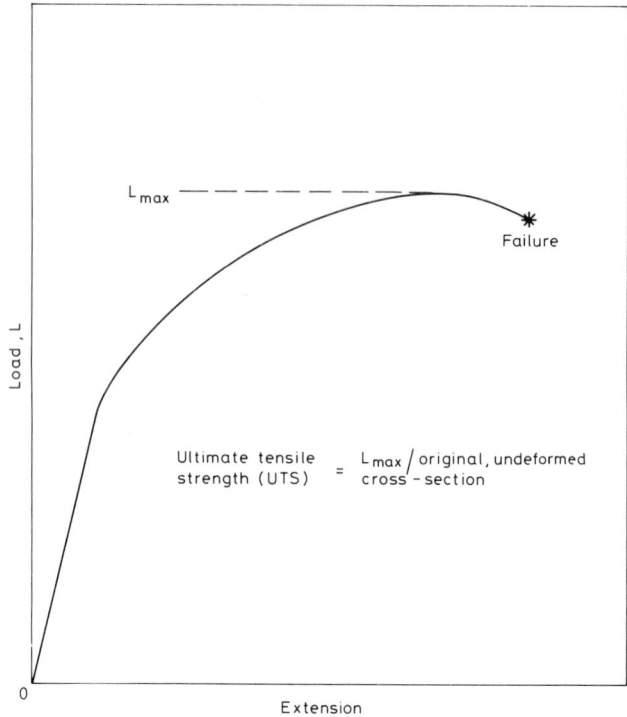

Fig. 11.4 If the applied load, rather than the stress, is plotted against the strain, the curve passes through a maximum which defines the UTS.

rate of application of the stresses and the total time under load. Thus, for example, if the time-rate is zero and the material is being subjected to constant stress for long periods, the conditions are those used to study creep. Prolonged application of periodically varying stresses is employed in the investigation of fatigue.

When a solder joint is stressed by some mechanically or thermally induced deformation it yields under both tension and shear. Within a small volume of deformed material around a chosen point, in which the stresses may be considered to be uniform, the state of stress may be described by three mutually perpendicular tensile (or compressive) stresses, σ_{xx}, σ_{yy} and σ_{zz} and six shear stresses $\sigma_{xy} = \sigma_{yx}$, $\sigma_{yz} = \sigma_{zy}$ and $\sigma_{zx} = \sigma_{xz}$, as shown in Figure 11.5. The first subscript specifies the direction of the normal to the plane where the stress is acting and the second the direction of the force which gives rise to the stress.

Resulting from the tensile and shear stresses, both tensile and shear strains arise. The tensile strain ε_{xx} resulting from the tensile stress σ_{xx}, for example, is measured as the elongation δ_x in the x direction divided by the initial length x_o of the specimen or workpiece, in that direction. Thus $\varepsilon_{xx} = \delta_x/x_o$.

Referring to a cube of material, shown in Figure 11.6(a), the shear strain ε_{zy}, for example, due to the shear stress σ_{zy} that has displaced the upper surface of the cube along the y direction, is defined as the tangent of the shear angle ψ. The

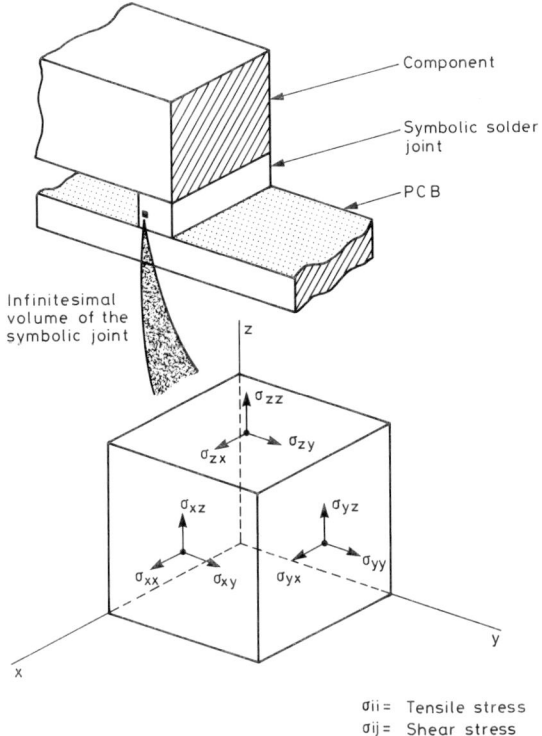

σ_{ii} = Tensile stress
σ_{ij} = Shear stress

Fig. 11.5 Definition of the three tensile stresses σ_{ii} and the six shear stresses, σ_{ij}. The first subscript specifies the direction of the normal to the plane where the stress is acting, and the second the direction of the force which gives rise to the stress.

shear modulus (or rigidity modulus) μ is the ratio of the shear stress to the shear angle:

$$\sigma = \mu \tan \psi$$

For isotropic systems the rigidity modulus μ, Young's modulus E, and Poisson's ratio v, are inter-related:

$$\mu = \frac{E}{2(1+v)}$$

The deformation shown in Figure 11.6(a) is characterised by the fact that any plane of the cube parallel to the z axis is displaced rigidly along a straight line, in this case the y direction, and the separation of any pair of such 'shearing planes' remains constant, i.e., the cube does not extend or contract along the z axis. This is referred to as 'simple shear'. However, as shown in Figure 11.6(b), the same deformed shape can result by displacing the cube in both the y and z directions simultaneously. Since σ_{yz} is equal to σ_{zy}, it can be seen that this is the mode of deformation actually experienced in practical situations. This model is known as 'pure' shear. There are three principal axes that do not change their directions in

384 The Solder Fillet

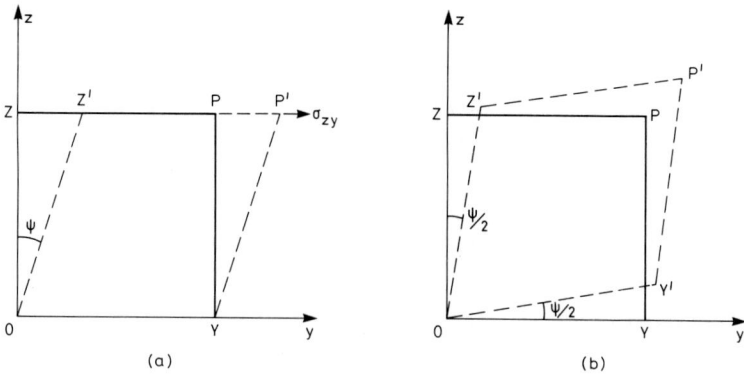

Fig. 11.6 Defining the shear angle ψ of a cube of material, (a) in simple shear and (b) in pure shear.

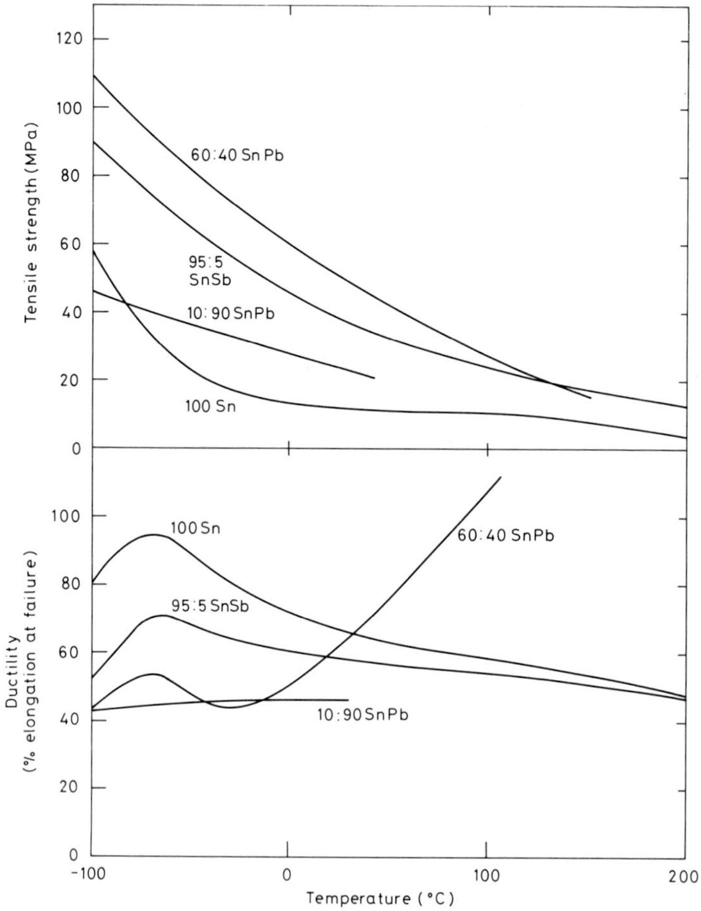

Fig. 11.7 The tensile strength and ductility of some solder alloys over their working temperature range.[399]

the course of the deformation, ZY, OP and the x axis. This constancy of the principal axes is the main feature differentiating pure shear from simple shear.

The mechanical strength properties of solder are very low in terms of commonly used engineering materials. However, in electronic assembly, solder is always used in conjunction with two surfaces to be joined and hence the strength of the bulk alloy is not entirely relevant to the fitness for purpose required. The strength of a solder joint is very dependent on the type of joint and the nature of the materials being joined. Thus, test data on lap joints or plug and socket joints in shear, or bulk solder in tension give only limited indications of how a joint will behave between a component and a circuit board.

Another characteristic of almost all tin alloys is their relatively fast solid-state reactions with elements such as copper and gold, found on the terminations of electronic components, giving rise to intermetallic compounds. These interfacial layers are often brittle and therefore reduce mechanical strengths, notably the impact properties.

The mechanical properties of SnPb solder and solder joints of various configurations are scattered through the literature[399, 400] and it is not unusual to find a relatively wide range of values for a particular property. The tensile strength, modulus of elasticity, shear strength, Poisson's ratio, creep strength, fracture toughness, fatigue properties, etc. have all been subject to investigation as a function of, for example, the tin-lead ratio, temperature, frequency, strain rate and time. The influence of temperature on the tensile strength of some bulk solders is shown in Figure 11.7. All alloys show a decrease in strength as the temperature rises from about $-100°C$. The ductility of near-eutectic tin-lead solders increases as the temperature rises in the range $0-100°C$, whilst for high-tin solders and high-lead solders the ductility decreases slowly over that temperature range. These temperature variant effects arise from the different dominant physical mechanisms occurring when a stress is applied. The deformation behaviour of solder at different temperature ranges is summarised in Figure 11.8.

The influence of the rate of stress ramping during testing of both the tensile and the shear properties of solder alloys has been investigated and it is found that both the tensile and shear strengths increase with the speed of testing.[399] This effect on the shear strength is illustrated in Figure 11.9 as a function of test temperature and in Figure 11.10 for several solder alloy compositions.

The tensile properties of solder also depend crucially on the microstructure generated as the specimen or the joint cools.[401, 402] By adjusting the casting temperature and the cooling rate, the grain size can be made to vary considerably and it has been demonstrated, for example, on one composition of solder that the yield strength can vary by a factor of two, proportionally to the square of the mean grain size diameter, in the range $1-2$ μm.

The creep rupture strength (the maximum nominal stress that can be applied for a given time and temperature without causing rupture) and the stress relaxation (the gradual reduction of stresses in a joint having an initial fixed strain) in solder are very complex since the alloy properties are very time and temperature sensitive. Usually, creep rupture strength is measured by testing under constant stress, whilst stress relaxation is investigated by testing under constant strain. The creep rupture strength of solders has been studied extensively.[403-405] For example, in Figure 11.11 are data from creep measurements on bulk samples of 60:40 SnPb solder at $20°C$ and $100°C$; strength

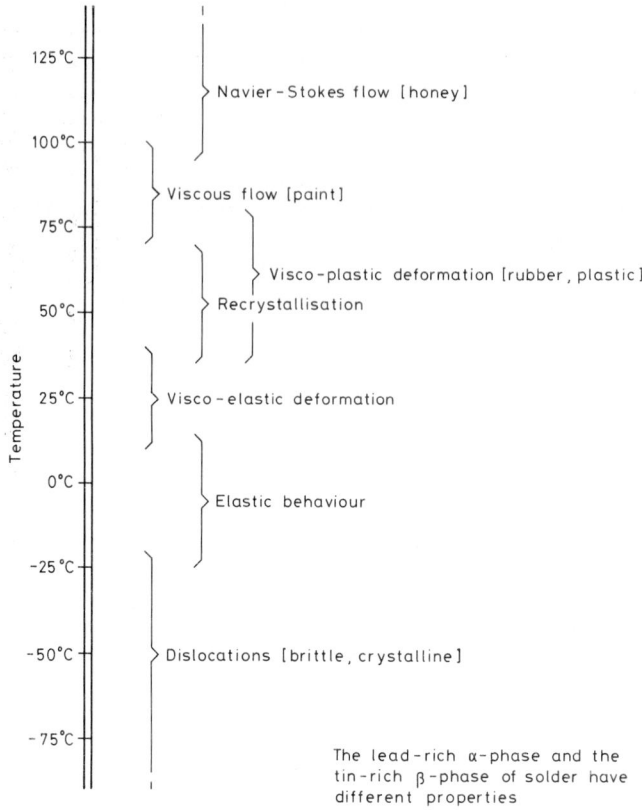

Fig. 11.8 The deformation behaviour of solder at different temperature ranges.

values drop quite significantly between the two temperatures. The creep rupture strength of solder joints shows similar behaviour and has also been shown to change with time after the making of the joint because of diffusion mechanisms. Some data[406] are shown in Figure 11.12, for accelerated ageing at 170°C, the testing however again being performed at 20°C and 100°C. The complexity of the diffusion mechanisms and concomitant microstructural changes occurring during ageing are illustrated here; at 20°C the creep rupture strength deteriorates with ageing whereas at 100°C it improves.

11.2.4 The Phase Diagram

The phase diagram of the tin-lead binary alloy system is shown in Figure 11.13. Such a diagram enables the determination of the composition of separate microstructural phases that can co-exist in a solder fillet with any given nominal tin-lead ratio. The phase diagram describes what is thermodynamically possible at equilibrium; it gives no indication as to the kinetics of attainment of the equilibrium microstructure.

The composition scale in the diagram is calibrated in percentage of tin by mass, corresponding to specified solder compositions. The phase diagram calibrated in

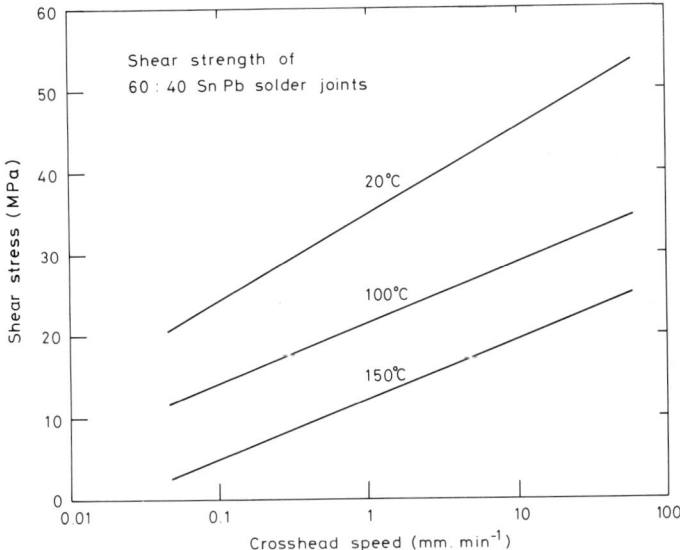

Fig. 11.9 Both the tensile and shear strengths of solder alloys and solder joints increase with the speed of testing and decrease in the rising temperature, as illustrated here for 60:40 SnPb solder joints in shear.[399]

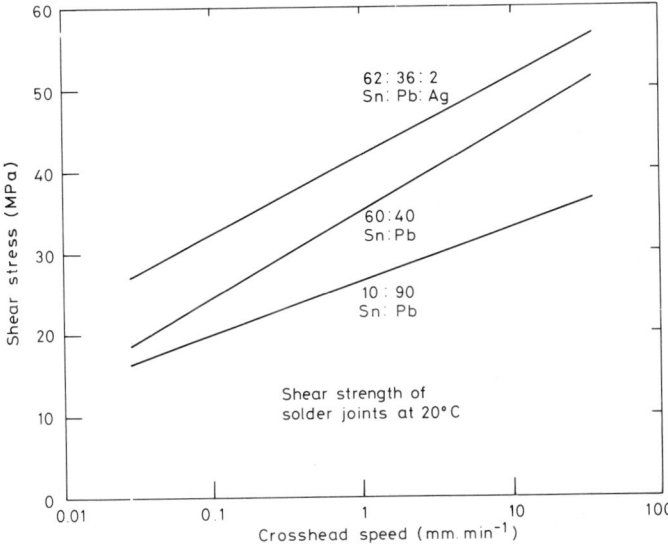

Fig. 11.10 The shear strength of some solder alloys at 20°C as a function of the speed of testing.[399]

atomic percentage was shown in Figure 6.11. Because the atomic weight of lead is greater than that of tin, the atomic percentage of tin is greater than its percentage by mass. In a SnPb binary alloy:

$$at\%Sn = 100/(0\cdot 43 + \frac{57}{wt\%Sn})$$

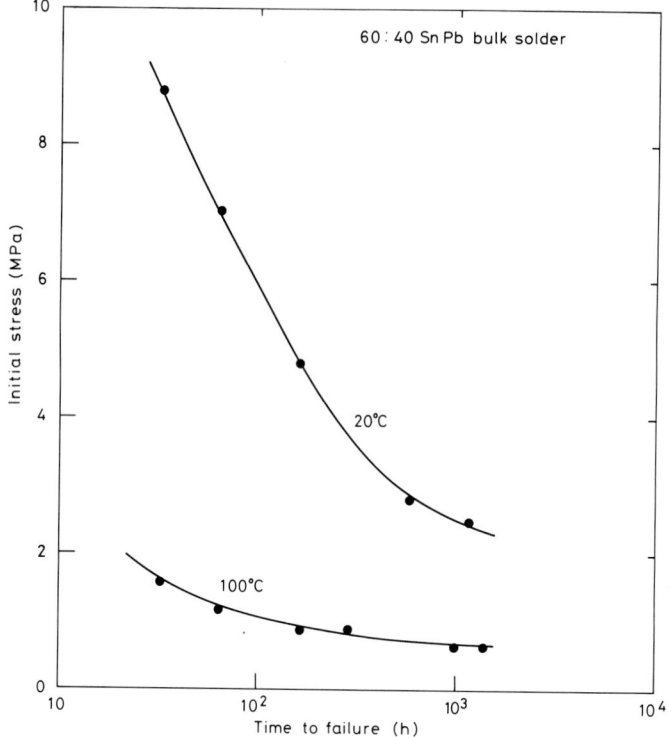

Fig. 11.11 Creep measurements on 60:40 SnPb solder at 20°C and 100°C.[404]

In Figure 11.13, point A is the melting point of pure lead (327°C) and point B is the melting point of pure tin (232°C). The line ACB joining these points across the composition range is called the liquidus, above which, in an equilibrium situation, only liquid exists. The point C is the lowest attainable melting point (183°C) which occurs at the eutectic composition of 61·9% Sn (74·0 at%Sn).

The line ADEB is called the solidus, below which only solid exists. In the two intermediate triangular regions, both solid and liquid exist together in equilibrium. (This is sometimes termed the pasty region.)

Consider, by way of example, a solder alloy of composition 40 Sn:60 Pb cooling from the liquid state. The liquid is homogeneous. When the temperature falls below 230°C (point a), liquid will only continue to exist in an increasingly tin-rich composition, defined by the part of the liquidus aC. Therefore at point a, crystals of lead begin to solidify within the liquid. However, these are not pure lead because tin is soluble in lead and the composition of the solidifying crystals is that given at point b. As cooling continues the liquid becomes more tin-rich following the line aC and the concomitant solidifying lead crystals absorb more tin, following the line bD.

When the temperature has fallen to 183°C, the liquid has attained the eutectic composition of point C and the lead rich crystals have the composition of point D. Crystallisation now continues without any reduction in temperature, during which time the liquid composition remains at C, whilst solidifying out are

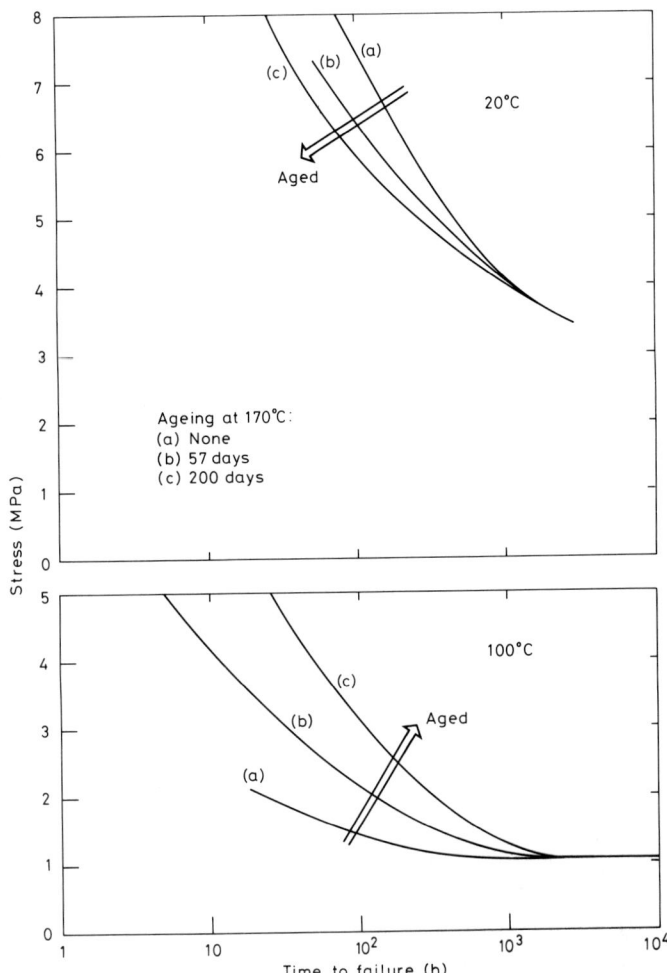

Fig. 11.12 Creep behaviour of solder joints as they are aged at 170°C. When tested at 20°C the creep rupture strength deteriorates with ageing, but when tested at 100°C ageing improves performance.[406]

more lead-rich crystals of composition D plus tin-rich crystals of composition E. These crystals, solidifying at the eutectic temperature, are very fine compared with the previously solidifed primary, lead-rich crystals. Below 183°C the solder is completely solid. Solidification has therefore occurred over a temperature range 230–183°C for this alloy. The microstructure described above is illustrated in Figure 11.14.

If the initial liquid solder composition has a high tin content between points C and B, then, as solidification occurs, the primary crystals are tin-rich.

If the solder composition is exactly that of the eutectic, there is not a temperature range of solidification but a single solidifying point of 183°C. The microstructure contains no primary crystals but only the very fine dispersion of lead-rich crystals of composition D and tin rich crystals of composition E.

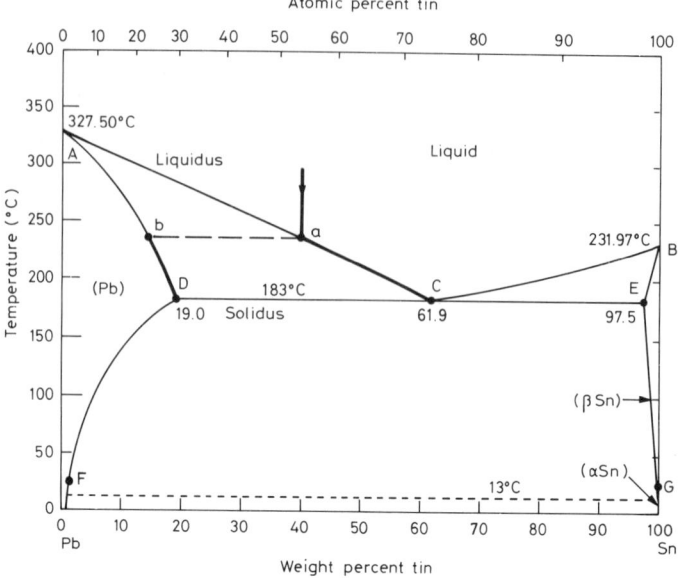

Fig. 11.13 The phase diagram of the tin-lead binary system.[221]

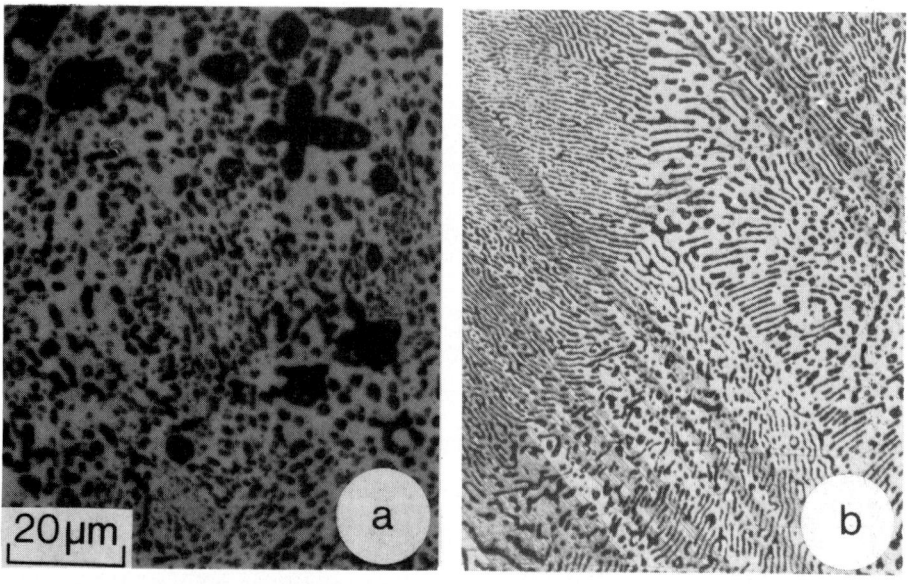

Fig. 11.14 Typical microstructures of tin-lead solder showing (a) the lead-rich primary crystals and (b) the lamellae structure of the eutectic composition.

11.2.5 Diffusion Reactions

Besides the basic tin-lead microstructure, the strength of soldered joints is also dependent on the degree to which diffusion of the alloying species, or of impurities, or of elements leached from the solderable surfaces can occur. These diffusion reactions occur both when the solder is in the liquid state and after solidification. There are three categories of diffusion reactions:

1 *Leaching*, refers to the dissolution of the substrate material into the liquid solder during soldering with a consequent change in the strength properties both of the solidified solder and the dissolved interface. This problem has been considered in the chapter on pastes (Section 6.4) where the choice of alloy can affect the leaching, and in the chapter on solderability (Section 10.4.7) where leaching and the consequent formation of an intermetallic layer are seen to influence wetting properties.

The degree of leaching affects the strength properties of the joint. The effects of soldering time on the strength of joints made on gold-palladium substrates by vapour phase soldering have been investigated. The joint strength decreases with longer soldering times due to dissolution of gold and palladium in the solder and subsequent poor wetting and poor adherence of the solder to the underlying film. The effect was shown to be avoided by use of a nickel barrier layer. Similarly, silver is leached from silver-palladium surfaces on chip component terminations, reducing the strength of the joints and eventually exposing unsolderable ceramic surfaces, so that no joint at all is made on solidification.

2 *Ageing*, refers to a solid state diffusion occurring after soldering, between the solder and the substrate, causing a thickening of the intermetallic layer and a change in the compositional profile of the elements across the interface. The phenomenon of ageing was considered in detail in the chapter on solderability (Sections 10.4.2 and 10.4.9) since the intermetallic layer at the interface between the substrate and the solderable coating is crucial to the wetting properties. The intermetallic layer is also crucial to the strength properties of the joint. Figure 11.15 shows how the strength of solder joints on gold-palladium declines during ageing after the soldering.[407] In this figure, below the dashed line fracture occurs mostly at the intermetallic layer, but above the dashed line fracture occurs in the bulk solder of the joint.

Additionally, SnPb solders experience a loss in strength during storage after solidification. This ageing, which can result in a 20% loss of strength, is caused by the precipitation of secondary tin from the saturated lead-rich phase. The presence of antimony retards this precipitation and thus helps to retain the solder strength during storage.

3 *Coarsening*, an ageing phenomenon caused by solid state diffusion, but referring specifically to changes (usually growth) of the scale of the morphology of the metallurgical microstructure. This mechanism is driven by the thermodynamic requirements to minimise the total free energy of the system, which leads to a minimising of the interfacial area of the separate phases. Since the surface/volume ratio of a finely dispersed microstructure is larger than that of a coarse microstructure, there is a tendency for the phases to coarsen, limited by temperature controlled diffusion of the species.

After solidification, the diffusion mechanisms that enable the equilibrium

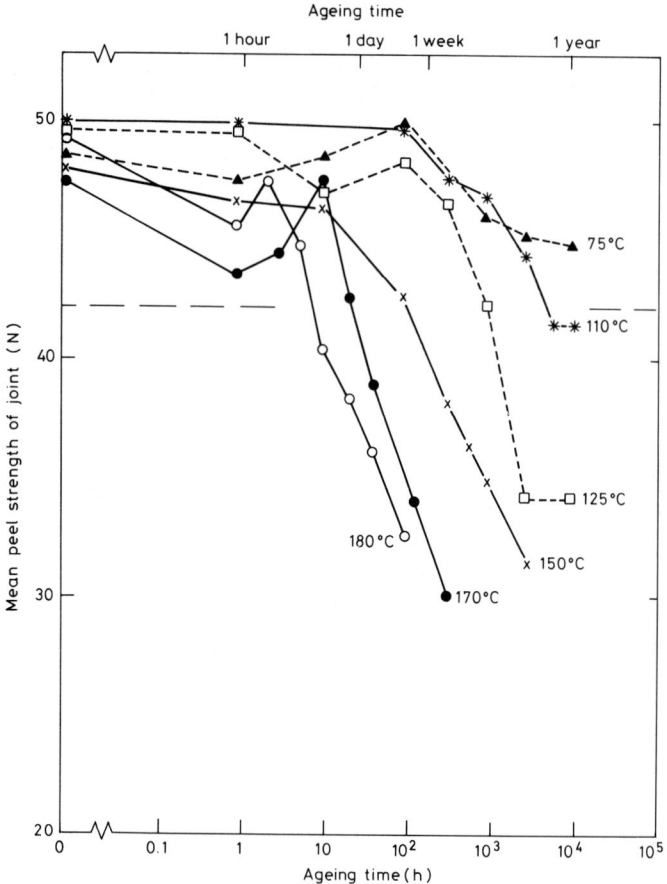

Fig. 11.15 Solder joint strength as a function of ageing time at different temperatures. These data are for gold-palladium metallisations. Above the dashed line, failure occurs mostly in the bulk solder, below the line it occurs mostly at the intermetallic layer.[407]

conditions defined by the phase diagram to be attained become greatly slowed. However, if cooling after solidification takes place sufficiently slowly, the compositions of the lead-rich and tin-rich regions alter, governed by the composition lines DF and EG respectively in Figure 11.13. The lead-rich crystals at solidification have the composition of point D, but at room temperature lead cannot contain more than 1·9 wt% of tin, defined by point F, so that tin precipitates out of the lead rich phase and similarly lead precipitates out of the tin-rich phase. Generally, this can only be observed in the larger primary crystals which, if cooling is sufficiently slow, are found to contain small included crystals of a different composition.

If cooling after solidification is fast, the microstructure extant at the solidifying temperature is frozen in. In these cases, changes occurring over a long time by solid state diffusion mechanisms will be more marked, because the initial state is more distant from the equilibrium, phase diagram state. Nevertheless, given sufficient time, a consistent microstructure will always be attained.

11.2.6 Cooling Curves

The description just given above is slightly simplified because when the solidification temperature of a particular composition is reached, during cooling, precipitation of that phase is not spontaneous, but requires some activation energy which is usually supplied thermally. This effect is manifested as an undercooling of the liquid until such time that the precipitating phase is nucleated, whence a return to the solidifying temperature.

The phenomenon of undercooling is seen when the temperature of a melt is monitored as it is losing heat at a constant rate. Such a cooling curve is shown in Figure 11.16. Provided no solidification occurs, the temperature of the cast decreases at a constant rate. During solidification, heat is liberated (the latent heat of solidification), causing a slowing of the cooling. However, in order to nucleate the solidification of crystals the temperature has to fall slightly below the solidification temperature, as shown,[408] both for the primary crystals and for the full solidification at 183°C.

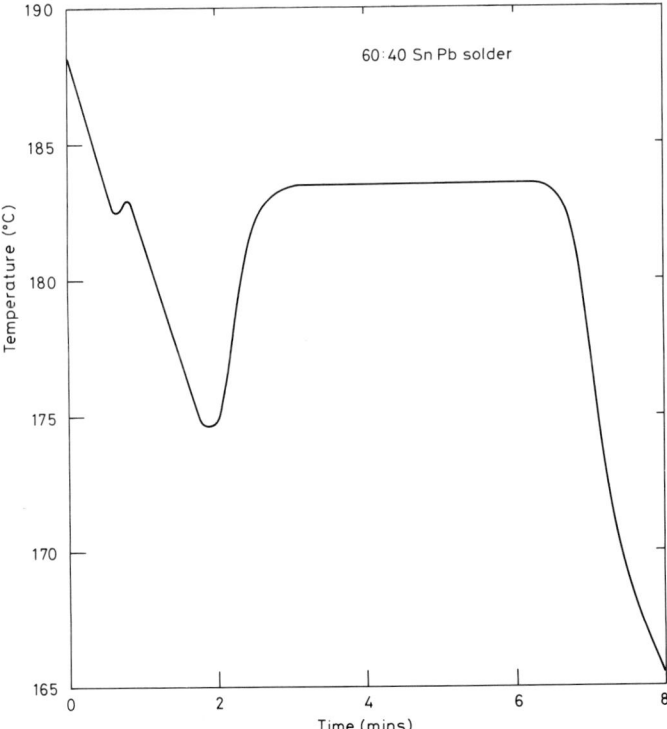

Fig. 11.16 A partial cooling curve for 60:40 SnPb solder, showing the undercooling required to nucleate the solidification of the primary crystals and the eutectic structure.[408]

The undercooling necessary to nucleate the eutectic structure is quite considerable (up to 8°C) and this is the reason why, in order to obtain a fully eutectic microstructure with no lead-rich primary crystallites present, the composition has to be slightly enhanced in tin. Thus, 63%Sn37%Pb rather than 61·9%Sn38·1%Pb is very often referred to as the eutectic solder. This loosely means that when the 63%Sn37%Pb is cooled and solidified in a solder joint a fully eutectic microstructure is obtained.

11.2.7 Alloying and Impurity Elements

Because it is economically unfeasible to use very high purity tin and lead to produce solder alloys, some impurity elements are always present. In addition to impurities, some alloys contain controlled additions of one or more metallic species in order to improve some aspect of performance.

The most notable addition used in solder alloys for surface mounting assembly is silver, to alleviate problems of excessive dissolution of silver from the solderable terminations of chip capacitors and resistors and ceramic chip carriers, as shown in Figure 6.9.

Mention was made of solder compositions when discussing pastes (Section 6.4) and also the effect of impurities of wettability (Section 10.3.2). Impurities and alloying additions also have an effect on the mechanical properties of surface

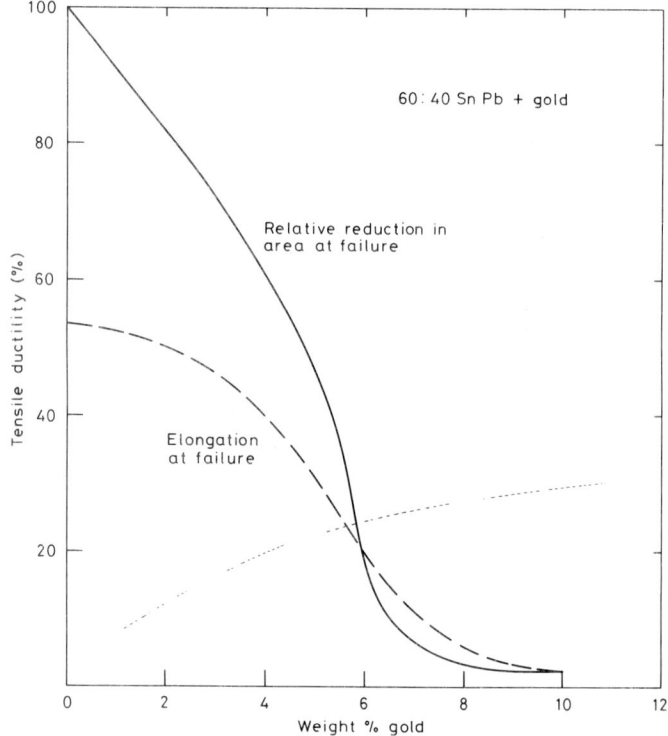

Fig. 11.17 Gold, as an impurity in solder, is very detrimental to its ductility because of the formation of brittle tin-gold intermetallic compounds, mainly $AuSn_4$.[408]

mounted solder joints both by modifying the intrinsic strength properties of the solder itself and the shape and interfacial properties of the fillet.

With regard to the equilibrium phase diagram, impurity additions up to the level corresponding to the ternary eutectic composition lead to a finely dispersed ternary phase in the microstructure and a lowering of the solidus temperature down to the ternary eutectic temperature. Higher levels of impurity result in larger crystals of the ternary phase in the microstructure and an increase in the liquidus temperature.

Some surface mounting components have gold or gold-platinum terminations. Unfortunately, gold has a very high dissolution rate and a high solubility (15 wt%) in molten, near-eutectic Sn-Pb solder at soldering temperatures, but only a limited solubility in the solid. Thus gold can dissolve from the termination into a restricted amount of solder, applied locally as a paste, and attain relatively high concentrations. As the solder cools, the tin-gold intermetallic compound precipitates, $AuSn_4$, form in the fillet. This is very brittle and has a marked effect on the strength of the solder joint. Figures 11.17 and 11.18 show how ductility and fracture toughness of solder both fall off dramatically at concentrations above about 4%Au.[408]

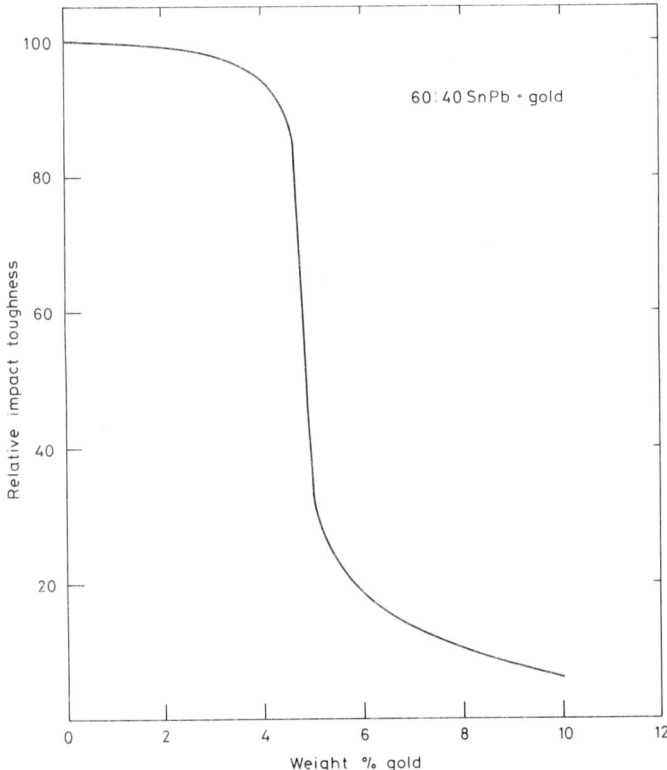

Fig. 11.18 The resistance of solder to impact deteriorates drastically when it is contaminated with gold.[408]

The copper-tin intermetallic phase Cu_6Sn_5 is probably slightly embrittling, whereas the silver-tin intermetallic Ag_3Sn is believed to be soft and ductile and to have no significant effect on mechanical properties.

Apart from possible embrittling effects, the influence of impurity levels on the liquidus temperature is also of importance for the soldering process as the presence of solid particles affects the flow properties of the solder and may lead to defects such as icicling and bridging. If the solidified particles formed in the solder fillet during solidification are less dense than the surrounding liquid solder, they float to the surface, giving rise to a grittiness of appearance of the solidified fillet.

As a guide, it can be said that any impurity must be kept below the level corresponding to the eutectic of the ternary tin-lead-impurity phase diagram, if solder joint properties are not to be degraded by a primary intermetallic phase.

11.3 FATIGUE IN SOLDER JOINTS

Failures of solder joints in surface mounted assemblies are invariably associated with a cyclic fatigue mechanism.

All materials will crack when subjected to cyclic strain. The problem is generically termed fatigue, and the phenomenon of fatigue is responsible for the majority of mechanical failures in engineering components in service. It is caused by a persistent cycling stress at a lower magnitude than the normal monotonic fracture stress of the material. This low-stress cycling causes cracks which, in metals, nucleate at the surface and grow into the bulk until the residual coherent volume is small enough for the applied stress to exceed the fracture stress, whence normal tensile or shear failure occurs.

On a printed circuit board the components and the board material are relatively rigid compared with the solder. The solder, being the most compliant material in the joint, therefore encounters most movement as any strain is relieved and hence the solder is most susceptible to fatigue cracking. Cyclic strain arises through ambient temperature cycling of the board, through power cycling (on/off), or through mechanical flexing and vibration. The problem has been exacerbated considerably by the advent of surface mounting devices, since the quantity of solder is very small and there is little compliance between the component and the board. In particular, much attention has been given to the leadless devices such as chip capacitors and resistors, and ceramic chip carriers, for which only the solder fillet exists to relieve any strain.

11.3.1 Prediction of Fatigue Life

A typical fatigue stress or strain cycle is shown in Figure 11.19 which defines the maximum, minimum and mean stress or strain. The cycle in practical situations usually is not a sine wave and the mean stress is not necessarily zero.

The fatigue behaviour of a material cannot be predicted or described from fundamental principles and mechanical property data. However, since it is of great practical interest to be able to predict fatigue behaviour in service from short-term laboratory tests, various empirical equations relating fatigue life to material and testing parameters have been developed.[409]

For metals, it is generally found that, when test data of log strain range $\Delta \varepsilon$ are plotted against log fatigue life N_f over a limited range of $\Delta \varepsilon$, a straight line is obtained, indicating a relationship of the form:

$$\Delta \varepsilon = C\, N_f^{-c} \qquad (11.1)$$

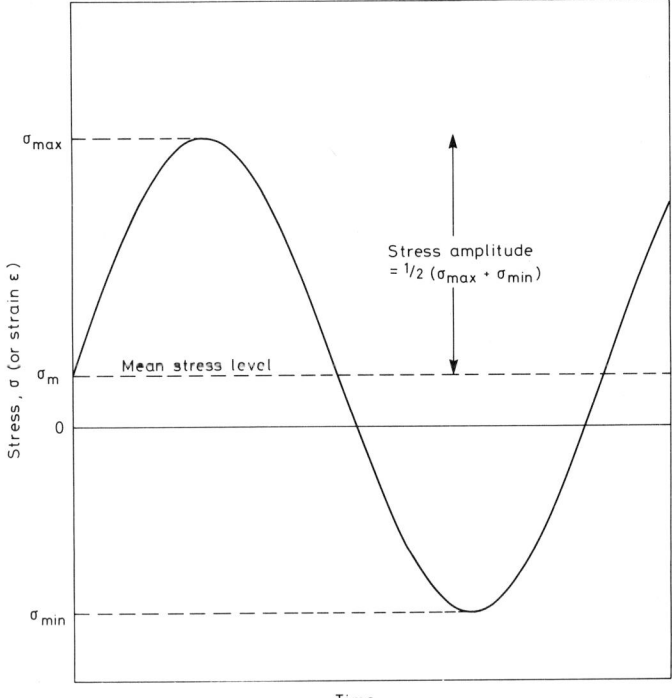

Fig. 11.19 A typical fatigue stress or strain cycling regime. The cycle need not be sinusoidal and the mean stress need not be zero.

where N_f is expressed as the number of cycles to failure.

Equation (11.1) is known as the Coffin-Manson law.[410]

When the whole range of $\Delta \varepsilon$ is considered, the curve may be of the form shown in Figure 11.20 with two distinct regions, one at low $\Delta \varepsilon$ where elastic strain predominates and one at high $\Delta \varepsilon$ where plastic strain predominates. Thus

$$\Delta \varepsilon = \Delta \varepsilon_e + \Delta \varepsilon_p = C_e N_f^{-c_e} + C_p N_f^{-c_p} \tag{11.2}$$

For solder, the yield strain $\Delta \varepsilon_e$ (i.e., the strain at which the solder begins to undergo plastic deformation) is very small, about 0·02%, so that usually $\Delta \varepsilon_e$ is negligible compared with $\Delta \varepsilon_p$. Some understanding of the constant C_p can be obtained by considering that, over the first half of a fatigue cycle, the test piece is subjected to pure tension so that, when $N_f = 0·5$, $\Delta \varepsilon$ is equal to the fracture ductility ε_f. Thus for solder

$$N_f = \frac{1}{2} \left(\frac{\varepsilon_f}{\Delta \varepsilon} \right)^{1/c} \tag{11.3}$$

In practice, it turns out that ε_f is not exactly equal to the fracture ductility, and so is called the fatigue ductility coefficient. It is usually determined empirically, corresponding to the value of $\Delta \varepsilon$ when $N_f = 0·5$.

Equation (11.3) is relevant to fatigue resulting from any cyclic strain, be it arising from mechanical flexing of a PCB, temperature cycling or power cycling.

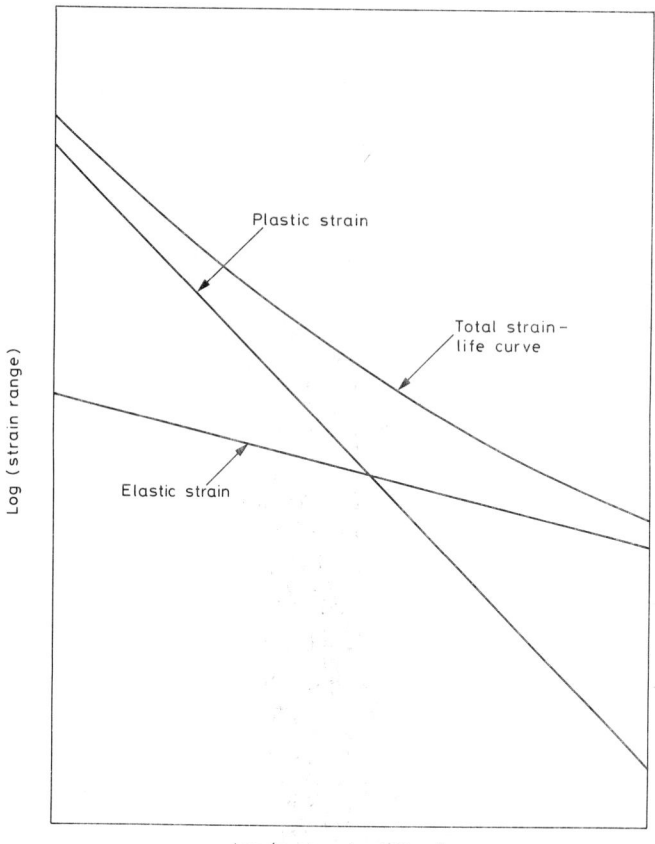

Fig. 11.20 In fatigue, log (lifetime) is proportional to log (cyclic strain range) in either the elastic or the plastic regime.

The imposed strain range $\Delta\varepsilon$ can usually be calculated from the test or environmental conditions.

The exponent c is another empirical constant and is the slope of the line when $\ln N_f$ and $\ln \Delta\varepsilon$ are plotted. For all metals, c lies in the range 0·45 to 0·60 over a wide range of conditions, but is found to increase at high temperatures and long hold times, to approach a value of 1.

In addition to a dependence on the strain range $\Delta\varepsilon$ and the ductility ε_f, the number of cycles to failure N_f has also been found to vary with temperature, cycling frequency, hold times during cycling and microstructural characteristics of the alloy tested. The effect of temperature can be accommodated by variations in ε_f and c but the effects of frequency, ν, and microstructure, m, are usually taken in by an additional factor, $f(\nu,m)$.

The effect of microstructure f(m) is empirical, covering effects of the original microstructure, such as grain size, interphase spacing and particle distribution, as well as possible deterioration in the microstructure during fatigue. The fatigue life can be affected drastically by the microstructure when significant plastic deformation is involved as is always the case for solders.

Regarding the effect on fatigue life of cyclic frequency v and hold time, a number of possible forms of the function $f(v)$ have been used to fit experimental data to the basic Coffin-Manson equation. Those specific to solder joints will be discussed in detail later.

11.3.2 Origin of Fatigue in Solder Joints

The problem of fatigue in solder joints lies in the fact that heat is generated in electronic components which makes them expand. Figure 11.21 demonstrates the fatigue caused by power cycling. The generated heat is dissipated through radiation, convection and conduction. That which is dissipated by conduction flows through the solder joint to the board, heating up both. The board material generally has a different coefficient of expansion from the component. Thus, the thermal expansion mismatch, together with the delay in the heating of the component and the board, causes a relative movement between the component and the board. Since both parts are rigid compared with the solder, the repeated movement, caused perhaps by a switching current, produces a cyclic stress and eventual fatigue failure. Thus, when a solder joint is being made and when it is functioning, it is subject to very complex states of stress and strain due to the significant thermal expansion mismatch between the surface mounted component and the substrate. Ceramic chip carriers and ceramic capacitors, for example, have temperature coefficients of expansion (TCEs) of 6–7 ppm.$°C^{-1}$, whilst for epoxy/glass laminates the range is 12–16 ppm.$°C^{-1}$.

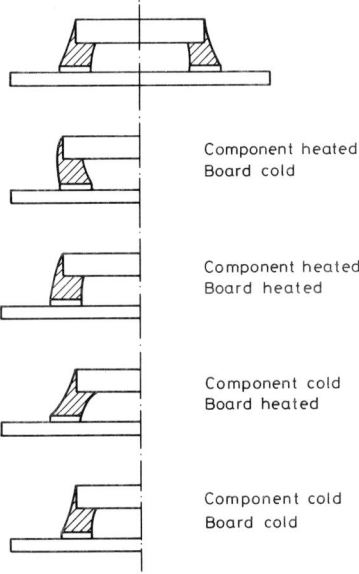

Fig. 11.21 Solder joint fatigue in surface mounted assemblies is generally caused by power cycling.

Because the component and the substrate are usually of different materials and because they have different time-variant heat inputs, there is not a steady state expansion mismatch, but a transient expansion mismatch during power-up and power-off. Figure 11.22 shows schematically how the temperature of a ceramic

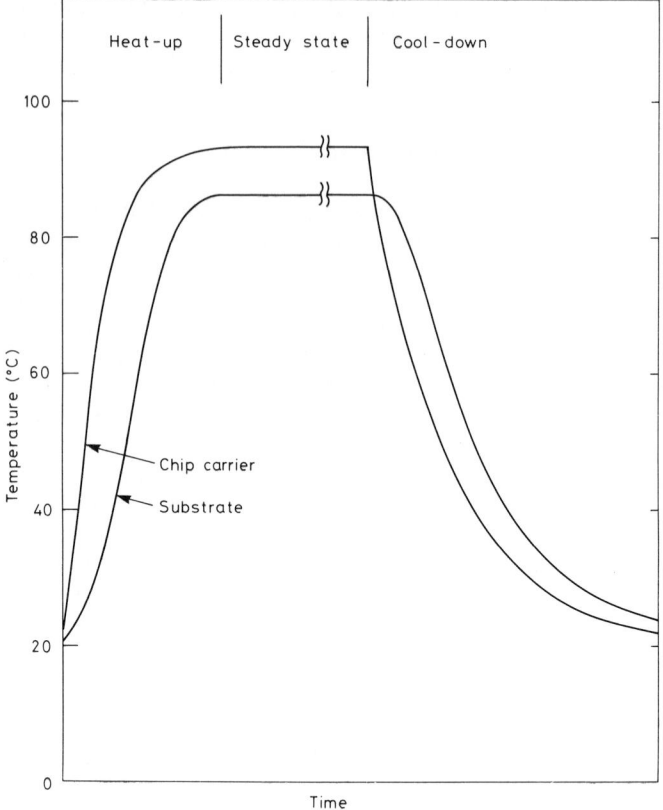

Fig. 11.22 Typical temperature variations of a LCCC and its interconnection substrate during power-up, steady state and power-off.[411]

chip carrier, for instance, and a substrate to which it is soldered, change due to the heat generated in the device.[411] By taking the temperature differential in combination with the difference in the TCEs, the differential expansion can be calculated as shown in Figure 11.23. Even if the substrate and the chip carrier are of the same material, there exists a strain in the solder joints during power-up and power-down. This then results in a low-cycle fatigue of the joints with a period, typically, of one cycle per day.

In practice, a solder joint on an SMD will deform by both shear and tensile strain as sketched in Figure 11.24. Generally, tensile strain is more detrimental to fatigue life than is shear strain,[412] as illustrated in Figure 11.25. Figure 11.5 showed a symbolic joint and the stresses acting upon it, where σ_{ii} denotes the tensile stresses and σ_{ij} the shear stresses. By integrating the stress components over the surface area of the joint, the resultant forces (F_x, F_y, F_z) and the moments (M_x, M_y, M_z) are obtained, as shown in Figure 11.26. Using strain gauges and holographic interferometry, these forces and moments have been measured[413] and have shown that the deformation is not simply within the solder as shown in Figure 11.21 but arises from a complex combination of several deformation modes, shown schematically in Figure 11.27. Furthermore, the

The Solder Fillet

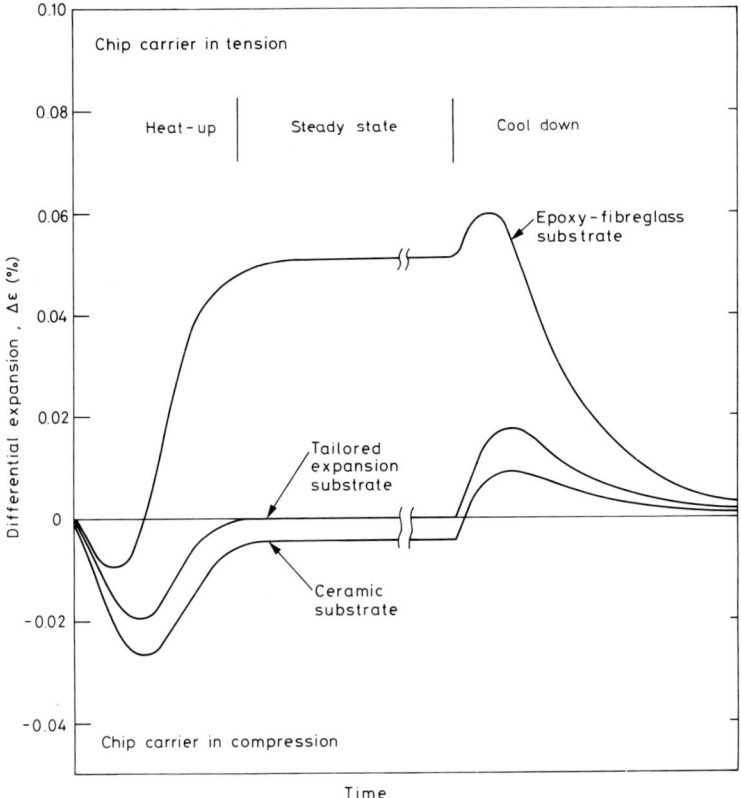

Fig. 11.23 The temperature variations shown in Figure 11.22, in combination with different TCEs of the LCCC and the substrate, give rise to cyclical strains between the chip carrier and its substrate.[411]

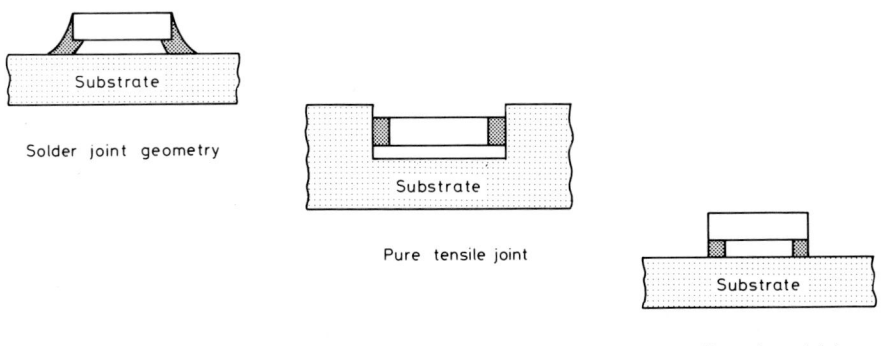

Fig. 11.24 In practice a solder joint on a surface mounted assembly deforms both in tension and in shear.

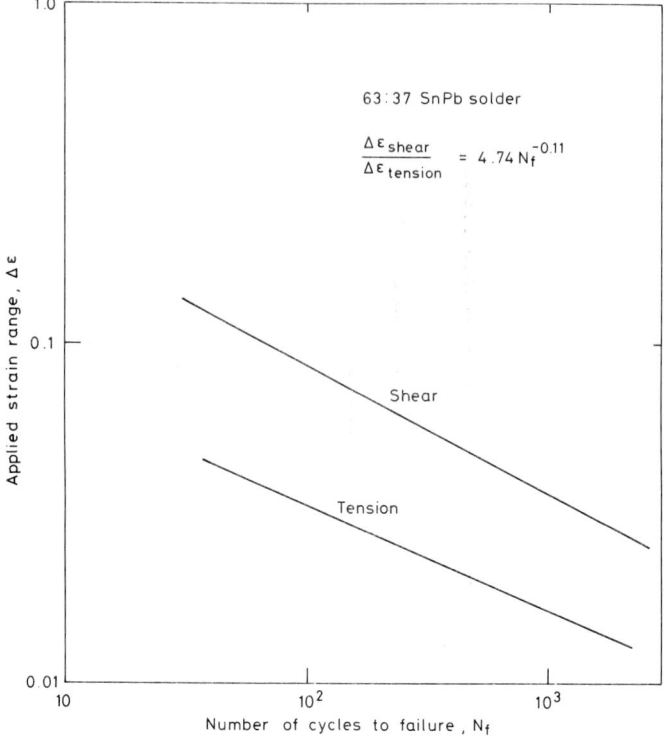

Fig. 11.25 In general, tensile strain is more detrimental to fatigue life than shear strain.[412]

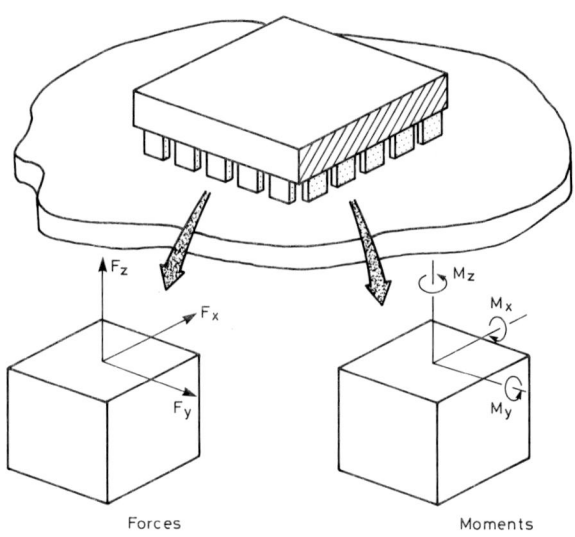

Fig. 11.26 The forces and the rotational moments associated with a solder joint in stress.

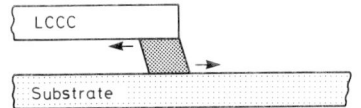

Mode A: In-plane displacement (simple shear)

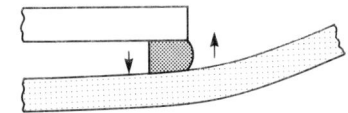

Mode B: Out-of-plane rotation

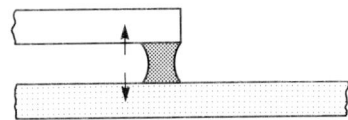

Mode C: Out-of-plane displacement

Fig. 11.27 In practice, deformation of a solder joint is a complex combination of several deformation modes.[413]

deformation of the joint during ambient temperature cycling is significantly different from that during power cycling. Thus power cycling cannot be correctly simulated by ambient temperature cycling.

During thermal chamber cycling of an LCCC on a PCB, in addition to out-of-plane bending of the PCB, there is an out-of-plane bending of the chip carrier,[414] as shown in Figure 11.28. In power cycling the main deflection is an out-of-plane bending of the board towards the chip carrier because of the temperature gradient through the thickness of the PCB. These results emphasise the fact that failure is by a combination of tensile and shear stresses.

11.3.3 Fatigue Mechanisms of Solder

Under stress, metals fail by cracking, initiated and then propagated until such time that the residual coherent cross-sectional area can no longer support the applied load. The mechanisms of the crack initiation and propagation vary with material, environmental conditions and temperature. The temperature regimes of failure mechanisms are loosely defined by the ratio of the operating or test temperature to the melting point of the material, on the kelvin temperature scale. Due to the low melting points of solder alloys, this ratio is very high, being about 65% at room temperature, 75% at 70°C and over 90% at 150°C, a commonly used temperature in fatigue testing of solders. Thus, the fatigue mechanisms operating at or above room temperature are similar to those prevailing in other engineering alloys at elevated temperatures. Whilst at 'low' temperatures, crack initiation usually occurs at free surfaces, and crack propagation is transgranular,

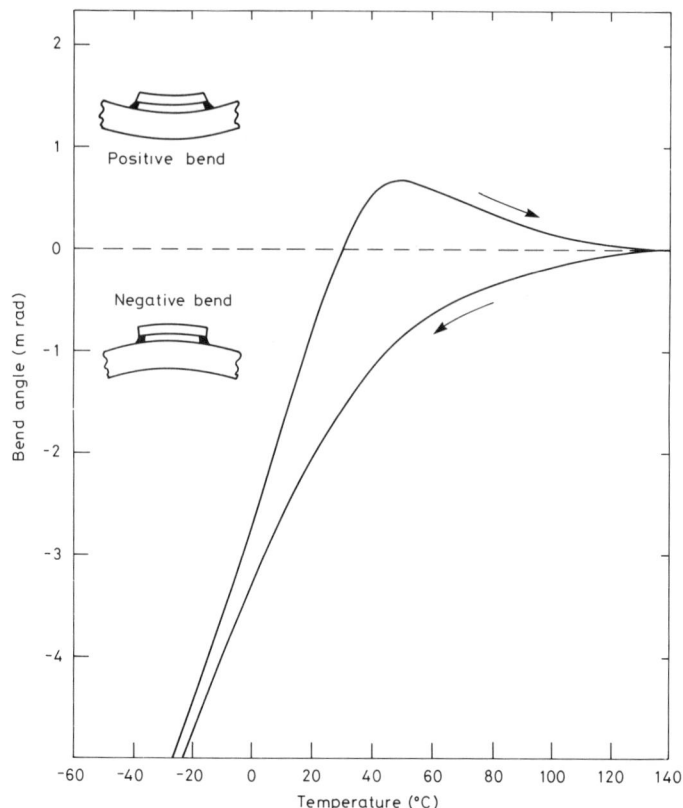

Fig. 11.28 Typical out-of-plane bending of a chip carrier and its substrate during isothermal heating-up and cooling-down.[413]

at 'high' temperatures initiation is not surface sensitive and propagation is intergranular. At low stress levels, cavities form where crystallographic slip bands meet the grain boundaries, while at higher stress levels deformation takes place by grain boundary sliding, as in creep.[415] Also 'high' temperature fatigue varies with cyclic frequency and hold-times because deformation occurs by diffusion controlled and hence time-dependent processes.[416]

The fatigue behaviour of solder alloys can be influenced by the environment, but not by affecting crack initiation as is the case at 'low' temperatures, but by hindering crack propagation. Thus, preventing oxidation significantly increases fatigue life and, for example, coating the surface with oil can increase the fatigue life by a factor of ten.[417] However, a corrosive flux residue left on test joints can cause a drastic reduction in fatigue life.

The basic susceptibility of solder joints to fatigue failure at ambient temperatures arises directly from the fact that such temperatures are very close, thermodynamically speaking, to the melting point of solder. The effectively 'high' temperature facilitates a relaxation of the stress during the hold times of the cyclic strain. In Figure 11.29(a) is drawn a stress-strain curve for a material in which very little relaxation occurs. The area under the curve is termed the fracture toughness of the material. The fatigue damage per strain cycle is equivalent to the area within the hysteresis loop defined by the cyclic strain range $\Delta\varepsilon$. The lifetime cycles-to-failure, N_f, of the material is given approximately by:

$N_f \approx$ fracture toughness/cyclic fatigue damage.

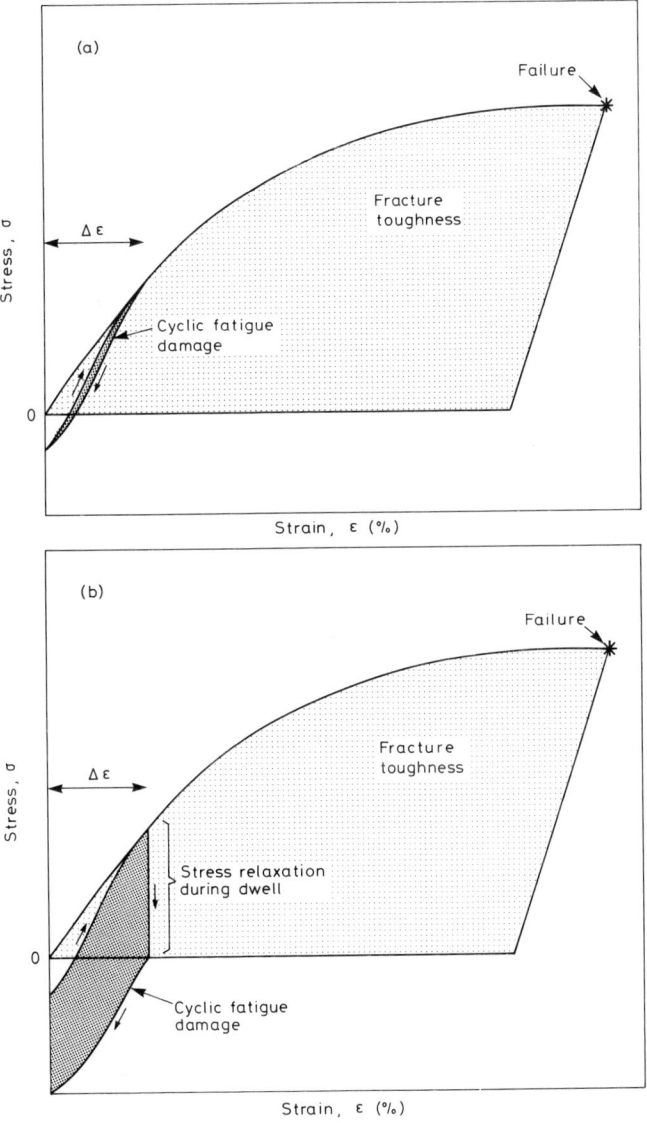

Fig. 11.29 The area under a stress-strain curve is termed the fracture toughness. The stress-strain area of a fatigue cycle is greatly increased if the material can relax its stress within each cycle. The relative fatigue lifetime is the ratio of the fracture toughness to the cyclic fatigue damage.

The situation pertaining to solder at very low cycling is shown in Figure 11.29(b), where there is sufficient time at the hold positions of each cycle for the stress in the solder to relax. The fatigue damage at each cycle is much greater and hence there is an earlier exhaustion of the ductility of the material. Besides demonstrating why solder has poor fatigue resistance, this figure also explains why N_f decreases as the cyclic frequency decreases. Since the stress relaxation is faster at higher temperatures, it also explains why N_f decreases as the temperature increases.

Measurements of the rate of stress relaxation[418, 419] in 60:40 SnPb solder as a function of its temperature are shown in Figure 11.30. It is found that the data can be expressed by an Arrhenius equation with an activation energy of 62 kJ.mol^{-1}.

The dependence of fatigue on the stress relaxation phenomenon means that the cyclic fatigue lifetime N_f is dependent on the strain rate, an effect shown in Figure 11.31 using a triangular cyclic strain at 20°C on 60:40 SnPb solder joints. The temperature effect of the stress relaxation on the fatigue properties is illustrated in Figure 11.32, again for 60:40 SnPb solder joints. An increase in the temperature from 20°C to 100°C results in a decrease in the fatigue lifetime cycles by a factor of 1000.

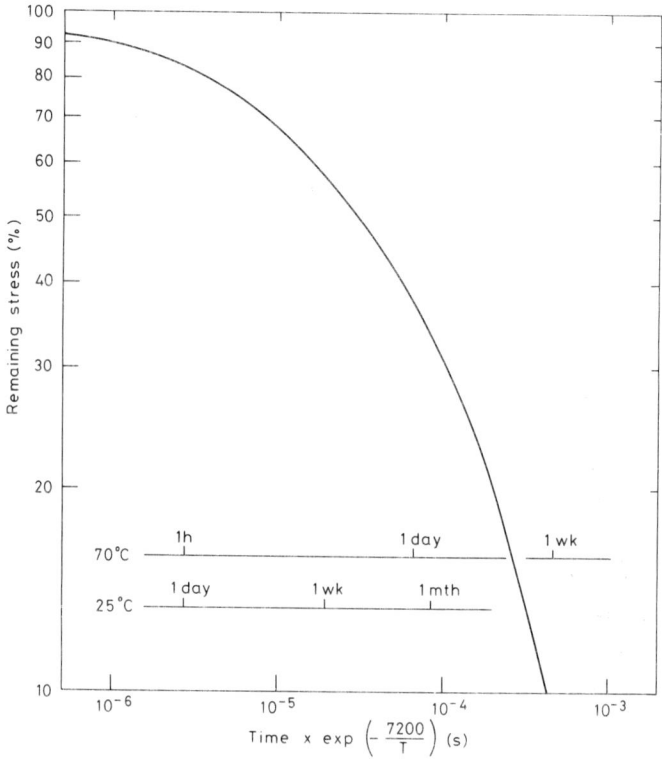

Fig. 11.30 Measurement of the rate of stress relaxation of 60:40 SnPb solder as a function of its temperature. The curve is universal for all temperatures since the abscissa incorporates the temperature T, in kelvin. (Data from ref. 418).

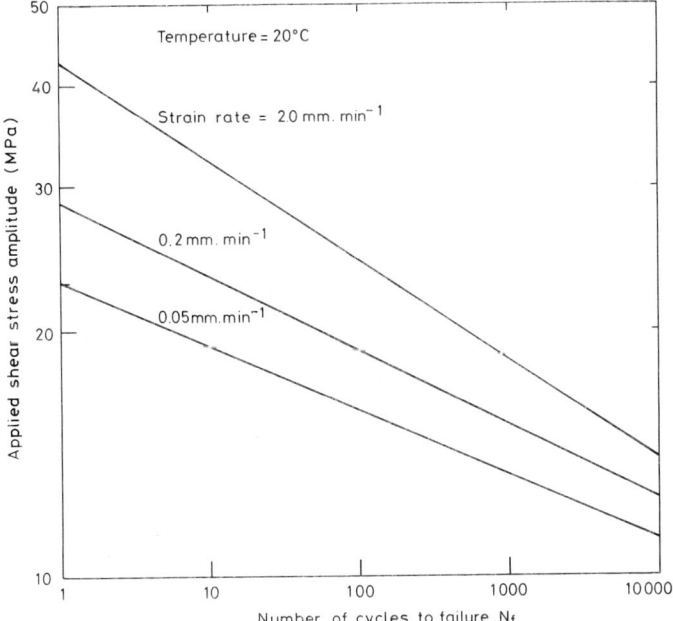

Fig. 11.31 The fatigue lifetime of 60:40 SnPb solder joints using a triangular cyclic strain imposed at different ramping rates.[401]

In addition, the mean stress level of the induced cyclic strain (defined in Figure 11.19) has a marked effect on the fatigue lifetime at low stress levels, as shown in Figure 11.33. Also shown in the figure is the comparative creep strength which shows a further decline at low stress levels. In these measurements the influence of varying the cyclic frequency (strain rate) has not been taken into account, but the general correlation between fatigue and creep is valid.

The relaxation of the solder occurs in fatigue by the same mechanisms as creep, during which plastic strain anneals and the solder becomes softer. The fractional creep occurring on each cycle of thermal fatigue can be calculated from the cyclic frequency, strain range, temperature and thermal cycle range, plus a knowledge of the mechanical property data of the bulk solder alloy. It is found that the fatigue lifetimes of solder joints correlate with the fractional creep per cycle, as shown in Figure 11.34. As can be seen, the longer the thermal cycle or the higher the temperature the more creep occurs in the solder per cycle and the shorter is the fatigue lifetime. All these data shown are for a 10% strain range ($\Delta \varepsilon = 0 \cdot 1$).

11.3.4 Fatigue Life of Leadless Ceramic Chip Carriers

Most effort on the investigation of fatigue in surface mounted assemblies has been directed at leadless ceramic chip carriers.[420–423] Intuitively, and from practical experience, these devices have the least compliance when soldered to a printed circuit board and hence are most likely to be susceptible to fatigue failures. On the other hand they are devices used in high reliability applications.

In Equation (11.3) the predicted lifetime N_f is dependent upon the strain range

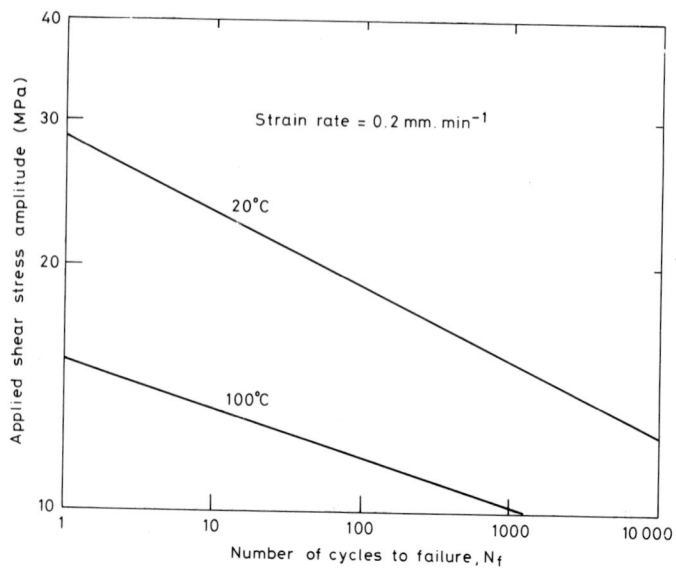

Fig. 11.32 The fatigue life of 60:40 SnPb solder joints is greatly reduced by increasing the temperature because more stress relaxation per cycle can occur at the elevated temperature.[406]

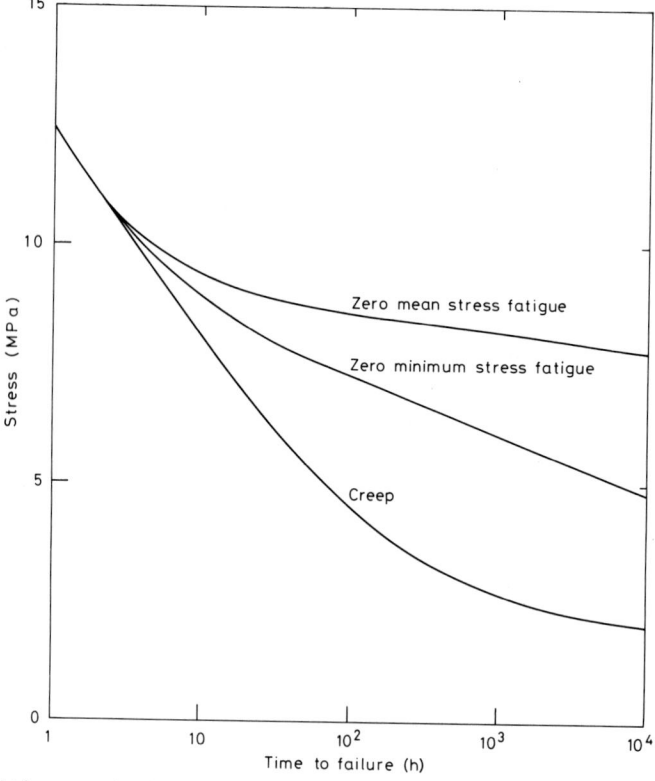

Fig. 11.33 At low stress levels, the mean stress level of the induced cyclic strain has a marked effect on fatigue lifetime.

$\Delta\varepsilon$, the fatigue ductility coefficient ε_f and the fatigue ductility exponent c. These will now be considered in turn for the leadless ceramic chip carrier soldered to a substrate.

The Strain Range $\Delta\varepsilon$

Referring to Figure 11.35, the shear strain range developed when a chip carrier soldered to a substrate is subjected to temperature cycling is[424]

$$\Delta\varepsilon = \frac{r\Delta T\Delta\alpha}{h} \qquad (11.4)$$

where, as shown in the figure, h is the height of the joint and r is its distance from the centre of the carrier. ΔT is the temperature cycle amplitude and $\Delta\alpha$ is the difference in expansion coefficient between substrate and carrier. Thus, the joint lifetime is lowest for the corner joints for which $r = r_{max} = \ell/\sqrt{2}$, and the lifetime is increased by increasing the stand-off h. For a given joint configuration and given materials properties, the strain range is proportional to the temperature range, and so, in Equations (11.1) and (11.3), for thermal cycling ΔT may be substituted for $\Delta\varepsilon$.

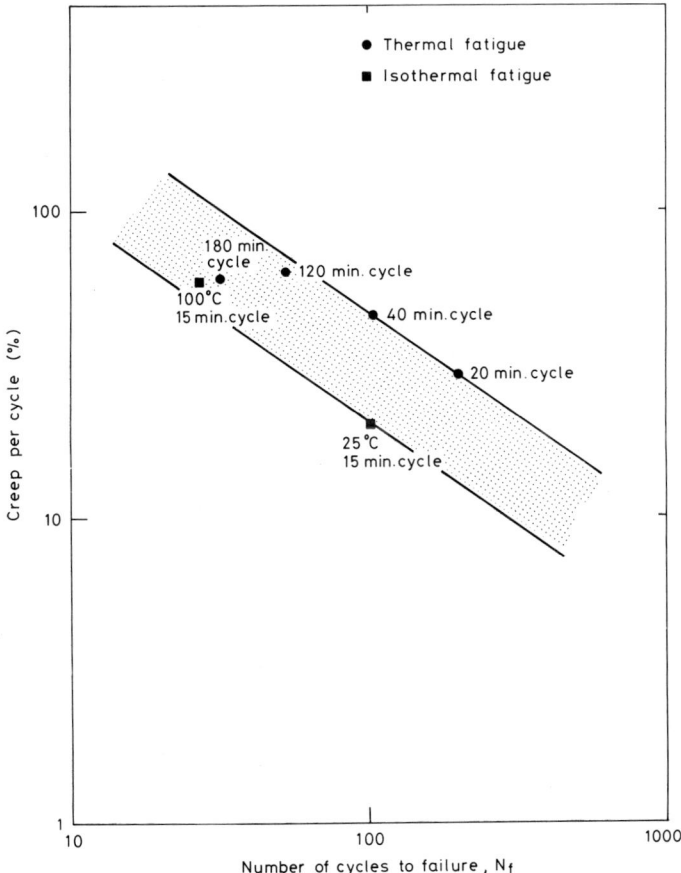

Fig. 11.34 The fatigue lifetimes of solder joints correlate with the fractional creep per cycle.

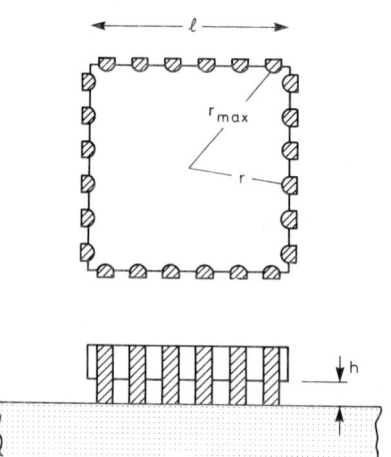

Fig. 11.35 Definition of the parameters ℓ, r and h for a LCCC.

The simple expression for $\Delta\varepsilon$, in Equation (11.4) is based on a number of assumptions:

1 The component and substrate are rigid.
2 The solder joint dimensions are small compared with the component size.
3 The solder volume remains constant.
4 The solder joint deforms uniformly.
5 The joint is a pure shear joint.

It is known that, in reality, these assumptions are, at least in part, invalid. Equation (11.4) is, at best, very much a simplification but it predicts, in agreement with observation, that fatigue life decreases with increasing size of chip carrier, increasing temperature cycle excursion and increasing TCE difference; the fatigue life increases with increasing stand-off height and increasing ductility of the solder alloy. Additional factors not accounted for by the above expression include the variation of the properties of the solder with varying temperature and varying strain rate,[425, 426] as well as the dependencies on the moduli of the chip carrier and substrate materials.

To take into account that the solder joint deforms by both shear and tensile strain and that the component and substrate deform to some extent, as well as the solder, an empirical factor b can be included:

$$\Delta\varepsilon = \frac{b r \Delta T \Delta\alpha}{h} \qquad (11.5)$$

This factor becomes important when using substrates with special compliant layers, for which b≈0. When a thermally matched substrate is used, such as copper-clad Invar, the factor $\Delta\alpha \approx 0$. Both these dramatically reduce $\Delta\varepsilon$ and hence increase the predicted lifetime N_f in thermal cycling tests.

However, the situation for electronic assemblies in service is not that of temperature cycling but of power cycling. In power cycling, as in ambient temperature cycling, the concern arises from the TCE mismatch giving rise to either a transient or an equilibrium strain, $\Delta\varepsilon$, between the component and the substrate:[427]

$$\Delta\varepsilon = \alpha_c(T_c - T_o) - \alpha_s(T_s - T_o)$$
$$= (\alpha_c - \alpha_s)(T_c - T_o) + \alpha_s(T_c - T_s) \tag{11.6}$$

where α_c and α_s are the linear TCEs for the chip carrier and the substrate respectively, T_c and T_s are the mean effective temperatures of the chip carrier and the substrate immediately below the chip carrier, and T_o is the power-off steady state temperature.

During ambient temperature variations and temperature cycling in an oven, $T_c = T_s$ so that Equation (11.6) becomes

$$\Delta\varepsilon = (\alpha_c - \alpha_s)(T_c - T_o)$$
$$= \Delta\alpha(T_c - T_o) \tag{11.7}$$

and the strain varies directly with the expansion mismatch. However, during power cycling, $T_c \neq T_s$ and hence the strain $\Delta\varepsilon$ cannot be reduced to zero by matching the expansion coefficient of the substrate to that of the chip carrier. Even if the steady-state $\Delta\varepsilon$ is zero, during the power-up and power-down thermal transients will give rise to strain. Only by close thermal coupling of the chip carrier to the substrate can $T_c \approx T_s$ and power cycling be simulated by temperature cycling. In the absence of such thermal coupling, temperature cycling can give very misleading data regarding fatigue life, especially in cases where a matched substrate is used and $\Delta\alpha \approx 0$, such that virtually no cracking occurs in thermal cycling tests. Furthermore, in power cycling, not only is there a transient temperature difference between the chip carrier and the substrate but also a temperature gradient in the substrate, resulting in a warpage that cannot be simulated in thermal cycling tests.

Fatigue Ductility Coefficient, ε_f

The ductility coefficient is a measure of the ductility of the solder under fatigue conditions. The simplest treatments put ε_f equal to the tensile fracture ductility ε'_f, which ranges from about 0·5 at 20°C to 0·8 at 100°C. Considering that the strain in the joint is mainly shear rather than tensile, other more complicated expressions have been derived, for example:[421]

$$\varepsilon_f = (1 + v)\ln\left(\frac{1}{1 - RA}\right) \tag{11.8}$$

where v is Poisson's ratio, a measure of the resistance to shear, and RA is the area reduction of the solder at a purely tensile fracture. Taking $v = 0·4$ for solder, ε_f then takes a value of about 1·0 when RA is 50% up to 3·2 when RA is 90%.

Elsewhere an alternative expression has been derived:[412]

$$\varepsilon_f = 0·99\{\ln(1 + \varepsilon'_f)\}^{0·75} \tag{11.9}$$

where ε'_f is the tensile fracture ductility, so that ε_f from this expression takes a value 0·5 at 20°C and 0·66 at 100°C.

Values for ε_f have been determined[428] by extrapolation of the results of fatigue measurements made on eutectic SnPb solder.[429, 430] By plotting log N_f against log $\Delta\varepsilon$, as shown in Figure 11.36, we have, from Equation (11.3)

$$\log N_f = -\frac{1}{c}\log \Delta\varepsilon + \left(\log \frac{1}{2} + \frac{1}{c}\log \varepsilon_f\right) \tag{11.10}$$

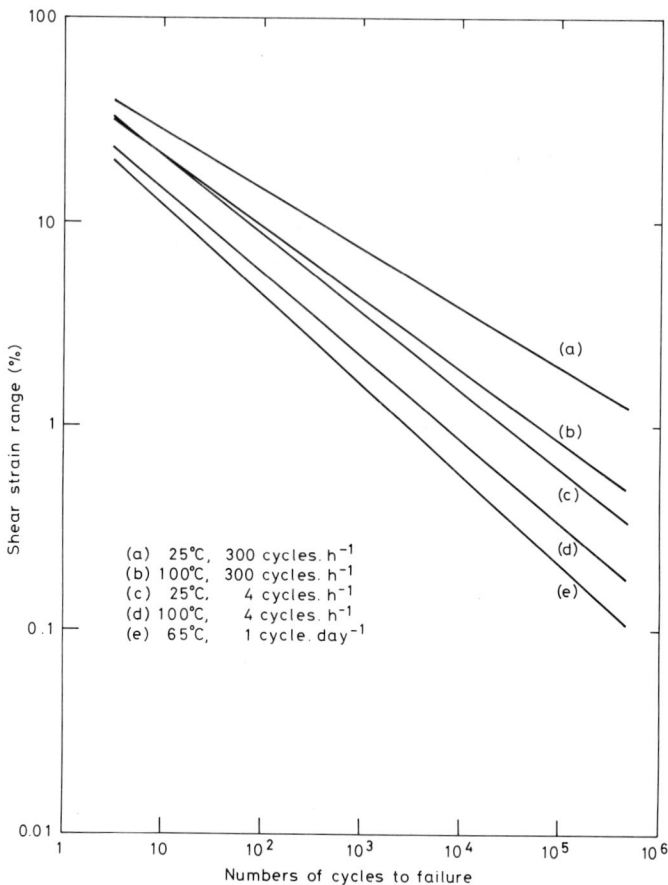

Fig. 11.36 Some experimental data of fatigue of eutectic SnPb solder.[429]

so that c can be obtained from the slope and ε_f from the intercept. For the temperatures of 25°C and 100°C, and frequencies of 300 cycles per hour, 4 cycles per hour and 1 cycle per day, the major change occurs in c, with only a very small variation in fatigue ductility. These data give $\varepsilon_f = 0.65 \pm 0.5$ for all conditions, in reasonable agreement with the analytical expression of Equation (11.9). These values of ε_f incorporate the effects of changes in microstructure during fatigue, so that, although the tensile ductility increases with temperature, the calculated fatigue ductility coefficient does not. Similarly, ε_f does not increase with decreasing cyclic frequency although fracture ductility increases with decreasing strain rate, because the increased time available for stress relaxation counteracts the increased ductility.

Fatigue Ductility Exponent, c

As mentioned above, values for the fatigue ductility exponent c can be obtained from fatigue data and, for the conditions of the measurements, it lies in the range 0.3 to 0.45. An approximate analytical expression is obtained for very low cycle power fatigue:[428]

$$c = 0 \cdot 442 + 6 \cdot 10^{-4}\overline{T}_s - 1 \cdot 74 \cdot 10^{-2} \ln(1 + \nu) \tag{11.11}$$

where $\overline{T}_s(°C)$ is the mean cyclic solder joint temperature and ν is the cyclic frequency in cycles per day.

11.3.5 Effect of Frequency and Hold Time

In general, the fatigue lifetime N_f should decrease with decreasing cyclic frequency and increasing hold times because more time is then available for diffusion controlled processes to respond to the applied strain. From Equation (11.3), therefore, decreasing cycling frequency must result in an increase of the exponent c or a decrease in the fatigue ductility coefficient ε_f. It has been demonstrated experimentally that, for constant temperature, mechanical fatigue of solder lap joints under shear, ε_f, is very little affected by the cycling frequency, the effect of which is accounted for by the exponent c. Thus, when the effect of frequency and hold time is not implicitly allowed for in the analytical expression of the Coffin-Manson relation, it will probably affect the exponent c, increasing its value as the cycling frequency is decreased and hold times are increased.

The effect of cycling frequency and hold times on the fatigue phenomenon is at least partly due to relaxation of the elastic strain. When a solder joint is strained, the elastic stress initially present decreases with time and the initial elastic strain is transformed to plastic strain damage. Thus, if the relaxation time is less than the time for half a fatigue cycle, all the elastic strain is absorbed as damage, reducing the fatigue life. However, if the relaxation is slow compared with the fatigue cycle, some or all of the elastic strain is recovered during the subsequent half-cycle. The relaxation phenomenon does not explain the full effect, since the fatigue lifetime continues to decrease as the cycling frequency is decreased, beyond that accounted for solely by relaxation of elastic stresses.

Experimental results have demonstrated an empirical relation for the effect of frequency, incorporating into the Coffin-Manson relation a term $f(\nu) = k_1 \nu^{1/3}$ where k_1 is a constant and ν the cycling frequency. Thus fatigue lives, under identical test or service conditions apart from the frequency of cycling, are related by:[431]

$$\frac{N_f(1)}{N_f(2)} = \left\{ \frac{\nu(1)}{\nu(2)} \right\}^{1/3}$$

Other possible forms of $f(\nu)$ have been discussed, such as incorporating the fatigue exponent c:

$$f(\nu) = \nu^{(1-k_2)c}$$

where k_2 is an empirical constant.

As shown in Equation (11.11), the frequency, for power cycling at low frequencies, can be incorporated into the fatigue ductility exponent. The validity of this relationship is demonstrated in Figure 11.37 in which the curve is the prediction of N_f from the Coffin-Manson equation as a function of the cyclic frequency under low cycle fatigue.[428] The three data points are obtained by power cycling or oven cycling but with the same mean temperature $\overline{T}_s$ and the same cyclic strain range $\Delta\varepsilon$.

It has been found empirically that the effect of hold time can be taken into

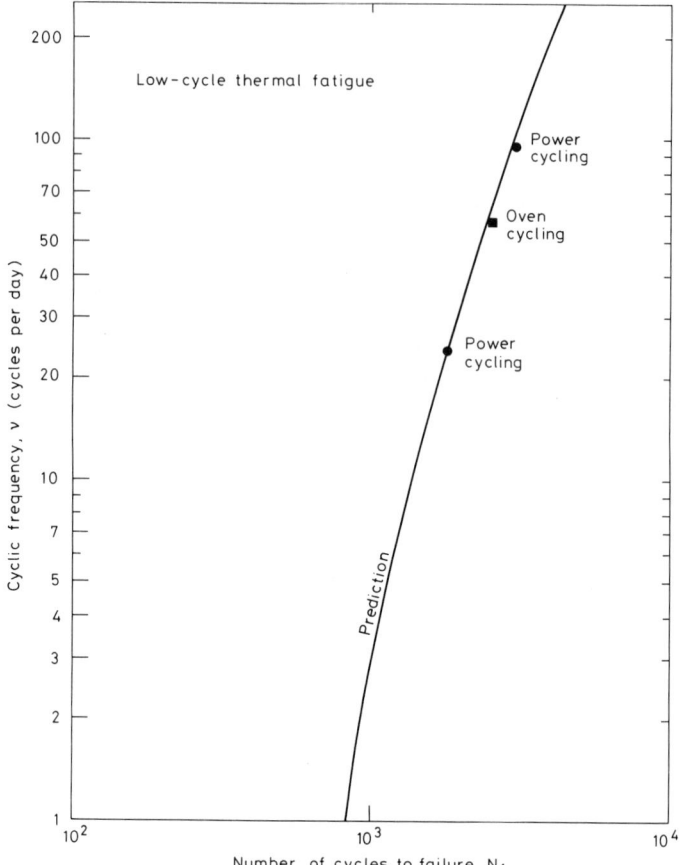

Fig. 11.37 Fatigue lifetime as a function of cycling frequency. The points are experimental measurement; the curve is a prediction from Equations (11.10) and (11.11). (Data from ref. 428).

account with reasonable accuracy by simply incorporating it into the cycle frequency. In other words, the difference in fatigue life measured using a sinusoidal stressing cycle and a square-wave stressing cycle is not great.

11.3.6 High Frequency Fatigue

In Figure 11.37, for low cycle thermal fatigue with cycling periods in the range 15–120 minutes, the model prediction and some relevant experimental test results agreed that the fatigue lifetime is reduced as the cycling frequency decreases because the longer available times at the extremes of the strain cycle allow a relaxation of the stress and therefore a greater fatigue damage per cycle.

At high frequencies, where very little stress relaxation can occur, the situation can be very different. High frequency fatigue testing of solder joints is carried out either to simulate vibration during service or as an accelerated form of test for low cycle thermal or mechanical fatigue in service. Whether or not confidence can be placed on high frequency mechanical fatigue as an accelerated measure of low cycle mechanical fatigue depends on the temperature of the service environment

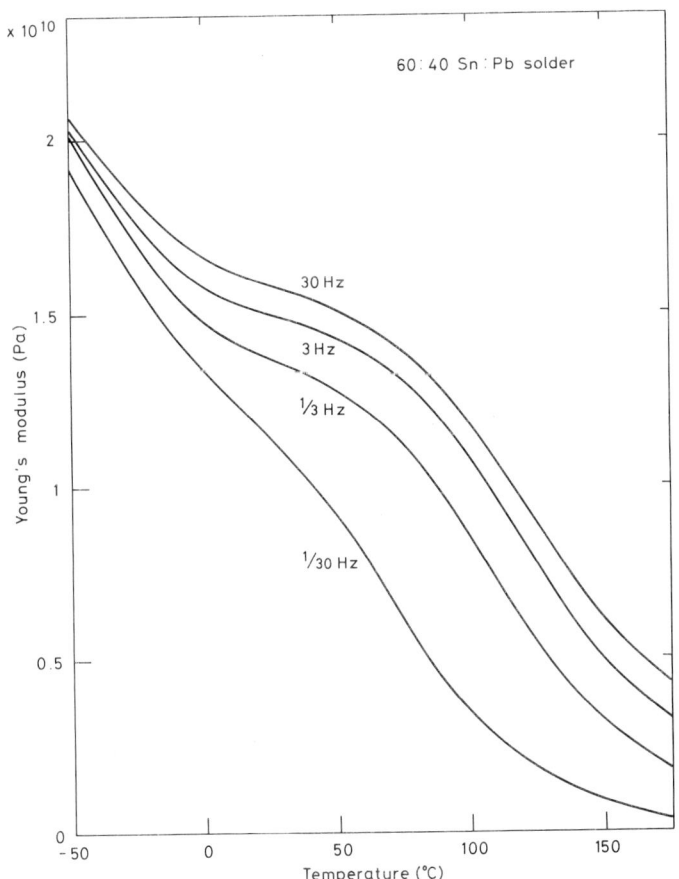

Fig. 11.38 The physical properties of solder alloys are dependent on the frequency of a cyclic stress: Young's modulus of 60:40 SnPb solder for example.[432]

and consequently on whether or not significant stress relaxation can occur.

Some physical properties of solder alloys are frequency dependent at these relatively high strain rates (frequencies $v \gtrsim 0.01$ Hz) and therefore a simple frequency-based acceleration factor for a test is not possible. For example, Figure 11.38 shows the frequency and temperature dependence of Young's modulus over the relevant ranges, for 60:40 SnPb solder.[432]

Mechanical flexing, at relatively high frequencies, of a surface mounted assembly is usually carried out with a full reversal bending about a zero-strain plane. Some typical results are shown in Figure 11.39 for solder joints on several types of 68 lead plastic chip carriers subjected to fully reversed, square wave mechanical cycling on a 1 metre bend radius, with cycling periods in the range 5 seconds to 20 minutes. The test temperature is 81°C. The curve is a prediction from the model used previously for low cycle fatigue (Equations (11.10) and (11.11)).

The experimental data are in good agreement with the prediction from the parametric model and it seems that mechanical fatigue can be used, with caution,

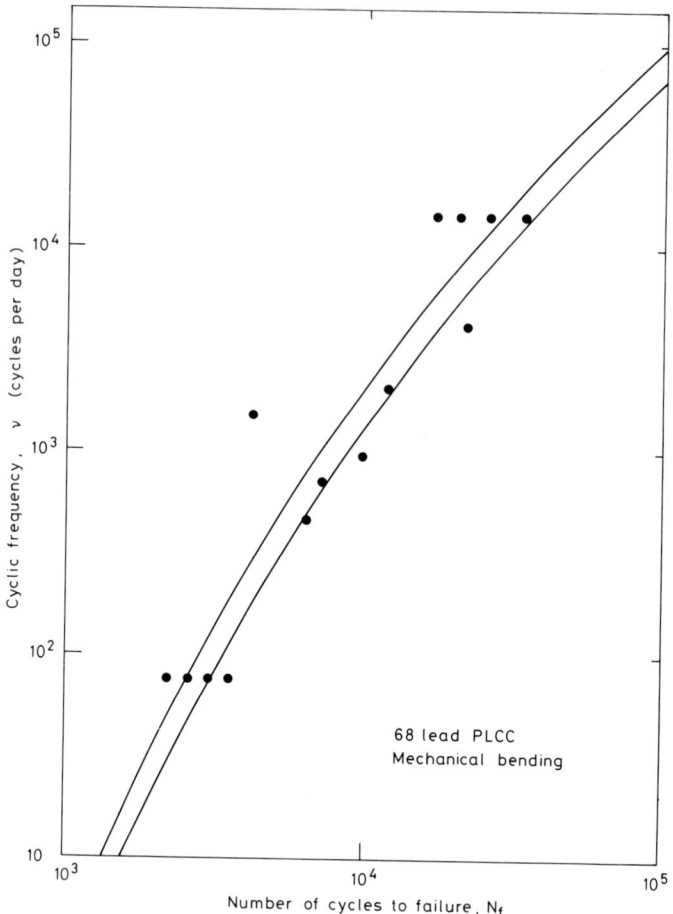

Fig. 11.39 Mechanical flexing with full reversal bending, of 68 lead PLCCs soldered to a substrate. The points are experimental and the curve band a prediction from Equations (11.10) and (11.11). (Data from ref. 427).

to assess the performance of systems at much lower frequencies. It is often the only means of judging the fatigue properties of solder joints on the leaded devices, such as in this case, for which, because of the compliancy of the leads, the thermal fatigue lifetime would be extremely long. Mechanical flexing enables both the strain range and the frequency to be increased.

High frequency thermal fatigue testing in an oven is not possible experimentally because of the thermal lag of the oven. As an alternative, two baths of inert liquid can be used, one hot and the other cold, with a rapid transfer mechanism of the test piece from one to the other. The use of a suitable pair of inert fluorocarbon fluids is recommended, since they are simply different boiling fractions of the same chemical family and hence fluid carry-over from one bath to the other has only a minimal time-variant effect on the heat transfer properties. This type of test is generally referred to as thermal shock and, by using long dwell times at each temperature, may be used at any cycling frequency less than about 0.1 Hz. Unfortunately, the mode of failure of surface mount solder joints when undergoing a cyclic thermal shock treatment is very different from that when undergoing thermal fatigue, even when the cycling frequencies are the same and are low enough to allow stress relaxation to occur. In thermal shock testing crack initiation occurs at places of stress concentration in the joint, usually at a point where solder, metallisation and the environment meet.

11.3.7 Effect of Solder Fillet Geometry

The geometry of the solder joint has a significant effect on the deformation induced by thermal differentials in the workpiece, and thus affects the fatigue life of the joint. Two factors are of prime importance. Increasing the stand-off decreases the shear strain for a given displacement of the ends of the solder joint and therefore increases the fatigue life. Extending the solder fillet beyond the side edges of the component metallisation increases the area of the joint and thus decreases the shear stress in the joint.

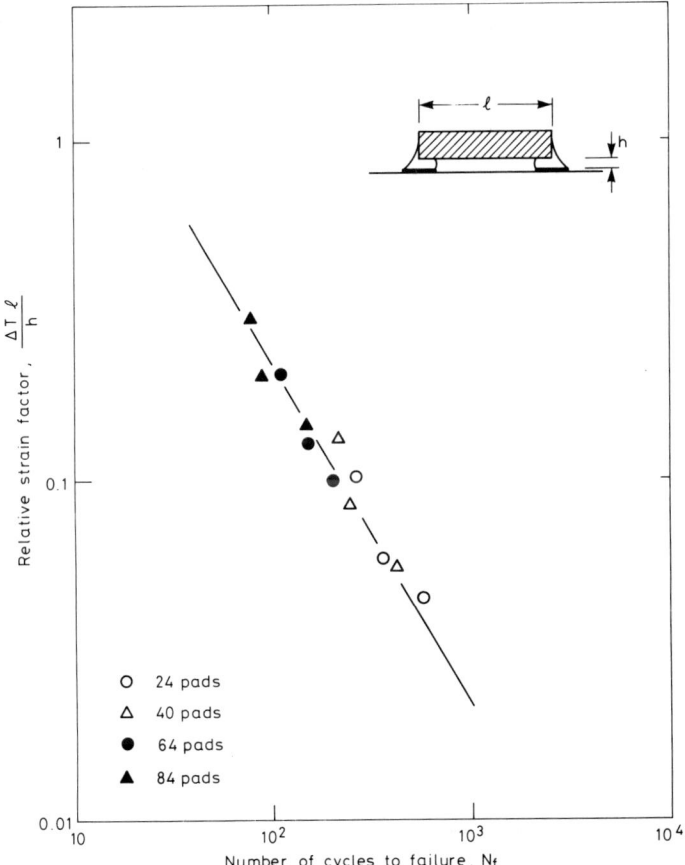

Fig. 11.40 Experimental data verifying Equation (11.4) for the strain range of a LCCC subjected to fatigue. Four component sizes and three stand-off heights are tested, but by normalising using the factor ℓ/h, the data lie on one straight line.[412]

The validity of the simple expression for $\Delta\varepsilon$ in Equation (11.4), with its inverse relationship to the stand-off height h, has been demonstrated[412] by varying the size l, the stand-off height h, and the temperature excursion ΔT, in thermal fatigue tests on four ceramic chip carrier sizes (24, 40, 64 and 84 pads), each with three values of h (80, 135 and 185 μm), as shown in Figure 11.40. By plotting $\log(\frac{l\Delta T}{h})$ against $\log N_f$, the data are normalised and a straight line, as predicted, is obtained.

The geometry of the solder joint also influences the heat flow between the component and the circuit board, and therefore affects the temperature gradients between a chip carrier and the substrate during power dissipation.[433] Obviously the thermal resistance of the joint increases, and hence the temperature differential increases as the stand-off height increases or the area of the joint decreases. Figure 11.41 shows this effect of stand-off height on the board temperature as a chip carrier is switched on with a constant power dissipation, and then switched off.

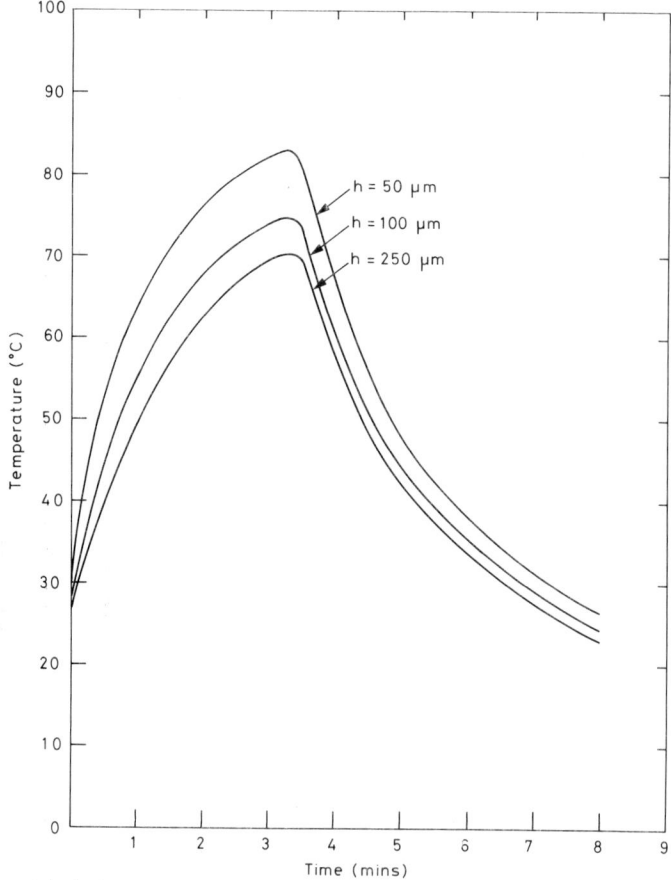

Fig. 11.41 The temperature of a substrate board as a soldered chip carrier is powered up and switched off, as a function of its solder stand-off height from the board.[433]

Extending the solderable pad on the printed circuit board outwards from the leadless device will also improve the fatigue properties of the joint. It is reported[434] that extending the fillets from 0 to 1 mm from the pads on 44-terminal LCCCs increased the fatigue life from 200 to 500 cycles in thermal cycling tests over the temperature range $-55°C$ to $+125°C$. By increasing the size of the fillet in the horizontal plane, the shear strength of the joint is improved as illustrated in Figure 11.42. In this figure is given the force parallel to the plane of the PCB necessary to fracture a 44-terminal LCCC from a PCB. The required force has only limited dependence on the stand-off height h, but increases by a factor of more than two when the pad extension is increased.

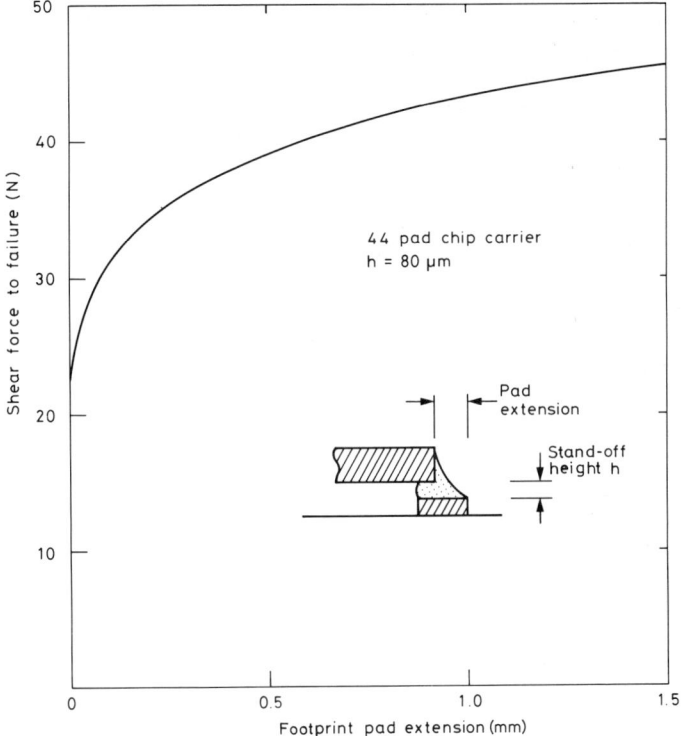

Fig. 11.42 The horizontal shear force required to fracture a 44 lead LCCC from its footprint as a function of the footprint pad extensions.[434]

This observation may not, however, be the whole story. The application of finite element analysis to the fatigue of SMT joints is rare and has been limited to simplified situations.[435] However, it seems that, in order to accommodate the linear stresses and the bending moment in the solder joint, it is advantageous to have a large joint area on the chip carrier and a smaller joint area on the PCB pad, while making a concave fillet.[436]

The shape of the fillet and the amount of solder therein are factors which need consideration in the context of low cycle fatigue.[437] Model experiments have shown that, whilst the breaking strength in shear, of a joint with standard pad sizes, increases with increasing solder mass in the joint, the shear stress at failure

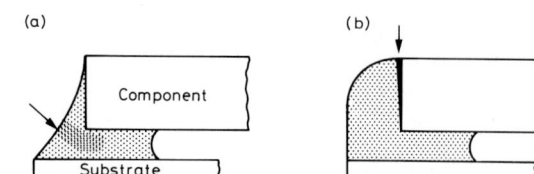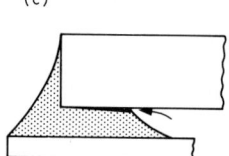

Fig. 11.43 Failure modes of leadless solder joints caused by thermal fatigue or thermal shock: (a) commonly observed fatigue zone observed in section, (b) if solder bulk is large the component shears upwards, (c) a stress concentrating notch.

(load per wetted area of joint) decreases slightly. Even if a fillet is formed to the end of a component the shear strength is not increased but the breaking load is.[438]

The solder fillet shape depends on a combination of the extent of the metallisation and the volume of solder; the type of cracking observed in joints on chip components varies with this geometry.[439]

In microsections of thermally fatigued joints is seen a band of stressed solder along a shear plane, as shown in Figure 11.43(a). Microcracks first appear at the outer surface of the solder and grow through the fatigued material. However, if the solder fillet is very bulky, as it may well be if produced by wave rather than reflow soldering as shown in Figure 11.43(b), the differential expansion of the component may be relieved by shear at the solder-to-end cap interface or, occurring in thermal or mechanical shock rather than fatigue, the cracking initiates at a sharp internal 'notch' effected by the solder and the underside of the component, shown in Figure 11.43(c).

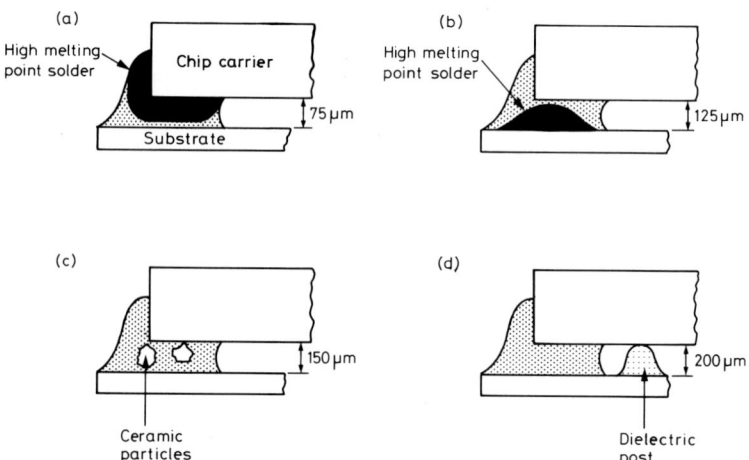

Fig. 11.44 Various methods for producing a controlled stand-off of a component from the board: (a) pre-tinning the component, (b) pre-tinning the board, (c) inert particles in the solder paste, and (d) polyester posts screen printed and fired on to the board.[424]

To some extent the profile of the solder fillet can be controlled by the dimensions of the metallisations and the substrate lands. A stand-off height of up to about 0·25 mm can be achieved by applying solder 'bumps' to the board under the component. (If the board is inverted during reflow the spacing can be increased.) Other, more controllable methods of accomplishing a specific stand-off have been elucidated in detail. These methods are:[424]

(i) pre-tinning the chip carrier with a high melting point solder;
(ii) pre-tinning the substrate with a high melting point solder;
(iii) incorporating ceramic granules or metal spheres in the solder paste;
(iv) screen printing dielectric corner posts on to the substrate.

These techniques are shown schematically in Figure 11.44.

Because of the very precise viscosity and flow requirement of the solder paste during screen printing, its loading with inert particles is not totally straightforward, and more care must be taken in the thermal reflow cycle to control solder balling of the loaded paste. Most promise is expressed in the use of dielectric post stand-offs, although this requires an extra printing and firing procedure. For some components it is possible to set the stand-off height mechanically by using the underside heat sinks that are available.

11.3.8 Effect of Substrate Material

In order to reduce the mechanical stresses resulting from differential thermal expansion when using leadless components, attempts have been made to match the expansion coefficient of the interconnection substrate to that of the component, as discussed in Section 3.5. A number of substrate materials have been considered by the industry and some typical expansion coefficients were

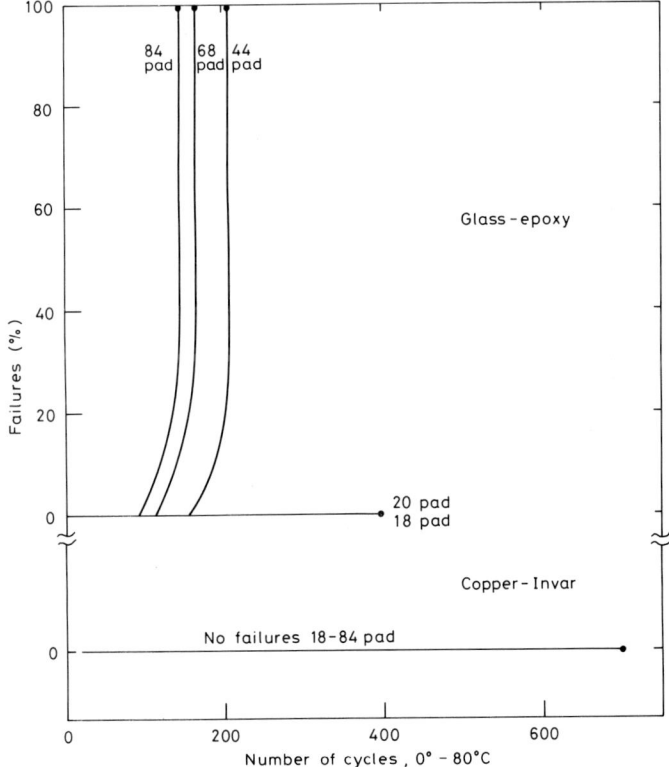

Fig. 11.45 Experimental data of fatigue failures of LCCCs mounted on FR-4 substrates and on copper-clad Invar substrates.[3]

shown in Figure 3.31 compared with that for a ceramic chip carrier. The beneficial effect of reducing $\Delta\varepsilon$ is demonstrated in Figure 11.45, presenting results of thermal cycling between 0°C and +80°C of a ceramic leadless chip carrier soldered to an epoxy-glass PCB and also to a copper-clad-Invar cored PCB, both with a minimal stand-off height.[3] The effect becomes more marked for larger LCCCs. Joints on all the 84, 68 and 44 pad packages have failed after about 200 thermal cycles when an epoxy-glass substrate is used. With the thermally matched substrate no failures are monitored even after 750 cycles.

Another approach to the problem has been to use a compliant substrate manufactured from epoxy-fibreglass laminate coated with an elastomer layer on both sides so that, although there may be TCE mismatch, the induced strains are taken up by the substrate, designed for the purpose, and not the solder joint.

Using the Coffin-Manson equation with the semi-empirical expressions given in Section 11.3.4 for the strain range $\Delta\varepsilon$, the fatigue ductility ε_f and the fatigue ductility exponent c, the importance of the choice of the TCE of the substrate material can be predicted as in Figure 11.46. The ordinate in the figure is the

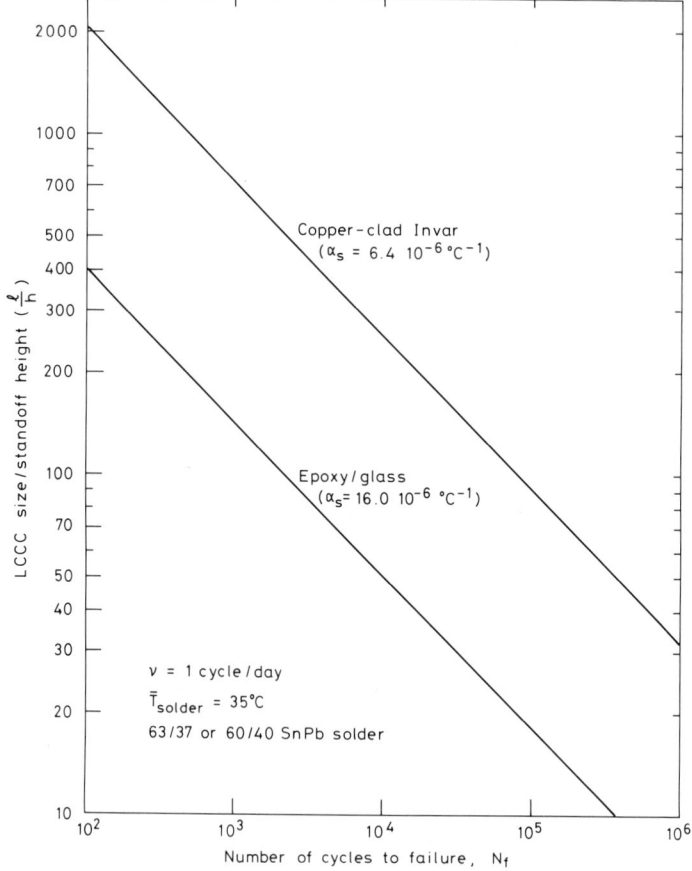

Fig. 11.46 Predictions of fatigue lifetime of LCCCs soldered with near-eutectic SnPb solder to FR-4 and to copper-clad Invar, cycling at 1 cycle a day.

parameter ℓ/h, the ratio of the size of a square LCCC to its solder stand-off height. Two substrate materials are shown: a copper-clad Invar substrate with a tailored TCE of $6 \cdot 4 \; 10^{-6} \; °C^{-1}$ and an FR-4 epoxy-glass laminate with a TCE of $16 \cdot 0 \; 10^{-6} \; °C^{-1}$.

Since these predictions are for power cycling, the frequency is taken, by way of illustration, as 1 cycle per day and the mean solder temperature (between power-off and power-on equilibrium) as $\bar{T}_s = 35°C$.

For a given LCCC and stand-off height it can be seen that the difference between an FR-4 and a copper-clad Invar substrate material represents a factor of 40 in lifetime.

11.3.9 Effect of the Solder Alloy

The strength properties of a solder alloy in a particular regime of operation of a solder joint are dependent upon the inherent physical properties of the bulk alloy, the microstructure which is a function of the cooling rate during solidification, and the composition and structure of the two interfaces of the joint. In the fatigue regime, tests on bulk alloys and on PCB joints have shown that, in general, Sn-Pb alloys exhibit very poor resistance to fatigue, and alternative alloys have been sought especially with a view to the fatigue resistance of large LCCCs soldered to epoxy-fibreglass substrates. These alternatives can be classified into three groupings:

(i) variations in the Sn-Pb ratio away from the eutectic composition;
(ii) minor additions to the basic Sn-Pb alloys;
(iii) alloys other than those based on Sn-Pb.

The fatigue behaviour of four Sn-Pb alloys with compositions 70:30, 63:37, 50:50 and 40:60 have been tested by cycling PCBs with through-hole component leads from $-65°C$ to $+150°C$ for 800 cycles with 30 minute hold-times, and only the eutectic 63:37 alloy showed any electrical failure, indicating the worst fatigue resistance.[440] The improvement in fatigue resistance for off-eutectic compositions is most marked at elevated temperatures and high strain ranges.

The solder alloy most commonly used for the assembly of surface mounting components is Sn62:Pb36:Ag2 which is a eutectic composition in which the silver hardens the tin phase. This alloy shows significantly better fatigue properties than the binary Sn-Pb alloy when tested in plated-through-hole joints, cycling between -40 and $+125°C$. The benefit of the ternary alloy is attributable to its much higher ductility and its resistance to coarsening of the microstructure.

As an alternative to Sn-Pb alloys, a number of other additions to tin have been tried and some of the resulting alloys are listed in Table 11.2. Unfortunately there is no consensus agreement between the test results from different laboratories, and this goes to emphasise the relevance and the effect of variation of the test procedures and parameters (strain rates, joint shape, joint microstructure, etc.).[441-443] For example, the fatigue performance of Sn50:In50 has been assessed as 'very good', and 'poor'; likewise Sn95:Sb5 has been judged to be 'excellent', and 'poor'. The uncertainty of data is particularly clear in the case of Pb50:In50 where fatigue properties have been variously described as 'very good', and 'even worse than the Sn:Pb eutectic composition'. Thus, the assessment in Table 11.2 represents a weighted opinion taken from the published literature.

Of the other binary tin alloys tried, the silver (3·5–5%) binaries give rise to very good fatigue resistance, but the bismuth eutectic composition Sn42:Bi58 has relatively poor fatigue properties.

Creep tests have shown that small changes in the solder composition give large changes in the lifetime and this may be valid also for fatigue. In surface mount joints the quantity of solder is very low and dissolution of the solderable pad materials can easily result in relatively high impurity levels.

Table 11.2
Comparative Fatigue Characteristics of Solder Alloys used in SMT[430]

Performance	Composition	Microstructure
Worst	63Sn:37Pb	2-phase, eutectic
	60Sn:40Pb	2-phase, near eutectic
Poor	62Sn:36Pb:2Ag	2-phase, Ag hardened Sn phase
	65Sn:35In	2-phase, variable composition
	42Sn:58Bi	2-phase, eutectic
	50Sn:50In	2-phase, variable composition
	50Sn:50Pb	2-phase, eutectic + pro-eutectic
Fair	99In:1Cu	2-phase, eutectic
	90Sn:10Pb	2-phase, mostly pro-eutectic Sn
	$99\frac{1}{4}$Sn:$\frac{3}{4}$Cu	2-phase, eutectic
Good	99Sn:1Sb	1 phase, solid solution
	50Pb:50In	1 phase, solid solution
	100Sn	1 phase
Excellent	96Sn:4Ag	1 phase, particle hardened
	95Sn:5Sb	1 phase, particle hardened

11.3.10 Effect of Joint Microstructure

Because of the low melting points of solder alloys (even at ambient temperatures the solder is operating at over ⅔ of its melting point, in kelvin), metallurgical changes occur in the joint over relatively short times. A reduction in the fatigue life of solder occurs as the joint ages, because of the continuing growth of intermetallic compounds at the joint interfaces, and this suggests a steepening of the Coffin-Manson line as the number of cycles increases which could have grievous consequences. This effect is illustrated in Figure 11.47 in which the fatigue life is demonstrated to decline significantly as joints are aged.

Microstructural variations can drastically affect fatigue life at temperatures where significant plastic flow occurs during fatigue (as is always the case for solders). A coarser phase structure results in a reduction in fatigue life. Considering a 63-37% eutectic SnPb solder, the as-solidified structure consists of alternate lamellae of tin and lead phases with the lamellar spacing x being related to the freezing rate R by $x^2R = $ constant. The tendency is for the microstructure to coarsen. By doing so the interfacial area between the two phases is reduced and the free energy is likewise reduced. Both plastic deformation and increasing temperature accelerate the diffusion necessary for the coarsening process.

That a coarsening of the initial phase structure does result in a reduction of fatigue life is illustrated in Figure 11.48, in which the lamellar spacing has been varied.[412] Thus x is required to be small, indicating that a fast quenching from the

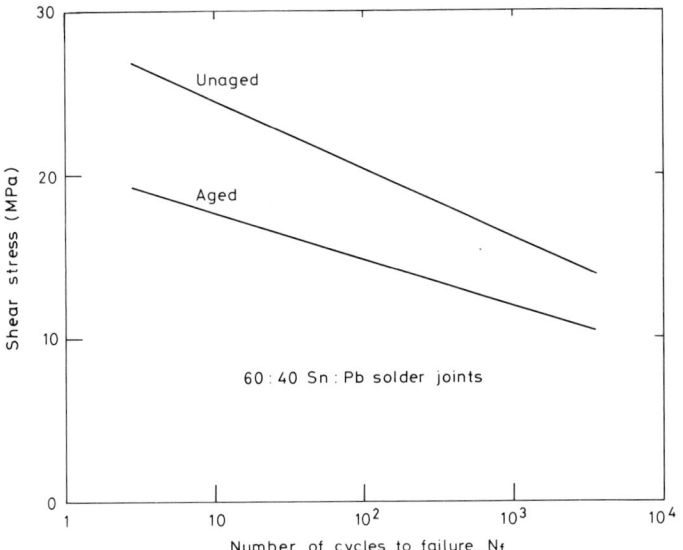

Fig. 11.47 The fatigue life of 60:40 SnPb solder joints declines significantly as the joints age.[406]

molten state is preferred to slow cooling when the joint is being made. This enables the joint at least to start its life with a fine structure.

The measurement of any change of structure is quantitative, in that the interphase spacing alters with the cooling rate of the molten solder, but also qualitative in that, for example, the lamellar structure of the SnPb eutectic composition changes to one with lead islands in a tin matrix. During fatigue, the microstructure coarsens and, whilst the tin-lead phase boundary is the dominant structural discontinuity before coarsening, after coarsening of the crystal structure both the tin-lead phase boundaries and the tin-tin grain boundaries are significant. Thus, to consider only phase spacing when describing the microstructure produced during thermal fatigue is not always sufficient.

Figure 11.48 shows a phase effect, but it has been suggested[429] that grain size is another factor affecting the fatigue life. An initially coarse grain structure, resulting from a slow solidification, apparently decreases the fatigue life, but it is not easy to isolate the effect of a coarse phase distribution from that of a coarse grain structure.

It is believed that plastic deformation of solder, such as occurs during cyclic fatigue, causes the tin-rich phase to go through a sequence of work-hardening, recrystallisation and grain growth whilst the lead-rich phase flows without any hardening or grain growth. Crack formation occurs by intergranular or phase boundary slip and the work hardening can also lead to transgranular cracking in the tin phase.

In an attempt to isolate the microstructural factors that contribute to good or poor fatigue properties of different solder alloy compositions, it has been found that poor properties correlate with a structure consisting of two phases in similar volume fractions, with widely differing mechanical properties. This usually means that a eutectic lamellar microstructure is formed on solidification which is unstable and tends to coarsening during thermal ageing and thermal cycling. The alloys with good fatigue properties, on the other hand, are usually a single phase, as seen in Table 11.2, sometimes hardened by fine precipitate particles, with a structure that is relatively stable during thermal ageing.

It is also probable that, in broad terms, high ductility is an essential property since this gives rise to a high fatigue ductility coefficient ε_f, in the Coffin-Manson relation.

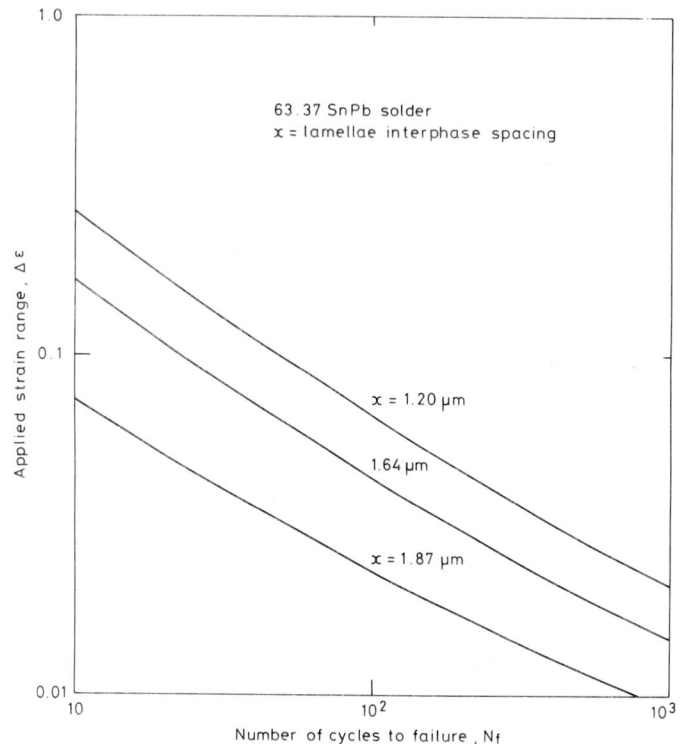

Fig. 11.48 The fatigue life of eutectic SnPb solder improves as the phase microstructure is made finer.[412]

11.3.11 Effect of Temperature

The effect of temperature on fatigue life is not included explicitly in the basic Coffin-Manson relationship, although the temperature range defines the strain range in thermal fatigue. Any temperature effect is included implicitly in the materials property ε_f and in the fatigue ductility exponent c. Thus, from an analysis of the data in Figure 11.36, for example, it is found from the intercepts that ε_f decreases, and from the slopes that c increases, with rising temperature, both of which changes imply a decrease in fatigue life. It is possible that these changes are due to oxidation of the surface contributing to the fatigue process.

Corrosive flux residues left on test joints can also cause a drastic reduction in fatigue life, and such corrosive effects will be very temperature dependent.

11.3.12 Effect of Test Conditions

The aim of all mechanical testing of engineering materials is to provide data for predictive behaviour of the material in service and it is therefore necessary either to simulate service conditions as closely as possible in the testing or to understand the relationship between accelerated test conditions and service conditions. Unfortunately, in complex metallurgical systems such as solder alloys the accelerated effects arising from increasing the temperature, increasing the strain rate, increasing the strain range, etc. are multi-parametric and inter-related so the use of accelerated procedures, even in a qualitative manner, must be regarded with some caution.

Both the monotonic and cycling mechanical properties of solder alloys are dependent on the strain rate and temperature of test. As an example, the yield stress of eutectic Sn-Pb solder at $-50°C$ is five times higher than at $+125°C$ so that, in thermal cycling between these temperatures, the temperature range is not simply giving rise to the test strain range, but is altering the physical properties of the alloy. Thus, if the temperature range, and hence the strain range, is halved one cannot simply assume an increased lifetime as predicted by the Coffin-Manson relation.

In addition, the limit of proportionality, i.e., the elastic limit, decreases by a factor of two from $-70°C$ to $+60°C$. The ultimate shear stress of solder joints is also proportional to the strain rate and inversely proportional to temperature, as shown in Figure 11.9 for 60:40 Sn-Pb solder and varies significantly from one solder alloy to another, as shown in Figure 11.10. Hence the fatigue failure of solder joints is directly proportional to the strain rate, as shown in Figure 11.31, and also, as shown in Figure 11.32, the fatigue lifetime of solder joints decreases enormously with increasing temperature.

With regard to the relevance of testing and measurement to the in-service behaviour of joints, the mean value of the cyclic stress used in fatigue testing dictates considerably the measured fatigue lifetime of solder joints as shown clearly in Figure 11.33.

11.3.13 Choice of Failure Criterion

It is interesting to note that the slope c, of the Coffin-Manson relation, but not the fatigue ductility coefficient ε_f, varies depending upon the choice of the criterion used to define solder joint failure.[401] This is shown in Figure 11.49 for the criteria of an observable crack in the joint and an electrical failure of a joint. Since most of the solder in a joint is redundant as far as its electrical continuity and electrical performance are concerned, an electrical breakdown of a joint is a much less severe go-no go criterion of a fatigue test failure.

Visual inspection for cracks and fractures in solder joints under accelerated fatigue tests, besides necessitating disruptions of the test at periodic intervals and being extremely time consuming and labour intensive, does not provide a definitive and repeatable failure mode. Neither does the periodic testing for a resistance increase in a continuity net which includes the solder joints under test. Its main shortcoming results from the possibility that a static electrical test might not give any indication even of fully developed fractures in the solder fillets because the fracture surfaces, acting like switch contacts, provide low resistance continuity for many cycles after fracture. This criterion can result in fatigue life observations which are more than an order of magnitude too long.

During both thermal cycling in an oven and power cycling of devices, electrical discontinuity typically occurs first for extremely short durations during the heat-

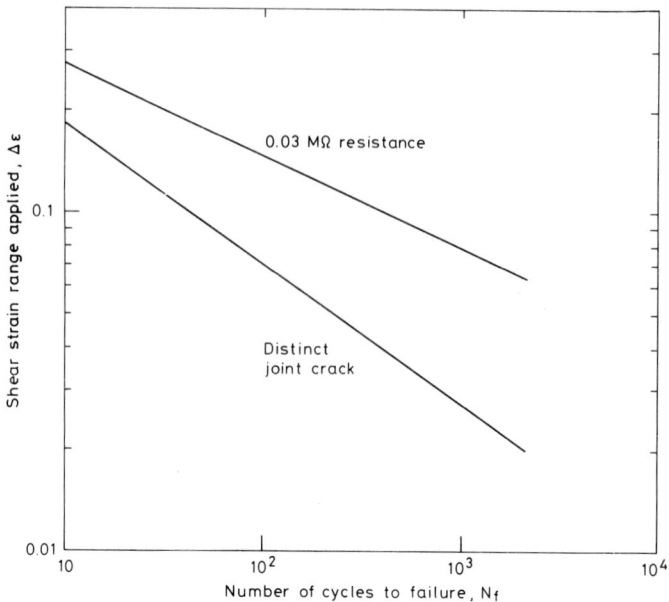

Fig. 11.49 The choice of a failure criterion in fatigue testing is important. Experimental data show that an observable crack in the solder gives a much earlier warning of fatigue damage than does increasing electrical resistance.[401]

up or cool-down transients near the steady state condition, as shown in Figure 11.50. As cycling continues, the discontinuity duration increases, but maybe thousands of cycles pass before the discontinuity is no longer intermittent. Thus, only continuous monitoring during cyclic testing can give an accurate indication of failure. This failure detection criterion not only closely corresponds to the failure indication in actual product operation, but it does not require test interruptions with time consuming and uncertain failure searches.

11.3.14 Validity of Fatigue-life Predictions

Effects occur within the temperature range commonly employed for thermal cycling tests such that distinct differences in plastic strain are induced at each extreme of a cycle which are unique to a particular temperature range and rate of cycling. The dependencies on strain rate and temperature arise because the rate of application of strain is usually so fast that stress relaxation cannot occur in the joints. Stress relaxation is a 'high temperature' process related to creep, converting elastic strain to plastic strain. Most models of thermal fatigue of LCCC solder joints ideally assume all the strain to be plastic, whereas the tests derived from the models usually do not allow time for full plastic strain to develop; actual practical events depend on the strain rate and dwell times at the cycle extremes. The temperature range and the mid-point of the thermal cycle are also crucial, as they determine the maximum, the minimum and the mean stresses of the cycle.

It is evident from these arguments that there must be strong reservations about the validity of simulating long slow changes in ambient temperature or even

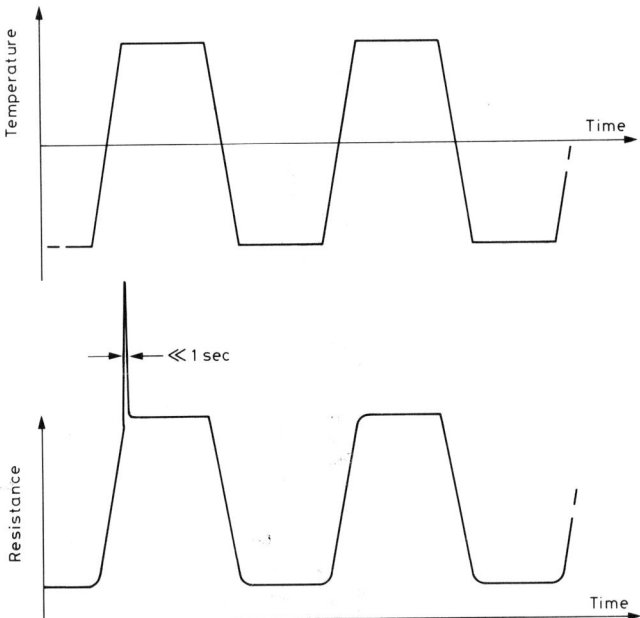

Fig. 11.50 If electrical resistance is monitored continuously during thermal cycling, a short, intermittent electrical discontinuity occurs first during the heat-up or cool-down transients.

rapidly switched power dissipation in surface mounted assemblies by rapid cycling of the ambient temperature.

Almost all the experimental work on LCCC solder joint reliability has been done by thermal cycling to meet the US military specification MIL-883, method 1011. This defines thermal shock/cycling as the stress a device undergoes when it is switched alternately from a bath of hot liquid to a bath of cold liquid, or when it is temperature cycled in an environmental chamber. The method specifies several testing conditions:

(i) condition A states that $+100°C$ is hot and $0°C$ is cold;
(ii) condition F states that $+200°C$ is hot and $-195°C$ is cold; while
(iii) condition B, which is that most commonly adopted, states that $+125°C$ is hot and $-55°C$ is cold.

Such a test as (iii), for example, gives one fatigue lifetime point taken at a strain range corresponding to $\Delta T = 180°C$, as shown in a Coffin-Manson plot in Figure 11.51. The specified service requirement might be for a temperature excursion $\Delta T = 40°C$ at a frequency of 1 cycle per hour for 40 years as, for example, operating calls at a telephone exchange. It has been pointed out[444] that, unfortunately, there are not sufficient experimental data available to confirm that the curve through that single point must follow line (a), given by taking empirical values for the fatigue ductility ε_f and the fatigue ductility exponent c. Very small differences, for example in alloy composition, might well reduce the predicted fatigue life from 100 years, given by line (a), to 15 years, given by line (b) in the figure. It is known too that solder joints undergo metallurgical changes with time due to diffusion mechanisms, and it is not unlikely that the real fatigue

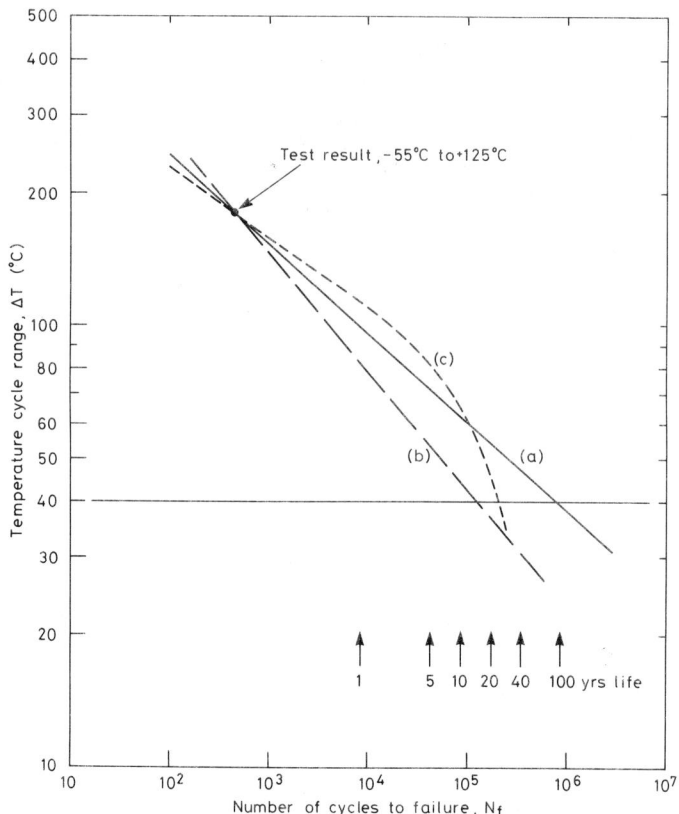

Fig. 11.51 Extrapolation of accelerated fatigue testing to predict service life is fraught with uncertainty. Small errors in the values assumed for ε_f and c can result in catastrophic errors of predicted lifetime. Additionally the Coffin-Manson line is unlikely to be straight.

lifetime curve could be more like (c) in the figure, with serious consequences. Because the fatigue lifetime appears logarithmically in the Coffin-Manson relation, minor changes in the measurement have great influence on the extrapolated predicted lifetime. Furthermore, in service, solder joints are subjected to a succession of temperature cycles of widely varying amplitude and frequency, and methods of summing the cumulative effects are oversimplified since, for example, low amplitude cycles followed by high amplitude cycles are more deleterious than high amplitude cycles followed by low amplitude cycles.

Another factor to be taken into account when predicting fatigue lifetimes is the spread of values measured for the lifetime in the accelerated test regime, and precisely the significance of the predicted lifetime in terms of joint survival rate.

Thus, any predicted solder joint lifetime, as shown in Figure 11.52, has a degree of confidence that is determined by the precision of the measurement made under accelerated conditions.[444]

In summary, inappropriate cyclic testing can lead to the identification of erroneous failure modes, and misleading predictions of lifetime of surface mounted assemblies. The concerns for the test procedures are:

(i) Cyclic stress levels are too high: this can result in failures in the metallisation pads on the solder joints which are stress induced and thus not representative of the strain-induced solder joint fatigue failures.
(ii) Cycling frequencies are too high: this can produce results that are unrepresentative of functional service behaviour, owing to the dwell times being too short for stress relaxation of the solder, and hence incorporating disproportionate influences of second-order effects.
(iii) Temperature extremes are too high or too low: this can result in material behaviour that cannot be related to service conditions.
(iv) Test method is inappropriate: oven cycling of assemblies with matched TCEs ($\Delta\alpha < 2$ ppm.°C^{-1}) does not cause sufficient thermal expansion mismatch to produce failures during reasonable test durations; mechanical cycling may be too fast to allow stress relaxation of the solder; thermal shock cycling gives rise to inappropriate cracking modes.

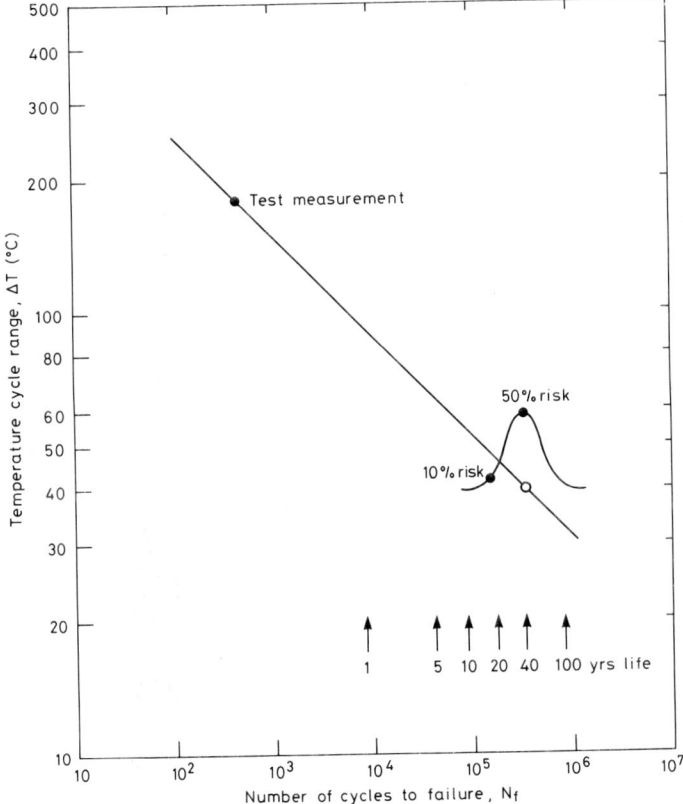

Fig. 11.52 Uncertainty in the accelerated fatigue test measurement and in the slope of the Coffin-Manson line gives rise to a range of predicted service lifetimes, occurring with different probabilities.[444]

Accelerated cyclic fatigue testing methods can be assessed as shown in Table 11.3.

Table 11.3
An Assessment of Accelerated Cyclic Fatigue Testing Methods[432]

Test	Information	Test Complexity	Test Duration
Functional power cycling	accurate	high	very long–long
Oven temperature cycling	good, with caution	moderate	long–moderate
Mechanical cyclic flexing	comparative, with caution	low	moderate–short
Thermal shock cycling	misleading	moderate–low	short

11.3.15 Fatigue Conditions of Solder Joints in Service

Predictive fatigue testing has been criticised on two grounds. First, in reality, cyclic heating of surface mounted assemblies results, more often than not, from power cycling rather than ambient temperature cycling. Secondly, the tests deal with relatively large strain ranges for hundreds of cycles over a relatively short time, whereas real conditions often impose much smaller strain ranges, for a much larger number of cycles, each with a longer hold time.

The types of cycles encountered in service conditions have been discussed. For example, a temperature excursion $\Delta T = 75°C$ and a cyclic frequency $\nu = 1 \cdot 5$ cycles per day corresponds to average switching conditions of consumer electronics. For telecommunications equipment two possibilities have been suggested: for indoor office equipment $\Delta T = 85°C$ (0 to 85°C) at $\nu = 10$ cycles per year, and for outdoor equipment $\Delta T = 125°C$ (-40 to $+85°C$) at $\nu = 25$ cycles per year.

As an example, considering telephone exchange equipment,[445] the temperature cycles experienced by a solder joint during its lifetime have been divided into four categories given in Table 11.4. During manufacture of the assembly, the solder cools from its solidifying temperature to room temperature, and this may be repeated during repairs. When the station is at work the PCB is assumed to be at 80°C during the day, and switched off at night, at 20°C. During operation a frequency of one cycle per hour is assumed.

Table 11.4
Possible Thermal Cycles of Solder Joints in a Telephone Exchange[445]

	T_{max} (°C)	T_{min} (°C)	ΔT (°C)	No. of Cycles
(a) Production and repairs	+183	+20	163	10
(b) Storage and transport	+80	−40	120	500
(c) Day/night, on/off	+80	+20	60	15000 (40 years)
(d) Operating calls	+100	+60	40	350000 (40 years)

Thus, the fatigue life starts with a large strain range but few cycles and finishes with a large number of cycles over a small strain range. Testing is carried out under the conditions of a large strain range for a few cycles, because testing at low strain ranges is impractical due to the very long times required. Thus extrapolation of the Coffin-Manson line from the test result is necessary. In Figure 11.53 are plotted the four fatigue requirements, given in Table 11.4, of each soldered joint on a strain-fatigue lifetime graph. Suppose that an accelerated test at $\Delta T = 180°C$ produces a fatigue life of 1000 cycles, as shown, then the Coffin-Manson line has been drawn with several possible values of the fatigue exponent c. For approval for service, the test prediction line must lie higher than all four points.

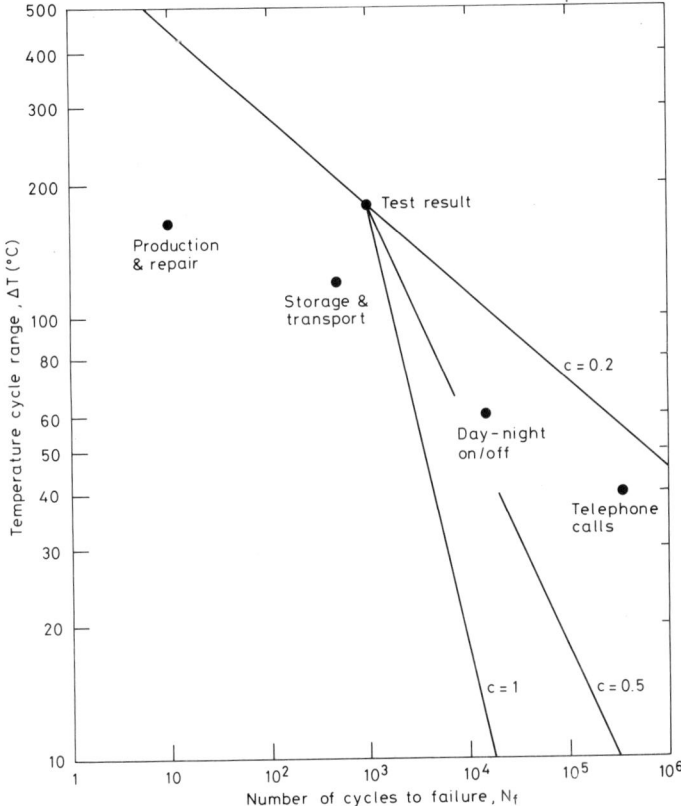

Fig. 11.53 An example of solder-joint fatigue life in service, in a telephone exchange. The test result and its extrapolation to service lifetimes must lie higher on the plot than all the fatigue regimes encountered during assembly, installation and service.[445]

The plot shows that many cycles with a small temperature excursion is more deleterious than a few cycles over a large temperature range. Since the value of c is not known and may indeed change with time, extrapolation from the test condition is very uncertain. The most conservative assumption, that $c = 1$, would be unduly restrictive; the test result would have to exceed 40,000 cycles without failure for specification (d) in Table 11.4 to be met.

It has been pointed out that, at low values of ΔT, the solder joint deformation comes more into the elastic regime than the plastic, which would cause the slope of the Coffin-Manson line to decrease to $0\cdot1-0\cdot2$ in the manner shown in Figure 11.20. Thus the fatigue performance would become relatively better at the low strain ranges encountered in service compared with the high strain ranges encountered under test conditions. The Coffin-Manson line would, under these circumstances, tend to level off horizontally. The value of ΔT (or $\Delta \varepsilon$) at which this happens depends on the nature of the solder joint geometry, the chip carrier material and the substrate material. For solder, the elastic limit is about $0\cdot02\%$ strain and this corresponds to between 5.10^5 and 10^6 cycles. Thus the Coffin-Manson line might not be expected to level off appreciably except at very low strain ranges and corresponding N_f values in excess of 10^5.

Chapter 12

POST-ASSEMBLY OPERATIONS

12.1 CLEANING

12.1.1 To Clean or Not to Clean?

The main function of cleaning an electronic assembly after soldering is to remove flux residues. Pure non-activated rosin flux should, by preference, be left unwashed on the board, but more usually an activated flux is required for good soldering and consequently washing is normally required. If water-soluble flux is used, washing is mandatory. Water-soluble fluxes are washed with water, which is cheap, but subsequent drying can be difficult and expensive.[446] Rosin fluxes are washed away with solvents that must be capable of removing both ionic and non-ionic contaminants from the assembly. The rosin itself is non-ionic but the activators are ionic and it is these activators, acids and salts, which cause the major problems. Being ionic they attract moisture which affects the surface resistivity of the board. It is this moisture problem which makes cleaning essential for any assembly to which a conformal coating is to be applied. The rosin, as well as oil and grease, is non-ionic, but such contaminants can all cause problems for automatic testing by contaminating the probes and also forming an insulating barrier, preventing the test facility from working properly.

The type of flux used in wave soldering or the type of flux in the solder paste is very important in deciding whether or not to clean. The less active fluxes obviously present less of a problem than the rosin activated and the synthetic activated fluxes. In some types of flux the rosin element is designed to harden and encapsulate the activators, rendering them harmless. With this type of flux it is essential that cleaning, if carried out, is done completely and not just sufficiently well to remove the flux but leave the activators, causing more problems than those solved.

The main objections to washing are based on the costs of investment and control. The most expensive feature of washing is the drying. The boards and the assembly of components must be suitable for washing, with no capillaries under the components or open via-holes that may trap the solvent. Also when using organic solvents the environmental and safety costs must be considered.

The advantages of washing, besides being a customer stipulation, are that the assembly is more commercially attractive, is easier to handle and insert into racks in equipment, can be more reliably tested and can be more easily inspected and repaired.

12.1.2 Effects of Contamination

The expense and the procedural problems of cleaning surface mounted assemblies are tolerated because the effects of contamination during service can be catastrophic.[447]

12.1.2.1 CORROSION

Although service failures through corrosion, induced by contaminants left on the assembly, are rare, the results of corrosion can be very serious and it is one of the most feared consequences of contamination. The most common type of corrosion is electrolytic, which occurs when a current passes through an electrolyte between two metallic electrodes, as illustrated in Figure 12.1. It is not necessary for there to be any external voltage applied to the assembly for this to occur since it is common for several different metals to be used in the construction of the assembly, with different electromotive forces. Thus, if two dissimilar metals in electrical contact are bridged by a drop of condensed moisture with a small quantity of ionic contaminant present, a short-circuited voltaic cell is formed and a current will flow. The voltage produced at such a cell is equal to the difference between the electromotive forces (emf) of the component metals, the values of which for some metals commonly occurring in printed circuit assemblies are given in Table 12.1. The values of the voltages are referenced to hydrogen as zero.

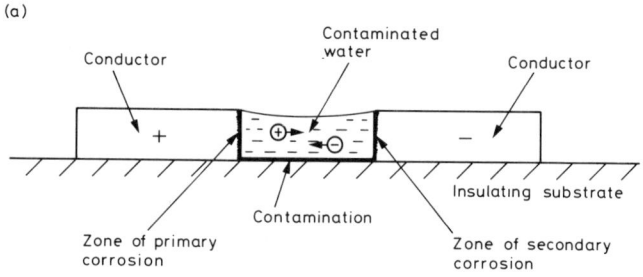

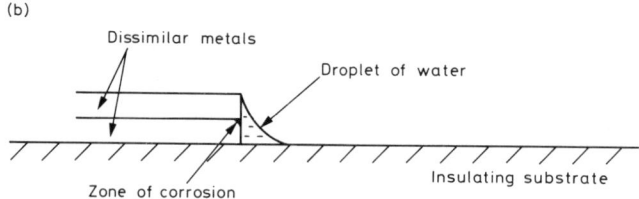

Fig. 12.1 Electrolytic corrosion can occur either (a) in the presence of an electric field between adjacent conductors and a film of moisture or (b) because of an induced voltage between dissimilar metals.

Table 12.1

Electromotive Series of Common Metals (emf at 25°C)

Elements		volts	Alloys	volts
Magnesium	Mg	−2·37	brasses	−0·75 to +0·30
Aluminium	Al	−1·66	stainless steels	−0·30 to +0·10
Zinc	Zn	−0·76	nickel-iron	−0·45 to +0·30
Iron	Fe	−0·45	steels	−0·40 to +0·30
Nickel	Ni	−0·26	bronzes	−0·10 to +0·30
Tin	Sn	−0·14		
Lead	Pb	−0·13		
Copper	Cu	+0·34		
Silver	Ag	+0·80		
Gold	Au	+1·69		

The avoidance of electrolytic corrosion is possible only by removal of all traces of moisture and ionic contamination from the circuit and then by protecting it from recontamination.

Occasionally solder joints may fail by stress corrosion. This is a phenomenon that gives rise to a detrimental effect on the properties of the electronic assembly as a result of a combination of stress and corrosion, the degree of the effect being greater than that expected from either the stress or the corrosion alone.

12.1.2.2 LEAKAGE CURRENTS

The problem of a lowering of the surface resistance of an insulating substrate between a pair of conductors as a result of contamination is much more frequently encountered than corrosion. This problem has become considerably more significant with the growth of microelectronics for two reasons.[448, 449] First, the conductor spacings have become greatly reduced and secondly semiconductor devices have progressed from low impedances between 10 and 10^4 ohms to high impedances between 10^6 and 10^{12} ohms in a continual striving to reduce power consumption. Leakage currents as low as 10^{-12} amperes are sometimes sufficient to cause malfunction of binary gates.

Leakage currents can arise by way of either ionic or non-ionic contaminants. Ionic products on the printed circuit assembly simply act directly as electron carriers to allow a current to flow. Non-ionic products are usually organic compounds such as fluxes used in the manufacture of the assembly, adsorbed on the surface, which can conduct in the presence of a mono-atomic layer of moisture adsorbed from the atmosphere. This type of effect is aggravated through the moisture being acidified by the dissolved carbon dioxide from the air, to form carbonic acid which is itself highly ionic.

Leakage currents can also arise from totally non-ionic phenomena such as external metallic particles (e.g. solder balls), a badly etched circuit, a solder sliver or a metallic whisker growing between conductors. A metallic whisker is a hair-like crystal that grows apparently spontaneously without need of an applied electric field. Typically whiskers have diameters of a few micrometres but may grow to lengths of a few millimetres. Pure tin, particularly electrodeposited tin, is prone to whisker generation whereas tin-lead solder with a tin content below 70% is normally exempt. It is thought that the growth mechanism is initiated by stress in the deposit. Whiskers grow typically at rates of 0·01 to 10 mm per year.

Whiskers are not seen as a result of surface contamination but are mentioned here as a possible source of leakage currents and because of their apparent similarity to dendrites. Dendrites are again metallic filaments that grow from the surface of a metal but by an electrolytic mechanism. It is necessary to have an electrolyte and a voltage for dendritic growth. The speed of growth might be as high as 0·1 mm per minute from the cathode. A similar growth occurs from the anode but at a slower rate. Dendritic growth has been observed with silver, copper, tin-lead, gold, gold-platinum and gold-palladium conductors.

The mechanism of growth of the dendrites is a very localised ionic contamination which, in the presence of moisture, creates a conduction path of extremely small dimensions. A very low current in such circumstances can then function as an electrolytic deposition cell. The anodic growth mechanism is thought to be different and based on the formation of hydroxide ions at that surface.

12.1.2.3 COATING DE-BONDING

Surface mounted assemblies are often protected by a conformal coating. Contamination at the interface between the board and the coating can prevent adequate adhesion and cause the coating to peel or flake away. If a protective coating is required, prior removal of all flux residues is essential.

Conformally coated circuit boards may exhibit a pattern of small white dots in the coating after humidity testing. This effect is sometimes called mealing and its appearance is a definite indication of ionic contamination of the board surface at the time the coating was applied. It is usually a cause for product rejection.

12.1.2.4 WHITE RESIDUES

There are many defects which masquerade under the generic name of white residues, the most common of which being associated with polymerised rosin. If the polymerisation is widespread, one of the milder cleaners may be ineffective in totally removing the residues, resulting in a faint dull whitish film which is non-ionic in nature but is a cosmetic defect and is usually unacceptable. The effect is most commonly observed when using PCBs with a matt dry-film solder resist because such surfaces more readily retain the residues.

Sometimes white residues are ionic in nature, occurring when flux activators or salts formed as a result of reactions with flux activators, are left on the board surface.

12.1.2.5 INSULATING CONTACT SURFACES

A fairly obvious problem that can occur if a board is left uncleaned, or is inadequately cleaned, is that an insulating deposit might be left on contact surfaces such as edge connector conductors and circuit test points. Since rosin is an insulator, if it is left on the contact surfaces a very high resistance can result at a contact.

12.1.3 Solubility

Rosin used for the basis of flux for electronic assembly is not a precisely defined material, being a product from the steam distillation of sap from pine trees. However, for the most part, the rosin solids are mild organic acids, mainly abietic acid plus 10–15% of the pimaric acids. After soldering, the materials to be cleaned include the rosin and its activator, their breakdown products and the metal oxide reaction residues, in addition to any airborne contamination, particulate matter or finger oils that result from handling.

Table 12.2 shows how contaminants to be removed can be classified as ionic (polar) or non-ionic which generally require different types of solvent. Ionic solvents include water, alcohols and acetone whereas solvents for non-ionic contaminants are most typically the chlorinated or fluorinated hydrocarbons.

Table 12.2

Typical Contaminants

Ionic	Non-ionic
etching residues	rosin
plating salts	oil
fingerprints	grease
perspiration	make-up
	hand lotion
	hair oil
	particulates

12.1.3.1 DISSOLUTION

The mechanism of dissolution is the most important one for cleaning electronic assemblies. The contamination becomes dissolved in the liquid which becomes contaminated and must be removed, usually by successive rinses of the same liquid at purer levels each time. With this practice the ideal of perfect cleanliness is only approached asymptotically as the rinsing liquid becomes purer.

The problem of solute saturation of the liquid can arise. Solvents can dissolve only a given amount of a particular contaminant. Certain by-products from the soldering process have a very low solubility, i.e., they are virtually insoluble in water and alcohol, whilst others have a very low solubility in the organic solvents. For example, sodium chloride, a contaminant from fingerprints, has a very high solubility in cold water, some 300 g.l^{-1}, whereas in alcohol it saturates at only a few ppm and in fluorinated solvents it is insoluble. Other examples are the organometallic and inorganic salts of lead and tin produced by reaction of the solder with the flux chemicals. Lead palmitate, for instance, an organometallic reaction product of lead oxide and palmitic acid flux activator, has a hot water solubility of only $0 \cdot 07$ g.l^{-1} and is totally insoluble in alcohol and the organic solvents. Tin chloride, an ubiquitous reaction product when using halide activated fluxes, is, in common with all chlorides, very soluble in water but only slightly soluble in alcohol and insoluble in the fluorinated solvents. On the other hand, abietic acid, a principal constituent of rosin fluxes, saturates in water at less than 1 ppm but is soluble in alcohol.

It is clear that no single solvent can be expected to remove, by dissolution alone, all the contaminants from a soldered electronic assembly.[450] Apart from the non-ionic rosin itself, water better dissolves the contaminants found after soldering electronic assemblies than do the alcohol-organic solvent mixtures. A combination cleaning cycle is occasionally employed, using first an alcohol-organic solvent mixture, followed by a water wash in order to utilise the very different dissolution characteristics of both the solvent types.

12.1.3.2 SOLUBILISATION

Solubilisation describes a means whereby an insoluble substance is rendered soluble. One special case of solubilisation by chemical modification using soaps is called saponification, but other methods can be used.

Saponification is used mainly to render rosin soluble in water, for aqueous cleaning of electronic assemblies, almost all other contaminants being readily soluble in pure water. Alkaline amines, with a pH in the range 11–12, are usually used as saponifiers, and the rosin is converted to a water-soluble soap. The reaction is exothermic and considerable energy arises locally in the region where it is required for the cleaning, helping to dissolve the rosin soap in the solution. Nevertheless, some mechanical spraying is generally required in addition. Saponification is a chemical process and therefore requires an excess of the saponifier to be present to ensure a complete conversion of rosin to soap. Also, because it is a chemical process, the physical conditions in which the reaction takes place, notably the temperature, may also be critical.

Saponification is usually by a chemical reaction but may also be by a physical bonding using long chain molecules, one end of which has a strongly hydrophobic radical and the other end a strongly hydrophilic radical. At very low concentrations in water the distribution of these molecules is random but as soon as the concentration reaches a certain critical value the hydrophobic ends of the molecules are attracted together so that, once a spherical group forms, water is excluded from the centre. This effect is shown schematically in Figure 12.2.

The spherical groups are called micelles, and have the properties of significantly reducing the surface tension of water and of being able to solubilise hydrophobic organic substances such as oils and greases.[451] This occurs because the outer surface of a micelle is extremely hydrophilic. When it comes into contact with an immersed surface the micelle unwinds so that more of its hydrophilic ends have more contact with the water. This is illustrated in Figure 12.3. An extremely strong ionic attraction between the water and the hydrophilic ends of the molecules pulls the hydrophobic substance apart and small droplets of it become completely surrounded to form a stabilised micelle. The solubilisation is then as permanent as the micelle-forming molecules. The mechanism works only if the adsorption bond between the contamination and the solid surface is weaker than both the ionic attraction between the water and the soap molecules and the attraction between the soap molecules and the contamination.

Another approach to solubilisation is to render the cleaning solution slightly acid in order to convert insoluble oxides on metal surfaces into soluble salts.

12.1.3.3 FLUX SOLUBILITY

For cleaning surface mounted assemblies particularly, it is important to maximise the solubility of contaminants in the solvent since only a small amount of the solvent is likely to contact the contaminant if it is lying under components. Also the better the solubility characteristics the faster will the contaminant be removed and the shorter the required cleaning cycle.

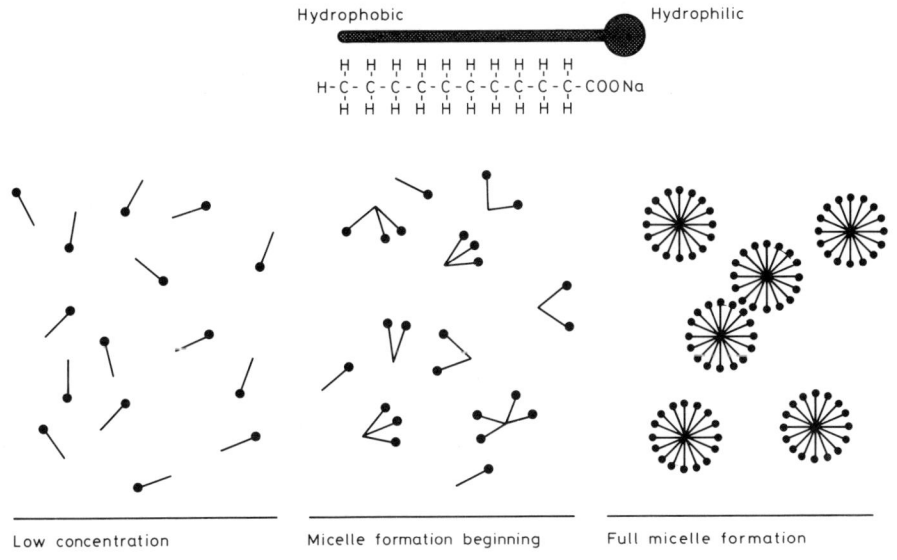

Fig. 12.2 The formation of spherical micelles of saponifier molecules.

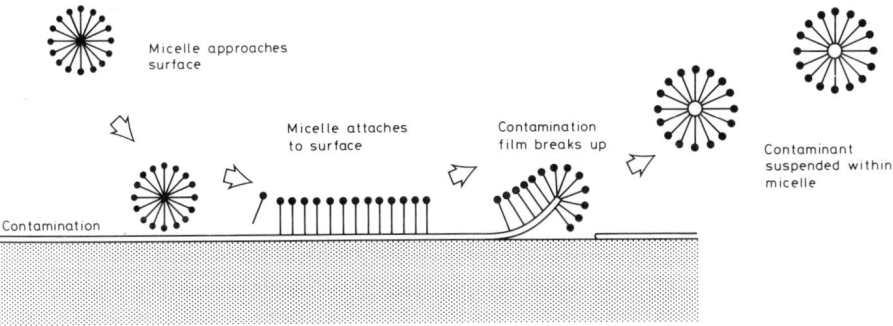

Fig. 12.3 The manner in which a micelle breaks up and removes contamination from a surface.

The solubilities of solvents for rosin flux are often compared in the literature on fluxes on the scale of kauri-butanol values which refers to the specific solubility of a standard solution of kauri gum in butyl alcohol, under certain conditions. However, the solubility of relevant contaminants may be very different and in particular the kauri-butanol values are not relevant to the dissolution of rosin. Kauri gum is a hard fossil rosin that is exuded from trees that have long since decayed. It is insoluble in esters and only partially soluble in hydrocarbons and ketones, whereas flux rosin is soluble in all these three classes of solvent. The kauri-butanol test was originally conceived for hydrocarbon solvents which are not used in defluxing solvents, and it is therefore inappropriate in this context.

The rosin used in fluxes for electronic soldering is a natural product from trees and consists of a mixture of related monocarboxylic acids. These acids all have twenty carbon atoms including three fused carbon rings. Abietic acid is the principal one, which reacts with metal surfaces to form metal abietates.

12.1.3.4 METAL ABIETATES

During soldering with a rosin flux the surfaces of the solder and the terminations can react with the carboxylic acids in the flux to form a class of compounds called metal soaps which are virtually insoluble in water and the fluoro-chlorocarbon solvents. Whilst a whole range of compounds form between tin, lead, copper or iron and the acids in the rosin, the tin and lead abietates have received special attention.[452] The solubility of these two contaminants in four solvents is shown in Table 12.3.

Table 12.3

Solubilities of Metal Abietates at Room Temperature (mg/g of saturated solution)

	Tin Abietate	Lead Abietate
Water	3·2	0·7
Dimethyl-formamide	207·9	99·0
Chlorocarbon blend	275·5	17·1
Fluoro-chlorocarbon blend	40·8	2·2

The lead abietate is considerably less soluble than the tin abietate in all cases. The fluorocarbon fluids are much less useful in removing these residues, which are usually yellow-brown in colour,[453] than are the chlorocarbon fluids.

12.1.3.5 SOLUBILITY PARAMETER THEORY

Solubility parameter theory is applied in a wide range of technologies. It assumes that the solubility of a molecule is determined by its chemical structure and the total solubility parameter S_t comprises three partial solubility parameters: a non-ionic (dispersion) term S_d, an ionic (or polar) term S_p and a hydrogen bonding term S_h. These terms are summed in quadrature:

$$S_t = \sqrt{(S_d^2 + S_p^2 + S_h^2)}$$

The solubility parameters of a great number of pure solvents are tabulated.[454] Solubility parameters of solvent mixtures can be summed in proportion to their volume concentrations:

$$S_t(\text{mixture}) = \sum_i V^i S_t^i$$

where V^i are the volume fractions of the components and S_t^i their respective total solubility parameters. The solubility parameters of solids, such as the constituents of rosin, are not so readily available as those of solvents, but methods exist for calculating values in such cases.

The general principle of the solubility parameter theory is that maximum solubility of a solute in a solvent occurs when their three respective partial solubility parameters are matched. As a measure of how good the match is, the sum of the squares of the differences between each pair of parameters is used:

$$\Delta S = \sqrt{(S_d^i - S_d^j)^2 + (S_p^i - S_p^j)^2 + (S_h^i - S_h^j)^2} \qquad (12.1)$$

where the superscript i refers to the solvent and the superscript j to the solute. The lower the value of ΔS, the better the solvent-solute match and the better the solubility.

The use of Equation (12.1) has been tested[455] for the solubility of pure abietic acid in a wide range of solvents and solvent blends and also for the solubility of rosin flux. The data on abietic acid are shown in Figure 12.4, plotting the measured saturation solubility of abietic acid in the solvent as a function of ΔS. It is seen that the solubility does indeed increase as the solubility parameters of the solvent and the solute become closer in value, i.e., ΔS becomes smaller.

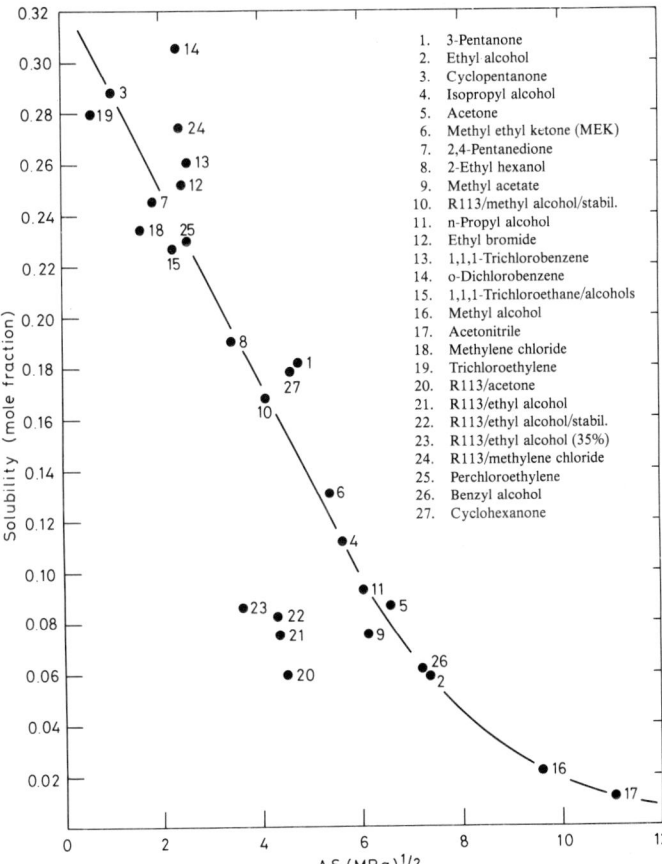

Fig. 12.4 The saturation solubility of abietic acid in a range of solvents and solvent blends as a function of the difference in the solubility parameters between solute and solvent.[455]

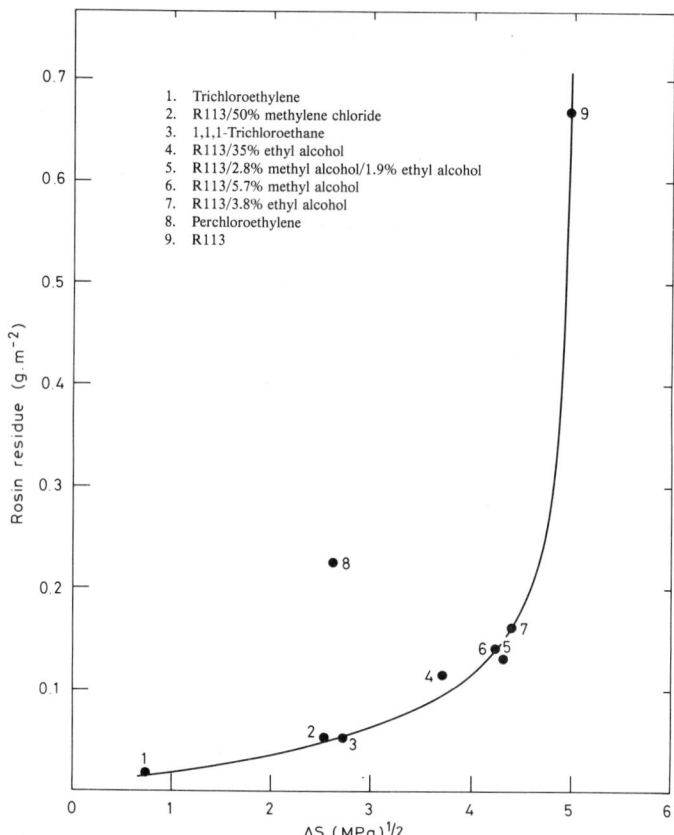

Fig. 12.5 The amount of rosin flux left on a test coupon after 2 minutes ultrasonic cleaning in a range of solvents as a function of the difference in the solubility parameters.[456]

The data on RA fluxes are given in Figure 12.5. In this case the available measurements[456] are of the weight of flux, per unit area, left on test coupons after two minutes of ultrasonic cleaning in the solvent. Again the correlation with the solubility parameter is very good, showing that the theory can be used successfully to select the best solvents or solvent blends for the purpose.

In general, good solvents for ionic contaminants must have high values for the polar and hydrogen bonding partial solubility parameters, S_p and S_h. Thus, for example, the respective values for methanol are 12·3 and 22·3 MPa$^{\frac{1}{2}}$, contrasting with 1.1.2 trichloro-trifluoroethane for which they are 4·3 and zero MPa$^{\frac{1}{2}}$. This also explains why the chlorocarbon fluids have superior ionic solubility to the chloro-fluorocarbon fluids.

12.1.3.6 SOLVENT TEMPERATURE

Solvents can normally be used at any temperature between room temperature and their normal boiling point. Whilst cold cleaning may be adequate for some circuit board cleaning it is normal to use organic solvents at or near their boiling points (in the hot liquid or the vapour state) and to operate aqueous systems at about 70°C.

In vapour cleaning facilities, choice of a higher boiling point solvent, and hence a higher vapour temperature gives rise to a greater volume of solvent condensing on the workpiece for a given initial temperature of the workpiece. This is because the latent heats of vaporisation of all the organic solvent alcohol blends are quite similar. Superheated vapours have been tried as a means to increase the effectiveness of vapour condensation cleaning.

The solubility of a solute in a solvent increases almost universally with increasing temperature—a phenomenon common in daily experience. The quantitative effect of temperature on solubility is often quite dramatic. Thus the higher the temperature, the greater the rate of dissolution of contaminants in the washing solvent.

In the case of cleaning rosin-based fluxes, provided the flux residue has not charred or polymerised significantly during the assembly operation, at about 85°C the rosin begins to soften as its viscosity becomes much reduced. The cleaning process is aided by this softening and so there is an added advantage in moving to this temperature.

The physical properties of a solvent that are mainly responsible for its ability to ingress into pockets and capillaries, and to wet and flow around the assembly, are its surface tension, its viscosity and its density.[457] In measurement reference tables, these properties are usually compared at 25°C, but since these properties, especially viscosity, are temperature dependent, this can be quite misleading.

12.1.3.7 SURFACE WETTING BY SOLVENTS

The physical criteria that determine whether or not a solid surface is wet by a solvent are of course exactly those that determine whether a solid surface is wet by molten solder. The concepts of the Young-Dupré equation (Section 10.2.3) and the contact angle apply equally. The condition that is most desirable for cleaning is one in which the contact angle θ is close to zero and the solvent spreads spontaneously. In such a situation, referring to Figure 10.3, the spreading coefficient is

$$\sigma_{SLV} = \gamma_{SV} - \gamma_{SL} - \gamma_{LV} \tag{12.2}$$

and if $\sigma_{SLV} > 0$ spontaneous spreading will occur. Unfortunately the value of σ_{SLV} is usually difficult to determine from Equation (12.2) because γ_{SV} and γ_{SL} are normally unknown. It is more useful, in the context of solvents, to quote a critical surface free energy γ_C, that can be measured experimentally, where

$$\gamma_C = \gamma_{SV} - \gamma_{SL} \tag{12.3}$$

Then, if the surface tension of the solvent γ_{LV} is less than γ_C, $\sigma_{SLV} > 0$, and complete wetting will occur. If γ_{LV} is greater than γ_C, the wetting will be only partial or non-existent. Values of γ_C for some materials found in assembled PCBs are given in Table 12.4 and some values of γ_{LV} for relevant solvents in Table 12.5. By combining the data in these two tables it can be seen that complete wetting occurs of virtually all the materials listed by all the organic solvents, but not by water.

Table 12.4

Critical Surface Free Energies, γ_C, of Materials Found in Electronic Assemblies

	$\gamma_C (mN.m^{-1})$
polytetrafluoroethylene (PTFE)	18·5
polypropylene	29
polyethylene	31
polystyrene	30–36
polyvinylchloride (PVC)	39
polycarbonate	42
polyamide resin	52
urea formaldehyde resin	61
glass (fused silica)	∼300
ceramics	∼1000

Table 12.5

Typical Surface Tensions of Solvents, γ_{LV}, at 25°C

	$\gamma_{LV}(mN.m^{-1})$
1.1.2 trichlorotrifluoroethane (R113)	17·7
ethyl alcohol	22·3
1.1.1 trichloroethane	25·1
water	72

Even at 25°C, $\gamma_{LV} < \gamma_C$ for the organic solvents, and hence total wetting would occur. Thus one organic solvent does not wet better than another since any decrease in γ_{LV} below γ_C cannot result in a benefit being gained over the condition of total wetting that already exists.

12.1.3.8 CAPILLARY PENETRATION

The clearance of an electronic component over the substrate on a surface mounted assembly may be as little as 60 μm. The penetration of the cleaning solvent into such spaces is extremely important and the ability of the solvent to do so depends on its surface tension, viscosity and density. The movement of the solvent into tight spaces is analogous to the common phenomena of capillary rise and capillary flow. One must consider not only the equilibrium penetration situation but the factors that influence the kinetics of solvent flow through such a capillary.[458]

If liquid is in contact with one end of a capillary-like gap, the liquid is pulled into that gap. The driving force for this capillary action is the pressure differential that exists across the curved meniscus surface of the liquid initially at the entrance to the capillary and then when the liquid is in the capillary. As an approximation, the meniscus can be assumed spherical between two flat surfaces, as shown schematically in Figure 12.6. The pressure differential across the meniscus (Equation 10.1) is, ignoring any hydrostatic pressure from the height of liquid at the entrance to the capillary:

$$\Delta P = \gamma_{LV} \left(\frac{1}{R_1} + \frac{1}{R_2} \right)$$

where the radii R_1 and R_2 in the directions of the principal axes are ∞ and $a/\cos\theta$, where a is the radius of the capillary and θ the contact angle. Thus

$$\Delta P = \frac{\gamma_{LV}\cos\theta}{a}$$

and the equilibrium solvent *penetration* is directly proportional to its surface tension. Provided the surfaces of the capillary are wet by the solvent such that θ is close to zero, the depth of penetration of water into a capillary is some three times further than it is for the chlorinated and fluorinated solvents, since its surface tension is about three times the value.

The *rate* at which the liquid flows into a capillary is a function not only of the surface tension but the viscosity. The rate of flow of liquid through a horizontal capillary of radius a and length ℓ is

$$R = \frac{\pi a^3 \gamma_{LV}\cos\theta}{8\ell\eta} \, m^3.s^{-1} \qquad (12.4)$$

where η (Pa.s) is the viscosity of the liquid. As the viscosity increases so the rate of flow decreases. For a given capillary size the penetration rate is proportional to $\gamma_{LV}\cos\theta/\eta$, evaluated at the relevant temperature. Some typical values are given in Table 12.6.

Table 12.6
Capillary Flow Parameters of Cleaning Fluids

	Temp. °C	θ deg	γ_{LV} mN.m^{-1}	η mPa.s	$\frac{\gamma_{LV}\cos\theta}{\eta}$ m.s^{-1}
Fluorocarbon-alcohol blend	25	~0	17·4	0·66	26·4
Chlorocarbon-alcohol blend	25	~0	25·2	0·80	31·5
Water	25	60	72·0	0·89	40·4
Fluorocarbon-alcohol blend	40	~0	14·7	0·52	28·3
Chlorocarbon-alcohol blend	73	~0	17·0	0·42	40·5
Water	70	45	64·4	0·40	113·8

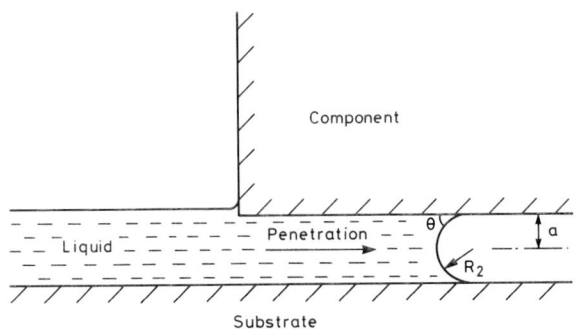

Fig. 12.6 Capillary penetration of liquid, at equilibrium, under a surface mounted component.

Regarding the effect of surface tension, solvent penetration in both aspects of rate and degree is enhanced by a high value whereas the wetting of surfaces is better with a low value. Thus the surface tension should be low enough to allow spontaneous wetting and spreading, while remembering that solvent penetration is aided by having a high surface tension and low viscosity.

12.1.3.9 SHEAR STRESS CLEANING

One aspect of the efficacy of a cleaning action, in addition to solubility, is the shearing force of the moving liquid across the workpiece surface. This effect is particularly important for cleaning under components on surface mounted assemblies where the fluid circulation is critical.[459] Consider, in Figure 12.7, fluid flowing through the narrow gap between the underside of a component and the substrate. The rate of volume flowing into the gap by capillary action alone is given by Equation 12.4, but, if the liquid is forced into the gap, by the use of a spray, for example, the subsequent flow rate is not determined by the surface tension but the applied pressure differential Δp across the ends of the gap. In these circumstances the velocity of liquid through the gap is

$$v = \frac{\Delta p h^2}{12 \ell \eta} \quad [\text{m.s}^{-1}] \tag{12.5}$$

where h is the height of the gap and, as before, ℓ is the length of the gap and η is the viscosity of the liquid. The drag force at the liquid-solid interfaces (equal to the shear stress on the solid surface) is then

$$F_{\text{drag}} = k\varrho A v^2 \quad [\text{N}] \tag{12.6}$$

where k is a dimensionless drag coefficient which has to be determined experimentally (except for judging comparative effects), ϱ is the density of the liquid, and A the projected area of a contaminant, such as a solder ball, facing the flowing liquid.

Once capillary action has drawn the cleaning fluid into the gap, Δp goes to zero unless an external pressure is applied by using a spray or forced movement of the liquid.

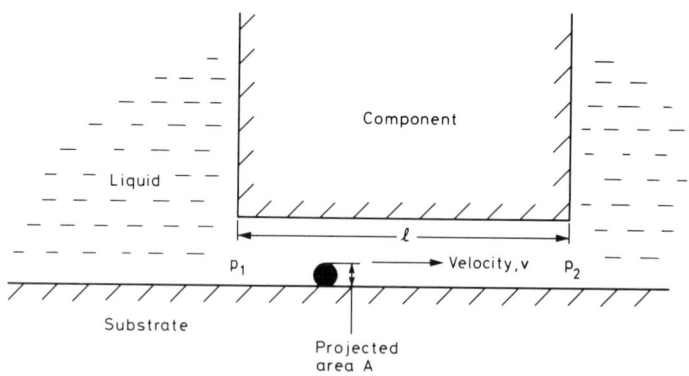

Fig. 12.7 The velocity v of a liquid flowing through a capillary is dependent on the difference in pressure p_1 and p_2 at either end of the capillary, and the viscosity of the liquid. The shearing force for cleaning within the capillary depends on the density of the liquid, its velocity and the projected area of the contaminant.

It is clear from Equations (12.5) and (12.6) that the pressure exerted on, for example, a solder ball under a component, and therefore the ease with which it is removed, increases with increasing density of the liquid, increases inversely with the square of the viscosity and increases with the fourth power of the stand-off height. Thus the stand-off height has a quite dramatic significance in the effectiveness of the cleaning under components by a mechanism of shear stress of the flowing liquid.[460]

12.1.4 Cleaning with Organic Solvents

Organic solvents are commonly employed to remove flux residues if a rosin flux has been used for the soldering. These are typically chlorinated or fluorinated hydrocarbon liquids mixed with ethyl alcohol or isopropyl alcohol.

In a study of ten activated (RA) and mildly activated rosin (RMA) flux types in combination with seven commercially available cleaning solvents designed for flux removal, it was found that RMA fluxes do not necessarily result in less contaminated boards after organic solvent cleaning and, more importantly, the cleanliness results obtained were more dependent on the flux type than on the solvent. Thus the choice of solvent is not usually very critical whereas the choice of flux is, if low contamination levels are a necessity.

Residues of fluxes containing organic acids are more difficult to remove than plain rosin fluxes, especially at soldering temperatures in excess of 250°C when the organic acids tend to decompose, giving rise to the white residues that are insoluble in the solvent blends.

12.1.4.1 AZEOTROPIC SYSTEMS

Organic cleaning fluids used for printed circuit assemblies are invariably a mixture of two or more solvents. This is because, in general, an ionic solvent is required to remove ionic contaminants and a non-ionic solvent is required to remove non-ionic contaminants.

If the fluid mixture has exactly the same composition in the liquid and the vapour when it is at its boiling point, it is described as an azeotrope. This means that it can be distilled without significantly changing composition and is therefore easily purified for recovery and re-use. Azeotropic systems are ideal for vapour cleaning operations. Since they are readily re-used, azeotropic cleaning fluids should contain stabilisers and anti-oxidants to prevent decomposition and a build-up of acids over long-term use.

Non-azeotropic mixtures are generally used below their boiling points, and again they are mixtures of ionic and non-ionic solvents. Unlike the azeotropes these systems fractionate when they are boiled, i.e., the vapour becomes richer in one component than the other. Because of this, the constituent solvents are depleted at a non-uniform rate and the relative concentrations can vary over wide ranges. Such systems are not easily reclaimed by distillation, since they have to be analysed and reblended after distillation. Some non-azeotropic systems can present a serious hazard by becoming enriched in alcohol or another flammable solvent, and continuous monitoring of composition is highly recommended. In a non-azeotropic system, the lowest available boiling azeotropic composition of the solvents will boil first. Thus the vapour will be extremely close to the azeotropic composition whilst the liquid in the sump will become enriched with the highest

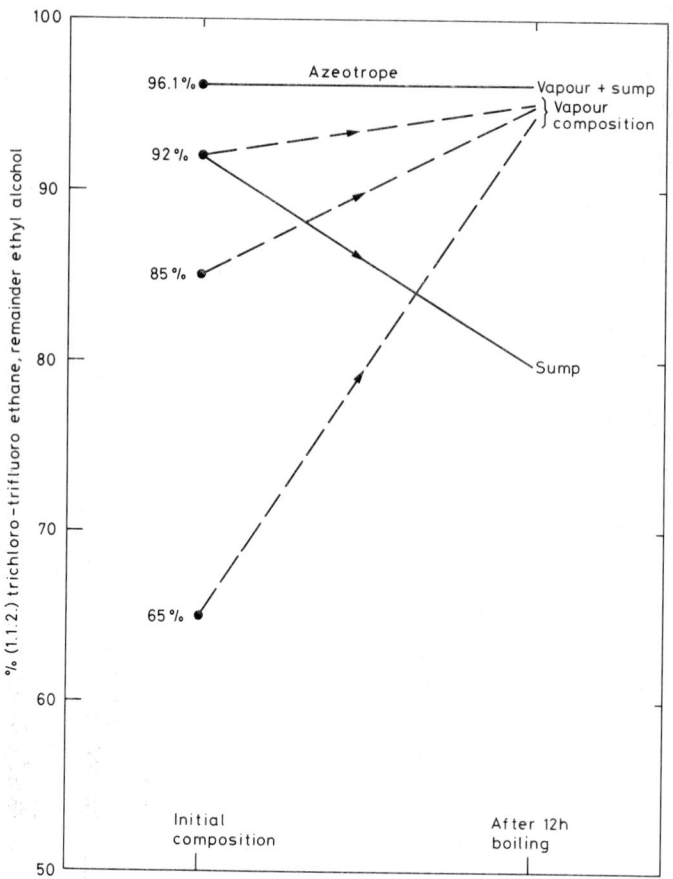

Fig. 12.8 Measured vapour compositions and sump compositions of mixtures of 1.1.2 trichloro-trifluoro-ethane and ethyl alcohol after 12 hours boiling.

boiling point component. This effect is shown in Figure 12.8 for a mixture of a fluorocarbon solvent (1.1.2 trichloro-trifluoro-ethane) and ethyl alcohol. This mixture has an azeotropic composition of 96·1%:3·9% by volume. Addition of excess alcohol up to a ratio of 85%:15% has very little effect on the composition of the vapour whilst the sump becomes further enhanced in the alcohol.

12.1.4.2 CHLORINATED SOLVENTS

Chlorinated solvents as a family can be said to be fairly volatile, fairly toxic, medium cost, highly efficient degreasers but poor ionic contamination dissolvers, aggressive to synthetic polymers, non-flammable, readily available, easily distilled and therefore tend to be chemically unstable.

In Table 12.7 are listed the five most used chlorinated solvents, although carbon tetrachloride and trichloro-ethylene are not popular industrially because of their toxicity. The boiling point of perchloro-ethylene is too high to allow its use generally for cleaning electronic assemblies by vapour degreasing but is frequently used as a cold solvent. 1.1.1-trichloro-ethane is the least aggressive

Table 12.7
Common Chlorinated Solvents

		Boiling Point (°C)	Density at 25°C (g.cm^{-3})	Surface Tension at 25°C (mN.m^{-1})	MAC ppm
Carbon tetrachloride	CCl$_4$	76·5	1·594	26·8	10
Trichloro-ethylene	Cl$_2$C=CHCl	87·2	1·322	32·0	50
Perchloro-ethylene	Cl$_2$C=CCl$_2$	121·0	1·619	32·3	100
1.1.1 trichloro-ethane	Cl$_3$C–CH$_3$	74·1	1·456	25·9	200
Dichloro-methane	CH$_2$Cl$_2$	40·0	1·317	28·2	200

towards plastics and is commonly used in the electronics assembly industry. Dichloro-methane has the lowest boiling point of the common chlorinated solvents and is a useful constituent of many proprietary cleaning fluids.

Also given in Table 12.7 are the maximum allowable concentrations (MAC), which are the peak concentrations of vapour in air, expressed in parts per million, to which a person may be exposed for a maximum of 40 hours per week. Sometimes the threshold limit value (TLV) is used instead of the MAC; this is the average rather than the peak concentration. It has the disadvantage that high peaks are permissible provided the average is below the limit.

As mentioned above, chlorinated solvents are relatively poor removers of ionic contaminants and for this reason are frequently blended with alcohols. The mildest of the chlorinated solvents, perchloro-ethylene, if blended with alcohol, produces a range of excellent cold flux removers. However, even if the ratio of the mixture is about 1:1 the boiling point is depressed only some 10–20°C and so the mixture remains inappropriate for hot use in a vapour reflux system.

Chlorinated solvents are relatively unstable chemically, decomposing under certain circumstances in the presence of water, to undesirable chemicals such as phosgene and hydrochloric acid. The decomposition needs to be catalysed, but in a soldering facility this can be provided by, for example, tin at temperatures in excess of 150°C, a lit cigarette, or any reactive metal such as aluminium or zinc in the presence of flux residues. For this reason, chlorinated solvents are usually stabilised with an amine, which has the property of allowing an accumulation of acidic contamination to occur before the solvent starts to decompose. Suppliers of the solvents offer a simple test from which an imminent onset of decomposition can be identified.

12.1.4.3 FLUORINATED SOLVENTS

Fluorinated solvents are generally much milder than chlorinated solvents, which makes them safer to use with delicate components and with board and component markings. Care, however, must be taken to ensure that the chosen fluid has adequate solvency for the task. As a class, fluorinated solvents have higher evaporation rates, lower boiling points and are generally less toxic than the chlorinated solvents. Fluorinated solvents are customarily used in vapour cleaning systems.

Unlike chlorinated solvents whose chemical names are in common usage, the fluorinated solvents are invariably sold under a trade name, and classified according to an internationally designated numbering system developed for use of the chemicals as refrigerants. Three such solvents given in Table 12.8 are used in the electronics assembly industry, usually mixed with alcohol. However, R-113, (1.1.2 trichloro-trifluoro-ethane) is far and away the most commonly used. Its usefulness arises from a low boiling point but a high vapour density of 7·38 mg.cm^{-3}, more than six times that of air, which prevents undue evaporation of liquid from a deep vapour cleaning chamber. R-11 is less toxic and cheaper but its low boiling point renders it inconvenient in use. R-112 is more aggressive than R-113 but, having a melting point of 26°C, it is solid at room temperature which makes it difficult to use.

For use in post-soldering cleaning of electronic assemblies, R-113 is mixed with one or more alcohols in order to improve the dissolution of ionic contaminants, and stabilised against decomposition, in a similar manner to the chlorinated solvents. The most effective alcohol is methyl alcohol, being the most ionic, but in view of the toxicity of its vapours, either ethyl alcohol or iso-propyl alcohol are normally used. Some properties of azeotropic mixtures of R-113 and ionic solvents are given in Table 12.9. If used in the liquid state rather than the vapour state, it is possible and advantageous to increase the alcohol content to improve the solubility of ionic contaminants; commercial blends with up to 35% alcohol are available.

Table 12.8

Common Fluorinated Solvents

			Boiling Point (°C)	Density at 25°C (g.cm^{-3})	Surface Tension at 25°C (mN.m^{-1})	MAC ppm
Trichloro-fluoro-methane	CFCl$_3$	R-11	23·8	1·467	18	500
1.2 Difluoro-tetrachloro-ethane	FCl$_2$C−CCl$_2$F	R-112	92·8	1·634	23	1000
1.1.2 Trichloro-trifluoro-ethane	FCl$_2$C−CF$_2$Cl	R-113	47·6	1·565	18·7	1000

Table 12.9

Common R-113 Azeotropes

Additive	Concentration (volume %)	Boiling Point (°C)	Density at 25°C (g.cm^{-3})	Surface Tension at 25°C (mN.m^{-1})	MAC ppm
None	0	47·6	1·565	18·7	1000
Methyl alcohol	6·1	39·6	1·488	19·3	300
Ethyl alcohol	3·9	44·5	1·500	19·0	900
Iso-propyl alcohol	2·8	46·4	1·517	18·7	800
Acetone	12	43·6	1·406	18·7	900
{ Methyl alcohol { Ethyl alcohol	0·2 4·3	44·5	1·505	19·0	800

12.1.5 Aqueous Cleaning

The fluxes and contaminants on an assembled circuit board can be removed using washing fluids based on water. There are two different types of process here. First, the flux may be a water soluble product, synthesised from organic acids and not involving rosin, whose residues are intended to be readily removed by water. Secondly, the flux may be rosin based and the aqueous washing fluid is a straight replacement for the organic solvent. Rosin, which is non-ionic, will not dissolve in water, but can be water-washed if it is saponified by suitable chemicals.

12.1.5.1 WATER-SOLUBLE FLUX REMOVAL

The flux vehicle, the activator and the wetting agents of a water-soluble flux should be readily soluble and washable in water. However, the solubility of the thermally modified products arising from an excessive preheat or solder temperature may be less certain. This is because this type of flux usually contains organic chemicals which have a tendency to polymerise under the influence of heat.

The process of washing simply involves copious amounts of water in a combination of spraying, washing, rinsing and drying, with the number of cycles being determined by the purity of the successive rinse waters. The final rinse

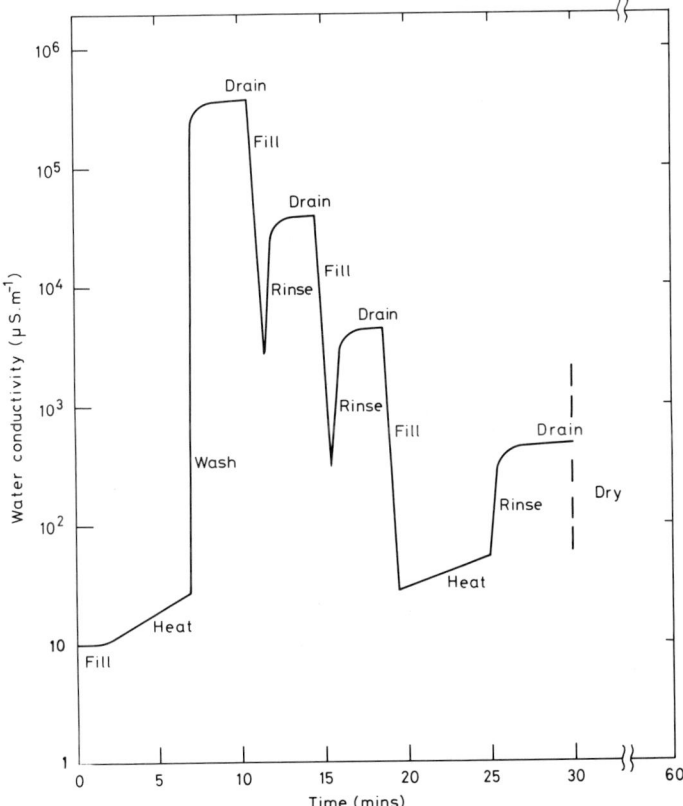

Fig. 12.9 Typical water conductivity in a dishwasher-type cleaner during an operating cycle of 1 hot wash, 2 cold rinses and 1 hot rinse, followed by evaporative drying, all water being fed into the machine at a conductivity of 10 $\mu S.m^{-1}$.

water is always blown off the board rather than left to dry, so that no water drops remain which would otherwise deposit locally any residual salts dissolved. The conductivity of the water after each successive rinse can be automatically monitored to ensure that the cleaning is progressing as specified. A typical control recording of a batch 'dishwasher' system is shown in Figure 12.9. In this instance, the water is fed into each rinse cycle with a conductivity of 10 $\mu S.m^{-1}$, each cycle reducing the conductivity of the rinse water by a factor of about ten.[461]

Large quantities of water are required which must usually be purified before use. The methods of purification and of the disposal of waste water have been discussed in detail.[461] Both batch and in-line facilities are available.

Because some of the residues are only sparingly soluble in water and require protracted times to clean off surfaces, various chemicals added to the water have been used with variable results. These include ammonia and mild caustic solutions.

Large volumes of water and relatively long immersion times are required and consequently care must be taken to ensure that moisture and ionic residues are not trapped in capillaries and under components. A well-designed air blowing system for drying after the final rinse overcomes this problem. There is no guarantee however that the water under components is fully replaced at each washing stage.

12.1.5.2 AQUEOUS REMOVAL OF ROSIN FLUX

Rosin fluxes and their residues cannot be removed by water unless saponified. The end product of the saponification reaction is a rosin soap which is readily soluble in water. Since water is additionally a highly efficient ionic cleaner and much superior to any solvents in this respect, a properly designed water cleaning operation can produce very clean parts. The saponifier itself is of course highly ionic in nature and can be a serious contamination problem unless copious rinsing follows the washing. Again, care must be taken to prevent both the cleaner and rinse water from becoming trapped in capillaries and under components on the circuit board.

12.1.6 Cleaning Techniques

12.1.6.1 LIQUID SOLVENT CLEANING

Cleaning with liquid solvents is usually carried out with mixtures of chlorinated solvents and alcohol in in-line systems. Several techniques are available, the most popular being wave cleaning, brush cleaning and dip cleaning. 'Cold' cleaning is used where the factors of cost and throughput outweigh the need for the highest standards of board cleanliness. It is clear that on surface mounted assemblies the flow of solvent under components and around the solder fillets is quite limited, because of the small capillaries involved. Because of this, cold cleaning is rarely chosen.

Wave cleaning is very similar in concept to a wave fluxer used prior to wave soldering. A broad flat standing wave of the cleaning solution is produced by an immersed pump, over which the soldered assembly passes while it is still warm from the soldering operation. Fairly aggressive solvents are normally used because of the short time for which the assembly is in contact and the solvent must also have a low evaporation rate because of the open nature of the cleaning wave.

Brush cleaning is similar to the wave method but with a brush, rotating counter to the direction of the passage of the work, installed in the centre of the wave. The brush adds an extra mechanical scrubbing action, but of course has no effect under the components.

Dip cleaning consists of immersing the assembly in successively cleaner baths of the same solvent. It is a relatively inefficient and tedious process, usually restricted to small batches. Such a system can be improved by using automatic replenishment of the solvent in the baths, with a cascade concept as shown in Figure 12.10. This type of system can be used for water-based cleaning as well as solvent use.

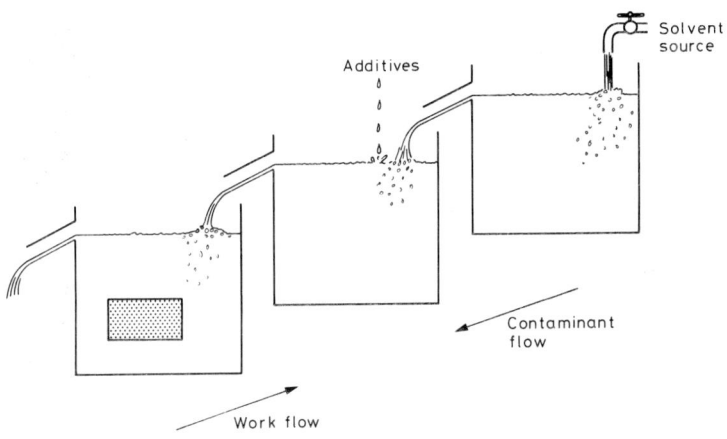

Fig. 12.10 Schematic representation of successive dip cleaning.

12.1.6.2 VAPOUR SOLVENT CLEANING

Most high reliability circuit board cleaning is performed in a vapour degreaser, usually, but not necessarily, employing an azeotropic solvent blend. The solvent is continuously boiled in a heated sump from which the generated vapours rise to an upper cooled zone where they condense. This condensate should be returned to a second sump rather than directly to the primary sump. In this way the solvent is being distilled continuously and the clean sump fed continuously by the cleaned distillate, thereby diluting any contaminants in the clean sump and overflowing them into the dirty boiling sump. At equilibrium, therefore, a vapour degreaser will continuously recycle the solvent from the boiling sump into the vapour layer, from the vapour layer back into the clean sump and finally, from the clean sump, it overflows back into the boiling sump. Figure 12.11 is a schematic of such a system. A typical cleaning cycle combining vapour and liquid solvent cleaning would be as follows: Immersion of the workpiece is first made in the solvent vapour; since the board is colder than the vapour, some solvent becomes dirty so it is advisable to perform this step above the boiling tank to avoid dripping into

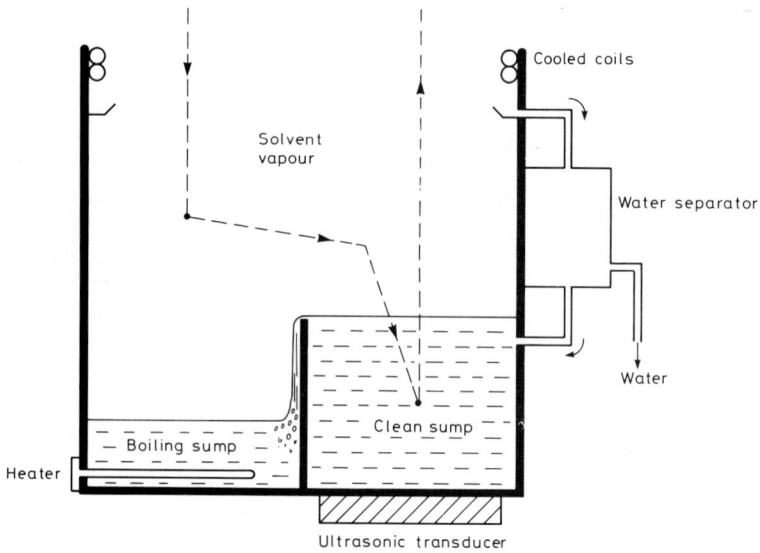

Fig. 12.11 A typical system for solvent vapour cleaning.

the clean solvent. The second step is immersion in the cold solvent which promotes further cleaning and also cools the board. Ultrasonic energy can additionally be used here to improve the effectiveness. As a third step, the board is removed to the vapour layer where there is a further condensation of pure solvent which acts as the final rinse. Obviously azeotropic mixtures are preferable for such a system. Generally, fluorinated solvents are used because they will not damage the plastic components of the assembly, even at the elevated temperature of the solvent vapour.

12.1.6.3 SPRAY SOLVENT CLEANING

High pressure spraying of assemblies has been used extensively with aqueous systems but only recently with solvent systems. The speed and efficiency of solvent cleaning under surface mounted components can be greatly improved by the use of such sprays. Ideally, needle jets are directed across the leading edge of each individual component. Difficult flux residues arising from flux overheating or ageing can be removed in this way by increasing the spray pressure.

The effect of varying parameters has been demonstrated in a model test sample obtained by vapour phase reflowing, under standard conditions, a small deposit of RMA solder paste deposited between two 3 mm thick glass slides incorporating spacers between, and held together using spring clips. After reflow, the flux residue is clearly visible as a halo around the solder. This assembly simulates flux residues sandwiched between a large surface mounted component and a substrate. The cleaning of the flux residue can be assessed visually, through the glass. In this work[462] the solvent used throughout is the stabilised azeotrope of R-113 and methyl alcohol sprayed at 45° to one open side of the assembly through a fan nozzle with an orifice diameter of 1·4 mm and a dispersion of 15°. Figure 12.12 shows an effect of spray pressure on the efficiency of cleaning and Figure 12.13 the effect of the clearance (i.e., the stand-off) on the ease of cleaning. As shown in Section 12.1.3.9, the effectiveness of pressurised cleaning increases as the fourth power of the stand-off so that increasing the clearance by a factor of 3 theoretically reduces the cleaning time by some 80 times.

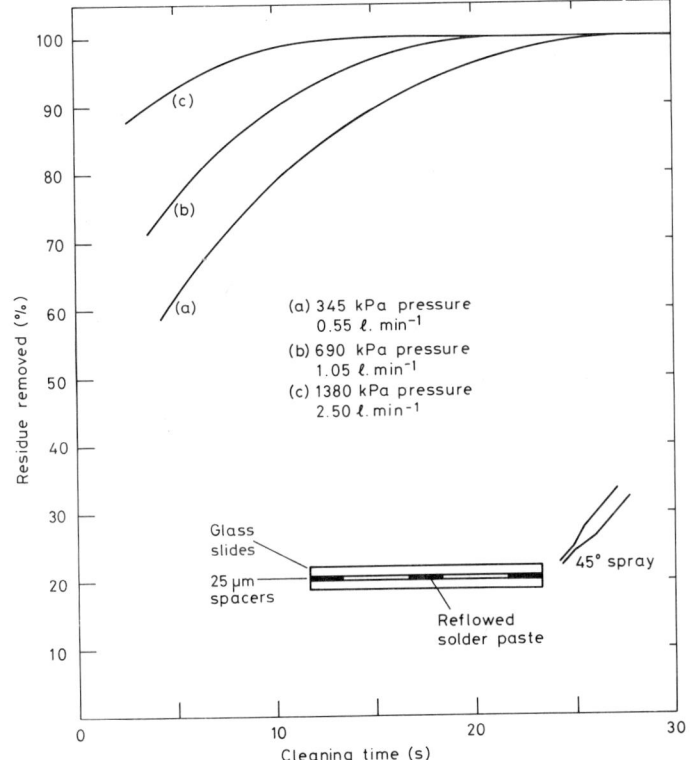

Fig. 12.12 The effect of pressure on the speed of cleaning flux residues within a 25 μm capillary using a solvent spray. The solvent is the azeotrope of 1.1.2 trichloro-trifluoro-ethane and methyl alcohol. The cleaning was done immediately after reflow of the solder paste.[462]

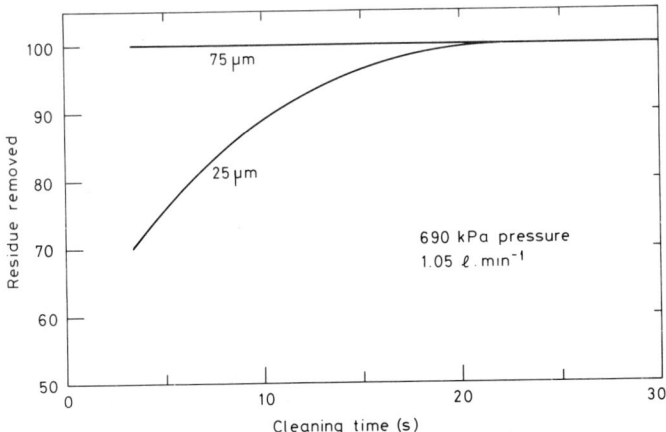

Fig. 12.13 Solvent spray cleaning as in Figure 12.12: the effect of reducing from 75 μm to 25 μm, the component stand-off height from the substrate.[462]

12.1.6.4 ULTRASONIC AGITATION

The effectiveness of cleaning in a liquid, by the wave or dip methods, can be greatly enhanced by the injection of ultrasonic energy to the liquid. The action is to introduce a mechanical oscillation into the liquid solvent of such an amplitude and frequency that the liquid is unable to flow fast enough to follow it. This causes a reduction in pressure in the liquid which, if it falls below the vapour pressure of the solvent, gives rise to spontaneous generation of cavities or bubbles at the points where the local pressure is low, as shown schematically in Figure 12.14. Upon the reverse mechanical cycle, the local pressure rises again and the bubbles implode, the adiabatic compression causing a significant temperature rise. The energy is dissipated as mechanical shock waves which are extremely efficient at dislodging and carrying away surface contaminants. In a properly designed system operating at a suitable frequency, the dirt particles themselves help to nucleate the ultrasonic energy in the liquid, so that the energy is localised intensely exactly where it is most required.

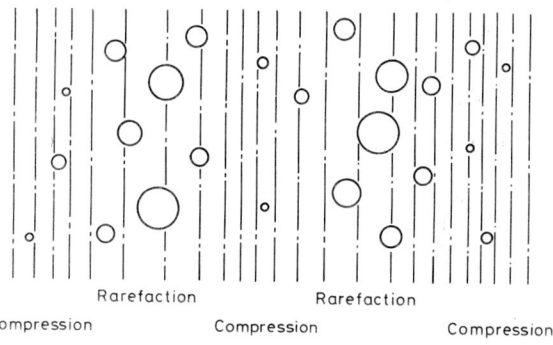

Fig. 12.14 Ultrasonic energy applied to a liquid gives rise to spontaneous generation of cavities and bubbles where the local pressure is low.

The best frequency range for cleaning electronic assemblies is 25–75 kHz and, although the optimum has to be achieved by some trial and error, the exact frequency is not critical.[463] A reasonable starting condition for cleaning immersed electronic assemblies is 40 kHz at an intensity of 1·5 to 2 times that required to induce cavitation in the solvent. Many organic solvents can have significant amounts of gas dissolved which means that the cavitation mechanism is less effective as the cavities will tend to implode on to gas bubbles. In such cases, a new batch of solvent must be degassed by heating it to near its boiling point and, as it cools, subjecting it to ultrasonic cavitation for at least an hour.

In the past there has always been some doubt in using ultrasonic agitation to assist cleaning electronic assemblies because of possible damage or a reduction in reliability of the components themselves. Such events as a lifted solder pad on a PCB and a broken chip-to-carrier bond have been reported, but never in controlled tests to determine whether such problems would have occurred anyway, perhaps at a much more inconvenient time, in service for instance.

The cleaning efficiency—with ultrasonic agitation—has, however, been compared with other methods, in controlled tests and, for surface mounted assemblies, was found to give great benefit. This is because the spaces between

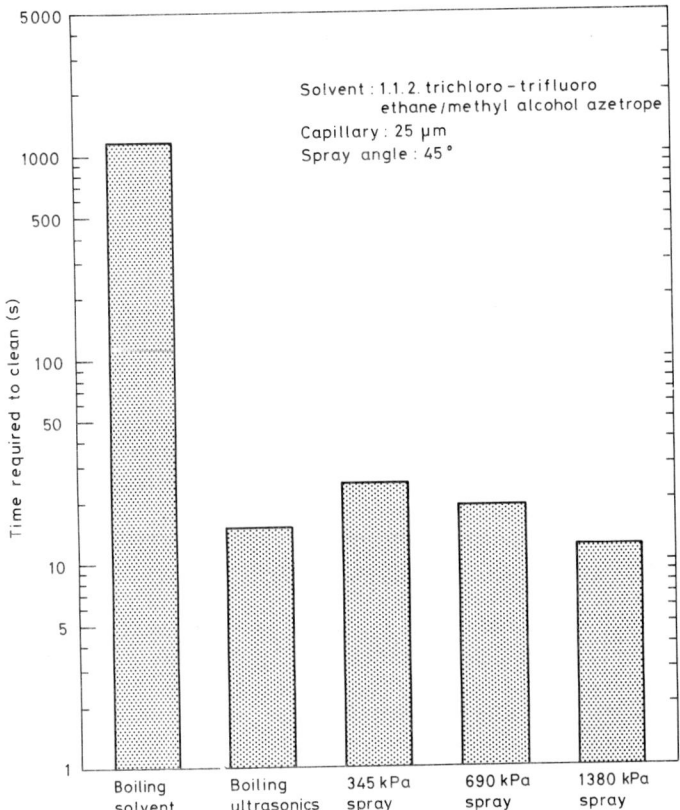

Fig. 12.15 Comparison of the cleaning times to remove flux residues from a capillary using immersion in the boiling liquid with and without ultrasonics, and using a pressure spray.[462]

and under the components are so small that the solvents require additional energy both to dissolve and to wash away the contaminants.

Using the same model test samples described in Section 12.1.6.3, Figure 12.15 gives comparative data for cleaning times with the same solvent as used for the spray data in Figure 12.12. The azeotropic liquid is at its boiling point of 39·6°C and the testpiece is immersed in the liquid. The cleaning time is equivalent to that using a high pressure spray but some 100 times less than that required without the aid of ultrasonic power.

Ultrasonic agitation can be incorporated into wave cleaning as well as dip cleaning processes. It is convenient to use wave systems for in-line facilities, and ultrasonic transducers can be placed on the wave tank such that cavitation is generated on the surface of the workpiece as it progresses through the surface of the wave.

12.1.6.5 WATER CLEANING

There are available a number of batch, conveyorised batch, and in-line machines for water washing but all operate using a combination of high pressure spraying, immersion, rinsing and drying. Their principles of operation and their relative efficiencies have been discussed at length.[464] Basically the mechanisms of cleaning are similar to those of a conventional dishwasher and the degree of cleanliness is dependent upon the number of rinses and the effectiveness of the hot air dryer in removing droplets of water before they evaporate.

12.1.7 Cleaning Considerations Specific to Surface Mounted Assemblies

Because of the very high component packing density, the very small spaces between components and the substrate, the use of adhesives and the use of solder pastes, there exist some very real problems for the cleaning and contamination control of surface mounted assemblies, that do not pose so great a concern for conventional assemblies.

Frequently on surface mounted boards conductors pass under components, not only reducing the effective stand-off of the component but also providing capillary tracks more likely to retain fluids. In some instances small holes have been drilled through the substrate under components to allow for better circulation of cleaning fluids under the components. This is obviously only cost effective in exceptional circumstances.

Assemblies to be wave soldered must utilise adhesives in the process. First, the adhesive will be effective only if applied to a clean dry surface and it may be necessary to check the surface cleanliness and perform a cleaning cycle on the bare boards before assembly. After curing and soldering, the adhesive is responsible for more problems in the cleaning procedure. First, it reduces the space between the component and the substrate and this restricts the flow of solvent beneath the component. Secondly, the profile of the cured blob of adhesive is ill-defined and may well present pockets and capillaries in which fluids can become trapped. Thirdly, the solvent may attack chemically the cured adhesive, giving rise to further contamination that cannot be continually diluted by successive rinses. Fourthly, the adhesive may crack during the soldering because of the thermal shock but this may not be an inspectable or a rejectable condition because the adhesive at that stage is redundant. Such cracks, however, can trap the solvent and give rise to ionic contamination.

Epoxy adhesives offer better resistance to solvent attack than the cyanoacrylate 'superglues'. An alternative approach to the relationship between the adhesive and the cleaner is to use a combination in which the adhesive completely dissolves in the cleaning fluid, a possibility since it is no longer required for mechanical fixing or strength. Such a process involves a gum-type adhesive and aqueous cleaning. It has the disadvantage of requiring inordinately long dissolution, and therefore cleaning, times.

Flux residues that have charred or hardened by polymerisation or decomposition can be virtually impossible to remove. In excess heating reflow methods such as infra-red reflow, local hot-spots on the board surface can result in charring of the flux, whereas in asymptotic heating methods, such as vapour phase reflow, the relatively long times required at temperature can lead to flux polymerisation.

Surface mounting components are required to tolerate total immersion in molten solder and consequently use packages whose materials and design are more robust than leaded components. This has the consequence that it is usually possible to employ stronger solvents for surface mounted assemblies than conventional assemblies. Whilst fluorinated solvent mixtures are most commonly used in vapour phase cleaning, more highly efficient flux removers based on chlorinated solvent mixtures can be used for surface mounted assembly, albeit with penalties of health and fire precaution requirements. One such suitable azeotrope is 64% trichloro-ethylene plus 36% methyl alcohol which boils at 60·2°C.

Aqueous cleaning with saponifiers offers some advantageous possibilities in the cleaning of surface mounted assemblies, because of the ability of water to penetrate into narrow capillaries. However, although water may penetrate to the places where contaminants lie, it is not sure that the latter will dissolve or be washed away. For this to happen, sufficient water has to circulate under the components and rinse away the contaminants. If water penetration without circulation occurs, the problem may even be exacerbated by spreading the contaminant more evenly between conductors and increasing the risk of electrical leakage. In controlled tests, no significantly different cleanliness results of surfaces under surface mounted components have been achieved using either organic solvents or saponified water.

The aqueous route does, however, have a greater potential for use in high pressure jets which, besides being efficient at cleaning flux residues from beneath components, are also effective at removing solder balls. The removal of solder balls that are visible can be undertaken during visual inspection with a suitable tool, but those under components are difficult to see and even more difficult to remove. Potentially they present a very serious problem. Whilst ultrasonic cleaning, if permissible, may succeed, probably the more efficient method involves the use of high-pressure spraying with multi-directional jets directed under the components. In this respect, saponified water cleaning offers a potential advantage.

12.1.8 Measurement of Cleanliness

12.1.8.1 TECHNIQUES FOR CONTAMINATION ASSESSMENT

In order to maintain control over the cleaning process and to ensure that the final cleanliness is satisfactory, it is necessary to monitor and measure the degree of cleanliness. There now exist surface analytical instruments involving such techniques as Auger electron spectroscopy, X-ray photoelectron spectroscopy and secondary ion mass spectrometry that are capable of analysing in a quantitative manner the chemical composition of the outermost few atom layers on a surface, either of small identified areas some tens of nanometres in dimension or average measurements of areas some millimetres in dimension. Such techniques are available in research and development facilities but are too expensive and are inappropriate for production tests. The simpler quality assurance techniques used routinely, however, are restrictive in three ways:

(i) they monitor only ionic contaminants;
(ii) they measure the contamination washed off the assembly, not that left residing on it;
(iii) they measure an average contamination for the assembly, whereas corrosion, for example, is dependent upon very local concentrations of ionic contaminants.

The principle of the instruments used to assess contamination levels is to measure the conductivity of a given volume of water or other solvent after it has extracted the soluble ions from the surface of the assembly.[465] This measurement is compared with the initial conductivity of the test liquid. The test may be either static or dynamic.

Thus, the simplest method is to immerse a sample circuit board into pure water under given conditions and monitor the change in electrical conductivity of the water. There are two reasons why this method is not ideal. First, the relatively high surface tension of pure water (72 mN.m^{-1}) means it is difficult for the water to circulate through all the small paths under components; the water may well gain access but does not necessarily circulate. The second reason is that many of the non-ionic contaminants found on assembled circuit boards are not soluble in water and the ionic contamination may be prevented from coming into contact with the water by these other contaminants.

Both these problems are resolved, at least in part, by using a mixture of water and alcohol. A 50% mixture with iso-propyl alcohol, for example, reduces the surface tension to about 25 mN.m^{-1}, encouraging much greater circulation around the components and at the same time providing solubility for the non-ionic residues such as the rosin, thus ensuring that the ionic flux activators, for example, are also dissolved.[466]

In a test instrument the water-alcohol solution may be pumped at a constant rate through a recirculating loop containing an ion exchange column to remove all traces of ions from the solvent before it enters the test tank. Without the sample in the tank the recirculating liquid soon reaches a high purity level. With a soldered assembly in the tank, the conductivity of the liquid rises in proportion to the ionic contamination being washed off it. The data over a period of time are integrated to yield a measure of the total amount of ionic contamination removed from the assembly. An alternative approach is simply to allow the dissolved contamination in the circulating fluid to accumulate in the liquid by by-passing the ion extractor once the workpiece is introduced. In this method no integration of data is necessary. The instrument is easily calibrated by injecting a known amount of sodium chloride into the tank as a well-defined contamination level.

12.1.8.2 LEVELS OF IONIC CONTAMINANTS

Ionic contamination is measured as a function of the change of conductivity of a standard solution. It is usually expressed in terms of the amount of sodium chloride that would give the same change in electrical conductivity.[467] Thus the often seen unit is 'equivalent micrograms of sodium chloride', (μg. eq NaCl). If a flat object has a measured amount of known contaminant evenly distributed over its whole surface, its numerical value can be expressed as weight per unit area, usually μg.cm^{-2}. However this supposes that the nature of the contamination is known and that its distribution is uniform over a flat surface. It is necessary to define the surface area over which the weight of ionic contamination is measured, in order to express the contamination level as μg. eq NaCl.cm^{-2}. It is contentious as to the area that should be taken—to include or exclude the surface area of components in particular. The surface ionic contamination per unit area of component packages is only about one tenth that retained on the substrate and it is therefore advisable, as a safety factor, to incorporate just the area of the bare board into the quantification of the measurements.

The resistivity (or more correctly the specific resistance) of a substance, in this case the rinse liquid, has dimensions ohm.metre (Ω.m) and is the resistance between opposite faces of a cube of the substance. The conductivity K_e (or more correctly the specific conductance) is the inverse of the resistivity with units of 1/(ohm.metre) or siemens/metre ($\Omega^{-1}.m^{-1}$ or $S.m^{-1}$).

In a dilute solution such as the test rinse water-alcohol solvent, the ionic contaminants dissolved are fully dissociated, and the conductivity is then proportional to the number of ions present, because each ion has its own specific conductance (or specific resistance). Thus the conductivity of the solution equals the sum of the products of specific conductance and concentration for each type of ion, i, dissolved:

$$K_e = \sum_i \text{(specific conductance)}_i \text{(concentration)}_i$$

Carefully deionised water, at 25°C, has a resistivity of about $0 \cdot 17$ MΩ.m or a conductivity of 6 μS.m^{-1} resulting from H$^+$ and OH$^-$ ions.

Figure 12.16 shows the relative effects of additions of NaCl and other salts and acids on the conductivity of water. The abscissa is specified in terms of ion concentration, i.e., gram-equivalent per cubic metre of liquid. (The equivalent weight of a compound is its molecular weight/ionic valency, and the gram-

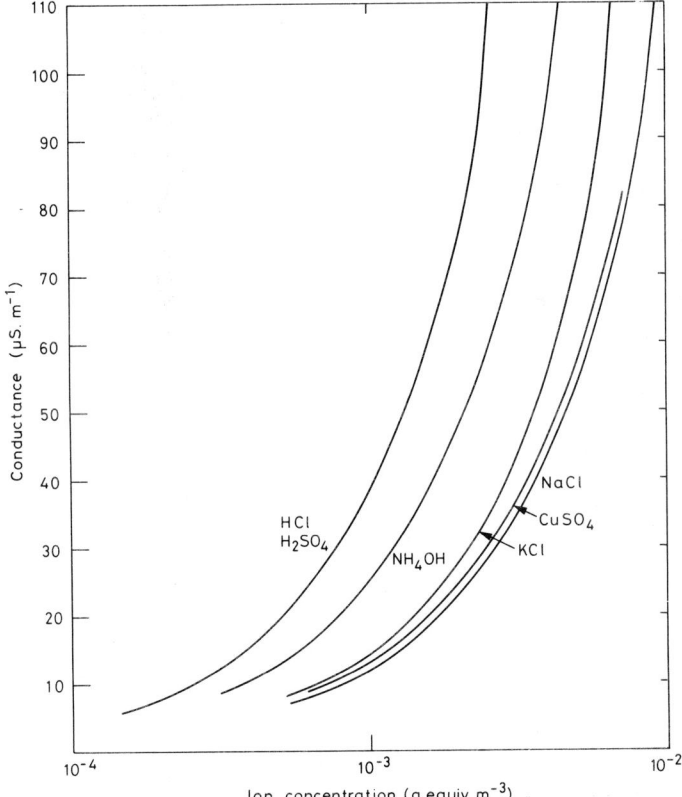

Fig. 12.16 The effects of additions of NaCl and other salts and acids on the conductivity of water. Because the curves are of the same shape the relative effects can be represented by a single number, as given in Table 12.10.

Table 12.10

Concentrations that have the Same Ionic Conductivity, of Common PCB Contaminants, Referred to Sodium Chloride

		Relative Concentration ($\mu g \cdot m^{-3}$)
Carbonic acid	H_2CO_3	0·16
Hydrochloric acid	HCl	0·18
Sulphuric acid	H_2SO_4	0·24
Malonic acid	$HOOC \cdot CH_2 \cdot COOH$	0·27
Succinic acid	$HOOC \cdot CH_2 \cdot CH_2 \cdot COOH$	0·31
Calcium hydroxide	$Ca(OH)_2$	0·31
Sodium hydroxide	NaOH	0·35
Propionic acid	$CH_3 \cdot CH_2 \cdot COOH$	0·40
Ammonium chloride	NH_4Cl	0·72
Calcium chloride	$CaCl_2$	0·88
Sodium carbonate	Na_2CO_3	0·96
Sodium chloride	NaCl	1·000
Potassium chloride	KCl	1·02
Copper (II) chloride	$CuCl_2$	1·12
Sodium succinate	$NaOOC \cdot CH_2 \cdot CH_2 \cdot COONa$	1·61
Copper sulphate	$CuSO_4$	1·28
Lead chloride	$PbCl_2$	2·04

equivalent is that weight in grams.) The equivalent weight of sodium chloride is 58·4 so that, if, for example, a solution has a concentration of 0·5 g.m^{-3} of sodium chloride, its ion concentration is 8·6 10^{-3} gram equivalent per cubic metre and hence, from Figure 12.16, the conductance is about 95 $\mu S \cdot m^{-1}$.

Since the curves have an approximately common shape the relative effects remain nearly constant over the relevant range of ion concentrations. Thus the relative effect of a particular ionic contaminant can be expressed in terms of the effect of NaCl, by a factor that accounts for the relative translation of the curves in Figure 12.10 and the equivalent weights of the solutes. Some approximate comparative values are given in Table 12.10 for a range of compounds that are known common contaminants on assembled PCBs. Thus if, for example, a conductivity is measured that is equivalent to 1 $\mu g \cdot cm^{-2}$ NaCl, but the contaminant is actually sodium hydroxide, the real concentration is only 0·35 $\mu g \cdot cm^{-2}$ NaOH. The smaller the number in Table 12.10, the more dangerous is the contaminant, weight for weight.

12.1.8.3 LEVELS OF NON-IONIC CONTAMINANTS

A standard extraction method for ionic contaminants exists because water is a universal solvent for such contaminants. No such universal solvent exists for non-ionic materials, and even if it did it would probably also dissolve some of the materials of the electronic assembly itself. Thus, it is not easy to judge contamination levels of ion-ionic residues using a solvent extraction method. As mentioned before, the surface analytical techniques of Auger electron spectroscopy, X-ray photoelectron spectroscopy and secondary ion mass spectroscopy are capable of making a fingerprint identification of residues with an approximate value of concentration. However, these techniques are impractical for routine quality control.

Several extraction procedures have been attempted for various classes of contaminants, for example water-soluble and alcohol-soluble. In some cases the procedure is quite specific to one contaminant, e.g., rosin. One such test, which is extremely sensitive, is a particularly valuable non-destructive procedure for checking for rosin residues left under surface mounted components. The procedure consists of running a few drops of reagent grade iso-propyl alcohol from an eye dropper across the surface area in question, allowing it to drip on to a filter paper. After a few minutes it is dry, and to it is applied a drop of a syrup made by dissolving 300 g of sucrose in 1 litre of distilled water. Then after a few minutes, one drop of reagent grade concentrated sulphuric acid is applied to the same spot on the filter paper. If rosin has been present in the iso-propyl alcohol, the filter paper turns pink or red, with sufficient sensitivity to detect a few micrograms.

Most of the common non-ionic contaminants such as sugars, polyalcohols and the constituents of rosin can be selectively measured using ion chromatography of water-alcohol extraction solutions. The principle of ion chromatography for a non-ionic solute depends on finding the correct agent which will react with it to convert it into an ionic product which will hydrolyse. For example, rosin can be saponified in sodium hydroxide, the resultant soap being ionogenic. Contamination testers for specific substances are available which perform this operation automatically to give directly a concentration measure of the contaminant.

12.1.8.4 INSULATION RESISTANCE TESTS

The measurements of cleanliness hitherto described depend on a change in property of a solvent used to remove any contaminant on the board. An insulation resistance test attempts to measure the effect of contamination remaining on the board. However, there are numerous reasons, other than surface contamination, why poor electrical resistance can occur. Conversely there are numerous contaminants which do not adversely affect the electrical resistance but manifest themselves in other unacceptable ways. For example, a silicone oil contamination may actually improve the surface insulation but would prevent adhesion of a conformal coating. Insulation resistance tests are complementary to the extraction ionic contamination tests and cannot be used as a replacement.

It is the surface electrical insulation that is of paramount importance for surface mounted assemblies. The usual way of measuring it is with an interleaved double comb pattern such as shown in Figure 12.17, having uniformly spaced copper conductors normally produced on a separate board that undergoes all the processes of the assembled PCBs in the same batch. The surface resistivity is calculated from the measured resistance, the separation and the total overlapping lengths between the two electrodes.

The measured value of resistance between the combs is frequently dependent upon the atmospheric temperature and humidity, in addition to the surface residual contamination, and care must be exercised in the interpretation of the acquired data. More fundamentally, the resistivity across the comb pattern on a test panel in a batch of assemblies may have little relation to that under a surface mounted component that is much more difficult to clean.

If a surface resistance test is called for in a specification, care must also be

Fig. 12.17 A standard conductor comb pattern on a PCB for monitoring the surface resistivity.

taken to avoid voltage induced changes in the materials at the field strengths required to produce easily measured current levels. The resistivity calculated from current-voltage data may be significantly voltage-dependent because of migration of ions within the surface region of the laminate, under high potential stress, causing an artificially low resistivity at high voltages. Modern circuitry, unprotected by any coating, normally has a voltage gradient between printed conductors of 10–20 V.mm^{-1}, and test conditions should therefore be limited to around 100–200 V.mm^{-1}, sufficient to allow a reasonable safety margin.

12.1.8.5 PERMISSIBLE IONIC CONTAMINATION

From practical experience of failure mechanisms on high reliability printed circuit boards due to contamination, it is suggested in some specifications that an acceptable upper limit for surface ionic contamination be set at 1 μg.cm^{-2} eq NaCl as measured using a water-alcohol extraction technique with a calibrated instrument. For surface mounted assemblies this level is better set at 0.5 μg.cm^{-2} except if the circuitry is of low impedance or if a lower reliability is more acceptable than the problems associated with more stringent cleaning.

The higher cleanliness level is required for surface mounted boards for three reasons. First, conductor spacings are usually much smaller so that the surface resistivity must be equivalently greater to retain the same inter-track resistance. Secondly, surface mounted assemblies are of modern design and therefore tend to operate at low power with high impedances and hence the leakage currents tolerable are necessarily much lower. Thirdly, the cleaning quality under surface mounted components is more difficult to ensure than under inserted components, and it is therefore necessary to have a wider safety margin on average contamination levels.

12.2 INSPECTION OF SOLDERING QUALITY

In the ideal situation, after the board has been assembled with electronic components, no inspection of the soldering quality should be required since every manufacturing operation and materials property should be within a defined specification. However, each of these parameters has a gaussian distribution of probability of possessing a certain value and, in a multi-parametric manufacturing route such as exists here, there must always remain a finite chance that the tails of these distributions will conspire to produce an end result that is out of specification. It is therefore customary to make some kind of inspection of the solder joints in addition to subjecting the assembly to a functional test.

The inspection and assessment of solder joints on a printed circuit assembly is probably the most difficult task required of the quality assurance department. Whilst there are becoming available instruments designed to inspect solder joints automatically against fixed criteria, most inspection continues to be carried out visually.

12.2.1 Visual Inspection

12.2.1.1 CLASSIFICATION OF DEFECTS

It is generally agreed that there are three classes of defects arising from the soldering process of a printed circuit assembly. These are major, minor and cosmetic. The major defects are those which prevent, or are likely to prevent at some time during future service, the assembly from functioning to its specified performance. Minor defects are those which probably will not reduce materially the functional ability of the assembly for its intended purpose. Cosmetic defects are those which may affect a perceived appearance requirement but do not reduce the performance of the assembly during its service lifetime.[468]

Major defects must be repaired, whilst the repair of minor defects depends on the specified requirements dictated by the end use of the assembly, be it ultra-high reliability space or medical use, high reliability telecommunications or process control use on the one hand, or consumer electronics such as televisions or pagers on the other hand. The identification of cosmetic defects, whilst not requiring obligatory repair, should invite an investigation of their cause.

Major defects commonly occurring on surface mounted assemblies are:

(i) *No solder:* no continuous solder connection between the substrate and the component metallisation. This is commonly called a skip in wave soldering, arising from a shadowing effect of the solderable surfaces by a component body as the board passes through the solder wave (Section 5.5.4). The defect may arise also because either of the two surfaces to be joined has poor solderability or because the circumstances might be conducive to the problem of tombstoning (Section 10.7.2) when one termination of a chip component is wet before the other. Non-wetting of solderable surfaces can also arise from contamination or a fault in the application of the solder paste or of the solder resist.

(ii) *De-wetting:* a lack of solder adhesion to either the substrate or the component metallisation, which may result in an open circuit. De-wetting, as discussed in Section 10.3.6, is the withdrawal of molten solder from a solderable surface which initially gave the appearance of being wet. The phenomenon is characterised by irregularly shaped areas of thicker solidified droplets of solder, between which is a thin coating of solder. De-wetting can arise if the solder coating is too thin. If too much heat is applied during the formation of the joint, the intermetallic compound grows through to the surface. It is this change in the surface energy of the solder-substrate surface that normally gives rise to the energy-beneficial change in the surface structure of the molten solder. Another cause of de-wetting is particulate contamination on the surface to be soldered. Initially, the hydrostatic pressure of the molten solder covers over small non-wettable areas but then, as the solder film spreads and becomes thinner, it de-wets from these areas.

(iii) *Solder bridges:* unwanted continuous solder paths between two or more conductors that should not be short circuited in the electronic design. Bridges are an indication of incorrect process conditions in a wave soldering machine or poor design of the board layout and footprints. The conditions giving rise to solder bridges during wave soldering and the means by which bridges can be minimised have been discussed in Sections 5.7.1 and 5.10.2.1.
(iv) *Cracking* in the solder joint: which may cause an open circuit or an intermittent open circuit during operational or environmental stress. Cracked solder joints are relatively rare on newly soldered assemblies, but can arise during the cooling of the solidified solder if components are constrained, for example, by the mounting adhesive.
(v) *Solder balls:* solder paste particles or agglomerations of particles that have not become amalgamated within the solder fillet during reflow and are thus left stranded as isolated solder balls. The methods and the paste specifications by which the problem can be minimised were discussed in Section 6.10.
(vi) *Solder residues:* besides the specific cases of solder balls and solder bridges, other causes of excess solder can be classified as major faults if they create electrical shorts. In such cases the solder in question has formed a web or a skin over a part of the surface of the board, which arises during wave soldering from excessive oxidation of the wave at the exit point of the board, usually associated with faulty adjustment of the fluxing station.
(vii) *Disturbed solder*: arises if the two component parts of the joint move with respect to each other during the time that the solder is solidifying. The surface of the solidified solder is not smooth and the thin oxide formed has become folded into the surface. The mechanical properties of such joints are usually considerably reduced.
(viii) *Whiskers and dendrites:* needle shaped metallic growths extending out from conductors which may eventually cause electrical shorting. This was discussed in Section 12.1.2.2.
(ix) *Misalignment of components:* a displacement sideways or rotationally of a component with respect to its footprint on the substrate which may affect the acceptability of the solder fillet or the electrical characteristics in relation to the neighbouring components. Misalignment can be caused by a placement machine that is out of specification, or a subsequent movement prior to or during the soldering.

Minor defects commonly occurring on surface mounted assemblies are:
(i) *Excess solder:* more than the optimum amount of solder in the fillet which thus obscures the joint surfaces and prevents proper inspection. The degree to which this condition is acceptable or not is variable, but some guidelines for various components are given in the next section.
(ii) *Solder residues:* excess solder not necessarily associated with the joint fillets, for example icicles or slivers of solder left as the board leaves a solder wave, arising from too high a viscosity of the solder, or arising from solder plating overhang at the edge of conductors.
(iii) *Insufficient solder:* less than the optimum amount of solder, specifically there not being enough solder to form a well defined fillet. Again the

degree to which this condition is acceptable depends on the fitness-for-purpose criterion.
(iv) *Minor de-wetting or non-wetting:* if only a small portion of the surfaces intended to be joined exhibits these faults, the proportion of joint area that is allowed to be not wet or to have de-wet is defined by an agreed specification.
(v) *Pinholes:* small holes in the surface of the solder that are usually an external indication of voids within the solder fillet. The voids themselves arise from gases generated in the vicinity of the molten solder during the soldering operation, but they cannot, of course, be detected visually.
(vi) *Contamination:* Contamination may be a major fault if it prevents the circuit from functioning at its full performance, but more usually it interferes only with a secondary operation such as the adherence of a conformal coating. The most common fault found during visual inspection is a cloudy white film that can arise from a variety of causes, but often as a result of polymerised flux residues (Section 12.1.2.4).

The difficulty that an inspector of solder joints is faced with is to judge the state of the interior of a solder joint from its external appearance. The inspection criteria must be such that their discrimination ensures that good joints are not rejected and bad joints are not accepted. Over-zealous rejection of good joints results in excess rework costs, reinspection costs and loss of solder joint strength, as well as production delay. On the other hand, acceptance of bad joints leads to poor reliability and excessive repair costs at later levels of higher assembly or in service. The criteria used for visual inspection are quite subjective and rely to a large extent on human judgement.

Solder joints on surface mounted devices present some specific problems, notably of leadless ceramic chip carriers and of 'J' leaded devices for which the critical part of the solder joint is mostly hidden under the component package. The size and proximity of the solder joints dictate that some magnification must be used and consequently the sheer number and density of joints severely limits the throughput. Additionally, great care must be taken to ensure that all inspection aids, as well as inspectors, are rendered completely free of electrostatic charges.

For visual inspection, a magnification of 5–30 X is required. A microscope, especially one with a zoom lens, is the most useful inspection tool, allowing the inspector to view the solder joints from different angles and at different magnifications. Microscropes have the disadvantage of causing inspector fatigue rather quickly. This is overcome to some extent by using a video camera to raster scan the assembly while viewing and inspecting on a high definition video screen. This is not a trivial operation for a surface mounted assembly because a wide range of viewing angles is required to inspect each joint fully.

Visual inspection is very much limited by operator judgement. There are several reports showing the dependence of rejection levels on the specific operator and on the time of day, using control assemblies inserted several times into inspector's batches of boards.

The criteria on which the acceptability of solder joints is based are not absolute or even agreed in specifications. This is because a full appraisal of failure modes of solder joints on surface-mounted assemblies has not yet been made.[469, 470] Also the end use of the product and the value added by the soldering determine the investment in inspection time and equipment that can be made. To a large extent, each company develops its own workmanship standards for its specific products.

Whatever the degree of acceptance used, all solder joints should have evidence of wetting and adherence at all the surfaces, indicating the presence of a metallurgical bond with metallic continuity and strength. The solder surface should demonstrate a smooth transition from the component terminal to the solderable land on the board and the fillet should appear to be concave without any sharp radii or crevices. The inspector must be convinced that in an acceptable joint good wetting has occurred, there is the correct amount of solder and the fillet surface is sound and smooth, these three aspects being judged against agreed acceptance criteria.

The defects encountered on boards that have been wave soldered, vapour phase reflow soldered and infra-red reflow soldered are of the same types but their relative occurrences are different. Vapour phase soldering tends to produce more tombstoning whilst reflow soldering in general tends to produce more voids and pin-holes than wave soldering, but fewer solder bridges or insufficient solder joints. All categories of defect are, of course, process dependent and will increase whenever the process gets out of specification.

12.2.1.2 CHIP COMPONENT SOLDER FILLETS

The amount of solder and the shape of the fillets attaching chip components show quite large variations. Multilayer chip capacitors have metallisation on the faces around the entire ends of the components, whereas chip resistors usually have only three-face metallisation with none on their sides. This difference gives rise to different solder fillet profiles, as does the choice between wave and reflow soldering. With wave soldering the joints usually have a larger solder volume, and the fillet shape depends on its position relative to the component and the conveyor direction. Because the board has been soldered upside down with a large reservoir volume of solder available, the solder fillet may be convex but still have wet perfectly. On a reflowed solder paste board, however, because a limited amount of molten solder has had to rise against gravity, a convex fillet is an indication of poor wetting and is therefore a rejectable condition. In the case of reflow soldering a concave fillet is required, with the meniscus having risen at least one third of the height up the metallisation.

A major defect to which all leadless components potentially succumb is misplacement. This is the misalignment of a component on its solderable footprint, which can occur at one of a number of operations: the original placement, curing of the glue, the wave or reflow process, etc. For chip components, in the case of the solder lands being wider than the component, a relatively large sideways shift of up to half the component width is acceptable, provided due consideration is given to the mechanical properties of the joint and the minimum insulation distances from neighbouring components and conductors. If the width of the solder lands is less than the width of the components, these lands must have a fillet across their full width. Conversely, if the width of the component is less than the width of the land, the component must have a fillet across its full width, even if misaligned.

The component may also be misplaced along the line of its length. This may be to such an extent that the overlap of one end of the component is minimal or even non-existent, producing either an open circuit or a joint of doubtful integrity. For a wave soldered joint on a chip component an overlap down to zero should produce an acceptable fillet, whereas for reflow soldering an overlap of at least 0·25 mm is normally necessary.

An assessment of misaligned chip components that have suffered a swing about an axis normal to the board or a combined X-Y displacement can be made in terms of these guidelines for the two separate displacements in the directions of the width and of the length of the component.

The chip components are susceptible to lifting and tombstoning, particularly during solder reflow by vapour phase heat transfer, as listed under major defects: no solder.

12.2.1.3 LEADED COMPONENT SOLDER FILLETS

Figure 12.18 illustrates five common lead formats found on surface mounting IC packages. The quadpack (albatross wing) lead is not conducive to good solder fillet shape and high first-pass yields. The SO (gull-wing) lead is better as it does not lie flat on its pad and hence allows solder to flow under it to form a fillet. The PLCC (inset J-lead) allows a well defined solder fillet to form and generally gives high success but, being tucked in beneath the component body, is difficult to inspect. Preferred, for reasons of both thermal management and inspectability, are the outboard J-lead and the butt lead. The butt lead should be placed on a round pad to give it some self-alignment ability during soldering.

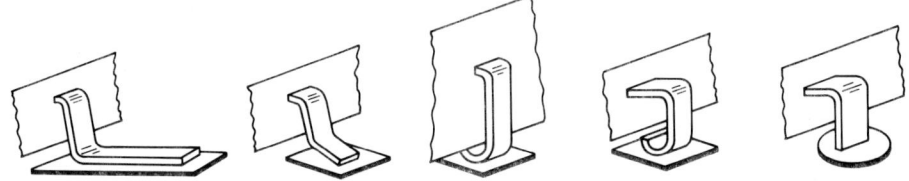

Fig. 12.18 Leads on surface mounting IC packages: the quadpack albatross wing, the SO gull wing, the PLCC inset J-lead, the outboard J-lead and the butt lead.

The basic criteria of acceptance by visual inspection, of solder fillets on these various leads are essentially the same. The foot of the lead must exhibit a visible fillet rising from the land to the top surface of the lead and the heel fillet must be continuous between the heel of the lead and the land.

In the case of the discrete components such as the SOT-23, good joints are obtained only if the feet of the leads are positioned entirely within the solder lands and the solder fillet should rise up the lead on all sides. It has been shown for the discrete components that a large amount of solder is preferable to a small amount, regarding the reliability in thermal cycling. With smaller amounts of solder, however, the visual assessment of the joint is easier because of the opportunity offered to view the wetting of the sides of the terminations. Unlike the requirement for inserted components to see clearly the outline of the lead within the solder fillet, to ensure that the wire is of adequate length, there is no necessity for this criterion for surface mounted components which all come with leads of predetermined form and length.

Besides the misalignment of the component on its footprint by a displacement in the X direction, in the Y direction, by rotation or through any combination of these, multileaded components must also be inspected for displacement of individual leads normal to the plane of the board. One or more leads can be inadvertently bent during assembly, giving rise to a mechanically weak or even non-existent joint because the gap between the foot of the lead and the land is too great.

12.2.1.4 LEADLESS CHIP CARRIER SOLDER FILLETS

The ceramic chip carriers with solderable castellations are the most difficult to inspect because much of the joint is obscured from view by the component package. Some solder joints are sketched in Figure 12.19 as a guide to the acceptability of the inspected fillets. Shown are the preferred fillet geometries and the acceptable extremes of maximum and minimum amounts of solder plus two unacceptable joints with insufficient and excessive solder present.

In the preferred joint type shown in Figure 12.19(a), the castellation is completely filled with solder, there is no visible metallisation and no visible evidence of intermetallic compound formation. (If the metallisation is gold, it should have been pre-tinned prior to placement.) The concave surface of the fillet virtually guarantees that the solder has flowed under the chip carrier by capillary action.

Figure 12.19(b) details the maximum allowable size of fillet. Again the castellation is completely filled and there is no visible sign of the metallisation or of any intermetallic compound. However, in this case the fillet has a convex surface and evidence of full wetting to the land and to the castellation must be carefully sought by inspection. Also evidence of solder flow under the component must be seen.

The minimum allowable fillet size is sketched in Figure 12.19(c). All of the surfaces of the metallised castellation are fully covered with solder and there is evidence that the solder has flowed across the land and merged with its coating. The fillet is fairly thin but does extend under the component and does cover all the solderable areas.

Figure 12.19(d) is an example in which the minimum acceptable fillet has not been achieved. The important consideration here is that the fillet is too short, extending only a little distance beyond the chip carrier. In some instances the fillet may be unacceptably small even if the solder has apparently flowed across the full area of the coating on the land. This usually occurs if a means exists whereby the molten solder can flow away from the fillet, for example along a track or down a via-hole, as discussed in Section 4.3.4.

Figure 12.19(e) is an example of an unacceptable joint in which the maximum allowable size of solder fillet has been exceeded. This situation is often the result of poor solder flow or non-wetting of the component metallisation. The joint is bulbous, and if wetting is poor the solder may not have flowed under the component.

12.2.2 X-ray Inspection of Solder Joints

X-ray inspection techniques can detect solder defects that are not in line of sight for a visual inspection process. In principle the assembled circuit board is imaged using X-rays transmitted through the board rather than using visual light reflected from the surface. The technique can therefore additionally examine both the parts of soldered connections hidden from view and the internal

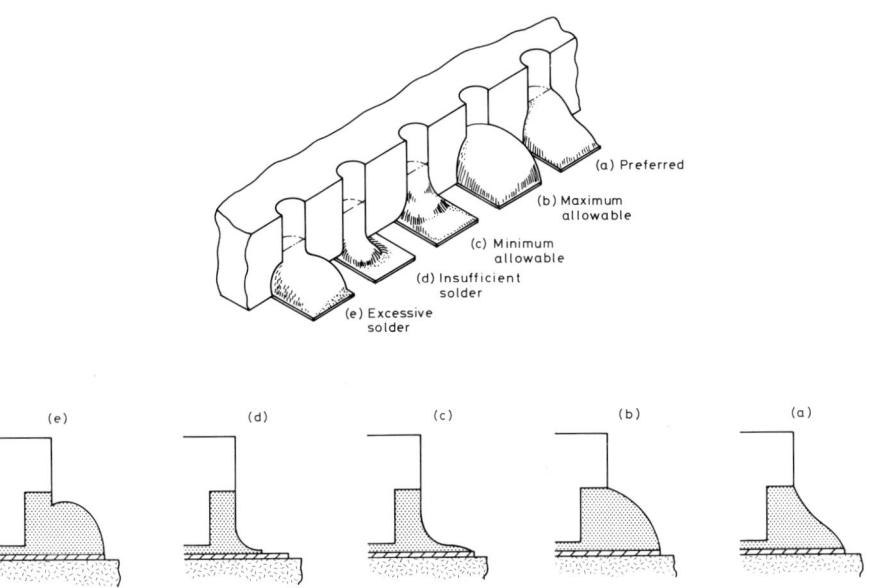

Fig. 12.19 Sketches of five solder fillets on a LCCC indicating their acceptability.[2]

structure of the solder fillet.[471]

X-ray inspection of PCBs has had a long use for detection of multilayer misregistration. Instruments are now available for automatically handling fully assembled surface mounted boards and, additionally, can be made fully automatic not only in the handling but in the quality assessment, by comparing the X-ray image with a pre-set quality acceptance criterion with no human monitoring. More usually at present, however, a real-time video display of the X-ray image is presented for operator assessment, transformed to a colour image which greatly facilitates the operator's task. Interpretation of an X-ray image requires considerable operator experience because it is a 2-dimensional representation of a 3-dimensional assembly. Some X-ray inspection systems are fitted with gimbals for the PCB mounting to enable the joint to be X-rayed for faults from every angle.

The very high solder-joint density on surface mounted boards calls for speed of inspection which, in the case of X-ray inspection, calls for a powerful X-ray source. Generally this is achieved by using a highly focused electron beam impinging on the X-ray target. The spatial resolution of any feature identified in the X-ray image depends on the contrast arising from differences in X-ray absorption between the various materials, but also on the amount of the materials through which the X-rays have penetrated. With a 100 kV X-ray source a focused spot size of about 15 μm can readily be achieved. Whilst the X-ray image of, for example, a 15 μm gold wire in plastic could be resolved by such a spot, a void within a solder fillet would probably be imaged only if it were in excess of 50 μm or so. Nevertheless, real-time X-ray inspection is a very powerful technique for pinpointing solder balls, bridges, skips or splashes as well as debonded wires within IC packages.

Operator inspection and assessment of a video display of the X-ray image cannot be an in-line facility. For fully automated in-line X-ray inspection against pre-set acceptance criteria it is necessary to define an angle of tilt to the X-ray beam for each joint, to enable adequate defect detection to be achieved. Any over-shadowing components must be avoided. Since the axis of board rotation does not intersect the joint under inspection, x, y and z re-alignments are required after tilting to return the joint to the X-ray focus. Plus the two possible orthogonal angles of rotation, each image taken requires five focal adjustments programmed into the controlling software. Additionally, the automated processing of the 2-dimensional image against acceptance criteria is very complex.

A new fully automatic X-ray inspection technique which simplifies both the programming and the alignment hardware is showing great promise for all types of surface-mounted and mixed-technology boards. The instrument uses X-ray laminography in which the area under inspection is held in a horizontal plane but the X-ray source and the detector are configured to produce a high-resolution image of a laminar horizontal slice through the joint, stepped through the joint at pre-determined depths. For a single-sided surface mounted board intended for consumer electronics, two slices give an adequate indication of the solder fillet shape, voids, solder balls, etc. to be automatically judged against pre-set criteria. For double-sided surface mounted boards for consumer electronics, four slices are required. For military quality boards the number of slices is usually doubled. Typically the instrument* operates with a 20 μm spot size, making five slice images per second of an area approximately 1 cm square, including all the movement time between adjacent areas. Because significantly less time is required to align the imaged plane at the focus of the X-ray beam, compared with the 2-dimensional transmission X-ray system, the inspection speed of this 3-dimensional laminography technique is actually faster. Additionally, the images are much more susceptible to computer interpretation and hence the efficiency of inspection is superior.

12.2.3 Laser Inspection of Solder Joints

The advantages of fully automated robotic inspection and assessment of solder joints may be summarised as:

(i) objectivity of assessment;
(ii) consistency between inspectors;
(iii) reliability (not subject to tiredness);
(iv) product quality level set according to end use;
(v) potentially low inspection cost;
(vi) statistical analysis of faults, leading to process optimisation.

There are now available fully automated instruments which, in some instances, have gained approval for their use as a replacement for human visual inspection. One such injects a pulse of heat from an infra-red laser into the solder joint while monitoring the temperature rise and fall.[472] The output data are measurements, not pictures, and so the instrument, once set up, requires no operator interpretation to produce an accept or reject decision.

It is, in principle, possible to devise a non-destructive test of integrity of a piece of material or an engineering component by injecting a given quantity of heat

---

* Manufactured by Four Pi Systems, San Diego, California.

into it and measuring its temperature rise and decay by infra-red emission. This rise and decay, the thermal signature, will be affected by, and therefore may be expected to diagnose, the structural integrity and homogeneity of the region being inspected. This principle has recently found application in the quality control and inspection of the joints in the soldering of electronic components on printed circuit boards, using an infra-red laser to inspect in turn all the solder joints.

The principle of one laser inspection instrument* is as follows,[473] referring to Figure 12.20. Each solder joint in succession is exposed to a laser pulse of infra-red radiation focused to a spot size compatible with the joint size being assessed. The wavelength of the radiation is 1·06 μm from a YAG laser; the pulse length can be varied from 10 to 200 ms, but typically 30 ms is used. The assembly under test is mounted on a motor driven horizontal X-Y table under the control of a

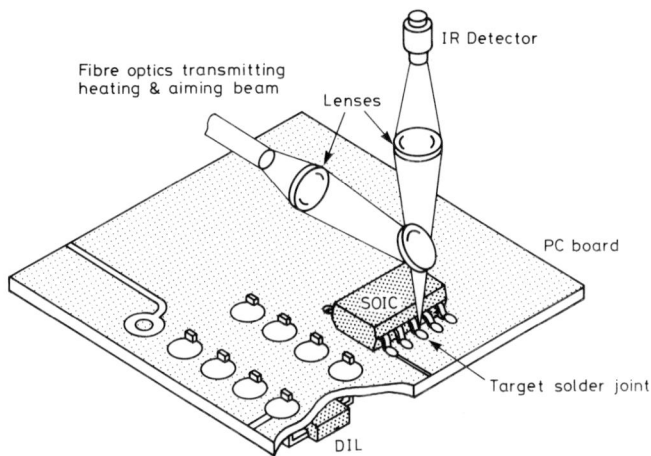

Fig. 12.20 Schematic of the principle of operation of an infra-red laser solder-joint inspection instrument.

microprocessor. The co-ordinates of each solder joint are entered into the program, either manually on the first board of the batch or from CAD software. The laser pulse, plus the X-Y platen movement time, typically allows an average inspection rate of 10 joints/second.

The infra-red radiation is absorbed by the solder and the component termination. The temperature of the joint rises by a few degrees and, at the end of the injected pulse, it decays back to the ambient temperature. During this temperature rise and fall the joint radiates heat which is monitored by an infra-red detector. This is an indium-antimony device held at 77°K, with a response time less than 1 μs. Its spectral sensitivity extends from 6 μm but is limited at the high energy end by a blockage filter to eliminate detection of any incident radiation.

A laser radiating in the visible spectrum is also used for alignment and initial set-up, the two lasers being brought into coincidence with a single glass fibre. The angle of incidence of the infra-red radiation can be pre-selected for each solder joint, to avoid awkward obstructions to the laser path.

The rise and decay of detected radiation (the thermal signature of the joint) is compared with its 'ideal' counterpart, and any differences are indicative of either

---

* The Laser-Inspect, manufactured by Vanzetti Systems, Stoughton, Massachusetts.

a difference in the thermal mass or a difference in the surface absorption of the solder joints. Various types of thermal signature are shown schematically in Figure 12.21, showing how it is possible to distinguish various types of joint that are all nominally identical.[474]

The effects of laser power and pulse time on the thermal signatures and the discrimination between good and bad joints have been investigated at length.[475]

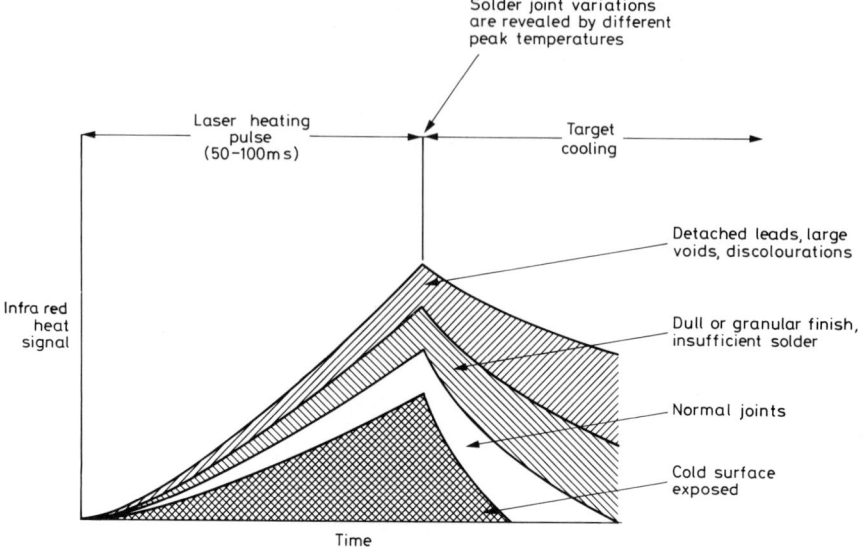

Fig. 12.21 Some types of thermal signature obtained by infra-red laser inspection of soldered joints.[474]

The longer the thermal pulse, the greater the chance of detecting faults below the solder surface, but with a reduction in discrimination because of heat flow out of the solder into the board. A short pulse gives good discrimination of surface faults but little information of hidden defects. In this mode the instrument is equivalent to the human eye but with the significant advantages of objectivity and reliability.

For laser radiation at 1·06 μm wavelength, the solder and component lead typically have an absorption coefficient of 200–500 μm^{-1} and hence an attenuation length of only some ten to twenty atomic layers. The incident radiation therefore can be considered as absorbed at the surface and the subsequent emission then involves the problem of conductive heat flow. If the laser is heating the joint from time zero to time t_o with a heat energy per unit time and per unit area of H (J.m^{-2}.s^{-1}), then the temperature rise ΔT at a depth z into the solder fillet is:[476]

$$\Delta T = \frac{2H}{K}\left[\left(\frac{\varkappa t}{\pi}\right)^{\frac{1}{2}} \exp\left(\frac{-z^2}{4\varkappa t}\right) - \frac{z}{2}\operatorname{erfc}\frac{z}{2\sqrt{\varkappa t}}\right] \qquad \text{for } 0 < t < t_o$$

and*

$$\Delta T = \frac{2H}{K}\varkappa^{\frac{1}{2}}\left[t^{\frac{1}{2}}\operatorname{ierfc}\left(\frac{z}{2\sqrt{\varkappa t}}\right) - (t-t_o)^{\frac{1}{2}}\operatorname{ierfc}\left(\frac{z}{2\sqrt{\varkappa(t-t_o)}}\right)\right] \qquad \text{for } t > t_o$$

---

* See footnote about ierfc(x) on page 284.

where $K(J.s^{-1}.m^{-1}.K^{-1})$ is the thermal conductivity and $\varkappa(m^2.s^{-1})$ is the thermal diffusivity. Thus, at the surface of the joint ($z=0$) the temperature rise caused by a laser pulse of t_o seconds duration is

$$\Delta T = \frac{2H}{K}\sqrt{\frac{\varkappa t}{\pi}} \qquad z=0, \quad 0<t<t_o$$

$$\Delta T = \frac{2H}{K}\sqrt{\frac{\varkappa t}{\pi}}\left[1-\sqrt{\frac{t-t_o}{t}}\right] \qquad z=0 \quad t>t_o$$

For near eutectic solder, the values of K and $\varkappa$ are respectively 50 J.m^{-1}.s^{-1}.K^{-1} and $3 \cdot 4\ 10^{-5}$ m^2.s^{-1}. These parameters give rise to the shape of the signatures shown in Figure 12.24. By changing either the surface properties so that the heat input H changes, or the internal properties (with a void for example) so that the thermal conductivity K and thermal diffusivity $\varkappa$ change, a different value of ΔT at the surface at time $t=t_o$ is obtained while the shape of the signature remains common.

The distance that the thermal pulse travels in time t is given approximately by the diffusion distance $\sqrt{\varkappa t}$ so that, for example, within a 30 ms exposure to the thermal pulse, the heat front could have travelled 1 mm into the solder fillet.

This laser-induced thermal signature inspection is best suited to large production runs of one board type, so that the acceptability or rejectability of the thermal signature of each joint on the board can be determined and learnt by the data-processing computer.

12.2.4 Scanning Acoustic Microscopy

Because of the huge potential market for instruments that can automatically inspect the quality of solder joints on surface mounted PCBs, techniques have been developed, or are being developed, that operate across the electromagnetic spectrum from the infra-red to X-rays. Yet another option is to employ acoustic waves.

In scanning acoustic microscopy an image of a sample, in this case a solder joint, is produced, whose contrast depends on the different elastic properties within the sample. The technique can be configured for seeing through several millimetres of material. Acoustic waves are produced by a piezoelectric transducer and focused to a convergent beam that is rastered across the sample. Detection of the reflected echo signal enables a direct modulation of the brightness of a TV monitor, and hence a picture of the structure of the sample. The lateral resolution attainable depends upon the loss characteristics and the frequency of the acoustic radiation used. Typical values are 30 μm at 50 MHz and 1·5 μm at 1 GHz. However, for imaging to depths required for full solder joint inspection, the lower part of the frequency spectrum is required. Lower frequency waves give better penetration through the board, its components and joints but higher frequency waves give better resolution.[477]

The technique is able to image voids and cracks within solder fillets as well as microstructural inclusions, such as intermetallics. It will detect solder balls, solder webs and bridges on both the inspection side and the reverse side of a board. A disadvantage is that, since the acoustic waves must be coupled from the lens into the sample, a liquid coupling medium, such as water, alcohol or an inert fluorocarbon fluid, must be used.

The technique has been modified[478] by using a laser to raster the side of the board opposite to the acoustic wave generation. The scanning laser acoustic microscope (SLAM) operates in an acoustic frequency range 10–200 MHz, with the laser raster scanning at some 30 times a second. The laser produces an image that is modulated by the sound waves, identifying occlusions, voids, delaminations and so on, within the board.

12.2.5 Evaluation of Solder Joint Inspection Methods

A number of factors must be considered when a particular inspection system is being evaluated, in respect of the type and end-usage of the assembly as well as the specification to which the product is being made.

(i) *Speed of inspection:* This includes not only the rate that solder joints can be inspected but also any time required for introducing the assembly to the inspection equipment, such as for alignment.

(ii) *Discrimination:* This is the ability of the system to distinguish acceptable joints from rejectable joints, against a pre-determined 'go/no-go' criterion. Whilst the discrimination should be as high as possible, it is usually increased only at the expense of other factors, most notably the speed of inspection. In addition, not all inspection systems can detect all defects. For example, criteria concerning voids in solder fillets cannot be evaluated using visual inspection by eye.

(iii) *Specification of criteria:* The quality of every product must always be compromised by its cost of manufacture, and hence an inspection technique must be able to have its criteria of acceptance tailored to meet the specified requirements.

(iv) *Versatility:* The system must be readily adaptable to handle different types and sizes of assembly with rapid re-programming of software.

(v) *Objectivity:* The human element in inspection is best kept to a minimum since it is very difficult to maintain consistency between different inspectors or, for one inspector, between different times. A fully automated system with unchangeable criteria set by management would be ideal.

(vi) *Manipulation:* An assembly may have to be inspected from a variety of angles. The less it needs to be handled the better, and a fully automatic untouched system greatly reduces the chance of mechanical or electrostatic damage.

(vii) *Field of view:* This should be as wide as possible to allow inspection of as many joints as possible without having to move the assembly.

(viii) *Depth of field:* When using an optical system for evaluating a solder joint, it is much easier if the depth of field is adequate for the entire joint to be viewed at once.

(ix) *Operator fatigue:* Operator fatigue can play a major part in determining the reproducibility of systems which rely on human judgement for an assessment of a solder joint. Fatigue is generally enhanced by flickering displays, sitting in unusual positions, etc.

(x) *Capital expense:* The cost of a piece of inspection equipment amortised over a given period may make a not insignificant contribution to the cost of each assembly.

(xi) *Operating costs:* This includes not only an inspector's time to perform and document his observations but also the cost of expendable supplies such as X-ray film, the maintenance of the equipment and the plant area and overheads for the instrument.

The above criteria are assessed in Table 12.11 for the available inspection systems of a naked eye, a conventional microscope, a wide scan viewer, a video display, X-ray imaging and infra-red laser inspection, compared with the ideal inspection system.

The average naked eye is incapable of detecting defects in solder joints on surface mounted assemblies. This remains essentially true even if enhanced with a 3 X illuminated ring magnifier.

The primary drawback of the conventional microscope is the relatively high operator fatigue factor and the level of operator training required. Parts manipulation is easy, allowing the inspector to view the entire solder joint rapidly and to view it from a variety of angles from all sides, with a minimum of effort. Zoom lens microscopes are especially convenient for taking a closer look at suspected defects.

The wide scan viewer presents a magnified image on a screen which is much easier to look at for long times than it is to use a microscope. However, the poor depth of field and the inability to view the solder joint from a variety of angles makes doubtful its application to surface mounted assemblies.

A video display also presents the inspector with a large screen image. The monitor used needs to have a high definition to detect many of the faults encountered and, again, such systems are limited by a relatively poor depth of field.

An X-ray imaging system is expensive in capital investment and relatively costly in operation. Considerable operator training is required if the real-time video images are to be assessed visually. However, X-ray inspection machines can be made automatic in their assessment of assembly quality, by comparing the

Table 12.11

Inspection System	Speed of inspection	Discrimination	Specification of criteria	Versatility	Objectivity	Manipulation	Field of view	Depth of view	Operator fatigue	Capital expense	Operating costs	Comments
Ideal system	Very fast	Very high	Very easy	Very high	Very high	Very easy	Very large	Very deep	Very low	Very low	Very low	Does not exist
Naked eye	Fast	Very low	Medium	Very high	Low	Very easy	Very large	Very deep	Low	Very low	Low	Cheap but very poor detection efficiency
Conventional microscope	Medium	High	Medium	High	Low	Easy	Medium	Medium	High	Low	Low	Suitable for most assemblies
Wide scan viewer	Medium	Medium	Medium	High	Low	Hard	Large	Very shallow	Low	Medium	Low	Generally impractical because of poor depth of field
Video	Fast	Low	Medium	High	Low	Medium	Medium	Shallow	Low	Medium	Low	Difficult to locate some types of defect
X-ray imaging	Fast	High	Easy	Medium	High	Hard	Large	N/A	Medium	High	Medium	Good for critical assemblies, internal faults.
Infra-red laser	Medium	High	Medium	Low	Very high	Medium	N/A	N/A	Very low	Very high	Low	Fully automated and objective. Needs programming and needs to learn for each board type.

image with a stored image and pre-set acceptance criteria. The great advantage of an X-ray inspection technique is the ability to detect hidden faults such as solder balls or splashes under components, voids in solder fillets and debonded wires in chip-on-board assemblies.

Inspection systems utilising an infra-red laser to monitor the thermal response of a solder joint are relatively new and the interpretation of the acquired signal in terms of specific faults is not yet fully understood or quantified. The inspection speeds and the multi-angle access to the joints gives the technique the opportunity to replace the human eye, but its capability of detecting hidden internal faults is rather contentious. The greatest benefit of an infra-red laser system is its potential for full automation and hence its objectivity of joint assessment. The system even has the potential to use a co-axial laser to make instant solder rework repairs on faulty joints (Section 9.5.6).

12.3 POST ASSEMBLY TESTING

Electrical testing may begin long before the assembly procedure is complete. The bare board is often tested for continuity and isolation before being populated with components. Additionally, the individual components can each be tested for functionality as they are placed on to the board and any misalignment and omission of the soldering quality are assessed visually or by an X-ray or laser technique. Thus the post assembly electrical testing of a circuit board is the last step in a testing régime designed to ensure top reliability in service.

12.3.1 A Testing Strategy

There are many types of automatic test equipment (ATE) that can test very complex, high density surface mounted assemblies.[479] They can be divided into three groups, based on complementary test philosophies: functional, in-circuit and environmental. The functional testers test the fitness for purpose of the whole assembly, and address the assembly through its edge connectors and/or customised test fixtures. Instruments for this type of post assembly testing need, in principle, be no different from those used for functional testing of insertion assemblies except that the programming may have to be changed to account for faster switching or processing times within a circuit. Automatic fault isolation and automatic learning software generation are available. The main problem with functional testers is that full fault diagnostics are difficult and can only be accomplished on digital assemblies.

The in-circuit test systems provide the same capabilities as the functional testers, but they can isolate failures on both digital and analogue assemblies. In these test facilities, the assembly is placed on a bed-of-nails array of probes that interface points on the assembly with the test station. The bed-of-nails needs to be designed to coincide with the printed circuit design for the surface mounted components. The in-circuit testers for insertion assemblies have a probe array on 2·54 mm centres which is rarely appropriate for surface mounted assemblies. In that respect it should be noted that bed of nails testers require contact points which take some of the valuable space the designer is trying to save by using the surface mounting production route.[480] Whereas with insertion assemblies the bed of nails contacts the board surface that is unpopulated with components, on surface mounted assemblies the test probes have to compete with components for probe test points. For surface mounted assemblies the probe sizes have been reduced, but this is limited by their rigidity and tendency to slide on the surface of the assembly.

Environmental screening of assemblies stresses the board by burn-in, temperature cycling and power cycling in order to precipitate latent defects.

The boundaries between the three strategies of functional, in circuit and environmental testing is blurred in that, for example, a functional tester can be used for in-circuit testing, or an environmental test may substitute for a functional test.

Manual programming of ATE for complex assemblies is an impractically enormous task and universal software packages are normally used with some operator interactions, to generate patterns of an input stimulus and the expected response.

12.3.2 Test Efficiency

Each test technique addresses specific defect classes of the fault spectrum. The efficiency with which the test equipment executes this task is the ratio of the faults found to the number of all discernible faults, and is a measure of the closeness of match between what defects the tester can identify and what defects actually exist in the fault spectrum. The test cost is related to the efficiency and there is a point of diminishing returns beyond which it makes no sense to improve tester performance.

The performance of a tester can be compared with the ideal only for particular types of defect. These can be divided into three basic groups concerned with the components, the assembly and the operation. The fault spectrum of components might include wrong component value, inoperative devices and out-of-tolerance components. The assembly faults are normally restricted to shorts and open circuits. The operational faults might include PCB pattern errors, temperature drift effects, etc. An in-circuit tester can readily detect and diagnose assembly faults and faults in components that are not too complex, but not be able to identify pattern faults, because it examines components individually or in subsets and does not functionalise the board in its entirety. Thus, the efficiency of an in-circuit tester might be as in curve (a) of Figure 12.22. A second tester may have a complementary test efficiency as in curve (b) if it is designed to specifically test the functioning of the completed assembly. A combination of these two testers gives a good overall test efficiency.

12.3.3 In-circuit Testing

Whilst functional testers provide the best active testing of assemblies, they are not well suited to finding process based faults and can be expensive to program to component level diagnostics. In-circuit testers can locate solder and assembly problems and provide active testing with diagnostics to component level. In-circuit testers are well suited to the typical spectrum of surface mounted assembly failures and are therefore the more popular type.

Figure 12.23 illustrates some of the problems encountered when attempting to use a traditional bed-of-nails fixture for testing surface mounted assemblies.[481] With the traditional fixture the test probes are supported and isolated through a carriage plate on a 2·54 mm square array. For surface mounted assemblies

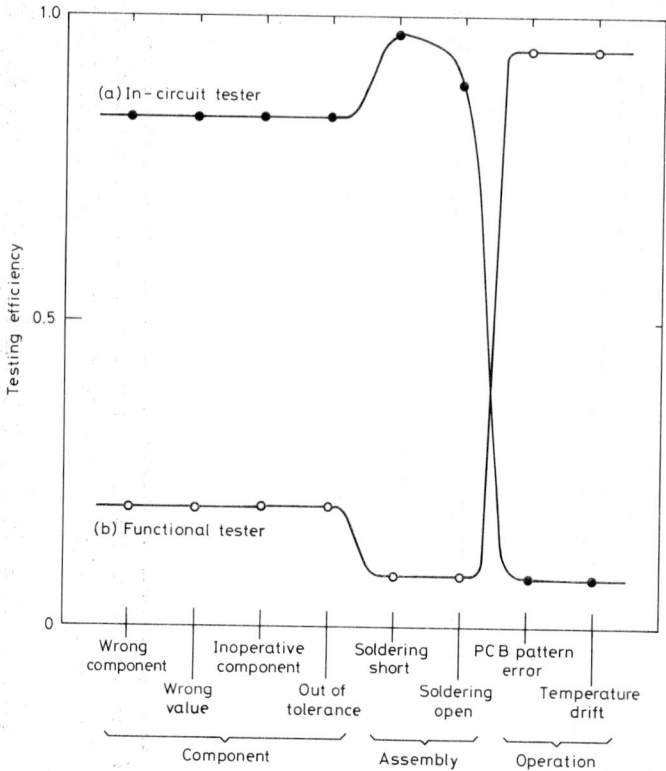

Fig. 12.22 Typical testing efficiencies of (a) an in-circuit tester and (b) a functional tester.

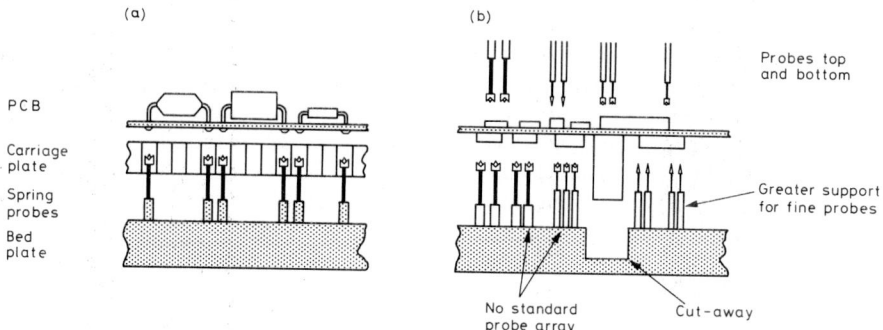

Fig. 12.23 Bed-of-nails test fixtures for (a) insertion mounted assemblies on a 2·54 mm pitch and (b) surface mounted assemblies with components on both sides of the board and with terminations on no fixed grid pattern.

components may well be on both sides of the board so that access to test points is more difficult and there is no requirement in such assemblies for a square array of component terminations. Additionally the pitch of the terminations is likely to be at least a factor of two less.[482]

In the case of conventional PCBs all component leads are accessible from the underside of the board and connection is made directly by test probes. The height of the lead ends is well controlled and there is little to restrict access. Testability of surface mounted assemblies is not only affected by devices that do not permit probing but by several other factors. It is not advisable to probe directly discrete or flatpack solder joints because probes striking an open or dry joint can cause the joint to appear good during test if the pressure of the probe is sufficient to make connection. Probes can also skid off a sloping solder surface resulting in poor contact, no contact or a short circuit to the next probe.[483] These points are illustrated in Figure 12.24. The solutions to these problems of probing are:

(i) to use specially designed test points on the circuit board wherever possible[484] and
(ii) to use dedicated multipinned probes for specific devices such as SOICs, PLCCs and LCCCs.

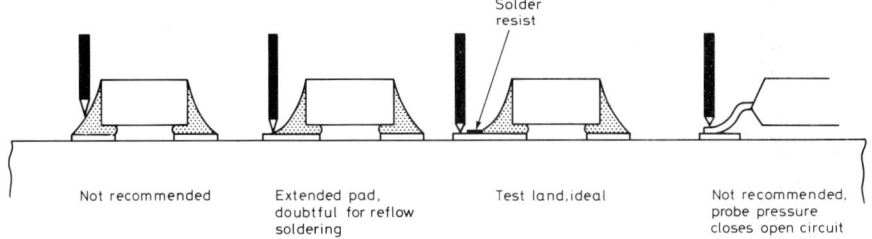

Fig. 12.24 Some problems encountered with probing surface mounted boards.

12.4 REWORK AND REPAIR

12.4.1 Reworking Solder Fillets

Inspection and test of surface mounted assemblies will identify faults that arise from three sources: board defects, component defects and processing defects. The processing defects are those arising from the placement (misalignment, wrong component, etc.) and from the soldering. Soldering defects are deviations from the perceived ideal that either give an immediate functional failure or a finite chance of becoming a functional failure in the future. It is the evaluation of the degree of that chance that is the subject of much subjective conjecture. Both inspectors and their customers tend, quite naturally, to err on the conservative side which leads to a much higher proportion of reworked solder joints than is really necessary for reliable circuit functioning over the full lifetime of the circuit. Furthermore, the usual visual inspection procedures identify only those faults that have some surface manifestation, and therefore reworking, besides being applied to many joints that do not need to be reworked, also misses some joints that might benefit from rework. This irony is illustrated in the Venn diagram in Figure 12.25.

Reworking of individual solder joints on surface mounted assemblies is a highly skilled operation requiring excellent eyesight and a steady hand. In some assemblies, manual access to solder joints is impossible, even with a miniaturised soldering iron, because of the close proximity of components and the fact that for many solder joints the bulk of the solder is obscured by the component body. The aim, therefore, of surface mounting must be towards defect-free soldering.

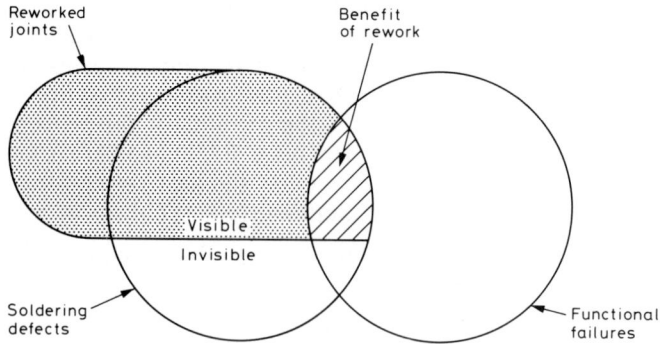

Fig. 12.25 A Venn diagram showing that, of all functional failures in an assembly, only a proportion are due to soldering defects and only a proportion of these are visible and therefore identified for rework.[4]

12.4.2 Replacing Components

When a functional fault has been isolated to one specific component, either during post-assembly testing or during service, it becomes necessary to remove that component from the printed circuit board and replace it with another. When repairing insertion-mounted boards there is normally sufficient access to the solder joints to apply a heat source in the form of a soldering iron plus a tool to remove the solder, such as a solder-wick or a suction vacuum, so that each terminal can be desoldered in turn. This is not the case for surface mounted components, whose terminals must all be desoldered simultaneously. A number of enabling techniques are available.[485]

12.4.2.1 HEATED COLLET

One technique of applying heat locally to one component is by way of a heated collet that is of the dimensions to contact only the soldered joints of one component. The work is normally carried out at a repair station that is equipped with a micro-manipulator and in-built microscope for aligning the assembly under the collet. The collet and the technique for its use are discussed in Section 9.3.3 as a small-scale assembly technique. This is illustrated in Figure 9.4. The collet and the central pin can be raised and lowered separately, usually pneumatically. On some facilities the central pin is hollow and a vacuum can be applied to lift the component when the solder has melted, whilst on others the removal of the component relies on a frictional fit of the collet over the component. Thus desoldering starts with the collet push-fitted over the component, then heating of the collet until the solder melts, followed by lifting of the collet and the enclosed component. To make a replacement, dabs of solder paste must first be applied to the footprint using a syringe, then the component inside the collet is lowered into contact, placed precisely in position and a slight downward pressure is applied by the central pin. The heating element inside the collet walls is then turned on to heat up all the leads and pads simultaneously. The heat is applied until the solder reflows and then the central pin maintains pressure while the solder solidifies.

Hand held heated collets are also available at a fraction of the cost. These probes are heated like a soldering iron but the tool consists of two halves of a collet (for example, two Ls for chip carriers) that can be applied, scissors fashion, to the component. As for the pneumatic work stations, several collet sizes and shapes are required. The hand held probe can be used with only one hand, using the scissors movement to grip the component and lift it clear, once the solder has melted. The heating can conveniently be switched by foot.

All heated collets, whether pneumatically or manually applied, need a working space around the component and cannot be used if the components are extremely close. Therefore the designer of the assembly must make the decision between sacrificing component density or completely sacrificing assembled boards if a fault is identified.

12.4.2.2 HOT GAS

In Section 9.4 the use of hot air or a hot gas as a heat transfer fluid for reflow of solder pastes was mentioned. A directed jet of hot gas can also be used very successfully for localised heating for the removal and replacement of faulty components. Heat can be supplied to places which are otherwise difficult to reach and this allows a greater freedom of design. In hot air rework stations, a variety of jets and shrouds can be used to direct the heat only where it is required. Both the temperature and flow rate of the gas may be adjusted, the optimum setting being based mainly on trial and error.

The heat transfer from a gas is relatively low, but this is not a serious problem for rework. The speed of operation can be increased by increasing the gas flow at the risk of blowing the component out of position if no mechanical restraint is incorporated, or by increasing the gas temperature at the risk of damaging or charring the board and component.

Various modifications of a basic hot air rework station are available. For example, the underside of the substrate can be cooled with a stream of cold air to help protect it from damage. The removal and replacement of components is normally achieved using a vacuum probe.

Whereas the heating with a hot collet is by direct conduction into the metallic parts of the joint, a hot gas jet invariably heats also the substrate around the joint. When the solder has melted and the component is being lifted clear there is a very real danger, therefore, that the footprint pad may lift because of the solder surface tension between the component and the pad. To alleviate the problem, the component is usually given a small sideways movement before lifting.

Whilst the gas used is normally air, nitrogen, nitrogen plus hydrogen, and argon have also been used successfully, offering the advantage of prevention of oxidation which occurs when air is used.

12.5 PROTECTIVE COATINGS

The cleanliness of the surface mounted assembly can to some extent be retained during usage, with a protective coating. Such a coating aims to prevent deterioration of an established reliability by protecting the circuit in such a way that any future contamination does not influence its performance. The board is protected from, for example, moisture and handling contaminants. Besides allowing the board to function satisfactorily in hostile environments or after handling with bare hands, the coating increases the general robustness of the circuit by protecting both the components and their joints.[486]

Coatings can be particularly beneficial to surface mounted assemblies because of their very fine conductive patterns, which are sensitive to dust particles as well as small changes in surface resistance and voltage breakdown. Before application of a coating, a high quality washing is essential.

Many types of protective coatings are used, based on acrylics, epoxies, polyimides, polyurethanes and silicones,[487] which may be applied in a variety of ways. There are essentially three families of suitable coatings, those that conform to the surface topography of the circuit board, those that do not but nevertheless are surface coatings, and thirdly those that use a bulk potting procedure.

12.5.1 Protective Concepts

12.5.1.1 CONFORMAL COATINGS

A conformal coating is usually less than 0·1 mm thick and is one whose topography conforms to the underlying structure.[488] The coating is applied as a relatively low viscosity liquid, covering and wetting separately all of the free surfaces, with virtually no capillaries where the coating and substrate surfaces meet. After curing,[489] the coating is expected to be absolutely homogeneous and completely free from inclusions, bubbles or other signs of defect. An autoclave test should not reveal any blistering, peeling or measling (the result of the formation of minute blisters under the coating due to moisture trapped by contaminant at the interface).

12.5.1.2 NON-CONFORMAL SURFACE COATINGS

If the applied coating has a thickness of more than 0·1 mm over the whole of the surface of the board and an average thickness of about 0·5 mm, it generally does not conform to the underlying structure because its viscosity and surface tension are such that it forms a meniscus, closing over all gaps that are below about 0·3 mm in size. When using such a coating the assembly must be placed in a vacuum chamber while the coating is still liquid, to remove all the air from beneath components and from blind cavities.

The application of a non-conformal coating is considerably more difficult than with the more fluid conformal coatings, but such a coating is much more rugged and offers a lot more mechanical protection.

12.5.1.3 POTTED COATINGS

In situations where extra weight on the assembly is not a restriction but great ruggedness is a requirement, an assembly can be immersed using a cast of liquid resin that is subsequently cured to a certain hardness. As the resin cures it undergoes some shrinkage, creating stresses on the assembly which might reduce its reliability. Generally, the faster the polymerisation, the greater the stresses; it may be necessary to cure at room temperature for up to two weeks to avoid this problem.

The potting is normally done by pouring the liquid resin over the assembly in a mould and again it is beneficial to subject it to a vacuum to remove all the air bubbles.

12.5.2 Methods of Application

Exceptionally, protective coatings can be applied by evaporation in a vacuum system, but for most surface mounted assemblies this is impracticable and the usual route is via a liquid. There are four main techniques to applying the coating to the assembly: dipping, spraying, brushing and injection.

The easiest method to use is dipping, and good thickness control is possible except for a frequent slight overthickness along one edge due to draining from the trailing edge of the assembly. It is particularly important to control and maintain a specified temperature and viscosity of the coating liquid. The immersion rate should be as slow as practicable, in order to minimise the entrapment of air. The withdrawal process is critical: the slower the withdrawal speed, the thinner the coating. The dipping system suffers from two potentially serious drawbacks. First, sufficient coating liquid must be prepared to fill the available tank and may therefore by largely wasted if only a few boards need coating, and secondly it is possible for contaminants from one immersed board to pollute the tank of liquid coating, remain undetected and subsequently spoil the following boards.

The most economic way of applying surface coatings is by spraying. Neither of the two problems just mentioned arises, but the process of obtaining a thin, uniform coating is very difficult. During the spraying, the assembly must be fully rotated in three dimensions to ensure that all the blind surfaces are reached. The most difficult areas are of course under large components with small stand-offs and therefore, if at all possible, via-holes should be avoided in these areas and only tracks which are pre-coated with solder resist should be taken through such areas.

Brushing application of protective coatings is usually restricted to special small batches. The technique is manual, using a high quality house-painter's brush. To reduce the risk of contamination from one board to the next, both the brush and the coating reservoir should be changed frequently.

As an alternative to potting an assembly by pouring the resin over it in a mould, an injection moulding process can be employed, to avoid the need for subsequent evacuation to remove included air.

The evaporative deposition of a conformal coating requires an enormously greater capital investment in equipment but offers a number of quite significant technical advantages. The polymer coating is deposited at 25°C in a chamber pumped to about 10 Pa pressure. The advantages to be gained are that the coating has a much more uniform thickness with a very low porosity; it also has a dielectric strength increase of a factor 2–3, a surface resistivity and volume resistivity increase of a factor 100, and is particularly useful for high frequency circuits.

12.5.3 Types of Coatings

The coatings available for application in the form of a liquid are either of a single or a double component type.

The single component types are cured by one of several mechanisms that all rely on contact with the air. Thus the curing takes place from the outer surface and this reduces the curing of the underlying coating. It is therefore essential that single component systems are used only as very thin coatings, if necessary applying several thin coatings rather than one thicker coating.

There is a wide variety of two-component systems which obviously have the requirement of a mixing process before use. In some cases the ratio of the two constituents and the degree of mixing is critical to their properties. Modern epoxy-based conformal coatings contain flexibilisers to minimise stresses induced in the coating and the assembly as the epoxy hardens during cure.

Phenolic Lacquers are the cheapest form of coating, but are limited by their electrical characteristics at high impedances and frequencies. They represent a general type of coating for non-demanding conditions in the temperature range -55 to $+125°C$. They are easy to apply and can be removed by readily available solvents such as xylene and toluene. They have a good appearance and are easily repairable. The *acrylic lacquers* are used for the temperature range -60 to $+135°C$ where excellent electrical properties are required. Thin coatings can be 'soldered-through' or the coating can be readily removed using solvents such as toluene and methyl-ethyl-ketone (MEK). They are relatively sensitive to chemical and thermal attack but have a good shiny appearance and are easily repairable. *Urethane lacquers* are particularly good where resistance to moisture, solvents and abrasion is required, and can be used for applications in the temperature range -55 to $+125°C$. Thin coatings can be 'soldered-through' but otherwise the coating must be removed mechanically. These lacquers are relatively difficult to apply and the appearance is usually not particularly appealing. Where higher temperature operation is necessary, *silicone lacquers* can be used in the range -55 to $+260°C$. These coatings also have excellent dielectric and resistance properties. They are easy to apply, have a good appearance and can be repaired. *Silicone rubber* coatings can also be used in the temperature range -55 to $+260°C$ and have good abrasive properties. They are flexible, transparent and moderately difficult to remove, but require a primer coat in order to obtain optimum adhesion. *Polystyrene* coatings are recommended for use where the highest dielectric properties are required. *Epoxy* coatings are useful in the temperature range -60 to $+200°C$ and are general coatings for use where excellent electrical and solvent resistance properties are required. Thin coatings can be 'soldered through' but removal is difficult. They can be patched and have a good appearance but, being a two-component system, are more difficult to apply. *Paraxylylene* is a vacuum-deposited polymer for use in a temperature range -65 to $+145°C$. This type of coating is truly conformal, penetrating all crevices and coating all surfaces with a layer of constant thickness. The protection afforded against humidity, solvents and abrasion is excellent. Application, however, requires specialised equipment and the coating cannot be repaired using conventional techniques.

The two most widely used coatings for high-technology surface mounted boards are conformal, based on epoxy and polyurethane resins.

Chapter 13

QUALITY AND RELIABILITY

13.1 INTRODUCTION

Quality means the conformance to the requirements that satisfy the customer, in other words the achievement of electrical and physical characteristics to meet an inspection specification. Reliability of a component or a system is the ability it has to perform a given function under stated conditions for a desired duration. Thus reliability is essentially the maintenance of quality over the working life of the item.

High quality is associated with a high standard of materials and manufacturing processes. This implies close and constant inspection and control. However, it is apparent in the electronic assembly industry that the improved output arising from stringent quality practices actually pays for the cost of maintaining those practices, through a reduction in repair both during manufacture and during service.[490]

Whilst good quality does have some impact on short-term reliability, enabling low-risk guarantees to be offered by manufacturers of electronic assemblies, the long-term reliability is more dependent upon statistically governed physical processes occurring within the manufacturing materials than on the quality of manufacture.[491] Thus, simplistically, short-term reliability is influenced by the manufacturer but long-term reliability is more influenced by the designer.

One assessment of reliability, used especially in the semiconductor industry, is to select individual components, assemblies or complete systems and test to specified, usually standardised, conditions. The number of units that fail to meet the conditions of the tests are compared with the total number of units tested, and a failure rate is given.[492] As experience is gained in the correlation between the test results and the service reliability of a process or technology, the failure rate becomes a confident prediction of the reliability of the product.

13.2 RELIABILITY BEHAVIOUR

The reliability (or failure rate) of almost all manufactured goods passes through three regimes during the lifetime of the product. For electronic components and assemblies these three regimes are quite distinct and form, as shown in Figure 13.1, the well-known bathtub curve. The early high failure rate is associated with a period of infant mortality during the debugging and burn-in of the manufactured product. During the main period of life expectancy the failure

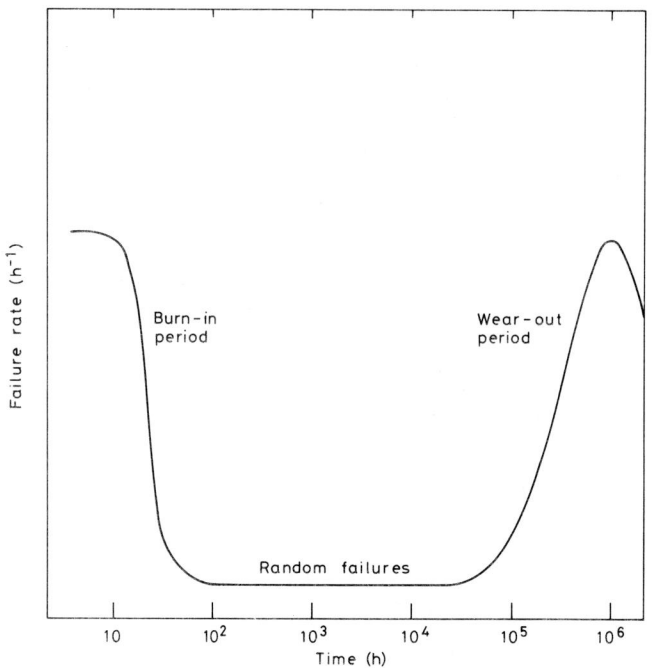

Fig. 13.1 The bathtub reliability curve.

rate is low and is only non-zero because of statistically random events occurring within components. At the end of the design life there is a sharply rising failure rate as components wear out.

Most of the infant mortality can be associated with poor quality control during manufacture of the assembly, while most of the mid-life mortality can be attributed to component manufacture. It must be noted that the time scale in Figure 13.1 is logarithmic and therefore that, although the failure rate in mid life is much lower than during infancy, the number of failures (i.e., the area under the curve on a linear time scale) can be considerably higher.

The design of an accelerated ageing test is aimed to predict, over a short-term assessment, the onset of the wear-out period under the conditions of service of the component or the assembly.

13.3 RELIABILITY FUNCTIONS

The bathtub reliability curve can be described in terms of some mathematical function p(t) which is the probability of failure as a function of time t. The cumulative failure function Q(t), which is the probability that an item will have failed by time t, is then

$$Q(t) = \int_0^t p(t)\,dt \tag{13.1}$$

Consequently, the reliability function R(t) is the probability of survival after a time t, given by

$$R(t) = 1 - Q(t) \tag{13.2}$$

where the functions are normalised so that, clearly, $Q(\infty) = 1$.

The rate of failure, or hazard rate $H(t)$ is the rate of increase of the cumulative failure function as a proportion of the reliability function. Thus

$$H(t) = \frac{1}{R(t)} \frac{dQ(t)}{dt}$$

$$= \frac{p(t)}{R(t)} \tag{13.3}$$

It is required to predict the usable service life of an electronic assembly and the mean time between its functional failures. To do this it is necessary to have, for each type of component in an electronic assembly, a measure of both the mean random failure rate and the mean wear-out failure rate. From these can be calculated the probability that each component, and therefore the whole assembly will survive for a given time, assuming that inspection and burn-in have enabled a satisfactory replacement of all the infant mortalities.

13.3.1 Random Failures

The occurrence of random failures contributes a constant failure rate, which gives rise to the predominantly flat bottom to the bathtub curve. Thus, for random failures, the failure rate

$$H(t) = K$$

a constant, and consequently

$$Q(t) = 1 - e^{-Kt}$$

and $p(t) = K e^{-Kt}$ \hfill (13.4)

Since the failure rate in the random failure regime is independent of the time, predictive statistical analysis can be carried out by testing a sufficiently large number of components to failure within that appropriate regime. Thus, if N components that survived infant mortality are tested for t hours and produce F failures, then the mean time to failure (MTTF) per component is simply

$$\text{MTTF} = \frac{Nt}{F} \text{ hours}$$

and the mean failure rate for a component is the inverse of the MTTF:

$$K_c = \frac{F}{Nt} \text{ per hour}$$

If an assembly or a system comprises N components and its survival depends upon every one of the components, the system failure rate K_s is

$$K_s = N K_c$$

and the mean time to failure of the system is only $1/N$ of the mean time to failure of each component.

If the probability of component or interconnection failure is small during the lifetime of the assembly, the statistical analysis of random failures can be described by the Poisson distribution, to give the likelihood that a given assembly will have zero, one, two . . ., failures during its lifetime. The Poisson distribution gives the probability that n failures will occur within a defined assembly provided these events are individually independent and that the number occurring in one assembly does not influence the number occurring in any other assembly. Thus, in this case it applies to low frequency random failures within a large number of components or solder joints.

The probability of occurrence of a failure, f, is equal to the total number of defects occurring within the batch over their lifetime, divided by the total number of components at risk. For one assembled board with N components and joints, the average number of defects occurring during its lifetime is fN. The probability that no defects will occur on a board during that time is

$$P_{(no\ defects)} = e^{-fN} \qquad (13.5)$$

and the probability that n defects will occur on a board during that time is

$$P_{(n\ defects)} = e^{-fN}\frac{(fN)^n}{n!} \qquad (13.6)$$

Figure 13.2 gives the percentage of boards of a batch that remain defect-free during their lifetime, given by Equation (13.5). It depends on the reliability function of the components and the number of joints per board.

Equation (13.6) gives the proportion of the batch that will experience one defect (n = 1), two defects (n = 2), etc. Thus, if a board type comprises 500 components and joints, whose individual reliability R over the lifetime of the assembly is 99·9%, then N = 500, f = 0·001 and hence approximately 60% of the assemblies should experience no faults, 30% one fault, 8% two faults, 1% three faults and so on.

13.3.2 Wear-out Failures

Whereas the random failures are not dependent upon the passage of time, it is clear from Figure 13.1 that the failure rate from wear-out is very dependent on time. Therefore the only sure way of measuring the wear-out failure function is to hold the components on test until they fail. For most components the onset of wear-out failure should be in excess of 10 years, sometimes in excess of 40 years, so that obviously this approach to reliability testing is not viable.

Two other approaches are used to study wear-out reliability. In the first, a representative sample of a component batch is subjected to an accelerated ageing process which attempts, in a controlled manner, to compress the time scale of Figure 13.1 by accelerating the mechanisms that are believed to contribute to the failures. The acceleration is achieved by increasing the temperature, the humidity, the electronic performance requirement, or a combination of two or all of these. Such accelerated failure data are useful for comparing component performance but it is doubtful that the relationship between accelerated and real-time reliability data is a simple linear one.

The second approach is to determine the criteria of acceptance and rejection relating to certain properties of components that are known to contribute to their unreliability. Such criteria might be, for example, the interdiffusion of two species across an interface within the IC micro-architecture, the interdiffusion across a solderable coating-substrate interface, or the rate of ingress of moisture

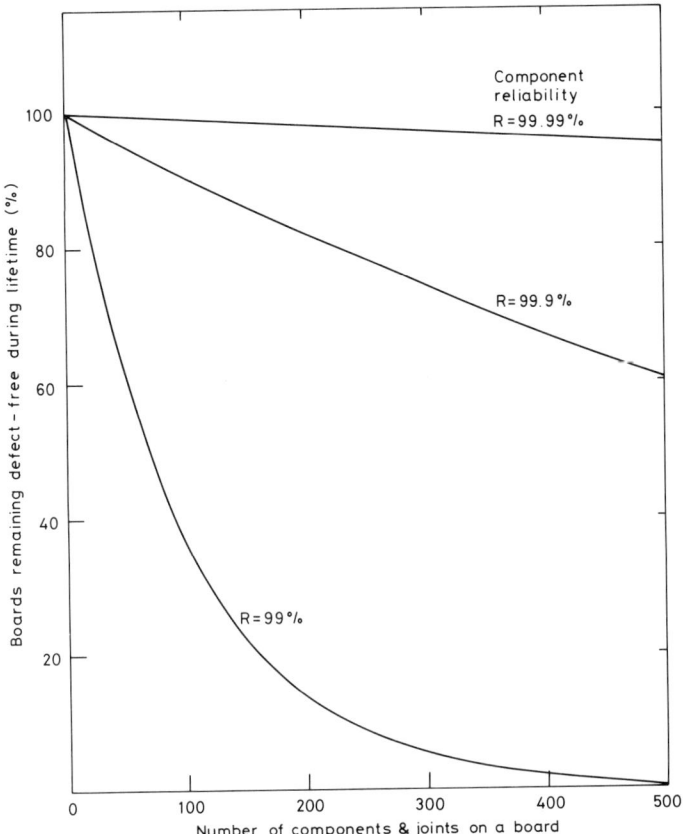

Fig. 13.2 The number of PCB assemblies that remain defect-free during their lifetime depends on the number of components and joints on the board as well as their reliability functions.

through a hermetically sealed encapsulation, and so on.

In both approaches, an understanding of the solid-state mechanisms that occur within components, both active and passive, and within solder joints is required.

As shown in Figure 13.1, wear-out failures are characterised by a rising failure rate which reaches a maximum before declining as the population available for failure diminishes. The failure rate in this regime is simply a statistical distribution about a mean value and hence is of the form of the Gaussian or normal probability curve:

$$p(x)dx = \frac{1}{\sigma\sqrt{\pi}} \exp-\frac{(x-\bar{x})^2}{2\sigma^2} dx \qquad (13.7)$$

being the probability that the value of the variable (in this instance log time) will be between x and x + dx.

Here $\bar{x}$ is the mean value of x, i.e., at the peak in the distribution, and σ is the standard deviation of the distribution. Within a batch of components, the better the manufacturing quality control, the more alike they will be, so the more likely they are to fail close to the same time, and the smaller will be the standard deviation σ of the distribution p(x). Referring to Figure 13.1, the parameter x in Equation (13.7) is actually log time:

x = log t

whence dx = $dt/t \ln_e 10$

so that

$$p(t)dt = \frac{1}{\sigma\sqrt{2\pi}\, t \ln_e 10} \exp -\frac{(\log t - \overline{\log t})^2}{2\sigma^2} dt \qquad (13.8)$$

where $\overline{\log t}$ gives the mean time to failure, $\bar{t}$. If only wear-out failures are considered, this is the time for half the components to fail, as clarified in Figure 13.3. The bell-shaped Gaussian distribution extends from zero to infinity, symmetrically on a logt scale, about $\overline{\log t}$. The points of inflexion on the curve, where the slope of the bell curve is steepest, occur at ±σ, one standard deviation on either side of the mean. The probability that failure occurs within ±σ of the mean is 68·3%, increasing to 95·4% within ±2σ and to 99·7% within ±3σ. These figures arise from the area under the Gaussian distribution, so that the probability that failure occurs within ±logt of the mean is

$$p(\bar{x}\pm x) = \int_{\bar{x}-x}^{\bar{x}+x} p(t)\, dt$$

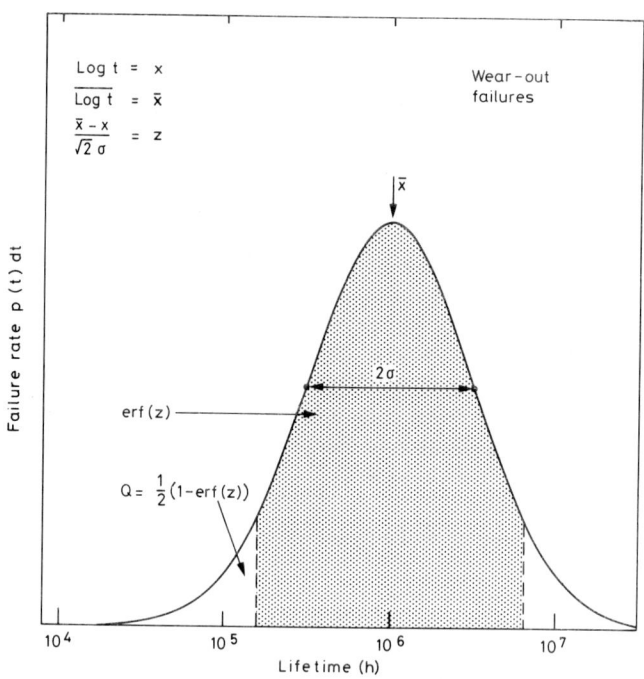

Fig. 13.3 The Gaussian distribution of wear-out failures.

Quality and Reliability

remembering that $x = \log t$. This can be integrated by substituting the variable

$$z = \frac{\bar{x} - x}{\sqrt{2}\sigma}$$

whence

$$p(\bar{x} \pm x) = \frac{2}{\sqrt{\pi}} \int_0^z e^{-z^2} dz \qquad (13.9)$$

This is called the error function, erf(z), and is tabulated, or can be calculated as the sum of an infinite series:

$$\mathrm{erf}(z) = \frac{2}{\sqrt{\pi}} \left(z - \frac{z^3}{3(1!)} + \frac{z^5}{5(2!)} - \frac{z^7}{7(3!)} + \frac{z^9}{9(4!)} - \frac{z^{11}}{11(5!)} + \ldots \right)$$

The cumulative failure function, $Q(t)$, defined in Equation (13.1), is the integral of the Gaussian distribution from $x = 0$ up to x. Thus, clearly

$$Q(t) = \tfrac{1}{2}[1 - \mathrm{erf}(z)] \qquad (13.10)$$

The form of this failure function is shown in Figure 13.4. In order to simplify the curve fitting of this function to measurements of component failure, 'probability graph paper' is available on which the ordinate is calibrated with a

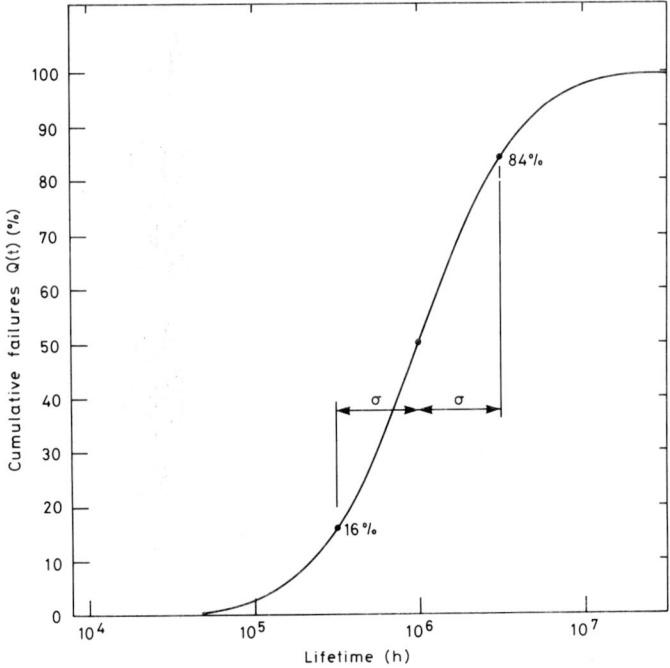

Fig. 13.4 The integral of the Gaussian distribution: the cumulative failure function $Q(t) = \tfrac{1}{2}[1 - \mathrm{erf}(z)]$.

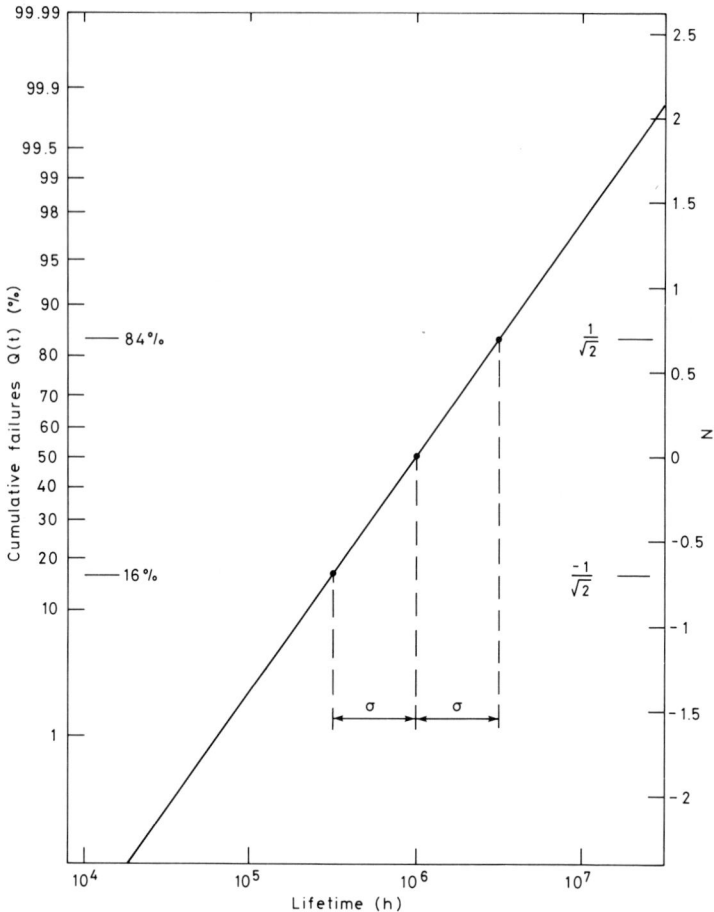

Fig. 13.5 The cumulative failure function is transformed to a straight line by plotting it on 'probability graph paper' whose ordinate is calibrated in linear z.

linear z scale as shown in Figure 13.5. Then the failure function is transformed to a straight line with a slope of $1/\sigma$. When $t = \bar{t}$, $z = 0$ and $Q(t) = 50\%$. When t is given by $\overline{\log t} \pm \sigma$, $z = \pm 1/\sqrt{2}$ and $Q(t) = 84\%$ and 16% respectively. If the sample of measurements form part of a Gaussian distribution then they will fall on a straight line when plotted on probability graph paper.

13.4 ACCELERATED ASSESSMENT OF RELIABILITY

The mean lifetime in service, of a type of component, can be assessed by accelerating the mechanisms of ageing and failure in a controlled fashion. The changes of three parameters are often used to compress the time scale of the ageing mechanisms that lead to failure; these are temperature overstress, electrical overstress and humidity overstress.

13.4.1 Ageing Mechanisms

Changes in physical properties of electronic components arise through diffusion of species. The mean diffusion distance travelled by an atom in a time t, in an isotropic material, is $(Dt)^{1/2}$ where D is the relevant diffusivity. This is strongly temperature dependent, of the form

$$D = D_0 \exp \frac{-Q}{kT} \qquad (13.11)$$

where Q is an activation energy, k is Boltzmann's constant, T is the temperature in kelvin and D_0 is the diffusion coefficient.

The simplest case, and the one that has therefore been subject to most study, is that of the thick film resistor.[32, 493, 494] This shows a very consistent resistance change ΔR as a function of time and temperature:

$$\Delta R = \text{const. } \sqrt{Dt}$$

Besides the effect of temperature, there are also detrimental effects on the active interfaces of devices arising from contact with the environment, and in particular the ingress of moisture. Although the advent of silicon planar technology effectively removed the active *pn* junctions from the exposed surface, the junctions are nevertheless vulnerable to the effects of ionic contaminants introduced into the insulating oxide layer, which can cause junction leakage or gain degradation by extending the depletion layer at the surface.[495] For plastic encapsulated devices, epoxies have good moisture impermeability as well as good adhesion and sealing to the metal leads, and thus inhibit the ingress of moisture to the IC chip. Epoxies are, however, less pure than the silicone alternatives, and contain additional impurities in the fillers of the transfer moulding compounds. Consequently, when moisture does penetrate to the chip, it will have leached impurities from the epoxy and is free to form corrosion cells at the chip surface.

13.4.2 Thermal Acceleration

The properties of electronic components and assemblies invariably degrade faster as the temperature is increased because most physical processes follow an Arrhenius-type reaction rate dependence on the temperature, that arises from Equation (13.11):

$$R_T = \text{const. } \exp \frac{-Q_T}{kT} \qquad (13.12)$$

where R_T is the rate of the reaction or rate of degradation of the physical property, Q_T is an activation energy for the thermal process, T the temperature in kelvin and k is Boltzmann's constant ($1 \cdot 38 \; 10^{-23}$ J.K^{-1}).

Different mechanisms have different activation energies, but Equation (13.12) is found to be valid for all the temperature dependent deterioration mechanisms that lead to failure in electronic components and their interconnections. As an observation, an activation energy of about $1 \cdot 5 \; 10^{-19}$ J (0·9 eV) is relevant to failure mechanisms for electronic components, which corresponds to a halving of life for every 6°C rise in temperature around room temperature and for every 10°C rise in temperature around 150°C. Clearly this has two consequences; a serious deterioration in component lifetime occurs if the thermal management of the assembly is not addressed satisfactorily, and secondly very large acceleration factors can be achieved with relatively small increases of test temperature over service temperature.

The desired life of electronic components is usually about 20 years and so, to make reliability measurements over sensibly short periods, of less than 1000 hours, a time compression factor of around 200 is sought. From Equation (13.12) it is easy to show that this factor can be achieved by, for instance, increasing a service temperature of 70°C to a test temperature of 140°C, the lifetime at the higher temperature being predicted to be 1/200 of that at the lower temperature. If, as in this case, the activation energy Q_T is known or assumed, the reliability after 20 years can be predicted from the cumulative failures at just one temperature, namely 140°C. If, however, the activation energy is not known, failures must be induced at a series of temperatures and the time taken, at each temperature, to record, say, 1%, 5%, 10% cumulative failures, or whatever failure level is deemed to be acceptable. Then, by plotting these times against the inverse temperature, a straight line Arrhenius plot should be obtained that can be

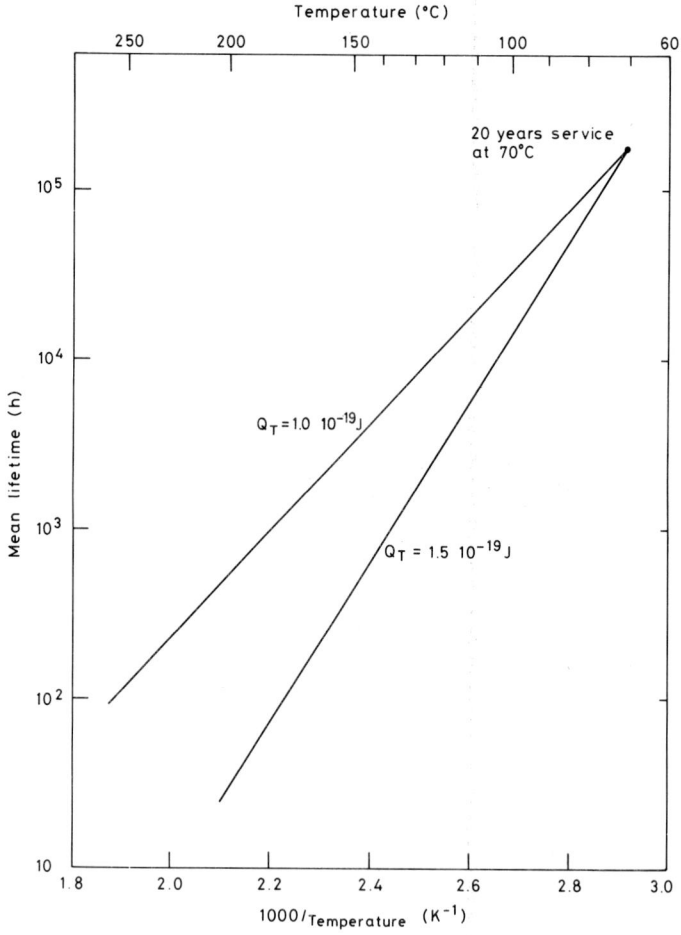

Fig. 13.6 An Arrhenius plot of temperature-accelerated failure data can be used to predict long-term service life. The extrapolation depends on the activation energy Q_T used.

extrapolated to predict the lifetime at the service working temperature, as illustrated in Figure 13.6. The lines shown in this figure are applicable to a predicted service lifetime of 20 years at 70°C, with activation energies for failure mechanisms, of $1 \cdot 5 \, 10^{-19}$ J and $1 \cdot 0 \, 10^{-19}$ J.

Since the test data at elevated temperatures are normally taken within the range 100-1000 hours lifetime, the extrapolation of the data to times that may be 200 times greater can give rise to considerable uncertainty concerning the predicted lifetime.

Another precaution to be taken when using this type of accelerated ageing is to ensure that all the test temperatures are within the limits of validity for the component under test. As the temperature is increased, so the mechanism of deterioration may change from that which is responsible for wear-out failures at normal service temperatures.

The acceleration factors for thermal overstress are found empirically[496] and are normally quoted assuming an ambient temperature that is the worst-case operating temperature of 70°C. This ensures that the thermal overstress test encompasses even worst-case service operation. As a rule-of-thumb, for purely thermal overstress, a value of $Q_T \sim 1 \cdot 5 \, 10^{-19}$ J can be taken to give adequate long-term service reliabilities at an ambient temperature of 70°C. The acceleration factor is then given by

$$\text{acceleration factor} = \exp \frac{Q_T}{k} \left[\frac{1}{T_{amb}} - \frac{1}{T_{test}} \right] \quad (13.13)$$

represented by the $1 \cdot 5 \, 10^{-19}$ J line in Figure 13.6.

13.4.3 Temperature-sensitive Parameters

If temperature overstress is to be used to judge the reliability of a component in an accelerated manner, it is worth considering which properties of the component or assembly under test are temperature sensitive. This is especially true if the component or assembly is powered up during the test.

Chip resistors, as mentioned in Section 2.7.2, have a very low temperature coefficient, below 0·02% per degree celsius, and more typically 0·005% C^{-1}. Thus, a 50°C temperature excess should only give rise to a 1% change in resistance. In fact, the quoted values for temperature coefficients are not constant, but average values, usually measured over the range 20–80°C. For some sensitive ink systems the resistance variation with temperature can be quite marked within the range 20–80°C and even more marked at higher temperatures.

The effect of normal and excessive operating temperatures on capacitance values was discussed in Section 2.7.1. The changes in value occurring are very dependent upon the type of capacitor construction used and the nature of the materials. For an NPO ceramic chip capacitor, its value over the temperature excursion 20°C to 120°C typically increases by only 0·1%. For an X7R ceramic chip capacitor the corresponding change might be expected to be 5–10%, or for a tantalum chip capacitor 2–5%.

In active components the temperature makes its effect known through solid state diffusion of species within the semiconductor chip. All diffusion processes can be described by Equation (13.11) for the diffusivity of a species.

The distance that one species diffuses into another in time t is then given statistically by the diffusion distance $(Dt)^{\frac{1}{2}}$. The more basic properties of semiconductors are the resistivities of the intrinsic and doped layers which depend on the concentrations and diffusivities of the carrier species. The diffusivities depend exponentially upon the inverse temperature in the manner of Equation (13.11). As a result, the forward current at a junction increases as the temperature increases, with an activation energy $Q \approx E_g - eV$, where E_g is the energy gap of the semiconductor, V is the forward voltage and e the electronic charge.

13.4.4 Electrical Acceleration

A second method of accelerating the ageing time scale for reliability testing is to overstress the component or assembly electrically. This can be achieved by increasing either the currents or the voltages in a controlled manner, to values in excess of their service operation levels.

Current induced failures are generally associated with an electromigration phenomenon which is a movement of metal atoms within the device architecture that is induced at high current densities. As ICs have become smaller and more complex, with finer geometries, current densities have increased. Measurable electromigration proceeds when the current density is in excess of about 10^9 A.m^{-2}, at a rate that is proportional to the square of the current density.[497] The rate is also temperature dependent, being of the form

$$R_J = \text{const.} J^2 . \exp \frac{-Q_J}{kT} \tag{13.14}$$

where J is the current density and Q_J, as before, is an activation energy. Figure 13.7 illustrates this effect with data for aluminium conductor tracks at 180°C, relating the life until electromigration failure with the current density.[498] Extrapolation of this relationship to lower current densities enables some prediction of device lifetime to be made.

The acceleration of device failure by use of an overvoltage is not widely used because the functional dependence of the device on voltage can be quite complex, and therefore neither representative mechanically of service failure, nor easy to extrapolate to more realistic conditions.

13.4.5 Damp Heat Acceleration

The accelerated ingress of moisture to an active semiconductor device is used as a means of predicting long-term behaviour from short-term tests. The diffusion of moisture through ceramic packages is insignificant in terms of device deterioration, but the permeation of moisture with the associated leaching of ionic contaminants from plastic IC packages has been recognised as a problem for a long time.[499] An enormous amount of experimental work[500-502] has been carried out to study in detail the controlling parameters of the moisture diffusion and to study the mechanisms of failure of ICs related to moisture.

It has been found empirically that, over almost the full range of humidity, the semiconductor lifetime varies inversely as the exponential of the square of the relative humidity.[503] Thus the rate of degradation due to humidity, R_H, can be expressed as

$$R_H = \text{const.} \exp (A . Q_H . RH^2)$$

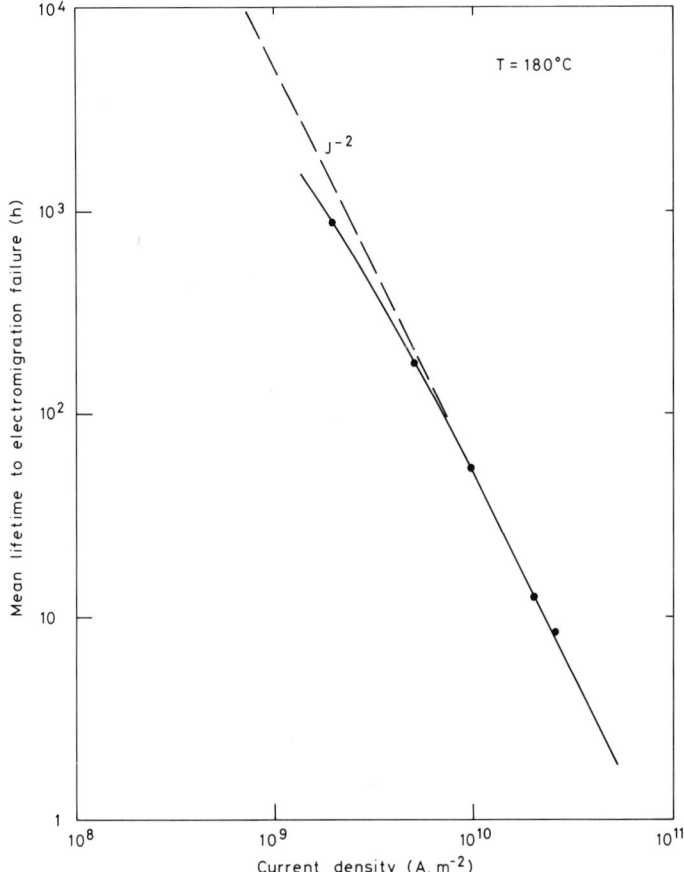

Fig. 13.7 Mean life of conductor tracks at 180°C subjected to electrical overstress, showing the J^{-2} dependence of electromigration.[498]

where A is a constant and Q_H is an activation energy.

The RH^2 relationship arises from the form of the adsorption isotherms[504, 505] and the direct link between moisture adsorption and corrosion failure of semiconductor components.

Because the diffusion of the moisture through a plastic encapsulation is affected greatly by the temperature, it is customary to accelerate the ageing effect by combining high humidity with a high temperature. The rate of degradation for accelerated ageing by damp heat stress is then of the form

$$R_{TH} = \text{const.} \exp\left[A \cdot Q_H \left(RH_{test}^2 - RH_{amb}^2 \right) + \frac{Q_T}{k}\left(\frac{1}{T_{amb}} - \frac{1}{T_{test}}\right) \right] \quad (13.15)$$

Using both heat and moisture, acceleration factors for ageing can be very large, but care must be taken to remain within the validity of the test. It is essential that RH < 100% and that saturation and water precipitation are avoided. Also, under severe conditions the properties of the encapsulant and the adhesives may be

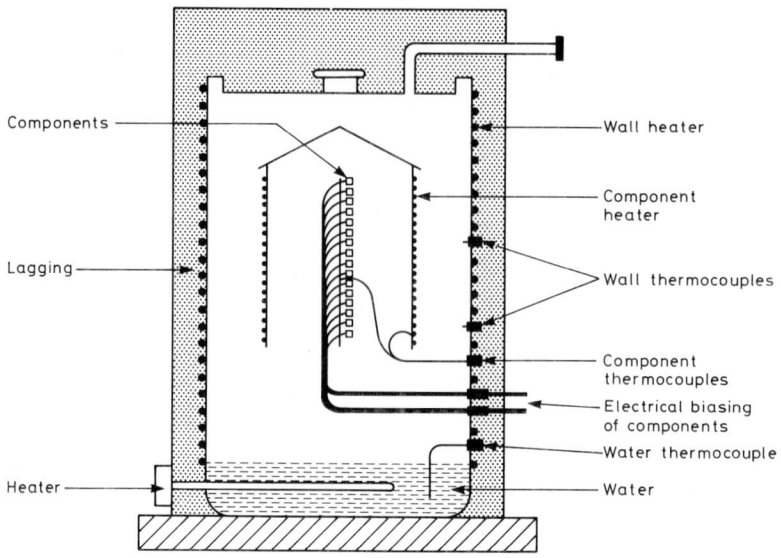

Fig. 13.8 Sketch of a damp heat autoclave for accelerated temperature-humidity testing of components and assemblies.[501]

irreversible, for example the semiconductor may debond from its base—a situation unlikely to occur in service.

Figure 13.8 is a sketch of a typical damp heat autoclave for accelerated ageing of components.[501] The relative humidity within the autoclave is the ratio of the saturated vapour pressure of moisture at the water temperature to that at the component temperature. The temperature can be increased above 100°C by pressuring the autoclave with nitrogen or argon.

It is again clear that the extrapolation to service lives of 20 years leaves a large margin of uncertainty. The acceleration factor for damp heat overstress is less clear than for dry heat overstress.[506] In the latter case the ambient service temperature can be taken as a worst case 70°C. For the damp heat case the activation energy and the ambient relative humidity can vary depending on the type of component and the failure mechanism involved. If RH is expressed in per cent, a mean value for the product $A.Q_H$ in Equation (13.15) is found[507] to be 5.10^{-4}. It is normal practice to employ worst-case values to set the acceleration factor, in order to embrace all the conceivable failure mechanisms. A commonly used damp heat ageing test condition is 85°C, 85% RH and the calculated test times required to simulate service conditions are shown in Figure 13.9.

13.5 PRACTICAL RELIABILITY

One advantage of using surface-mounting components to replace DIL and other insertion mounting components, that is often stated, is increased system reliability. This implies that, in service, the surface mounted assemblies will have a lower failure rate. It is prudent therefore to examine the available evidence that supports this significant claim.

The reliability of an electronic assembly is a function of the intrinsic reliability

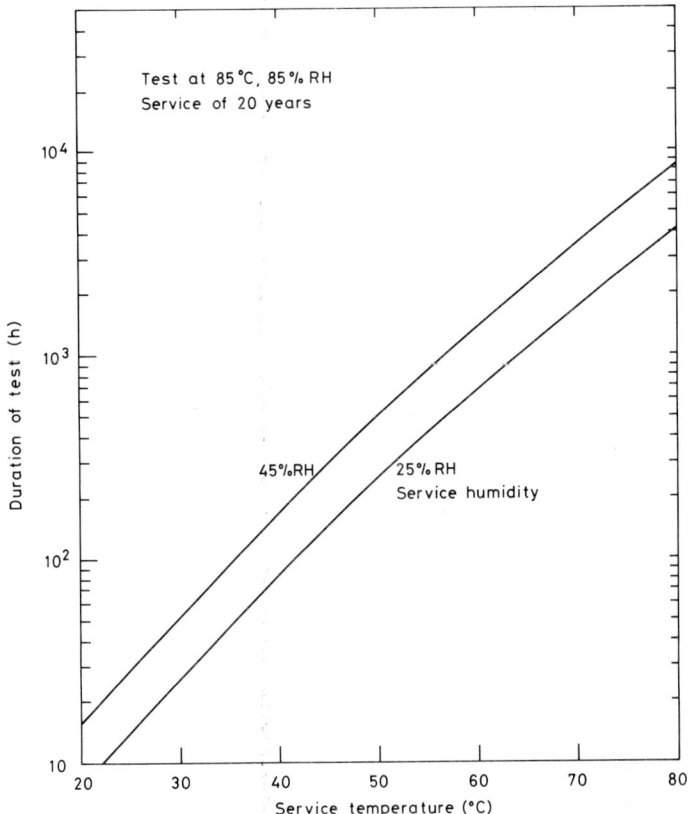

Fig. 13.9 The calculated test times using Equation (13.15) for accelerated ageing at 85°C, 85%RH to simulate long-term service conditions. The service conditions illustrated are in the temperature range 20–80°C and either 25% or 45% RH.

of the components used and the reliability of the attachment and interconnection system. It is conceivable that the attachment process will have a detrimental effect on the reliability of the components, for example, by a too-high solder immersion or reflow temperature. The quoted reliability figures for components do not take such possibilities into account. Again, a major difference between the reliability of a surface mounted assembly and a conventional assembly is the dependence of reliability upon the compatibility of the substrate-component combination for the service conditions. A large leadless chip carrier soldered to an FR-4 laminate PCB will have a tragically low life expectancy even under the more benign service conditions, even though the component reliability may be higher than its DIL equivalent.

13.5.1 Component Reliability

Individual, unattached components are tested both in real time and in accelerated ageing tests and reliability is expressed in terms of a typical failure rate per hour of service operation. The accelerated ageing tests employed are

commonly three in number:
- (i) a dry heat overstress at, for example, 125°C;
- (ii) a damp heat overstress at 85°C, 85% RH;
- (iii) a temperature cycling between −65°C and +150°C or −55°C to +125°C.

In all cases the component is of course powered up. From the results of these tests, using suitable activation energies, acceleration factors can be determined and failure rates, under normal service conditions, computed. For example, some typical predicted failure rates are given in Table 13.1 for 55°C operation. (For active devices this is the junction temperature; for passive components the isothermal temperature.)

Component reliability is frequently expressed in units of 'failure in time standard' (FITS), defined as one failure per 10^9 component hours. For high reliability equipment, failure rates of less than about 50 FITS are required, relaxed to 100–1000 FITS for consumer electronics.

Table 13.1

Typical High Quality Component Failure Rates, 55°C[3]

	Failures per Hour	FITS
Chip resistors	$<10^{-9}$	1
Chip capacitors	$<10^{-9}$	1
SOTs	$1\ 10^{-8}$	10
SOICs	$1\ 10^{-8}$	10
PLCCs	$5\ 10^{-8}$	50

13.5.2 Assembly Reliability

The reliability of an electronic assembly is a function not only of the reliability of the individual components and the number of components, but of the assembly process and the interconnection circuits. As discussed fully in Chapter 11, the primary long-term failure mode of surface mounted assemblies is low cycle fatigue of the solder joints, brought about by thermally induced or mechanically induced cyclic strains. The methods of test and the survival rates of assemblies have been presented in Section 11.3.

The reliability of a surface mounted assembly containing leadless devices depends upon the compatibility of the components and the interconnection substrate. This is a complex issue, depending mainly on the physical size of the component and the difference between the temperature coefficients of expansion of the component and the substrate.

13.5.3 Zero-hour Quality

The zero-hour quality of a batch of components is the measured quality determined by tests carried out by the purchaser of the components at the time of their delivery. The acceptance tests are carried out statistically, ensuring that each batch meets specified acceptance quality levels (AQL). These levels are set by the component manufacturer, being realistic and practical, taking into account the demands of the market where the product needs to be able to sell competitively. The manufacturer attempts to remove all the components that will fail in infancy. In practice, the actual defect levels are considerably lower than the quoted AQLs, but the advertised AQL is a compromise between promotional needs and a safeguard against rejection of a delivery by the customer with the consequent loss of customer confidence. With fully automated component manufacture, many of the larger suppliers realistically aim for zero defects at delivery, and perhaps establish an AQL as low as 5 in a million. This defect level must include the possibility of inoperative components, components out of tolerance electrically or physically, and components whose markings, if required, are illegible.

Chapter 14

ECONOMICS AND TRENDS

14.1 SMT GROWTH

Surface mounting technology has evolved from the two different technologies of hybrids (surface mounting on ceramic) and conventional PCBs (insertion mounting on organic substrates). However, in a very dynamic industry SMT is expected to continue to evolve until it bears little resemblance to either.

14.1.1 Growth of Infrastructure

One of the main factors that will drive the acceptance of surface mounting manufacturing is the growth of the industrial infrastructure in the technology. By infrastructure is meant the combination of many factors which, in total, can be summed up as the confidence that the step to SMT will be successful. These factors include the availability of a distribution network for components, reliability data, the dissemination of technical literature, the standardisation of components and quality inspection criteria, and so on. The infrastructure will be strengthened in particular by the growing belief that adoption of the surface mounting manufacturing process will result in cost levelling in the present and cost savings in the future.

The growth of an infrastructure necessary to support an industry usually follows a curve of the form shown in Figure 14.1. Growth is slow at first, restricted by the natural reluctance to take an unnecessary financial risk. As the technology is accepted, growth is rapid until it begins to saturate as nearly all the necessary back-up expertise and organisation is in place. The activity then continues at a relatively steady level until the technology begins to decline due to the advent of superseding innovation.

The infrastructure for surface mounting has come mainly from the hybrid microelectronics industry which has been mounting electronic components on surfaces of substrates for 30 years. For example, the robotic pick-and-place machines are based on designs for assembling hybrid circuits but modified to accommodate much larger substrates and a greater variety of component packages. These larger machines, however, are being manufactured mainly by the same companies who have historically manufactured automatic insertion equipment.[508] This is a logical progression because SMT is having a far greater impact on the insertion mounting industry than it is on the hybrid industry.

The technical support for surface mounting manufacture has, up to the

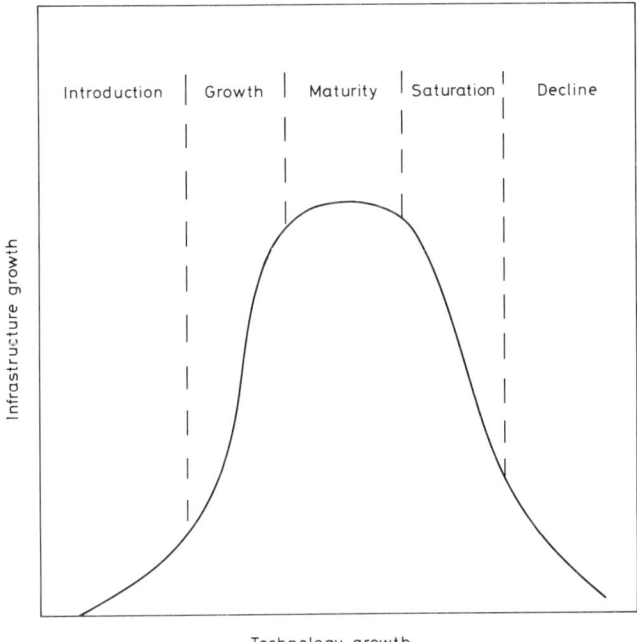

Fig. 14.1 The infrastructure activity growth and decline curve, in support of a technology.

present, come mainly from the hybrid assembly industry, but this is changing as more companies are developing their own expertise in surface mounting assembly. The growth of the surface mounting industry is fostering a new group of technical personnel who understand its specific needs. The infrastructure of surface mounting includes a pool of such expertise that is now in a position to develop training courses with hands-on opportunities.

14.1.2 Growth of Technology

Because the technology of surface mounting is relatively new, there are many issues that remain unresolved. Component availability, supply and costing form part of the industrial infrastructure base, but uncertainties of reliability and processing form part of the technology base. Factors that will promote the acceptance of this manufacturing technology are perceived as:

1. The ability of the processing and the component packaging formats to provide improvements in system performance through increased circuit density, faster switching speeds, and faster access times.
2. The growing demand for active devices with improved operating characteristics, having higher lead counts in smaller, lighter weight packages.
3. Technical solutions to the problems specific to SMT, such as compatibility of thermal coefficients of expansion between components and substrates, thermal management of devices and assemblies, the requirement of zero-defect soldering, and the issues arising from testing surface mounted assemblies.

Like the infrastructure, most of the components and materials used in surface mounting have come from the hybrid industry, but this is changing as insertion mounting components and materials are being adapted.[509] Additionally, new components and materials are being developed specifically for use in SMT. This trend is expected to continue such that the cost of surface mounting components will fall below the cost of comparable insertion mounting components.[510]

The printed circuit board, the most popular substrate interconnection for surface mounting components, will undergo considerable change. The copper foil now used to clad organic laminates will become thinner to accommodate the need for finer, sub-100 μm, conductor tracks and spaces. The foil will therefore have to be improved because most currently available electro-deposited copper is too irregular to etch evenly to provide fine lines without pinholes.

Polymer thick film circuits may well make significant inroads into the conventional copper/organic laminate PCB market. This will reduce the use of discrete resistors as more resistors and cross-overs are printed directly on to the laminate substrate. The growth of polymer thick film will be determined partly by developments in the equipment available; if discrete components can be placed more economically than film resistors can be screened and cured, the discrete components will continue to be used.

The new technologies of PCB production, such as polymer thick film, will work on almost any substrate. It is therefore expected that substrate materials other than epoxy-fibreglass will increase their market share, developed for specific purposes.

Technological advancements are expected in component development.[511] At present the trend is towards greater use of tape automated bonding (TAB). This is particularly useful for high reliability in high lead-count devices. The ease of handling devices using the TAB format is not an insignificant factor.

The equipment used in surface mounting is benefiting from a continuous programme of technical development. The speed of operation of placement equipment is now capable of meeting the demanded requirements, but the reliability of pick-up and of placement can be aided by the use of pattern recognition techniques. Such techniques are also available for inspection of populated boards automatically. It is perhaps in the area of robotic solder joint inspection that most potential for technological development exists. The human eye is an extremely discriminating inspection tool that is not readily replaced by a machine.

14.2 ECONOMICS OF SMT

The ability to reduce production costs of electronic assemblies by using surface mounting processes and components without unreasonable risk depends on a number of factors:

—type of product to be manufactured
—the market being served
—quality and reliability requirements demanded by the market
—product mix
—product life cycle
—production volume
—degree of complexity of board assembly

—requirement for higher densities
—degree of labour intensity of current assembly methods
—degree of automation needed in equipment and processes
—availability of components
—availability of resources.

The cost differential between a PCB assembled conventionally and by surface mounting is not based solely on the manufacturing costs. Each of the following has some impact on the economic decision for the manufacturing route:

—reliability
—testability
—repairability in-process
—repairability in service
—materials handling
—component inventory
—capital support equipment.

In examining cost savings possible with surface mounting, a global viewpoint must be taken.[512] Very often immediate manufacturing savings are realised by turning from insertion mounting to surface mounting but these savings may be small or even non-existent compared with the effect of the sub-assembly on the entire system. Redesigning an entire system to take advantage of the reduced size of surface mounted boards can be quite costly. However, it can result in a lower production cost, a more compact system with a higher performance, or a product with a greater customer appeal.

14.2.1 Manufacturing Costs

The savings realised by choosing a surface mounting production route over other design and manufacturing technologies vary with the design and the type of ICs required by the design. Indeed, in some cases the manufacturing costs may well be more for the surface mounted assembly. In such instances it is necessary to determine what, if any, benefits to the full system are to be realised. If there are none, then surface mounting should not be implemented.

If high lead-count ICs are part of the design, a higher degree of automation can be achieved with surface mounting than would be possible using insertion mounting. Equipment that will insert DIL packages with over 40 leads or QUIP or QUAD leaded packages is not readily available and it takes only one high lead-count package to make the assembly unmanufacturable using automated insertion technology exclusively. Such designs can be evaluated to determine any economic justification of re-designing the board for surface mounting.

At present the price of surface mounting components is not significantly less and very often more than that of their insertion mounting equivalents. Any premium is partially offset by the fact that no lead cutting or forming is required. Also, as components for surface mounting can be handled using an array of bulk feeders or tapes, the cost of sequencing the components, required for automatic insertion equipment, is avoided. Although DIL packages can be fed directly from their shipping package tubes into the insertion equipment, the tubes must be changed much more frequently than on pick-and-place machines, because of the larger component size, and this means that fewer insertion machines can be supervised by each operator.

Probably the greatest deterrent to the growth of surface mounting manufacturing is the initial cost of implementing the technology.[513] Most firms have a large commitment to and investment in insertion mounting equipment. The processes are well developed and management feels confident with the entire manufacturing set-up. To change to surface mounting assembly requires a substantial capital investment at one time. There is very little scope for a gradual implementation of the new technology—the screen printers, the placement machines, reflow equipment and rework stations are required together rather than step by step. To change over to surface mounting is expensive and time consuming. Not only is there the capital cost to consider but also the opportunity cost. Thus it is easier for a newly organised company to begin manufacturing surface mounting assemblies than it is for an established company to change.

The established companies which are successfully making the commitment to surface mounting are doing so mainly as an expansion or upgrading programme and not as a replacement. The insertion equipment is kept in full production until the end of its life and only when replacement is called for is surface mounting capacity increased.

Experience has shown that successful implementation of surface mounting technology is driven by technical reasons rather than economic. As semiconductor integration becomes more complex the lead-counts are increasing accordingly. These are incompatible with the DIL format and a surface mounting format is a technical necessity. To redesign an existing circuit simply in order to make use of a surface mounting assembly process without any additional technical functions being incorporated in the circuitry is likely to meet with economic doom.

The costs of surface placement machines and insertion machines are comparable but the payback on the former is faster because of a generally higher placement rate, provided of course that the production requirement is there. Another potential saving arises with the possibility of manufacturing arrays of substrates simultaneously because of their smaller size when using the surface mounting format. Operations such as screen printing and component placement usually benefit from array processing and a consequent reduction in handling.

14.2.2 Assembled System Costs

The manufacturing saving or the cost premium of each PCB, occurring through implementation of a surface mounting assembly route, cannot be considered in isolation. For example, one advantage of surface mounting often highlighted is the greater board density and consequent smaller assemblies. In systems whose size is not dictated by the size of the electronic circuit boards the higher density of surface mounting is neither an advantage nor a disadvantage. A television is one example; the size is determined by the picture size requirement.

Even if the cost of the assembled PCB is not reduced by the use of surface mounting techniques, there may be other economic reasons for implementing surface mounting. The greater potential component density may allow the number of boards per system to be reduced by repartitioning the electronic circuits on the boards. This reduces the cost of materials and labour even if the space saving is inconsequential.

By reducing the circuit board sizes or by consolidating more components to a board, savings may arise in a simplified cooling system, a smaller mother-board interconnection system, a smaller power supply, and so on.

Other benefits accrue with the reduction of the physical size of the system. Smaller systems are less costly to store and to ship. Also, they are often more attractive to the customer.

14.2.3 Increased Sales Potential

The cost savings realised by making products more saleable are less quantifiable than the direct savings generated during production, but they are no less real. Higher sales raise the production rate and, since surface mounting is a volume orientated technology, there are economics of scale; the cost per unit falls as the volume of products produced increases.

For some items the questions of size and weight are paramount: pocket calculators, paging devices, pacemakers, watches, etc. For calculators the size reduction may be limited by the useful size of the keyboard and so the potential miniaturisation is taken advantage of by an increase in the functions, or a reduction in the thickness.

Small size is not the only saleable advantage of surface mounting. In some products high speed operation is important, and here surface mounting is normally superior because of shorter signal paths and lower inductances.

14.2.4 The Economic Decision

At the end of the day, manufacturing industry must make a profit and, for the most part, the decision whether to use surface mounting will be based on economic arguments rather than scientific or technological ones. If it can be shown that an assembly can be manufactured at a lower cost using surface mounting, then the technology will be implemented. Which incurred expenses are to be included in that costing are to some extent arguable. There are several methods for making the cost determinations but the net present value (NPV) method seems to give usable results. The calculation takes into account the time value of money and discounts future income. Precise NPV calculations also take into account when money is spent for equipment and technological development. For example, money spent for future process development has a lesser value since it is assumed to be invested at a specified interest rate until it is spent. Likewise, money that is earned in the future is discounted at the same interest rate because the money spent to make that income possible could have been invested elsewhere and would be earning interest or dividends from the time that it was invested.

Thus, if N_i is the net cash flow for each year, from $i=1$ to $i=n$, where n is the lifetime of the product production, and if C is the capital expenditure and k the cost of that capital, then

$$\mathrm{NPV} = \sum_{i=1}^{i=n} \frac{N_i}{(k+1)^i} - C \qquad (14.1)$$

The investment is a sound one if NPV is positive.

In the present context of adoption of surface mount technology, N, rather than the net cash flow, is really the difference in manufacturing cost between the two technologies concerned. The manufacturing cost differential including labour, materials and overheads, is computed for the number of units anticipated to be manufactured during the year i ($1<i<n$).

The value of k used in Equation (14.1) can be either the cost of capital if it were borrowed or the internal rate of return that could be achieved if the money were invested in an alternative internal programme. The capital expenditure, C, includes the cost of the equipment used in surface mounting less the cost of equipment if conventional assembly is used, plus the cost of developing surface mount technology. The development cost may be discounted if it will take an extended period of time. Any scrap value for the equipment should also be discounted from the capital cost.

Experience is showing that there are economies to be realised by judicious implementation of surface mounting.[514] However, most companies are not scrapping their existing manufacturing plant but augmenting it with a surface mounting capability.

14.2.5 Introducing SMT

Justification for the introduction of surface mounting cannot be made on the basis of operational improvements alone. Rarely can it even be justified as a tactical medium-term investment. The introduction of SMT must be part of a valid business strategy aimed to give a competitive edge.

The impact of the introduction of surface mounting technology to a business is felt in every department.[515] The design and development department requires CAD systems with new capabilities; it has to overcome the challenge posed by poor component standardisation, and design from the start with inspection and test in mind. Materials management has to develop new procurement methods and meet new goods inspection, storage and stock control requirements. The process development sector has to select, purchase and install new production lines for prototype and volume production, gaining new expertise on placement, reflow soldering and cleaning. The manufacturing and test department must progress rapidly up the learning curve and aim for zero defects. It is important to support these methodologies with people who can plan and control the successful introduction of all aspects of the new technology. It is the strategic and organisational requirements together with the costs and risks that need to be appraised. The introduction process may be considered in three phases: first, detailed planning is required to make all levels of management ready for the change, secondly the change is implemented, and finally the process of continuous optimisation is set up.

There is a range of possible approaches to introducing SMT.[516] These span from the apparently minimum risk approach of redesigning an existing board to an apparently high risk approach of introducing a new product that is entirely surface mounted. In fact, the latter strategy is more likely to derive a successful economic marketing edge. Another approach is to contract out the surface mount design or the assembly to specialised centres of expertise.[517] There is no universally correct approach to the problem and the most advantageous method will depend upon the circumstances of the company. It remains to be said that SMT can bring great benefits, but to date most introductions of SMT cost more, bring fewer short-term benefits and take a lot longer than originally envisaged.

The strategic issue of SMT implementation must be thoroughly resolved before significant investment is made in the technology.

14.3 TRENDS

14.3.1 Surface Mounting Components

Components for surface mounting are available in a wide variety of formats which, for mainly historical reasons, have made different penetrations into different markets. The military products tend to use the leadless ceramic chip carriers; the consumer industry in the United States uses almost exclusively J-leaded plastic packages, whilst the Japanese consumer products use gull-wing leaded devices. So entrenched are these national divisions that in the United States a range of J-leaded SOIC packages is becoming available.

What is clear is that, worldwide, the penetration of surface mounting formats at the expense of insertion components into the electronics assembly industry is growing year by year, and is predicted to continue to do so.[518] This is shown in Figure 14.2.

The growth of ICs over the decade 1980-1990, worldwide, rises from about 10^{10} pieces to some $2 \cdot 5 \; 10^{10}$ pieces with, as shown in the figure, the proportion of surface mounting devices rising from near zero% to around 35-40% of the

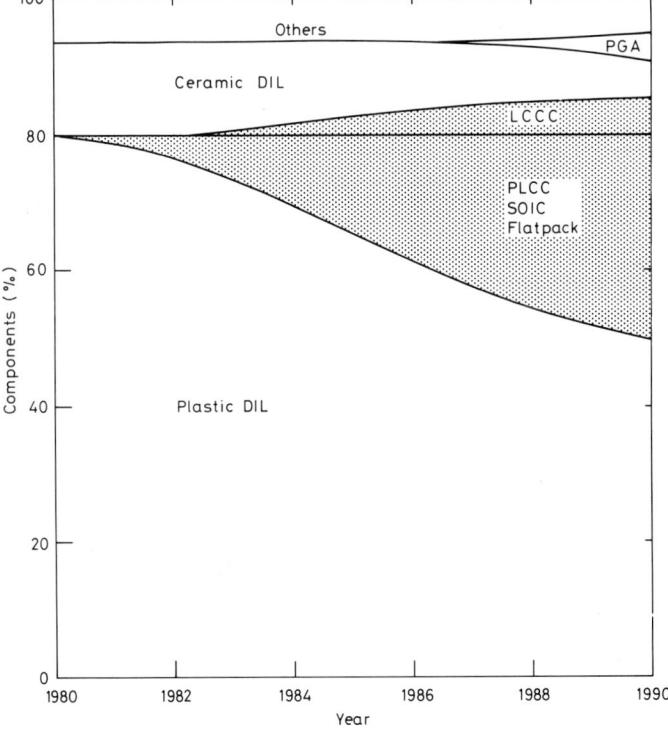

Fig. 14.2   IC packaging trends 1980–1990.[3]

market over the same period. Some predictions are even as high as 50% by 1990.[519] Figure 14.3 compares other component types as well as ICs, over the five year period 1985-1990. Besides these five main component types there are the trimmer capacitors and resistors, inductors, connectors, displays and so on which contribute to a total worldwide market of some 25 10^{10} components in 1985 of which 4·5 10^{10} or 18% were surface mounted. By 1990 the market is predicted to have increased to around 42·5 10^{10} components of which some 17 10^{10} or 40% will be surface mounted.

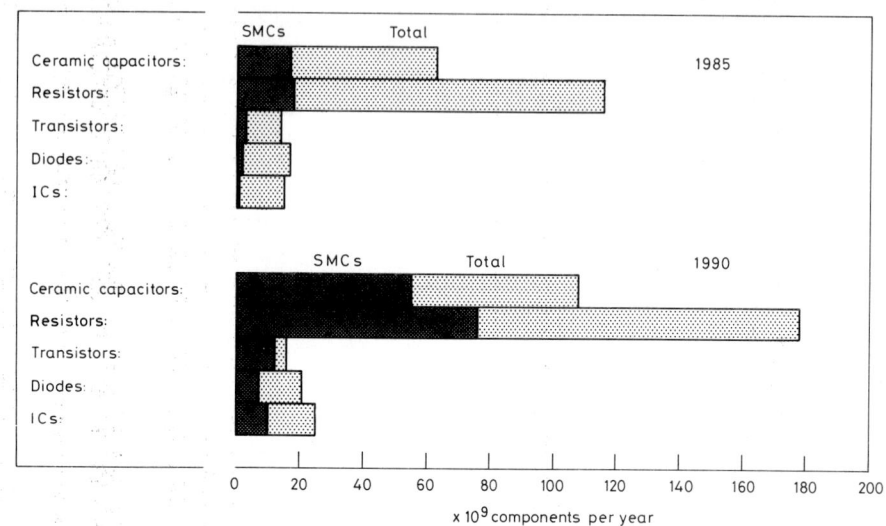

Fig. 14.3 Worldwide component use in 1985 and prediction for 1990, including the SMT penetration into the market.[519]

These surface mounting penetration figures of 18% in 1985 and 40% in 1990 arise from different degrees of SMT usage in different parts of the world. The three major manufacturing areas with large commitments to surface mounting are Western Europe, Japan and the United States of America. Figure 14.4 gives a prediction for the change of use of SMT between 1985 and 1990 for these three.

14.3.2 Surface Mounting Technology in Europe

The development of SMT in Europe has been driven from expertise in hybrid microelectronics manufacturing. Two of the major plastic packaging styles, the SOT-23 for diodes and transistors and the SOIC range for integrated circuits, have European origins. These were designed for use with thin and thick film hybrid circuits.

Japan was first to exploit surface mounting technology in cameras, calculators and other consumer electronics applications. Europe did not immediately follow the Japanese lead, largely because the types of mass produced products in which the technology has found success were not manufactured on a large scale in Europe. Consequently Japan built an enormous lead in surface mounting technology.

Economics and Trends   515

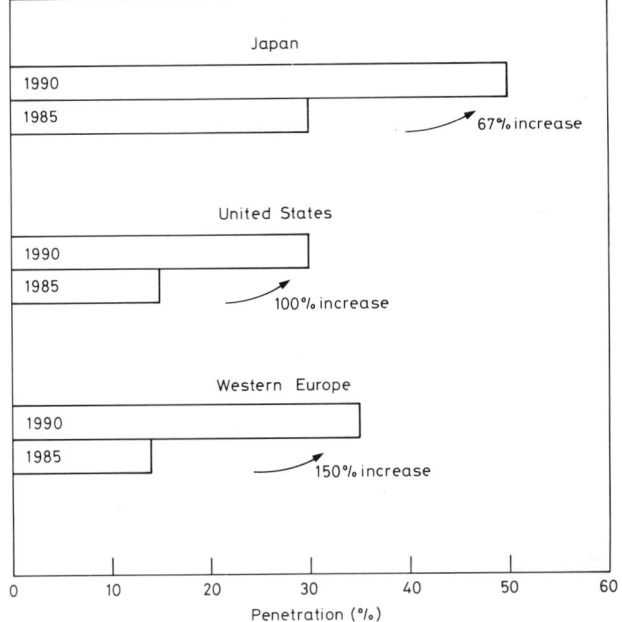

Fig. 14.4 SMT penetration in the three major manufacturing regions.[519]

Europe has now responded[520] and the European market for surface mounting components is planned to double over the years 1985-1989 and treble over the years 1985 to 1991.[521] The main areas of activity in Europe are in small appliances and TV assemblies, radio communications (especially cellular radio),[522] main exchange telephone equipment and data processing. Examination of the European electronics industry does indicate that it is unlikely to emulate Japan in surface mounting technology. For products where neither speed of operation nor space are paramount there is less incentive to make the move to SMT. Whereas in Japan the production of, for example, leaded capacitors is predicted to begin to fall before 1995, a continued growth of all leaded components is expected in Europe.

14.3.3 Surface Mounting Technology in Japan

In Japan during the decade 1975-85 the need to maintain leadership in consumer electronic products in the world market was overwhelming. The major concerns were miniaturisation, reliability and cost.[523] Therefore the Japanese started SMT in consumer products unlike the USA where SMT started in military applications. Although surface mounting components were included in consumer products from 1970, it was a slim-line radio marketed by Matsushita Electric Industry in 1975 that is credited with the innovative use of SMT in a mass produced consumer product. However, the lack of infrastructure and multiple sourcing of components and solder pastes meant that further breakthrough into the market by SMT was slow, and the expansive stage began only around 1980. The availability and use of chip resistors and capacitors started to accelerate

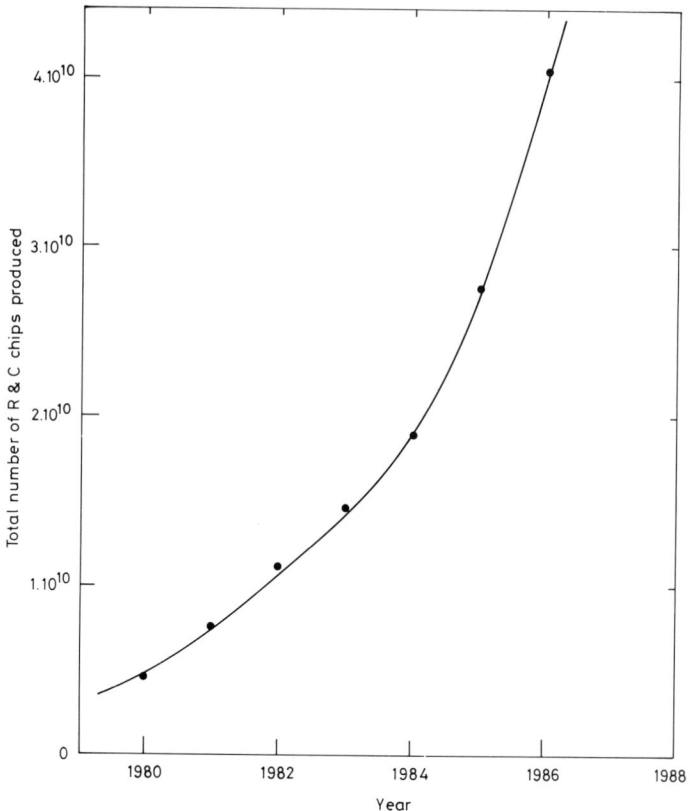

Fig. 14.5 The growth in availability and use of chip resistors and capacitors over the period 1980–1986.[524]

tremendously,[524] as shown in Figure 14.5. Although the cost of chip components still remained higher than the cost of conventional components, as shown in Figure 14.6, the proponents of SMT placed a greater emphasis on reduction of the total assembly and system costs by the introduction of automated assembly processes. In 1986 about 30% of capacitors and 20% of resistors made in Japan were in chip form. DIL still dominates the IC packaging options, but it is predicted that in 1990 over 50% of all discrete components and 40% of ICs will be in surface mounting format.[525, 526] This more or less follows the predicted components usage worldwide.

14.3.4 Surface Mounting Technology in USA

In the United States the impetus for surface mounting arose from military rather than consumer requirements. The leadless ceramic chip carrier was developed for military use in the late 1970s. However, its high cost and required use of special substrates precluded its widespread adoption.

With the recent growth of leaded components surface mounted on organic substrates, there remains a strong requirement in the United States for specialised low-volume types of components in surface mounting formats, in order to avoid,

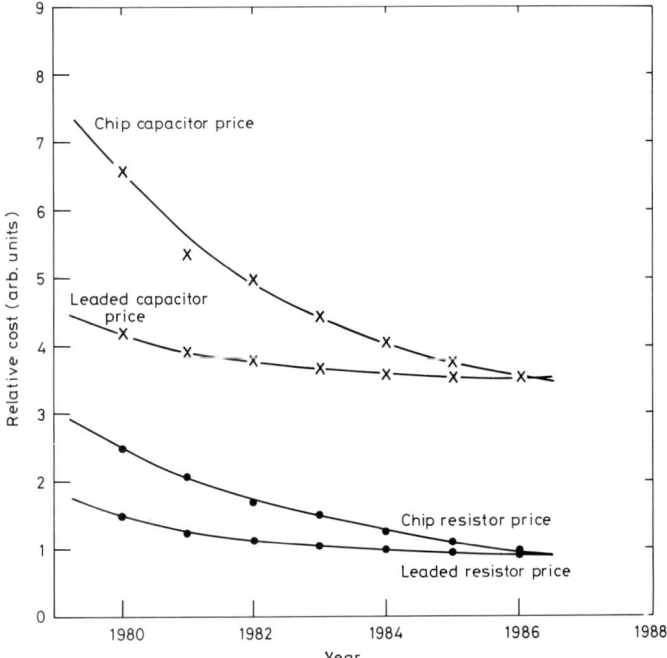

Fig. 14.6 The relative costs in Japan of chip resistors and capacitors, and leaded resistors and capacitors.[523]

even in small batch, high technology assemblies, the cost inefficiencies inherent in manufacturing mixed technology boards.

The greatest use of surface mounting manufacturing processes in the United States, however, will be the communications, computer and automotive industries where large volume component demand is strongest. In the communications industry particularly there is a tremendous drive to achieve equipment portability, lower weight and smaller overall size. The expectation is that certain components, however, will always remain in through-hole leaded format, including crystals, battery clips, audio-type transformers, relays and so on.

The mainframe computer manufacturers are being driven towards surface mounting assembly because of the VLSI technology and the need for higher speeds. The desk-top and micro-computer manufacturers, however, are being driven to the new technology by very strong price competition as well as the demands to achieve smaller, more compact equipment size. The forecast is that, by 1992, between 70 and 100% of computer memory boards will be surface mounted, but with much less penetration in the production of logic and analogue boards.

The third large area of surface mounting acceptance in the United States is the automotive industry, being driven by the need to reduce the size and weight of the electronics as well as to increase the available functions in an extremely cost-sensitive market. High volumes of production are able to justify the capital investment of the new technology.[527]

The industrial electronics market is not, in general, being driven by demands for miniaturisation, high density or speed, but more by high reliability, cost sensitivity, and long product life. The trend in this market is to evaluate the use of surface mounting on one product at a time.[528]

In the military sector, the major applications for the use of surface mounting are those in which the density factor is beneficial. This is particularly apparent in avionics equipment, communication equipment, 'intelligent' weaponry and such like.[529]

14.4 THE CHALLENGE OF SURFACE MOUNTING TECHNOLOGY

Because the technology of surface mounting is relatively new there are many issues that remain unresolved relating to reliability, processing, cost and availability.[530] Potential users and component suppliers have to evaluate the complex issues associated with the adoption of the new manufacturing process.

Probably the major deterrent to acceptance of SMT is the entrenched position of through-hole technology combined with a commonly experienced resistance to change. The argument is made in terms of the economics, not only in the cost of machinery but in the cost of redesigning and retooling printed circuit board layouts.

The lack of industry-wide standards is a consequence of a lack of widespread expertise in surface mount design and implementation, together with a lack of long-term reliability data. The need for standards is concerned not only with component packaging but with the unresolved issues of testing, inspection and reworking of assemblies.

Surface mounting presents a whole new set of challenges to the manufacturer. The transition is from bar solders and liquid fluxes to solder pastes and adhesives, from wave soldering to reflow soldering. New cleaning, testing and inspection techniques have to be acquired. Probably the greatest challenge of all is the goal of zero-defect assembly, a requirement brought about by the need to avoid, if at all possible, inspection and reworking of the soldered surface mounted assembly.

Appendix

UNITS AND SYMBOLS

Throughout the book, SI units have been used together with a rationalised range of symbols. Within each chapter any symbol used is almost invariably unique in its assignation, and from chapter to chapter a consistency has been applied wherever possible, subject to the limited number of symbols available. A comprehensive list of symbols follows, after some useful conversion factors and fundamental constants.

A.1 SI Base and Derived Units

There are seven base units of the 'Système International d'Unités' whose names and symbols are given in Table A.1.

Table A.1

The SI Base Units

Quantity	Name	Symbol
Length	metre	m
Mass	kilogram	kg
Time	second	s
Electric current	ampère	A
Thermodynamic temperature	kelvin	K
Amount of substance	mole	mol
Luminous intensity	candela	cd

In addition, there are two supplementary units of plane angle and solid angle, namely the radian (rad) and the steradian (sr) respectively, there being 2π radians defining a circle and 4π steradians defining a sphere.

All other derived units can be expressed algebraically in terms of the base and supplementary units, but a number have been given special names and symbols in order to simplify their use. The SI derived units with special names, relevant to this book, are given in Table A.2.

Table A.2

The SI Derived Units with Special Names

Quantity	Name	Symbol	Expression in Other Derived Units	Expression in Base Units
Force	newton	N		$m.kg.s^{-2}$
Pressure, stress	pascal	Pa	$N.m^{-2}$	$m^{-1}.kg.s^{-2}$
Energy work, quantity of heat	joule	J	$N.m$	$m^2.kg.s^{-2}$
Power, radiant flux	watt	W	$J.s^{-1}$	$m^2.kg.s^{-3}$
Quantity of electricity, electric charge	coulomb	C		$s.A$
Electric potential, potential difference, emf	volt	V	$W.A^{-1}$	$m^2.kg.s^{-3}.A^{-1}$
Capacitance	farad	F	$C.V^{-1}$	$m^{-2}.kg^{-1}.s^4.A^2$
Electrical resistance	ohm	Ω	$V.A^{-1}$	$m^2.kg.s^{-3}.A^{-2}$
Conductance	siemens	S	$A.V^{-1}$	$m^{-2}.kg^{-1}.s^3.A^2$
Magnetic flux	weber	Wb	$V.s$	$m^2.kg.s^{-2}.A^{-1}$
Inductance	henry	H	$Wb.A^{-1}$	$m^2.kg.s^{-2}.A^{-2}$
Frequency	hertz	Hz		s^{-1}

The units for all other physical quantities are expressed in terms of combinations of base and derived units, some examples of which are shown in Table A.3.

Table A.3

Examples of SI Derived Units

Quantity	Name	Symbol	Expression in Base Units
Surface tension	newton per metre	$N.m^{-1}$	$kg.s^{-2}$
	joule per square metre	$J.m^{-2}$	$kg.s^{-2}$
Dynamic viscosity	pascal second	Pa.s	$m^{-1}.kg.s^{-1}$
Moment of force	newton metre	N.m	$m^2.kg.s^{-2}$
Heat flux density	watt per square metre	$W.m^{-2}$	$kg.s^{-3}$
Heat capacity	joule per kelvin	$J.K^{-1}$	$m^2.kg.s^{-2}.K^{-1}$
Specific heat capacity	joule per kilogram kelvin	$J.kg^{-1}.K^{-1}$	$m^2.s^{-2}.K^{-1}$
Thermal conductivity	watt per metre kelvin	$W.m^{-1}.K^{-1}$	$m.kg.s^{-3}.K^{-1}$
Permittivity	farad per metre	$F.m^{-1}$	$m^{-3}.kg^{-1}.s^4.A^2$
Permeability	henry per metre	$H.m^{-1}$	$m.kg.s^{-2}.A^{-2}$
Angular velocity	radian per second	$rad.s^{-1}$	
Radiant intensity	watt per steradian	$W.sr^{-1}$	

A.2 SI Prefixes

Prefixes have been adopted for use with SI units, to form decimal multiples and submultiples. They are used throughout the book, as given in Table A.4.

Table A.4
The SI Prefixes

Factor	Prefix	Symbol	Factor	Prefix	Symbol
10^{12}	tera	T	10^{-1}	deci	d
10^{9}	giga	G	10^{-2}	centi	c
10^{6}	mega	M	10^{-3}	milli	m
10^{3}	kilo	k	10^{-6}	micro	μ
10^{2}	hecto	h	10^{-9}	nano	n
10	deca	da	10^{-12}	pico	p
			10^{-15}	femto	f
			10^{-18}	atto	a

A.3 Conversion Factors

Many non-SI units are unfortunately in common usage throughout manufacturing industry. These are based mainly on the British (or Imperial) system with the yard, pound, second, degree Fahrenheit as the main base units. For convenience some relevant conversion factors are given in Table A.5.

Table A.5
Some Conversions between Non-SI and SI Units

Length	1 in	=	25·4 mm
	1 yd	=	0·9144 m
	1 mil, thou	=	25·4 μm
	1 ångstrom (Å)	=	0·1 nm
Area	1 sq in	=	645·16 mm^2
	1 sq ft	=	0·09290 m^2
	1 sq yd	=	0·83613 m^2
Volume	1 cu in	=	16·3871 cm^3
	1 cu ft	=	0·02832 m^3
	1 gal (UK)	=	4·5461 dm^3
	1 gal (US)	=	3·7854 dm^3
Velocity	1 ft/s	=	0·3048 m.s^{-1}
Mass	1 oz	=	28·3495 g
	1 lb	=	0·45359 kg
Density (ϱ)	1 lb/cu in	=	27·6799 g.cm^{-3}
	1 lb/cu ft	=	16·0185 kg.m^{-3}
Force	1 kgf	=	9·80665 N
	1 lbf	=	4·44822 N
Pressure	1 lbf/sq in (psi)	=	6·89476 kPa
	1 mbar	=	100·0 Pa
Viscosity, dynamic (η)	1 cP (centipoise)	=	1·0 mPa.s
Viscosity, kinematic	1 cSt (centistokes)	=	1·0 mm^2.s^{-1}
Energy	1 cal	=	4·1868 J
	1 Btu	=	1·05506 kJ
	1 kW.h	=	3·6 MJ
	1 therm	=	105·506 MJ
	1 eV	=	1·60219 $\times 10^{-19}$ J

Power	1 hp (horsepower)	=	745·70 W
Surface tension (γ)	1 dyn/cm	=	$1·0$ mN.m^{-1}
Temperature range (ΔT)	1 °F	=	0·55556 K
Heat flux (Q/A)	1 Btu/h/sq ft	=	3·15459 W.m^{-2}
Specific heat capacity (c)	1 Btu/lb/°F	=	4·1868 J.g^{-1}.K^{-1}
	1 cal/g/°C	=	4·1868 J.g^{-1}.K^{-1}
Thermal conductivity (K)	1 Btu/h/ft/°F	=	1·73073 W.m^{-1}.K^{-1}
	1 kcal/h/m/°C	=	1·163 W.m^{-1}.K^{-1}
Heat transfer coefficient (h)	1 Btu/h/ft^2/°F	=	5·67826 W.m^{-2}.K^{-1}
	1 kcal/h/m^2/°C	=	1·163 W.m^{-2}.K^{-1}

A.4 Fundamental Constants

The SI units form a rationalised set covering all the physical properties of matter. They are rationalised in that, for example, 1 W = 1 A × 1 V exactly and in perpetuity. However, the absolute values of these units, as manifested in some physical phenomenon, depend on fundamental universal properties which are expressed in terms of fundamental physical constants. The values of these constants expressed in SI units are known with greater precision than generally required.

There are three classes of fundamental constants, based on

(i) particular physical objects, such as the mass of an electron at rest;
(ii) characteristics of physical phenomena, such as the gravitational coupling constant;
(iii) universal constants entering into universal physical laws, such as Planck's constant.

Some recommended 'best values' of fundamental physical constants and conversion factors,[531] relevant to this book, are given in Table A.6.

Table A.6

Fundamental Physical Constants

Quantity	Symbol	Value	
Speed of light in vacuum	c	2·997924	10^8 m.s^{-1}
Planck's constant	h	6·626176	10^{-34} J.s
Elementary charge	e	1·602189	10^{-19} C
Avogadro's constant	N_A	6·022045	10^{23} mol^{-1}
Gravitational constant	G	6·6720	10^{-11} N.m^2.kg^{-2}
Molar gas constant	R	8·31441	J.mol^{-1}.K^{-1}
Boltzmann's constant	k (R/N_A)	1·380662	10^{-23} J.K^{-1}
First radiation constant	k_1 ($2\pi hc^2$)	3·741832	10^{-16} W.m^2
Second radiation constant	k_2 (hc/k)	1·438786	10^{-2} m.K
Stefan-Boltzmann constant	ϱ ($2\pi^5 k^4/15h^3c^2$)	5·67032	10^{-8} W.m^{-2}.K^{-4}
Permeability of vacuum	μ_0	4π (exactly)	10^{-7} H.m^{-1}
Permittivity of vacuum	ε_0 ($1/\mu_0 c^2$)	8·854188	10^{-12} F.m^{-1}
Mass of electron at rest	m_e	9·109534	10^{-31} kg
Mass of proton at rest	m_p	1·672649	10^{-27} kg

Appendix 523

A.5 Symbols Used

Chapter 2

A	[m]	maximum width of component body
B	[m]	maximum length of component body
C	[m]	typical height of component
C_{20}	[F]	capacitor value at 20°C
C_{80}	[F]	capacitor value at 80°C
K	[s^{-1}]	constant of rate of degradation due to temperature
K'	[s^{-1}]	constant of rate of degradation due to temperature and humidity
L	[m]	maximum length of a component
Q_H	[J]	activation energy for humidity degradation
Q_T	[J]	activation energy for temperature degradation
R_T	[s^{-1}]	rate of degradation due to temperature
R_{TH}	[s^{-1}]	rate of degradation due to temperature and humidity
RH	[%]	relative humidity
T	[K]	temperature
TC	[K^{-1}]	temperature coefficient of a capacitor
W	[m]	maximum width of a component
k	[$J.K^{-1}$]	Boltzmann's constant
p	[m]	lead pitch
w	[m]	lead width
θ_{JA}	[$K.W^{-1}$]	thermal resistance between active junction and ambient
θ_{JC}	[$K.W^{-1}$]	thermal resistance between active junction and case

Chapter 3

A	[m^2]	cross-sectional area
E	[Pa]	elastic modulus
K	[$J.s^{-1}.m^{-1}.K^{-1}$]	thermal conductivity
K_n	[$J.s^{-1}.m^{-1}.K^{-1}$]	thermal convectivity with natural convection
K_p	[$J.s^{-1}.m^{-1}.K^{-1}$]	thermal convectivity with forced convection
Q	[J]	quantity of heat
R_θ	[$K.W^{-1}$]	thermal resistance
T	[K]	temperature
T_f	[K]	temperature of cooling fluid
V	[m^3]	volume
c	[$J.kg^{-1}.K^{-1}$]	specific heat
d	[m]	thickness
h	[$W.m^{-2}.K^{-1}$]	heat transfer coefficient
ℓ	[m]	length
n	–	number
r	[$\Omega.m$]	electrical resistivity
t	[s]	time
v	–	Poisson's ratio
w	[m]	width

524 Appendix

x	[m]	spatial coordinate
y	[m]	spatial coordinate
z	[m]	spatial coordinate
Ψ	[deg]	spreading angle
α	[K^{-1}]	temperature coefficient of expansion
δ	[m]	thickness of a boundary layer
ε	–	emissivity
ε_y	–	yield strain
ϱ	[kg.m^{-3}]	density
σ	[W.m^{-2}.K^{-4}]	Stefan-Boltzmann constant
σ_y	[Pa]	yield stress

Chapter 4

A	[m]	maximum footprint width
B	[m]	maximum footprint length
C	[m]	inner footprint dimension
L	[m]	LCCC size
N	–	number of solderable terminations in a row on one component
W	[m]	component width
a	[m]	pad width
b	[m]	pad length
d	[m]	pad separation on same footprint
f	[m]	pad separation on adjacent footprints
g	[m]	corner separation of pads on a quad-footprint
ℓ	[m]	length of component
p	[m]	pad pitch
q	[m]	nominal position of pad
Δq	[m]	positional uncertainty of substrate
r	[m]	pad extension beyond component termination
s	[m]	overlap of component and pad
t	[m]	nominal position of component
Δt	[m]	positional uncertainty of component
u	[m]	separation of adjacent components
Φ	[deg]	rotational placement accuracy

Chapter 5

C	[J.K^{-1}]	heat capacity
Q	[J]	heat energy
T	[K]	temperature
ΔT	[K]	temperature change
V	[m^3]	volume
W	[W]	power
X_1	–	} position identifiers
X_2	–	
c	[J.kg^{-1}.K^{-1}]	specific heat

Appendix 525

| t | [s] | time |
| ϱ | [kg.m^{-3}] | density |

Chapter 6

A	[m^2]	area
A$_o$	–	fractional open area of a mesh
D	[m]	dissolution from a surface
E$_A$	[J]	activation energy for oxidation
E$_D$	[J]	activation energy for dissolution
F	[N]	force
L	[K]	liquidus temperature
M	[N.m]	viscometer torque
S	[K]	solidus temperature
T	[K]	temperature
V	[m^3]	volume of reflowed solder
a	[m]	size of mesh opening
b	[m]	diameter of mesh wire
p	[m]	radius of inner cylinder of viscometer
q	[m]	radius of outer cylinder of viscometer
h$_e$	[m]	emulsion thickness of a screen
h$_m$	[m]	mesh thickness of a screen
h$_r$	[m]	reflowed solder deposit height
h$_w$	[m]	wet paste print thickness
k	[J.K^{-1}]	Boltzmann's constant
ℓ	[m]	length of inner cylinder of viscometer
r	[m]	radius of a circular solderable pad
t	[s]	time
v	[m.s^{-1}]	velocity
w	[kg.m^{-2}]	weight increase per unit area due to oxidation
x	[m]	distance
η	[Pa.s]	dynamic viscosity
Ω	[rad.s^{-1}]	rotational velocity

Chapter 7

A	[m^2]	area of thermal emission source
B	[m^2]	area of thermal receptor
E$_{ph}$	[J]	energy of a photon
G	[W]	total irradiance
J	[W]	total radiosity
Q	[J]	heat energy
Q$_r$	[J]	radiative heat energy
Q$_{cc}$	[J]	conductive plus convective heat energy
R	–	infra-red character of a furnace
T	[K]	temperature, in kelvin
V	[m^3]	volume
W	[W]	total emissive power
W$_b$	[W]	emissive power of a blackbody

W_{12}	[W]	net exchange of radiation between bodies 1 and 2
W_n	[W]	net heat input
W_l	[W]	heat loss
W	[W.m^{-1}]	spectral emissive power
c	[m.s^{-1}]	speed of electromagnetic radiation
c	[J.kg^{-1}.K^{-1}]	specific heat
h	[J.s]	Planck's constant
h	[W.m^{-2}.K^{-1}]	heat transfer coefficient
h_r	[W.m^{-2}.K^{-1}]	radiative heat transfer coefficient
k	[J.K^{-1}]	Boltzmann's constant
k_1	[W.m^2]	first radiation constant ($2\pi hc^2$)
k_2	[m.K]	second radiation constant (hc/k)
k_3	[m.K]	Wien's radiation constant
t	[s]	time
t_h	[s]	heating time
z	[m]	depth
α	–	absorptivity
α_λ	–	spectral absorptivity
ε	–	emissivity
ε_λ	–	spectral emissivity
ε_θ	–	directional emissivity
θ	[deg]	angle from the normal to the radiant source
λ	[m]	wavelength of electromagnetic radiation
λ_{max}	[m]	wavelength of maximum of W_λ distribution
λ_s	[J.kg^{-1}]	latent heat of fusion of solder
μ_λ	[m^{-1}]	linear absorption coefficient
ν	[Hz]	frequency
ϱ	–	reflectivity
ϱ_λ	–	spectral reflectivity
σ	[W.m^{-2}.K^{-4}]	Stefan-Bolzmann constant
τ	–	transmissivity
τ_λ	–	spectral transmissivity
φ	[deg]	angle in the plane of the radiant source

Chapter 8

A	[m^2]	surface area
C	[J.K^{-1}]	heat capacity
E	[J]	energy
K	[W.m^{-1}.K^{-1}]	thermal conductivity
K_L	[W.m^{-1}.K^{-1}]	thermal conductivity of a liquid film
K_t	[W.m^{-1}.K^{-1}]	thermal conductivity of cooling tube material
M	[kg.m^{-2}.s^{-1}]	mass flow of a condensing liquid
Q	[J]	heat energy
R_a	–	Rayleigh number
T	[K]	temperature
T_c	[K]	temperature of condensation
T_f	[K]	temperature of fluid
T_o	[K]	temperature at start of heat transfer

T_s	[K]	temperature of a solid surface
$T(z)$	[K]	temperature at distance z
ΔT_{crit}	[K]	critical temperature excess
V	[m^3]	volume
a	[m]	width of a plate
b	[m]	length of a plate
c	[J.kg^{-1}.K^{-1}]	specific heat
c_L	[J.kg^{-1}.K^{-1}]	specific heat of a liquid
d	[m]	distance between droplets
g	[m.s^{-2}]	acceleration due to gravity
h	[W.m^{-2}.K^{-1}]	heat transfer coefficient
h_c	[W.m^{-2}.K^{-1}]	heat transfer coefficient of vapour on cooled tube
h_e	[W.m^{-2}.K^{-1}]	effective heat transfer coefficient
h_w	[W.m^{-2}.K^{-1}]	heat transfer coefficient of water in a tube
ℓ	[m]	semi-thickness of a plate
m	–	integer, 0, 1, 2 ...
n	–	integer, 0, 1, 2 ...
p	[m]	length of a cooling tube
r	[m]	radius of a tube
r_i	[m]	inner radius
r_o	[m]	outer radius
t	[s]	time
t_L	[s]	time during which latent heat is transferred
t_o	[s]	characteristic time of an exponential curve
x	[m]	spatial coordinate
y	[m]	spatial coordinate
z	[m]	depth/distance
β	–	thermal diffusion parameter, $\varkappa t/\ell^2$
γ	[N.m^{-1}]	surface tension of a liquid
δ	[m]	thickness of a liquid film
θ	[deg]	angle of inclination from the horizontal
ξ	–	normalised distance parameter, z/ℓ
$\varkappa$	[m^2.s^{-1}]	thermal diffusivity
λ_f	[J.kg^{-1}]	latent heat of condensation of a fluid
λ_s	[J.kg^{-1}]	latent heat of fusion of a solid
η_L	[Pa.s]	viscosity of a liquid
ϱ	[kg.m^{-3}]	density
ϱ_L	[kg.m^{-3}]	density of a liquid
ϱ_V	[kg.m^{-3}]	density of a vapour

Chapter 9

A_{nm}	[Hz]	Einstein coefficient of spontaneous emission
B_{nm}	[J^{-1}.m^3.s^{-2}]	Einstein coefficient of induced emission
B_{mn}	[J^{-1}.m^3.s^{-2}]	Einstein coefficient of absorbed emission
E_m	[J]	lower atomic energy level
E_n	[J]	higher atomic energy level
E_{ph}	[J]	energy of a photon
K	[J.m^{-1}.s^{-1}.K^{-1}]	thermal conductivity
N_m	–	population of atomic state m

528   *Appendix*

N_n	–	population of excited atomic state n
P	[W]	total radiated power
P_r	[W.m^{-2}]	power density at retina of eye
P_c	[W.m^{-2}]	power density at cornea of eye
Q	[J]	heat energy
T	[K]	temperature
ΔT	[K]	temperature change
T_s	[K]	temperature of solder-side of a substrate
V_i	[m^3]	volume of item i
V_s	[m^3]	volume of solder
a	–	} algebraic coefficients
b	–	
a	[m]	diameter of light source
b	[m]	radius of locally heated spot
c	[J.kg^{-1}.K^{-1}]	specific heat
c_i	[J.kg^{-1}.K^{-1}]	specific heat of item i
c	[m.s^{-1}]	velocity of light
d_e	[m]	diameter of pupil of the eye
d_r	[m]	diameter of image on the retina
f	[m]	focal length of eye
h	[W.m^{-2}.K^{-1}]	heat transfer coefficient
h	[J.s]	Planck's constant
k	[J.K^{-1}]	Boltzmann's constant
m	–	lower atomic level
n	–	higher atomic level
r	[m]	distance of light source from eye
t	[s]	time
w	[cm^{-1}]	wavenumber
x	–	} algebraic operators
y	–	
Λ	[J.m^{-3}.Hz^{-1}]	energy density of an electric field
Φ	[W.m^{-2}]	heat flux
ζ	–	dimensionless parameter
λ_s	[J.kg^{-1}]	latent heat of fusion of solder
λ	[m]	wavelength of electromagnetic radiation
ν	[Hz]	frequency of electromagnetic radiation
ν_{nm}	[Hz]	frequency of radiation characterised by atomic levels n and m
ϱ	[kg.m^{-3}]	density
ϱ_i	[kg.m^{-3}]	density of item i
ϱ_s	[kg.m^{-3}]	density of solder

Chapter 10

A_1	[m^2]	de-wetting model: base area of each solder droplet
A_2	[m^2]	de-wetting model: cap area of each solder droplet
A	[N.s]	} areas under wetting balance curve (Figure 10.34)
B	[N.s]	

Appendix

C	–	a solderable coating
C_i	[kg.m^{-3}]	concentration of species i
D	[m]	diameter of a sphere of solder
D	[m^2.s^{-1}]	diffusivity of one species in another
D_o	[m^2.s^{-1}]	diffusion coefficient
D_i^j	[m^2.s^{-1}]	diffusivity of element i in phase j
F	[J]	total surface free energy
F_i	[N]	
F^*	[N]	} defined forces measured using wetting balance
F'	[N]	
H	–	radius of constructed circle forming an index of solderability (Figure 10.34)
J_i^j	[kg.m^{-2}.s^{-1}]	flux of element i through phase j
M	[kg.m^{-1}]	component mass, per unit width of component
N	[m^{-2}]	number of solder droplets per unit area in model of de-wetting
Q	[J.mol^{-1}]	activation energy for intermetallic growth
R	[J.mol^{-1}.K^{-1}]	the Gas Constant
R	[m]	radius of a liquid sphere
R_1	[m]	} principal radii of a curve liquid surface
R_2	[m]	
R_a	[m]	average amplitude of roughness features on a surface
S	–	a metal substrate wetted by coating C
S	[N.m^{-1}.s^{-2}]	solderability index due to Schouten
SC	–	intermetallic compound formed at interface between substrate and coating
SS	–	fractional free surface of solder, forming a coating
ΔSS	–	difference in fractional free surface of solder before and after de-wetting
T	[K]	temperature, in kelvin
T_i	[N]	rotational movement, per unit width, of a tombstoning component
V	[m^3]	a volume of solder
W	[kg]	a mass of solder
W_R	–	surface roughness area ratio
X_i^b	–	bulk atomic concentration of species i
X_i^s	–	surface atomic concentration of species i
a	[m]	length of unwettable specimen immersing in solder (Figure 10.38)
b	[m]	radius of a circular solder cap
c	[m]	circumference or perimeter of a sample dipped in solder
d	[m]	thickness of a solder coating
f	–	fractional non-wettable area on a solderable surface (de-wetting model)
f	[m]	height of chip component (tombstoning model)
g	[m.s^{-2}]	acceleration due to gravity
h	[m]	height of a circular solder cap
k	[m^{-1}]	constant (Section 10.5.3)
ℓ	[m]	length of chip component (tombstoning model)

m	[m]	height of a meniscus rise
δp	[Pa]	pressure differential across a curved liquid surface
q	[m]	distance between centre of gravity of chip component and line of rotation during tombstoning
r	[m]	pad extension beyond component metallisation
r	[m]	maximum radius of curvature of sessile drop
s	[m]	overlap of chip metallisation and pad
t	[s]	time
t_i	[s]	} times defining points on wetting balance curve
t^*	[s]	
v	[m^3]	volume of a sample immersed in solder
v_1	[m.s^{-1}]	velocity of S–SC interface in the z direction
v_2	[m.s^{-1}]	velocity of SC–C interface in the z direction
x	[m]	horizontal coordinate of a point in a liquid
δx	[s]	small movement of a liquid surface
x	[s]	} displacements of wetting balance curve from a defined point (Figure 10.35)
y	[N]	
z	[m]	vertical coordinate of a point in a liquid
Δz	[m]	difference in height between two points in a liquid
z_o	[m]	thickness of intermetallic compound layer
z_s	[m]	width of supersaturated zone of molten solder, during cooling
Φ	[deg]	angle of rise from horizontal, of tombstoning component
Ψ	[deg]	$\tan^{-1}(f/\ell)$ in tombstoning chip
Ω	[deg]	$\tan^{-1}(r - f\sin\Phi)/f\cos\Phi$ in Figure 10.49
α	[deg]	angle to horizontal of plate immersed in a liquid
α_m	[deg]	maximum slope of a surface feature on a rough substrate
β	[deg]	angle of a liquid surface to the horizontal
γ	[J.m^{-2}]	surface tension (shorthand for γ_{LV})
γ_i	[J.m^{-2}]	surface tension of surface i
γ_{LV}	[J.m^{-2}]	surface tension of liquid-vapour interface
γ_{SL}	[J.m^{-2}]	surface tension of solid-liquid interface
γ_{SV}	[J.m^{-2}])	surface tension of solid-vapour interface
δ	[deg]	} angles defining the motion of a test PCB in the rotary dip test, Figure 10.45
ε	[deg]	
η	[deg]	
ζ	–	$r^2 g \varrho / \gamma$, (Equation 10.4)
θ	[deg]	contact angle
θ_R	[deg]	observed contact angle on a rough surface
λ_a	[m]	average wavelength of roughness features on a surface
ξ	[deg]	reflection angle of light from a solder meniscus (Figure 10.30)
ϱ	[kg.m^{-3}]	density
σ_{SLV}	[J.m^{-2}]	spreading energy
φ	[deg]	angle between normal to a surface and the vertical
ϕ	[deg]	angle defining a surface area of a liquid (Figure 10.1)

Appendix

χ	[deg]	angle of Schouten's tangent to the wetting balance curve
ψ	[deg]	angle defining a surface area of a liquid (Figure 10.1)

Chapter 11

C	–	Coffin-Manson constant
C_e	–	Coffin-Manson constant in elastic régime
C_p	–	Coffin-Manson constant in plastic régime
E	[Pa]	Young's modulus
L	[kg]	load
N_f	–	fatigue life: number of cycles to failure
R	[K^{-1}]	freezing rate
RA	–	fractional reduction in area at a purely tensile failure
T	[K]	temperature
ΔT	[K]	temperature cycle amplitude
T_c	[K]	temperature of ceramic chip carrier
T_o	[K]	power-off, steady state temperature
T_s	[K]	temperature of substrate
b	–	empirical factor in model of LCCC strain range (Equation 11.5)
c	–	fatigue ductility exponent
c_e	–	fatigue ductility exponent in elastic régime
c_p	–	fatigue ductility exponent in plastic régime
h	[m]	stand-off height of a solder joint on an LCCC
k_1	–	⎫ constants suggested for incorporating frequency
k_2	–	⎭ effects into the Coffin-Manson equation
ℓ	[m]	size of a square LCCC
m	–	microstructure parameter
r	[m]	distance of solder joint from centre of LCCC
v	–	Poisson's ratio
x	[m]	separation of microstructural lamellae in eutectic microstructure
x_o	[m]	initial length of a test specimen
α_c	[K^{-1}]	TCE of ceramic chip carrier
α_s	[K^{-1}]	TCE of substrate
$\Delta\alpha$	[K^{-1}]	difference in TCE of LCCC and substrate
δ_i	[m]	elongations
ε_{ii}	–	tensile strains
ε_{ij}	–	shear strains
$\Delta\varepsilon$	–	strain range
$\Delta\varepsilon_e$	–	elastic strain range
$\Delta\varepsilon_p$	–	plastic strain range
ε_f	–	fatigue ductility coefficient
ε_f'	–	tensile fracture ductility coefficient
θ	[deg]	angle defining the slope of the elastic stress-strain curve: $\tan\theta = E$
μ	[Pa]	shear (rigidity) modulus
ν	[Hz]	cyclic frequency

532   Appendix

σ_e	[Pa]	elastic limit of stress
σ_{ii}	[Pa]	tensile stresses
σ_{ij}	[Pa]	shear stresses
σ_y	[Pa]	yield stress
ψ	[deg]	shear angle

Chapter 12

A	[m^2]	projected area of a particulate contaminant in a flowing liquid
F_{drag}	[N]	drag force at a liquid-solid interface
H	[J.m^{-2}.s^{-1}]	heat energy per unit area, per unit time
K_e	[S.m^{-1}]	electrical conductivity
K	[J.s^{-1}.m^{-1}.K^{-1}]	thermal conductivity
ΔP	[Pa]	pressure differential across a liquid surface
R_1	[m]	} principal radii of a curved liquid surface
R_2	[m]	
R	[m^3.s^{-1}]	rate of flow of a liquid
S_d	[Pa$^{\frac{1}{2}}$]	non-ionic (dispersion) solubility parameter
S_h	[Pa$^{\frac{1}{2}}$]	hydrogen bonding parameter
S_p	[Pa$^{\frac{1}{2}}$]	ionic (polar) solubility parameter
S_t	[Pa$^{\frac{1}{2}}$]	total solubility parameter
ΔSS	[Pa$^{\frac{1}{2}}$]	difference in solubility parameters of solvent and solute
ΔT	[K]	temperature change
V	–	volume fractions of a solvent mixture
a	[m]	radius of a capillary
c	[J.kg^{-1}.K^{-1}]	specific heat
h	[m]	height of a gap beneath a component
k	–	dimensionless drag coefficient
ℓ	[m]	length of gap beneath a component
Δp	[Pa]	pressure differential in a liquid across a gap
t	[s]	time
t_o	[s]	length of a laser heating pulse
v	[m.s^{-1}]	velocity of a flowing liquid
z	[m]	depth from surface of a solder fillet
γ_C	[N.m^{-1}]	critical surface free energy
γ_{LV}	[N.m^{-1}]	surface tension of a liquid (liquid-vapour interface)
γ_{SL}	[N.m^{-1}]	interfacial tension between solid and liquid
γ_{SV}	[N.m^{-1}]	surface tension of a solder-vapour interface
η	[Pa.s]	viscosity of a liquid
θ	[deg]	contact angle
κ	[m^2.s^{-1}]	thermal diffusivity
ϱ	[kg.m^{-3}]	density
σ_{SLV}	[N.m^{-1}]	spreading coefficient

Appendix

Chapter 13

Symbol	Units	Description
A	$[J^{-1}]$	constant defining humidity degradation
D	$[m^2.s^{-1}]$	diffusivity
D_o	$[m^2.s^{-1}]$	diffusion coefficient
E_g	$[J]$	energy gap of a semiconductor
F	–	number of failures in a test batch
H(t)	$[s^{-1}]$	hazard rate
J	$[A.m^{-2}]$	current density
K	$[s^{-1}]$	constant random failure rate
K_c	$[s^{-1}]$	mean failure rate of a component
K_s	$[s^{-1}]$	mean failure rate of a system
MTTF	$[s]$	mean time to failure
N	–	number of components in a test batch
P	–	probability of a given number of failures occurring
Q(t)	–	probability of failure before time t
Q	$[J]$	activation energy
Q_H	$[J]$	activation energy for humidity accelerated degradation
Q_J	$[J]$	activation energy for electrically accelerated degradation
Q_T	$[J]$	activation energy for thermally accelerated degradation
R(t)	–	probability of survival after time t
ΔR	$[\Omega]$	change in electrical resistance
R_H	$[s^{-1}]$	rate of degradation of a property due to humidity overstress
R_J	$[s^{-1}]$	rate of degradation of a property due to electrical overstress
R_T	$[s^{-1}]$	rate of degradation of a property due to thermal overstress
R_{TH}	$[s^{-1}]$	rate of degradation of a property due to temperature and humidity
T	$[K]$	temperature
T_{amb}	$[K]$	ambient service temperature
T_{test}	$[K]$	temperature of accelerated ageing test
V	$[V]$	forward voltage of an active junction
e	$[C]$	electronic charge
f	–	probability of component failure
k	$[J.K^{-1}]$	Boltzmann's constant
n	–	number of component failures
p(t)	–	probability of failure at time t
p(x)	–	probability of failure at x (= log t)
t	$[s]$	time
x	–	log time
$\bar{x}$	–	mean value of x
z	–	$(\bar{x} - x)\sqrt{2}\sigma$
σ	–	standard deviation of x

Chapter 14

C	[£]	capital expenditure
N_i	[£]	net cash flow in year i
i	–	integer, 1, 2, 3 ...
k	[%]	cost of capital
n	–	lifetime, in years, of product production

REFERENCES

1. Sage, M. G., 'The Fourth Electronic Revolution', *Circuit World*, **Vol. 11**, No. 4, pp. 52–53 (1985).
2. Caswell, G., 'Surface Mount Technology', International Society for Hybrid Microelectronics (ISHM), Silver Spring, Maryland (1984).
3. Mullen, J., 'How to use Surface Mount Technology', Texas Instruments, Dallas, Texas (1984).
4. Klein Wassink, R. J., 'Soldering in Electronics', Electrochemical Publications, Ayr, Scotland (1984).
5. Dettmer, R., 'Solder Side Up – The Move to Surface Mounting', *Electronics & Power*, pp. 117–120, February (1984).
6. Prasad, P., 'Critical Issues in Implementation of SMD Technology', Proceedings IPC Fall Meeting, Denver, Colorado, Technical Paper IPC-TP-491 (1983).
7. Reiner, M., 'VLSI Packaging', *Hybrid Circuits*, No. 6, pp. 9–13 (1985).
8. Marston, P., 'Future Semiconductor Packaging Trends', *Electronic Engineering*, pp. 145–154, September (1982).
9. Noach, K., 'The Small Outline Integrated Circuit (SOIC)', Seminar 'The Choice of Chip Carriers and PCBs', Noresund, Norway. Elektronikkindustriens Komponent-forum (1984).
10. Hollander, D., 'Discrete SMCs Come of Age', *Electronic Packaging & Production*, **Vol. 27**, No. 1, pp. 75–76 (1987).
11. Houldsworth, J., 'SMD Handles Higher Power', *Electronic Packaging & Production*, **Vol. 27**, No. 1, pp. 77–78 (1987).
12. Val, C., Kersuzan, G. and Dreyfus-Alain, B., 'The Chip Carrier, Universal Chip Support for High Level Integration', *Hybrid Circuits*, No. 1, pp. 4–10 (1982).
13. Reynolds, M. J., 'Plastic Chip Carriers', *Electronic Production*, **Vol. 13**, No. 9, pp. 53–55 (1984).
14. Blackshaw, M. F. and Dance, F. J., 'The Design, Manufacture and Assembly of High Pin Count Plastic Pin Grid Array Packages', Proceedings 35th IEEE Electronic Components Conference, Washington DC, pp. 199–205 (1985).
15. Love, G. F., 'Packages for High-density Surface Mount Applications', *Circuit World*, **Vol. 13**, No. 3, pp. 36–39 (1987).
16. White, M. L., Serpiello, J. W., Striny, K. M. and Rosenzweig, W., 'The Use of Silicone RTV Rubber for Alpha Particle Protection on Silicon Integrated Circuits', Proceedings IEEE 19th International Reliability Physics Symposium, Orlando, Florida pp. 43–47 (1981).
17. Eastman, K., 'Ceramic Quad Package Meets High-density SMT Needs', *Electronic Packaging & Production*, **Vol. 27**, No. 1, pp. 70–71 (1987).
18. Diletti, H., Hoffmann, R., Sele, G. and Rees, J., 'Resistor Chips for Thick Film Hybrids', *Hybrid Circuits*, No. 6, pp. 18–19 (1985).
19. Wheeler, J. M., 'Thick Film Conductors and Resistors on Dielectrics for High Reliability Applications', *Hybrid Circuits*, No. 8, pp. 24–27 (1985).
20. Bancroft, R. C., 'A Survey of Laser Trimming', *Hybrid Circuits*, No. 5, pp. 18–20 (1984).
21. Spitz, S. L., 'Trimmed for Precision', *Electronic Packaging & Production*, **Vol. 25**, No. 10, pp. 48–56 (1985).
22. Hauschild, F. D., 'TCRs of Thick Film Resistors Produced in High Volume', *Hybrid Circuits*, No. 2, pp. 28–31 (1983).
23. Reynolds, M.J., 'Surface Mounting Connectors', *Electronic Production*, **Vol. 13**, No. 2, pp. 24–25 (1984).
24. Yates, R., 'Surface Mounting: Don't Forget The Connector', *Electronic Production*, **Vol. 14**, No. 4, pp. 41–45 (1985).
25. Brearley, D., 'The Connector/PCB Interface: Key to Success in Surface Mounting of Connectors', *Microelectronics Journal*, **Vol. 17**, No. 3, pp. 14–20 (1986).

26 Weaver, H., 'Surface-mount Enhances High-density Connector System', *Electronic Packaging & Production*, **Vol. 27**, No. 1, pp. 130–132 (1987).
27 Ginsberg, G., 'Latest Developments on Surface Mounted Connectors for Printed Wiring Boards', Proceedings IPC Fall Meeting, San Francisco, California, Technical Paper IPC-TC-514 (1984).
28 Patel, D. and Wiltshire, B., 'The Rôle of Connectors in Surface Mount Technology', *IPC Technical Review*, **Vol. 26**, No. 4, pp. 10–13 (1985).
29 Coleman, M., 'Ageing Mechanisms and Stability in Thick Film Resistors', *Hybrid Circuits*, No. 4, pp. 36–41 (1984).
30 Fu, S-L., Liang, M-S., Shiromatsu, T. and Wu, T-S., 'Electrical Characteristics of Polymer Thick Film Resistors', *IEEE Transactions on Components, Hybrids and Manufacturing Technology*, **Vol. CHMT-4**, pp. 283–288 (1981).
31 Hendricks, A.H.C., 'Atmospheric Corrosion Tests on Pd/Ag Thick Film Conductors', *Hybrid Circuits*, No. 11, pp. 69–71 (1986).
32 Sinnadurai, F.N. and Wilson, K.J., 'The Ageing Behaviour of Commercial Thick-Film Resistors', *IEEE Transactions on Components, Hybrids and Manufacturing Technology*, **Vol. CHMT-5**, pp. 308–317 (1982).
33 Morten, B. and Prudenziati, M., 'Thermal Ageing of Thick-film Resistors', *Hybrid Circuits*, No. 3, pp. 24–26 (1983).
34 Itoh, K., 'Heat Management Faces Demand of High Thermal Density', *Electronic Packaging & Production*, **Vol. 27**, No. 1, pp. 136–139 (1987).
35 Markstein, H.W., 'Surface-mount Substrates: The Key in Going Leadless', *Electronic Production*, **Vol. 13**, No. 3, pp. 23–30, 53 (1984).
36 Angstenberger, A., 'Base, Substrate Materials for the Construction of Electronic Assemblies; Special Application of Surface Mount Technology PCBs', *Circuit World*, **Vol. 12**, No. 4, pp. 44–47 (1986).
37 Fishman, D. and Cooper, N., 'Mounting Leadless Chip Carriers on to PCBs', *Hybrid Circuits*, No. 1, pp. 38–43 (1982).
38 Fisher, J.R., 'Cast Lead Process and Surface Mount Reliability', *IPC Technical Review*, **Vol. 26**, No. 5, pp. 13–18 (1985).
39 Lacruche, B., 'Thermal Problems in Micro-assemblies', *Hybrid Circuits*, No. 7, pp. 12–16 (1985).
40 Carslaw, H.S. and Jaeger, J.C., 'Conduction of Heat in Solids', Oxford University Press, Second Edition (1960).
41 Fukuoka, Y. and Ishizuka, M., 'Transient Temperature Rise for Multi-chip Packages', *Hybrid Circuits*, No. 3, pp. 52–57 (1983).
42 Seraphim, D.P., 'Chip-Module Package Interfaces', *IEEE Transactions on Components, Hybrids and Manufacturing Technology*, **Vol. CHMT-1**, pp. 305–309 (1978).
43 Leonida, G., 'Handbook of Printed Circuit Design, Manufacture, Components and Assembly', Electrochemical Publications, Ayr, Scotland (1981).
44 Scarlett, J.A. 'An Introduction to Printed Circuit Board Technology', Electrochemical Publications, Ayr, Scotland (1984).
45 Smith, C.A., 'The Glass Transition Temperature and its Measurement in Epoxy Glass Laminates', *Circuit World*, **Vol. 12**, No. 2, pp. 62–64 (1986).
46 Electronic Production-BPA(UK) Joint Report, 'Know your Laminates—Present and Future', *Electronic Production*, **Vol. 13**, No. 10, pp. 49–56 (1984).
47 Sanjana, Z.N., Valentich, J. and Marchetti, J.R., 'Thermal Expansion of Circuit Board Materials for Leadless Chip Carriers', Proceedings IPC Fall Meeting, Denver, Colorado, Technical Paper IPC-TP-488 (1983).
48 Märtens, A., 'Ultra-thin Copper Foils: A Solution for High Density Circuit Boards', Proceedings 3rd Printed Circuit World Convention, Washington DC, Paper WC III–50.
49 Sculpher, M.R., 'Advanced Solder Masks', *Circuit World*, **Vol. 12**, No. 3, pp. 9–11 (1986).
50 Tilsley, G.M. and Roos, L., 'New High-resolution Dry Film Photoresist', *Circuit World*, **Vol. 13**, No. 1, pp. 72–77 (1986).
51 Cassada, K., 'The Application of Dry Film Solder Mask Resists', *Electri-onics*, **Vol. 32**, No. 10, pp. 28–30 (1986).
52 Smith-Vargo, L., 'Focus of Photoimageable Solder Masks', *Electronic Packaging & Production*, **Vol. 26**, No. 6, pp. 87–89 (1986).
53 Elliott, D.A., 'Hot Air Levelling of Printed Wiring Boards', Proceedings NEPCON-East 77, Philadelphia, Pennsylvania, pp. 325–332 (1977).

54 Masciana, G.C., 'Quality Aspects of Hot-air Soldering', *Evaluation Engineering*, p.46, September (1983).
55 Emerson, A., 'Using Hot Air Levelling with SMOBC', *Electronic Production*, **Vol. 14**, No. 9, pp. 37–41 (1985).
56 Elliott, D.A., 'Hot Air Levelling Excels for Surface-mount Boards,' *Electronic Packaging & Production*, **Vol. 26**, No. 2, pp. 91–95 (1986).
57 Keeler, R., 'SMOBC Clears the Way to Fine-line Circuitry', *Electronic Packaging & Production*, **Vol. 26**, No. 10, pp. 29–31 (1986).
58 Val, C. and Humbert, N., 'Alumina With a Thermal Conductivity Close to Beryllia', *Hybrid Circuits*, No. 2, pp. 45–50 (1983).
59 Storbeck, I., Balke, H. and Wolf, M., 'Substrate Bowing of Multilayer Thick Film Circuits', *Hybrid Circuits*, No. 11, pp. 21–23 (1986).
60 Wall, C.I., 'Screenprinting in the PCB Industry', *Circuit World*, **Vol. 12**, No.3, pp. 36–40 (1986).
61 Hargrave, C.E., 'Thick Film Screen Techniques', *Hybrid Circuits*, No. 2, pp. 21–27 (1983).
62 Pound, R., 'Thick-film Inks Mark Hybrids with Today's Circuits', *Electronic Packaging & Production*, **Vol. 27**, No. 1, pp. 111–112 (1987).
63 Pitt, K.E.G., 'An Introduction to Thick-film Component Technology', Mackintosh (1981).
64 Stein, S.J., Huang, C. and Bless, P., 'Pd/Ag Conductors Show Promise', *Electronics Manufacture and Test*, pp. 29–31, December (1986).
65 Björklund, G., 'Thick-film and Associated Technologies', in 'Handbook of Microelectronics Packaging and Interconnection Technologies', Ed. Sinnadurai, F.N., Electrochemical Publications, Ayr, Scotland (1985).
66 Sinnadurai, F.N., Spencer, P.E. and Wilson, K.J., 'Some Observations on the Accelerated Ageing of Thick Film Resistors', Proceedings ISHM European Hybrid Microelectronics Conference, Ghent, Belgium, pp. 113–121 (1979).
67 Sinnadurai, F.N., 'Reliability of Microelectronics Packaging and Interconnection' in 'Handbook of Microelectronics Packaging & Interconnection Technologies', Ed. Sinnadurai, F.N., Electrochemical Publications, Ayr (1985).
68 Brierley, C.J., Pedder, D.J. and McCarthy, J.P., 'The Characterisation of Novel PWB Substrate Materials for Leadless Ceramic Chip Carrier Attachment', *Hybrid Circuits*, No. 10, pp. 5–8 (1986).
69 Menozzi, G., 'Advanced PWB Materials for Surface Mounted Devices in Aerospace Applications', *Hybrid Circuits*, No. 11, pp. 72–81 (1986).
70 Reimann, W.G. and Gates, L.E., 'Polyimide-quartz Fabric Printed Circuit Boards for Mounting Leadless Chip Carriers', Proceedings IPC Fall Meeting, San Diego, California, Technical Paper IPC-TP-443 (1982).
71 Hartman, H., 'Substrates for Leadless Chip Carriers', *Electronic Production*, **Vol. 14**, No. 5, pp. 18–24 (1985).
72 Wicher, D.P. and Hatfield, W.B., 'Porcelain Steel Technology: a Bona Fide Alternative', Proceedings ISHM International Microelectronics Conference, Minneapolis, Minnesota, pp. 176–187 (1978).
73 Spector, M., 'Porcelain Coated Steel Substrates for High Density Component Interconnection', *Insulation/Circuits*, **Vol. 25**, No. 1, pp. 15–17 (1979).
74 Stein, S.J., Huang, C. and Gelb, A.S., 'Thick Film Materials on Porcelain Enamelled Steel Substrates', Proceedings 29th IEEE Electronic Components Conference, Cherry Hill, New Jersey, pp. 121–125 (1979).
75 Stein, S.J., Huang, C. and Gelb, A.S., 'Comparison of Enamelled Steel Substrate Properties for Thick Film Use', Proceedings ISHM European Hybrid Microelectronics Conference, Ghent, Belgium, pp. 525–538 (1979).
76 Hugh, S.C., 'Thermal Design Criteria for Porcelainised Steel vs. Alumina Substrates', Proceedings ISHM International Microelectronics Conference, New York, pp. 322–326 (1980).
77 Anon, 'Clad Metal May Allow New Packaging Concept for Leadless Chip Carriers', *Insulation/Circuits*, **Vol. 28**, No. 3, pp. 38–39 (1982).
78 Langridge, C., 'Copper-Molybdenum Metal Cores for PCBs', *Electronic Production*, **Vol. 14**, No. 7, pp. 35–38 (1985).
79 Leibowitz, J., Winters, W. and Kolkin, J., 'Graphite Layers in SMT Boards Control Thermal Mismatch', *Electronic Packaging & Production*, **Vol. 25**, No. 6, pp. 86–88 (1985).
80 Jones, C., Yeager, R. and Gray, F., 'Constraining Core Technique for Surface Mounted Components', *IPC Technical Review*, **Vol. 25**, No. 10, pp. 13–18 (1984).
81 Dance, F.J. and Wallace, J.L., 'Clad Metal Circuit Board Substrates for Direct Mounting of

Ceramic Chip Carriers', *Electronic Packaging & Production*, **Vol. 22**, No. 1, pp. 228–237 (1982).
82 Gray, F., Cartwright, L. and Lindblom, S., 'Copper-clad Invar as a Constraining Core for Reliable LCC Applications: Design and Fabrication', *IPC Technical Review*, **Vol. 27**, No. 1, pp. 14–19 (1986).
83 Hanson, J.R. and Hauser, J.L., 'New Board Overcomes TCE Problem', *Electronic Packaging & Production*, **Vol. 26**, No. 11, pp. 48–51 (1986).
84 Delagi, R., 'Designing with Clad Materials', *Machine Design*, pp. 79–83, November 20 (1980).
85 Lindblom, S.H., 'Thermophysical Properties of Copper-clad Invar', Proceedings IPC Fall Meeting, San Diego, California, Technical Paper IPC-TP-617 (1986).
86 House, J.R., 'Multiwire—An Alternative to Etched Tracks', *Electronic Production*, **Vol. 14**, No. 6, pp. 28–31 (1985).
87 Kozima, F., Fukutomi, N., Iwasaki, Y. and Kida, A., 'Multiwire Boards for VLSI Packaging', Proceedings 3rd Printed Circuit World Convention, Washington DC, Paper WC III–32.
88 Chadzynski, P.Z., 'Multiwire Boards and Surface Mount', *Printed Circuit Fabrication*, **Vol. 9**, No. 2, pp. 36–46 (1986).
89 Buck, T.J., 'A Discrete-wired Solution for High Speed Surface-mount Packaging', *Electronic Packaging & Production*, **Vol. 26**, No. 6, pp. 136–139 (1985).
90 Pound, R., 'Discrete Wiring Supports Higher Signal Speed and Circuit Density', *Electronic Packaging & Production*, **Vol. 25**, No. 10, pp. 84–88 (1985).
91 Ginsberg, G.L., 'Packaging and Interconnecting Structures for Chip Carrier—Surface Mounting Technology', Proceedings 3rd Printed Circuit World Convention, Washington DC, Paper WC III–40.
92 El Refaie, M., 'Interconnect Substrate for Advanced Electronic Systems', *Electronic Engineering*, pp. 133–141, September (1982).
93 Schmidt, W., 'Denstrate Technology', *Circuit World*, **Vol. 13**, No. 2, pp. 22–25 (1987).
94 Martin, F.W., 'Polymer Thick-film Extends Options for Hybrid and PC Fabrication', *Circuits Manufacturing*, pp. 72–77, May (1977).
95 Roos-Kozel, B.L., Caiazzo, F.M. and Francis, J.S., 'Improved Polymer Inks Increase Applications', Proceedings ISHM International Symposium on Microelectronics, Anaheim, California, pp. 239–244 (1985).
96 Johnson, R.W., 'Polymer Thick-films: Technology and Materials', *Circuits Manufacturing*, pp. 54–60, July (1982).
97 Castelli, G. and Lovati, G., 'Integration of Polymer Thick Films with PCB Technology in the Telecommunication Field', Hybrid Circuits, No. 14, pp. 42–48 (1987).
98 Kabe, A. and Morooka, I., 'Polymer Thick Film Circuits in Japan', *Hybrid Circuits*, No. 6, pp. 24–26 (1985).
99 Wall, C.I., 'Polymer Thick Films: What They Are, How They Are Used and Where?' *Circuit World*, **Vol. 12**, No. 2, pp. 42–44 (1986).
100 Jones, B., 'Going Hybrid with PTF Resistors', *Electronic Packaging & Production*, **Vol. 27**, No. 1, pp. 118–121 (1987).
101 Martin, F.W. and Shahbazi, S., 'Polymer Thick Film for High Reliability Applications', *Hybrid Circuits*, No. 3, pp. 33–36 (1983).
102 Martin, F.W., 'The Use of Polymer Thick Film for Making PCBs', *Hybrid Circuits*, No. 1, pp. 22–23 (1982).
103 Heason, E.N., 'Adding Polymer Circuitry to PCBs', *Electronic Production*, **Vol. 16**, No. 4, pp. 19–27 (1987).
104 Green, W.J., 'Commercial Production of Polymer Thick Film Resistors', *Hybrid Circuits*, No. 5, pp. 15–17 (1984).
105 Green, W.J., 'Processing Polymer Thick Film Circuitry with Radiation Curing', *Hybrid Circuits*, No. 2, pp. 12–13 (1983).
106 Huang, C., Lo, E. and Stein, S.J., 'Infra-red Curing of Polymer Thick Film Resistors', Proceedings 'Circuit Expo '84'', Worcester, Massachusetts, pp. 79–83, October (1984).
107 Stein, S.J., Huang, C. and Cang, L., 'Base Metal Thick Film Conductors', *Solid State Technology*, **Vol. 24**, No. 1, pp. 73–79, 110, (1981).
108 St John, F., 'Solderable Polymer Thick Film Conductors for Surface Mount Technology', *Hybrid Circuits*, No. 6, pp. 20–23 (1985).
109 'SMD Technology', Siemens AG, Munich (1985).
110 Prasad, P., 'Designing Surface Mount Boards for Manufacturability', Proceedings SMART II, Los Angeles, pp. 184–185 (1986).
111 Jacob, G.W., 'Designing for Testability', *Hybrid Circuits*, No. 5, pp. 21–24 (1984).

References

112 Smith, M. and Cook S., 'Probe Problems with SMDs', *Electronic Production*, **Vol. 14**, No. 8, pp. 38–43 (1985).
113 Markstein, H.W., 'Making SMT Assemblies Testable', *Electronic Packaging & Production*, **Vol. 26**, No. 10, pp. 60–64 (1986).
114 Pierce, T.R., 'Surface Mounted Devices', Proceedings IPC Fall Meeting, San Francisco, California, Technical Paper IPC-TP-512 (1984).
115 Capillo, C.A., 'How to Design Reliability into Surface-mount Assemblies', *Electronic Packaging & Production*, **Vol. 25**, No. 7, pp. 74–80 (1985).
116 Hutchins, C.L., 'Surface Mount Land Design Criteria', Proceedings SMART II, Los Angeles, pp. 222–224 (1986).
117 Prasad, P., 'Guidelines for Surface Mount Land Pattern Design', Proceedings SMART II, Los Angeles, pp. 216–218 (1986).
118 Knight, W.G., 'Solder Masks Reappraised', *Circuit World*, **Vol. 12**, No. 1, pp. 19–21 (1985).
119 Fenner, M., 'Flatpack Soldering Considerations', *Surface Mount*, pp. 6–7, June/July (1986).
120 Peel, M.E., 'SMT-Ignored Problems', *Microelectronics Journal*, **Vol. 17**, No. 3, pp. 36–41 (1986).
121 Pawling, J.F., 'Surface Mounted Assemblies', Electrochemical Publications, Ayr, Scotland (1987).
122 Keeler, R., 'Real-World Constraints on Printed Circuit CAD Performance', *Electronic Packaging & Production*, **Vol. 27**, No. 2, pp. 88–90 (1987).
123 Summers, S.C., 'The Evolution and Growth of Computer Aided Design', *Electri-onics*, **Vol. 32**, No. 9, pp. 28–30 (1986).
124 Boscha, B., 'Pick-and-Place Machines Boost Productivity', *Electronic Packaging & Production*, **Vol. 26**, No. 12, pp. 66–67 (1986).
125 Stark, A.M., 'SMD Pick-and-Place Machines—An Assessment', *Electronic Production*, **Vol. 13**, No. 10, pp. 73–79 (1984).
126 Page Walton, J., 'Equipment for Assembling Boards with SM Devices', *Electronic Production*, **Vol. 14**, No. 4, pp. 11–16 (1985).
127 Spitz, S.L., 'SMA Promotes Automation and Boosts Machine Vision', *Electronic Packaging & Production*, **Vol. 26**, No. 1, pp. 82–87 (1986).
128 Field, J., Payne, J. and Cullen, C., 'SMD Placement Using Machine Vision', *Electronic Packaging & Production*, **Vol. 26**, No. 1, pp. 128–129 (1986).
129 De Jong, J.J., 'A Machine for In-line Assembly of Large SMT Boards', *Electronic Packaging & Production*, **Vol. 26**, No. 8, pp. 62–64 (1986).
130 Dance, F.J., Blackshaw, M.F. and Goldman, P., 'Chip-on-Board Has Designs on High-density Packaging', *Electronic Packaging & Production*, **Vol. 25**, No. 10, pp. 70–75 (1985).
131 Fuchs, E., 'Chip-on-Board: an Economical Packaging Solution', *Electronic Packaging & Production*, **Vol. 25**, No. 1, pp. 182–185 (1985).
132 Keeler, R., 'Chip-on-Board Alters The Landscape of PC Boards', *Electronic Packaging & Production*, **Vol. 25**, No. 7, pp. 62–67 (1985).
133 Landis, R., 'Alternative Bonding Methods for Chip-on-Board Technology', Proceedings 36th IEEE Electronic Components Conference, Seattle, Washington, pp. 53–58 (1986).
134 Johnson, K.I., 'Microjoining Developments for the Electronics Industry', *Hybrid Circuits*, No. 2, pp. 5–11 (1983).
135 Weston, A.D., 'Wire Bonding as a Technique for Semiconductor Device Assembly', *Hybrid Circuits*, No. 7, pp. 26–35 (1985).
136 Johnston K.I., Scott, M.H. and Dawes, C.J., 'Development of Al Ball/Wedge Wire Welding', Proceedings ISHM International Microelectronics Conference, New York, pp. 134–139 (1980).
137 Johnson, K.I., Scott, M.H. and Edson, D.A., 'Ultrasonic Wire Welding: Ball/Wedge Wire Welding', *Solid State Technology*, **Vol. 20**, No. 4, pp. 91–95 (1977).
138 Rodwell, R. and Worrall, D.A., 'Quality Control in Ultrasonic Wire Bonding', *Hybrid Circuits*, No. 7, pp. 67–72 (1985).
139 Pitt, V.A. and Needes, C.R.S., 'Thermosonic Gold Wire Bonding To Copper Conductors', *IEEE Transactions on Components, Hybrids and Manufacturing Technology*, **Vol. CHMT-5**, pp. 435–440 (1982).
140 Ginsberg, G.L., 'Chip and Wire Technology: The Ultimate in Surface Mounting', *Electronic Packaging & Production*, **Vol. 25**, No. 8, pp. 78–83 (1985).
141 Suwa, M., 'Aluminium Ball Bonding Having the Reliability of Gold Wire', *Nikkei Electronics*, June 11 (1984).
142 Atsumi, K., 'Ball Bonding Techniques for Copper Wire', Proceedings 36th IEEE Electronic Components Conference, Seattle, Washington, pp. 312–317 (1986).

143 Okuda, T., 'Copper Wire Bonding Technology Ready For Practical Use' (Japanese), *Nikkei Microdevices*, pp. 89–100, September (1985).
144 Cheype, J.M., 'Development of TAB for VLSI Devices'. Proceedings 36th IEEE Electronic Components Conference, Seattle, Washington, pp. 65–69 (1986).
145 Liljestrand, L-G., 'Bond Strengths of Inner and Outer Leads on TAB Devices', *Hybrid Circuits*, No. 10, pp. 42–48 (1986).
146 Totta, P.A., 'Flip-chip Solder Terminals', Proceedings 21st IEEE Electronics Components Conference, Washington DC, pp. 275–284 (1971).
147 Ginsberg, G.L., 'Chip-on-Board Profits from TAB and Flip-chip Technology', *Electronic Packaging & Production*, Vol. 25, No. 9, pp. 140–143 (1985).
148 Granger, J.J., Basset, J.C., Vimont, N. and Viret, P., 'Wire Bonded Chip Encapsulation Using Resin Droplets', Proceedings ISHM 4th European Hybrid Microelectronics Conference, Copenhagen, Denmark, pp. 146–154 (1983).
149 Strauss, R., 'Machine Soldering: The State of the Art', *Brazing & Soldering*, No. 2, pp. 38–41 (1982).
150 Burke, J., 'Soldering Techniques for Surface Mounting', *Electronic Production*, Vol. 14, No. 3, pp. 37–39 (1985).
151 Billing, B.G., 'PC Board Soldering Systems', *Electronic Packaging & Production*, Vol. 23, No. 1, pp. 52–57 (1983).
152 Elliott, D.A., 'Wavesoldering of Discrete Surface Mounted Devices', Proceedings 3rd Printed Circuit World Convention, Washington DC, Paper WCIII–64.
153 Denda, S. and Ikegami, A., 'Automatic Assembly and Chip Components for Thin Consumer Electronic Products in Japan', Proceedings ISHM International Microelectronics Symposium, Los Angeles, California, pp. 240–246 (1979).
154 Klein Wassink, R.J. and Vledder, H.J., 'The Attachment of Leadless Electronic Components to Printed Boards', *Hybrid Circuits*, No. 3, pp. 28–32 (1983).
155 Keeler, R., 'For SMD Wave Soldering, Adhesives Still Hold Fast', *Electronic Packaging & Production*, Vol. 25, No. 6, pp. 118–121 (1985).
156 Grant, S., 'Adhesives for Surface Mounting', *Electronic Production*, Vol. 13, No. 10, pp. 69–71 (1984).
157 Seah, M.P., Howie, F.H. and Lea, C., 'Blowholing in PTH Solder Fillets, Part 3: Moisture and the PCB', *Circuit World*, Vol. 12, No. 4, pp. 26–33 (1986).
158 Grant, S. and Wigham, J., 'Use of Adhesives in Surface Mounting', *Hybrid Circuits*, No. 8, pp. 15–16 (1985).
159 Rubin, W., 'Some Recent Advances in Flux Technology', *Brazing & Soldering*, No. 2, pp. 24–28 (1982).
160 Adamson, C., 'Wave Soldering Machine Design', *Electronic Production*, Vol. 15, No. 6, pp. 19–25 (1986).
161 Bud, P.J., 'Mass Production Techniques Using The Principle of Wave Soldering', *Welding Journal*, Vol. 52, pp. 431–439 (1973).
162 Manko, H., 'Understanding the Solder Wave and Its Effects on Solder Joints', *Insulation/Circuits*, Vol. 24, No. 1, pp. 45–49 (1978).
163 Bud, P.J., 'Wave Soldering Technique for the Soldering and Tinning of Coils, Windings and Wire', *Insulation/Circuits*, Vol. 20, No. 4, pp. 37–40 (1974), Vol. 20, No. 5, pp. 65–68 (1974).
164 Bernard, C.D., 'Horizontal Versus Inclined Conveyor Wavesoldering', *Circuits Manufacturing*, Vol. 17, No. 9, pp. 53–55 (1977).
165 Comerford, M.F. and O'Rourke, H.T., 'An Analysis of Some Important Parameters in Automatic Soldering', Proceedings NEPCON-East 79, Boston, Massachusetts, pp. 197–208 (1979).
166 Carr, R., 'Solder Waves—A Review', *Electronic Production*, Vol. 13, No. 10, pp. 26–30 (1984).
167 Page Walton, J., 'Soldering Equipment for Surface Mounted Components', *Electronic Production*, Vol. 13, No. 10, pp. 101–103 (1984).
168 Boey, W.K. and Walker, R.J., 'Wave Soldering of Surface Mount Components', *Circuit World*, Vol. 12, No. 3, pp. 25–26 (1986).
169 Down, W.H., 'Reliable Wavesoldering of Chip Components', *Electronic Packaging & Production*, Vol. 21, No. 7, pp. 171–173 (1981).
170 Turner, C. and Matthews, R., 'Wavesoldering Techniques—A Re-evaluation', *Electronic Production*, Vol. 16, No. 5, pp. 23–28 (1987).
171 Fishman, D. and Cooper, N., 'Mounting Leadless Chip Carriers on Printed Circuit Cards', Proceedings International Electronic Packaging Symposium, pp. 45–59 (1981).

References

172 Brierley, C.J. and McCarthy, J.P., 'A Practical Comparison of Wave Soldering Equipment Designs for Surface Mounted Components', *Brazing & Soldering*, No. 10, pp. 7–10 (1986).
173 Strauss, R., 'Soldering of Surface-mounted Devices: New Tasks and Their Solutions', *Brazing & Soldering*, No. 10, pp. 21–23 (1986).
174 Karpel, S., 'Mass-soldering Equipment for The Electronics Industry: 1: Wave Soldering', *Tin and its Uses*, No. 127, pp. 1–6 (1981).
175 Hedges, R., 'Know Your Soldering Machine Maintenance', *Electronic Production*, Vol. 14, No. 9, pp. 44–45 (1985).
176 Stayner, R.A., 'Soldering with Oil', *Electronic Packaging & Production*, Vol. 14, No. 3, pp. 103–106 (1974).
177 Boynton, K.G., 'Oil in Wave Soldering', *Electronic Production*, Vol. 2, No. 11, pp. 60–62, 64 (1973).
178 Klein Wassink, R.J. and Verguld, M.M.F., 'Bridge-free Wave Soldering of Surface Mounted Devices', *Brazing & Soldering*, No. 9, pp. 24–27 (1985).
179 Comerford, M.F., 'Removing Solder Bridges from Printed Circuit Boards', *Insulation/Circuits*, Vol. 27, No. 4 pp. 31–34 (1981).
180 Comerford, M.F., 'Debridging of High Density PWAs by the GBS System', Proceedings NEPCON-West 82, Anaheim, California, pp. 342–353 (1982).
181 Lambert, L., 'Airknives and Solder Defect Control', *Electronic Production*, Vol. 13, No. 5, pp. 58–60 (1984).
182 Verguld, M.M.F., 'Wave Soldering: The Efficiency of Debridging with a Hot air Knife, *Circuit World*, Vol. 13, No. 2, pp. 14–18 (1987).
183 Elliott, D.A., 'Wavesoldering of Surface Mount Devices', Proceedings of Circuit Expo 84, Worcester, Massachusetts, pp. 125–127 (1984).
184 Hamilton, P.G.B., 'Surface Mount Assembly Problems: Elimination of PC Solder Resist', *Circuit World*, Vol. 13, No. 3, pp. 72–73 (1987).
185 Weinhold, M., 'Designing for Zero-defect Soldering', *Circuit World*, Vol. 12, No. 4, pp. 5–8 (1986).
186 Lin, K.M. and Kacker, R.N., 'Optimising the Wave Soldering Process', *Electronic Packaging & Production*, Vol. 26, No. 2, pp. 108–115 (1986).
187 Elliott, D.A., 'Trouble-shooting Solder Defects, *Electronics Manufacture & Test*, pp. 21–22, December (1986).
188 Pascoe, G., 'Fault Finding in the Wave Soldering Process', *Brazing & Soldering*, No. 1, pp. 23–26 (1981).
189 Leonida, G., 'Quality Control of Solder Joints', *Brazing & Soldering*, No. 1, pp. 19–20 (1981).
190 Leonida, G., 'Quality Control of Solder Joints', *Brazing & Soldering*, No. 2, pp. 42–47 (1982).
191 Davy, J.G., 'A Comprehensive List of Wave Solder Defects and Their Probable Causes', *Brazing & Soldering*, No. 9, pp. 50–59 (1985).
192 Davy, J.G., 'Wave Solder Defects', *Electronic Packaging & Production*, Vol. 26, No. 2, pp. 120–123 (1986).
193 Lea, C. and Howie, F.H., 'Blowholing in PTH Solder Fillets, Part 1: Assessment of the Problem', *Circuit World*, Vol. 12, No. 4, pp. 14–19 (1986).
194 Howie, F.H. and Lea, C., 'Blowholing in PTH Solder Fillets, Part 2: The Nature, Origin and Evolution of the Gas', *Circuit World*, Vol. 12, No. 4, pp. 20–25 (1986).
195 Lea, C., Seah, M.P. and Howie, F.H., 'Blowholing in PTH Solder Fillets, Part 4: The Plated Copper Barrel', *Circuit World*, Vol. 13, No. 1, pp. 28–34 (1986).
196 Lea, C. and Howie, F.H., 'Blowholing in PTH Solder Fillets, Part 5: The Rôle of the Electroless Copper', *Circuit World*, Vol. 13, No. 1, pp. 35–42 (1986).
197 Lea, C. and Howie, F.H. 'Blowholing in PTH Solder Fillets, Part 6: The Laminate, The Drilling and The Hole-wall Preparation', *Circuit World*, Vol. 13, No. 1, pp. 43–50 (1986).
198 Howie, F.H., Tilbrook, D. and Lea, C., 'Blowholing in PTH Solder Fillets, Part 7: Optimising the Soldering', *Circuit World*, Vol. 13, No. 2, pp. 42–45 (1987).
199 Lea, C., Howie, F.H. and Seah, M.P., 'Blowholing in PTH Solder Fillets, Part 8: The Scientific Framework Leading to Recommendations for its Elimination', *Circuit World*, Vol. 13, No. 3, pp. 11–20 (1987).
200 Zado, F.M., 'The Theory and Practice of High Efficiency Soldering with Non-corrosive Rosin Soldering Fluxes', Proceedings 2nd Printed Circuit World Convention, Munich, Germany, Vol. 1, pp. 154–161 (1981).
201 Elliott, D.A., 'Overview of Surface Mounted Device Wavesoldering Problems and Solutions', *Electri-onics*, Vol. 31, No. 3, pp. 37–40 (1985).
202 Comerford, M.F. and Swift, D.R., 'An Analysis of some Key Variables in the Wave Soldering

of PCBs with Surface Mounted Components', Proceedings NEPCON-West 83, Anaheim, California, pp. 204–212 (1983).
203. Abbott, W.G. 'Wavesoldering Fineline Circuitry and SM Components', *Electronic Production*, **Vol. 14**, No. 4, pp. 25–29 (1985).
204. Lee, I.W.H., 'Screenable Solder Paste', *Welding Journal*, **Vol. 56**, No. 10, pp. 32–36 (1977).
205. Anjard, R.P., 'Solder Pastes for Electronics: A Status Report', *Tin and Its Uses*, No. 143, pp. 13–17 (1985).
206. Rubin, W., 'Solder Pastes for Surface Mounts', *Electronic Production*, **Vol. 15**, No. 3, pp. 74–80 (1986).
207. Anjard, R.P., 'Quality Solder Paste Systems for Use in Microelectronic Applications', *Solid State Technology*, **Vol. 26**, No. 10, pp. 183–189 (1983).
208. Daebler, D.H., 'Specifying Solder Paste Materials for Stencilling Applications on Thick Film Circuits', *Electronic Packaging & Production*, **Vol. 21**, No. 4, pp. 99–106 (1981).
209. Roos-Kozel, B.L., 'Parameters Affecting the Incidence of Pad Bridging in Surface Mounted Device Attachment', *Hybrid Circuits*, No. 6, pp. 5–8 (1985).
210. Glaessgen, R., 'Electronic Grade Solder Paste: Characteristics and Application', *Solid State Technology*, **Vol. 24**, No. 4, pp. 54–55 (1981).
211. Dixon, T., 'SMT Forces Solder Paste Improvement', *Electronic Packaging & Production*, **Vol. 24**, No. 8, pp. 122–125 (1984).
212. Anjard, R.P., 'Factors Affecting the Quality of Solder Powder Used in Thick Film Solder Paste Systems', Proceedings ISHM International Microelectronics Conference, Reno, Nevada, pp. 153–163 (1982).
213. Socolowski, N., 'Solder Pastes for Hybrid Applications', *Microelectronic Manufacturing & Testing*, pp. 24–26, October (1982).
214. Roos-Kozel, B., 'Reliable Solder Paste Quality Control', *Brazing and Soldering*, No. 10, pp. 35–37 (1986).
215. Smith-Vargo, L., 'Solders with Precision', *Electronic Packaging & Production*, **Vol. 26**, No. 4, pp. 92–93 (1986).
216. Anjard, R.P., 'Solder Paste For Surface Mount Technology', *Brazing & Soldering*, No. 9, pp. 19–23 (1985).
217. Taylor, B.E., Slutsky, J. and Larry, J.R., 'Technology of Electronic Grade Solder Pastes', *Solid State Technology*, **Vol. 24**, No. 9, pp. 127–135 (1981).
218. Bulwith, R.A. and MacKay, C.A., 'Silver Scavenging Inhibition of Some Silver Loaded Solders', *Welding Journal Research Supplement*, **Vol. 64**, pp. 86s–90s (1985).
219. Kay, P.J. and MacKay, C.A., 'The Growth of Intermetallic Compounds on Common Basis Materials Coated with Tin and Tin-Lead Alloys', *Transactions of the Institute of Metal Finishing*, **Vol. 54**, pp. 68–74 (1976).
220. Mather, J.C. and Hagge, J.K., 'Kinetics of Intermetallic Compound Formation and of Base Metal Dissolution into Molten Solder During Vapour Phase Soldering', *Brazing & Soldering*, No. 3, pp. 29–32 (1982).
221. Hansen, M., 'The Constitution of Binary Alloys', McGraw-Hill (1958).
222. Morrison, L., 'Low Temperature Solder Developments', *Electronic Production*, **Vol. 16**, No. 5, pp. 33–35 (1987).
223. International Tin Research Institute, 'Solder Alloy Data', ITRI Publication No. 656 (1986).
224. MacKay, C.A., 'Causes and Effects of Solder Contamination', *Electri-onics*, **Vol. 29**, No. 3, pp. 44–48, **Vol. 29**, No. 4, pp. 41–44, **Vol. 29**, No. 5, pp. 51–53 (1983).
225. Ackroyd, M.L., MacKay, C.A. and Thwaites, C.J., 'Effect of Certain Impurity Elements on the Wetting Properties of 60/40 SnPb Solders', *Metals Technology*, **Vol. 2**, pp. 73–85 (1975).
226. Raman, K.S., 'Influence of Metallic Additions on the Spreading Characteristics of Sn-Pb Solders', *Chemical Era*, **Vol. 13**, pp. 97–116 (1977).
227. Thwaites, C.J., 'Antimony in Soft Solders—A Review of the Effects and Its Use in the Soldering Industry', *Brazing & Soldering*, No. 11, pp. 22–26 (1986).
228. Kvurt, O.S. and Ginzburg, F., 'Effect of Sb and As on the Properties of SnPb Solder', *Soviet Journal of Non-ferrous Metals*, **Vol. 11**, No. 6, pp. 64–65 (1970).
229. Siattery, J.A. and White, C.E.T., 'A Primer on the Use of Solder Creams in Hybrid Assembly', *Electronic Packaging & Production*, **Vol. 21**, No. 10, pp. 146-156 (1981).
230. Smith-Vargo, L., 'Screen Printers Acquire Sophistication', *Electronic Packaging & Production*, **Vol. 27**, No. 3, pp. 74–76 (1987).
231. Partridge, S.A., 'The Rôle of the Stencil in High Definition Screen Printing', *Circuit World*, **Vol. 13**, No. 2, pp. 4–13 (1987).
232. Schoenthaler, D., 'Soldering Surface Mounted Chip Carriers to Printed Circuits', *Western Electric Engineer*, **Vol. 27**, No. 1, pp. 73–79 (1983).

References

233 Hwang, J.S. and Lee, N.C., 'A New Development in Solder Paste with Unique Rheology for Surface Mounting', Proceedings ISHM International Symposium on Microelectronics, Anaheim, California, pp. 23–30 (1985).
234 Anjard, R.P., 'The Rheology of Solder Pastes', Proceedings IPC Fall Meeting, San Diego, Technical Paper IPC-TP-614 (1986).
235 Rubin, W., 'Solder Creams and Vapour Phase Soldering in Hybrid Technology', *Hybrid Circuits*, No. 3, pp. 18–21 (1983).
236 Searle, G.F.C., 'A Simple Viscometer for Very Viscous Liquids', *Proceedings Cambridge Philosophical Society*, Vol. 16, pp. 600–606 (1912).
237 Ainsworth, P.A., 'Measurement of Tackiness of Solder Pastes', *Hybrid Circuits*, No. 11, pp. 27–29 (1986).
238 Condon, J., 'Wet Nitrogen Storage Streamlines SMT Assembly', *Electronic Packaging & Production*, Vol. 27, No. 1, pp. 64–65 (1987).
239 Peterson, N.C., 'Using Screened Solder Paste', *Circuits Manufacturing*, Vol. 22, No. 6, pp. 60–61 (1982).
240 Roos-Kozel, B., 'Designing Solder Paste Materials to Attach Surface Mounted Devices', *Solid State Technology*, Vol. 26, No. 10, pp. 173–178 (1983).
241 der Marderosian, A. and Gionet, V., 'The Effects of Entrapped Bubbles in Solder for The Attachment of Leadless Ceramic Chip Carriers', Proceedings 21st IEEE International Reliability Physics Symposium, Phoenix, Arizona, pp. 235–241 (1983).
242 Mahalingham, M., Nagarkar, M., Lofgran, L., Andrews, J., Olsen, D.R. and Berg, H.M., 'Thermal Effects of Die Bond Voids in Metal, Ceramic and Plastic Packages', Proceedings 34th IEEE Electronic Components Conference, New Orleans, Louisiana, pp. 469–477 (1984).
243 MacKay, C.A., 'Some Causes of Problems with Existing SMT Solder Creams and Possible Improvements', *Circuit World*, Vol. 13, No. 3, pp. 4–7 (1987).
244 Roos-Kozel, B., 'Advances in Solder Paste Technology to Eliminate Voids', Proceedings Circuit Expo '84, Worcester, Massachusetts pp. 49–53 (1984).
245 Chapman, A.J., 'Heat Transfer', Macmillan Publishing Company, New York, 3rd Edition (1974).
246 Kreith, F., 'Principles of Heat Transfer', International Textbook Company, Scanton, Pennsylvania, 2nd Edition (1965).
247 Flattery, D.K., 'Infra-red Reflow for the Solder Attachment of Surface Mounted Devices', *Hybrid Circuits*, No. 9, pp. 9–12 (1986).
248 Dow, S.J., 'Use of Radiant Infra-red in Soldering Surface Mounted Devices to PCBs', *Brazing & Soldering*, No. 8, pp. 16–19 (1985).
249 'Performance Data for Tubular Infra-red Lamps', Sylvania Bulletin 0-270, Sylvania Electric Products Incorporated, Salem, Massachusetts, USA.
250 Zubler, E.G. and Mosby, F.A., 'An Iodine Incandescent Lamp with Virtually 100% Lumen Maintenance', Proceedings National Technical Conference of Illuminating Engineering Society, San Francisco, California, pp. 7–11 (1959).
251 Schoenthaler, D., 'Solder Jointing and Fusing with Radiant Heating', Assembly & Joining Techniques, IPC, Illinois (1978).
252 Dow, S.J., 'Using Radiant Infra-red to Solder Surface Mounted Devices', *Electronic Production*, Vol. 14, No. 9, pp. 19–28 (1985).
253 Eckert, E.R.G. and Drake, R.M., 'Heat and Mass Transfer', McGraw-Hill, New York (1959).
254 King, S.R., 'SMT Module Yield Improvement with Infra-red Reflow', Proceedings NEPCON-West 86, Anaheim, California, pp. 995–998 (1986).
255 Cox, N.R., 'Reducing the Risk with Reflow', *Electronics Manufacture & Test*, pp. 53–56, July/August (1985).
256 Cox, N.R., 'Optimisation of Solder Reflow with Infra-Red', Proceedings ISHM International Symposium on Microelectronics, Anaheim, California, pp. 43–47 (1985).
257 Gardon, R., 'An Instrument for the Direct Measurement of Intense Thermal Radiation', *Review of Scientific Instruments*, Vol. 24, pp. 366–370 (1953).
258 Bartosz, T., 'Solder Fusing: Process and Equipment', Proceedings IPC Fall Meeting, Denver, Colorado, Technical Paper IPC-TP-486 (1983).
259 Weltha, M.D., 'A Hot Air Soldering Facility', Proceedings INTERNEPCON 76, Brighton, UK, pp. 112–118 (1967).
260 Scagnelli, G.J., D'Erchia, F.J. and Wittenberg, A.M., 'Forced Hot Air Fusing of Solder Plated Circuits', *Welding Journal*, Vol. 54, No. 10, pp. 718–724 (1975).
261 Lea, C. and Johns, K.W.E., 'Liquid Phase Soldering', *Brazing & Soldering*, No. 12, pp. 34–39 (1987).
262 McAdams, W.H., 'Heat Transmission', 2nd Edition, McGraw-Hill, New York (1942).

263 Mollendorf, J.C., 'The Applicability of Approximate and Exact Transient Heat Transfer Analysis to Heating Processes Used to Solder Multilayer Circuit Boards', *IEEE Transactions on Parts, Hybrids and Packaging*, **Vol. PHP-11**, No. 2, pp. 96–104 (1975).
264 Markstein, H.W., 'SMT Reflow: IR or Vapour Phase?' *Electronic Packaging & Production*, **Vol. 27**, No. 1, pp. 60–63 (1987).
265 Ahearn, J.F., Ursch, R.R., and Kilham, L.F., 'Cooling of Electronic Equipment by Means of Heavy Inert Vapours', Proceedings National Conference on Aero Electronics, pp. 403–409 (1957).
266 Ammann, H.H. and Oien, M.A., 'Solder Sliver Removal from Flex Circuits by Condensation Reflow', Proceedings NEPCON-West 78, Anaheim, California, **Vol. 1**, pp. 59–66 (1978).
267 Pfahl, R.C., Mollendorf, J.C. and Chu, T.Y., 'Condensation Soldering', *Welding Journal*, **Vol. 54**, pp. 22–25 (1975).
268 Ammann, H.H. and Farkass, I., 'On the Applicability of the Nusselt Correlation to Transient Condensation Heating', Proceedings NEPCON-West 77, Anaheim, California pp. 257–269 (1977).
269 Bosworth, R.C.L., 'Heat Transfer Phenomena', Associated General Publications, Sydney (1952).
270 Nusselt, W., 'The Surface Condensation of Water Vapour' (German), *Zeitschrift des Vereines Deutscher Ing.*, **Vol. 60**, pp. 541–546 and 569–575 (1916).
271 Kapitza, P.L. and Kapitza, S.P., 'Wave-like Flow of Thin Layers of a Viscous Liquid' (Russian), *Journal of Experimental and Theoretical Physics*, **Vol. 18**, pp. 19–29 (1948), **Vol. 19**, pp. 105–120 (1949).
272 Nimmo, B., 'Laminar Film Condensation on a Horizontal Surface', Stanford University, Technical Report 237–12 (1966).
273 Leppert, G. and Nimmo, B., 'Laminar Film Condensation on Surfaces Normal to Body or Inertial Forces', Transactions ASME: *Journal of Heat Transfer*, **Vol. 90**, pp. 178–179 (1968).
274 Gerstmann, J. and Griffith, P., 'The Effects of Surface Instabilities of Laminar Film Condensation', Technical Report 5050–36, MIT, Cambridge, Massachusetts (1975).
275 Chu, T.Y., Mollendorf, J.C. and Pfahl, R.C., 'Soldering Using Condensation Heat Transfer', Proceedings NEPCON-West 74, Anaheim, California, pp. 101–104 (1974).
276 Wenger, G.M. and Mahajan, R.L., 'Condensation Soldering', Assembly Joining Handbook, IPC, Illinois (1980).
277 Chu, T.Y., Pfahl, R.C. and Wenger, G.M., 'Condensation Soldering with Vapour Blanket Reduces Cost of This High Yield, High Quality Process', Proceedings NEPCON-West 76, Anaheim, California, pp. 259–264 (1976).
278 Mahajan, R.L., 'Determination and Control of Heater Surface Temperature in Condensation Soldering', Proceedings NEPCON-West 79, Anaheim, California, pp. 355–364 (1979).
279 Wright, A.W., Mahajan, R.L. and Wenger, G.M., 'Thermal and Soldering Characteristics of Condensation Heating Fluids', Proceedings NEPCON-West 85, Anaheim, California, p. 62 (Abstract only) (1985).
280 Morini, A., Macchi, E. and Giglioli, G., 'Experimental Results on the Thermal Stability of Some Fluorocarbons', Proceedings 15th Conference of the International Society of Energy Conversion Engineers, Seattle, Washington, published by the American Institute of Aeronautics and Astronautics, pp. 992–997 (1980).
281 Clayton, J.W., 'Fluorocarbon Toxicity and Biological Action', *Fluorine Chemistry Reviews*, **Vol. 1**, pp. 197–252 (1967).
282 Danielson, R.D., 'Vapour Phase Soldering with Perfluorinated Inert Fluids', Proceedings NEPCON-West 79, Anaheim, California, pp. 374–382 (1979).
283 Slinn, D.S.L., 'A New Vapour Phase Soldering Fluid Having Improved Safety and Thermal Stability Characteristics', Proceedings NEPCON-West 84, Anaheim, California, pp. 121–128 (1984).
284 Marhevka, J.S., Johnson, G.D., Hagen, D.F. and Danielson, R.D., 'Generation of PFIB Reference Sample and Determination by Gas Chromatography with Electron Capture and Flame Ionisation Detection', *Analytical Chemistry*, **Vol. 54**, pp. 2607–2611 (1982).
285 Turbini, L.J. and Zado, F.M., 'Chemical and Environmental Aspects of Condensation Reflow Soldering', *Electronic Packaging & Production*, **Vol. 20**, No. 1, pp. 49–59 (1980).
286 Morse, R., 'Economic and Practical Views on Vapour Phase', *Electronics Manufacture & Test*, pp. 25–26, December (1986).
287 Smith, M.S., 'Soldering Surface Mounted Components Using Vapour Phase In-line', Proceedings INTERNEPCON 84, Brighton, UK, pp. 1–8 (1984).
288 Lilienthal, P.F., Wenger, G.M. and Zado, F.M., 'Residue Removal Methods for Condensation

References

Soldering Systems', Proceedings NEPCON-West 77, Anaheim, California, pp. 270–273 (1977).

289 Elliott, D.A. and Bud, P.J., 'Reflowed Solder Coating for Printed Circuits', *Plating*, **Vol. 59**, pp. 288–294 (1972).

290 Schoenthaler, D., 'Solder Fusing with Heated Liquids', *Welding Journal Research Supplement*, **Vol. 53**, pp. 489s–509s (1974).

291 Briggs, S., 'Perfluoropolyethers—The Most Versatile Fluids for Vapour Phase Reflow Soldering', *Brazing & Soldering*, No. 7, pp. 6–12 (1984).

292 Becker, G., 'From Soldering Iron to Laser—A Review of Soldering Methods for Surface Mounting', *Hybrid Circuits*, No. 12, pp. 22–27 (1987).

293 Browne, L.T., 'Reflow Soldering of Hybrid Circuits', *Insulation/Circuits*, **Vol. 21**, No. 9, pp. 27–29 (1975).

294 Loeffler, J.R., 'Numerically Controlled Laser Soldering—Fast, Low Cost, No Rejects', *Assembly Engineering*, **Vol. 20**, No. 3, pp. 32–34 (1977).

295 Kujawa, T., 'Laser Soldering Boosts Productivity', *Lasers & Applications*, pp. 93–94, September (1982).

296 Dixon, T., 'Lasers Bring Precision to Electronics Manufacturing', *Electronic Packaging & Production*, **Vol. 24**, No. 3, pp. 98–104 (1984).

297 Lish, E.F., 'Lasers Tackle Tough Soldering Problems', *Electronic Packaging & Production*, **Vol. 24**, No. 6, pp. 154–161 (1984).

298 Yariv, A., 'Quantum Electronics', John Wiley & Sons, New York (1967).

299 Lish, E.F., 'Applications of Laser Microsoldering to Printed Wiring Assemblies', *IPC Technical Review*, **Vol. 26**, No. 7, pp. 10–20 (1985).

300 Burns, F. and Zyetz, C., 'Laser Microsoldering', *Electronic Packaging & Production*, **Vol. 21**, No. 5, pp. 109–120 (1981).

301 Bolam, C.F., 'The Laser and Microsoldering', Society Manufacturing Engineers, Technical Paper AD74–810.

302 Harry, J.E., 'Industrial Lasers and Their Applications', McGraw Hill, London (1974).

303 Lish, E.F., 'Considerations for Laser Soldering Surface Mounted Connectors', *Electri-onics*, **Vol. 30**, No. 6, pp. 28–32 (1984).

304 Dustoomian, A.S., 'Intelligence Comes to Laser Soldering', *Electronics*, pp. 75–77, 10 July (1986).

305 Carr, J.F., 'Laser Safety' in 'Lasers in Industry', Ed. Charschan, S.S., Van Nostrand Reinhold, New York (1972).

306 Smith-Vargo, L., 'Adhesives That Possess a Science All Their Own', *Electronic Packaging & Production*, **Vol. 26**, No. 8, pp. 48–49 (1986).

307 Piccoli, S., 'A Comparison of Electrical Performance of Conductive Epoxy and Soldered RF Connections in Microwave Hybrid Circuits', Proceedings ISHM International Microelectronics Symposium, Orlando, Florida, pp. 79–86 (1975).

308 Estes, R.H. and Kulesza, F.W., 'Surface Mount Technology—The Epoxy Alternative', Proceedings NEPCON-West 84, Anaheim, California, pp. 219–234 (1984).

309 Kulesza, F.W. and Estes, R.H., 'Conductive Epoxy Solves Surface Mount Problems', *Electronic Products*, pp. 83–88, March 5 (1984).

310 Pound, R., 'Conductive Epoxy is Tested for SMT Solder Replacement', *Electronic Packaging & Production*, **Vol. 25**, No. 2, pp. 86–90 (1985).

311 de Gennes, P.G., 'Wetting: Statics & Dynamics', *Reviews in Modern Physics*, **Vol. 57**, pp. 827–863 (1985).

312 Bashforth, F. and Adams, J.C., 'An Attempt to Test the Theories of Capillary Action', Cambridge University Press (1883).

313 Butler, J.N. and Bloom, B.H., 'A Curve Fitting Method for Calculating Interfacial Tension from the Shape of a Sessile Drop', *Surface Science*, **Vol. 4**, pp. 1–17 (1966).

314 Shanahan, M.E.R., 'Equilibrium of Liquid Drops on Thin Plates; Plate Rigidity and Stability Considerations', *Journal of Adhesion*, **Vol. 20**, pp. 261–274 (1987).

315 Klein-Wassink, R.J., 'Wetting of Solid Metal Surfaces by Molten Metals', *Journal of the Institute of Metals*, **Vol. 95**, pp. 38–43 (1967).

316 Wei-Heng Shih and Stroud, D., 'Theory of the Surface Tension of Liquid Metal Alloys', *Physics Review B*, **Vol. 32**, pp. 804–811 (1985).

317 Bailey, G.L.J. and Watkins, H.C., 'The Flow of Liquid Metals on Solid Metal Surfaces and its Relation to Soldering, Brazing and Hot-dip Coating', *Journal of the Institute of Metals*, **Vol. 80**, pp. 57–76 (1951).

318 Bondi, A., 'The Spreading of Liquid Metals on Solid Surfaces', *Chemical Reviews*, **Vol. 52**, pp. 417–458 (1953).

319 Deigham, R.A., 'Surface Tension of Solder Alloys', *Journal of Hybrid Microelectronics*, **Vol. 5**, No. 2, pp. 307–313 (1982).

320 Howie, F.H. and Hondros, E.D., 'The Surface Tension of Tin-Lead Alloys in Contact with Fluxes', *Journal of Materials Science*, **Vol. 17**, pp. 1434–1440 (1982).

321 Lea, C., 'The Physical Basis of Wettability in Metallic Systems', Proceedings 8th International Vacuum Congress, Cannes, *Supplement Le Vide, les Couches Minces*, **Vol. 201**, pp. 467–470 (1980).

322 Smith, G.C. and Lea, C., 'Wetting and Spreading of Liquid Metals: The Rôle of Surface Composition', *Surface & Interface Analysis*, **Vol. 9**, pp. 145–150 (1986).

323 Lea, C., 'Composition-depth Profiling using Auger Electron Spectroscopy', *Metal Science*, **Vol. 17**, pp. 357–367 (1983).

324 Smith, G.C., 'Surface Segregation and the Surface Energies of Liquid Metal Alloys', *Brazing & Soldering*, No. 13, pp. 6–9 (1987).

325 Wynblatt, P. and Ku, R.C., 'Surface Energy and Solute Strain Energy Effects in Surface Segregation', *Surface Science*, **Vol. 65**, pp. 511–531 (1977).

326 Seah, M.P., 'Quantitative Prediction of Surface Segregation', *Journal of Catalysis*, **Vol. 57**, pp. 450–457 (1979).

327 Eustathopoulos, N., 'Energetics of Solid-liquid Interfaces of Metals and Alloys', *International Metallurgical Reviews*, **Vol. 28**, pp. 189–210 (1983).

328 Wenzel, R.N., 'Resistance of Solid Surfaces to Wetting by Water', *Industrial & Engineering Chemistry*, **Vol. 28**, pp. 988–994 (1936).

329 Shuttleworth, R. and Bailey, G.L.J., 'The Spreading of a Liquid over a Rough Solid', *Discussions Faraday Society*, **Vol. 3**, pp. 16–22 (1948).

330 Hitchcock, S.J., Carroll, N.T. and Nicholas, M.G., 'Some Effects of Substrate Roughness on Wettability' *Journal of Materials Science*, **Vol. 16**, pp. 714–732 (1981).

331 Nicholas, M.G. and Crispin, R.M., 'Some Effects of Anisotropic Roughening on the Wetting of Metal Surfaces', *Journal of Materials Science*, **Vol. 21**, pp. 522–528 (1986).

332 Johnson, R.E. and Dettre, R.H., 'Wettability and Contact Angles', *Surface and Colloid Science*, **Vol. 2**, pp. 85–153 (1969).

333 Klein Wassink, R.J. and de Kluizenaar, E.E., 'De-wetting of Molten Solder from Copper', Proceedings Deutscher Verlag für Schweisstechnik Conference on 'Soldering and Welding in Electronics and Precision Mechanics', Munich, Germany, **Vol. DVS-71**, pp. 16–21 (1981).

334 Shipley, J.F., 'Influence of Flux, Substrate and Solder Composition on Solder Wetting', *Welding Journal Research Supplement*, **Vol. 54**, pp. 357s–362s (1975).

335 Stoneman, A.M. and MacKay, C.A., 'The Effect of Natural and Artificial Ageing on the Solderability of Tin and Tin-Lead Coated Wires', Proceedings INTERNEPCON 77, Brighton, UK, pp. 49–52 (1977).

336 Thwaites, C.J., 'Solderability of Coatings for Printed Circuits', *Transactions Institute of Metal Finishing*, **Vol. 43**, pp. 143–152 (1965).

337 Ackroyd, M.L. and MacKay, C.A., 'Solders, Solderable Finishes and Reflowed Solder Coatings', *Circuit World*, **Vol. 3**, No. 2, pp. 2–8 (1977).

338 Hagge, J.K. and Davis, G.J., 'Ageing, Solder Thickness and Solder Alloy Effects on Circuit Board Solderability', *Circuit World*, **Vol. 11**, No. 3, pp. 8–15 (1985).

339 Mather, J.C. and Hagge, J.K., 'Kinetics of Intermetallic Compound Formation and of Base Metal Dissolution into Molten Solder During Vapour Phase Soldering', *Brazing & Soldering*, No. 3, pp. 29–32 (1982).

340 Steen, H.A.H., 'Ageing of Component Leads and Printed Circuit Boards', Research Report IM-1716, Swedish Institute for Metals Research (1982).

341 Schmitt-Thomas, Kh.G., 'Status and Trends of Soft Soldering Techniques in Research, Development and Industrial Application, (German), Proceedings Deutscher Verlag für Schweisstechnik Conference on 'Soft Soldering in Research and Practice', Munich, **Vol. DVS-82**, pp. 1–12 (1983).

342 Dyson, B.F., Anthony, T.R. and Turnbull, D., 'Interstitial Diffusion of Copper in Tin', *Journal of Applied Physics*, **Vol. 38**, pp. 3408–3409 (1967).

343 Kay, P.J. and MacKay, C.A., 'The Growth of Intermetallic Compounds on Common Basis Materials Coated with Tin and Tin-Lead Alloys', *Transactions Institute of Metal Finishing*, **Vol. 54**, pp. 68–74 (1976).

344 Anon, 'Copper-tin Intermetallics', *Circuits Manufacturing*, **Vol. 20**, No. 9, pp. 56–64 (1980).

345 Le Fevre, B.G. and Barczykowski, R.A., 'Intermetallic Compound Growth on Tin and Solder Platings on Cu Alloys', *Wire Journal International*, **Vol. 18**, No. 1, pp. 66–71 (1985).

346 Unsworth, D.A. and MacKay, C.A., 'A Preliminary Report on Growth of Compound Layers

on Various Metal Bases Plated with Tin and Tin Alloys', *Transactions Institute of Metal Finishing*, **Vol. 51**, pp. 85–90 (1973).
347 Tu, K.N., 'Interdiffusion and Surface Reaction in Bimetallic Cu-Sn Thin Films', *Acta Metallurgica*, **Vol. 21**, pp. 347–354 (1973).
348 Revay, L., 'Interdiffusion and Formation of Intermetallic Compounds in Tin-Copper Alloy Surface Coatings', *Surface Technology*, **Vol. 5**, pp. 57–63 (1977).
349 Creydt, M., 'Diffusion at Galvanic Surfaces in Soft Soldering at Temperatures in the range 23 to 212°C' (German), Zurich Technical Highschool, PhD Thesis (1971).
350 Lubyova, Z., Fellner, P. and Matiasovsky, K., 'Diffusion in the Systems Iron-Tin and Copper-Tin', *Zeitschrift für Metallkunde*, **Vol. 66**, pp. 179–182 (1975).
351 London, J. and Ashall, D.W., 'Compound Growth and Fracture at Copper-Tin-Silver Solder Interfaces', *Brazing & Soldering*, No. 11, pp. 49–55 (1986).
352 Muckett, S.J., Warwick, M.E. and Davis, P.E., 'Thermal Ageing Effects between Thick-film Metallisations and Reflowed Solder Creams', *Plating & Surface Finishing*, **Vol. 73**, No. 1, pp. 40–50, January (1986).
353 Leibfried, W., 'Wetting, Leaching and Solder Adhesion of Copper Thick Film Conductors on Alumina and Multilayer Glasses', *Hybrid Circuits*, No. 10, pp. 24–37 (1986).
354 Ma, C.H. and Swalin, R.A., 'Study of Solute Diffusion in Liquid Tin', *Acta Metallurgica*, **Vol. 8**, pp. 388–395 (1960).
355 Brothers, E.W., 'Intermetallic Compound Formation in Soft Solders', *The Western Electric Engineer*, pp. 47–63, Spring/Summer (1981).
356 Shoji, Y., Uchida, S. and Ariga, T., 'Dissolution of Copper Cylinders in Molten Tin Under Dynamic Conditions', *Welding Journal Research Supplement*, **Vol. 60**, pp. 19s–24s (1981).
357 Knott, U. and Schmitt-Thomas, Kh.G., 'Influence of Crystal Size and Diffusion Layer Thickness in the Alloy Zone between Solder and Substrate' (German), Technical University of Munich, Research Report 4012 (1981).
358 Bader, W.G. and Baker, R.G., 'Solderability of Electro-deposited Solder and Tin Coatings After Extended Storage', *Plating*, **Vol. 60**, pp. 242–246 (1973).
359 Bernier, D., 'Effect of Ageing on the Solderability of Various Plated Surfaces', *Plating*, **Vol. 61**, pp. 842–845 (1974).
360 DeVore, J.A., 'The Mechanisms of Solderability and Solderability-related Failures', Proceedings 3rd Printed Circuit World Convention, Washington DC, Paper WCIII–43 (1984).
361 Warwick, M.E. and Muckett, S.J., 'Observations on the Growth and Impact of Intermetallic Compounds on Tin-coated Substrates', *Circuit World*, **Vol. 9**, No. 4, pp. 5–11 (1983).
362 Boggs, W.E., Kochik, R.M. and Pellissier, G.E., 'The Effect of Alloying Elements on the Oxidation of Tin', *Journal of Electrochemical Society*, **Vol. 110**, pp. 4–11 (1963).
363 Fidos, H. and Piekarski, K., 'Oxidation of Tin Coatings on Copper Wires', *Journal of the Institute of Metals*, **Vol. 101**, pp. 95–96 (1973).
364 Britton, S.C. and Bright, K., 'An Examination of Oxide Films on Tin and Tinplate', *Metallurgia*, **Vol. 56**, pp. 163–168 (1957).
365 de Kluizenaar, E.E., 'Surface Oxidation of Molten Soft Solder: An Auger Study', *Journal of Vacuum Science & Technology*, **Vol. A1**, pp. 1480–1485 (1983).
366 Tompkins, H.G., 'The Interaction of Some Atmospheric Gases with a Tin-Lead Alloy', *Journal of Electrochemical Society*, **Vol. 120**, pp. 651–654 (1972).
367 Ackroyd, M.L., 'A Survey of Accelerated Ageing Techniques for Solderable Substrates', Proceedings INTERNEPCON 76, Brighton, UK, pp. 214–235 (1976).
368 Wilson, G.C., 'Accelerated Ageing of PCBs with Specific Reference to Solderability', *IPC Technical Paper* IPC-TP-122, Institute for Interconnecting & Packaging Electronic Circuits, Lincolnwood, Illinois (1976).
369 Schoenthaler, D., 'Accelerated Ageing for Solderability Evaluations', Proceedings 3rd Printed Circuit World Convention, Washington DC, Paper WC111–03 (1984)
370 Davis, P.E., Warwick, M.E., Kay, P.J. and Muckett, S.J., 'Intermetallic Compound Growth and Solderability', *Plating and Surface Finishing*, **Vol. 70**, No. 8, pp. 49–53 (1983).
371 Mackay, C.A., 'A Comparison of Solderability Values as Measured by Different Solderability Test Methods', Proceedings INTERNEPCON 79, Brighton, UK, pp. 67–72 (1979).
372 Warwick, M.E., 'A Comparison of Solderability Tests for Wire', *Brazing & Soldering*, No. 2, pp. 48–52 (1982).
373 Allen, N. and George, W.R., 'Standardising Instruments for Soldering Assemblies', *Welding and Metal Fabrication*, **Vol. 47**, pp. 267–271 (1979).
374 Barranger, J., 'Setting Standards for Solderability', *Welding & Metal Fabrication*, **Vol. 48**, pp. 267–271, (1980).

375 Verguld, M.M.F., 'The Measurement of the Solderability of Metallic Pads on Thick-film Substrates', *Hybrid Circuits*, No. 10, pp. 16-19 (1986).
376 Albrecht, H-J., Scheel, W. and Freund, T., 'The Wetting Angle Measuring Unit: A New Optoelectronic Solderability Tester', *Brazing & Soldering*, No. 8, pp. 8-15 (1985).
377 ten Duis, J.A. and van der Meulen, E., 'Measurement of the Solderability of Components', *Philips Technical Review*, **Vol. 28**, pp. 362-364 (1967).
378 Allen, B.M., 'The Kinetics of Soldering', *Welding and Metal Fabrication*, pp. 47-50, January/February (1982).
379 Wallis, D.R., 'Trends in Testing for Wettability', *Brazing & Soldering*, No. 3, pp. 11-16 (1982).
380 Lin, K.M. and Harry, T.R., 'High Performance Instrument for Solder Wetting Studies', *The Western Electric Engineer*, pp. 11-19, Spring (1982).
381 Mackay, D., 'The Meniscograph: A Method of Solderability Measurement'. *Circuits Manufacturing*, **Vol. 13**, No. 7, pp. 52-56 (1973).
382 Peth, H., 'The Wetting Balance, a new Aid for Testing Soft Solder Fluxes' (German), *Journal of Materials Technology*, **Vol. 6**, pp. 367-376 (1975).
383 Schouten, G., 'A Figure of Merit for Solderability', *Philips Telecommunication Review*, **Vol. 38**, No. 3, pp. 131-138 (1980).
384 Schmitt-Thomas, Kh.G., Zahel, H.M. and Knott, U.C., 'The Wetting Balance for Solderability Testing' (German), *Elektronik Produktion und Prüftechnik*, **Vol. 11**, pp. 683-686 (1982).
385 Yoshida, H., Warwick, M.E. and Hawkins, S.P., 'The Assessment of the Solderability of Surface Mounted Devices using the Wetting Balance', *Brazing & Soldering*, No. 12, pp. 21-29 (1987).
386 Gunter, I.A., 'The Solderability Testing of Surface Mount Devices Using the GEC Meniscograph Solderability Tester', *Circuit World*, **Vol. 13**, No. 1, pp. 8-12 (1986).
387 Hartmann, H.J., 'Extended Application of the Wetting Balance for the Evaluation of Flux-free Soldering Processes' (German), Proceedings Deutscher Verlag für Schweisstechnik Conference, **DVS-40**, pp. 149-156 (1976).
388 Becker, G., 'Scanning the Solderability of a Surface', *Welding Journal Research Supplement*, **Vol. 60**, pp. 202s-206s (1981).
389 Podlesnykh, V.G. and Tkachev, M.A., 'Evaluating the Wetting of Materials by Solders Using Scanning Meniscography', *Welding Production*, **Vol. 31**, No. 7, pp. 49-52 (1984).
390 Allen, B.M., 'Solderability of Microcircuits and Leadless Components', Proceedings Deutscher Verlag für Schweisstechnik Conference on 'Soft Soldering in Research and Practice', **DVS-82**, pp. 198-205 (1983).
391 Gunter, I. 'Solderability Testing of SM Components', *Surface Mount*, pp. 6-7, September/October (1986).
392 Thwaites, C.J., 'A New Solderability Test Apparatus', *Electrical Manufacture*, **Vol. 8**, No. 5, pp. 18-19, 22-23 (1964).
393 Klein Wassink, R.J., 'Improving the Rotary-dip Solderability Test', *Welding & Metal Fabrication* **Vol. 49**, pp. 214-216 (1981).
394 Bakszt, M., 'Providing Solderability Retention by Means of Chemical Inhibitors', *Metal Finishing*, **Vol. 83**, pp. 35-38 (1985).
395 Lea, C., 'A Surface of Standard Solderability' (German), Proceedings Deutscher Verlag für Schweisstechnik Conference on 'Soft Soldering in Research and Practice', Munich, **Vol. DVS-82**, pp. 172-178 (1983).
396 Snow, A.C., 'Some Practical Observations on the Processes Involved in Surface Mounted Technology', *Circuit World*, **Vol. 11**, No. 2, pp. 12-13 (1985).
397 Klein Wassink, R.J. and Verguld, M.M.F., 'Drawbridging of Leadless Components', *Hybrid Circuits*, No. 9, pp. 18-24 (1986).
398 Hedges, E.S., 'Tin and its Alloys', Edward Arnold, London (1960).
399 Manko, H.H., 'Solders and Soldering', McGraw-Hill, New York, Second Edition (1979).
400 Hawkins, S.P., Thwaites, C.J. and Warwick, M.E., 'The Mechanical Properties of Soldered Joints to Surface Mounted Devices', *Brazing & Soldering*, No. 10, pp. 4-6 (1986).
401 Sinnadurai, N., Cooper, K. and Woodhouse, J., 'Assessing the Joints in Surface Mounted Assemblies', *Microelectronics Journal*, **Vol. 17**, No. 2, pp. 21-31 (1986).
402 Ahmed, M.M.I. and Langdon, T.G., 'The Effect of Grain Size on Ductility in the Superplastic Pb-Sn Eutectic', *Journal of Materials Science: Letters*, **Vol. 2**, pp. 337-340 (1983).
403 Baker, W.A., 'The Creep Properties of Soft Solders and Soft Soldered Joints', *Journal of the Institute of Metals*, **Vol. 65**, pp. 277-297 (1939).
404 Ainsworth, P.A., 'The Formation and Properties of Soft-soldered Joints, *Metals and Materials*, **Vol. 5**, pp. 374-379 (1971).

405 Solomon, H.D., 'Creep, Strain-rate Sensitivity and Low Cycle Fatigue of 60/40 Solder', *Brazing & Soldering*, No. 11, pp. 68–75 (1986).
406 Stone, K.R., Duckett, R., Muckett, S. and Warwick, M.E., 'Mechanical Properties of Solders and Soldered Joints', *Brazing & Soldering*, No. 4, pp. 20–27 (1983).
407 Keller, H.N., 'Temperature Ageing of External Connections for Condensation Soldering', *IEEE Transactions on Components, Hybrids and Manufacturing Technology*, **Vol. CHMT-2**, pp. 180–195 (1979).
408 Steen, H.A.H. and Becker, G., 'The Effect of Impurity Elements on the Soldering Properties of Eutectic and Near-eutectic Tin-Lead Solder', *Brazing & Soldering*, No. 11, pp. 4–11 (1986).
409 Tomkins, B., 'Fatigue: Introduction and Phenomenology', in 'Creep and Fatigue in High Temperature Alloys', Applied Science Publishers, London, pp. 73–110 (1981).
410 Manson, S.S., 'Thermal Stress and Low-cycle Fatigue', McGraw-Hill, New York (1966).
411 Engelmaier, W., 'Effects of Power Cycling on Leadless Chip Carrier Mounting Reliability and Technology', *Electronic Packaging & Production*, **Vol. 23**, No. 4, pp. 58–63 (1983).
412 Hagge, J.K., 'Predicting Fatigue Life of Leadless Chip Carriers using Manson-Coffin Equations', Proceedings International Electronics Packaging Symposium, pp. 199–208 (1982).
413 Hall, P.M., 'Forces, Moments and Displacements During Thermal Chamber Cycling of Leadless Ceramic Chip Carriers Soldered to Printed Boards', *IEEE Transactions on Components, Hybrids and Manufacturing Technology*, **Vol. CHMT-7**, pp. 314–327 (1984).
414 Lee, T.S.F., Wiltshire, B. and Culver, D., 'Joint Strength Analysis of Surface Mounted Components', Proceedings IPC Fall Meeting, Los Angeles, California, Technical Paper IPC-TP-566, (1985).
415 Engberg, G., Larsson, L.E., Nylén, M. and Steen, H., 'Low Cycle Fatigue of Soldered Joints', *Brazing & Soldering*, No. 11, pp. 62–65 (1986).
416 Dunn, B.D., 'The Resistance of Space-quality Solder Joints to Thermal Fatigue', *Circuit World*, **Vol. 6**, No. 1, pp. 16–27 (1979).
417 Honeycombe, R.W.K., 'The Plastic Deformation of Metals', Arnold, London (1968).
418 Baker, E., 'Stress Relaxation in Tin-Lead Solders', *Materials Science and Engineering*, **Vol. 38**, pp. 241–247 (1979).
419 Baker, E. and Kessler, T.J., 'The Influence of Temperature on Stress Relaxation in a Chill-cast Tin-Lead Solder', *IEEE Transactions on Parts, Hybrids and Packaging*, **Vol. PHP-9**, pp. 243–246 (1973).
420 Kinser, D.L., Vaughan, J.G. and Graff, S.M., 'Reliability of Soldered Joints in Thermal Cycling Environments', Proceedings ISHM International Microelectronics Conference, Vancouver, BC, pp. 4–7 (1976).
421 Taylor, J.R., Brierley, C.J. and Pedder, D.J., Proceedings INTERNEPCON 82, Brighton, UK, pp. 236–241 (1982).
422 Lake, J.K. and Wild, R.N., 'Some Factors Affecting Leadless Chip Carrier Solder Joint Fatigue Life', Proceedings 28th National SAMPE Symposium, pp. 1406–1414 (1983).
423 Lynch, J.T., 'Surface Mounting Reliability', *Hybrid Circuits*, No. 11, pp. 36–38 (1986).
424 Bilson, R.T., Hepher, M.R. and McCarthy, R.J., 'Controlling Solder Fillets on Leadless Chip Carriers', *Electronic Production*, **Vol. 13**, No. 1 pp. 24–28 (1984).
425 Brierley, C.J. and McCarthy, J.P., 'The Reliability of Surface Mounted Solder Joints Under PWB Cyclic Mechanical Stresses', *Circuit World*, **Vol. 12**, No. 3, pp. 16–19 (1986).
426 Lauer, L.D., 'Dynamic Mechanical Testing of Solder and Solder Joints', *Circuit World*, **Vol. 13**, No. 1, pp. 13–17 (1986).
427 Engelmaier, W., 'Fatigue Life of Leadless Chip Carrier Solder Joints During Power Cycling', *IEEE Transactions on Components, Hybrids and Manufacturing Technology*, **Vol. CHMT-6**, pp. 232–237 (1983).
428 Engelmaier, W., 'Functional Cycles and Surface Mounting Attachment Reliability', *Circuit World*, **Vol. 11**, No. 3, pp. 61–72 (1985).
429 Wild, R., 'Fatigue Properties of Solder Joints', *Welding Journal Research Supplement*, **Vol. 51**, pp. 521s–526s (1972).
430 Steen, H.A.H., 'Thermal Fatigue of Solder Joints with Special Reference to Surface Mounted Components', Research Report IM-1837, Swedish Institute for Metals Research (1983).
431 Norris, K.C. and Landzberg, A.A., 'Reliability of Controlled - collapse Interconnections', *IBM Journal of Research & Development*, pp. 266–271, May (1969).
432 Engelmaier, W., 'Test Method Considerations for SMT Solder Joint Reliability', *Brazing & Soldering*, No. 9, pp. 40–43 (1985).
433 Wright, E.A. and Wolverton, W.M., 'The Effect of the Solder Reflow Method and Joint Design on the Thermal Fatigue Life of Leadless Chip Carrier Solder Joints', Proceedings IEEE

34th Electronic Components Conference, New Orleans, Louisiana, pp. 149–155 (1984).
434 Taylor, J.R. and Pedder, D.J., 'Joint Strength and Thermal Fatigue in Chip Carrier Assembly', *Journal of Hybrid Microelectronics*, Vol. 5, pp. 209–214 (1982).
435 Shine, M.C., Fox, L.R. and Sofia, J.W., 'A Strain Range Partitioning Procedure for Solder Fatigue', *Brazing & Soldering*, No. 9, pp. 11–14 (1985).
436 Dierke, J.H., 'Increasing Solder Joint Reliability of Leaded Surface Mounted Components', *Brazing & Soldering*, No. 7, pp. 13–14 (1984).
437 Thwaites, C.J. and Duckett, R., 'Some Effects of Soldered Joint Geometry on Their Mechanical Strength', *Revue de la Soudure*, Vol. 4, pp. 196–201 (1976).
438 Thwaites, C.J., 'Some Metallurgical Studies Related to Surface Mounting of Electronic Components', *Circuit World*, Vol. 11, No. 1, pp. 8–12 (1984).
439 Schmitt-Thomas, Kh.G. and Wege, S., 'Behaviour of Solder Joints Under Thermal and Mechanical Stress', *Brazing & Soldering*, No. 11, pp. 27–33 (1986).
440 Bangs, E.R. and Beal, R.E., 'Effect of Low Frequency Thermal Cycling on the Crack Susceptibility of Soldered Joints', *Welding Journal Research Supplement*, Vol. 55, pp. 377s–383s. (1976).
441 Fox, L.R., Sofia, J.W. and Shine, M.C., 'Investigation of Solder Fatigue Acceleration Factors', *IEEE Transactions on Components, Hybrids & Manufacturing Technology*, Vol. CHMT-8, pp. 275–282 (1985).
442 Lau, J.H. and Rice, D.W., 'Solder Joint Fatigue in Surface Mount Technology: State of the Art', *Solid State Technology*, Vol. 28, No. 10, pp. 91–104 (1985).
443 DeVore, J.A., 'Fatigue Resistance of Solders', Proceedings NEPCON-West 82, Anaheim, California, pp. 409–414 (1982).
444 Becker, G., 'Creep and Fatigue Testing of Micro Solder Joints', Proceedings Symposium on Thermal Fatigue in Surface-mounted Electronic Components, Stockholm, Paper 2 (1983).
445 Becker, G., 'Testing and Results Related to the Mechanical Strength of Solder Joints', Proceedings IPC Fall Meeting, San Francisco, Technical Paper IPC-TP-288 (1979).
446 Page Walton, J., 'PCB Cleaning—Aqueous or Solvent', *Electronic Production*, Vol. 16, No. 1, pp. 12–18 (1987).
447 Spitz, S.L., 'Cleaning PCBs for Higher Quality', *Electronic Packaging & Production*, Vol. 25, No. 9, pp. 100–106 (1985).
448 Capillo, C.A., 'Surface Mounted Assemblies Create New Cleaning Challenges', *Electronic Packaging & Production*, Vol. 24, No. 8, pp. 76–81 (1984).
449 Johnson, P.G., 'Cleaning Surface Mounted Assemblies', *Electronic Production*, Vol. 14, No. 11, pp. 27–30 (1985).
450 Pascoe, G., 'Recent Advances in Removing Electronic Flux Residues', *Brazing & Soldering*, No. 6, pp. 30–31 (1984).
451 Ellis, B.N., 'The Theory of Cleaning, Part 1', *Circuit World*, Vol. 11, No. 4, pp. 37–39 (1985).
452 Lovering, D.G., 'The Nature of Insoluble Residues Formed on Printed Circuits Soldered with Rosin Fluxes', *Circuit World*, Vol. 11, No. 4, pp. 20–23 (1985).
453 Davy, J.G., 'Tan Residue from Cleaning Rosin Flux from PWAs with Fluorocarbon-methanol Azeotrope', Proceedings NEPCON-West 83, Anaheim, California, pp. 565–569 (1983).
454 Barton, A.F.M., 'Handbook of Solubility Parameters and Other Cohesion Parameters', CRC Press, Boca Raton, Florida (1983).
455 Cabelka, T.D. and Archer, W.L., 'Cleaning: What Really Counts', Proceedings ISHM International Microelectronics Symposium, Anaheim, California, pp. 520–528 (1985).
456 Turbini, L.J., Eagle, J.G. and Stark, T.J., 'A Comparison of Removal of Activated Rosin Flux by Selected Solvents', Proceedings IPC Fall Meeting, San Francisco, California, Technical Paper IPC-TP-305 (1979).
457 Ellis, B.N., 'The Theory of Cleaning, Part 2', *Circuit World*, Vol. 12, No. 1, pp. 8–10 (1985).
458 Comerford, M., 'Cleaning Printed Wiring Assemblies: The Effects of Surface Mounted Components', *Electri-onics*, Vol. 30, pp. 13–19 (1984).
459 Keeler, R., 'Post Solder Cleaning Meets its Match in SMT Geometry', *Electronic Packaging & Production*, Vol. 27, No. 1, pp. 86–89 (1987).
460 Musselman, R.P., 'Shear Stress Cleaning of Printed Wiring Boards and Assemblies', Proceedings NEPCON-East 85, Boston, Massachusetts, pp. 244–259 (1985).
461 Ellis, B.N., 'Removing Flux Using Batch Water Machines', *Electronic Production*, Vol. 16, No. 4, pp. 49–54 (1987).
462 Lemond, D.S., 'Key Process Design Factors for Efficient Fluorosolvent Spray Cleaning of Surface Mounted Assemblies', Proceedings IPC Fall Meeting, San Diego, California, Technical Paper IPC-TP-604 (1986).
463 Lambert, W., 'Ultrasonics Aid Cleaning', *Electronic Production*, Vol. 14, No. 10, pp. 121–125 (1985).

464 Page Walton, J., 'PCB Cleaning Systems', *Electronic Production*, **Vol. 15**, No. 6, pp. 10–15 (1986).
465 Ellis, B.N., 'Contamination Control: Quo Vadis?', *Circuit World*, **Vol. 12**, No. 4, pp. 40–43 (1986).
466 Ellis, B.N., 'Quality Control of Surfaces for High Reliability Electronics', *Circuit World*, **Vol. 13**, No. 2, pp. 28–32 (1987).
467 Ellis, B.N., 'Ionic Contamination Control of Circuits with Surface Mounted Components', *Brazing & Soldering*, No. 9, pp. 15–18 (1985).
468 Keller, J., 'Re-defining Solder Joint Acceptability', *Electronic Packaging & Production*, **Vol. 26**, No. 2, pp. 101–103 (1986).
469 Pound, R., 'Inspection Equipment Exposes Quality of Soldered Joints.' *Electronic Packaging & Production*, **Vol. 26**, No. 2, pp. 84–86 (1986).
470 Schoenbaum, G.L. and Socha, E., 'Selecting the Accept/Reject Criteria for an Automated Solder Process Inspection Machine', *Circuit World*, **Vol. 13**, No. 2, pp. 19–21 (1987).
471 Pound, R., 'Image Processing Boosts the Power of Non-destructive Testing', *Electronic Packaging & Production*, **Vol. 25**, No. 6, pp. 98–104 (1985).
472 Vanzetti, R., Traub, A.C. and Supino, L., 'The Ultimate in Automatic Inpsection', *Circuit World*, **Vol. 8**, No. 4, pp. 12–17 (1982).
473 Vanzetti, R., Traub, A.C. and Richard, A.A., 'Laser Inspection of Solder Joints', *Brazing & Soldering*, No. 2, pp. 34–37 (1982).
474 Vanzetti, R., 'Automatic Laser Inspection System for Solder Joint Integrity Evaluation', Proceedings 3rd Printed Circuit World Convention, Washington DC, Paper WC III–44 (1984).
475 Lea, C., Howie, F.H. and Seah, M.P., 'Automated Inspection of PCB Solder Joints: An Assessment of the Capability of the Vanzetti LI-6000 Infra-red Laser Inspection Instrument', *Brazing & Soldering*, No. 8, pp. 34–42 (1985).
476 Seah, M.P. and Lea, C., 'Certainty of Measurement Using an Automated Infra-red Laser Inspection Instrument for PCB Solder Joint Integrity'. *Journal of Physics E. (Scientific Instruments)*, **Vol. 18**, pp. 676–682 (1985).
477 Smith, G.C., 'The Scanning Acoustic Microscope—A New Tool for the Materials Scientist', *Materials Science and Technology*, **Vol. 2**, pp. 881–887 (1986).
478 Murray, J., 'Looking into the SMD Solder Joint, Deeply', *Circuits Manufacturing*, **Vol. 27**, No. 1, pp. 48–54 (1987).
479 Page Walton, J., 'Testing Surface Mounted Boards', *Electronic Production*, **Vol. 16**, No. 2, pp. 8–14 (1987).
480 Nicholas, P., 'SMA Technology—Current Practices and Criteria for Testability', *Circuit World*, **Vol. 13**, No. 2, pp. 33–36 (1987).
481 Maunder, C., Roberts, D. and Sinnadurai, N., 'Chip Carrier Based Systems and Their Testability', *Hybrid Circuits*, No. 5, pp. 29–36 (1984).
482 Lawrence, B., 'Fixing the Future', *Electronics Manufacture and Test*, pp. 66–69, February (1986).
483 Tygard, C.M., 'Probes Ease Testing of Hybrid and SMT Assemblies', *Electronic Packaging & Production*, **Vol. 26**, No. 2, pp. 205–210 (1986).
484 Turino, J., 'Testability Circuit Solves SMT Board Access Problems', *Electronic Packaging & Production*, **Vol. 26**, No. 1, pp. 110–114 (1986).
485 Wallgren, L., 'The Removal and Replacement of SMCs', *Electronic Packaging & Production*, **Vol. 26**, No. 1, pp. 102–104 (1986).
486 Smith, L.J., 'Coatings Cover High-volume PCBs', *Electronic Packaging & Production*, **Vol. 26**, No. 2, pp. 148–150 (1986).
487 Malloy, G.T. and Keister, F.Z., 'Silicone Conformal Coatings for Microcircuits', Proceedings ISHM International Microelectronics Symposium, Chicago, Illinois, pp. 440–449 (1981).
488 Markstein, H.W., 'Conformal Coatings Seal Out Adverse Environments', *Electronic Packaging & Production*, **Vol. 27**, No. 3, pp. 82–84 (1987).
489 Goldman, I.B. and Krajewski, A., 'An Ultra-violet Conformal Coating System: Materials and Processes', *Circuit World*, **Vol. 12**, No. 1, pp. 4–7 (1985).
490 Anon, 'American Manufacturers Strive for Quality, Japanese Style', *Business Week*, March (1979).
491 Moltoft, J., 'Reliability Assessment and Screening by Reliability Indicator Methods', Proceedings ELECTRONICA, Munich, Germany, November (1982).
492 George, A.H., 'Reliability and Quality Assurance: Basic Concepts and Their Application to Microcircuits', *Hybrid Circuits*, No. 5, pp. 12–14 (1984).
493 Ansell, M.P., 'Conduction Processes in Thick-film Resistors', *Electrocomponent Science & Technology*, **Vol. 3**, pp. 131–151 (1976).

494 Pranchov, R.B. and Campbell, D.S., 'Model for Reliability Prediction of Thick Film Resistors', *Electrocomponent Science & Technology*, **Vol. 11**, pp. 185–190 (1984).
495 Sinnadurai, F.N., 'Mechanisms and Modes of Failure in Silicon Planar Semiconductor Devices', British Post Office, March (1970).
496 Reynolds, F.H., 'Thermally Accelerated Ageing of Semiconductor Components', *Proceedings IEEE*, **Vol. 62**, pp. 212–222 (1974).
497 Black, J.R., 'Electromigration—A Brief Survey and Some Recent Results', *IEEE Transactions on Electron Devices*, **Vol. ED-16**, pp. 338–347 (1969).
498 Sim, S.P., 'Procurement Specification Requirements for Protection Against Electromigration Failures in Aluminium Metallisations', *Microelectronics and Reliability*, **Vol. 19**, pp. 207–218 (1979).
499 Harrison, J.C. and Rickard, E.F., 'Plastics for Long Life Microcircuit Encapsulation: Materials Properties and Possible Failure Mechanisms', Proceedings IEE Conference on Reliability in Electronics, pp. 129–136 (1969).
500 Peck, D.S., 'The Design and Evaluation of Reliable Plastic Encapsulated Devices', Proceedings IEEE 8th Annual Reliability Physics Symposium, pp. 81–93 (1970).
501 Sinnadurai, F.N., 'The Accelerated Ageing of Plastic Encapsulated Semiconductor Devices in Environments Containing a High Vapour Pressure of Water', *Microelectronics & Reliability*, **Vol. 13**, pp. 23–27 (1974).
502 Lawson, R.W., 'The Accelerated Testing of Plastic Encapsulated Semiconductor Components', Proceedings IEEE 12th Annual Reliability Physics Symposium, Las Vegas, Nevada, pp. 243–247 (1974).
503 Peck, D.S. and Zierdt, C.H., 'Temperature–Humidity Acceleration of Metal–electrolysis Failure in Semiconductor Devices', Proceedings IEEE 11th Annual Reliability Physics Symposium, Las Vegas, Nevada, pp. 146–152 (1973).
504 Brunauer, S., Emmett, P.H. and Teller, E., 'Adsorption of Gases in Multimolecular Layers', *Journal American Chemical Society*, **Vol. 60**, pp. 309–319 (1938).
505 Flood, E.A., Editor, 'The Solid-Gas Interface', Dekker, New York (1967).
506 Sbar, N.L. and Korakiewicz, R.P., 'New Acceleration Factors for Temperature-Humidity Bias Testing', Proceedings IEEE 16th Annual Reliability Physics Symposium, San Diego, California, pp. 161–178 (1978).
507 Sim, S.P. and Lawson, R.W., 'The Influence of Plastic Encapsulants and Passivation Layers on the Corrosion of Thin Aluminium Films Subjected to Humidity Stress', Proceedings IEEE 17th Annual Reliability Physics Symposium, San Francisco, California, pp. 103–112 (1979).
508 Derman, G., 'The Impact of Surface Mount Technology on Electronics Manufacturing', *Microelectronics Journal*, **Vol. 17**, No. 2, pp. 5–11 (1986).
509 Brown, D., Bracken, J. and Brasch, J., 'Expense and Availability Direct SMT Growth', *Electronic Packaging & Production*, **Vol. 26**, No. 1, pp. 94–97 (1986).
510 Barnwell, P.G., 'Future Trends and Developments in Circuit Technology', *Hybrid Circuits*, No. 9, pp. 13–14 (1986).
511 Mangin, C-H., 'Surface Mount Component Tug-of-war', *Electronic Packaging & Production*, **Vol. 27**, No. 1, pp. 80–81 (1987).
512 Tye, R., 'Managing Surface Mount?', *Surface Mount*, pp. 6–7, November/December (1986).
513 Maes, M., 'Automating Surface-mount Assembly in a European Plant', *Electronic Packaging & Production*, **Vol. 27**, No. 3, pp. 30–33 (1987).
514 Kalenik, S. and Anderson, B., 'Design Economics of Surface Mount PCBs', *Circuit World*, **Vol. 12**, No. 1, pp. 26–28 (1985).
515 Rowe, G.L., 'Surface Mount Technology: Setting up a Surface Mount Facility', *Hybrid Circuits*, No. 6, pp. 27–34 (1985).
516 Markstein, H.W., 'Choosing SMT and Setting up a Facility', *Electronic Packaging & Production*, **Vol. 26**, No. 1, pp. 74–78 (1986).
517 Brown, D. and Gallanda, P., 'The Contractors—Building Business via SMT', *Electronic Packaging & Production*, **Vol. 26**, No. 5, pp. 24–26 (1986).
518 Nakahara, H., 'Future Trends of Electronics in Asia', *Circuit World*, **Vol. 11**, No. 3, pp. 32–34 (1985).
519 SMD and SMT Trends, Marata Report (1986).
520 Pound, R., 'SMT is Poised for Penetration of European Manufacturing', *Electronic Packaging & Production*, **Vol. 26**, No. 1, pp. 66–70 (1986).
521 Donnelly, E., 'The Future of Electronics in Europe', Proceedings 3rd Printed Circuit World Convention, Washington DC, Paper WC III-56 (1984).
522 Taylor, K., 'Surface Mount Technology for High Reliability Telecomms Applications', *Circuit World*, **Vol. 12**, No. 3, pp. 27–29 (1986).

523 Nakahara, H., 'Surface Mount Technology in Japan—From Consumer Electronics to Supercomputer', *Circuit World*, **Vol. 13**, No. 3, pp. 25–28 (1987).
524 Nakahara, H., 'SMT Expands Options in Japan', *Electronic Packaging & Production*, **Vol. 26**, No. 1, pp. 58–62 (1986).
525 Cohen, C.L., 'Japan's Packaging Goes World Class', *Electronics*, **Vol. 58**, No. 45 pp. 26–31, (1985).
526 Nakahara, H., 'Japan's Swing to Chip-on-board', *Electronic Packaging & Production*, **Vol. 26**, No. 12, pp. 38–41 (1986).
527 Marques, L., 'Surface Mount Assembly in High Volume Manufacturing', *Circuit World*, **Vol. 12**, No. 1, pp. 58–60 (1985).
528 Lancaster, M., 'Why is the USA Behind in SMT?' *Electronic Packaging & Production*, **Vol. 26**, No. 5, pp. 28–30 (1986).
529 Lyman, J. 'Military Moves Headlong into Surface Mounting', *Electronics*, **Vol. 59**, No. 26, pp. 93–98 (1986).
530 Page Walton, J., 'Reacting to the Challenge of Surface Mounting', *Electronic Production*, **Vol. 13**, No. 9, pp. 29–33 (1984).
531 Cohen, E.R. and Taylor, B.N., 'Review of Best Values of Fundamental Constants', CODATA Bulletin No. 63 (1986).

Author Index

(see also separate Subject Index)

References in the text to an author's work can be traced back from here by means of the item numbers. **These are not page numbers but item numbers in the bibliography starting on p. 535.**

Abbott, W. G. 203
Ackroyd, M. L. 225, 337, 367
Adams, J. C. 312
Adamson, C. 160
Ahearn, J. F. 265
Ahmed, M. M. I. 402
Ainsworth, P. A. 237, 404
Albrecht, H.-J. 376
Allen, B. M. 378, 390
Allen, N. 373
Ammann, H. H. 266, 268
Anderson, B. 514
Andrews, J. 242
Angstenberger, A. 36
Anjard, R. P. 205, 207, 212, 216, 234
Ansell, M. P. 493
Anthony, T. R. 342
Archer, W. L. 455
Ariga, T. 346
Ashall, D. W. 351
Atsumi, K. 142

Bader, W. G. 358
Baker, E. 418, 419
Baker, R. G. 358
Baker, W. A. 403
Bailey, G. L. J. 317, 329
Bakszt, M. 394
Balke, H. 59
Bancroft, R. C. 20
Bangs, E. R. 440
Barczykowski, R. A. 345
Barnwell, P. G. 510
Barranger, J. 374
Barton, A. F. M. 454
Bartosz, T. 258
Bashforth, F. 312
Basset, J. C. 148
Beal, R. E. 440
Becker, G. 292, 388, 408, 444, 445
Berg, H. M. 242
Bernard, C. D. 164

Bernier, D. 359
Billing, B. G. 151
Bilson, R. T. 424
Björklund, G. 65
Black, J. R. 497
Blackshaw, M. F. 14, 130
Bless, P. 64
Bloom, B. H. 313
Boey, W. K. 168
Boggs, W. E. 362
Bolam, C. F. 301
Bondi, A. 318
Boscha, B. 124
Bosworth, R. C. L. 269
Boynton, K. G. 177
Bracken, J. 509
Brasch, J. 509
Brearley, D. 25
Brierley, C. J. 68, 172, 421, 425
Briggs, S. 291
Bright, K. 364
Britton, S. C. 364
Brothers, E. W. 355
Brown, D. 509, 517
Browne, L. T. 293
Brunauer, S. 504
Buck, T. J. 89
Bud, P. J. 161, 163, 289
Bulwith, R. A. 218
Burke, J. 150
Burns, F. 300
Butler, J. N. 313

Cabelka, T. D. 455
Caiazzo, F. M. 95
Campbell, D. S. 494
Capillo, C. A. 115, 448
Carr, J. F. 305
Carr, R. 166
Carroll, N. T. 330
Carslaw, H. S. 40
Cartwright, L. 82

Author Index

Cassada, K. 51
Castelli, G. 97
Caswell, G. 2
Chadzynski, P. Z. 88
Chapman, A. J. 245
Cheype, J. M. 144
Chu, T. Y. 267, 275, 277
Clayton, J. W. 281
Cohen, C. L. 525
Cohen, E. R. 531
Coleman, M. 29
Comerford, M. F. 165, 179, 180, 202, 458
Condon, J. 238
Cook, S. 112
Cooper, K. 401
Cooper, N. 37, 171
Cox, N. R. 255, 256
Creydt, M. 349
Crispin, R. M. 331
Cullen, C. 128
Culver, D. 414

Daebler, D. H. 208
Dance, F. J. 14, 81, 130
Danielson, R. D. 282, 284
Davis, G. J. 338
Davis, P. E. 352, 370
Davy, J. G. 191, 192, 453
Dawes, C. J. 136
de Gennes, P. G. 311
Deigham, R. A. 319
De Jong, J. J. 129
de Kluizenaar, E. E. 333, 365
Delagi, R. 84
Denda, S. 153
Derman, G. 508
der Marderosian, A. 241
D'Erchia, F. J. 260
Dettmer, R. 5
Dettre, R. H. 332
DeVore, J. A. 360, 443
Dierke, J. H. 436
Diletti, H. 18
Dixon, T. 211, 296
Donnelly, E. 521
Dow, S. J. 248, 252
Down, W. H. 169
Drake, R. M. 253
Dreyfus-Alain, B. 12
Duckett, R. 406, 437
Dunn, B. D. 416
Dustoomian, A. S. 304
Dyson, B. F. 342

Eagle, J. G. 456
Eastman, K. 17
Eckert, E. R. G. 253
Edson, D. A. 137
El Refaie, M. 92
Elliott, D. A. 53, 56, 152, 183, 187, 201, 289

Ellis, B. N. 451, 457, 461, 465, 466, 467
Emerson, A. 55
Emmett, P. H. 504
Engberg, G. 415
Engelmaier, W. 411, 427, 428, 432
Estes, R. H. 308, 309
Eustathopoulos, N. 327

Farkass, I. 268
Fellner, P. 350
Fenner, M. 119
Fidos, H. 363
Field, J. 128
Fisher, J. R. 38
Fishman, D. 37, 171
Flattery, D. K. 247
Flood, E. A. 505
Fox, L. R. 435, 441
Francis, J. S. 95
Freund, T. 376
Fu, S-L. 30
Fuchs, E. 131
Fukuoka, Y. 41
Fukutomi, N. 87

Gallanda, P. 517
Gardon, R. 257
Gates, L. E. 70
Gelb, A. S. 74, 75
George, A. H. 492
George, W. R. 373
Gerstmann, J. 274
Giglioli, G. 280
Ginsberg, G. 27, 91, 140, 147
Ginzburg, F. 228
Gionet, V. 241
Glaessgen, R. 210
Goldman, I. B. 489
Goldman, P. 130
Graff, S. M. 420
Granger, J. J. 148
Grant, S. 156, 158
Gray, F. 80, 82
Green, W. J. 104, 105
Griffith, P. 274
Gunter, I. A. 386, 391

Hagen, D. F. 284
Hagge, J. K. 220, 338, 339, 412
Hall, P. M. 413
Hamilton, P. G. B. 184
Hansen, M. 221
Hanson, J. R. 83
Hargrave, C. E. 61
Harrison, J. C. 499
Harry, J. E. 302
Harry, T. R. 380
Hartman, H. 71
Hartmann, H. J. 387
Hatfield, W. B. 72
Hauschild, F. D. 22

Hauser, J. L. 83
Hawkins, S. P. 385, 400
Heason, E. N. 103
Hedges, E. S. 398
Hedges, R. 175
Hendricks, A. H. C. 31
Hepher, M. R. 424
Hitchcock, S. J. 330
Hoffmann, R. 18
Hollander, D. 10
Hondros, E. D. 320
Honeycombe, R. W. K. 417
Houldsworth, J. 11
House, J. R. 86
Howie, F. H. 157, 193, 194, 195, 196, 197, 198, 199, 320, 475
Huang, C. 64, 74, 75, 106, 107
Hugh, S. C. 76
Humbert, N. 58
Hutchins, C. L. 116
Hwang, J. S. 233

Ikegami, A. 153
Ishizuka, M. 41
Itoh, K. 34
Iwasaki, Y. 87

Jacob, G. W. 111
Jaeger, J. C. 40
Johns, K. W. E. 261
Johnson, G. D. 284
Johnson, K. I. 134, 136, 137
Johnson, P. G. 449
Johnson, R. E. 332
Johnson, R. W. 96
Jones, B. 100
Jones, C. 80
Jong, J. J. de 129

Kabe, A. 98
Kacker, R. N. 186
Kalenik, S. 514
Kapitza, P. L. 271
Kapitza, S. P. 271
Karpel, S. 174
Kay, P. J. 219, 343, 370
Keeler, R. 57, 122, 132, 155, 459
Keister, F. Z. 487
Keller, H. N. 407
Keller, J. 468
Kersuzan, G. 12
Kessler, T. J. 419
Kida, A. 87
Kilham, L. F. 265
King, S. R. 254
Kinser, D. L. 420
Klein Wassink, R. J. 4, 154, 178, 315, 333, 393, 397
Knight, W. G. 118
Knott, U. C. 357, 384
Kochik, R. M. 362

Kolkin, J. 79
Korakiewicz, R. P. 506
Kozima, F. 87
Krajewski, A. 489
Kreith, F. 246
Ku, R. C. 325
Kujawa, T. 295
Kulesza, F. W. 308, 309
Kvurt, O. S. 228

Lacruche, B. 39
Lake, J. K. 422
Lambert, L. 181
Lambert, W. 463
Lancaster, M. 528
Landis, R. 133
Landzberg, A. A. 431
Langdon, T. G. 402
Langridge, C. 78
Larry, J. R. 217
Larsson, L. E. 415
Lau, J. H. 442
Lauer, L. D. 426
Lawrence, B. 482
Lawson, R. W. 502, 507
Lea, C. 157, 193, 194, 195, 196, 197, 198, 199, 261, 321, 322, 323, 395, 475, 476
Lee, I. W. H. 204
Lee, N. C. 233
Lee, T. S. F. 414
Le Fevre, B. G. 345
Leibfried, W. 353
Leibowitz, J. 79
Lemond, D. S. 462
Leonida, G. 43, 189, 190
Leppert, G. 273
Liang, M-S. 30
Lilienthal, P. F. 288
Liljestrand, L-G. 145
Lin, K. M. 186, 380
Lindblom, S. 82, 85
Lish, E. F. 297, 299, 303
Lo, E. 106, 107
Loeffler, J. R. 294
Lofgran, L. 242
London, J. 351
Lovati, G. 97
Love, G. F. 15
Lovering, D. G. 452
Lubyova, Z. 350
Lyman, J. 529
Lynch, J. T. 423

Ma, C. H. 354
McAdams, W. H. 262
McCarthy, J. P. 68, 172, 425
McCarthy, R. J. 424
Macchi, E. 280
MacKay, C. A. 218, 219, 224, 225, 243, 335, 337, 343, 346, 371
Mackay, D. 381

Author Index

Maes, M. 513
Mahajan, R. L. 276, 278, 279
Mahalingham, M. 242
Malloy, G. T. 487
Mangin, C.-H. 511
Manko, H. 162, 399
Manson, S. S. 410
Marchetti, J. R. 47
Marhevka, J. S. 284
Markstein, H. W. 35, 113, 264, 488, 516
Marques, L. 527
Marston, P. 8
Märtens, A. 48
Martin, F. W. 94, 101, 102
Masciana, G. C. 54
Mather, J. C. 220, 339
Matiasovsky, K. 350
Matthews, R. 170
Maunder, C. 481
Menozzi, G. 69
Mollendorf, J. C. 263, 267, 275
Moltoft, J. 491
Morini, A. 280
Morooka, I. 98
Morrison, L. 222
Morse, R. 286
Morten, B. 33
Mosby, F. A. 250
Muckett, S. J. 352, 361, 370, 406
Mullen, J. 3
Murray, J. 478
Musselman, R. P. 460

Nagarkar, M. 242
Nakahara, H. 518, 523, 524, 526
Needes, C. R. S. 139
Nicholas, M. G. 330, 331
Nicholas, P. 480
Nimmo, B. 272, 273
Noach, K. 9
Norris, K. C. 431
Nusselt, W. 270
Nylén, M. 415

Oien, M. A. 266
Okuda, T. 143
Olsen, D. R. 242
O'Rourke, H. T. 165

Page Walton, J. 126, 167, 446, 464, 479, 530
Partridge, S. A. 231
Pascoe, G. 188, 450
Patel, D. 28
Pawling, J. F. 121
Payne, J. 128
Peck, D. S. 500, 503
Pedder, D. J. 68, 421, 434
Peel, M. E. 120
Pellissier, G. E. 362
Peterson, N. C. 239

Peth, H. 382
Pfahl, R. C. 267, 275, 277
Piccoli, S. 307
Piekarski, K. 363
Pierce, T. R. 114
Pitt, K. E. G. 63
Pitt, V. A. 139
Podlesnykh, V. G. 389
Pound, R. 62, 90, 310, 469, 471, 520
Pranchov, R. B. 494
Prasad, P. 6, 110, 117
Prudenziati, M. 33

Raman, K. S. 226
Rees, J. 18
Reimann, W. G. 70
Reiner, M. 7
Revay, L. 348
Reynolds, F. H. 496
Reynolds, M. J. 13, 23
Rice, D. W. 442
Richard, A. A. 473
Rickard, E. F. 499
Roberts, D. 481
Rodwell, R. 138
Roos, L. 50
Roos-Kozel, B. L. 95, 209, 214, 240, 244
Rosenzweig, W. 16
Rowe, G. L. 515
Rubin, W. 159, 206, 235

Sage, M. G. 1
St. John, F. 108
Sanjana, Z. N. 47
Sbar, N. 506
Scagnelli, G. J. 260
Scarlett, J. A. 44
Scheel, W. 376
Schmidt, W. 93
Schmitt-Thomas, Kh. G. 341, 357, 384, 439
Schoenbaum, G. L. 470
Schoenthaler, D. 232, 251, 290, 369
Schouten, G. 383
Scott, M. H. 136, 137
Sculpher, M. R. 49
Seah, M. P. 157, 195, 199, 326, 475, 476
Searle, G. F. C. 236
Sele, G. 18
Seraphim, D. P. 42
Serpiello, J. W. 16
Shahbazi, S. 101
Shanahan, M. E. R. 314
Shine, M. C. 425, 441
Shipley, J. F. 334
Shiromatsu, T. 30
Shoji, Y. 356
Shuttleworth, R. 329
Sim, S. P. 498, 507
Sinnadurai, F. N. 32, 66, 67, 401, 481, 495, 501
Slattery, J. A. 229

Slinn, D. S. L. 283
Slutsky, J. 217
Smith, C. A. 45
Smith, G. C. 322, 324, 477
Smith, L. J. 486
Smith, M. 112, 287
Smith-Vargo, L. 52, 215, 230, 306
Snow, A. C. 396
Socha, E. 470
Socolowski, N. 213
Sofia, J. W. 435, 441
Solomon, H. D. 405
Spector, M. 73
Spencer, P. E. 66
Spitz, S. L. 21, 127, 447
Stark, A. M. 125
Stark, T. J. 456
Stayner, R. A. 176
Steen, H. A. H. 340, 408, 415, 430
Stein, S. J. 64, 74, 75, 106, 107
Stone, K. R. 406
Stoneman, A. M. 335
Storbeck, I. 59
Strauss, R. 149, 173
Striny, K. M. 16
Stroud, D. 316
Summers, S. C. 123
Supino, L. 472
Suwa, M. 141
Swalin, R. A. 354
Swift, D. R. 202

Taylor, B. E. 217
Taylor, B. N. 531
Taylor, J. R. 421, 434
Taylor, K. 522
Teller, E. 504
ten Duis, J. A. 377
Thwaites, C. J. 225, 227, 336, 392, 400, 427, 438
Tilbrook, D. 198
Tilsley, G. M. 50
Tkachev, M. A. 389
Tomkins, B. 409
Tompkins, H. G. 366
Totta, P. A. 146
Traub, A. C. 472, 473
Tu, K. N. 347
Turbini, L. J. 285, 456
Turino, J. 484
Turnbull, D. 342
Turner, C. 170
Tye, R. 512
Tygard, C. M. 483

Uchida, S. 356
Unsworth, D. A. 346
Ursch, R. R. 265

Val, C. 12, 58
Valentich, J. 47
van der Meulen, E. 377
Vanzetti, R. 472, 473, 474
Vaughan, J. G. 420
Verguld, M. M. F. 178, 182, 375, 397
Vimont, N. 148
Viret, P. 148
Vledder, H. J. 154

Walker, R. J. 168
Wall, C. I. 60, 99
Wallace, J. L. 81
Wallgren, L. 485
Wallis, D. R. 379
Warwick, M. E. 352, 361, 370, 372, 385, 400, 406
Watkins, H. C. 317
Weaver, H. 26
Wege, S. 439
Wei-Heng Shih 316
Weinhold, M. 185
Weltha, M. D. 259
Wenger, G. M. 276, 277, 279, 288
Wenzel, R. N. 328
Weston, A. D. 135
Wheeler, J. M. 19
White, C. E. T. 229
White, M. L. 16
Wicher, D. P. 72
Wigham, J. 158
Wild, R. N. 422, 429
Wilson, G. C. 368
Wilson, K. J. 32, 66
Wiltshire, B. 28, 414
Winters, W. 79
Wittenberg, A. M. 260
Wolf, M. 59
Wolverton, W. M. 433
Woodhouse, J. 401
Worrall, D. A. 138
Wright, A. W. 279
Wright, E. A. 433
Wu, T-S. 30
Wynblatt, P. 325

Yariv, A. 298
Yates, R. 24
Yeager, R. 80
Yoshida, H. 385

Zado, F. M. 200, 285, 288
Zahel, H. M. 384
Zierdt, C. H. 503
Zubler, E. G. 250
Zyetz, C. 300

Subject Index

(See also separate Author Index)

abietic acid
   solubility   443
abietates
   solubility   442
absorptivity
   of environment
      radiant heating   224–5
   workpiece characteristics
      radiant heating   219–20
accelerated ageing treatment   339–45, 492, 496–502
accelerated fatigue testing
   validity   430–34
acoustic microscopy
   inspection of solder joints   476–7
acrylic lacquers
   coatings   488
activators
   fluxes   162–3, 164, 165
adhesives
   application by placement machine   116
   assembly methods   303–7
   bonding prior to wave soldering   121–6, 147
   causing cleaning problems   460
   use in wave soldering   95–6
aerosol losses
   vapour phase soldering   264–5
ageing
   accelerated reliability assessment   496–502
   diffusion reactions   391
   effect on wettability   327–45
   thick film circuits   71
air
   environment
      humidity   344–5
      oxidation and corrosion   338–9, 340
      radiant heating   224–5
   fluid heat transfer   234
air knife
   wave soldering   145–6
alcohol
   cleaning solvents   452
alloys: soldering *see* solder alloys

alumina
   substrates   63, 64
aluminium
   wire bonding   118, 119
aluminium electrolytic capacitors   35
aluminium nitride
   substrates   64
antimony
   solder alloys   142, 169, 380
      effect on wetting   320, 321
aqueous cleaning   452–4, 459–60, 461
area emission sources
   radiant heating   211, 217–19, 226, 227, 231–2
area-of-spread test   348–51
assembly *see* manufacturing processes
assessment *see* testing
autoclave
   accelerated testing   502
automatic placement *see* machinery: component placement
automatic test equipment   480–1
azeotropic systems
   cleaning solvents   449–50, 452

back-wash zone
   wave soldering   134, 135
ball bonding   118–19
balling
   solder   468
      cleaning problems   461
   solder pastes   157–8, 186, 188, 193–4
      laser soldering   298
bare chips   20–22
   assembly methods   117–20
beam leads   120
'bed-of-nails' testing   480, 482
Bernoulli's law   139
beryllia
   substrates   63–4
bismuth
   solder alloys   142, 169, 379
blind vias   54–5
blister tape
   component packing   43

Subject Index

boiling
  fluid heat transfer 255–8
bonding
  bare chips 20–22, 117–20
  prior to wave soldering 121–6, 147
bridging
  wave soldering 143–4, 149, 151–3, 468
  use of air knife 145–6
brush cleaning 455
brush coating 487
bulk packing
  components 42, 115, 116
buried vias 54–5

capacitance
  inter-conductor 26
capacitors 30–35
  costs of chip v. leaded 517
  footprints 107
  growth in availability 516
  thick-film circuits 70
capillary penetration
  cleaning solvents 446–8
carbon dioxide gas laser 291–2, 296, 297
  board damage 295
  heating efficiency 293–4
  power levels and pulse duration 295
  safety precautions 302–3
ceramic chip capacitors 30–33
ceramic chip carriers *see* LCCC packages
ceramic packages 28–9
ceramic substrates
  matched TCE 63–71
chip carriers 13–17
chip components 29–38
  availability and use 516
  costs 517
  inspection of solder joints 470–1
chip-on-board assembly 117–20
'chips'
  defined 29
chlorinated solvents
  cleaning 450–1
chlorine
  corrosive effects 339
cleaning
  aided by flux 162
  post-assembly operations 435–66
    (*see also* Contents List, Ch. 12)
cleanliness
  measurement 461–6
coarsening
  diffusion reactions 391–2
coatings 485–8
  de-bonding 438
  rôle in wettability 327–8
coefficients
  ductility 411–12
  dynamic viscosity 181
  heat transfer *see* heat transfer coefficient
  TCE *see* temperature coefficient of expansion
Coffin-Manson law 396–7
collet soldering 282
  repair work 484–5
colophony *see* rosin
compatibility
  component/substrate
    thermal expansion 45–7, 82
compliant joints
  thermal mismatch 47
compliant substrates 82–3
  thermal mismatch 47
components 6–44
  (*see also* Contents List, Ch. 2)
  expansion compatibility with substrate 45–7, 82
  footprints 100–110
  overview 2
  placement 111–17
    accuracy 101–3
    machinery
      effect on design 93, 94
    preferred layouts 97, 98, 99
    reliability 503–4
    repair work 484–5
  shifting during soldering 371–7
  trends 515–16
  wave soldering defects 148–53
computer-aided design 110–11
computer-controlled placement 114, 116
  laser soldering 286, 297
condensation soldering *see* vapour phase soldering
condensing coils
  vapour phase soldering 246, 269, 270, 271
conduction
  heat dissipation 48–50
conductive adhesives 303–7
conductive heating, local 279–85
conductivity
  contamination assessment 462–4
conductivity, thermal *see* thermal conductivity
conductors
  fan-outs 109
  intermetallic growth 336–7
  length
    packaging options 25–6, 27
    track density 53
  routing
    design considerations 97, 99, 100
  thick film circuits 67–9
  track density 52–5
conformal coatings 486
contamination (*see also* cleaning; impurities)
  assessment 461–6
  effects 436–8
control systems
  radiant heating 232–3
convection

fluid loss
  vapour phase soldering 263–4
  heat dissipation 50–51
cooling
  growth of intermetallic compounds 331–2
  solder metallurgy 393–4
  substrates 48–52
cooling tubes
  condensation heat transfer 246, 269, 270, 271
copper
  ball bonding 119
  conductors 68, 69
  intermetallic growth with tin 334–6, 337–8
  solder alloys 380
  solderability standards 370–71
copper-clad Invar 75–81
copper foil
  organic laminates 59
corrosion
  effects 436–7
  of solderable surfaces 338–9
  vapour phase soldering 262–3
costs
  advantages of SMT 4
  economics of SMT 509–12
  PCBs 4
cracking
  solder joints 403–7, 468
creep
  fatigue mechanisms 407–9
creep rupture strength
  properties of solder 385–6
curing
  adhesives 116–17, 124–6, 306
  polymers
    thick films 84
cylindrical ceramic capacitors 33–4
cylindrical diodes 13

damp heat
  accelerated tests 500–2
defects
  classification
    solder joints 467–9
  wave soldering 148–53
dendrites
  contamination effects 438, 468
density
  conductors 52–5
  fluxes 128–30
design 93–111
  avoidance of bridging 151–3
  restricted by SMT 4
desoldering
  by laser 301
  repair work 484–5
de-wetting 323–4, 325, 467, 469
die bonding
  chip-on-board assembly 117
dielectric constant
  substrate materials 85, 86
dielectric layers
  capacitors 31, 32
dielectric pastes
  thick film capacitors 70
diffuse emission
  radiant heat 206
diffuse radiation
  heat transfer 206–8
diffusion
  fluid loss
    vapour phase soldering 263–4
  solder metallurgy 391–2
diffusivity, thermal 237–8
DIL packages
  board area requirement 22–5
  conductor lengths 25–6, 27
  disadvantages 8–9
  modification 18, 19
  thermal resistance 40
dimensions
  chip components 30
  flatpacks 17
  chip carriers 15, 16
  SOIC packages 11
  tantalum capacitors 35
diodes 13
dip cleaning 455
dip coating 487
dip testing
  solderability 346–8, 369–70
dissipation, thermal 48–52
dissolution rates
  of metals in solder 165–6, 167, 338
double-sided PCBs
  assembly variations 90–91, 92–3
drag-out
  fluid losses
    vapour phase soldering 265–6
drawbridge effect 373–7
drilling
  substrates
    interconnection density 55
droplets
  surface tension 309–11
dropwise condensation 241–2
dross
  formation on solder 142–3, 143–4
drying
  solder pastes 186–8, 189
dual wave soldering 139–40
ductility
  solder joints 411–13

economics of SMT 506–18
edge connectors 37–8
elastica curve 313, 314–15
electrical characteristics
  capacitor dielectrics 32

electrical performance
  packaging options 25–6
electrical testing 480–3
  accelerated 500, 501
  effect on design 93, 96–7
electrolytic capacitors 34–5
  footprints 107
electrolytic corrosion 436–7
electromagnetic spectrum 197
electromotive forces
  common metals 436–7
electroplating *see* plating
emissive power
  defined 198
emissivity
  defined 203
  directional 221
encapsulation
  bare chips 120
epoxy adhesives 303–4, 305
epoxy coatings 488
epoxy laminates
  and interconnection 57–8
  local heat conduction 283–5
etched mesh
  screen printing 176
etching
  stencils
    solder paste application 177–8
Europe
  development of SMT 514–15
evaporation
  fluid loss
    vapour phase soldering 263–4
expansion
  component/substrate compatibility
    45–7, 82
  porcelain-enamelled steel 73
  temperature coefficient *see* temperature
    coefficient of expansion

fabrication processes *see* manufacturing
  processes
failures
  reliability functions 490–6, 504
fan-outs
  conductors 109
fatigue
  solder joints 396–434
faults *see* defects
FC-70 248, 252–4
  condensing capacity of coils 270
  contamination by R113 267
  power requirements 258
FC-5311 252–4
  power requirements 258
feeding methods
  component placement 115, 116
fillets, solder *see* solder fillets
film condensation 241–5
filtration

fluid loss
  vapour phase soldering 266, 271–2
'fines'
  solder pastes 157, 158
flatpacks 17, 152
flexible layers
  compliant substrates 82–3
flip-chips 22, 120
'floating'
  component movement 372–3
fluid heat transfer
  reflow soldering 234–77
    (*see also* Contents List, Ch. 8)
fluorinated solvents
  cleaning 451–2
Fluorinert FC-70 *see* FC-70
fluorocarbons
  heat transfer measurements
    reflow soldering 239–41
  vapour phase soldering 248, 251–66
Flutec PP11 252
fluxes
  aqueous cleaning 452–4
  filtration
    vapour phase soldering 271–2
  need for 324–6
  need for board cleaning 435
  solder pastes 162–5
    causing voids 191, 192
  solubility 440–2
  vapour phase soldering 258–60
  wave soldering 126–30, 134
fluxless soldering
  laser soldering 296
foam fluxing 127–8
footprints
  design considerations 100–110, 151–2
fracture toughness 406
frequency
  cycling
    fatigue in solder joints 414–18
furnaces
  radiant heating 226–8, 229–32

Galden perfluoropolyethers 253
gas
  lasers *see* carbon dioxide gas laser
  soldering methods 285
  repair work 485
gate-arrays
  lead count 7
geometry
  of solder fillet
    effect on fatigue life 417–21
  of workpiece
    affecting radiant heating 223–4
glass cloth
  laminates 59
glass transition temperature
  resins 57, 58, 303, 304, 305
globule balance

solderability testing 363–8
glue *see* adhesives
gold
    impurities
        solder metallurgy 394, 395
    intermetallic growth with tin 336
    wire bonding 118, 119
gold-palladium
    ageing effects 391
    leaching 392
grid arrays 18, 23

hand assembly
    components 111
heat *see headings beginning* thermal . . .
heat transfer coefficient 208–9, 235–6
    evaluation 238–41
    liquid phase soldering 274
    vapour phase soldering 244, 245–6, 254
heated collet *see* collet soldering
heating (*see also* pre-heating)
    fluid heat transfer
        reflow soldering 234–77
        (*see also* Contents List, Ch. 8)
    soldering procedures
        effect on drying 186, 189
        localised 279–85
heating, radiant
    reflow soldering 196–233
        (*see also* Contents List, Ch. 7)
heating cycles
    infra-red heating 229–32
high frequency fatigue 414–17
holograms
    laser soldering 296
hot air knife
    wave soldering 145–6
hot air levelling 60, 61–2, 147
hot gas soldering 285
    repair work 485
hot-plate solder reflow 278–9
humidity
    accelerated ageing 344–5, 500–2
    affecting plastic packages 27–8
hybrid circuits
    evolution of SMT 506–7
hydrogen-nitrogen
    environment
        radiant heating 225
hysteresis
    wetting 322

icicles
    soldering defects 149
impurities (*see also* contamination)
    effect on wetting 319–20
    in solder baths 141–2
    in solder pastes 170–72
    solder metallurgy 394–6
in-circuit testing 481–3
indium

solder alloys 169, 170, 380
inductors 37
infra-red heating
    laser inspection of solder joints 474–6, 480
    reflow soldering 210–19, 225–33
inspection (*see also* testing)
    soldering quality 466–80
insulation resistance test 465–6
interconnections (*see also* conductors)
    types of substrate 55–87
intermetallic compounds
    growth 329–38
Invar
    effect on solder fatigue 422–3
    metal core substrates 75–81
ionic contamination
    assessment 462–4, 466
iron
    intermetallic growth with tin 336
irradiance
    defined 198

Japan
    development of SMT 515–16
jet wave soldering 138–9
joints
    conductive adhesives 307
    soldering *see* solder joints

kauri-butanol values 441
Kevlar
    matched TCE substrates 71, 72
Kirchoff's law 205
Kirkendall voids 118

Lambert's law 206, 207
laminates
    and interconnection 55–63
    matched TCE 71–2
lamps
    radiant heat sources 211–17
land grid arrays 18
lands *see* pads
Laplace equation 310
laser beams
    solderability testing 352
laser inspection
    solder joints 474–7, 480
laser soldering 286–303
laser trimming
    resistors 70
layout *see* design
LCCC packages 13–15
    fatigue life 407–13
    footprints 104
    inspection of solder joints 472–3
    thermal resistance 40, 41
leaching
    and solder alloy 165–6, 391
lead

oxidation   339
solder alloys, 141, 168, 169, 170, 380 (*see also* tin-lead solders)
    effect on wetting   317–20
leaded chip carriers *see* PLCC packages
leadframes   29
leadless ceramic chip carriers *see* LCCC packages
leads
    albatross wing   471
    butt   18, 471
    gull wing   11, 471
    J-lead   15, 471
    length
        packaging options   25–6
    packages
        DIL   8–9
        no. required   7–8
leakage current
    contamination effects   437–8
liquid phase soldering   273–7
    evaluation of heat transfer   238–41
liquid solvents
    cleaning techniques   454–5
liquids
    surface tension   309–11
liquidus
    defined   388
local conductive heating   279–85
LS-230   252–4
    power requirements   258

machinery
    component placement   111–17
        effect on design   93, 94
magazines
    component packing   44, 115, 116
Manhattan effect   373–7
manual assembly
    components   111
manufacturing processes   111–20
    assembly variations   88–93
    costs   509–10
    miscellaneous attachment methods   278–307
    organic laminates   60–63
    overview   2–3
masks
    soldering   108, 146–8
mealing   438
measurement *see* testing
melamine laminates   56, 59
MELF capacitors   33–4
MELF diodes   13
memories
    lead count   7
meniscus
    shapes   313–15
        solderability testing   352
meshes
    screen printing   173–4, 306

sieving
    solder powder   156
metal core substrates   72–82
metallisation *see* conductors
metallurgy
    solder alloys   168–70, 378–96
micelles
    saponification   440, 441
microprocessors
    lead count   7
microstructure
    solder joints
        effect on fatigue   424–6
miniaturisation
    components   6–9
mixed assembly   90, 91–3
modification
    through-hole components   18, 19
moisture
    affecting plastic packages   27–8
multilayer PCBs
    manufacturing processes   62–3
    track density   53
multiple-beam laser soldering   296
multiple-lead soldering   281–2
Multiwire connection   81–2

Nd:YAG laser   290, 291, 296, 297
    board damage   295
    heating efficiency   292–3
    power levels and pulse duration   295
    safety precautions   302–3
net present value
    economic decisions   511
nichrome filament lamps
    radiant heat sources   211, 217, 218, 226, 227
nickel
    intermetallic growth with tin   336
nitrogen
    environment
        radiant heating   225
nitrous oxide
    corrosive effects   339, 340
non-conformal coatings   486
non-ionic contamination
    assessment   464
nucleate boiling   255–8

oil
    wave soldering   143–5
organic solvents
    cleaning   449–52
organic substrates
    and interconnection   55–63
    matched TCE   71–2
    metal-cored   74–81
outgassing
    wave soldering   149
oxidation
    of solder   142–3

Subject Index

of solderable surfaces 338–9
oxide content
   solder powder 159–60

packages
   construction and materials 26–9
   effect on design 93
   trends 513–14
   types 9–19
packing
   of components 42–4
pads
   electrical testing 93, 96–7
   footprint design 100–110
palladium-silver
   conductors 67, 68, 69
panel plating 60
paraxylylene coatings 488
particles
   solder powder 154–8
passive components 29–38
   footprints 107–8
pastes, solder *see* solder pastes
pattern plating 60
PCBs *see* substrates
peanut oil test
   solder pastes 160
peel-back zone
   wave soldering 134, 135
perfluoro-isobutylene (PFIB) 261–2
performance
   advantages of SMT 4
phase diagram
   solder alloys 168, 386–90
phenolic lacquers
   coatings 490
phenolic laminates
   and interconnection 56–7
phosphorus
   in nickel coatings
      effect on intermetallic growth 337
pick-and-place machines 111–14
pick-up heads
   placement machines 116
pin grid arrays 18, 23
pin transfer
   adhesives 123
   solder pastes 179
pinholes
   solder defects 469
placement
   components 111–17
      accuracy 101–3
   machinery
      effect on design 93, 94
Planck's law 199–201
plastic capacitors 34
plastic leaded chip carriers *see* PLCC
   packages
plastic packages 27–8
plating

organic laminates 60–63
PLCC packages 15–17
   board area requirement 22–5
   footprints 105
   inspection of solder joints 471–2
   thermal resistance 40
polyester laminates 56, 59
polyimide adhesives 303
polyimide laminates
   and interconnection 58
polymers
   thick film circuits 83–4
      future trends 508
polystyrene coatings 488
pool boiling 255
porcelain-enamelled steel
   metal-core substrates 72–4, 75
potted coatings 486
powder, solder 154–62
power
   operating requirements
      fluid heat transfer 255–8
power cycling
   fatigue in solder joints 399–403, 410
preheating
   prior to wave soldering 130–32
   solder pastes
      effect on voids 193
primary fluid
   vapour phase soldering 251–66
printed circuit boards *see* substrates
printing, screen *see* screen printing
process control
   radiant heating 232–3
   vapour phase soldering 272–3
production processes *see* manufacturing
   processes
protective coatings *see* coatings

quad-in-line packages 22
quad packs 17, 152
quality control *see* inspection; reliability;
   testing
quartz
   matched TCE substrates 71–2

R113 249, 266, 267
   cleaning solvents 452
   condensing capacity of coils 270
radiant heating
   reflow soldering 196–233
      (*see also* Contents List, Ch. 7)
radiation
   heat dissipation 51
radiosity
   defined 198
random failures
   reliability functions 491–2
Rayleigh number 245
reflectivity
   workpiece characteristics

radiant heating 219
reflow soldering 196–277
  (see also Contents List, Ch. 7 & 8)
  assembly variations 89
  by hot plate 278–9
  design constraints 94, 95
  reflow temperature 166–8
  reliability 489–505
    (see also Contents List, Ch. 13)
  placement machine functions 117
  temperature-dependent effects 39
repair work 483–5
  adhesive assembly 304
  by laser 301
resins
  fluxes 162–4
  use in substrates 55–9
resistance
  conductors 25
resistance soldering 283
resistivity
  conductors 65, 67, 68
  polymers
    affecting by curing temperature 84
resistors 36–7
  costs of chip v. leaded 517
  footprints 107
  growth in availability 516
  thick film circuits 69–70
resists, solder
  PCB manufacture 61
rework
  solder fillets 483–4
rheology
  solder pastes 179–86
rosin
  fluxes 162–4
    contamination assessment 464
    filtration 271–2
    solubility 258–60, 438–49
  wave soldering 144
rotary dip testing 369–70
roughness, surface see surface roughness

safety precautions
  laser soldering 302–3
sales potential 511
saponification
  cleaning methods 440, 441, 454, 461
saturated vapour density 344
scanning acoustic microscopy
  inspection of solder joints 476–7
scanning mode wetting balance 361–2
screen printing
  adhesives 122–3
  solder pastes 172–7
screenability
  solder pastes 183
secondary fluid
  vapour phase soldering 266, 267, 269–71

semiconductor packages see packages
sequential placement machines 111–14
sessile drops
  surface tension 309–11
shadowing
  wave soldering 137–8
shear strength
  solder alloys 171, 380–6
shear stress
  cleaning solvents 448–9
  fatigue in solder joints 399–402
short circuits
  conductive adhesives 304–5
silicone coatings 488
silicone laminates 56, 59
silver
  and tin-lead solders 141, 165, 166, 168–9, 170
  conductors 67, 68, 69
  intermetallic growth with tin 336
  solder alloys 380
simultaneous placement machines 111, 113, 114
single-lead soldering 279–81
single-sided PCBs
  assembly variations 88–9, 91–2
size (see also dimensions)
  advantages of SMT 4, 510–11
skipped joints
  soldering 137–8, 149, 151, 467
slump
  solder pastes 183–4
Small Outline Diode packages see SOD packages
Small Outline Integrated Circuit packages see SOIC packages
Small Outline Transistor packages see SOT packages
SMDs
  defined 9
snap-off
  screen printing 176
sockets 18–19
SOD packages 13
  footprints 106
sodium chloride
  use in contamination assessment 462–4, 465
SOIC packages 9–11
  and solder bridging 151–2
  board area requirement 22–5
  footprints 104, 105
  thermal resistance 40
solder
  reflow properties 254–5
  strength properties 380–6
solder alloys 140–43, 165–72
    (see also tin-lead solders)
  effect on solder fatigue 423–4
  effect on wetting 316–20
  purity 170

Subject Index

solder balls *see* balling: solder
solder bridges *see* bridging
solder dip test 346–8
solder fillets 378–434
  (*see also* solder joints; Contents List, Ch.11)
  inspection 470–2
  rework 483–4
  voids
    solder pastes 186–7, 190–93
solder joints (*see also* solder fillets)
  inspection 466–80
  laser soldering 298–9
  thermal expansion compatibility 46, 47
  wave soldering 135
solder masks 108, 146–8
solder pastes 154–95
  (*see also* Contents List, Ch. 6)
  application prior to placement 116
solder resists
  PCB manufacture 61
solder robbers 152, 153
solderability 308–77
  (*see also* Contents List, Ch. 10)
  capacitors 33
soldering (*see also* laser soldering; reflow soldering; wave soldering)
  components must withstand 37
  effect on design 93, 94–6
  miscellaneous attachment methods 278–303
  overview 2–3
solids content
  solder powder 160–62
solidus
  defined 388
solubilisation
  cleaning methods 440
solubility
  of contaminants 438–49
  rosin flux
    vapour phase soldering 258–60
solvents (*see also* solubility)
  cleaning techniques 454–9
  organic
    cleaning 449–52
  solder pastes
    effect on voids 192–3
SOT packages 11–13
  footprints 102, 106
spectrum, electromagnetic 197
spray coating 487
spray fluxing 128, 129
spray solvent cleaning 456–7
squeegees
  screen printing 176
stand-off height
  effect of solder paste 160–61
  effect on cleaning 456–7
  effect on solder fatigue 417–21
standards
  lack of 518
  solderability 370–71
Stefan-Boltzmann law 203–5
stencils
  solder paste application 177–8
Stonehenge effect 373–7
storage
  after drying
    solder pastes 187, 190
  and ageing 328, 339–40
  solder pastes 188–9
strain range
  fatigue life of LCCCs 407–11
stress (*see also* fatigue)
  properties of solder 380–6
stress relaxation
  fatigue mechanisms 404–7, 413–14
substrates 45–87
  (*see also* Contents List, Ch. 3)
  assembly variations 88–93
  costs 4
  future trends 508
  materials 2
    effect on solder fatigue 421–3
  positioning
    placement machines 115
  use of board area 6–8, 22–5
surface composition
  soldering wettability 317–20
surface roughness
  soldering wettability 320–22
vapour phase heaters
  effect on heat transfer 268
workpiece characteristics
  radiant heating 221
surface tension
  cleaning solvents 446
  soldering wettability 309–11, 316–17
'swimming'
  component movement 372–3
syringe dispensing
  adhesives 122, 123, 306
  solder pastes 178–9

TAB (tape automated bonding) 20–22, 119, 508
tackiness
  solder pastes 184–6
tantalum capacitors 34–5
  footprints 107
tape automated bonding 20–22, 119, 510
tape packing
  components 42–3, 115, 116
telecommunications equipment
  solder fatigue predictions 432–4
temperature (*see also* headings beginning thermal . . .)
  influencing solder fatigue 426
  influencing solder tensile strength 384–5

influencing viscosity
  solder pastes 179–80
  of cleaning solvents 444–5
  of reflow
    solder alloys 166–8
  parameters used in accelerated tests 499–500
temperature coefficient of expansion 41
  epoxy laminates 58
  matching 46–7
    ceramic substrates 63–71
    laminates 71–2
    metal-cored substrates 77–80
    substrate materials compared 86
temperature coefficients
  capacitors 31–2, 33
tensile stress
  fatigue in solder joints 399–403
  properties of solder 380–6
testing (see also inspection)
  cleanliness 461–6
  conditions
    effect on solder fatigue 426–7
  electrical see electrical testing
  liquid phase soldering 276–7
  reliability assessment 496–502
  solder fatigue
    validity 428–32
  solder joints
    high frequency fatigue 414–16
  solder pastes 159–60
    slump 183–4
    solder balling 194
    tackiness 185–6
    viscosity 181–2
  solderability 345–70
    accelerated ageing 340–44, 345
thermal acceleration
  reliability assessment 497–9
thermal characteristics
  SM components 38–41
thermal conductivity
  fluid heat transfer 236–7
  metal-cored substrates 77, 79–81
  local heating 283–5
  substrate materials compared 85
thermal degradation
  fluorocarbons 260–63
  workpiece characteristics
    radiant heating 221–3
thermal diffusivity 237–8
thermal expansion
  component/substrate compatibility 45–7, 82
  fatigue in solder joints 399
thermal management
  substrates 47–52
thermal mounting 110
thermal resistance 39–40, 41
thermal shock testing 416
thermal signature

laser inspection of solder joints 474–7
thermal vias 54–5
thermocompression bonding 20, 118
thermocouples
  process control
    radiant heating 232, 233
thermosonic bonding 20, 118
thick film circuits
  ceramic substrates 66–71
  polymers 83–4
    future trends 508
thickness
  deposits
    solder pastes 175
thin film circuits
  ceramic substrates 65–6
tin
  intermetallic growth 333–7, 337–8
  metallurgy 376–7
  oxidation 339
tin-antimony solders 169
tin-bismuth solders 169
tin-indium solders 169, 170
tin-lead-silver solders 165, 170
tin-lead solders 141, 168
  effect on wetting 317–20
  intermetallic growth 333–7, 338
  metallurgy 379–86
  phase diagram 168, 386–90
tin-silver solders 168–9
tombstone effect 373–7
toxicity
  fluorocarbons 260–62
trimming
  resistors 70
tungsten filament lamps
  radiant heat sources 211–17, 218, 226, 227

ultimate tensile strength (UTS) 381
ultrasonic agitation
  cleaning techniques 458–9
ultrasonic bonding 20, 118
undercooling
  solder metallurgy 393
United States
  development of SMT 516–18
urethane lacquers
  coatings 488

vapour phase soldering 241–7, 248–73
  compared with liquid phase 274–6
  evaluation of heat transfer 238–41
vapour solvent cleaning 455–6
ventilation
  fluid loss
    vapour phase soldering 265
vias
  interconnection density 53–5
  PCB manufacture 62–3
viscosity

adhesives   125
  solder pastes   179–82, 183
visual inspection
  solder joints   467–72, 478, 479
voids
  solder joints   186–7, 190–93

washing *see* cleaning
water
  aqueous cleaning   452–4, 459–60, 461
  water-soluble fluxes   164–5
  aqueous cleaning   453–4
wave cleaning   454–5
wave fluxing   128
wave soldering   121–53
    (*see also* Contents List, Ch. 5)
  assembly variations   89
  design constraint   94, 95
wear-out failures
  reliability functions   492–6
wedge bonding   118–19
wettability
  soldering   308–45

wetting
  aided by flux   162, 163
  by cleaning solvents   445–6
wetting balance
  solderability testing   353–62
    compared with globule balance   368
wetting force   354–5
whiskers
  contamination effects   437–8, 468
white residues   438
Wien's law   201–2
wire bonding
  bare chips   20, 117–19

X-ray inspection
  laminography   474
  solder joints   474–5, 479–80

YAG laser *see* Nd:YAG laser
Young-Dupré equation   312

zero-hour quality   504–5

"(0202) 674333 *YOUR* HOTLINE FOR A PLACE ON THE BOARD..."

Surface Electronics – The leader in Surface Mount Technology. Our Custom Services are Design, Assembly and Test. Surface Electronics is the credible, experienced and reliable manufacturer.

Surface Electronics ltd

Patrick House, West Quay Road, Poole, Dorset BH15 1JF, UK
Tel: (0202) 674333   Fax: (0202) 678028   Telex: 41184 SURFEL G

IF YOU WANT TO STAY ON TOP

YOU WON'T GET A BETTER DEAL

THAN FROM

manix
SURFACE MOUNTING SYSTEMS

In surface mounted technology our technical expertise and unrivalled product range make us leaders of the pack

AUTOVEYOR LTD, School Street, Loughborough, Leics. Tel: (0509) 237003. Fax: (0509) 237004.

SURFACE MOUNT DESIGN
For Surface Mount Technology Consult THE Design Team with Experience

The design opportunities and manufacturing constraints that Surface Mount Components bring have led to new concepts and challenges in the layout and design of printed circuit boards.

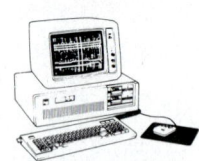

Our experienced team of designers produce complete designs, artwork and manufacturing drawings using Racal Redac CAD and Gerber Digitising equipment.

Coupled with our mechanical design and electro/mech packaging expertise we can offer selective design facilities or a complete product design service.

Watford Design Services Limited
Kebbell House Delta Gain Carpenders Park
Watford WD1 5EF Herts

For further information or assistance on your Surface Mount, Mixed or Conventional Technology pcb design project please call....

MEMBER OF S.M.A.R.T.   Len Lawrence on 01-421-2377   Fax 01-421-2950

A Cookson Group Company

Fry's Metals
NEW FLOWCREAM
for Surface Mount Assembly

✷ Perfected screenability
✷ Fine tracking – No problem
✷ Extended screen life
✷ Excellent reflow properties
✷ Flux residues removable to Specification MIL-P-28809
✷ Suitable for vapour phase, infra red, or other reflow methods
✷ Both screenable and syringe dispensable types

FRY'S METALS LTD
Tandem Works, Merton Abbey, London SW19 2DP.
Telephone 01-648 7020. Telex 265732.

S34

Protonique

$$R = \rho \frac{1}{A} \text{ ohms}$$

- Measure automatically up to 96 insulation resistances in the range 10^6 to 10^{13} ohms, individually or in 1 to 48 groups.
- Record the results on magnetic discs during climatic testing from 1 to 64 days, with any cycle.
- Display and print out all anomalous results during the test.
- Analyse each group automatically using sophisticated statistical techniques, as soon as the test is finished.
- Correlate the results from any standard or non-standard test pattern.

All this — and more — with **Irma la Douce**, the **Protonique Insulohmeter** IRMA-1, the only computerised Insulation Resistance Meter and Analyser.

For further information, please contact us or your local agent.

Protonique S.A.

P.O. Box 78
CH-1032 Romanel-sur-Lausanne
Telephone: (+ 41 21) 38 23 34
Telex: 454 144 PRTN CH
Telefax: (+ 41 21) 38 24 11 (Gp.3/2a)

SURFACE MOUNT PROBLEMS?
HOW AND WITH WHAT?

- **SMPS** have the answer to all your surface mount problems
- From prototype to production
- With the price range to suit your budget £445–£4800
- Save time and money, contact us now

SURFACE MOUNTED PRODUCTION SYSTEMS LTD

Unit 5, Sandbank Industrial Estate, Dunoon, Argyll PA23 8PB.
Tel: 0369 5116   Telex: 778261   Fax: 0369 4234.

A COMPLETE RANGE OF PRODUCTS TO MEET YOUR NEEDS.

Whether your requirements be manual systems, re-work, experimental work or full-production on any scale, Link Hampson have brought together a range of products to meet your needs.

In line with our group policy we have sourced a choice of product which is manufactured by companies who are both well-known and respected world-wide in their field. Our twenty years expertise in the market enable us to provide you, the customer, with the best equipment at the best possible price. Quality products that cover all your needs.

FOR A COPY OF OUR BROCHURE RING US ON
NEWBURY (0635) 44796
LINK HAMPSON LIMITED, BONE LANE, NEWBURY, BERKSHIRE RG14 5TD

SMT needs VISION

VISION SYSTEMS GIVE
IMPROVED QUALITY AND **RELIABILITY.**
CRS ARE THE **EYES** IN THE WORLD OF
ENGINEERING.
WITH OUR SKILLS AND EXPERIENCE IN
ELECTRONICS, COMPUTING & MECHANICS,
WE HAVE THE TECHNICAL ABILITY TO
INSPECT, MEASURE & REWORK
AND TO IMPROVE YOUR **PRODUCTIVITY**

COMPUTER RECOGNITION SYSTEMS LTD
UNIT 10, THE BUSINESS CENTRE, MOLLY MILLARS LANE,
WOKINGHAM, BERKS.

Tel: (0734) 792077 · Telex: 846028 · Fax: 774734

SURFACE MOUNT SPECIALISTS.

ASSEMBLY
The EPE 20/20 is a high density automatic pick and place system with a placement rate of up to 4000/hour. The optional 'Hot Head' also allows you to reflow expensive 2 and 4 sided devices during the placement operation.

SOLDERING
The Detrex Vapour Phase soldering systems offer the ultimate in high production reflow. They can be integrated with pick and place and cleaning systems for total automation.

CLEANING
SMC cleaning both batch and in line. Detrex's new compact SMT machine, only 10 feet long is a spray/irnmersion/spray cleaner, available with ultrasonics if required.

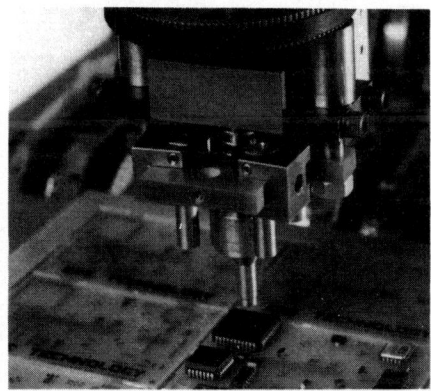

REWORK
Electrautom offer a variety of systems for SMD removal and repair. Equipment incorporates the latest technology for safe component handling. A full range of hand tools are also available.

CONSULT THE SPECIALISTS - YOUR AUTOMATIC CHOICE

Electrautom Ltd

Etom Buildings, Quarry Wood, Aylesford, Kent ME20 7NA.
Telephone: 0622 70188 Telex: 965021 Etom G.

TE TURNER ELECTRONICS LTD

Turner House, Roke Close,
Kenley, Surrey CR2 5NL
Tel: 01-668 0821 Telex: 893446 Fax: 01-668 2782

Specialist in Surface Mount Assembly Equipment

- Pick and Place Machine
- IR Reflow Machine
- Soldering Machine
- Vapour Phase Machine
- Adhesive Dispensing Equipment
- Rework Stations

SURFACE MOUNT BAREBOARD TEST SPECIALISTS

Sub-Contract testing of SMD boards on both sides simultaneously with our Topside Access Fixture System is available.

Component pad pitches down to 0·025" are possible.

Five years of experience and knowledge of testing unloaded PCB's of all types.

Prices available on application.

Cirtron Systems

14/15 Highdown Road, Sydenham Estate,
Leamington Spa, CV31 1XT
Telephone: 881718/312413   Telex: 31497

Services Include:
Hot Air Levelling. Automatic Optical Inspection of inner layers and Track Repair Service.

SOLDERING IN ELECTRONICS
by R. J. Klein Wassink

Pages—470+xviii; Tables—57; Figures—301; References/Bibliography—686; Size—23 × 15 cm.*
ISBN 0 901150 14 2*

The book 'Soldering in Electronics' has been written to cover soldering as a technique for mass production of electronic connection, although many items treated will be useful for other branches of soldering technology.

Chapter 1 deals with the interaction of the various aspects which determine the result of the soldering operation, namely the quality of the soldered joints. It is explained that the desired goals can only be achieved if soldering is treated as a coherent system, all separate aspects of which are to be well prepared, often long before the actual moment of the soldering operation. This way of thinking has many implications for the other chapters, such as on solderable coatings, solderability assessment, and joint design.

The book provides a combination of background knowledge and information for practice. In doing so it helps to clarify many points of basic discussion among soldering experts, such as wetting and dewetting, thermal influences, solderability and methods of its assessment, metallurgy of alloys and coatings, and process conditions. On the other hand it contains a wealth of detailed and practical information on soldering alloys, fluxes, coatings, soldering methods, pattern design and criteria for joint inspection. A separate chapter is devoted to the thermal aspects of soldering, a subject that is seldom discussed in literature despite its utmost importance for soldering.

It is this combination of theory and practice which makes the book indispensable to a broad group of readers with a responsibility for electronics manufacturing, both in R&D and production, such as design engineers, process engineers, production engineers, quality engineers, metallurgists and chemists.

At all relevant places in the text, literature references are included for those who wish to further their knowledge on the items discussed. Comprising over 400 references (listed in alphabetical author order) on soldering, the book provides an excellent survey of the existing literature. Moreover, a bibliography is offered of all books on soldering ever published, as well as a list of soldering items in several languages.

May be ordered direct from
ELECTROCHEMICAL PUBLICATIONS LTD,
8 BARNS STREET, AYR KA7 1XA, SCOTLAND. TEL: (0292) 263281. FAX: (0292) 284719

CLEANING AND CONTAMINATION OF ELECTRONICS COMPONENTS AND ASSEMBLIES
by B. N. Ellis

Pages—365 + xxi; Tables—17; Figures—99; References—159; Size—23 × 15 cm.
ISBN 0 901150 20 7

This book is a practical guide for all persons who are in any way concerned with cleaning or contamination control of either components or assembled circuits. It is also useful to those specifying components and, in particular, printed circuits. It equally covers the theoretical side to a sufficient extent to allow the average engineer or technician who is not a specialist in the field to understand the mechanisms involved in contamination and cleaning.

As a reference book, the text is divided into seven parts, logically divided into some thirty chapters illustrated by photographs, line drawings, graphs and tables. Each chapter has its own reference list. The introductory section comprises an historical background to cleaning in the electronics industry, a very complete chapter of definitions of all the terms employed and in the particular context, a short chapter on units employed, a theoretical treatise on the mechanics, physics and chemistry of cleaning (in simple terms) and one on the cost of cleaning.

The second part comprises some seven chapters cataloguing the diverse ways that contamination can occur in the electronics industry, whereas the third part describes, over three chapters, what effects contamination can have during the various manufacturing processes of components and assemblies and over the whole of their subsequent lives. Part 4 will be considered as being the most important by some production engineers because its four chapters describe all the currently used methods of cleaning and flux removal for the small, medium or large user with considerable detail on the products usually employed.

The fifth section deals with ionic contamination control. The first two chapters discuss respectively the American military specifications and the new British DEF standards. The third one gives a general view of the different instruments commercially available for measuring or detecting ionic contamination. The last chapter of this section gives an insight into some aspects of the theory of ionic contamination measurement and the solutions used for it. The next part treats the detection and measurement of contamination by other methods, with particular emphasis on non-ionic contaminants. Insulation resistance measurement is discussed in a separate chapter of this section. The last part, divided into three chapters, relates to the particular problems imposed by the use of surface mounted components and solder creams and pastes.

It is felt that this book will become a valuable reference work for the bookshelves of all companies involved in any aspect of electronics, particularly component, printed or hybrid circuit manufacturers or assemblers.

May be ordered direct from
ELECTROCHEMICAL PUBLICATIONS LTD,
8 BARNS STREET, AYR KA7 1XA, SCOTLAND. TEL. (0292) 263281. FAX. (0292) 284719